Physiology, Biophysics, and Biomedical Engineering

Series in Medical Physics and Biomedical Engineering

Series Editors: John G Webster, Slavik Tabakov, Kwan-Hoong Ng

Other recent books in the series:

Proton Therapy Physics
Harald Paganetti (Ed)

Correction Techniques in Emission Tomography
Mohammad Dawood, Xiaoyi Jiang, Klaus Schäfers (Eds)

Practical Biomedical Signal Analysis Using MATLAB®
K J Blinowska and J Żygierewicz

Physics for Diagnostic Radiology, Third Edition
P P Dendy and B Heaton (Eds)

Nuclear Medicine Physics
J J Pedroso de Lima (Ed)

Handbook of Photonics for Biomedical Science
Valery V Tuchin (Ed)

Handbook of Anatomical Models for Radiation Dosimetry
Xie George Xu and Keith F Eckerman (Eds)

Fundamentals of MRI: An Interactive Learning Approach
Elizabeth Berry and Andrew J Bulpitt

Handbook of Optical Sensing of Glucose in Biological Fluids and Tissues
Valery V Tuchin (Ed)

Intelligent and Adaptive Systems in Medicine
Oliver C L Haas and Keith J Burnham

An Introduction to Radiation Protection in Medicine
Jamie V Trapp and Tomas Kron (Eds)

A Practical Approach to Medical Image Processing
Elizabeth Berry

Biomolecular Action of Ionizing Radiation
Shirley Lehnert

An Introduction to Rehabilitation Engineering
R A Cooper, H Ohnabe, and D A Hobson

Series in Medical Physics and Biomedical Engineering

Physiology, Biophysics, and Biomedical Engineering

Andrew W. Wood

With contributions by

Anthony Bartel, Peter Cadusch, Joseph Ciorciari, David Crewther, Per Line, John Patterson, Mark Schier, and Bruce Thompson

CRC Press is an imprint of the
Taylor & Francis Group, an **informa** business
A TAYLOR & FRANCIS BOOK

Published in association with the IOMP.

Front cover: The Australian performance artist Stelarc operating his Exoskeleton – a six-legged, pneumatically powered walking machine controlled by arm movements. The left arm also has enhanced functionality, including additional flexion and rotation. Stelarc has completed a number of projects extending the capability of the human body, described at www.stelarc.org

Credits: EXOSKELETON; Cyborg Frictions, Dampfzentrale, Bern 1998; Photographer- Dominik Landwehr; STELARC

CRC Press
Taylor & Francis Group
6000 Broken Sound Parkway NW, Suite 300
Boca Raton, FL 33487-2742

Printed in the United States of America on acid-free paper
Version Date: 20111227

International Standard Book Number: 978-1-4200-6513-8 (Hardback)

Visit the Taylor & Francis Web site at
http://www.taylorandfrancis.com

and the CRC Press Web site at
http://www.crcpress.com

To Professor Richard Silberstein, who was the main force behind establishing Medical Biophysics at Swinburne University of Technology – for his challenging ideas and mentorship

. . . and to our families for their patience.

Contents

Part VII Systems Integration

About the Series

The Series in Medical Physics and Biomedical Engineering describes the applications of physical sciences, engineering, and mathematics in medicine and clinical research.

The series seeks (but is not restricted to) publications in the following topics:

- Artificial organs
- Assistive technology
- Bioinformatics
- Bioinstrumentation
- Biomaterials
- Biomechanics
- Biomedical engineering
- Clinical engineering
- Imaging
- Implants
- Medical computing and mathematics
- Medical/surgical devices
- Patient monitoring
- Physiological measurement
- Prosthetics
- Radiation protection, health physics, and dosimetry
- Regulatory issues
- Rehabilitation engineering
- Sports medicine
- Systems physiology
- Telemedicine
- Tissue engineering
- Treatment

The Series in Medical Physics and Biomedical Engineering is an international series that meets the need for up-to-date texts in this rapidly developing field. Books in the series range in level from introductory graduate textbooks and practical handbooks to more advanced expositions of current research.

The Series in Medical Physics and Biomedical Engineering is the official book series of the International Organization for Medical Physics (IOMP).

The International Organization for Medical Physics

The IOMP, founded in 1963, is a scientific, educational, and professional organization of 76 national adhering organizations, more than 16,500 individual members, several corporate members, and four international regional organizations.

The IOMP is administered by a Council, which includes delegates from each of the Adhering National Organizations. Regular meetings of Council are held electronically as well as every three years at the World Congress on Medical Physics and Biomedical Engineering. The President and other Officers form the Executive Committee, and there are also committees covering the main areas of activity, including Education and Training, Scientific, Professional Relations, and Publications.

Objectives

- To contribute to the advancement of medical physics in all its aspects
- To organize international cooperation in medical physics, especially in developing countries
- To encourage and advise on the formation of national organizations of medical physics in those countries that lack such organizations

Activities

Official journals of the IOMP are *Physics in Medicine and Biology*, *Medical Physics*, and *Physiological Measurement*. The IOMP publishes a bulletin *Medical Physics World* twice a year, which is distributed to all members.

A World Congress on Medical Physics and Biomedical Engineering is held every three years in cooperation with International Federation for Medical and Biological Engineering through the International Union for Physics and Engineering Sciences in Medicine. A regionally based International Conference on Medical Physics is held between World Congresses. The IOMP also sponsors international conferences, workshops, and courses. IOMP representatives contribute to various international committees and working groups.

The IOMP has several programs to assist medical physicists in developing countries. The joint IOMP Library program supports 69 active libraries in 42 developing countries, and the Used Equipment Programme coordinates equipment donations. The Travel Assistance Programme provides a limited number of grants to enable physicists to attend the World Congresses.

The IOMP website is being developed to include a scientific database of international standards in medical physics and a virtual education and resource center.

Information on the activities of the IOMP can be found on its website at www.iomp.org.

Preface

The aim of this book is to show that many aspects of human physiology lend themselves to numerical analysis. Many ways of monitoring physiological function also rely on an understanding of physics and engineering to appreciate fully how they operate. The book arises out of an undergraduate course in medical biophysics and a postgraduate course in biomedical instrumentation in which the authors were involved for many years. Although the emphasis is on numerical analysis only, a basic knowledge of mathematics is assumed, and every effort is made to supplement mathematical formulae with qualitative explanations and illustrations to encourage an intuitive grasp on the processes involved.

Most of the chapters have a range of numerical tutorial problems with, in most cases, worked solutions. These are based on examination questions at the middle and senior undergraduate level.

For some of the material, the computational package MATLAB® offers a convenient way to gain insight into some of the more advanced mathematical analysis of physiological or of clinical monitoring systems. Suitable MATLAB code is provided where this may aid understanding.

I acknowledge the help of colleagues in the preparation of this book. Particular chapters have been authored as follows: Anthony Bartel, Per Line, Peter Cadusch, Joseph Ciorciari, David Crewther, John Patterson, Mark Schier, and Bruce Thompson. In addition, others have been associated with teaching the course over many years. These include Peter Alabaster, David Liley, Ric Roberts, and David Simpson. I also acknowledge a debt of gratitude to Professor Richard Silberstein, who first designed and set up the medical biophysics course at Swinburne University in 1976 under the guidance of Stan Rackham, the then head of physics, and Dr. Ken Clarke, the then director of physical sciences at the Peter MacCallum Cancer Centre in Melbourne.

Additional resources, such as supplementary questions, worked solutions and suggestions for laboratory exercises can be found at http://www.crcpress.com/product/isbn/9781420065138.

Introduction

This book is about biophysics, physiology, and biomedical engineering. It brings together a number of disciplines to help understand biological phenomena and the instrumentation that can be used to monitor these phenomena. It is primarily about those biological phenomena that can be described by the physical phenomena of electricity, pressure and flow, and the adaptation of the physics of these phenomena to the special conditions and constraints of biological systems. In general, it covers a variety of areas of relevance to medicine and biology in which the methods of physics and engineering have been successfully applied. The book will concentrate mainly on human biological systems, although some of the principles apply equally as well to all living systems, including all animals, plants and bacteria.

It is helpful at the outset to have some definitions:

Physiology is the study of biological structure in relation to function.

Biophysics is the study of biological phenomena in terms of physical principles.

Biomedical engineering is the study of biological systems and the development of therapeutic technologies and devices in terms of engineering principles.

Further specializations: *cellular biophysics* covers such areas as bioenergetics (energy transformation at the cellular level), cell signaling (how information is carried within the cell), membrane transport (channels, receptors and transporters involved in cell permeation), and electrophysiology (electricity generated within cells and associated electrical phenomena). *Systems biophysics/bioengineering* covers the study of control theory and information theory to understand and analyze complex biological systems.

Brief Historical Introduction

From earliest recorded history, people have been fascinated by the way the body works and the nature of life. Animal and even human sacrifice in ancient times was associated with cleansing, and the blood in particular was regarded as having life in it. The heart was regarded not only as the seat of emotion, but also the organ of thinking and of reasoning. The ancient Egyptians did not regard the brain as being at all important, removing the brain at death, but preserving the heart. The Roman physician Galen in the 1st century AD identified the brain with thinking but still the heart with emotions. Medieval physicians regarded life as being distinct from the physical body, although from the Greek physician onward the vital force was considered to be made up of four elements or humors: air, water, earth, and fire. These were identified with four bodily fluids: blood, phlegm, black bile, and yellow bile, respectively. The terms sanguine, phlegmatic, melancholic (literally black bile) and choleric are still applied to temperaments. Medieval medicine was directed at stabilizing these humors.

The 18th century and the rise of the Enlightenment saw a deeper interest in human anatomy by careful dissection and with this the coupling of structure with function, the beginning of physiology. In 1791, Galvani first observed the twitching of dissected frog legs when contacted by wires of dissimilar metals and when connected to sparks from static electricity. He is regarded as the father of bioelectricity, even though he erroneously thought that "animal electricity" was released from fluids within the muscle by the action of the spark. A few years later, Volta (around 1800) constructed the first battery from electrodes of zinc and silver (later zinc and copper) and was able to show that current flowed whether he connected this to frogs' legs or to saline-soaked paper. He thus showed that it was the current that activated the muscles and not the release of "animal electricity."

The 19th century saw remarkable advances in the measurement of bioelectric events: from 1842, it was known that the heartbeat was accompanied by electrical current and the first electrocardiogram (ECG) was recorded directly from a frog heart in 1878. The first ECG measured from a human via the skin occurred in 1902, by Einthoven, using the recently developed string galvanometer (taking its name from Galvani), which had high enough sensitivity to measure millivolts. The invention of the thermionic valve and particularly the triode in 1913 provided a means to amplify these small voltages and to display them on a cathode ray oscilloscope (CRO). This led to the first recording of electrical activity associated with the brain, the electroencephalogram (EEG) by Berger in 1924. The 1930s saw the invention of the voltage clamp, which unlocked the secrets of the processes underlying nerve impulses. It is now possible to measure the current through individual ion channels using the patch clamp, which is a modification of the voltage clamp.

In parallel with the developments in bioelectricity, understanding of the fluid flow behaviors of both the heart and lungs has moved forward a long way since Harvey in 1628 described the circulation of blood. The work of E. H. Starling at the beginning of the 20th century gave insights into how the heart and capillaries regulate overall circulation. In the 1950s, the application of ultrasonics to medicine gave rise to new methods for measuring blood flow and imaging the heart in motion. Other areas of general medical imaging have made great strides since the first shadowy X-ray of the hand in 1895 and the first image of an orange using magnetic resonance imaging in 1978.

Integrated circuitry and miniaturization have led to medical monitoring instruments being transformed from the cumbersome devices of half a century ago, needing careful calibration and tweaking by technicians, to the easy-to-use systems of today. Miniaturization is continuing, with concentration on applications from the areas of nanotechnology, microelectromechanical systems (MEMS), and microphotonics. As computing power increases, the ability to extract hidden diagnostic data from recorded signals, the modeling of complex interacting systems, and the ability to store and transmit huge amounts of data has given rise to the subdisciplines of bioinformatics, telemedicine, and robotic surgery.

Overview of the Book

The first two chapters are a general introduction to the physiological systems to be encountered in the later chapters. This is primarily to give those majoring in the physical sciences a brief glimpse of the principles of organization and of the interdependence of cellular and organ systems. It gives references to the many physiology texts that can be used as further reading. The second chapter, on the other hand, introduces some of the basic electronic and electrochemical principles relevant to sensing biological signals and optimizing their recording. Readers with a physical sciences background may skim over material already familiar to them, as may those with a biological sciences background wish to do the same in Chapter 1. A third chapter introduces the reader to the specialized methods used to record electrical events from biological tissue, including single cells.

The next section (Part II: Chapters 4–7) concerns mainly the specialized molecules within cells that are concerned with transport of material (particularly ions) into and out of cells, the molecules involved in cell-to-cell signaling, and the proteins involved in the ability of some cells to contract and to generate tension.

The following part (Part III: Chapters 8–11) covers the electrical and mechanical properties of the heart, the special properties of blood, and circulation. It also covers the principles of monitoring cardiac and circulatory function. Some aspects of imaging will be discussed further in Chapter 23.

Part IV (Chapters 12–14) covers those organ systems (principally the lungs and kidneys) in which the interrelationship of pressures and flows, and their measurement, are particularly important. This is seen in the latter chapters covering heart–lung machines (cardiopulmonary bypass, neonatal monitoring, and intensive care).

Part V (Chapters 15–19) is concerned with the organization and function of the nervous system, including the sensory inputs. Much of this builds on concepts encountered in Part I.

Part VI (Chapters 20–24) deals mainly with systems monitoring: the recording and interpretation of electrical signals; the various imaging systems, both at the organ and whole body level, and at the cellular level (microscopy). The study of magnetic fields both as a way of inducing stimulation and monitoring electrical currents is also included here.

Part VII (Chapters 25–27) covers integrative aspects, including modeling, biomechanics, and some exciting emerging technologies. This last chapter can never be fully up-to-date but deals with the progressive miniaturization of sensors and actuators in biomedical engineering (nanotechnology).

There are many relevant areas which have been omitted or dealt with inadequately in comparison to other texts. In particular, many may find the emphasis on bioelectrical rather than biomechanical phenomena somewhat of an imbalance. This arises from the particular environment that existed when the course at Swinburne University of Technology was developed. There is a danger of a book of this type becoming overlong and trying to cover too many areas at insufficient depth. References to further reading are included at the end of each chapter, and it is hoped that the interested reader will find sufficient information in this book to at least get started.

Contributors

Principal Author

Andrew W. Wood
Brain and Psychological Sciences Research Center
Swinburne University of Technology
Hawthorn, Victoria, Australia

Assisted by

Joseph Ciorciari
Brain and Psychological Sciences Research Center
Swinburne University of Technology
Hawthorn, Victoria, Australia

Contributors

Peter C. Alabaster
Faculty of Engineering and Industrial Sciences
Swinburne University of Technology
Hawthorn, Victoria, Australia

Anthony Bartel
Faculty of Life and Social Sciences
Swinburne University of Technology
Hawthorn, Victoria, Australia

Peter J. Cadusch
Faculty of Engineering and Industrial Sciences
Swinburne University of Technology
Hawthorn, Victoria, Australia

David P. Crewther
Brain and Psychological Sciences Research Center
Swinburne University of Technology
Hawthorn, Victoria, Australia

Per Line
Faculty of Life and Social Sciences
Swinburne University of Technology
Hawthorn, Victoria, Australia

John Patterson
Faculty of Life and Social Sciences
Swinburne University of Technology
Hawthorn, Victoria, Australia

Mark A. Schier
Faculty of Life and Social Sciences
Swinburne University of Technology
Hawthorn, Victoria, Australia

Bruce R. Thompson
Department of Respiratory Medicine
Alfred Hospital
Prahran, Victoria, Australia

Part I

Introduction

1

Introduction to Physiological Systems

Andrew W. Wood

CONTENTS

1.1 Cell

The basic unit of biological tissue is the cell. The boundary of the cell is the plasma (or cell) membrane (see Figure 1.1), consisting of a bilayer mainly of phospholipids (which are types of fat) with embedded proteins (giving the cell surface recognition and permeability control properties, which will be explained further in Chapters 3 and 4). The central region is the nucleus (also surrounded by a lipid bilayer membrane), which, in humans, contains the genetic "blueprint" consisting of some 30,000 genes organized into 46 chromosomes. The rest of the cell (the cytoplasm) is far from being a simple fluid—it is a matrix of microtubules, filaments, and cavernous structures (the latter is known as the endoplasmic reticulum, or ER). There are also discrete structures (organelles) within the cytoplasm, such as mitochondria (the "powerhouse" of the cell), lysosomes, vacuoles, ribosomes, Golgi complex, and secretory granules. The nucleus is also associated with characteristic organelles, such as centrosomes and nucleolus.

Cells attach to each other in a variety of ways (see Chapter 5), but the "glue" is essentially a group of specific molecules (cell adhesion molecules, or CAMs) located on the cell surface. Certain enzymes can break down these CAMs, leading to disaggregation. These disaggregated cells can be grown in specialized nutrient fluids ("media"), and the ability to do this is the basis of emerging tissue engineering technologies. Cells can attach to surfaces particularly if these surfaces are negatively charged and "wettable." New instruments such as atomic force microscopes and laser tweezers (see Chapter 27) can be used to measure adhesion forces. The cell membrane is remarkably tough (see Figure 1.2), with typical values of elastic (Young's) modulus of 10^3 Pa.

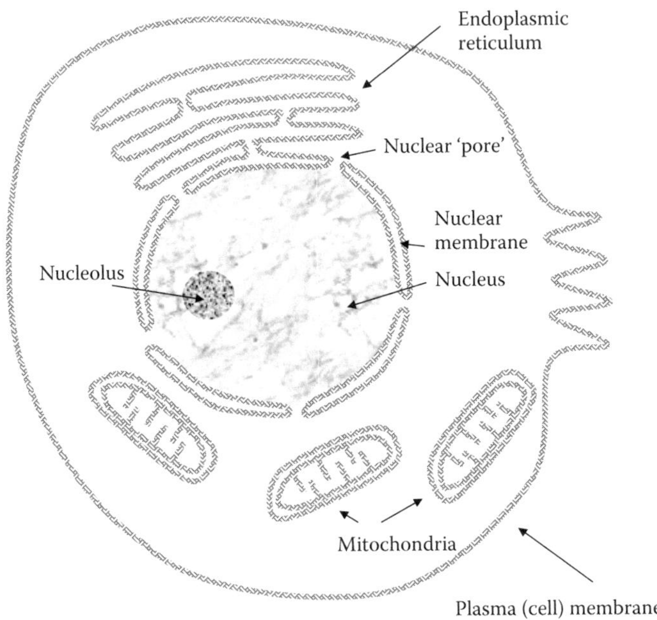

FIGURE 1.1
A very diagrammatic representation of a single cell. The cell is typically 10 μm–20 μm in diameter, with the plasma membrane forming the outer border. Parts of this can be formed into finger-like projections or "brush border." The mitochondria are a few micrometers long and consist of outer and inner membranes, the latter forming partitions or cristae. The nucleus is surrounded by a nuclear double membrane that is covered with pores. The nucleus, which contains the chromosomes attached to filaments, also has a specific region, the nucleolus, consisting of a large accumulation of RNA. The ER is a complex interconnected cavernous space formed from membrane material, which also interconnects with the nuclear membrane. The Golgi apparatus (which is not shown) is similar in appearance to the ER but performs a different function.

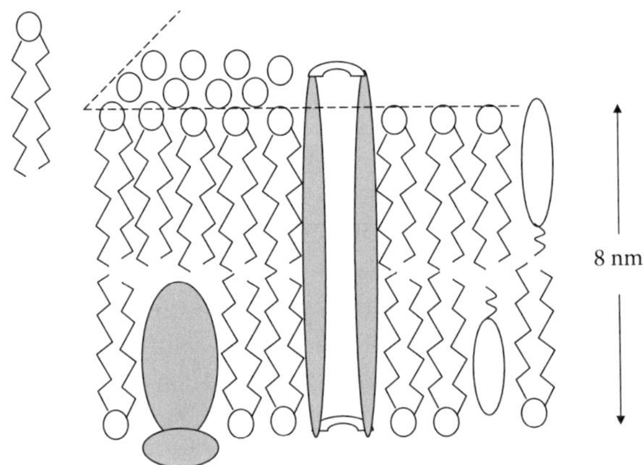

FIGURE 1.2
A phospholipid bilayer membrane. Two types of protein are shown shaded, one embedded in the lower layer and extending into the interior of the cell, the second traversing the membrane (in this case, a hollow "pore" or channel). A single phospholipid molecule is shown on the left, consisting of a polar hydrophilic head (containing the phosphate group) and two hydrophobic hydrocarbon chains, which form the interior of the membrane. Cholesterol, which has similar hydrophilic and hydrophobic parts, is also contained within the membrane (shown as an open oval) but in lower concentrations than the phospholipids (of which there are several types).

Plant cells are similar to animal cells except that their outer membranes are in contact with more structured walls consisting of cellulose in a matrix of sugars with proteins rich in amino acids such as glycine, proline, and hydroxyproline. The cytosol (or interior fluid space) of plant cells also contains chloroplasts (chlorophyll vesicles) and a large fluid-filled compartment or vacuole. Bacterial cell walls are also composed of protein structures and tend not to have nuclear membranes. Replication of cells is via duplication of genetic material, followed by each copy moving to opposite poles of the cell and then the nucleus and the cytoplasm forming two separate structures (progeny) (Figure 1.3).

Replication of viruses is somewhat different in that a virus consists only of nuclear material and needs to invade a host cell (animal, plant, fungus, or bacteria) to do this, often producing hundreds if not thousands of copies of itself.

Cultured cells grown in a plastic dish (Petri dish) often spontaneously develop into confluent monolayers (joined together in a single layer to occupy the entire surface) as they would in tissue. This property can be used in growing layers of skin cells to treat burns victims. In multicellular organisms, the main cell layers are the following: the epithelium (the skin and the lining of tubes and canals connecting to the outside world, such as the gut, the lung airspaces, and the kidney tubules), the endothelium (the tubular structures entirely contained within the organism, such as the arteries, veins, capillaries, and lymphatics), and the remainder (the connective tissue, fat, bones, nerves, and muscles). There is much interest in the therapeutic potential of stem cells because these cells (which are produced mainly in bone marrow) are known as "uncommitted cells" and have the capacity to grow into the cells determined to a great extent by their environment (nerve cells in the brain, heart muscle cells in the heart, and so on). During embryonic development, this process of differentiation

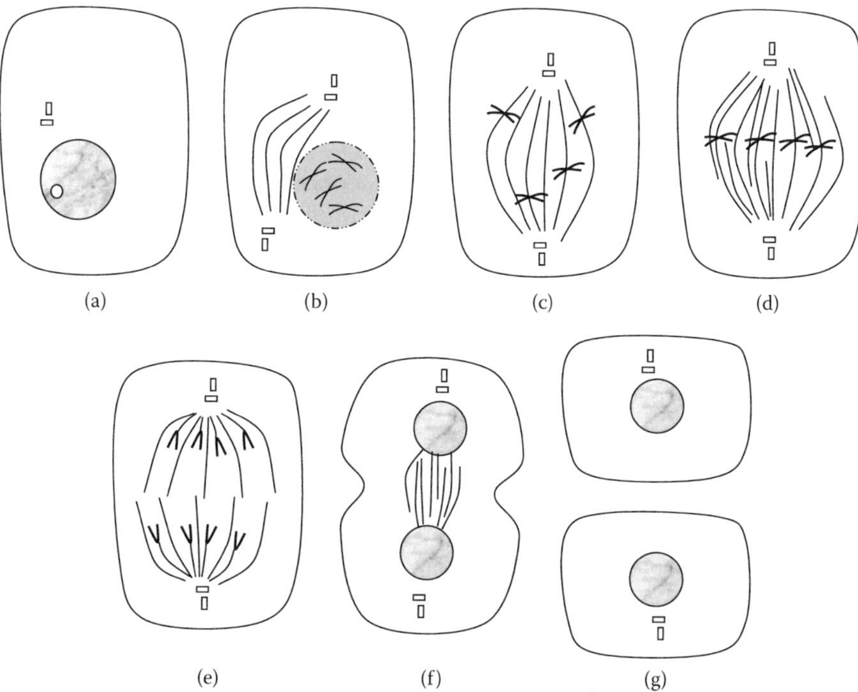

FIGURE 1.3
Mitosis. Phases: (a) Interphase; (b) early prophase; (c) late prophase; (d) metaphase; (e) anaphase; (f) telophase; (g) progeny cells following cytokinesis.

of stem cells into various forms follows a plan that is hardly understood, but which stem cell technology aims to emulate. The differentiation of cells into discrete organs during fetal life is an astounding process that cannot be described in detail here but is covered in greater detail in the references listed in the further reading section at the end of this chapter.

1.2 Systems of the Human Body

As separate organs develop during fetal life, so do the separate functions of the body, some (such as respiration and circulation) remaining in a seminonfunctional state until birth. In this book, the various systems will be studied in the following order: nerves, muscles, the heart and circulation, respiration, kidneys and other major organs, and the brain and the central nervous system (CNS). The essence of a systems approach is that each system can be studied as an entity, even though in reality, the systems are interdependent. To an extent, this arises from the development of classical physiology, in which isolated organs or "preparations" (such as nerves, muscles, and the heart) were studied in the laboratory. Much of the native function (such as conduction of impulses, contraction, and the spontaneous heartbeat) is preserved in these isolated preparations. In other experiments, organ systems have been effectively isolated within the body of an experimental animal. An example is the classic Startling "heart–lung preparation," wherein the heart was allowed to continue to beat (supported by the lungs, on a respirator) but the blood was diverted to an external circuit and all nervous connections to the heart were severed. In fact, much of what we know about the basic mechanisms of organ function have been inferred from experiments in which modifications of the structure (e.g., cutting of nerves, tying-off of blood vessels) or the addition of agents (e.g., hormones, inhibitors, electric stimuli) has led to changes in function. There follows a description, in the crudest of terms, of the various physiological systems described in this book, to enable those coming from a physical sciences background to move quickly onto the subsequent chapters. Nevertheless, such readers are strongly advised to consult the further reading section references at the end of this chapter.

1.2.1 Nerves

A nerve cell has an elongated cylindrical region of membrane (or axon) which can be up to 0.5 m in length and is typically 20 μm in diameter (a micrometer is a thousandth of a millimeter). Information is carried as a *nervous impulse*, which is a temporary (a millisecond or so) reversal in voltage (or *potential*) across the membrane. This potential reversal is propagated along the axon at a constant speed of several meters per second. At the axon terminal, the potential reversal causes the release of specific chemicals (*transmitters*) that are then able to interact with specific receptors on adjacent nerve or muscle cells, causing a potential reversal in them if certain conditions are satisfied.

1.2.2 Muscles

These are specialized cells containing two types of aligned interdigitating contractile proteins (*actin* and *myosin*) that have some of the properties of a linear motor, that is, they move in a preferred direction relative to each other, given a source of energy (in this case, adenosine triphosphate, or ATP) and an environment rich in calcium ions. This preferred movement is the basis for cell (and organ) contraction or the development of tension (if shortening is prevented). The environment of actin and myosin can change from calcium depleted (muscle relaxed) to calcium rich (muscle contracted) through the electrical stimulation of the nerve or the addition of chemical transmitters to receptors on the muscle cell surface (which is the normal method for muscle activation).

1.2.3 Heart and Circulation

The heart is basically a pulsatile pump. In humans, there are in fact two pumps in parallel, supplying the lungs and the rest of the body, respectively. In fluid mechanical terms, the pressure generated by the pumps is determined by the resistance of the circuit given that the flows produced by both pumps must be the same. A fluid mechanical version of Ohm's law can be applied. The resistance of small blood vessels is determined by blood viscosity and $(1/r^4)$, where r is the vessel radius. Inasmuch as the capillaries represent a large number of resistances in parallel, their overall resistance is actually rather less than that of the small arteries (*arterioles*), which do the bulk of the regulation of the peripheral resistance, via nerve connections to circular muscles surrounding these vessels.

The pumping action of the heart is brought about by a rhythmic contraction of the muscle that the heart consists of (cardiac muscle), commencing near the top and then moving down to the bottom in a synchronized manner. This rhythmic contraction is intrinsic to heart muscle cells, and disaggregated cells will continue to beat in isolation. The contraction is orchestrated by the "pacemaker" region of the heart, and alterations in the heart rate are brought about by the influence of nervous inputs from higher centers.

1.2.4 Respiration

The lungs fill and expel air because of the contractions of muscles in the chest wall and the diaphragm. Figure 1.4 shows a crude analog of the way the lungs inflate and deflate. The nervous impulses driving the rhythmic breathing movements are generated in a region of the brainstem (unlike in the heart, where the rhythm is endogenous) and will cease in the case of severe brain injury. The transfer of oxygen into and carbon dioxide from the blood is brought about by distributing the air over a large surface area in the numerous air sacs (*alveoli*) in which there is only a 2-μm diffusion path length from the air/cell interface and the blood contained in alveolar capillaries.

1.2.5 Kidneys

The kidneys (or renal system) essentially maintain the salt concentrations in the blood and other tissue fluid at their correct level. They do this via a remarkable combination of filtration, reabsorption (of essential materials), and secretion (of noxious substances). Around 99% of water that is allowed through the filter is reabsorbed.

1.2.6 Brain and Central Nervous System

The brain can be considered the control center of the body. It is responsible for memory, perception, thoughts (cognition), emotions, and behavior, and it consists of 10^{11} nerve cells (neurons) and 10^{13} connections. The central nervous system (CNS) is comprised of the brain and the spinal cord. The nervous system rapidly responds (at the millisecond level) to changes in the body's environment and to various activities via nerve impulses. These

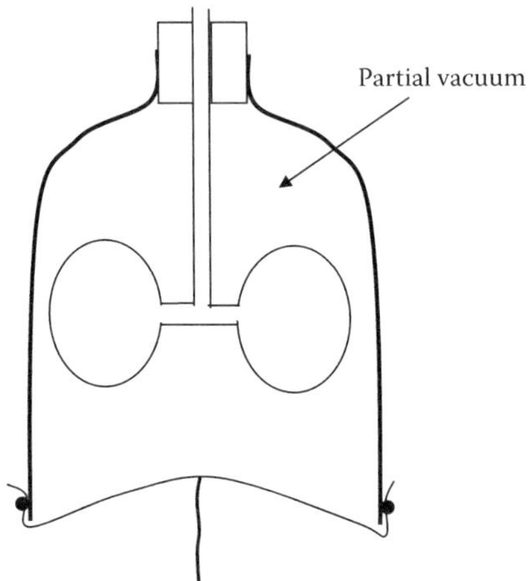

FIGURE 1.4
A representation of the mechanics of breathing. A large rigid jar has its bottom removed and a T-shaped tube attached to two balloons as shown. A rubber diaphragm is stretched over the gap at the bottom while expelling some of the air from the space. The two balloons inflate and deflate as the string attached to the rubber diaphragm is pulled down or allowed to return to its original position. Respiratory movements can be thought of essentially as muscles in the chest causing the chest diaphragm to move down. The elastic nature of the thoracic tissue causes the restoration during exhalation.

networks of cells have the capacity to "learn" through the process of neuroplasticity. These synaptic changes lead to the formation of memory.

1.3 Bioenergetics of Metabolism and Movement

In many physiological phenomena, we need to distinguish between those processes that require a source of energy (such as the action of picking up a 10-kg weight) and those that do not (such as letting an arm fall loosely by one's side). A cell can pump ions such as Na^+ and K^+ against a concentration gradient and also against an electrical potential difference that would tend to move the ions in the opposite direction. Both the difference in concentration and the difference in potential are associated with differences in energy, and this difference can be equated to the amount required from an energy source. Ultimately, this energy is derived from the food we eat and the oxygen in the air we breathe undergoing chemical reactions known as metabolism. The energy content (or calorific value) of foods is supplied on the packaging as kilojoules per serving or per 100 g. Another unit, the Calorie, is related to the kilojoules as 1 Cal = 4.2 kJ (here the *Calorie*, with a capital C, is distinguished from the *calorie*, small *c*, where the former is 1000 times the latter). A more useful measure is the number of kilojoules per mole (1 mole or mol is the molecular weight (MW) of the pure substance in grams). For example, 1 mol of glucose (which, having the molecular formula of $C_6H_{12}O_6$ and thus a MW of 72 + 12 + 96 = 180, would weigh 180 g) has a calorific value of 1700 kJ. The cells of the body are able to metabolize glucose (as noted below) in a process known as respiration:

$$C_6H_{12}O_6 + 6O_2 = 6CO_2 + 6H_2O\ (+\ 1700\ \text{kJ}).$$

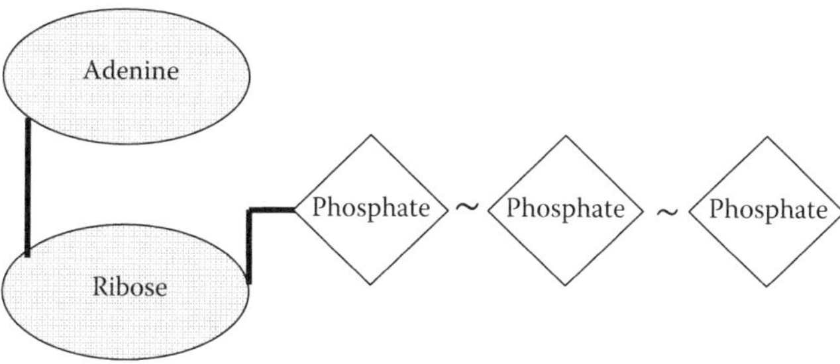

FIGURE 1.5
Adenosine triphosphate has three phosphate radicals, the last two of which are joined to the molecule by so-called high-energy bonds (shown as ~). When the last bond is broken (when ATP combines with water to form ADP), around 36 kJ per mole of ATP becomes available to cause changes in, for example, the structure of proteins involved in the catalysis of this reaction (ATPase). These ATPases have various forms, but in membranes, they are associated with specialized channels that pump Na^+ and K^+ against their electrochemical gradients. ATP synthases carry out the reverse function, using the energy from chemical gradients to turn ADP into ATP.

The liberated energy is available, at least partially, to power the energy-requiring (endergonic) processes within the cell. For example, as will be discussed in Chapter 4, the energy required to pump 3 mol of Na^+ and 2 mol of K^+ across a cell membrane is around 30 kJ, which is similar to the 29 kJ/mol liberated in the following reaction:

$$\text{ATP} = \text{adenosine diphosphate (ADP)} + \text{inorganic phosphate } (P_i).$$

This reaction is shown diagrammatically in Figure 1.5. This is an extremely important reaction and underlies many of the energy-requiring processes in biology, such as active ion transport, muscle contraction, and so on. In fact, ion transport is estimated to consume approximately 80% of metabolic energy during sleep, with much of the remainder appearing as the heat required to maintain body temperature at 37°C, which is around 17°C above ambient temperature. This basal metabolic rate (BMR) has been extensively studied and has median values of around 60 W for adults.

1.4 Homeostasis and Control

One of the remarkable properties of the human body is its ability to maintain equilibrium, despite changes in the external environment (humans can function adequately in climates ranging from arctic to equatorial and from the depths of seas to the highest mountaintops and to the surface of the moon). The regulation of body temperature (thermoregulation), blood pressure, blood salt (NaCl), and sugar (glucose) content, to name a few, tends to be tightly controlled within highly defined limits and in response to particular demands. Many of these parameters are controlled via a feedback mechanism, mostly negative feedback, although some mechanisms are more complicated, involving the interaction of many components.

A simple negative feedback mechanism is often used to regain homeostatic balance, as evidenced in the response of the pupil of the eye to light. The retina works best within a narrow range of light intensities, so the pupil diameter tends to get smaller as light levels go up and larger in low light levels. This adjustment is not immediate, as the common experience of coming out of a dark cinema into bright sunshine attests (see Figure 1.6).

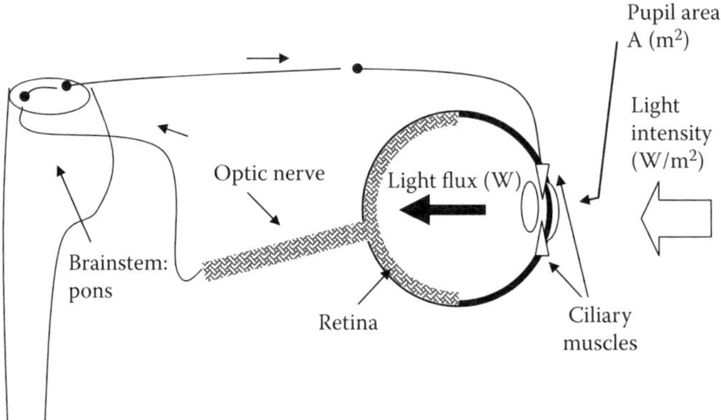

FIGURE 1.6
The pupillary reflex of the eye. The area A of the pupil determines how much light falls on the retina inasmuch as the light flux is the product of the area and the light intensity (in W/m^2) falling on the pupil. As the light flux on the retina (measured in W) increases, information in the form of nervous impulse rate is transferred to the brainstem. In a specialized region (the Edinger–Westphal nucleus), this information serves to increase the rate of nervous impulses to the ciliary muscles, which constrict the pupil as they contract (inasmuch as this consists of a circular or sphincter muscle). This in turn reduces the light flux by reducing A. Light flux on the retina is maintained at an optimal level.

The essential nature of negative feedback is that the value of a certain parameter (in this case, the number of rate of light energy (measured in watts) impinging on the retina) is monitored, and a certain control step (in this case, the diameter of the pupil) is brought into play such that the value of the parameter is either decreased or increased to coincide with a target value. This is expanded in Chapter 25.

1.5 Estimation

It is important, when dealing with biophysical or physiological phenomena, to be able to have in mind an approximate value of physical parameters in order to avoid errors in numerical analysis, particularly in view of the mixed units of constants and other coefficients obtained from the literature. For example, knowing that the human heart is roughly the size of a closed fist, then measuring the amount of water displaced from a filled container when the fist is plunged in gives a total heart volume of 400 mL. The heart consists of four chambers, so ignoring the amount of muscle (which can be done because two of the chambers are filled when the other two are empty) gives a maximum volume of 100 mL for each chamber. If the chamber were to empty 75% each beat, then 75 mL is expelled per beat, or 75×60 mL/min, given that the heart beats about once per second. This indicates that the flow rate into the major artery leaving the heart is around 4.5 L/min. In fact, this amount is nearer to 5–6 L/min in a resting human, which is coincidentally numerically similar to the volume of the blood in the circulation (blood volume, 6 L) and also the amount of air taken into the lungs in a minute (minute inspired volume).

The DuBois formula is useful for estimating body surface area (BSA), knowing a person's height and weight. The body mass index or BMI is important in evaluating whether a person is under- or overweight. A target BMI is in the range of 20–25, but this varies with age. The formulas are as follows:

$$\text{BSA (m)}^2 = 0.20247 \times \text{height (m)}^{0.725} \times \text{weight (kg)}^{0.425}$$

$$\text{BMI (kg/m}^2) = \text{weight (kg)/height (m)}^2.$$

Finally, those coming in to physiology from the physical sciences and used to being able to make precise measurements often are initially unable to appreciate the concept to biological variability. Normal or standard medical and biological parameters are usually quoted as being within a range, in other words, a mean plus or minus a standard error. If one thinks of the heights of individuals within a lecture group of, say, 70, then values may range from 1.5 m or less to more than 2 m. If the number of people with height in a certain range (say, 1.7 m–1.75 m tall, a range of 0.05 m) is taken, then the greatest number per range will be around 1.75 m, and, there will be progressively less in the ranges on either side. This bell-shaped or Gaussian distribution of heights is fairly common for biological measurements. The standard deviation (SD) is the average difference squared from the mean value. It is commonly denoted as σ. For a Gaussian distribution, 68% of readings will be within 1 SD from the mean, and 95.5% will be within 2 SDs. The mean value is often quoted with a "standard error of the mean" or SEM, where SEM = $\sigma/\sqrt{n}$. Thus, in a class of 100, where all except four people are in the range 1.5 m–2 m tall, then $s = 0.25/2$ and the SEM (or SE) will be $0.25/(2 \times 10) = 0.01$ m.

Tutorial Problems

1. Estimate the number of capillaries in the body, given that an average capillary is 20 μm in diameter, the blood velocity in them is approximately 2 mm/s, and blood is ejected from the heart at 6 L/min (100 mL/s).
2. Estimate the number of muscle fibers in the upper arm by measuring the circumference and then assuming that each fiber is 10 μm in radius and that muscle makes up 80% of the cross-sectional area of the upper arm.

Bibliography

Boron WF, Boulpaep EL. 2009. *Medical Physiology*, 2nd Edition. Elsevier Saunders, Philadelphia, PA, USA.

Hall JE. 2011. *Guyton and Hall Textbook of Medical Physiology*, 12th Edition. Elsevier-Saunders, Philadelphia, PA, USA.

Johnson LR. 2004. *Essential Medical Physiology*, 3rd Edition. Elsevier Academic Press, Waltham, MA. USA.

Marieb EN, Hoehn K. 2010. *Human Anatomy & Physiology*, 8th Edition. Pearson Benjamin Cummings, San Francisco, CA, USA.
Martini FH, Ober WC. 2005. *Fundamentals of Anatomy & Physiology*, 7th Edition. Prentice Hall, Upper Saddle River, NJ, USA.
Sherwood L. 2009. *Human Physiology: from Cells to Systems*, 7th Edition. Thomson Brooks/Cole, Belmont, CA, USA.

Second-hand copies of older texts can be obtained at modest cost from on-line sources. Look for:

Davson H, Segal MB. 1975–1976. *Introduction to Physiology*, Vols. 1–5. Academic Press, London with Grune & Stratton, New York.
Schmidt RF, Thews G. 1989. *Human Physiology*, 2nd Edition, Springer-Verlag, Berlin.

2

Fundamentals of Electrical Circuits for Biomedicine

Anthony Bartel and Peter C. Alabaster

CONTENTS

2.1 Amplifiers

2.1.1 What Is an Amplifier?

An amplifier is an electronic device that is used to increase the magnitude of an electrical signal. It can generate larger voltages and currents at its output than those applied to its input.

2.1.2 Amplifier Applications

Audio amplification: An audio amplifier receives a small input signal of millivolts from a microphone and increases this to volts, which may then be used to drive speakers.

Biomedical signal amplification: Attaching leads to a human head allows researchers to detect and measure small signals generated by the brain. These signals have an amplitude of a few microvolts. An amplifier increases the amplitude to volts, which can then be used as an input to oscilloscopes, chart recorders, or other digital displays.

Chemical potential amplification: A pH electrode in a solution generates a direct current (DC) potential of a few hundred millivolts. An amplifier increases this to volts. The magnitude of the signal can then be displayed on a digital multimeter or other display.

2.1.3 Construction

Amplifiers can be made from components such as transistors, field effect transistors, or valves. They can be made from operational amplifier chips. Operational amplifiers are circuits that require the attachment of no more than two external resistors to create a simple voltage amplifier circuit. These will be discussed later in the chapter.

2.1.4 Power Supply

An amplifier must be supplied with an external source of electrical power to enable it to generate output power greater than the power supplied to the input. This may be any source of DC power such as a battery.

2.1.5 The "Black Box" Model of an Amplifier

Like most electronic devices, an amplifier can be used without the user having any idea of its internal construction. It can be described as a "black box" with certain properties. Consider the amplifier "black box" to have two input terminals and two output terminals (Figure 2.1).

There are four different types of amplifier:

1. Voltage amplifier: These have fixed voltage amplification.
2. Current amplifier: These have fixed current amplification.
3. Transconductance amplifiers: These have a fixed relationship between input voltage and output current.
4. Transresistance amplifiers: These have a fixed relationship between input current and output voltage.

FIGURE 2.1
The amplifier "black box."

2.1.6 Voltage Amplifiers

Most amplifiers are voltage amplifiers. This will be the only type considered in this chapter. Such an amplifier can be specified by its voltage gain or voltage amplification. It is usually given the symbol G_v or A_v, where

$$G_v = V_o/V_i.$$

V_o is the output voltage, and V_i is the (unamplified) input voltage.

For example, if an amplifier has a voltage gain of 10, it will produce an open circuit voltage 10 times larger than the input voltage. (The concepts of open circuit and closed circuit will be discussed later in the chapter.) Therefore, $G_v = 10$.

2.1.7 Inverting Amplifiers

An amplifier supplied with a positive input voltage may produce a negative output voltage (and vice versa). This is called an inverting amplifier. For example, consider an amplifier with an input of +2 V that generates an output of –6 V:

$$G_v = V_o/V_i = -6/2 = -3.$$

Note that an inverting amplifier has *negative gain.*

If an inverting amplifier has a gain of –5 and has –2 V applied to the input, the output voltage will be 10.

2.1.8 Amplification of Alternating-Current Signals

The input signal to an amplifier may be an alternating voltage. In this case, we cannot specify the gain by a simple number. We need two quantities. The gain can be described in *polar form* by the change in magnitude of a signal and the change in phase of a signal.

For example, a gain of 8/60° indicates that the amplitude of the output voltage is eight times that of the input, and the phase of the output leads the input by 60°.

Note that the magnitude and phase response of an amplifier are functions of frequency. For example, an amplifier may have a gain of 20 at a frequency of 100 Hz and have a gain of only 3 at a frequency of 10 kHz. The phase response of this amplifier may be 0° at 100 Hz and –90° at 10 kHz.

The gain can also be described in *Cartesian form* using two components: A component that is in phase with the input wave (this is specified by a real number) and a component in quadrature (at 90° to the input wave) (this is specified by an imaginary number).

For example, a gain of $3 + 4\,j$ (where $j = \sqrt{-1}$) specifies an output wave made up from two components. These are as follows: (1) a component in phase with the input that has three times the magnitude of the input and (2) a quadrature component with four times the amplitude of the input signal.

The Cartesian form is preferred because both the quantities that are needed to describe the amplifier are present in a single complex number. The operation of the amplifier can be described over a range of frequencies by a single complex function of frequency. If the amplifier is described using polar form, then two separate functions of frequency are required; one for phase and one for gain.

2.1.9 Gain Expressed in Decibels

The magnitude of gain is often expressed in decibels. The symbol for the decibel is *dB*.

$$\text{Gain in dB} = 20 \log_{10} (V_o/V_i).$$

Consider an amplifier with a gain of 8/60°. The phase information is ignored, and only the change in magnitude is considered. $V_o/V_i = 8$; therefore, the gain in dB is

$$20 \log 8 = 18 \text{ dB}.$$

2.1.10 Attenuation

If the magnitude of an output signal is less than the magnitude of the corresponding input signal, the signal is attenuated. If an amplifier has a gain of less than 1, it is an attenuator.

Consider an amplifier with an input of 5 V and an output of 0.4 V.

$$G_v = V_o/V_i = 0.4/5 = 0.08.$$

This is expressed in decibels as

$$20 \log 0.08 = -21.9 \text{ dB}.$$

Note that if the signal is attenuated, the gain in decibels will be negative.

2.1.11 Amplifier Output

The output of most amplifiers is taken from one terminal and is measured with respect to earth or ground potential. The triangle is often used to symbolize the amplifier (Figure 2.2).

2.1.12 Types of Input

An amplifier might have a single input or alternatively have two input terminals.

2.1.12.1 *Single-Input Types*

The input signal is applied between the amplifier input and the ground or earth potential (Figure 2.3).

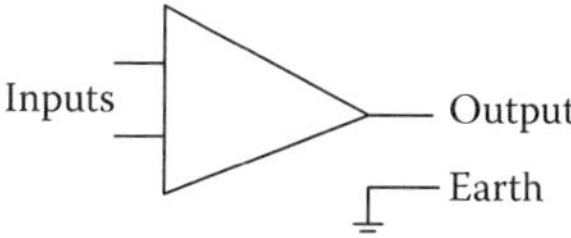

FIGURE 2.2
Amplifier diagram with the triangular amplifier symbol.

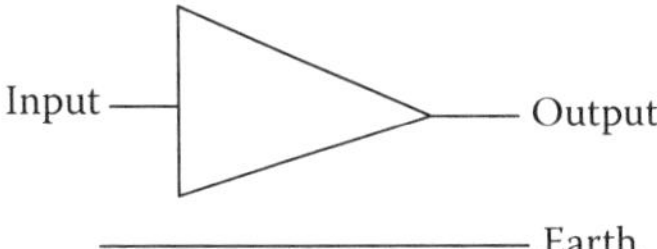

FIGURE 2.3
Amplifier with signal applied between the input and Earth.

2.1.12.2 Two Input Terminals (Differential Input)

The signal to be amplified is applied between the two input terminals. One input is called the *inverting input* and is given the symbol V_-. The other is called the *noninverting input* and is referred to by the symbol V_+.

Note that the positive and negative signs used with the symbols indicate which input is inverting and which is noninverting. They do not refer to the polarity of voltages that can be applied to the inputs. Both input terminals can have positive or negative voltages applied.

The gain of the amplifier is given by

$$G_v = V_o/(V_+ - V_-).$$

Consider an amplifier with

$$V_+ = 2\ V$$
$$V_- = 4\ V$$
$$V_o = 10\ V.$$

Therefore, the gain is –5.

The amplifier that amplifies the potential difference between the two inputs is called a *differential amplifier.*

Note that the word "differential" is used in instrumentation and engineering to indicate a subtraction or differencing process. This is different from the meaning in calculus where it refers to the rate of change.

2.1.13 Common-Mode Input

Signals that are common to both inputs are called *common-mode signals.* For example, if V_+ is 7 V and V_- is 5 V, then the differential input is $V_+ - V_- = 2$ V.

The common-mode input voltage is the average value of the two input signals. In this example, $(7 + 5)/2 = 6$ V.

2.1.14 Amplifier Circuits

An amplifier is used to amplify small voltages from a signal source connected to an amplifier input and produce higher-voltage signals across a load connected to the amplifier output (Figure 2.4).

FIGURE 2.4
Amplifier schematic diagram with input signal source and output load.

2.1.15 Signal Source Model

Electrical signals may be generated in many ways: mechanically, chemically, thermally, or optically. Signals can originate from sources that are complex, but all sources can be described by a simple circuit model. The model has only two components: an internal source voltage V_s in series with an output resistance R_s (Figure 2.5).

The characteristics of the source voltage depend on the type of source, such as a pH electrode, a nerve, or a thermocouple.

The source resistance R_s also depends on the type of source. The resistance of a pH electrode is of the order of 100 MΩ, a nerve potential has a source resistance of the order of 100 Ω, and a thermocouple has a source resistance of around 0.001 Ω.

When current is drawn from a source, some of the source voltage is dropped across the source resistance. The output voltage from the source V_i will therefore be less than the source voltage V_s.

2.1.16 Amplifier Model

An amplifier can be modeled using only three components (Figure 2.6):

An input impedance or input resistance R_i: The input circuit is modeled as a single impedance (usually resistance) connected across the input terminals.

An output impedance or resistance R_o: The output circuit of an amplifier is modeled as a voltage source V_a in series with a single impedance (usually resistance).

Open circuit voltage gain G_{vo}: The amplifier is modeled as a voltage source V_a, which can be determined by multiplying the input voltage V_i by the open circuit voltage gain G_{vo} of the amplifier.

2.1.17 Load

The load connected to the output of an amplifier may be a meter, a speaker, another amplifier, an analogue-to-digital converter, or other devices. Any load can be modeled as a single impedance (usually resistance) connected to the output of the amplifier (Figure 2.7).

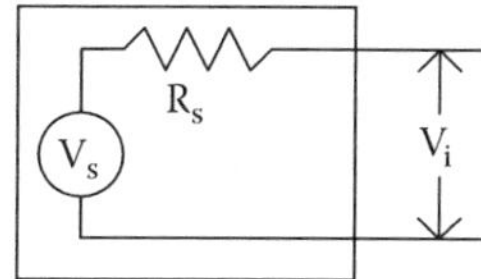

FIGURE 2.5
A signal source model.

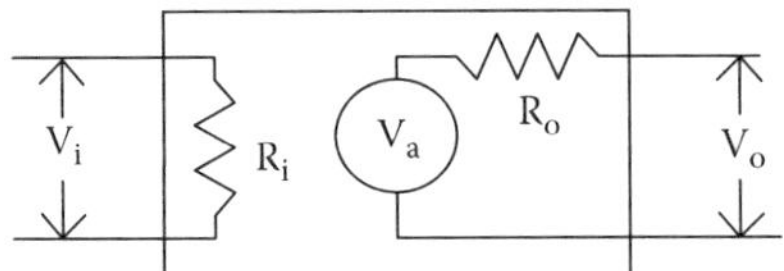

FIGURE 2.6
An amplifier model.

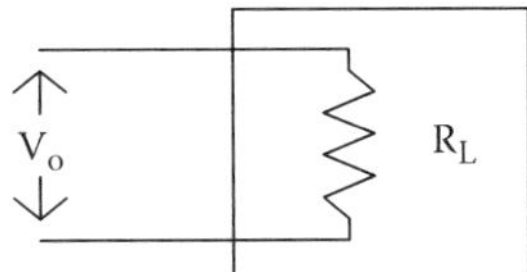

FIGURE 2.7
A model of the amplifier load.

When the amplifier is connected to the signal source, the signal source will experience a load according to the value of R_L.

2.1.18 Complete Model

When the amplifier is connected to the signal source, the signal source will experience a load according to the value of R_i just as the amplifier experiences the load R_L (Figure 2.8).

2.1.19 Gain–Impedance Relationship

A voltage amplifier also amplifies input current and input power, but the amplification of current and power is not fixed. These depend on both the input resistance of the amplifier (which is fixed for a particular amplifier) and the load resistance (which depends on the load connected to the amplifier output).

The voltage gain, current gain, input impedance, and output impedance of an amplifier are related by the gain–impedance equation.

At the amplifier input,

$$V_i = I_i R_i.$$

At the amplifier output,

$$V_o = I_o R_L.$$

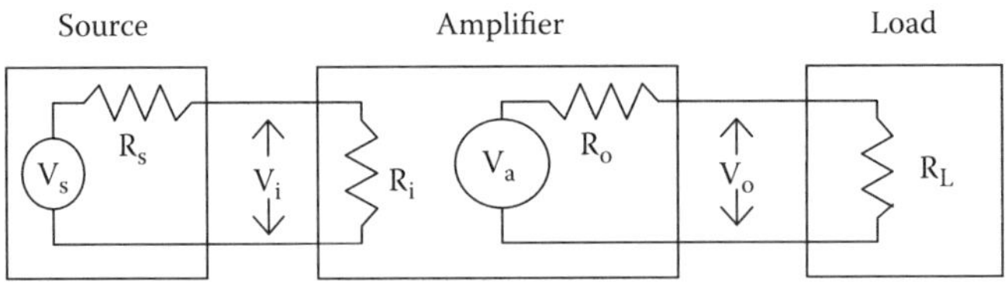

Source Amplifier Load

FIGURE 2.8
The complete amplifier model with input source and output load.

Dividing the output equation by the input equation gives

$$V_o/V_i = I_oR_L/I_iR_i.$$

This is the *gain–impedance relationship*. It states that

$$\text{Voltage gain} = \text{Current gain} \times \text{Impedance gain}.$$

2.1.20 Circuit Analysis

Consider the following: a signal source with a voltage of 1.4 V and a resistance of 6800 Ω is connected to an amplifier. The amplifier has an input resistance of 10 kΩ, an open circuit voltage gain of 4.3, and an output resistance of 2.2 kΩ. A load resistance of 3.3 kΩ is connected to the amplifier output.

1. Determine the total resistance across the signal source. This is the series combination of the internal source resistance R_s plus the amplifier input resistance R_i.
2. Determine the signal current flowing into the amplifier I_i. Divide the source voltage V_s by the resistance across the source.
3. Determine the signal voltage across the amplifier input terminals V_i. This is the product of the signal current flowing into the amplifier I_i by the amplifier input resistance R_i.
4. Determine the amplifier source voltage V_a. Multiply the input signal voltage V_i by the open circuit voltage gain of the amplifier G_{vo}.
5. Determine the total resistive load of the output circuit. This is the series combination of the amplifier output resistance R_o plus the load resistance R_L.
6. Determine the current flowing in the output circuit I_o. Divide the amplifier source voltage V_a by the resistance of the output circuit.
7. Determine the amplifier output voltage across the load resistance V_L. This is the product of the output current I_o and the output load resistance R_L.
8. Determine the loaded voltage gain of the amplifier G_{vL} by dividing the output signal voltage V_o by the input signal voltage V_i.
9. Determine the current gain of the amplifier G_i by dividing the output current I_o by the input current I_i.
10. Determine the input power delivered to the amplifier P_i from the product of the input voltage V_i and the input current I_i.
11. Determine the output power delivered to the load P_L from the product of the output voltage V_L and output current I_o.
12. Determine the power gain of the amplifier G_p by dividing the output power to the load by the input power to the amplifier.

2.1.21 Operational Amplifier Gain

Operational amplifiers are high-gain differential-input amplifiers. The gain of an operational amplifier is large and independent of frequency at low frequencies (the TL071 amplifier has a DC gain of 200,000). As the frequency is increased, a break point (or pole) is reached where the gain of the amplifier starts to decrease with frequency (for the TL701 amplifier, this occurs at 20 Hz). The gain continues to decrease with increasing frequency until it reaches unity gain (the TL071 amplifier has a unity gain frequency of 3 MHz). This decrease in gain with increasing frequency is often called a *gain roll-off with frequency.*

2.1.22 Bode Diagrams

The gain of an operational amplifier can be shown by a Bode diagram. The amplifier gain and frequency are both plotted on logarithmic scales. Gain is measured in decibels on a logarithmic scale. For example, a gain decrease of 10 is expressed as –20 dB, and a gain decrease of 100 is expressed as –40 dB. Frequency is measured in decades. For example, a frequency increase of 10 is expressed as +1 decade, whereas a frequency increase of 100 is expressed as +2 decades.

The advantage of this log–log representation is that changes in power law relationships between gain and frequency are shown as changes in the slope. A gain that is directly proportional to frequency produces a slope of +1 on a Bode plot. A gain that is inversely proportional to frequency produces a slope of –1 on a Bode plot (Figure 2.9). A gain that is directly proportional to frequency squared produces a slope of +2 on a Bode plot. A gain that is directly proportional to frequency cubed produces a slope of +3 on a Bode plot.

A Bode diagram also breaks up a gain curve into several straight-line segments, with slopes of zero, ±1, ±2, ±3, etc. (A slope of +1 is called +20 dB gain change per decade change in frequency. A slope of –1 is called –20 dB/decade. A slope of +2 is called +40 dB/decade. A slope of +3 is called +60 dB/decade.)

2.1.23 Operation Amplifier Circuit Gain

The very high and frequency-dependent gain of an operational amplifier makes it unsuitable for direct use. It can, however, be turned into a much more useful low-gain amplifier with a constant gain over wide band of frequencies, by using it in a feedback circuit. A feedback circuit sends back some fraction of the amplifier output to the amplifier input. This fed back signal is then added to the input signal to the amplifier circuit. A simple (inverting) operational amplifier circuit that amplifies 10 times is shown in Figure 2.10.

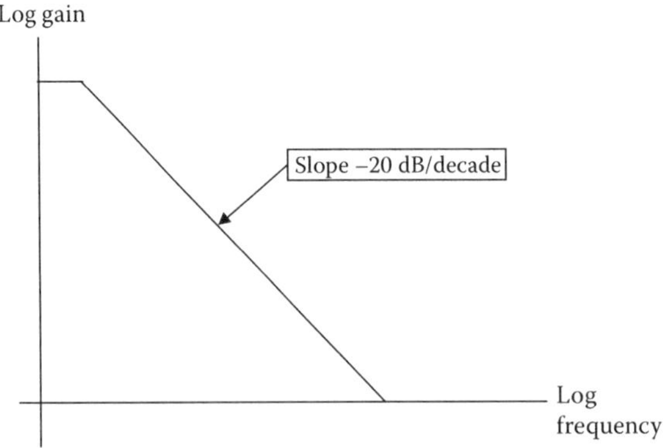

FIGURE 2.9
The frequency response of an operational amplifier.

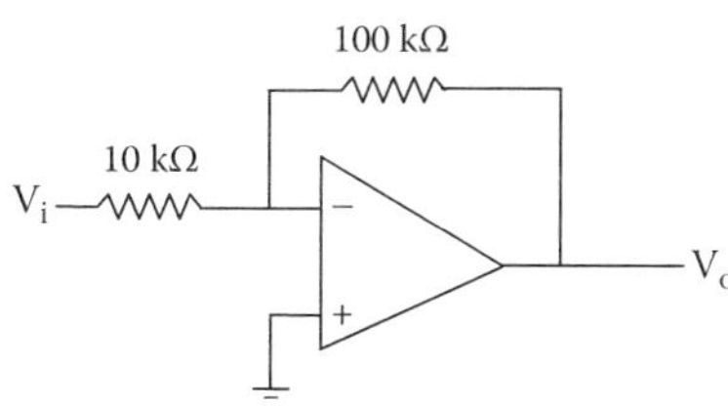

FIGURE 2.10
An op-amp circuit with a gain of –10.

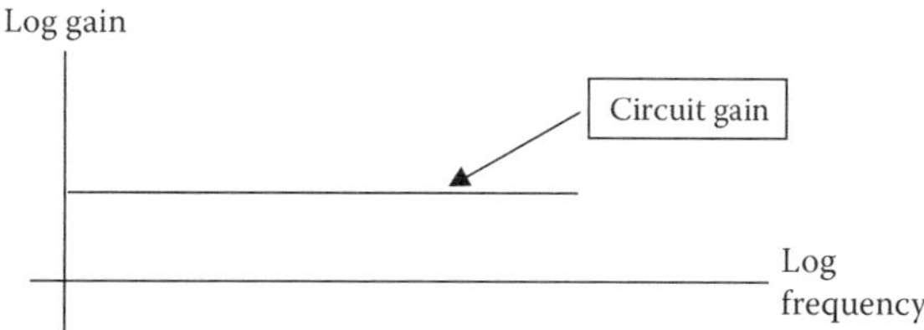

FIGURE 2.11
The frequency response of an op-amp circuit.

The circuit gain is much lower than the operational amplifier gain, and it is constant over a wide band of frequencies (Figure 2.11).

We must now distinguish between these two different types of gain. The gain of the operational amplifier is called the *open-loop gain*. The gain of the amplifying circuit is called the *closed-loop gain*. The Bode diagram in Figure 2.12 shows both of these.

We can now consider a new idea: the gain of the feedback loop. This may also be called the *control-loop gain* or the *loop gain*. It is found by dividing the operational amplifier open-loop gain by the circuit closed-loop gain.

The loop gain is easy to determine if we plot both the amplifier open-loop gain and the circuit closed-loop gain on the same Bode diagram. A Bode diagram uses a logarithmic

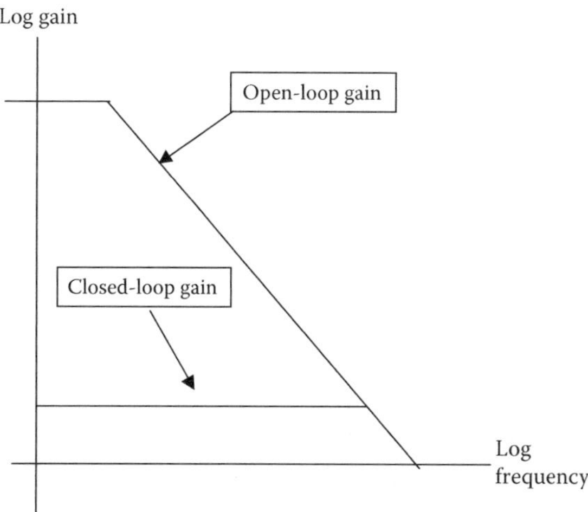

FIGURE 2.12
The frequency response of an op-amp (open loop) and op-amp circuit (closed loop).

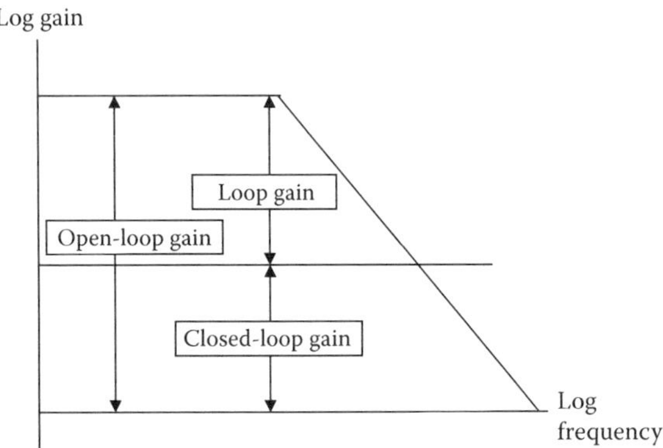

FIGURE 2.13
A frequency response showing loop gain.

scale, so division is transformed into subtraction. We can simply subtract the closed-loop gain from the open-loop gain on the diagram to determine the loop gain. This is shown in Figure 2.13.

2.1.24 Types of Feedback

Usually, the purpose of the feedback loop is to reduce the circuit closed-loop gain. The fed back signal opposes or cancels most of the input signal to the operational amplifier. This type of feedback is called *negative feedback*.

It is also possible to feed back output signals that constructively add to the input signal. This increases the closed-loop gain of the circuit, so that it becomes greater than the open-loop gain of the amplifier. This is called *positive feedback*. For example, an amplifier without feedback and with a gain of 10 could be used in a circuit with a gain of 100 by using positive feedback.

2.1.25 Extra Phase Shift

In an ideal inverting amplifier circuit, the phase difference between the input and output signals is 180°. In a practical circuit, however, there may be an additional phase shift around the feedback loop. This extra phase shift can be produced by both the operational amplifier and the feedback network.

The extra phase shift increases with frequency and adds to the 180° of the inverting amplifying circuit. If the total phase shift in this loop reaches 360° at a frequency for which the loop gain is unity, the feedback becomes effectively positive, resulting in a self-sustaining oscillation at that frequency.

2.1.26 Phase Shift Produced by a Roll-Off in the Open-Loop Gain of an Amplifier at High Frequencies

It can be shown that the phase response of an operational amplifier is determined from the rate of change of its open-loop gain with frequency.

If its gain is changing at –20 dB per decade change in frequency, the additional phase shift produced in the amplifier is –90°. The minus sign in the gain change means that the

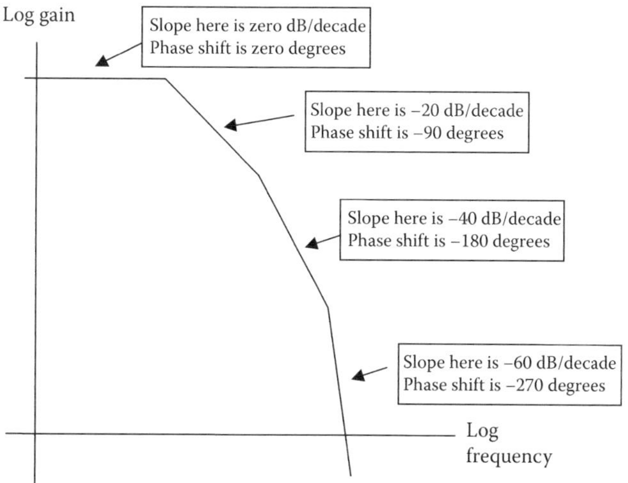

FIGURE 2.14
Phase shift produced by a roll-off in the open-loop gain of an amplifier at high frequencies.

gain decreases as the frequency increases. The minus sign in the phase indicates that the amplifier output lags the input.

If its gain is changing at –40 dB per decade, the additional phase shift produced is –180°. If the gain is changing –60 dB per decade, the additional phase shift produced is –270°. This is shown in the Bode diagram in Figure 2.14.

2.1.27 Phase Shift Produced by Frequency Response of Feedback Network

The feedback network sets the closed-loop gain of the circuit. It can also be shown that the phase response of a feedback network is determined from the rate of change of its gain with frequency.

If the amplifier closed-loop gain increases at +40 dB per decade, the phase shift in the feedback network is +180°. If the amplifier closed-loop gain increases at +20 dB per decade, the phase shift in the feedback network is +90°.

If we have an amplifier with constant closed-loop gain (a flat frequency response), the change in gain with frequency is zero; hence, the phase shift in the feedback network is also zero.

If the amplifier closed-loop gain decreases at –20 dB per decade, the phase shift in the feedback network is –90°. If the amplifier closed-loop gain decreases at –40 dB per decade, the phase shift in the feedback network is –180°. This is shown in the Bode diagram in Figure 2.15.

2.1.28 Loop Gain

We can determine the feedback or control-loop gain from Bode plots of the operational amplifier open-loop gain and the circuit closed-loop gain. This is shown in Figure 2.16.

It can be seen that in this example, the loop gain is constant at low frequencies and then rolls off at –20 dB per decade down to unity gain. The phase shift of the control loop is determined by the rate of change of loop gain with frequency. At low frequencies, the extra

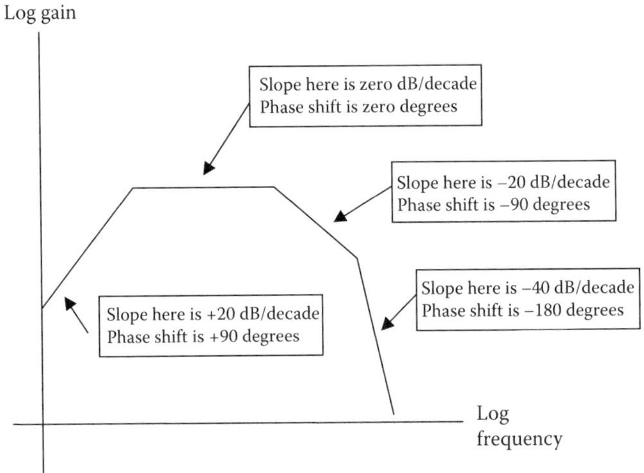

FIGURE 2.15
Phase shift produced by frequency response of the feedback network.

phase shift in the loop is zero. At frequencies greater than the break point (or pole), the loop gain decreases at –20 dB per decade. This produces an extra phase shift in the loop of –90°. The total phase shift in the loop is –270° because of the inverting input. This is not positive feedback. The system is stable and will not oscillate.

Consider what would happen if the circuit closed-loop gain has been set to a lower value. This is shown on the Bode diagram in Figure 2.17

It can be seen that in this example, the loop gain is constant at low frequencies. At frequencies above the first pole, the loop gain decreases at –20 dB per decade. This produces an extra phase shift in the loop of –90°. At frequencies above the second pole, the loop gain decreases at –40 dB per decade. This produces an extra phase shift in the loop of –180°. Positive feedback occurs because of the additional –180° produced by the inverting input. The system is unstable and will oscillate.

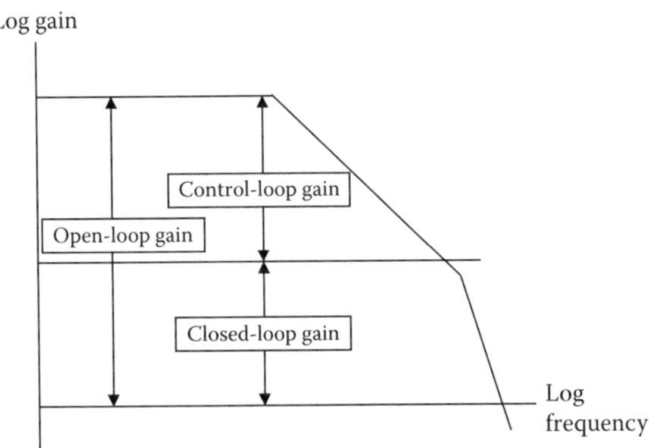

FIGURE 2.16
Control-loop gain for circuit with high closed-loop gain.

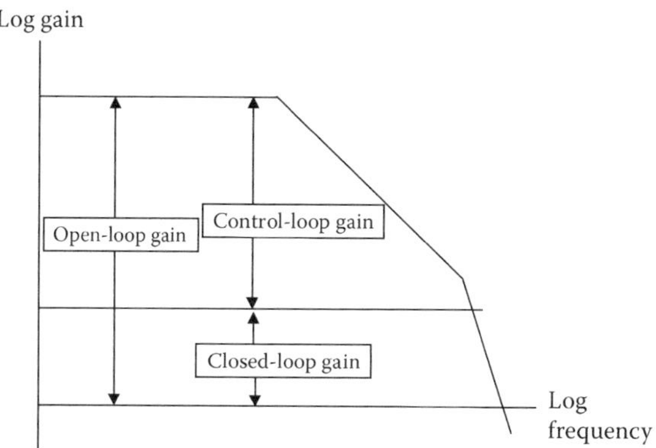

FIGURE 2.17
Control-loop gain for a circuit with lower closed-loop gain.

2.1.29 Stability Criterion

For an inverting amplifier circuit to be stable, the extra phase change in the loop must not reach 180° with a loop gain greater than 1.

We can now state this in another way. For an inverting amplifier circuit to be stable, at the point where the open-loop and closed-loop gain lines intersect on a Bode diagram, their slope difference must not be equal to or greater than 40 dB per decade.

2.1.30 Design of Operational Amplifiers

Operational amplifiers are designed to have a dominant pole (or resistor capacitor (RC) time constant) in their internal circuit, which controls their open-loop gain roll-off. The open-loop gain rolls off at –20 dB per decade down to unity gain (zero dB). This type of design is called a phase compensated amplifier. Some amplifiers are uncompensated and can have an open-loop gain roll-off of –40 dB or greater (Figure 2.18).

It can be seen that uncompensated amplifiers are stable if the circuit closed-loop gain is large enough but will oscillate if the closed-loop gain is set too low. Phase compensated amplifiers are stable for all closed-loop gain settings. Compensated operational amplifiers

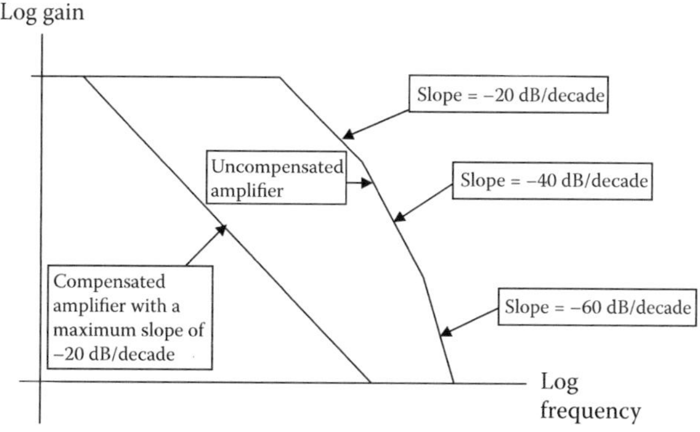

FIGURE 2.18
Frequency response of compensated and uncompensated amplifiers.

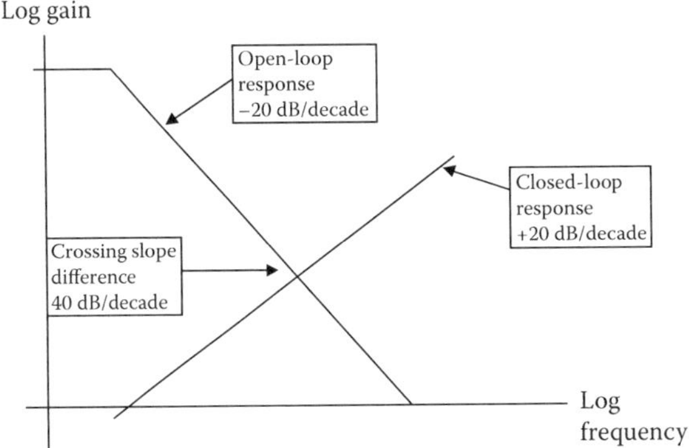

FIGURE 2.19
Frequency response of an RC differentiating circuit constructed using a phase-compensated operational amplifier.

may oscillate if extra phase shift is produced in the feedback network. Differentiating circuits are a problem; they have a +20 dB per decade increase in gain with frequency (Figure 2.19).

2.1.31 Oscillation of Amplifiers with Capacitive Loads

Operational amplifiers have an output resistance. This can be found in data books. The TL071 operational amplifier has an output resistance of about 250 Ω.

An external capacitive load connected to the amplifier output, together with the amplifier output resistance, forms a low-pass filter (Figure 2.20).

The result of this extra RC time constant (or extra pole) is an increase in the slope of the open-loop response at high frequencies and an increase in the amplifier open-loop phase shift at higher frequencies.

2.1.32 A Bode Plot Showing the Effect of the Extra Pole on the Open-Loop Gain of a Compensated Amplifier

If a 2nF load is connected to the amplifier output (Figure 2.21), then

$$\tau = RC = (2\times 10^{-9})(250) = 5\times 10^{-7}\,\text{s}$$
$$\omega = 1/\tau = 2\times 10^{6}\ \text{rad/s}$$
$$f = \omega/2\pi = 230\ \text{kHz.}$$

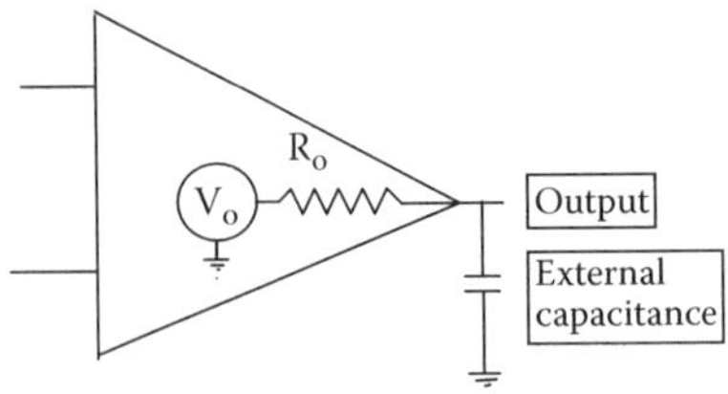

FIGURE 2.20
Capacitive load on an amplifier.

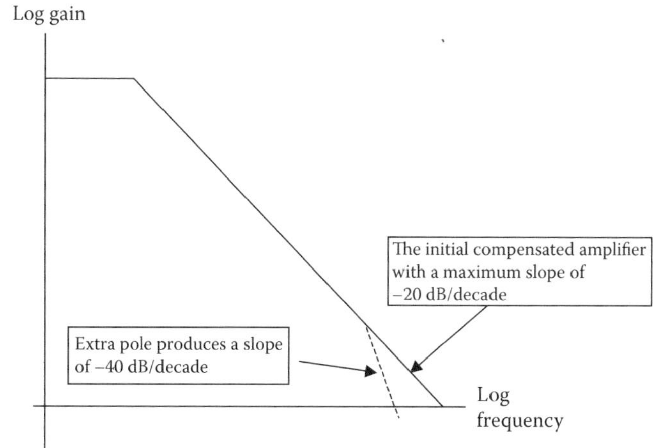

FIGURE 2.21
Frequency response showing the effect of the extra pole on the open-loop gain of a compensated amplifier.

The extra pole has caused the open-loop gain of the amplifier to roll-off at −40 dB per decade at frequencies greater than 320 kHz.

It is difficult to make a well-compensated operational amplifier oscillate in a simple inverting amplifier circuit. Differentiator circuits or high-pass filter circuits may oscillate. Uncompensated operational amplifier circuits may oscillate if the circuit closed-loop gain has been set too low.

Oscillation may also occur in equipment with feedback paths around several interconnected operational amplifiers. Each amplifier contributes to phase delay in the loop. A common example of this problem is driven right-leg electrocardiographic systems. The patient is not grounded. The common mode signal for the input instrumentation amplifier is inverted by a second amplifier and applied to the right-leg electrode. This feedback signal cancels the common mode pick-up on the patient and sets the patient common mode signal to zero for safety reasons.

The capacitive load of the patient on the inverting amplifier output is sometimes sufficient to push the system into oscillation. This oscillation is at a high frequency and will be blocked by the low-pass responses of later stages. It does, however, change the behavior of the input amplifiers and can produce artifacts in the recorded electrocardiogram. It may also be intermittent and may start or stop depending on the position of the patient and the connecting cables.

2.2 Active Filters

2.2.1 Introduction

A filter is a frequency-selective device that passes a specified band of frequencies from its input to its output. Frequencies outside this band are attenuated. Filters may be either passive or active. Passive filters are made only from resistors, capacitors, and inductors. They are not connected to an external power supply. Active filters also contain amplifiers and require external power.

There are four types of filter: low pass, high pass, band pass, and notch or band reject filters. This section describes a low-pass filter, that is, a filter that allows low frequencies to pass through it while attenuating high frequencies.

2.2.2 Passive Filters

A simple passive low-pass filter is the RC integrator circuit shown in Figure 2.22.

This can be considered as a voltage divider wherein some of the input voltage drops across the output and some is lost in the resistor. At low frequencies, the capacitor impedance is greater than the resistance, so most of the input voltage is dropped across the capacitor. At high frequencies, the capacitor impedance is small, and most of the input voltage is dropped across the resistor. The frequency at which the magnitude of the capacitive reactance equals the resistance is called the *break point* of the filter. Frequencies greater than the break point frequency are attenuated.

The break point occurs when $\frac{1}{\omega C} = R$; hence, $\omega = \frac{1}{RC}$ and $f = \frac{1}{2\pi RC}$.

For the component values in Figure 2.22, f = 1.6 kHz.

The amplitude response of this filter is shown on the Bode plot in Figure 2.23.

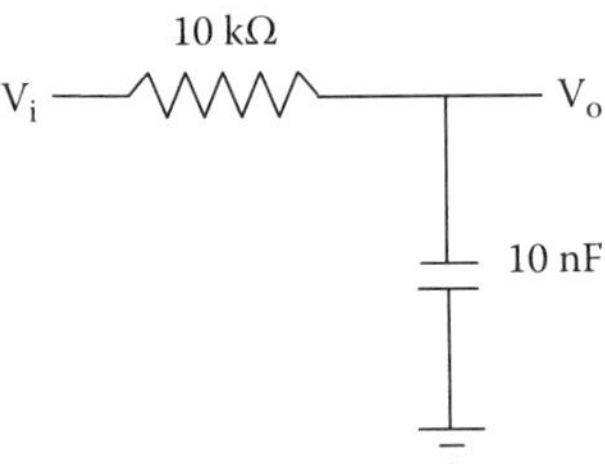

FIGURE 2.22
A low-pass passive filter.

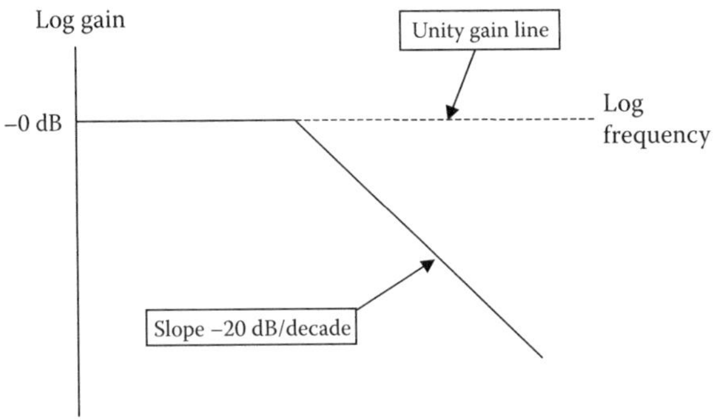

FIGURE 2.23
Frequency response of a low-pass filter.

The disadvantages of this type of passive filter are the following:

1. They can only attenuate input signals; they cannot be amplified.
2. Some passive filters require inductors. These may be large and may also pick up stray magnetic fields that inject 50-Hz signals into the filter.
3. They may load a high-impedance source connected to the filter or be loaded by a low-impedance output load to the filter output.
4. Their characteristics are hard wired into them. Their properties cannot be easily changed. They cannot be externally tuned.
5. They cannot be built with a sharp cutoff.

2.2.3 Active Filters

Active filters can be made in several different ways. These include Sallen–Key (or VCVS) filters, multiple feedback filters, state variable filters, and biquad filters. This section describes only the Sallen–Key low-pass filter. Details of other filter types, including high pass and band pass, can be found in many texts such as *Active Filters: Theory and Design* by S. Pactitis.

If we add a unity gain buffer to the previous passive filter example, we get a first-order low-pass Sallen–Key active-filter circuit, as shown in Figure 2.24.

We can easily add some gain to this design, as shown in Figure 2.25.

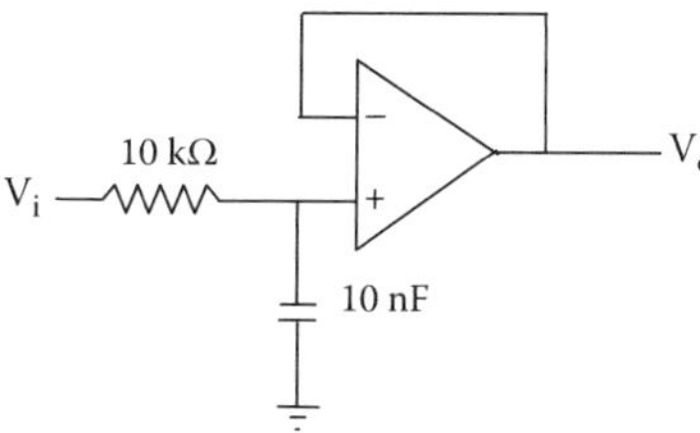

FIGURE 2.24
A low-pass active filter.

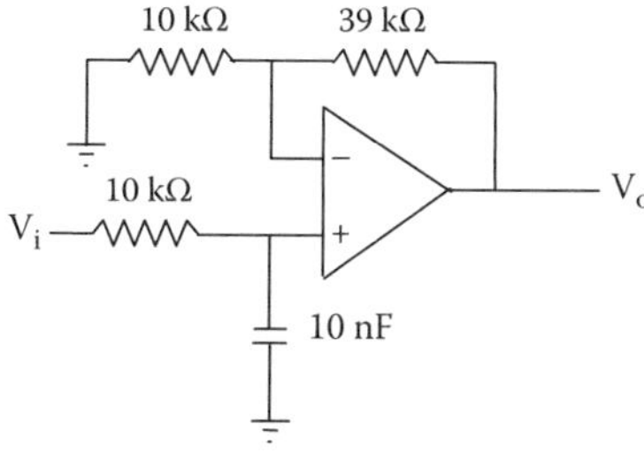

FIGURE 2.25
A low-pass filter with gain.

We now have the complete first-order low-pass Sallen–Key active-filter circuit. This circuit is noninverting and has a gain of (10 k + 39 k)/10 k = 4.9 at low frequencies. The break point is at 1.6 kHz, where the gain starts to roll-off at –20 dB/decade.

Active filters have some disadvantages. They require a power supply, and they are limited by their amplifier properties. These limitations include maximum input signal amplitude and high-frequency response. They also need high-tolerance low-drift components in some designs.

2.2.4 Theory

The first-order filter described is an example of a first-order system, a system described by a first-order differential equation. These systems are characterized in the time domain with a time constant, and in the frequency domain, with a break point frequency ω. These are related by $\omega = \frac{1}{\tau}$.

We shall now consider the design of a second-order filter. Second-order systems are described by a second-order differential equation. These systems are characterized in the frequency domain by a resonant frequency ω_0 and a damping factor d.

2.2.4.1 Importance of the Damping Factor

The damping factor has a range of zero to +2.

If the damping factor d is set at zero, the closed-loop gain becomes infinite at the break point frequency, and the filter circuit will oscillate. This can be a problem when working with narrow-band pass high-Q active filters. The damping is set low to make the frequency response sharp, but this renders the circuit unstable, and it may spontaneously burst into oscillation.

Note that phase shift oscillator circuits are very similar to active-filter circuits. It depends on the damping in the circuit, as to which of these you end up with.

If the damping factor $d < \sqrt{2}$, the filter is underdamped. The frequency response has a gain peak at the break point and a steep roll-off just after the break point. Figure 2.26

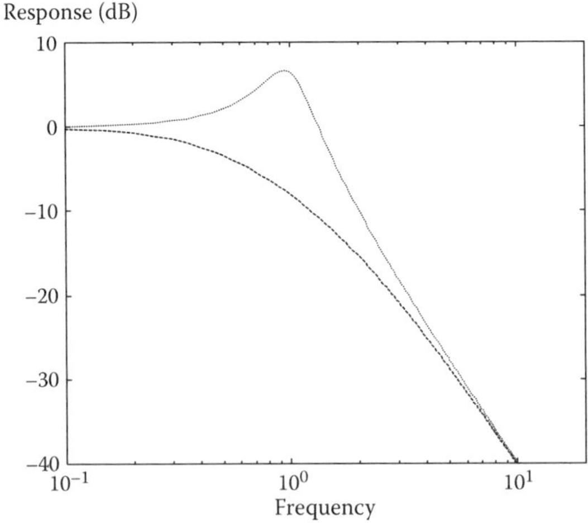

FIGURE 2.26
Response of underdamped and overdamped low-pass filters.

shows an example of an underdamped filter response. This type of filter is called a *Chebyshev filter.*

d = 0.766 produced a 3-dB peak Chebyshev filter
d = 0.886 produces a 2-dB peak Chebyshev filter
d = 1.045 produces a 1-dB peak Chebyshev filter

If the damping factor $d = \sqrt{2}$, the filter is critically damped. The frequency response does not peak. The response is flat to the break point and rolls off more slowly just after the break point. This type of filter is called a *Butterworth filter.*

If the damping factor $d > \sqrt{2}$, the filter is overdamped. The filter response rolls off very slowly from below the break point. The next diagram shows an example of an overdamped filter response. For example, if d = 1.73, we get a Bessel Filter. This has a poor gain amplitude response but the best possible phase response. If d = 2.0, the filter is equivalent to two RC first-order filters connected in series.

2.2.5 Amplitude Response of Underdamped and Overdamped Low-Pass Filters

The dotted line in Figure 2.26 shows an underdamped response, and the dashed line shows an overdamped response. (Note that this graph of response (dB) versus frequency has been normalized to a unity gain with a resonant frequency of 1 Hz.) Both responses asymptote to a roll-off of –40 dB/decade.

2.2.6 Phase Response of a Filter

The output of a filter is delayed in time behind the input. The ideal filter delays all frequencies by the same amount, so that output frequencies have the same phase relationship they had in the input signal. This is often called a *linear phase filter* because a constant delay means the phase shift produced is directly proportional to frequency.

Practical filters are dispersive, with different frequencies being delayed by different amounts. This produces a distortion of output signals. The most dispersive filters are Chebyshev filters. Chebyshev filters have a good amplitude response but a poor phase response. The least dispersive filters are Bessel filters. Bessel filters have a poor amplitude response but an excellent phase response.

2.2.7 Sallen–Key Second-Order Filters

A Sallen–Key second-order low-pass active-filter circuit is shown in Figure 2.27.

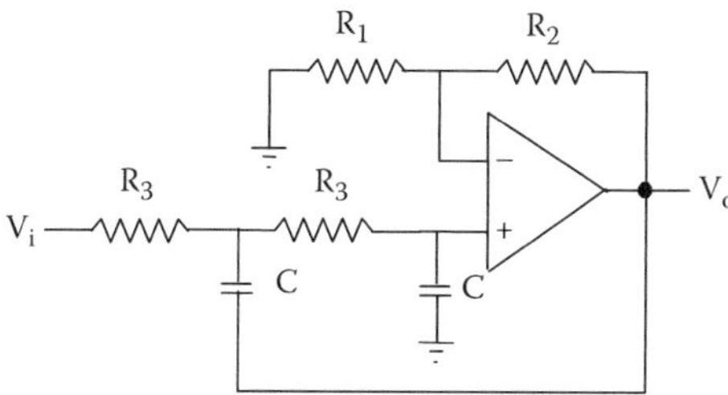

FIGURE 2.27
A second-order low-pass active filter.

2.2.7.1 Circuit Properties

The break-point frequency is set by the two R_3 resistors and the two capacitors. The frequency is given by $\omega = \frac{1}{R_3C}$.

The damping factor and gain are set by the values chosen for the resistors R_1 and R_2. The damping factor is given by $d = 2 - \frac{R_2}{R_1}$.

The gain of this filter is given by $G = 3 - d$.

2.2.8 Cascading Active-Filter Circuits

Active-filter circuits can be connected in series to make higher-order filters. The output of the first is connected to the input of the second. The output of the second is connected to the input of the third, and so on. The technical term for this arrangement is *cascading the filters.*

Two second-order filters connected in series will produce a fourth-order filter circuit. The combined amplitude response is the product of the two second-order responses. For example, if, at a frequency of 2 kHz, the first second-order filter circuit filter had a gain of 0.8 and the second second-order filter circuit had a gain of 1.4, the overall gain of the fourth-order combination at 2 kHz would be $0.8 \times 1.4 = 1.12$.

The combined fourth-order response can be easily found on a Bode plot by adding the two second-order responses together. This is shown in Figure 2.28.

The dashed line shows the overdamped response of a second-order filter, and the dotted line shows the underdamped response of the second second-order filter. The bold line shows the combined response of the fourth-order filter made by cascading the two second-order filters.

The combination produces a fairly flat amplitude response within the passband of the filter, with a sharp cutoff of frequencies outside the passband. Note: Both second-order responses asymptote to a roll-off of –40 dB/decade. The combined response of the fourth-order filter asymptotes to a roll-off of –80 dB/decade.

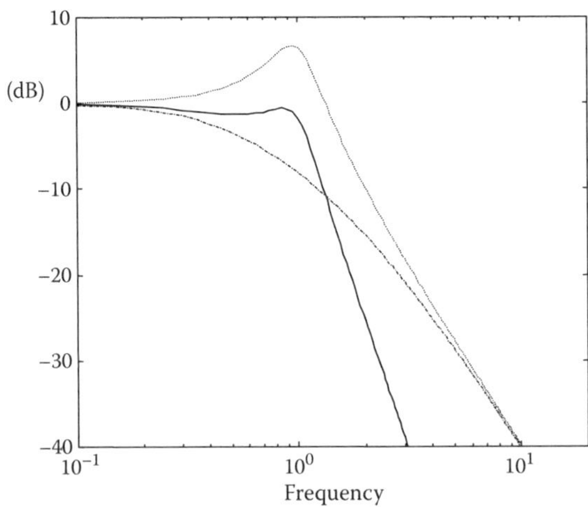

FIGURE 2.28
Response of two cascaded second-order filters (bold line).

2.3 Measuring Instruments and Systems

2.3.1 Need for Measuring Instruments

Sometimes, we do not need an instrument to determine the magnitude of a quantity. Our own senses may be sufficient to obtain the information required. For example, we can estimate length, weight, temperature, and color directly.

There are three limitations to human senses:

The first of these is that the information obtained is of low accuracy. For example, we could feel that the temperature of a cup of water was very hot, hot, warm, cold, or freezing. However, we could not accurately determine the temperature.

The second of these is the limited range of measurand. For example, if an object has a mass between 100 kg and 1 g, you could lift it and estimate its weight. However, you could not lift heavier objects and could not distinguish between masses of less than 1 g.

Finally, there are physical quantities that our human senses cannot respond to at all. We cannot see X-rays or feel magnetic fields.

Thus, an observation or measuring instrument is used when we require more information about a quantity than our senses alone can provide. The instrument acts as an extension of our senses, enabling us to notice things that we would not normally experience. We can use an instrument to measure a temperature accurately, weigh a mass of 50 mg, or detect X-rays.

A measuring instrument is an extension of our senses. It enhances our ability to perceive our environment. It enables us to be more aware of our environment.

2.3.2 Nature of Measuring Instruments

A measuring instrument acts as a *link* between the measurand under investigation and ourselves. We can consider this link as being made up from three parts:

Measurand ⟶ Instrument ⟶ Measurer

The arrows indicate a flow of information from the measurand to the instrument and from the instrument to the person. The role of the instrument is thus to transform information from the measurand into a form suitable for human perception.

2.3.3 Measurement Systems

A measurement system is made up of three components called a *sensor*, a *signal-processing device*, and a *display device*.

2.3.3.1 Sensor

A sensor is a device that transforms information about the magnitude of the measurand into a suitable output form. Usually, the output of a sensor is an electrical signal (a voltage or a current).

The reason electrical outputs are chosen is that most of our signal-processing technology is electrical. There is no reason in principle requiring the sensor to have an electrical output. You could, for example, build a sensor with a pneumatic of hydraulic output.

2.3.3.2 Signal-Processing Device

A signal-processing device transforms an input signal into a form suitable to drive a display device.

2.3.3.3 Display Device

A display device transforms an input signal into an output form that can be perceived by human senses. The output may be sound, patterns of light and dark, or touch.

We can now expand the previous diagram into

Measurand ⟶ Sensor ⟶ Signal Processing ⟶ Display ⟶ Measurer

The arrows indicate the flow of information through the system: from the measurand to the sensor, from the sensor to the signal-processing device, from the signal-processing device to the display device, and from the display device to the person.

2.3.4 Sensors

2.3.4.1 Sensitivity

The word *sensitivity* is not well defined when used with sensors. The sensitivity of a sensor can be measured in two ways:

1. By obtaining the magnitude of a response to a known stimulus: A large response caused by a small stimulus is a highly sensitive sensor. For example, we could define the sensitivity of a thermocouple in volts out per degree change in temperature.
2. By obtaining the magnitude of the smallest amount of some quantity that can be detected by the sensor: A sensitive sensor can detect much smaller amounts of something than a less sensitive sensor. For example, we could define the sensitivity of a specific ion electrode as the smallest ion concentration that could be detected by the electrode.

To avoid confusion, it is better to use different terms for these two different meanings of sensitivity. These are called *responsivity* and *detectivity*.

- The responsivity of a sensor is the ratio of the response obtained to the stimulus applied to the sensor.
- The detectivity of a sensor is a measure of the smallest amount of a quantity that can be detected by the sensor.

2.3.4.2 Responsivity

The responsivity of a sensor is the ratio of the magnitude of output signal from the sensor divided by the magnitude of the input stimulus to the sensor.

Responsivity = Output Response/Input Stimulus.

For example, a resistance thermometer has a responsivity measured in ohms/degree.

This simple definition of sensitivity is sufficient to specify the sensitivity of a linear sensor, where the output response is directly proportional to the input stimulus.

Some sensors have a nonlinear transfer characteristic. The response obtained is not directly proportional to the magnitude of the input stimulus. When this occurs, we have to redefine what is meant by responsivity. For example, if we filled a conductivity cell with electrolyte and measured its resistance as a function of the electrolyte concentration, we would get a curve like the one shown in Figure 2.29.

In cases like this, we must always specify an *operating point* to quote a responsivity value.

The responsivity can be measured in three ways.

2.3.4.2.1 *Absolute or Static Responsivity*

In this case, this is the ratio:

Responsivity = R/C at the operating point.

2.3.4.2.2 *Dynamic or Small Signal Responsivity*

If we want to know how a small change in concentration will affect the resistance of the cell, then we define the sensitivity as

Responsivity = dR/dC at the operating point.

2.3.4.2.3 *Relative Responsivity*

The previous definitions do not give us a feeling for the relative amount of change in the response and the stimulus.

For example, if we have a change $\delta R = 1\ K\Omega$, when the resistance at the operating point R is $10\ K\Omega$, then the charge is relatively large and is easy to detect. If, however, $\delta R = 1\ K\Omega$, when R is $10\ M\Omega$, then the same charge is relatively much smaller and is harder to detect.

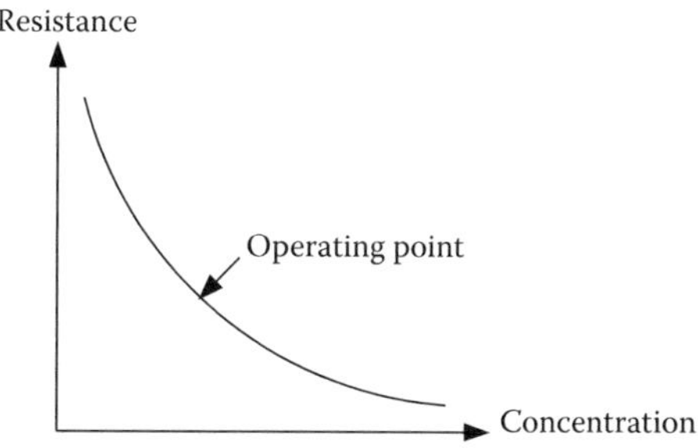

FIGURE 2.29
A responsivity curve for a conductivity cell.

A useful quantity is $\delta R/R$. This is a dimensionless number which tells us the relative change in R. (For example, if we had $\delta R = 1\ K\Omega$, when R is $10\ K\Omega$, then $\delta R/R$ is a 10% change.)

Similarly, a change in concentration would be expressed as $\delta C/C$.

The relative responsivity is given as

Responsivity = Relative Change in Response/Relative Change in Stimulus.

In this case, this is

$$\text{Responsivity} = (\delta R/R)/(\delta C/C) \text{ at the operating point.}$$

This is a dimensionless ratio.

2.3.4.3 Detectivity

The word *sensitivity* is sometimes used to mean the smallest amount of a quantity that can be detected with a sensor.

The smallest amount of signal that can be picked up from the sensor will be determined by the noise from the sensor. If the signal power from the sensor is much greater than the noise power, the signal is easily detected. If the signal power from the sensor is much less than the noise power, the signal is not easily detected. The signal can be hidden in the noise. Advanced techniques can extract a small signal from a large amount of noise. The smallest amount of signal that can be detected will depend upon the techniques used.

However, we can compare different sensors if we know the noise power they generate. We define the detectivity of a sensor as

$$\text{Detectivity} = 1/(\text{Noise Equivalent Power}).$$

We can use this definition to compare the properties of different sensors.

2.3.4.4 Active and Passive Sensing

2.3.4.4.1 *Passive Sensing*

A passive sensor transforms the measurand directly into an electrical signal, for example, a thermocouple or a pH electrode. This is shown in the diagram below:

Measurand ⟶ Sensor ⟶ Signal Processing

2.3.4.4.2 *Active Sensing*

In some measurement situations, the variable to be detected by the sensor is not present in the system under investigation. It must be generated by a signal generator and applied to the system. This is shown in block diagram form below:

Signal Generator ⟶ Measurand ⟶ Sensor ⟶ Signal Processing

Active sensing means stimulating the system under investigation and measuring the response to this stimulus. The magnitude of the measurand determines how much of the signal from the generator reaches the sensor, for example, a β-gauge paper thickness measurement or the absorption of light in a spectrophotometer.

2.3.4.5 Differential Sensing

Many physical quantities can be easily measured with a single sensor. However, in some measurement applications, it may be better to use two sensors. One of these sensors is used to sample the measurand, whereas the other is a reference sensor.

For example, consider the problem of measuring strain with a 120-Ω strain gauge, which is commonly formed from a copper track on a Mylar substrate (or similar). When we stretch the gauge, its resistance increases. The amount of strain is proportional to the resistance charge. In most cases, the resistance change is so small, say, 0.1 Ω. This small change in resistance creates two problems:

1. It is difficult to accurately measure a change in resistance from 120.0 Ω to 120.1 Ω. The sensor produces a large offset output.
2. The resistance of the strain gauge is temperature dependent. A change in resistance of 120.0 Ω to 120.1 Ω could also be caused by changing the temperature of the sensor.

Both these difficulties can be overcome by using a second strain gauge as a reference sensor in a bridge circuit (Figure 2.30).

In this type of circuit only the difference in resistance between the sample strain gauge and reference gauge sensors is detected. There is no offset signal. If the temperature changes, the resistance of both sensors should change by the same amount. This change should generate no differential signal.

Note: The word *differential* is used in instrumentation and engineering to indicate a subtraction or differencing process, for example, differential gears or differential inputs on oscilloscopes. *Differential* here does not mean taking a rate of change. It has nothing to do with calculus.

The signal that an amplifier receives is a differential signal. It is determined by the difference in resistance of the two sensors.

Signals that are common to both sensors are called *common mode signals*. In this case, if the resistances of both sensors increase by the same amount, this is a common mode signal. Common mode signals do not affect the detector.

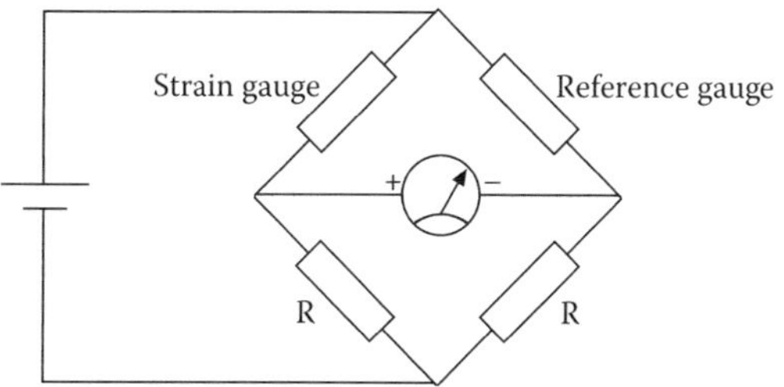

FIGURE 2.30
A strain gauge bridge circuit.

Another example of this principle is the problem of measuring the radiation absorbed by a chemical sample that is dissolved in a solvent. We could use a single sensing photo cell as shown below:

Light Source ⟶ Sample ⟶ Photo Cell

This system will measure the absorbence of the solvent as well as the absorbence of the dissolved material.

To overcome this problem, we use a double-beam system with two sample cells, one filled with the chemical plus solvent and the other filled with pure solvent. The beam from the monochromator is split into two parts, and two detectors could be used as shown below:

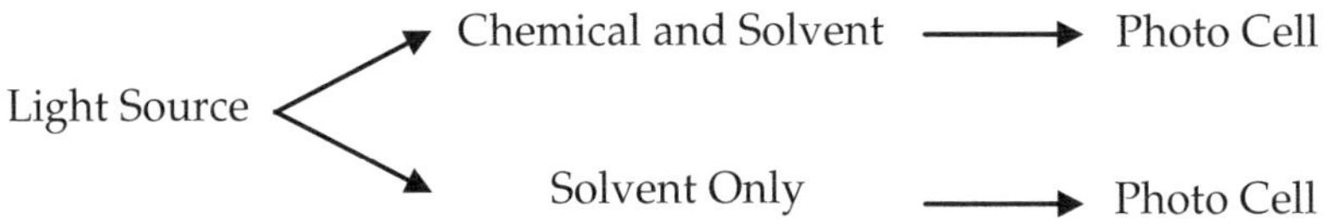

With this method, the difference in absorbence between the two solutions can be detected and common mode signals from the photo cells can be rejected.

2.4 Noise

The output of a measuring instrument will include both the desired output signal and unwanted signals called *noise*.

Noise can be classified as follows:

1. Random and nonrandom noise
2. Intrinsic and extrinsic noise

2.4.1 Random and Nonrandom Noise

Random noise is not predictable and causes random errors.

Nonrandom noise is predictable and is a source of systematic error, therefore, it can be removed by systematic processes.

2.4.2 Intrinsic and Extrinsic Noise

Intrinsic noise is produced by the measuring instrument, whereas extrinsic noise (interference) is produced by the environment.

2.4.3 Fundamental Limitations in Measurement Imposed by the Laws of Physics

If it were possible to work with a perfectly well defined uniform measurand and to eliminate all outside interference from a measuring instrument, a small amount of random error remains, caused by the intrinsic random noise of the instrument.

The amount of random error in an instrument can be reduced by improving the design or using higher-quality components in its construction. It can be shown, however, that there is a fundamental lower limit to the amount of random error in an instrument.

This limit can never be improved upon and is a consequence of the laws of physics. It cannot be overcome by building a better instrument.

There are two causes of this limit.

2.4.3.1 Thermal Noise

The thermal energy in a system of atoms or molecules is not uniformly distributed. The distribution of energy fluctuates with time; sometimes an atom has high energy, and at other times it has low energy.

The thermal energy of a system is determined by the absolute temperature T, and it can be shown that the kinetic energy of each degree of freedom of the system is kT/2. The constant "k" is called Boltzmann's constant and has the value of 1.4×10^{-23} J/K.

Examples of this thermal noise are the Brownian motion of small particles, low-level acoustic noise, and the electrical noise of resistors.

2.4.3.2 Quantum Noise

Matter, charge, and radiation do not come in continuously variable amounts. They are quantized. Quanta arrive at random intervals and cause random fluctuations in physical phenomena such as electron current and radiation intensity. (Shot noise in amplifiers is an example of quantum noise.)

The noise limit of an instrument may be determined by either thermal or quantum noise depending on the instrument.

Consider the case of a photomultiplier tube detecting a weak beam of green light with a wavelength of 500 nm. The thermal energy of the system at room temperature is 4×10^{-21} J. The photon energy is 4×10^{-19} J. The instrument is quantum noise limited since the quantum energy is 100 times the thermal energy.

2.4.4 Methods of Eliminating Interference

Interference between electronic circuits can be reduced using a variety of methods such as shielding, grounding, and appropriate cabling.

Noise cannot be entirely eliminated, but it may be reduced to such a level where it no longer causes interference. A unique solution to the problem of noise reduction may not exist. Designing equipment that does not generate noise is as important as designing equipment that is not susceptible to noise.

2.4.4.1 Shielding

Shielding serves a dual purpose by protecting a circuit from environmental noise or unwanted signals and protecting the environment from the circuit's signals. Shielding is mostly used to block electrostatic or "E" fields (Faraday shield).

For a shield to be effective, there must be no current flowing through it. This may be achieved by connecting the reference to only one point on the shield, but current flowing in the shield can produce secondary fields, reducing the shield's effectiveness.

A ground loop occurs when an electrical circuit is grounded at more than one point. Voltages can be induced in the circuit if the two earth points are not at the same potential. This can be caused by currents flowing in an earth line causing a potential drop. If a shielded cable has a potential difference at each end of the shield, a current flows in the shield, inducing noise in the conductor.

Ground loop voltages can be reduced by using two balanced signal transmission lines. The same amount of pickup will be induced in both lines. The pickup is thus a common mode signal that can be removed with a differential amplifier at the receiver end.

2.4.4.2 Active Shielding

Active shielding is where equal and opposite fields are generated to cancel out the interference fields. However, a mismatch of the electrical parameters generating the opposite field will result in addition to the noise.

An example of active shielding is where a "driven shield" on a cable can be used to minimize the capacitive load of the cable. A length of coaxial cable can have a capacitance of several hundred picofarads. This can impose a load on high-frequency signals from a high-impedance source. This load can be reduced by using a coaxial cable that contains a central signal wire surrounded by an inner shield and then an outer shield.

The inner shield conductor is held at the same potential as the central signal wire by connecting the signal wire to the input of a unity gain buffer amplifier, and the inner shield to the amplifier output. Since the inner signal wire and the shield conductor around it are at the same potential, there is no current between them. There is no signal loss from the central wire. The outer shield is earthed to shield the signal lines from pickup.

2.4.4.3 Cabling

Cables are the longest elements of a measurement system and therefore act as antennae that can both pick up and radiate noise. Coupling occurring between a field and a cable is known as *crosstalk*. The challenge of protecting a conducting cable from noise is inextricably linked to shielding and earthing, as described briefly in Sections 2.4.4.1 and 2.4.4.2 and in more detail in Section 2.4.4.4.

2.4.4.4 Electrical Interference

When two conductors are in close proximity, the electric and magnetic fields surrounding one of them can induce an unwanted voltage in the other. This unwanted signal can contaminate and corrupt an electrical signal present in the second conductor.

As stated previously, an unwanted signal in a system is called *interference*. In particular, the unwanted induced voltage on the second conductor is called *electrical interference*.

2.4.4.4.1 Electric Field Interference

Consider the general case of a voltage (V_1) on a conductor, producing a second voltage (V_2) on a nearby conductor.

The magnitude of the induced voltage on the second conductor will depend on two factors: the impedance of the coupling capacitance between the two conductors (Z_c) and the impedance of the second conductor to ground (Z_g) (Figure 2.31).

V_1 is the voltage on the first conductor, V_2 is the induced voltage on the second conductor, Z_c is the coupling impedance between the two conductors, and Z_g is the impedance between the second conductor and ground.

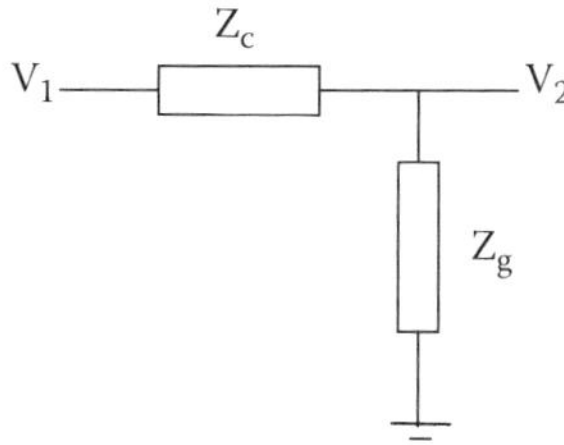

FIGURE 2.31
An electric field interference model.

This circuit is a voltage divider:

$$\frac{V_1}{V_2} = \frac{\text{Impedance of conductor 1 to ground}}{\text{Impedance of conductor 2 to ground}} = \frac{Z_c + Z_g}{Z_g} = 1 + \frac{Z_c}{Z_g}.$$

Electric field coupling between two conductors is determined by the electrical capacitance between them. The coupling impedance (Z_c) between the two conductors is thus the capacitive reactance (X_c) between them.

The impedance of the second conductor to ground is the parallel combination of the resistance of the second conductor to ground (R) and the capacitive reactance (X_g) between the second conductor and ground (Figure 2.32).

V_1 is the voltage on the first conductor, V_2 is the induced voltage on the second conductor, C_c is the coupling capacitance between the two conductors, R is the resistance between the second conductor and ground, and C_g is the capacitance between the second conductor and ground.

The impedance of conductor 2 to ground Z_g is given by

$$\frac{1}{Z_g} = \frac{1}{R} + \frac{1}{\left(\dfrac{1}{j\omega C_g}\right)} = \frac{1}{R} + j\omega C_g = \frac{1 + j\omega RC_g}{R}.$$

Hence,

$$Z_g = \frac{R}{1 + j\omega RC_g}.$$

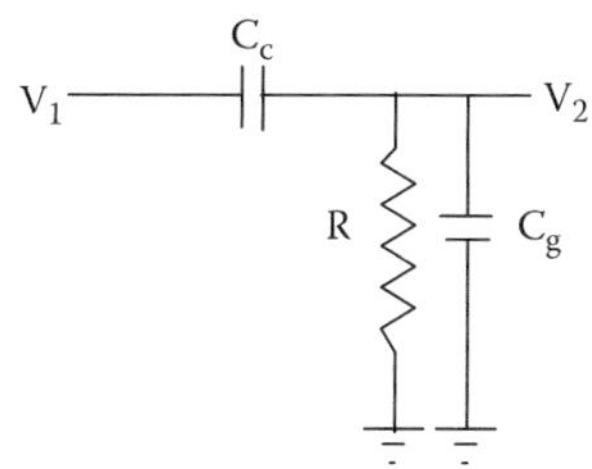

FIGURE 2.32
An electric field interference model.

Hence,

$$\frac{V_1}{V_2} = 1 + \frac{\left(\dfrac{1}{j\omega C_c}\right)}{\left(\dfrac{R}{1 + j\omega RC_g}\right)}$$

$$\frac{V_1}{V_2} = 1 + \left(\frac{1}{j\omega C_c}\right)\left(\frac{1 + j\omega RC_g}{R}\right)$$

$$\frac{V_1}{V_2} = 1 + \frac{1 + j\omega RC_g}{j\omega RC_c}$$

$$\frac{V_1}{V_2} = \frac{j\omega RC_c + 1 + j\omega RC_g}{j\omega RC_c}$$

$$\frac{V_1}{V_2} = \frac{1 + j\omega R\left(C_c + C_g\right)}{j\omega RC_c}.$$

And therefore,

$$\frac{V_2}{V_1} = \frac{j\omega RC_c}{1 + j\omega R(C_c + C_g)}.$$

Consider the behavior of this circuit at low and high frequencies.

At low frequencies, $j\omega R(C_c + C_g)$ is less than 1, and this ratio approximates to $\frac{V_2}{V_1} = j\omega RC_c$.

The amount of interference is directly proportional to frequency.

At high frequencies, $j\omega R(C_c + C_g)$ is greater than 1, and this ratio approximates to a constant

$$\frac{V_2}{V_1} = \frac{C_c}{C_c + C_g}.$$

The transition between the low frequency and high frequency approximations occurs when $1 = \omega R(C_c + C_g)$.

Hence,

$$\omega = \frac{1}{R\left(C_c + C_g\right)}.$$

2.4.4.4.2 Bode Plot

The noise voltage is directly proportional to frequency up to a break point frequency, $\omega = 1/R(C_c + C_g)$; at higher frequencies, it is set to a constant value (Figure 2.33).

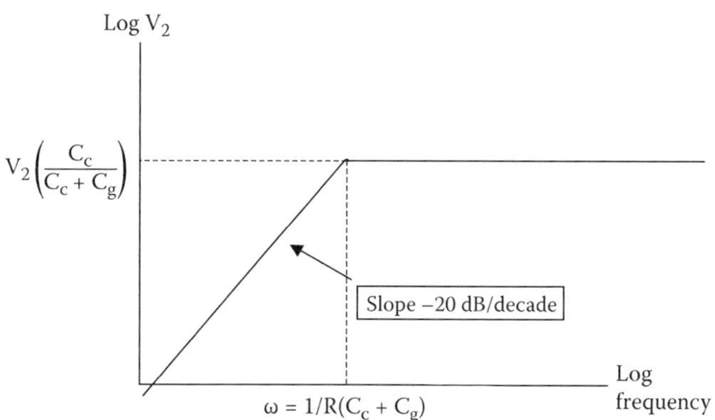

FIGURE 2.33
The frequency response of the electric field interference.

2.4.4.4.3 Coupling Capacitance

The capacitance between two conductors is determined by their size and their separation. Two large objects close together will have a large coupling capacitance. Two small objects further apart will have a smaller coupling capacitance.

Considering the specific case of two metal plates fairly close together, the capacitance between them can be shown to be

$$C = \kappa \frac{\varepsilon_0 A}{d},$$

where κ is the dielectric constant of the material between the plates, ε_0 is the permittivity of free space = 8.85×10^{-12} $C^2/N.m^2$, A is the area of the plates in square meters, and d is the separation between the plates in meters.

Dry air has a dielectric constant of 1.00059. If there is air between the plates, we can usually neglect κ in the above equation and write it as

$$C = \frac{\varepsilon_0 A}{d}.$$

2.4.4.4.4 Some Sample Calculations

1. The oscilloscope has an input impedance of 1 MΩ in parallel with 20 pF. The 50-Ω coaxial cable used in this laboratory has a capacitance of 100 pF/m. Assume, for example, that the capacitance of the cable and end leads is 130 pF. This is in parallel with the oscilloscope input of 20 pF, producing a total capacitance of 150 pF from the detector plate to ground. We will also assume that C_c is 35 pF:

$$\omega = \frac{1}{R(C_c + C_g)} = 5400 \text{ rad/s}.$$

2. At 100 Hz, the oscilloscope input resistance is less than the input capacitive reactance. The oscilloscope can be considered as a simple 1-MΩ resistance.

   At $f = 100$ Hz, $\omega = 2\pi f = 628$ rad/s, $\frac{V_2}{V_1} = j\omega RC_c$ rad/s, and hence, $V_2 = j\omega RC_c V_1$.

   The magnitude of V_2 is independent of the phase, so we can ignore the "j" in this equation if we just want the amplitude of the wave. If we assume a coupling capacitance of 35 pF, we get

   $$V_2 = j\omega RC_c V_1 = (628)(1 \times 10^6)(35 \times 10^{-12})(10) = 220 \text{ mV}.$$

3. At 10 kHz, the oscilloscope input capacitive reactance is less than the input resistance. The oscilloscope and cable can be considered as a 150-pF capacitor.

   $$\frac{V_2}{V_1} = \frac{C_c}{C_c + C_g}$$

   $$V_2 = \left(\frac{C_c}{C_c + C_g}\right) V_1 = \left(\frac{35 \times 10^{-12}}{35 \times 10^{-12} + 150 \times 10^{-12}}\right)(10) = 1.89 \text{ volts.}$$

2.4.4.5 Measurement of Voltage Induced on a Person

2.4.4.5.1 Sample Calculation

Consider a specific case of a person standing near a 230-volt root mean squared (RMS) (650 V peak to peak) mains cable. Assume that the capacitance between the person and the active wire is 5 pF. The person touches an oscilloscope input. The oscilloscope has an input resistance of 1 MΩ. What is the voltage on the person's body?

The impedance of the capacitor is $X_c = \frac{1}{j\omega C}$. If we ignore the phase, the magnitude of this impedance is $X_c = \frac{1}{\omega C} = \frac{1}{2\pi f C}$.

We substitute in a frequency of 50 Hz and a capacitance of 5×10^{-12} F.

The impedance of the coupling capacitor is 637 MΩ at 50 Hz.

The ratio of the oscilloscope input resistance to the total circuit impedance is $\frac{1}{650}$; hence, the voltage observed on the oscilloscope would be $\frac{650}{638} = 1.02$ V peak to peak.

2.4.4.5.2 Worked Example

Two signal lines are nearby. Line 1 has a 10-kHz, 5-volt RMS signal on it. Some of this is coupled into line 2. The coupling capacitance between the two signal lines is 10 pF. Line 2 is driven by a 600-Ω source and terminates at a 2.5-kΩ load. The capacitance of line 2 to ground is 75 pF.

The maximum amount of pick-up voltage on line 2 is as follows:

At high frequencies, the dominant impedance between line 2 and ground is the 75-pF capacitance. The circuit is a simple capacitance voltage divider. Let the capacitive reactance of the coupling capacitance be X_c and the capacitive reactance of the capacitance between line 2 and ground X_g; then,

$$\frac{V_2}{V_1} = \frac{X_g}{X_c + X_g}.$$

The coupling capacitance is C_c, and the capacitance between lines 2 and ground is C_g; then, $Xc = \frac{1}{j\omega C_c}$ and $X_g = \frac{1}{j\omega C_g}$, which we substitute into the above equation and get

$$\frac{V_2}{V_1} = \frac{C_c}{C_c + C_g}.$$

Substituting the capacitance values, we get V = 0.588 V.
Above what frequency does the maximum occur?
The transfer function has been shown to be

$$\frac{V_2}{V_1} = \frac{j\omega RC_c}{1 + j\omega R(C_c + C_g)}.$$

The transition between the low-frequency and high-frequency approximations occurs when $1 = \omega R\,(C_c + C_g)$.

Hence,

$$\omega = \frac{1}{R\left(C_c + C_g\right)}.$$

R, the resistance to ground of line 2, can be determined. This is the parallel combination of the source resistance on line 2 and the load resistance on line 2.

$$\frac{1}{R} = \frac{1}{R_S} + \frac{1}{R_L}$$

Substituting the values, the resistance to ground is 484 Ω.
Hence,

$$\omega = \frac{1}{R\left(C_c + C_g\right)}$$

and, substituting the values, f = 3.87 MHz.

2.4.4.5.3 *Capacitive Loading*

An oscilloscope has an input resistance of 1 MΩ and an input capacitance of 25 pF. It is connected to a circuit through a 1-m length of a 50-Ω cable. The capacitance between the inner and outer conductors of the cable is 100 pF.

What would the total capacitive load be on the circuit to which the oscilloscope is connected?

The total capacitive load is the parallel combination of the oscilloscope input capacitance and the cable capacitance. That is 125 pF.

What is the reactive load on the circuit at a frequency of 20 MHz?

At 20 Mhz, $\omega = 2\pi f = 1.256 \times 10^8$ rad/s.

$$X = \frac{1}{j\omega C} = 63.7\ \Omega.$$

At what frequency does the capacitive load of the oscilloscope become greater than the resistive load?

This occurs when $|X_c| = R$; hence, $\omega = \dfrac{1}{RC}$.

Substituting the values, we get f = 1.27 kHz.

Bibliography

Carter B, Mancini R. 2009. *Op-Amps for Everyone*. Newnes, Massachusetts.

Morrison R. 1992. *Noise and Other Interfering Signals*. J. Wiley & Sons, University of Michigan.

Pactitis SA. 2007. *Active Filters: Theory and Design*. CRC Press, Boca Raton, FL.

Placko D (ed.). 2007. *Fundamentals of Instrumentation and Measurement*. ISTE, London.

3

Properties of Electrodes

Andrew W. Wood

CONTENTS

3.1 Electrode Equilibrium Potentials

When a metal dips into a solution containing its own ions (for example, copper metal in a solution of $CuSO_4$), a chemical reaction takes place. Initially, this may consist of a slight dissolving of copper atoms from the electrode surface to form copper ions in solution. This causes a gain of positive charge to a normally balanced, or neutral, solution. This gives rise to a potential difference (p.d.) between the solution (positive) and the electrode metal (negative). This p.d. prevents any further dissolution of the electrode and sets up an equilibrium in the reaction:

$$Cu \leftrightarrow Cu^{2+} + 2e^-,$$

where e^- represents electrons released into the metal electrode. Please note that the reaction proceeds in both directions simultaneously, but at equilibrium, the left-to-right reaction is just as likely as the right-to-left (if the electrode is connected to an external source of current, such as a battery, then this is a different story—see Section 3.2). This arrangement (a metal dipping into a solution of its own ions) is known as an *electrode of the first kind*. This reaction can be represented in general as

$$M^{z+} + ze^- \leftrightarrow M.$$

Here, M represents a metal whose valence is z. The conditions for equilibrium can be studied by considering the electrochemical potentials ($\mu^{\sim}$) of the three components involved:

$$\mu^{\sim}_{(\text{metal ions in solution})} + z\mu^{\sim}_{(\text{electrons in metal})} = \mu^{\sim}_{(\text{metal atoms in electrode})}.$$

Now, since we can write that

$$\mu^{\sim} = \mu + zFV,$$

where V is the voltage (or potential) relative to an arbitrary zero (but see discussion below regarding the Stockholm convention), we can derive the following (for a derivation, see Atkins and de Paula [2009]):

$$V_M - V_S \equiv V_{MS} = V^0_{MS} + (RT/(zF)) \ln a_{(M)}.$$

Here, V^0_{MS} is termed the *standard half-cell potential*. V_{MS} is the p.d. between the metal and the solution, relative to the solution. The symbol $a_{(M)}$ refers to the *activity* of metal ions in solution, which is directly proportional to the metal ion concentration [M]. We can write

$$a_{(M)} = \gamma\,[M],$$

where γ is the activity coefficient of the ion in solution. In general, γ varies with [M], and tends toward 1, as [M] tends toward 0 (i.e., dilute solutions). Biological solutions can be treated as being dilute, even though γ has a value of around 0.5–0.8 for ions in cytoplasmic fluid. An approximate form for the half-cell potential equation is thus

$$V_{MS} = V^0_{MS} + 25 \ln [M] (\text{in mV}).$$

V^0_{MS} values depend on the metal, the type of solution the metal is dipped into, and what precise reaction takes place preferentially at the electrode (at a platinum electrode, for example, the reaction $H^+ + e^- \leftrightarrow \frac{1}{2}H_2$ (gas) takes precedence over a reaction involving Pt ions). Here are some examples of V^0_{MS} values—please note that by convention, the V^0_{MS} of the H_2 reaction just referred to is taken as 0 mV:

Reaction	V^0_{MS}(V)	Reaction	V^0_{MS}(V)
$Li^+ + e^- \leftrightarrow Li$	−3.045	$H^+ + e^- \leftrightarrow \frac{1}{2}H_2$	0.0
$K^+ + e^- \leftrightarrow K$	−2.925	$Cu^{2+} + 2e^- \leftrightarrow Cu$	+0.153
$Na^+ + e^- \leftrightarrow Na$	−2.714	$Ag^+ + e^- \leftrightarrow Ag$	+0.799
$Al^{3+} + 3e^- \leftrightarrow Al$	−1.66	$Cl^- \leftrightarrow \frac{1}{2}Cl_2 + e^-$	+1.36
$Zn^{2+} + 2e^- \leftrightarrow Zn$	−0.763	$F^- \leftrightarrow \frac{1}{2}F_2 + e^-$	+2.87
$Fe^{2+} + 2e^- \leftrightarrow Fe$	−0.44	In the case of these last two, the reaction would typically take place at a platinum electrode	

We have referred to half-cell potentials. In order to measure a p.d., two electrodes, each contributing a half-cell potential, have to be employed. If two identical metal electrodes are immersed into the same solution, the p.d. will be zero (except for small random variations of a few microvolts, due to statistical effects). If, on the other hand, copper in $CuSO_4$ and zinc in $ZnSO_4$ are joined via a porous membrane and the two solutions are the same concentration, the p.d. will be the difference between the standard half-cell potentials (in this case 153 – (–763) mV or a little less than 1 V). In fact, the porous plug itself will contribute a small p.d. as a result of Cu^{2+} and Zn^{2+} having differing ion mobilities.

There is shorthand for representing such electrode systems. The rules are as follows:

1. The symbol | denotes a metal/solution interface.
2. The symbol || denotes a liquid junction.

The Cu/Zn cell described above (wet cell) would be represented as

$$Cu \mid Cu^{2+}(aq, a_{Cu} = 0.1\ M) \parallel Zn^{2+}(aq, a_{Zn} = 0.1\ M) \mid Zn.$$

To avoid ambiguities, the following (Stockholm) convention is adopted:

1. The p.d. is given as the right-hand electrode relative to the left.
2. The V^0 for a given electrode is defined with the H_2 electrode on the left side.

3.1.1 Electrodes of the Second Kind

An *electrode of the second kind* is a metal in contact with its own *sparingly soluble* salt. A prime example is the silver–silver chloride or Ag/AgCl electrode. Here, Ag is the metal and AgCl is the salt, which is only very slightly soluble in water and will preferentially form a chocolate-colored coating on silver if NaCl solutions are electrolyzed. For example, if a current of a few microamperes is passed between two silver wires in saline solution (9 g NaCl/L), one of the wires will gain a brown coat (the positive wire, or anode) whereas the other wire (the cathode) will remain bright. Tiny gas bubbles will come off the wire while the current is passed, and a distinct chlorine smell may be noticed.

3.1.2 Anode

The electrode reactions are as follows:

$$Ag \rightarrow Ag^+ + e^-$$

$$Ag^+ + Cl^- \leftrightarrow AgCl.$$

The amount of Ag^+ in solution is governed by the so-called solubility product for AgCl. This dictates that the product $[Ag^+][Cl^-]$ for the solution cannot exceed $1.8 \times 10^{-10}\ M^2$. Since $[Cl^-] = 0.15$ M (the normal strength of saline), this implies that once the solution concentration of Ag^+ exceeds 10^{-9} M, the AgCl will form a precipitate on the electrode.

3.1.3 Cathode

Since there is so little Ag^+ in the solution, the above reaction does not go in reverse. The following is more likely:

$$Na^+ + e^- \rightarrow Na$$

then

$$Na + H_2O \rightarrow Na^+ + OH^- + \tfrac{1}{2}H_2.$$

Note that the anode reaction is readily reversible (if the current is reversed, the AgCl decomposes, eventually exposing the silver wire). This is a good property for an electrode because it means that the products of the electrode reaction quickly diffuse away from the electrode and do not cause polarization effects, which would make the electrode unstable. Note also that metal ions are often toxic to biological systems. Because in this case the $[Ag^+]$ is limited to nanomolar amounts, this minimizes the risk. We can show that the Ag/AgCl electrode acts like a chloride electrode, with a V^0_{MS} value of 222 mV, that is,

$$V_{MS} = 222 - 25 \ln [Cl^-] \text{ (mV)}.$$

Because of the stability of Ag/AgCl electrodes, they are often used as standard or reference electrodes (in pH meters, for example). They form the basis of electrocardiography (ECG) and electroencephalography (EEG) electrodes and of the microelectrodes used in cellular biophysics research. Another form of reference electrode is the calomel electrode, which has mercuric chloride (calomel) in contact with metallic mercury. Although these are used in some pH meters and gas analyzers, they are not so common in clinical biophysics.

3.2 Electrodes Not in Equilibrium (Overpotentials)

The half-cell potentials just referred to will only have the values predicted by the equations given above if the net current across the metal/solution interface is zero. In practice, if the two electrodes form an electrolytic cell (loosely called a battery), a current equal (and opposite) to the amount of current the cell is producing will have to be applied from an external source to make this net current zero. Under any other circumstance (for example, if the electrodes are used to stimulate tissue), the *p.d. between the electrodes and the p.d. at zero net current will be different by an amount known as the total overpotential.* These overpotentials are due to a number of sources; the two main ones are listed below:

1. *Charge-transfer overpotential*: predominates at low currents.
2. *Diffusion or concentration overpotential*: predominates at high currents.

In addition to these, there are voltage drops due to the passage of current through the bulk of the electrolyte and perhaps the connecting wires, if these have a significant amount of resistance.

We will now examine the two main overpotentials more closely.

3.2.1 Charge-Transfer Overpotential

This is due to the electrode reaction taking place at the metal/solution interface and is smaller for reversible (or so-called nonpolarizable) electrodes than for nonreversible electrodes. The value varies with the amount of net current I and depends on the detailed shape of the energy barriers involved in going from solution to metal or vice versa. The net current density J_{SM} (= I/A, where A is the surface area of the electrode) is related to η, the charge transfer overpotential, by the following expression (named the Butler–Volmer equation; for derivation, see Atkins and de Paula [2009]):

$$J^+_{MS} = j^+_e\{\exp(-zF\alpha\eta/(RT)) - \exp(+zF(1-\alpha)\eta/(RT))\}.$$

Here, j^+_e is the exchange current density (the one-way or unidirectional current across the metal/surface interface at equilibrium), α is the "symmetry factor," which is a measure of the fractional distance across the interfacial region where the peak of the barrier occurs, and MS refers to the direction metal to solution. Because α often has a value of 0.5, the expression simplifies to

$$J^+_{MS} = j^+_e 2\sinh(zF\eta/(2RT)).$$

If η is likely to be small (i.e., J^+_{MS} is small), we can approximate further to

$$J^+_{MS} = j^+_e(zF\eta/(RT)) \text{ or } \eta = 25\, J^+_{MS}/j^+_e(\text{mV}).$$

Thus, if we want the charge transfer overpotential to be less than 1 mV, we need to keep J^+_{MS} below $25 \times j^+_e$. The quoted value for j^+_e for a hydrogen reaction at a platinum electrode is 0.8 mA/cm^2; thus, for a 1-cm^2 area of electrode, the current needs to be kept below 20 mA.

3.2.2 Diffusion Overpotential

When a current is withdrawn from an electrode, the charge has to be provided by mobile ions in the solution, diffusing toward the electrode (which acts like a sink for charge). In effect, the region close to the electrode becomes depleted of metal ions if the electrode acts as a sink (this electrode will be gaining mass due to deposition) or will have an excess of metal ions if the electrode is dissolving. There is thus a zone of diffusion between bulk solution and the region adjacent to the electrode. This zone is known as the Nernst or diffusion layer and is around a few millimeters thick. Fick's first law of diffusion (see Chapter 5), coupled with the Nernst formula, can be applied to this region, yielding the following formula for diffusion overpotential η_c:

$$\eta_c = \{RT/(zF)\}\ln\{1 - \lambda J'/(cDzF)\}.$$

Here, λ is the thickness of the diffusion layer, J' the applied current density, c the ion concentration, and D the diffusion coefficient in the diffusion layer. When the concentration falls to zero at the electrode face, the current J' cannot increase any further. This is known as the limiting current density $J'_L = zFcD/\lambda$. The above expression can be written as

$$\eta_c = \{RT/(zF)\}\ln\{1 - (J'/J'_L)\}.$$

Note that if $J' = 0$, then $\eta_c = 0$, and if $J' \to J'_L$, then $\eta_c \to \infty$.

3.3 Impedances Associated with Solution–Electrode Interface

3.3.1 Direct-Current Resistances

These are found simply by dividing the above overpotential expressions by the net current density J. (We will now simplify J^{+}_{MS} and J′ to J.)

1. *Charge-transfer resistance*

$$R_t = \eta/(JA) = (RT/(zFA))\sinh^{-1}(J/(2j_e))/(J/(2j_e))$$

Here, A denotes the surface area of the membrane. Note that when J = 0, $\sinh^{-1}(x)/x \rightarrow 1$; thus, $R_t \rightarrow RT/(j_e zFA)$. When $J \rightarrow \infty$, $R_t \rightarrow 0$.

Also note that electrodes with large j_e and A values have smaller R_t values.

2. *Diffusion (or concentration) resistance*

$$R_c = \partial\eta_c/(\partial JA) = \{RT/(zFA)\}\partial \ln\{1 - (J/J_L)\}/\partial J = \{RT/(zFA)\}/(J_L - J)$$

In this case, when J = 0, $R_c = (RT/(J_L zFA))$, and when $J \rightarrow J_L$, $R_c \rightarrow \infty$.

Note that at low current densities, the total resistance is

$$R = (RT/(zFA))\{1/j_e + 1/J_L\}.$$

Remember that j_e is a unidirectional flux and J_L is a net flux.

3.3.2 Alternating-Current Impedances

Because charge builds up in both the layer immediately adjacent to the electrode (called the Helmholtz layer) and the Nernst layer, we can associate separate capacitance values with both these regions. For biological applications, the most important one of these is that associated with the Nernst layer. This, together with the alternating-current diffusion resistance, is known as the (series) Warburg impedance. Experimentally, it is found that the real (resistive) part of this impedance is numerically similar to the imaginary (capacitive) part and that both vary with the inverse square root of frequency. The Warburg law is thus written as

$$Z = \alpha/\sqrt{\omega} + j\alpha/\sqrt{\omega}.$$

Here, $\omega = 2\pi f$. Also, because the imaginary part of impedance is $1/(\omega C)$, that is, $\alpha/\sqrt{\omega} = 1/(\omega C)$, we can easily show that C is proportional to $1/\sqrt{\omega}$ as well.

The Warburg impedance predominates at frequencies below 10 kHz. Above this frequency, the impedance associated with the Helmholtz layer (charge-transfer process) becomes more important.

3.4 Factors in Optimizing the Recording of Biological Signals

Amplifier input impedance needs to be at least 100 times the value of the electrode impedance. For macroelectrodes, 1 MΩ is quite adequate. The input impedance of most cathode ray oscilloscopes is around this figure. For microelectrodes, the input impedance has to be at least 100 MΩ. The voltage that the input terminals of the amplifier "see" is given by

$$v_i = v_s\, R_i/(R_i + R_r),$$

where v_s is the voltage of the biological source (heart, nerve membrane, etc.), R_i is the input impedance of the amplifier, and R_r is the resistance of the electrodes.

Pickup from the 50-Hz mains supply is always a problem in recording tiny biological signals. Even the use of battery-powered equipment may not solve the problem because of pickup from mains-driven equipment nearby. The amount of 50-Hz contamination can be reduced by using a notch filter (which filters in a narrow band around 50 Hz), but because this can cause phase distortion, it is not the optimal solution. A differential amplifier with a common mode rejection ratio (M) of 10^5 or better is a good way to go. If the common mode signal v_c (in this case, the 50-Hz contamination) has a value of, say, 1 mV at the input, the value at the output is $v_o = Av_c/M$, or 0.01 mV for a gain (A) of 1000. If the resistances (impedances) of the two electrodes R_{r1} and R_{r2} are not matched, the value of the pickup at the output will be greater than this. In fact, it is given by

$$v_o = Av_c[(R_{r1} - R_{r2})/R_i + (1/M)].$$

3.5 Macroelectrodes

Electrodes for measuring from the skin surface are of two broad types: disposable and nondisposable. Electrodes of the first type are mainly used for the measurement of potentials associated with the heartbeat (electrocardiogram or ECG) and consist of an adhesive electrolyte gel in contact with a conducting foil, which is then connected to the monitoring instrument via lead wires using modified alligator clips. Other designs have electrolyte-soaked plastic foam in contact with a metal-coated button, which then has a press connector linked to the lead wire. Adhesive dressing tape is then used to secure the electrode to the skin. The disposable nature of these electrodes reduces the risk of cross-infection. The nondisposable types are often used for measuring brain electrical activity (electroencephalogram or EEG). In some designs, with around 130 electrodes held in an array within a stretch fabric "bathing cap," the electrodes consist of tin or stainless steel disks, with a central perforation to allow electrolyte gel to be injected (using a blunt needle) underneath the electrodes once the bathing cap is in place. The metal electrodes are held in plastic rings to provide a recess to contain the gel (see Figure 3.1).

In both ECG and EEG, proper preparation of the skin is paramount. Removal of skin oils by the use of an alcohol swab and the consequential application of abrasive paste to remove the dead outer skin layer both serve to reduce the electrode resistance, which is essential for good recordings (see previous section). With the "bathing cap" and other

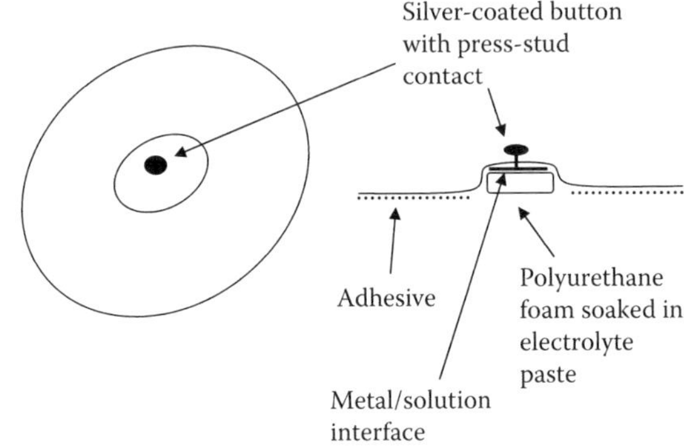

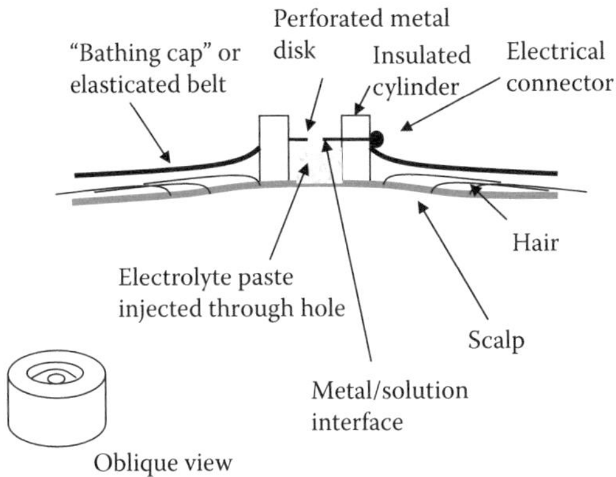

FIGURE 3.1
Disposable electrodes for ECG (top) and EEG (bottom) showing the metal/solution interfaces in either case.

scalp electrodes, it is also important to part the hair away from the region below the electrode to ensure good skin contact.

3.6 Microelectrodes

In order to record potentials from individual cells or from extracellular fluid spaces, electrical contact has to be made with areas of a micrometer or so. The work of Ling and Gerard in 1949 (quoted in Aidley 1998) showed how it was possible to use a thin glass capillary drawn out in a small electric heater to produce tips that were a micrometer or less in diameter (Figure 3.2). These glass capillaries are then filled with a strong electrolyte solution (usually 3 M KCl) to provide electrical contact with the cell contents or the extracellular space. A silver wire coated with AgCl then makes an electrical connection with the wires leading to the input preamplifier. Because the resistance (commonly referred to as electrode impedance) is dominated by the final section of the microelectrode, an estimate can be made of its value. For example, for $L = 10$ mm, $r = 10^{-6}$ m, and $\rho = 1/30\ \Omega$m, the impedance (given by $\rho L/(\pi r^2)$) is of the order of $1 \times 10^8\ \Omega$. Following the principles mentioned in Section 3.4, the input impedance needs to be around 100 times this figure. Field-effect transistors or other high-impedance components

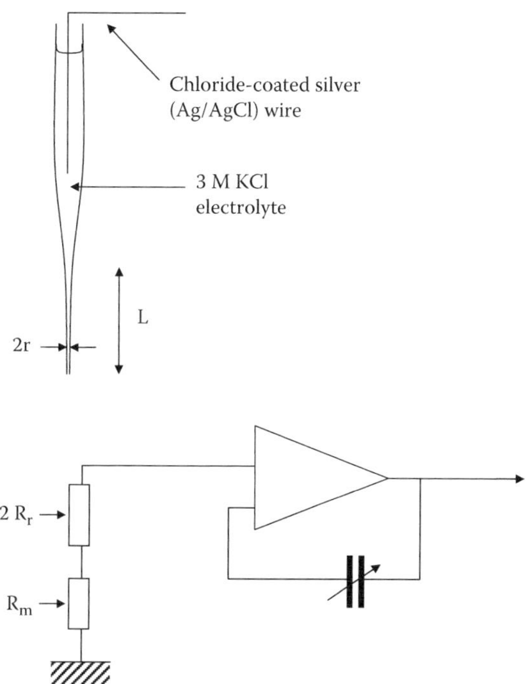

FIGURE 3.2
Top: a glass microelectrode. Typically, the tip is around 1 mm in diameter (2r), small enough to give minimal disruption of cell function when the cell is impaled by this tip. A second microelectrode is placed in the external solution in order to measure the transmembrane potential. In general, these electrodes need to be balanced with regard to their tip resistance and tip potential. Bottom: capacitance artifacts can be minimized by the use of a feedback amplifier, as shown here. R_r, R_m are resistances of electrode and membrane respectively.

need to be at the input stage of the preamplifier to prevent signal attenuation. The capacitance associated with the ability of the glass to hold charge tends to distort the signals measured (even if this amounts to 10^{-12} F (1 pF), the time constant is still several milliseconds, which is similar in duration to a nerve action potential). A feedback amplifier, with a variable capacitor in the feedback path, is used to compensate for this signal distortion.

Filling these glass microelectrodes with electrolyte solution also requires some skill. Basically, the microelectrodes are placed tip down in a heated flask containing KCl solution, and a vacuum is applied. When the vacuum is released, the air in the microelectrodes can be seen emerging as a series of small bubbles. Prior to use in experiments, the electrode impedance and spontaneous tip potentials are routinely measured to ensure that these are within acceptable ranges and give balanced pairs. Electrodes consisting of fine metal wires (insulated except for the final few micrometers) have also been used where the high-frequency components, such as the nerve firing rate (and not the steady transmembrane potentials), need to be monitored. In addition, microfabrication methods have been used to make arrays of silicon microelectrodes, which can be implanted in the brain to stimulate or record from specific regions. The electrodes are typically several hundreds of micrometers apart, so they are intercellular rather than intracellular. In experimental neuroscience, microelectrode arrays (MEAs) consisting of gold tracks etched onto a glass or plastic substrate have been used in the study of neurons in culture or of thin slices of rat brain, where function can be maintained for several days under appropriate treatment (Figure 3.3).

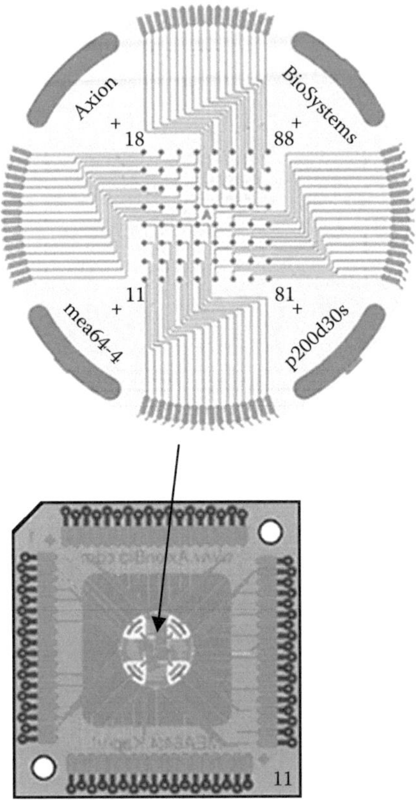

FIGURE 3.3
An MEA. The top diagram shows 64 gold electrodes, each 30 μm in diameter, in an 8 × 8 array (total size of array: 1.6 mm × 1.6 mm). Individual neurons can be grown in cell culture on this region, and then pairs of electrodes can be selected to stimulate and record from these cultured neurons. Alternatively, thin slices of brain (~0.2 mm in thickness, typically from the rat hippocampus, associated with memory formation) can be placed in this region. The array forms part of a larger plate (shown in the bottom diagram) that connects to the stimulating and recording electronics. The central region is submerged in physiological fluid to maintain the cells in a vital state. The electrode leads are on the reverse side of the plate, so they can be insulated from each other. (Courtesy of Axion BioSystems, Atlanta, GA, USA.)

Tutorial Questions

1. Explain what is meant by electrodes of the first and second kind and give examples of both.

   Estimate the p.d. in the following systems at 37°C:

   (a) Ag | AgCl | KCl (0.01 M) || KCl (0.003 M) | AgCl | Ag

   (b) Ag | AgCl | KCl (0.01 M) || KCl (0.01 M) | Hg_2Cl_2 | Hg

   Standard electrode potentials:

   $AgCl + e^- = Ag + Cl^-$; 0.222 V

   $Hg_2Cl_2 + 2e^- = 2Hg + 2Cl^-$; 0.268 V

   Assume that liquid junction potentials are negligible.

2. (a) If an electrode is not in equilibrium, explain the two major mechanisms thought to be responsible for the overpotential.

   (b) Given that the Butler–Volmer equation holds in an electrode of nickel placed in a dilute solution of KCl (such that a hydrogen reaction is occurring), what is the net current density from solution to metal if the exchange current density is 6×10^{-6} mA cm^{-2} and the overpotential has a measured value of 7.3 mV?

3. (a) Define "overpotential." What processes contribute to this overpotential?

   (b) What net current will produce an overpotential of 50 mV in a Pb electrode with an area of 0.5 cm^2? Assume that the electrode reaction is

$$H^+ + e^- = \tfrac{1}{2}H,$$

   in which the exchange current at this electrode is 5×10^{-12} mA/cm and the symmetry factor is 0.58. How is exchange current linked with electrode reversibility?

   (c) What is the limiting current that can be drawn from an Ag electrode dipped into a 1-mM solution of $AgNO_3$ if the area of the electrode is 0.1 cm^{-2}? (Take the diffusion coefficient for Ag^+ to be 2×10^{-5} cm^2/s and the Nernst layer thickness to be 0.4 mm.)

4. Explain what η, α, and j_e^+ are in the Butler–Volmer equation

$$J^+_{S\to M} = j_e^+\left\{\exp(-zF\alpha\eta/RT) - \exp(zF(1-\alpha)\eta/RT)\right\},$$

   and how are they measured?

   A current of 2 μA is passed through a Pt electrode with an area of 2 cm^2. Calculate the overpotential given that the process at the electrode is

$$H^+ + e^- = \tfrac{1}{2}H$$

   and that the j_e^+ for this process is 0.8×10^{-3} mA cm^{-2}. Estimate the charge-transfer resistance for this electrode. How does the charge-transfer overpotential differ from the diffusion (concentration) overpotential?

5. (a) Describe qualitatively the processes that give rise to charge-transfer and concentration (or diffusion) overpotentials.

   (b) A divalent metal electrode is placed in a solution of its own ions, and a current is passed between it and a second, similar electrode. The overpotential is measured between the first electrode and a third, reference electrode, and the results are shown in the table below.

Overpotential (mV)	**Net Current $J^+_{S\to M}$ (μA/cm^2)**	**ln($J^+_{S\to M}$)**
12.5	0.84	–0.18
25	1.92	0.65
50	5.75	1.75
75	15.8	2.76
100	44.7	3.80

Calculate the exchange current density for the electrode and the symmetry factor α, explaining what this factor is.

(c) Given that the limiting current density is 1.5 mA/cm^2, is the concentration overpotential important at these measured current densities? (Show reasoning.)

6. A differential amplifier has the following characteristics:
   - Input resistance to ground: $10^7\ \Omega$
   - Common mode rejection: 10^5
   - Gain: 10^3

   In the course of recording a bioelectric event, a 10-V (p–p) 50-Hz interference signal is induced between the differential inputs and ground.

   What will be the output voltage (p–p) of this interference signal given the following conditions?

   (a) Both recording electrodes have a resistance of $10^3\ \Omega$.

   (b) One electrode has a resistance of $10^4\ \Omega$ while the other has a resistance of $1.1 \times 10^4\ \Omega$.

7. (a) A current of 200 mA is drawn from an electrode constructed from a 2 cm × 2 cm square of platinum dipped into a solution containing H^+ ions. At this current, the overpotential is measured to be 200 mV. Assuming that in the energy diagram for the hydrogen reaction at this electrode the maximum value is midway between the inner and outer Helmholtz planes, calculate the magnitude of the exchange current at this interface. How does a particular exchange current influence the behavior of an electrode?

   (b) Distinguish between charge-transfer and diffusion overpotentials. Explain the term "limiting current density." If in the above system the limiting current density is 80 mA/cm^2, calculate the total direct-current resistance if the current drawn is still 200 mA. If the H^+ concentration were increased 10-fold, what effect would this have on resistance?

Bibliography

Aidley DJ. 1998. *The Physiology of Excitable Cells*, 4th Edition. Cambridge University Press, Cambridge, UK.

Atkins PW, de Paula J. 2010. *Physical Chemistry*, 9th Edition. Oxford University Press, Oxford.

Part II

Mainly Molecules

4

Molecular Biophysics

Andrew W. Wood

CONTENTS

4.1 Structure and Function of Some Important Biomolecules

In the previous chapter, we saw some of the techniques used to probe and monitor electrical and other signals from the human body and from single cells. Our understanding of the basic mechanisms behind physiological phenomena is aided by advances, particularly in the last half century, in knowledge of the structure and synthesis of the basic molecular building blocks of the cell. Electron microscopy has allowed the visualization of basic cellular components such as mitochondria, chromosomes, and the cell membrane, but the practical limit of resolution (around 1 nm) prevents the determination of the structure of individual molecules (although this can be inferred from molecular formulas and from knowledge of which parts of the molecule are happy to be close to water (termed *hydrophilic*) and those that do not (termed *hydrophobic*). More recently, atomic force microscopy (discussed further in Chapter 27) has been used to visualize the atomic details at membrane surfaces. However, the most important technique used to determine molecular structure has been X-ray crystallography. Because the wavelengths of X-rays (0.01 nm–10 nm) can be less than interatomic distances, details at this scale can be inferred. The particular molecule to be studied has first to be extracted from the cell and then crystallized, which is, in itself, no mean feat. In crystalline form, the molecule is in a regular lattice, and any repeating feature will give rise to characteristic "spots" in the crystallogram. The basic principles of crystallography are summarized in Figure 4.1. This relies on the work of the Nobel Prize–winning father-and-son team of W. and W. L. Bragg.

Because the amount of detail requires highly monochromatic (single wavelength) and intense X-ray beams, most of the significant work carried out at present uses X-rays from a synchrotron (a football-field-sized installation that derives X-rays from electrons accelerated to near the velocity of light being allowed to impinge on an X-ray-producing target). An example of a recent determination of the structure of the cell membrane-bound

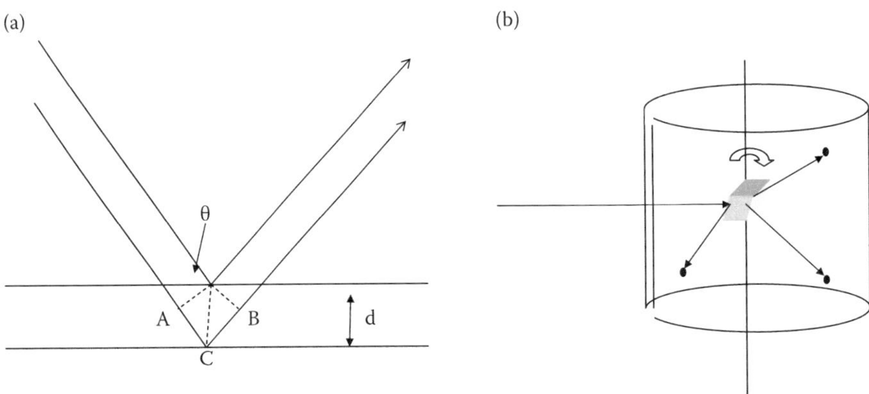

FIGURE 4.1
(a) The extra path length for the lower X-ray beam is 2d sin θ. If this is a whole number of wavelengths, then the emerging beams interfere constructively (Bragg condition). (b) Single crystal X-ray diffraction. As the crystal is rotated, the Bragg condition is satisfied for particular orientations and the cylindrical X-ray film is darkened at certain spots.

calcium ion pump is shown diagrammatically in Figure 4.2. Because the pump acts on Ca^{2+} without being chemically altered (the chemical reaction ATP = ADP + P_i does not cause modification in the pump's chemical make-up), it acts as an enzyme and, for this reason, is also known as a Ca^{2+}-dependent ATPase (the -ase ending denoting that it is an enzyme acting, in this case, on ATP).

Biological molecules are organic; that is, they are based on carbon (C) (although silicon (Si)-based life forms have been postulated as having predated carbon-based forms). Other atomic components of large biomolecules are hydrogen (H), oxygen (O), nitrogen (N), phosphorus (P), and sulfur (S). There are other atoms that are associated with biomolecules but do not strictly form part of their structure. These include, among others, calcium, iron, copper, and magnesium. Sodium, potassium, chloride, and, of course, water are the main constituents of tissue fluid. There now follows a brief description of the structures of the more important biomolecules.

4.1.1 DNA and RNA

The structure of deoxyribose nucleic acid (DNA) was published in 1953 by J. D. Watson and F. Crick. Together with X-ray crystallographers M. Wilkins and M. Perutz, they were awarded the Nobel Prize in 1962. DNA is the molecule out of which chromosomes are formed in the cell nucleus and which contains the genetic code in 40,000 or so genes in the human. The basic structure is a double helix of four subunits (bases) attached to a sugar-like backbone (Figure 4.3). The four bases are adenosine (A), cytosine (C), thiamine (T), and guanine (G). Because of the chemical composition of these four, A will always link with T and G with C (the easy way to remember this is that the curvy letters (G and C) and the straight letters (A and T) go together). When a particular gene (which is a stretch of DNA) is *transcribed*, the two strands of the helix are parted, and a complementary pair of DNA-like strands is formed. The enzyme RNA transcriptase is responsible for this process. The subtle difference is that instead of C, a different base, uracil (U), is substituted. The molecule is known as ribonucleic acid, or RNA. This form of RNA, the messenger RNA (or

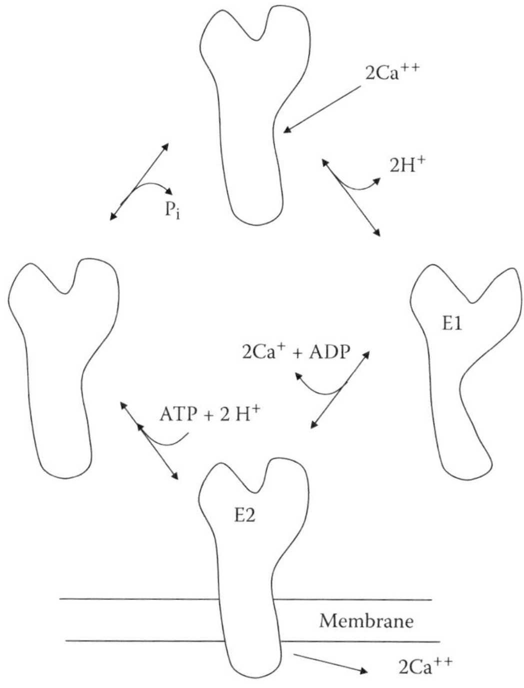

FIGURE 4.2
A diagrammatic representation of the calcium pump. As ATP is converted to ADP, the molecule undergoes a conformational change, allowing Ca^{2+} to enter a "tunnel." As H^+ exchanges for Ca^{2+}, the molecule returns to its original shape, releasing Ca^{2+} into the external solution. Note that the pump molecule has two conformations, E1 and E2, which prevent Ca^{2+} from returning to the cytoplasm, where the concentration is lower. (Redrawn from Olesen et al. [2007]. See this reference for full details of structure.)

mRNA), effectively carries the genetic code from the relevant gene out of the nucleus and into the endoplasmic reticulum (ER) of the cytoplasm. Here, the second part of converting the genetic "blueprint" into a protein product, translation, is brought about in specialized structures known as *ribosomes* (attached to the wall of the ER). Here, the genetic code is read off, but in order to appreciate this process, some knowledge of protein structure is now required.

4.1.2 Proteins

Enzymes and structural features such as microtubules, connective strands, and filaments are all constructed from proteins. Proteins also make up the large proportion of hair, cartilage, tendons, and also bone and teeth. Proteins are formed from sequences of some 24 amino acids, which have the generic structure H_2N-CHR-COOH, where R (the residue) is given in Table 4.1 with the three-letter and one-letter codes for each amino acid. Because small numbers of amino acids can be linked or converted to form peptides, proteins are sometimes referred to as *polypeptides*. Because each amino acid residue has specific hydrophobic or hydrophilic properties (and because Cys tend to form strong cross-bridges), a long polypeptide strand will tend to coil or convolute in order to minimize the interaction energy between each amino acid and its environment. This will give rise to a specific

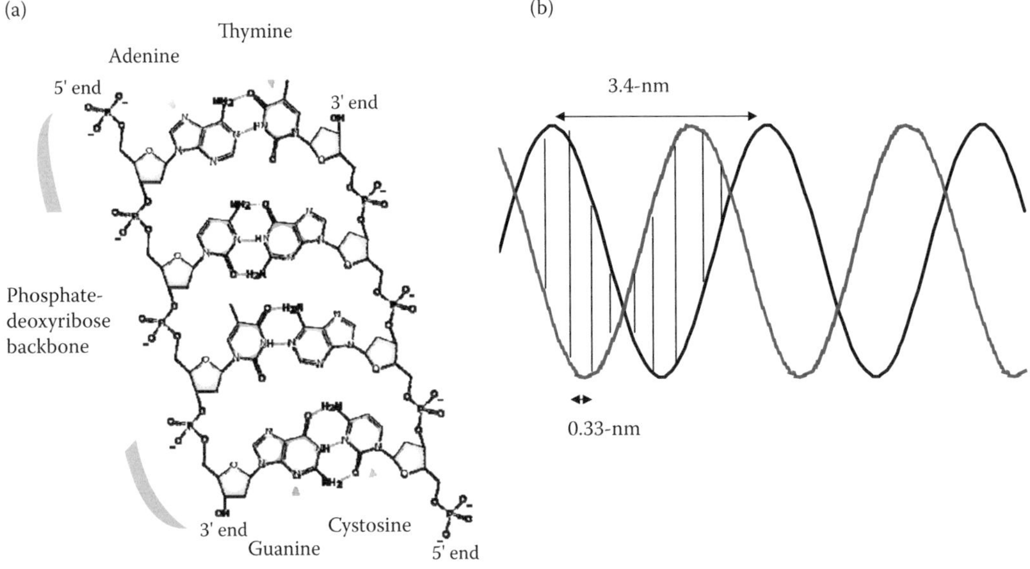

FIGURE 4.3
(a) DNA structure: the two base pairs—adenine and thymine, and guanine and cytosine—and the associated sugar–phosphate "backbone." (Courtesy of Madeleine Price Ball.) (b) DNA double-stranded helix showing the 0.33-nm distance between adjacent base pairs and the 3.4-nm "pitch." The radius is 1 nm.

tertiary structure and one that can be revealed by X-ray crystallography. Particular repeating sequences of amino acids give rise to a helical structure known as α-helices, which form the basis of structure of hair, teeth, and so on. Another form is the β-strand (or β-pleated sheet), which is often interspersed with α-helices in large proteins. Other proteins tend to form into a ball-like structure (globular proteins) of which probably the best known is hemoglobin, the oxygen-carrying molecule contained within red blood cells. The calcium pump shown in Figure 4.2 is globular protein, but one that is contained within the cell wall and has a number of α-helical regions (shown). Note that the protein may consist of several polypeptide subunits, which are produced in conjunction with separate genes. Much of what we know about the function of proteins (as opposed to structure) comes from the study of mutant microorganisms, in which particular subunits of a complex protein may be absent.

A particularly important set of proteins is the ribosomes, which, as we have seen, are responsible for the translation of strands of RNA to specific amino acid sequences. We are now in a position to revisit the process that we have begun to describe in the previous section. The ribosome has two main sections: the lower section, in which the mRNA is held, and the upper section, which is divided into two, each holding another form of RNA, transfer or tRNA, which consists of three bases and which can link to specific amino acids. The lower section contains exactly six bases of the mRNA, to which two complementary tRNAs attach. The right-hand tRNA has, in addition to the specific amino acid for this particular tRNA, the amino acids that have already been assembled in this process. In the final stage of the cycle, the new amino acid is attached to the end of the completed chain, and then the whole structure moves three bases to the right along the mRNA and the process repeats (Figure 4.4).

TABLE 4.1
List of Amino Acids and Associated Codons

Name	Three-Letter Abbreviation	One-Letter Abbreviation	Codon(s)
Alanine	Ala	A	GCU, GCC, GCA, GCG
Arginine	Arg	R	CGU, CGC, CGA, CGG, AGA, AGG
Asparagine	Asn	N	AAU, AAC
Aspartic acid	Asp	D	GAU, GAC
Cysteine	Cys	C	UGU, UGC
Glutamine	Gln	Q	CAA, CAG
Glutamic acid	Glu	E	GAA, GAG
Glycine	Gly	G	GGU, GGC, GGA, GGG
Histidine	His	H	CAU, CAC
Isoleucine	Ile	I	AUU, AUC, AUA
Leucine	Leu	L	UUA, UUG, CUU, CUC, CUA, CUG
Lysine	Lys	K	AAA, AAG
Methionine	Met	M	AUG
Phenylalanine	Phe	F	UUU, UUC
Proline	Pro	P	CCU, CCC, CCA, CCG
Serine	Ser	S	UCU, UCC, UCA, UCG, AGU, AGC
Threonine	Thr	T	ACU, ACC, ACA, ACG
Tryptophan	Trp	W	UGG
Tyrosine	Tyr	Y	UAU, UAC
Valine	Val	V	GUU, GUC, GUA, GUG
	START		AUG
	STOP		UAA, UGA, UAG

Note: The 20 amino acids with their three- and one-letter abbreviations, plus the groups of three RNA bases that code for that particular amino acid (codon). Note that in general, more than one codon codes for a specific amino acid. There are also specific codons that signal the start and stop of a particular codon sequence (or gene) which codes for a particular protein, consisting of a finite sequence of amino acids. There are in fact two more amino acids (selenocysteine and pyrrolysine) that are incorporated into proteins via another mechanism.

4.1.3 Cell Membrane: Phospholipid Bilayers

The membranes of the cell wall, the ER, and the nucleus all share a similar structure, the lipid (or phospholipid) bilayer. This basic structure of the membrane, identified from electron microscope images and other experimental techniques, is about 8 nm thick. Phospholipids have two distinct regions: a hydrophilic (water-loving) end containing the phosphate group and a hydrophobic (water-hating) part consisting of hydrocarbon chains that orient themselves to be away from water. In fact, solutions of phospholipids will form themselves quite spontaneously into small spheres, with the hydrophobic ends pointing inward, or even into spheres consisting of lipid bilayers, with the inner layer having hydrophilic regions next to the entrapped water. These structures are known as *micelles* and are useful for studying biochemical reactions in quasicellular environments. That actual cells have a similar lipid bilayer structure has been known for more than 50 years. In early experiments (by Gorter and Grendel [1925]), lipids were extracted from red blood cells by mixing these with solvent acetone and then spreading the lipids on a water

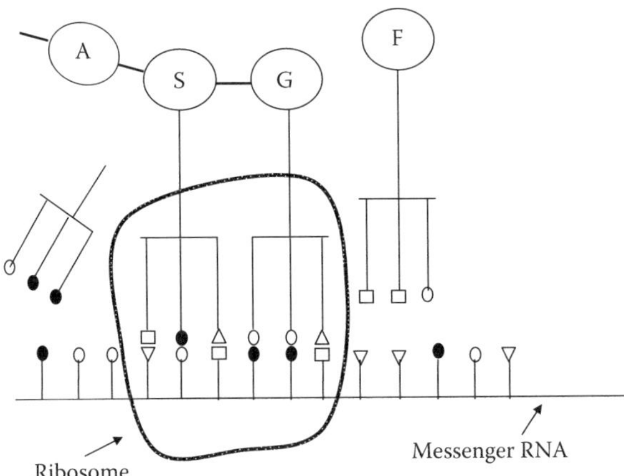

FIGURE 4.4
Ribosomal translation. Four transfer RNAs (tRNAs) are shown: with anticodons specific for A, S, G, and F. The messenger RNA (mRNA) feeds through the lower portion of the ribosome, which takes in two tRNAs at a time. The amino acid F (phenylalanine) is the next to be added to the protein sequence at top. The amino acid A has just been released from its specific tRNA. Whereas the S and G tRNAs are held in the ribosome, the respective amino acids S and G are chemically joined.

surface with the use of a device known as a Langmuir trough. Lipid spread on a water surface in this way forms a monolayer with the hydrophobic regions in the air. The movable dam of the trough has a sensitive force transducer on it that registers the minimum area that the monolayer can occupy before buckling starts. Knowing the number of red cells per unit volume and the cell diameter, one will not find it hard to estimate the total membrane area before lipids have been extracted and then compare this to the minimum area in the Langmuir trough. The membrane area turned out to be half of this minimum area, hence a bilayer. Subsequent electron microscope techniques, in which it is assumed that the osmium stain attaches to the hydrophilic regions, show a characteristic "tramline" pattern that would be expected from a bilayer.

There are several different types of phospholipid present in the membrane, and different types of cell have different compositions. Even the two sides of the bilayer have different compositions. Phosphatidyl choline, phosphatidyl ethanolamine sphingomyelin, and phosphatidyl serine are common phospholipids, each around 4 nm long and 0.3 nm across. Cholesterol is also an essential membrane component, having similar hydrophobic and hydrophilic regions to phospholipids. These membrane molecules are not held rigidly in place; they diffuse around laterally and even flip from one side to the other (see Figure 4.5).

It is also known from analysis that membranes contain protein, about 40% by weight. The original Davson and Danielli (Danielli and Davson 1935) model had the protein spread out over both surfaces in the bilayer, but the commonly accepted picture is now the Singer and Nicholson (1972) model, in which the protein is interspersed within the lipid bilayer and, in some cases, goes all the way from one side to the other as shown in Figure 1.2. For example, most ion pumps (such as the active sodium–potassium exchange pump) do this, as indeed do the passive voltage-gated sodium and potassium channels in nerve membranes. Strands of glycoprotein also extend from the external surface. These are involved with some of the recognition, or immune response, properties of the cell.

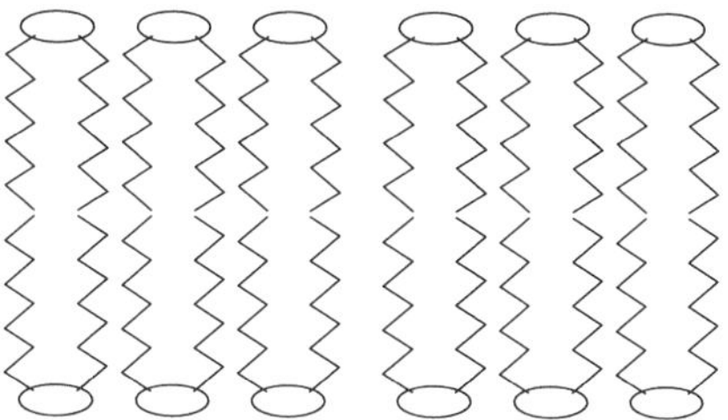

FIGURE 4.5
Representation of a lipid bilayer showing the two monolayers of phospholipid with their hydrophobic (hydrocarbon chain) ends facing each other. The ovals represent the polar (hydrophilic) end groups, which vary according to the species of phospholipid. For example, phosphatidyl choline (lecithin) has a polar part as follows:

```
CH2O-CO-R1
|
CHO-CO-R2
|
|    O
|    ||
CH2O-P-CCH2.CH2.N+(CH3)3
     |
     O-
```

where R_1 and R_2 are fatty acyl side chains such as $(CH_2)_{14}$-CH_3.

4.2 Origin of Bioelectricity

As mentioned in Chapter 1, the ability of certain fish to produce electric shocks has been known since ancient times. The familiar ECG, or electrocardiograph, measures the potential differences (PDs; around 1 mV in magnitude) on the skin surface due to the electrical activity of the heart, and PDs of smaller magnitude can be measured from the scalp (electroencephalogram, or EEG) and from contracting muscles (electromyogram, or EMG). This electrical activity is due to the passage along nerves and muscles of electrical impulses, each a few milliseconds in duration and consisting of a PD reversal across the membrane of around 100 mV. All cells have the property of polarization; that is, they will normally have a PD across the membrane of 50 mV–100 mV, outside positive. This standing, or resting, PD is equivalent to that of a rechargeable battery: some energy has to be put in to maintain the PD at this level. In the case of the cell, this is brought about by the activity of the sodium (Na^+), or most commonly the sodium–potassium (K^+), pump, which pumps three Na^+ ions for two K^+ ions in. This is "fuelled" by the Na/K ATPase enzyme activity of the pump, breaking down ATP and releasing 30 kJ/mol ATP, as we have seen in Chapter 1. The Na/K pump has similar features to the Ca^{2+} pump shown in Figure 4.2 (these will be described in greater detail in the next chapter). However, the tendency for positive charge to be transferred to the outside surface of the membrane gives rise to the PD (the process is actually a little more complicated, as we will see). The cell membrane can be thought of as a charged capacitor, with the polar end groups on the surface equivalent to parallel plates, and the lipid interior as the dielectric. The relative permittivity of a membrane lipid is around 5–7. The nervous impulse can be thought of as a sudden capacitor discharge, which is due to short-circuiting channels suddenly opening up in the dielectric, but again, this is an oversimplification. The pulse of 300 V produced by the electric fish *torpedo* is due to thousands of cells in series producing impulses simultaneously.

4.3 Molecular Motors

Recently, there has been a lot of interest in the locomotion of small bacteria and unicellular organisms called *flagellates*, which use a powered propeller-like tail. The hope is that it may be possible to replicate the molecules responsible for this locomotion artificially, to power the targeted delivery of therapeutic drugs to specific tissue. It appears that some of these motors are driven by an Na^+ pump or, in the case of Figure 4.6, an H^+ pump. There is a ring of specialized protein structures within the lipid bilayer (stator) and a corresponding ring on the tail or flagellum (rotor). The active transport out of the cell through the clefts, coupled with the rectification of Brownian motion (Brownian ratchet) between the stator and rotor, gives rise to the torque on the flagellum, which can rotate up to 3000 rps.

Another class of molecular motor is related to the molecules actin and myosin, which are the basis of muscular contraction. Myosin has a "head" which "cocks" when associated ATP breaks down to ADP. Normally, the head is attracted to particular sites on the actin, but as the head rotates during the "cocking" phase, the head is released. As the head returns to its initial position, it attaches further down the actin, and as the cycle repeats, the head continues to "walk" down the actin.

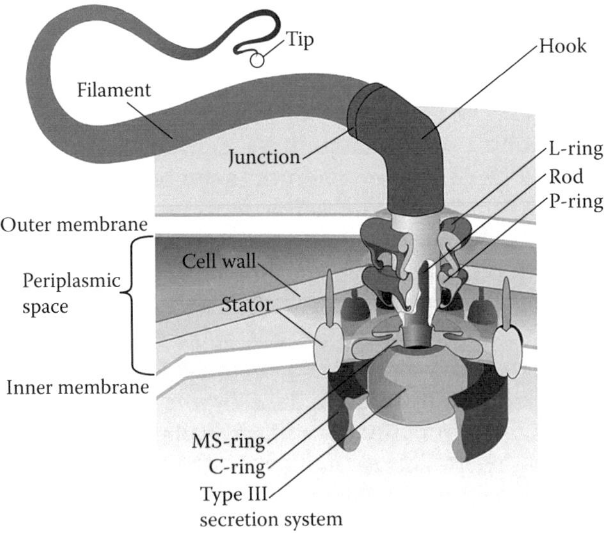

FIGURE 4.6
Diagram of a bacterial flagellum, which has a rotary molecular motor. The filament is free to rotate around the hook junction to the rod. The motive force is provided by a proton (H^+) pump in the lower portion of the structure, driving H^+ ions across the inner membrane. Rotational speeds of up to 1000 rpm can be attained. (Courtesy of LadyofHats. http://en.wikipedia.org/wiki/File:Flagellum_base_diagram_en.svg.)

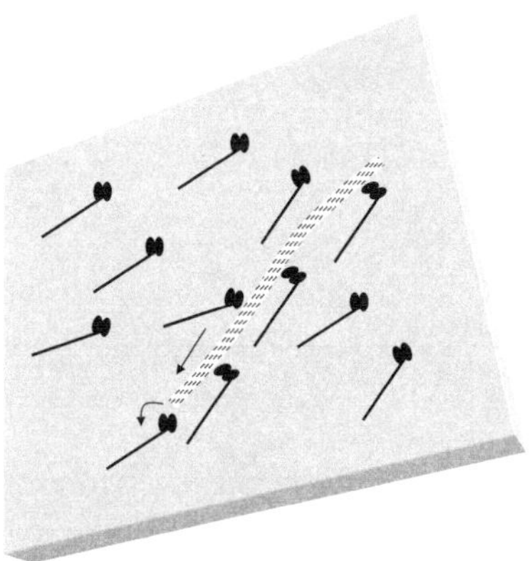

FIGURE 4.7
The muscle protein myosin (black shapes) immobilized on a coated glass slide. The light-colored rod represents an actin molecule, which moves in the direction as indicated when ATP is added to the solution above the slide. The "heads" of the myosin "cock" as shown to give the forward movement, which can be visualized by adding a fluorescent marker to the actin.

This can be demonstrated on a microscope slide in which myosin molecules have been laid down to form tracks. If fluorescent forms of actin are then added to the slide in a small amount of Ca^{2+}-containing water, plus ATP, they can be observed to move at a velocity of around 2 micrometers per second (Figure 4.7).

There are other classes of biomolecule that demonstrate similar properties. These are the dyneins and kinesins (which move in preferred, but opposite, directions along cellular microtubules), which are responsible for the motion of cilia (the hairlike projections from the cells lining the airways and gut, responsible, among other things, for the movement of mucus) and for the movement of certain materials within, for example, nerve cells.

Another class of motor, powered by nucleotide triphosphate (NTP) rather than ATP, is the helicases, responsible for the "unzipping" of DNA and RNA strands and the movement of RNA strands through membranes pores.

4.4 Structure of Water

At the level of ion channels (where the width is around 0.4 nm at the narrowest point of the channel), the amount of water associated with the particular ion and the degree of structure of this water become critically important. Water molecules exhibit an amazing variety of transient structure due to the particular conformation of the atoms within the molecule and the small excess of positive charge on the hydrogen atoms (and the corresponding negative charge on the oxygen). The isolated H_2O molecule is shown in Figure 4.8, with the way that these molecules join when water freezes to become ice. Figure 4.9c shows how the intense positive charge surrounding cations (a cation is an ion attracted to a negative electrode, or cathode) such as Na^+ and K^+ attracts a "hydrated shell," making the effective radius much greater than the ionic radius. Inasmuch as the electric field (E) surrounding a

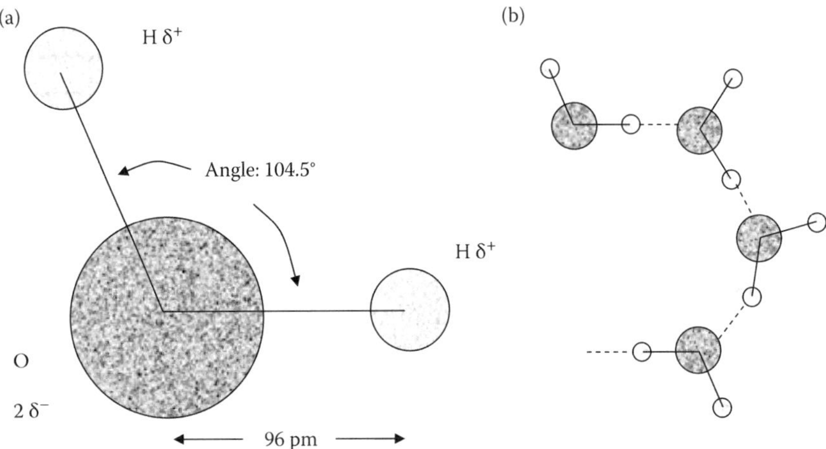

FIGURE 4.8
Structure of water. (a) A single water molecule, showing the approximate dimensions and the slight excess of positive charge on the H molecules and balancing negative charge on the O molecule. (b) Image shows how hydrogen bonding can create tetrahedral structures ("structured water").

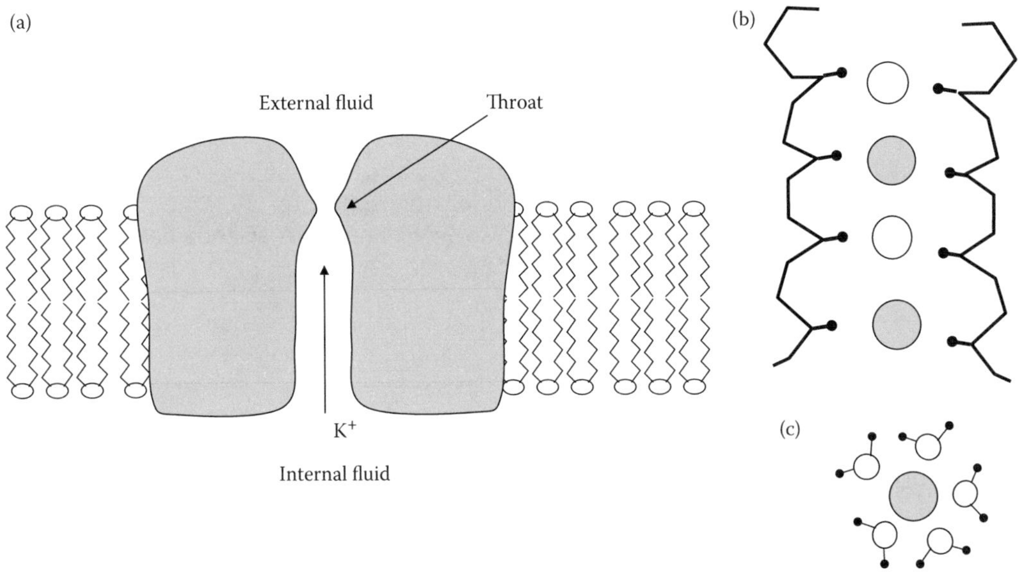

FIGURE 4.9
The K^+ channel. (a) Channel as a tube of protein straddling the membrane. The "throat" is around 0.4 nm in diameter and consists of four subunits, as shown in panel b. This shows how the K^+ ion must lose its water of hydration (shown in panel c) to move through the throat. The "naked" K^+ ions occupy either positions 1 and 3 or 2 and 4 as they "shuttle" through. Carboxyl groups on the amino acids lining the throat mimic the water of hydration molecules; hence, it is favorable for the K^+ ions to occupy these positions. Movement through the channel is controlled by a combination of concentration and voltage gradients across the membrane and, in some cases, the operation of a voltage-sensitive "gate" (not shown).

point source of charge varies as 1/r, the value for E is greater for a smaller ion such as Na^+ (atomic weight (AW) = 23) compared to K^+ (AW = 39). For this reason, Na^+ ions tend to diffuse more slowly than K^+, which is counterintuitive. For anions (an ion that is attracted to an anode and thus negatively charged), the hydration shell is formed of water molecules, with H atoms oriented inward.

Recently, the work of MacKinnon (2003) and others (MacKinnon won a Nobel Prize for his work in 2003) have established that in order for K^+ to permeate the K^+ channel, each individual ion has to lose its water of hydration (Figure 4.9b). In fact, each ion has to be accompanied by a second ion, so that they can occupy two of four positions within the "throat" (the other two positions being taken up by oxygen atoms of water molecules).

Water permeates cells via separate channels, or aquaporins, the study of which earned Agre the Nobel Prize the same year (Agre 2003).

At body temperature and within cells, water is not totally disordered but retains some of the ice-like characteristics, forming "flickering clusters." Certain classes of components within cells tend to promote structure, whereas others are "structure breakers." This leads to variations in water properties from location to location within the cell. Some have argued that this can explain at least some of the unique behaviors of cell physiology (Pollack 2001).

Bibliography

Agre P. 2003. "Peter Agre-Nobel Lecture." http://www.nobelprize.org/nobel_prizes/chemistry/laureates/2003/agre-lecture.html (accessed Oct. 3, 2011).

Danielli JF, Davson H. 1935. A contribution to the theory of permeability of thin films. *J. Cell. Comp. Phys.* 5:495–508.

Gorter E, Grendel F. 1925. On bimolecular layers of lipids on the chromocytes of the blood. *J. Exp. Med.* 41:439–443.

MacKinnon R. 2003. "Roderick MacKinnon-Nobel Lecture." http://www.nobelprize.org/nobel_prizes/chemistry/laureates/2003/mackinnon-lecture.html (accessed Oct. 3, 2011).

Olesen C, Picard M, Winther A-ML, Gyrup C, Morth J, Oxvig C, Moller JV, Nissen P. 2007. The structural basis of calcium transport by the calcium pump. *Nature* 450:1036–1042.

Pollack GH. 2001. *Cells, Gels and the Engines of Life: A New, Unifying Approach to Cell Function.* Ebner & Sons, Seattle.

Robertson RN. 1993. *The Lively Membranes.* Cambridge University Press, Cambridge.

Singer SJ, Nicholson GL. 1972. The fluid mosaic model of the structure of cell membranes. *Science* 175:720–731.

Watson JD, Crick FHC. 1953. A structure for deoxyribose nucleic acid. *Nature* 171:737–738.

5

Membrane Biophysics

Andrew W. Wood

CONTENTS

5.1 Membrane Structure and Function

5.1.1 Membrane Structure

As described in the previous chapter, the cell or plasma membrane is a phospholipid bilayer and is more than being just a bag to enclose the structures such as the nucleus and mitochondria within the cell. It is strongly selective in the materials it allows to cross, and certain materials are assisted by active transport (such as ion pumps). It naturally has a potential difference (PD) of around 1/10 V, with the inside negative with respect to the outside. Although some oil-soluble materials are able to permeate the lipid bilayer by directly dissolving within the lipid "tails," many substances permeate the membrane via specific cylindrical channels or pores. For example, there are many forms of potassium (K^+) channel, and water moves through specific channels known as the aquaporins. As we saw in the previous chapter, channels tend also to be "gated"; that is, they open in response to favorable PD changes or the association of particular transmitter substances with parts of the channel.

5.1.2 Membrane Function

The fluids on the inside and outside of the cell are essentially asymmetrical. The sodium chloride concentration of extracellular fluid is high (and the potassium chloride concentration low) in the same way it is in seawater. Within the cell cytoplasm, this situation is reversed, with the K^+ concentration ratio around 50 or more (see Figure 5.1a). Because each ion can be considered in isolation, there would be no reason that the extra K^+ would not diffuse out of the cell were it not for the PD across the membrane (the positive charge on the outside would repel K^+ ions as they tried to leave. If, however, the active ion pump were to be inhibited and the PD fell to zero, K^+ ions would flow out and Na^+ in to even up the concentrations. The cell would cease to function and would die. Nerve cells in fact use this asymmetry as a basis for the origin of the nervous impulse: here, there are momentary changes in the relative permeability of the membrane to Na^+ and K^+ ions such that for a few milliseconds, Na^+ rushes in then K^+ out. This is accompanied by a brief reversal of PD across the membrane, and it is this effect that is propagated along the nerve axon. A similar process couples the excitation to the contraction of muscle. These phenomena will be dealt with in greater detail in later chapters.

In multicellular organisms, the way individual cells join onto others is obviously very important. We will consider the three basic layers that organisms can be considered as being made up from: epithelium, endothelium and the remainder. In humans, the first of these is not only the skin but also forms the lining of many internal organs having connection with the outside world, such as the alveoli of the lungs, the tubules of the kidney, the internal surface of the gullet, stomach and intestines, and so on. The endothelium, which in some ways is similar to the epithelium, lines the blood and lymph vessels and would not normally be exposed to the outside world. The remainder consists of connective tissue,

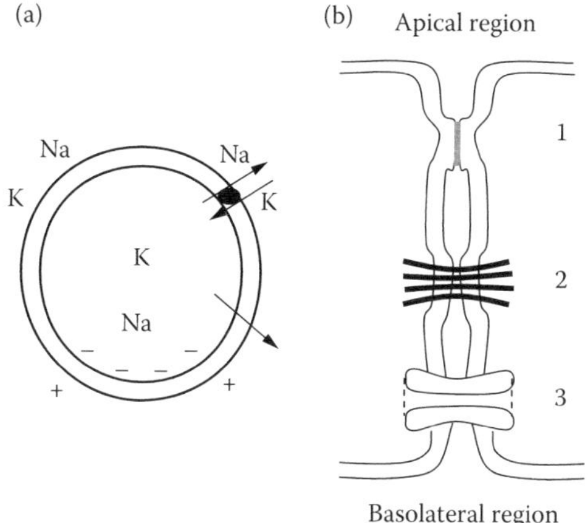

FIGURE 5.1
A simple representation of the cell is shown in (a), with high K^+ concentration inside and Na^+ concentration outside of the cell. The Na^+/K^+ pump preserves this asymmetry, but a small amount of K^+ may attempt to 'leak' out, giving rise to a positive charge on the outside of the cell. Two cells are shown in contact in (b) as they would be in an epithelial layer, for example. The three types of junction (shown 1, 2 and 3) are the tight junction, the desmosome and the gap junction respectively. Tight junctions have fused lipid bilayers—and consequently very high electrical resistance. These stop molecules getting between cells or from one cell to the next; desomosomes are like 'spot welds' between cells and consist of protein filaments; gap junctions consist of tube-like 'connexons' forming water-filled channels (whose diameter can be varied) allowing cell-to-cell communication.

nerves, muscle, and bone cells and does not form layers in quite the same ways as the other two types. Epithelial layers, in particular, are very interesting because they form into flat sheets with a relatively high electrical resistance (and hence low ion permeability) across them. Junctions between cells have been extensively studied, and some of these types are shown in Figure 5.1b.

The tight junctions form effective seals right around the top parts of epithelial cells, which force ions and water to take a cytoplasmic route from one side of the layer to the other. In fact, the ion pumps are concentrated toward the lower (in Figure 5.1a) or basolateral regions such that the Na^+ ions are pumped from the cytoplasm into the narrow clefts between the cells. The apical surface of the cell (usually characterized by countless tiny protrusions or microvilli) is quite permeable to Na^+, which is able to enter the cell by simple diffusion. This one-way movement of Na^+ causes a PD to be set up across the layer as a whole of around 50 mV–100 mV. As an example, a PD of this magnitude can be measured between the inside (or lumen) and outside (or serosal) surface or the kidney tubules. In fact, as we go along these tubules, the PD changes quite markedly (and even reverses along one section), allowing several separate processes to occur in different regions. Gap junctions are common in heart muscles, and these allow the easy passage of substances from cell to cell. In fact, the heart muscle functions as a complete unit (or syncytium) rather than a collection of individual cells, as in the case of skeletal muscle.

Many secretory processes have a strong link to membrane properties. An important distinction is between those glands, such as sweat and tear glands, which secrete across the epithelial layer, called exocrine glands, and those which secrete into the bloodstream (such as the pancreas), termed endocrine glands. Endocrine secretions, or hormones, are manufactured within the cytoplasm of cells then transported to the outside of these cells within small vesicles (with a bilayer structure) which initially fuse with the cell membrane and then disgorge their contents into the extracellular fluid. This process is known as exocytosis. The secretions then find their way into the bloodstream across capillary epithelia (usually, but not always). Exocrine secretory processes are somewhat similar, but secretions are usually into a blind tube or duct.

5.1.3 Mechanisms

Passage of substances across cell membranes is classified as either passive or active. These classifications are divided further:

5.1.3.1 Passive Mechanisms

These make use of existing physicochemical forces across the membrane—they do not need an energy supply and given sufficient time would move toward an equilibrium state. An appropriate way of studying these equilibria and movement toward them involves defining what is termed as the "electrochemical potential" of a chemical solution. This has terms relating to the concentration and pressure on the solution as well as its electrical potential. The unit of the electrochemical potential is joules per mole or J/mol.

5.1.3.2 Simple Diffusion

This involves a substance flowing from a more concentrated to a less concentrated region and the bigger the difference or gradient in concentration the greater the flow. In a mixture, each substance follows its own gradient, except if the substance is ionic, in which case the need to maintain electroneutrality adds an extra constraint. Simple diffusion is described mathematically by Fick's laws, which will be described soon. Across a membrane, in addition to the concentration difference, the degree to which the membrane allows the substance through (the permeability) determines the rate of flow, or flux as it is usually termed. (In fact, as in electricity, a flux is a flow or current per unit area: units are something like $mol/m^2/s$.) Diffusion is a random process at the molecular level: molecules tend to move into regions where they collide with fewer other molecules.

5.1.3.3 Electrodiffusion

If the inside and outside of the cell are at different electrical potentials and the substance is charged (i.e., is an ion) it will respond to the gradient in potential (which is the same as electric field). The resultant flow of charge is the same as an electrical current and Ohm's Law applies, although rather than resistance, its inverse, conductance is usually what is considered. Because a current of ions across a membrane will in general be due both to concentration and electrical gradients, the conductance is actually the electrical current divided by the electrochemical rather than electrical potential difference (this will be expanded in Section 5.5.1).

5.1.3.4 Osmotic Flow

This is just a specialized form of simple diffusion, but in this case the water is doing the diffusing across the membrane and the solute (or dissolved particle species), which is larger in diameter than water, is unable to permeate the membrane. The number of water molecules per unit volume (which is another way of expressing concentration) on the side where there is less solute will be greater, so water flows down its concentration gradient into the region with the higher solute concentration (which incidentally will make the solute concentration slowly less because of the diluting effect). For a rigid container this water flow can be prevented by applying an opposing hydrostatic pressure. It is this pressure that is known as osmotic pressure, and an appropriate unit is the pascal (Pa). In cell systems, most of the solutes that make up the osmotic pressure of the intracellular and extracellular fluids actually do get through the membrane, but relatively slowly.

5.1.3.5 Hydraulic Flow or Filtration

This is perhaps more important in plant rather than animal cells, but particularly in cell layers such as capillary endothelium, where there is distension of cells because of elevated fluid pressure, flow across the membrane can occur because of this. In general, the flow is of both solute and solvent species (i.e., NaCl and water) and is termed bulk flow. The membrane acts like a sieve, and the larger species do not permeate, but for the smaller species, the rate of flow depends on the pressure difference rather than the concentration difference.

5.1.3.6 Facilitated Diffusion

Here, some small carrier molecule within the membrane latches on to the substance in question (perhaps by chemical reaction), and the substrate–carrier complex then moves across the membrane (actually this movement is random, but the laws of probability give a preferred movement from the side with high substrate concentration to a lower one). The process is saturable; that is, the flow initially increases with increased substrate concentration but then levels off as the carriers all become occupied. Because the process just follows the laws of probability, there is no need for an external energy source.

5.1.3.7 Active Transport

The Na^+/K^+ pump was briefly discussed in Section 4.2 in the previous chapter. Some evidence of its structure and mechanism comes from isolating the protein subunits that still retain the enzyme activity associated with Na^+ and K^+. Having identified the molecular weight and the amino-acid sequence of the subunits, some guess can be made of detailed structure, but much is conjecture. Gross structure can be identified from electron microscopy and the study of mutant bacteria can further aid the understanding we have of its operation. The knob-like part inside the cell catalyses the breakdown of ATP to ADP but only in the presence of three Na^+ ions at specific sites. The energy released causes a conformational (or shape) change that exposes the Na^+ ions to the exterior. As they dissociate, two K^+ ions attach to sites specific for them as the molecule relaxes back into its original conformation, now exposing the K^+ ions to the interior.

5.2 Simple Diffusion and Electrodiffusion

5.2.1 Diffusive Processes

Fick's first law is empirical: if δc is the difference in concentration between two points δx apart, then the flow rate per unit area, or flux, J (in mol/m^2/s^1), in the x direction is proportional to the concentration gradient $\delta c/\delta x$, or dc/dx in the limit. The constant of proportionality D is known as the diffusion coefficient and the first law can be stated as

$$J = -D dc/dx$$

The minus sign indicates that if the concentration increases as x increases, then the flow is in the negative x direction.

In three dimensions, the equation becomes $\underline{J} = -\underline{\nabla} c$, where $\underline{\nabla}$ is the operator $\underline{i}d/dx + \underline{j}d/dy + \underline{k}d/dz$ and J becomes a vector.

Fick's second law gives the time dependence of changes and can be easily derived from the first law. Consider diffusion only in the x direction into and out of an imaginary slab (see Figure 5.2). The rate of change of concentration is related to the difference between the diffusive flux into and out of this slab. If S is the amount of the substance of interest contained in the slab, then the rate of change of concentration is given by $\delta S/(A.\ \delta x.\ \delta t)$ or $-\delta J/\delta x$. Substituting for J from the first law gives

$$\delta c/\delta t = \delta(-D\delta c/\delta x)/\delta x$$

or

$$\partial c/\partial t = D\partial^2 c/\partial x^2$$

and in three dimensions:

$$\partial c/\partial t = D\nabla^2 c.$$

Solution of this equation depends on the imposed boundary conditions and except for the simplest of arrangements gives fairly complicated analytical solutions. Numerical solutions, on the other hand, are much more convenient and three-dimensional computer packages exist to do this. It must be pointed out that the same equations describe the diffusion of heat by replacing c by temperature and D by κ, the thermometric conductivity (in m^2/s).

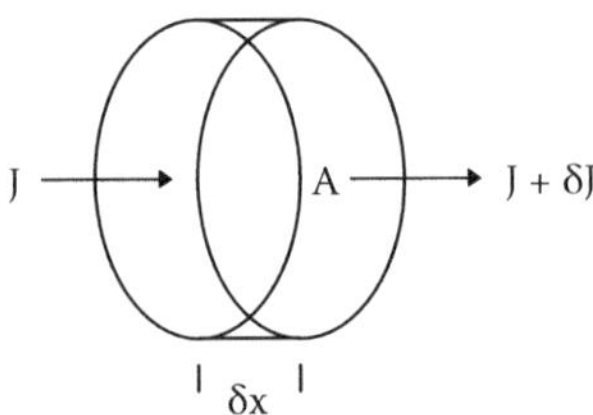

FIGURE 5.2
Diffusion. Imagine a small cylinder within a fluid, cross-sectional area A, δx thick. The flux rate of solute molecules into and out of the cylinder is J and J + δJ $mol/m^2/s$, respectively.

One elementary solution concerns a long cylinder, cross-sectional area A, with the diffusible material (amount S_0) deposited initially in the center (x = 0). The one-dimensional solution is

$$c(x,t) = \{S_0/[2A\sqrt{(\pi Dt)}]\}\exp(-x^2/(4Dt))$$

which is a series of bell-shaped curves. This expression can be used to estimate the average distance the particles travel in a given time. Considering only those traveling to the right

$$\langle x \rangle = \int xc(x,t)\, dt = \sqrt{(Dt/\pi)}$$

Given that D for potassium is around 2×10^{-9} m²/s, these ions will diffuse 25 microns (the width of an average cell) in 1 s.

WORKED EXAMPLE

Fluorescent acetylcholine is injected into a central portion of a nerve axon. Suppose that after 10,000 s the ratio of fluorescent intensity at the point of injection and a point 20 mm away is 2.0. Use the formula

$$c = S_0/(A\sqrt{(\pi Dt)})\exp(-x^2/(4Dt))$$

to calculate the diffusion coefficient D for acetylcholine (hint: take x = 0 as point of injection and consider ratio of concentrations c).

Answer

The ratio of concentrations at a particular instant in time is given by $1/(\exp(-x^2/(4Dt)))$, which in this case is 2. Taking logs, we get $\ln(2) = -(x^2/(4Dt))$, so substituting, we get $D = 0.02^2/(4 \times 10^4 \times \ln(2)) = 1.44 \times 10^{-8}$ m²/s.

Individual particles will of course execute a "random walk," but averaging over many particles will reveal a gradual "drift" of particles away from their original position at x = 0. This drift velocity v_d is related in rather a simple way to the flux J as follows: consider the slab in Figure 5.2 above above and suppose the distance dx between faces represents the distance on average that the particles drift, i.e., $dx = v_d dt$. The quantity of particles having crossed this face, area A, in dt seconds is thus S/(Adt), which is the same as the flux, J. The concentration in the slab is S/(Adx). We assume that a similar amount S left the slab in the same time across the other face. Comparing these equations shows that $J = v_d c$. A drift velocity of 25 microns/s thus corresponds to a flux of 2.5 μmol/m²/s or thereabouts within a cell.

The velocity of individual particles depends on temperature and to a certain extent concentration. The root mean square displacement along the x-axis has been shown by Einstein to be $\sqrt{(2Dt)}$, which is of the same order of magnitude as the distance moved from the origin in the case noted above. Individual particles can be thought of as heading off in random directions and from-time-to-time colliding with other particles in solution. If we assume that the average distance moved between collisions is d and the time between collisions (or time between jumps) is τ it can be shown by statistical theory that the diffusion coefficient $D = \frac{1}{2}d^2/\tau$ (Einstein–Smoluchowski), which is the same result as above. Einstein also showed that the RMS displacement was equal to $\sqrt{(2k\tau/f)}$ where k is Boltzmann's constant and f is the friction each particle experiences (in kg/s). Comparing these two equations gives $D = kT/f = RT/[Nf]$, where R is the gas constant and N Avogadro's number.

The same result can be derived by considering thermodynamic quantities such as the Gibbs free energy of the solution or more precisely the chemical potential μ (see below). For a gradient in μ in the x direction only (i.e., $\mu = f(x)$) the force acting on the particles is given by $F = -(d\mu/dx) = -d/dx(\mu^* + RT\ln c) = -(dc/dx)(RT/c)$. From Fick's first law this gives $F = JRT/(Dc) = v_dRT/D$. Treating the diffusion process as equivalent to spheres (the solute particles) moving through a viscous medium (the solvent: water) Stokes' law gives the frictional force acting as $6\pi\eta a v_d$ on each particle or $F' = 6\pi\eta a v_d N$ per mole, where η is water viscosity (10^{-3} Pa.s), a is particle effective radius and v_d the terminal or drift velocity of the particles. Equating the diffusive force F to the viscous force F′ gives

$$D = RT/(6\pi\eta aN) \qquad \text{or} \qquad D = kT/(6\pi\eta a) \qquad (\text{since } R/N = k)$$

This is known as the Stokes–Einstein relation. Note that because the molecular weight (M) varies with molecular volume, or a^3, we would expect that $D\sqrt[3]{M}$ would be constant for a range of molecules. In fact, for molecules of $M < 1000$, the relationship is nearer the quantity $D\sqrt{M}$ remaining constant.

For ions in solution, the situation is a little more complicated. In electrolysis situations where transfer of ions from solution to electrodes is possible (the electrode reactions may not in fact involve the deposition or release of the ion in question, but this can be ignored for the moment), the ions will travel at a speed determined by their size and the electric potential difference between the electrodes. In fact the electrical force $F_e = qE$, where q is the amount of charge involved) and E is the electric field (V/m) which will be given by the PD between the electrodes, E (V) and the distance between them d (m). The amount of charge transferred $q = zF = zNe$, where F is the Faraday constant, e the electronic charge, and z the valency of the ion. The mobility of the ion (u) is defined as the velocity per unit electric field (i.e., $u = v_d/E$), and because we can equate diffusive force $F = RTv_d/D$ to the electrical force $F_e = zFE$ and remembering the definition of u, we get

$$D = uRT/(|z|F),$$

which is known as the Einstein–Sutherland relation.

In the more general situation where an electrolyte solution is diffusing (i.e., Na^+ and Cl^- are diffusing from a high to a low concentration region) local electroneutrality has to prevail. The effective mobility (and hence diffusion coefficient) for this situation can be shown to be (see below in Section 5.3 for derivation)

$$1/u = (1/u^+ + 1/u^-)/2;$$

thus,

$$D = 2u^+u^-RT/((u^+ + u^-)|z|F).$$

WORKED EXAMPLE

1. The diffusion coefficient of a particular nonionic substance (nonelectrolyte) is 10^{-12} m^2/s, when measured in membrane lipid. If the lipid is 1000 times more viscous than water, what would you expect the diffusion coefficient to be for this substance in water?
2. If the lipid viscosity is 1 Pa.s at 37°C, estimate the molecular size of this substance.

Answers

1. Using the Stokes–Einstein equation ($D = RT/(6\pi\eta aN)$), we note that D is inversely related to viscosity η, thus the diffusion coefficient D will be 1000 times greater in water, hence 10^{-9} m^2/s.
2. Using the same equation, we can estimate particle radius $a = RT/(6\pi\eta ND) = 8.31 \times 310/(6 \times \pi \times 1.6 \times 10^{23} \times 10^{-12}) = 0.28$ nm.

TUTORIAL QUESTIONS: Nos. 1–3.

5.3 Electrochemical Equilibrium

We saw in the previous section that substances diffuse according to local concentration gradients. This movement of material implies work being done, because the difference in concentration can be regarded as a force acting on particles, which thus acquire a motion in a preferred direction. If two solutions are at identical temperature and pressure but differ in concentration, different levels of energy have to be assigned to them. The appropriate measure in this case is the Gibbs free energy. The Gibbs free energy per mole of a particular component in solution is termed the chemical potential, which is the sum of a standard chemical potential μ^0 for that particular species (glucose or urea, say) and a logarithmic term containing the concentration (or more accurately the activity) of the component in question. Just in the same way that electrical potential V is related to the work done in moving a charged particle from one position to another ($\int \underline{F} d\underline{s} = q(V_1 - V_2)$), we can write an analogous equation involving chemical potential ($\int \underline{F} d\underline{s} = \mu_1 - \mu_2$). The chemical potential is thus related to the amount of useful work that can be derived from the system and can help in deriving expressions for equilibrium between phases or at the boundary between two solutions (as is the case with a membrane). See Atkins and de Paula (2009) for some of the thermodynamic background to the concept of chemical potential.

$$\mu = \mu^0 + RT\ln(a) = \mu^0 + RT\ln(\gamma c) = \mu^0 + RT\ln(c)$$

for dilute solutions (R is the gas constant, T the absolute temperature, a the activity, γ the activity coefficient and c the concentration. ln is natural logarithm and μ^0 is the standard chemical potential for a particular material.)

Extension to ions—If the electrical potential varies from place to place, the ions in an electrolyte solution will also migrate in the direction of the electric field either toward or against the positive x direction, according to the sign of the charge, positive or negative. We have also seen that physiological solutions are predominantly Na^+, K^+, Cl^-, and so on. Remembering that if a charged particle is moved from a position at which the potential is V_1 to a position where it is V_2 the work done (see above) is $q(V_1 - V_2)$, we can associate qV_1 with the chemical potential μ_1 at position 1 and qV_2 with μ_2. Most physiology texts use the symbol E for electrical potential but we will use V, following physics/engineering practice. The relevant measure of energy of ions in electrolyte solutions is thus the electrochemical potential given as the sum of the chemical potential and an electrical term zFV, where z is the ion valence, F is the Faraday constant (96,500 Coulombs/mol) and V is the local electrical potential in volts. Note that F is numerically equal to the electronic charge e (1.6×10^{-19} Coulombs) multiplied by Avogadro's number (6.23×10^{23} ions/mol), thus zF is a measure of the total electrical charge, q, for each mole of that ion in solution. The product qV represents the work done in getting this charge to potential V, however this is accomplished. The electrochemical potential is usually denoted by the symbol $\tilde{\mu}$, where

$$\tilde{\mu} = \mu^0 + RT\ln(\gamma c) + zFV = \mu^0 + RT\ln(c) + zFV$$

for dilute solutions. This concept can in fact be extended: if we also allow work to be done on the solution by applying a pressure (as in osmosis: see below), an extra term can be added:

$$\tilde{\mu} = \mu^0 + RT \ln c + zFV + p\underline{v},$$

where $\underline{v}$ is partial molar volume.

5.3.1 Equilibria in Membrane Systems

5.3.1.1 Nernst Potentials

An approximately spherical cell surrounded by extracellular fluid is equivalent to a rectangular tank separated by a central barrier to diffusion. We will label the sides 1 and 2, respectively. Suppose the solution is KCl both sides of the membrane, with the concentration on side 1 greater than on side 2. Suppose also that the membrane itself is selective for K^+, in other words, K^+ does, but Cl^- does not permeate the membrane (Figure 5.3).

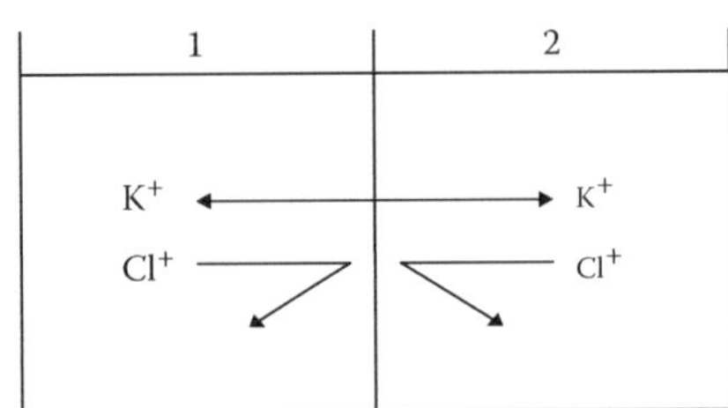

FIGURE 5.3
A potassium-selective membrane between two KCl solutions. Note that the solution on the left (side 1) has a higher value of $[K^+]$, so diffusion would tend to be from left to right, but because Cl^- is unable to follow, the membrane acquires a positive charge on the right hand side (side 2), so K^+ diffusion is prevented.

Thus, K^+ will come to an equilibrium when the electrochemical potential for this ion is the same on both sides of the membrane. In practical terms, K^+ will attempt to diffuse from side 1 to side 2, but as it does so, charges (due to the K^+ ions) will appear on the interface between the membrane and side 2. This small charge separation is sufficient to set up a sufficiently large potential difference between the two sides to prevent any further movement of charge. A typical potential difference of 50 mV between the two sides results from a mere 5×10^{-8} Coulombs excess charge, given the usual biological membrane capacitance of 1 μF/cm². Dividing this by F gives 5×10^{-13} mol approximately, which would be undetectable.

Equating the expressions for electrochemical potential for K^+ on the two sides 1 and 2 gives

$$\mu^0 + RT\ln(c_1) + zFV_1 = \mu^0 + RT\ln(c_2) + zFV_2.$$

And because μ^0 is always the same, we can write

$$RT\ln(c_1) + zFV_1 = RT\ln(c_2) + zFV_2$$

$$V_2 - V_1 = V_{21} = (RT/zF)\ln[c_1/c_2] = -V_{12}.$$

This is called the Nernst equilibrium formula, and V_{21} is referred to as the Nernst potential for K^+.

Note that at room temperature the term RT/F is $8.31 \times 300/96{,}500 = 0.025$ V, or 25 mV. Because $\ln(x) = 2.303 \log_{10}(x)$, we can also write this as

$$V_{21} = 58 \log_{10}[c_1/c_2] \text{ (mV)}.$$

This implies that a 10-fold difference in concentration between the two sides gives a PD of 58 mV, a 100-fold difference 116 mV and so on. At body temperature (310 K), a factor of 60 mV per decade is more appropriate. Note that for a divalent cation such as calcium, the Nernst potential is 29 mV for a 10-fold concentration difference (or decade change).

If the membrane referred to above is treated to make it permeable to chloride alone ($z = -1$), the Nernst potential is 58 mV per decade change, but the direction of the potential is reversed. If the PD across a membrane reflects the concentration difference of a particular ion this cannot by itself be taken as evidence if selective permeability to that ion (see below) especially if the nature of some of the other ions in solution is unknown, as in the case of the cytoplasm. However, as we will see in the case of nerve fiber cells, the use of the fact that when the membrane potential is the same as the Nernst potential for a particular ion (K^+ say) the net current of that ion is zero greatly simplifies analysis (because each ion current contributes to the total electrical current, which would be measured as flowing in an external circuit). In fact, the effective electrical force on each ion is $V_{12} - V_{Ni}$, where the second term represents the Nernst potential of the ith ion (note here that the driving force on the ion is not only the PD, V_{12}, as it would be in the case of a simple ohmic resistor, but also the force of the concentration gradient).

WORKED EXAMPLE

A cell is permeable only to K^+ ions. The membrane potential is held electronically at 60 mV, inside negative. The internal concentration is 10 times the external concentration for K^+. If the solution is at 20°C, which way does the K^+ flow? At what temperature is the net flux of K^+ zero?

Answer

The factor RT/F is 0.0252 at 200°C (T = 293 K), thus (RT/F)ln(10) = 58.1 mV, which is slightly less than 60 mV, so there is a residual negative charge on the inside, causing K^+ to flow in. If the temperature was increased to 302.5 K (29.5°C) there would be no net flux.

5.3.1.2 Donnan Equilibrium

Consider a slightly different situation to that given above in which a membrane separates two KCl solutions, but in this case, an extra anion or cation is added to one of the sides; for example, potassium isethionate (yielding isethionate⁻) is added to side 2 (Figure 5.4). The membrane is permeable to K^+ and Cl^- but not isethionate⁻. The question of whether the membrane is also permeable to water will be left open for now, but we will return to this later.

Because the concentration of Cl^- is greater on side 1, there is a tendency for it to diffuse through to side 2, but in order to preserve neutrality, it tends to drag K^+ with it. On side 2, therefore, both [K^+] and [Cl^-] increase whereas the concentrations on side 1 decrease. On the other hand, [isethionate⁻] cannot change. A PD appears across the membrane, which, at the final equilibrium state, can be predicted simultaneously from the K^+ and the Cl^- concentration gradients; in other words,

$$V_{21} = \frac{RT}{(+1)F}\ln\left(\frac{K_1}{K_2}\right) = \frac{RT}{(-1)F}\ln\left(\frac{Cl_1}{Cl_2}\right) = \left(\frac{RT}{(+1)F}\right)\ln\left(\frac{Cl_2}{Cl_1}\right)$$

or more simply,

$$\left(\frac{K_1}{K_2}\right) = \left(\frac{Cl_2}{Cl_1}\right).$$

Note that because the prime driving force is on the Cl^- ions, side 2 will go negative with respect to side 1. We noted above that this excess charge is infinitesimal and that both solutions will remain neutral with respect to positive and negative charge. This can be written as

$$[K_1] = [Cl_1]$$

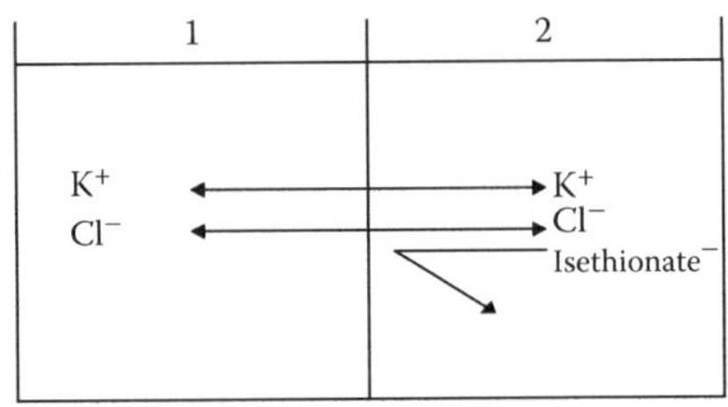

FIGURE 5.4
The Donnan system. An impermeable anion, isethionate, is present in the solution to the right of the membrane, and both K^+ and Cl^- are able to diffuse through the membrane. Because of the need to preserve electroneutrality, [Cl^-] will be less than [K^+] on the right (on the left, they will both be the same).

and

$$[K_2] = [Cl_2] + [\text{isethionate}^-].$$

The example shown below (based on Katz [1966]) gives a method for determining the final concentrations for an initial 100 mM KCl on side 1 and 50 mM KCl + 50 mM K^+ isethionate$^-$ on side 2. If the solutions contain cations and anions with valency greater than 1, we can generalize the ratio equations to

$$\left(\left[A_1^z\right]/\left[A_2^z\right]\right)^{1/z} = r,$$

where r is the Donnan ratio. Note that if A is an anion (z negative), the minus sign in the exponent gives the inversion shown above.

For example, in Figure 5.5, because the membrane is impermeable to isethionate, the amount will stay fixed at 50 units. Cl^- will tend to diffuse from 1 to 2, but solutions need to remain neutral, so

$$K_1 = Cl_1$$

$$K_2 = Cl_2 + 50$$

and when equilibrium is established,

$$K_1/K_2 = Cl_2/Cl_1 = r.$$

Substituting into the second equation gives

$$K_2 = K_1^2/K_2 + 50.$$

Total amount of K^+ will always be 200 units, so

$$\begin{aligned} K_2 &= (200 - K_2)^2/K_2 + 50 \\ K_2^2 &= (200 - K_2)^2 + 50K_2 \\ K_2^2 &= 40{,}000 - 400K_2 + K_2^2 + 50K_2 \\ K_2 &= 40{,}000/350 = 114 \text{ units}, \end{aligned}$$

Side 1	Side 2
K = 100 mM Cl = 100 mM	K = 100 Cl = 50 Isethionate = 50 mM

FIGURE 5.5
Numerical example of a Donnan system, with initially 100 mM KCl solution to the left of the membrane and 50 mM each of KCl and potassium isethionate to the right. Cl^- ions will tend to diffuse to the right because of the concentration gradient, but neutrality needs to be maintained throughout. The volumes on the two sides are the same.

and from the above,

$$K_1 = 86;\ Cl_1 = 86;\ Cl_2 = 86^2/114 = 64.$$

Note that $K_2 - Cl_2 = 50$, as required.

From the Nernst formula,

$$V_{21} = (RT/zF)\ \ln\ (K_1/K_2) = +\ 58\ \log\ (86/114) = [RT/((-1)F)]\ \ln\ (Cl_1/Cl_2)$$
$$= -58\ \log\ (86/64) = -7.2\ \text{mV}.$$

In addition, it should be noted that if ions in phase 1 are univalent,

$$z_R c_R = FV_{21}\,\Sigma c_i z_i^2/(RT)$$
$$= 2 \times (\text{Ionic strength in phase 1}) \times (\text{V in mV})/25,$$

giving a way of estimating the concentration of the impermeant species if its valence is known, or its valence if the concentration is known.

WORKED EXAMPLE

Hemoglobin is present in red blood cells at a concentration of 10 mM and with an effective valency of –8. When red blood cells are immersed in 100 mM KCl, show that the internal $[K^+]$ is 147.7 mM assuming KCl and hemoglobin are the only two components of significance inside the cell. What is the internal concentration of Cl^-? Estimate the membrane potential at 37°C.

Answer

The positive charge on the inside must balance, so $[K_i] = [Cl_i] + 8 \times 10$ (in mM).

$[K_i]$ must also satisfy the Donnan ratio $[K_i]/100 = 100/[Cl_i]$

If we put $x = [K_i]$ and eliminate $[Cl_i]$ we get

$$x = 10^4/x + 80;\ \text{or rearranging}\ x^2 - 80x - 10^4 = 0.$$

Using the quadratic formula, $x = (80 + \sqrt{(6400 + 4.10^4)})/2 = 147.7$, as expected. From the first equation above we get $[Cl_i] = 147.7 - 80 = 67.7$. We can estimate the membrane potential from the Nernst formula $V = (8.31 \times 310/96{,}500)\ln\ (147.7/100) = 10.5$ mV.

TUTORIAL QUESTIONS: Nos. 4–7 (parts).

5.3.1.3 Osmotic Equilibrium: van't Hoff Relationship for Dilute Solutions: Derivation

Consider the following system in which S is the solvent and I is the solute, which cannot cross the membrane (Figure 5.6). The membrane is freely permeable to the solvent, however. Let n_{s2}, n_{I2} be the number of moles of solvent and solute, respectively, on side 2.

Osmotic pressure $= \Pi = p_2 - p_1$ at osmotic equilibrium (i.e., no bulk flow).

The solvent species is the mobile one, so we consider that for equilibrium conditions.

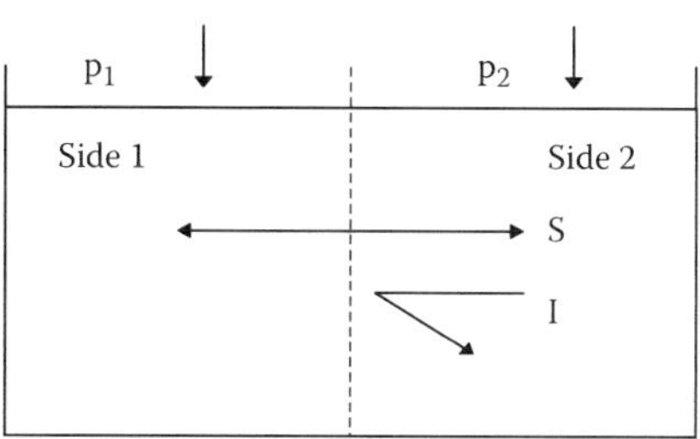

FIGURE 5.6
Osmosis. Substance I cannot cross the membrane, but S can. Because the I molecules are occupying locations that the solvent (water) could occupy, there is a concentration gradient for water from left to right, giving rise to a bulk flow of fluid. This can be prevented by adjusting the pressures p_1 and p_2. When flow stops, $p_2 - p_1$ is the osmotic pressure difference between the two sides.

$$\mu_{s1} = \mu^*_{s(p1)} + RT \ln x_{s1} = \mu^*_{s(p1)} \text{ because fraction} = 1 \text{ on side 1}$$

$$(\text{N.B., mole fraction } x_s = n_s/(n_s + n_1))$$

$$\mu_{s2} = \mu^*_{s(p2)} + RT \ln x_{s2.}$$

For equilibrium $\mu_{s1} = \mu_{s2}$ thus $RT \ln x_{s2} = \mu_{s(p1)} - \mu_{s(p2)} = \int_{p1}^{p2} \underline{v}dp$, where v is molar volume of pure solvent $= \underline{v}(p_1 - p_2) = -\underline{v}\Pi$ (this assumes the liquid is incompressible, and v is mean over pressure range). Thus

$$\Pi = -RT\ln x_s/\underline{v}.$$

Take solvent with single solute species, i.e., $x_{solv.} + x_{solu.} = 1$

$$\Pi = -RT \ln(1 - x_{solu.})/\underline{v}.$$

For several solute species, $x_{solv.} + \Sigma_i x_i = 1$

$$\Pi = -RT \ln(1 - \Sigma_i x_i)/\underline{v}.$$

Now if $\Sigma_i x_i$ small, $\ln(1 - \Sigma_i x_i) \approx -\Sigma_i x_i$

$$\Pi = +RT\,\Sigma_i x_i/v = +RT\,\Sigma_i n_i/(\Sigma_j n_j \underline{v}) \qquad \text{since } x_i = n_i/\Sigma_j n_j$$

(i.e., the summation in the numerator includes solute species only, and that in the denominator includes the solvent also).

In an ideal solution, each molecule occupies the same volume $\Sigma_j n_j \underline{v} = V$

$$\Pi = +RT\,\Sigma_i n_i/V$$

and remembering that $\Sigma_j n_i/V = c_i$, the concentration,

$$\Pi = RT\,\Sigma\, c_i$$

which is the van't Hoff relation.

WORKED EXAMPLE

1. A toad's urinary bladder is bathed on both sides by a physiological (Ringer's) solution containing the following (in mM): Na^+: 112; Cl^-: 116; K^+: 2; Ca^{2+}: 2; HCO_3^-: 2. A quantity of glucose (MW = 180) has been added at a concentration of 10 g/L to the inside solution. Calculate the difference in osmotic pressure (OP) in milli-osmol/L and in Pa at body temperature (37°C).
2. According to the literature, water flux in toad bladder in response to this OP is 1.5×10^{-6} kg/m²/s. Estimate the change in mass of a bladder containing 5 mL of glucose/Ringer after 30 min, assuming the bladder to be a hemisphere of 1 cm (0.01 m) radius. State whether this change is an increase or decrease.

Answers

1. The Ringer's solution has the same composition on both sides, so there is no osmotic imbalance due to the inorganic ions. The glucose, however does, and its concentration in the inside solution is 10/180 or 0.056 M (56 mmol/L). Using the van't Hoff relationship gives the osmotic pressure to be 56 milli-osmol/L, because glucose does not disassociate. To convert this to SI units, we note that 56 mmol/L is numerically the same as mol/m³. RTc is thus $8.36 \times 310 \times 56 = 14.5$ MPa.

   In Example 2, we note that the surface area of the bladder is $2\pi r^2 = 6.28 \times 10^{-4}$ m². Multiplying the flux 1.5×10^{-6} kg/m²/s by this area and by the time converted to seconds gives the amount in kg that the weight will change ($1.5 \times 10^{-6} \times 6.28 \times 10^{-4} \times 1800 = 1.7 \times 10^{-6}$ kg or 1.7 mg. It will be an increase, because the water is "drawn in" to the solution with the higher solute concentration.

TUTORIAL QUESTIONS: Nos. 8–12.

5.3.2 Nonequilibrium Situations: Ion Fluxes

5.3.2.1 Electrodiffusion Equation

This is an extension of Fick's first law of diffusion to include the drift velocity (or flux) of ions due to a potential difference between two points; that is,

$$J \text{ (Total Flux)} = J_d \text{ (Diffusive Flux)} + J_e \text{ (Electrical Flux)} = -D\,(dc/dx) + cv_e$$

where v_e is the drift velocity of ions.

Recall the definition of ion mobility as the (drift) velocity per unit electric field: ($u = v_e/(dV/dx)$, so we can substitute for v_e in terms of mobility. We can also substitute for D by using the Einstein–Sutherland formula. Note the – (sgn(z)) term to allow for the fact that as V increases, positive ions will drift in the direction of negative x and negative ions will do the reverse.

$$J = -(uRT/(|z|F))(dc/dx) - (\mathrm{sgn}(z))cu(dV/dx).$$

We can write this equation a little more neatly by defining a quantity $U = u/|z|$, thus

$$J = -(URT/F)(dc/dx) - zUc(dV/dx).$$

5.3.2.2 Diffusion Potentials

The diffusion or liquid junction potential formula can be derived by considering conditions of electroneutrality. Suppose we have a simple electrolyte A^+Y^- (e.g., KCl) and we consider the flux of A^+ and Y^- across the membrane:

$$J^+ = -(U^+RT/F)dc^+/dx - U^+c^+dV/dx \qquad \text{(the electrodiffusion equation)}$$

$$J^- = -(U^-RT/F)dc^-/dx + U^-c^-dV/dx \text{ because } z^- = -1.$$

$J^+ = J^-$; $c^+ = c^-$, so

$$-(U^+RT/F)dc/dx - U^+cdV/dx = -(U^-RT/F)dc/dx + U^-cdV/dx$$

$$dV/dx = (-U^+ + U^-)(RT/F)dc/dx/\{c(U^- + U^+)\}.$$

Integrating from side 1 to side 2 across the membrane,

$$V_2 - V_1 = V_{21} = (-U^+ + U^-)(RT/F)(\ln c_2 - \ln c_1)/(U^- + U^+)$$

$$= [(U^- - U^+)/(U^- + U^+)](RT/F)\ln (c_2/c_1) = [(U^+ - U^-)/(U^+ + U^-)](RT/F)\ln (c_1/c_2),$$

and because $U = u/|z|$,

$$V_{21} = [(u^- - u^+)/(u^- + u^+)](RT/F)\ln (c_2/c_1) = [(u^+ - u^-)/(u^+ + u^-)](RT/F)\ln (c_1/c_2).$$

Also, $m = u/(|z|F)$

$$V_{21} = [(m^- - m^+)/(m^- + m^+)](RT/F)\ln (c_2/c_1) = [(m^+ - m^-)/(m^+ + m^-)](RT/F)\ln (c_1/c_2).$$

WORKED EXAMPLE

An artificial cell is created and suspended in 150 mol/L NaCl. The internal concentrations of Na^+ and Cl^- are both 15 mM. If the potential across the membrane is 13 mV, inside negative, show that the ratio of Na^+ to Cl^- mobilities in this membrane (assuming that the inside contains only NaCl) is 0.64, approximately.

Answer

We can substitute $u^+ = 0.64\ u^-$ in the equation above, to see if V_{21} computes as 13 mV. We will make side 1 the inside. We remember, too, that RT/F has the value 26 mV at room temperature. Thus, $V_{21} = [(1 - 0.64)/(1 + 0.64)]\ 26 \ln (150/15)$ (in mV) $= 0.36 \times 26 \times 2.303/1.64 = 13.1$ mV.

TUTORIAL QUESTIONS: Nos. 4–6 (10–12).

5.4 Passive Transport through Membranes

5.4.1 Membrane Permeability

5.4.1.1 *Nonelectrolyte Substances*

We have seen previously that the movement of substances into and out of cells is controlled by the ability of that substance to diffuse either through the membrane lipid material or through water-filled pores (or channels) with a high specificity for particular ions. Because ions are also influenced by the transmembrane potential, we will deal with nonelectrolytes first. The permeability of a membrane to a particular substance is defined as the flux per unit difference in concentration across the membrane; that is,

$$P = J/(c_{in} - c_{out}) = J/\Delta c. \quad (5.1)$$

If we assume the membrane to be homogeneous, the permeability can obviously be related to the diffusion coefficient of that substance within the membrane material D_m.

In Figure 5.7, note that we have an abrupt change in concentration as we go from the aqueous inside and outside solutions to the membrane material itself. This is due to the partition effect arising from the relative solubilities of the substance in water (c_{in}, c_{out}) and membrane lipid material. Here, c'_{in} and c'_{out} refer to the concentrations of the substance just inside the membrane material at the boundaries.

In fact, the ratio of concentrations is constant, and we can write

$$c'_{in}/c_{in} = c'_{out}/c_{out} = k,$$

where k is known as the (oil/water) partition coefficient.

If we apply Fick's first law, $J = -D\, dc/dx$, to the diffusion of the substance within the membrane, we get $J = -D_m(c'_{in} - c'_{out})/t$, where t is the membrane thickness. Expressing this in terms of aqueous concentrations $J = -D_m k(c_{in} - c_{out})/t$ or $J = D_m k\Delta c/t$. Comparing this with our definition of permeability we get

$$P = D_m k/t \quad (5.2)$$

(D_m is diffusion coefficient for the substance *in the membrane*)

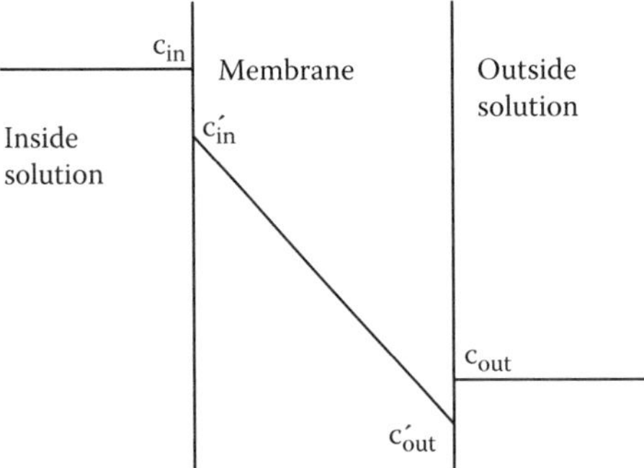

FIGURE 5.7
The concentration of a substance within the membrane c′ will in general be different from the concentration in solution c, but the ratios c′/c will be the same at both interfaces. The concentration profile within the membrane is unknown but can be assumed to be linear. The same assumption is made for PD within the membrane in the Goldman equation.

Extension to ions—If the membrane potential is zero, we can substitute for D_m using the Einstein–Sutherland formula; that is,

$$P = uRTk/(|z|Ft) \qquad \text{(again the mobility u is within the membrane).}$$

However, for a nonzero membrane potential the flux J will be affected by the electric potential gradient as well as the concentration gradient. Because P is defined in terms of the latter, the effects of the electric potential on J have to be allowed for. In order to do this we have to consider the electrodiffusion equation, dealt with earlier.

What we aim to do is to derive a method for estimating ionic permeability P in situations where the transmembrane potential V and the concentrations of the particular ion in question in the solutions bathing each side of the membrane (c_{in} and c_{out}) can be measured. The starting point is the electrodiffusion equation discussed in Section 5.3.2.1. We will need to make certain assumptions about the membrane. The most useful analysis has been that of D Goldman, whose simplifying assumption is that the voltage changes linearly with distance within the membrane (i.e., the electric field is constant). Thus $dV/dx = \Delta V/t$, where ΔV is the transmembrane potential ($V_{out} - V_{in}$ or V_{oi}) and t the membrane thickness. Initially, we rearrange the equation to make the concentration gradient within the membrane the subject:

$$\frac{dc}{dx} = -\frac{JF}{URT} - \frac{zF}{RT}\cdot\frac{dV}{dx}c(x) = -\frac{JF}{URT} - \frac{zF}{RT}\cdot\frac{\Delta V}{t}c(x).$$

This is of the form $dy/dx = -A - By$, which integrates to give $y = K_1 + K_2\exp(kx)$; where $K_1 = -A/B$; $k = -B$ and K_2 needs to be determined by examining boundary conditions. These are when $x = 0$, $c(x) = c'_{in}$ and when $x = t$, $c(t) = c'_{out}$.

Thus, $c'_{in} = K_1 + K_2\exp(0) = K_1 + K_2 = -A/B + K_2$. Thus, $K_2 = c'_{in} + Jt/(Uz\Delta V)$.

Hence, $c'_{out} = [c'_{in} + Jt/(Uz\Delta V)]\exp(-zF\Delta V/(RT)) - Jt/(Uz\Delta V)$.

For simplicity we will use Ψ for the factor $(zF\Delta V/(RT))$.

$$c'_{out} = [c'_{in} + Jt/(Uz\Delta V)]\exp(-\Psi) - Jt/(Uz\Delta V).$$

Rearranging:

$$c'_{out} - c'_{in}\exp(-\Psi) = [Jt/(Uz\Delta V)]\exp(-\Psi) - Jt/(Uz\Delta V).$$

Hence

$$J = -\frac{Uz\Delta V}{t}\left[\frac{c'_{out} - c'_{in}\exp(-\Psi)}{1-\exp(-\psi)}\right],$$

and because $P = uRTk/(|z|Ft) = URTk/(Ft) = Uz\Delta V/(\psi t)$, we can substitute for the factor $Uz\Delta V/t$ to give

$$J = -\frac{P\Psi}{k}\left[\frac{c'_{out} - c'_{in}\cdot\exp(-\Psi)}{1-\exp(-\psi)}\right] = -P\Psi\left[\frac{c_{out} - c_{in}\cdot\exp(-\Psi)}{1-\exp(-\psi)}\right], \tag{5.3}$$

recalling the definition of k (N.B., the concentrations are now referring to those in the bathing solutions, not those inside the membrane). The other assumption that has to be made (in addition to the constant field assumption) is that ΔV measured *inside* the membrane (between the two inner surfaces) is the same as that measured in the bathing solutions (at the outer surfaces of the membrane). There will probably be quite large voltage jumps in going from solution to membrane and vice versa, but the assumption usually is that they will cancel out.

We are now in a position to estimate ion permeability from

$$P = -J(1 - e^{-\Psi})/\{(c_{out} - c_{in}e^{-\Psi})\Psi\}. \tag{5.4}$$

WORKED EXAMPLE

The diffusion coefficient of a particular nonionic substance (nonelectrolyte) is 10^{-12} m^2/s, when measured in membrane lipid. Estimate the permeability coefficient of a cell to this substance, if the membrane is 10 nm thick and the lipid/water partition coefficient is 100.

If the substance is present in the bathing solution at a concentration of 1 mM and 0.1 mM in the cytoplasm, estimate the net flux of the substance in $mol/m^2/s$.

If the substance were an electrolyte, what would the net flux be if the univalent ion in question were to be assisted across the membrane by a potential difference of 25 mV as well as the concentration difference?

Answer

Using the relationship $P = D_m k/t$ we get $10^{-12} \times 100/10^{-8} = 10^{-2}$ m/s for membrane permeability. The net flux for the nonelectrolyte case is $J = P\Delta c = 10^{-2} \times (1 - 0.1) = 9 \times 10^{-3}$ $mol/m^2/s$ (note that the concentrations in mM are the same numerically is mol/m^3). For an electrolyte, the following equation must be used:

$$J = P\,(c_{out} - c_{in}e^{-\Psi})\Psi\}/(1 - e^{-\Psi}),$$

where

$$\Psi = zFV/RT = 1,$$

in this case (because RT/F = 25 mV).

$J = 10^{-2} \times 1 \times (1 - 0.1 \times e^{-1})/(1 - e^{-1}) = 1.5 \times 10^{-2}$ $mol/m^2/s$, rather more than in the nonelectrolyte case.

5.4.2 Tracer Fluxes

The usual method of estimating permeability experimentally is to use a marker or tracer to estimate flux. Tracers have generally been radioactive or fluorescent forms of the ion in question (such as ^{24}Na for Na$^+$ fluxes, ^{42}K for K$^+$, etc.). These effectively measure *one-way* or *unidirectional* fluxes, denoted j_{out} and j_{in}, because at the start of an experiment the tracer is present only on one side (i.e., the effective concentration on the side the tracer is *going to* is zero). The quantity J in the formula above is a *net* flux such that $J = j_{out} - j_{in}$.

If a tracer is used to estimate the flux of K$^+$ into a cell (j_{in}) the internal concentration c_{in} is effectively zero, so the formula above becomes (note the minus signs cancel)

$$P = j_{in}(1 - e^{-\Psi})/(c_{out}\Psi).$$

This can be written as $P = j_{in}/(c_{out}f)$, where f is a "field factor" $\Psi/(1 - e^{-\Psi})$. Remember ΔV is defined (implicitly) as $V_{out} - V_{in}$, so effectively Ψ is positive for the case of cation influx and negative for anion influx. For efflux (j_{out}) this is reversed.

WORKED EXAMPLE

This is an example of a muscle cell, adapted from Katz (1966), (the fluxes are unidirectional, obtained from radioactive tracer experiments). It is required to estimate membrane permeability to potassium P_K (see Figure 5.8).

Using influx data, Ψ is given by 90/25 and effectively $c_{in} = 0$. $c_{out} = 2.5$ mM = 2.5 mol/m^3. Using the formula above,

$$P_K = j_{in}(1 - e^{-\Psi})/(c_{out}\Psi) = 54 \times 10^{-9}(1 - \exp(-90/25))/(2.5 \times 90/25) = 5.8 \times 10^{-9} \text{ m/s}.$$

If we were to use the efflux data, the formula becomes

$$P_K = j_{out}(1 - e^{-\Psi})/(c_{in}e^{-\Psi}\Psi) = 88 \times 10^{-9}(1 - \exp(-90/25))/(140 \times .027 \times 90/25)$$
$$= 6.2 \times 10^{-9} \text{ m/s}.$$

In other words, they are not too dissimilar. For reasons we will discover later, the same exercise done with Na$^+$ data gives widely differing results.

Note that if we were to use the full formula (Equation 5.4) with J = 88 – 54 nmol/m^2/s = 34 nmol/m^2/s we would still get an answer for P_K of around 6×10^{-9} m/s.

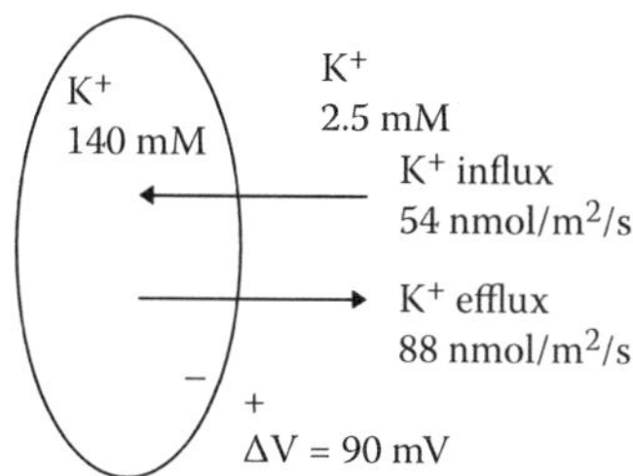

FIGURE 5.8
A muscle cell in which unidirectional K$^+$ fluxes have been determined using radioactive tracers. The transmembrane potential has also been measured, and the values of internal and external [K$^+$] are also known. The K$^+$ permeability of the membrane can be estimated two ways, from both the influx and the efflux data. In fact the net flux $J = j_{out} - j_{in}$ will also be in accordance with the equation discussed in the text, with an average value for P_K.

5.4.3 Electrodiffusive Processes

In Section 5.3.2.1, we looked at the equation for electrodiffusive flux and then used this to derive an expression for diffusion potentials. In bulk solution, the same condition as noted in the diffusion potential example applies; that is, $J^+ = J^-$; $c^+ = c^-$, so if we now eliminate the dV/dx term, we get

$$J = -U^+ \frac{RT}{F} \cdot \frac{dc}{dx} - U^+ \frac{c(-U^+ + U^-)}{c(U^- + U^+)} \cdot \frac{RT}{F} \cdot \frac{dc}{dx}$$

or

$$J = -\frac{2U^+U^-RT}{F(U^+ + U^-)} \cdot \frac{dc}{dx}$$

and comparison of this with Fick's first law gives

$D = 2u^+u^-RT/(F(u^+ + u^-)|z|)$, because $U = u/|z|$, and $z = \pm 1$ in the case of NaCl.

This formula was stated without derivation in Section 5.2.

5.4.4 Estimation of Membrane Potentials from Equilibrium Concentrations

Integration of the electrodiffusion equation across a membrane in general involves making some assumptions about the state of ions within the membrane (see Goldman's derivation above).

5.4.4.1 Hodgkin–Katz Equation

However, if we restrict ourselves to univalent cations as being the only types of ion which permeate the membrane and we assume that the net flux across the membrane is zero ($\Sigma J_i = 0$) we get

$$\Sigma J_i = 0 = -\Sigma\,(U_iRT/F)(dc_i/dx) - \Sigma z_i\, U_i\, c_i\,(dV/dx) \quad \text{and} \quad z_i = 1;$$

thus

$$\frac{dV}{dx} = -\frac{\Sigma(U_iRT/F)}{\{\Sigma c_iU_i\}} \cdot \frac{dc_i}{dx} = -\frac{RT}{F} \cdot \frac{d(\Sigma U_ic_i)}{dx} \cdot \frac{1}{\Sigma c_iU_i}.$$

$$\begin{aligned} V_2 - V_1 = V_{21} &= -(RT/F)\ln(\Sigma U_ic_{i2})/(\Sigma U_ic_{i1})) \\ &= \frac{RT}{F} \cdot \ln\left[\frac{P_{Na}[Na]_1 + P_K[K]_1}{P_{Na}[Na]_2 + P_K[K]_2}\right] \end{aligned} \tag{5.5}$$

because $P_i = u_iRT/k_i$.

This is the Hodgkin–Katz equation: even this approach assumes a partition k_i between membrane material and aqueous solution, even though the ions are probably contained in water-filled channels within the membrane and go through essentially in single file. The concept of concentration gradient within the membrane is thus rather nebulous.

5.4.4.2 Goldman Equation

Starting with Equation 5.3 above, if we again assume that the net flux of all ions across the membrane is zero, we get

$$\Sigma J_i = -\Sigma P_i \Psi \left[\frac{c_{out(i)} - c_{in(i)} \cdot \exp(-\Psi)}{1 - \exp(-\psi)} \right].$$

If we restrict ourselves to univalent cations and anions ($z_i = \pm 1$), we can thus write

$$\Psi = \Phi \text{ for } z = +1;\ \Psi = -\Phi \text{ for } z = -1 \text{ (i.e., } \Phi = \Delta VF/RT);$$

then summing cations and anions separately (and allowing for the fact that the total cation flow should equal total anion flow),

$$-\Sigma P_i \Phi \left[\frac{c^+_{out(i)} - c^+_{in(i)} \cdot \exp(-\Phi)}{1 - \exp(-\Phi)} \right] = -\Sigma P_i (-\Phi) \left[\frac{c^-_{out(i)} - c^-_{in(i)} \cdot \exp(+\Phi)}{1 - \exp(+\Phi)} \right].$$

If we multiply the first expression by the factor $\exp(+\Phi)$ (both numerator and denominator), we get

$$-\Sigma P^+_i \Phi \left[\frac{c^+_{out(i)} \exp(-\Phi) - c^+_{in(i)}}{\exp(+\Phi) - 1} \right] = -\Sigma P^-_i (-\Phi) \left[\frac{c^-_{out(i)} - c^-_{in(i)} \cdot \exp(+\Phi)}{1 - \exp(+\Phi)} \right],$$

i.e.,

$$+\Sigma P^+_i \Phi \left[\frac{c^+_{in(i)} - c^+_{out(i)} \cdot \exp(+\Phi)}{1 - \exp(+\Phi)} \right] = -\Sigma P^-_i \Phi \left[\frac{c^-_{in(i)} - c^-_{out(i)} \exp(+\Phi)}{1 - \exp(+\Phi)} \right].$$

Eliminating the factor $\Phi/(1 - \exp(+\Phi))$ and rearranging,

$$\Sigma P^+_i \left\{ c^+_{in(i)} - c^+_{out(i)} \cdot \exp(+\Phi) \right\} = -\Sigma P^-_i \left\{ c^-_{out(i)} - c^-_{in(i)} \exp(+\Phi) \right\}$$

$$\exp(+\Phi) = \left\{ \frac{\Sigma P^+_i \cdot c^+_{in(i)} + \Sigma P^-_i \cdot c^-_{out(i)}}{\Sigma P^+_i \cdot c^+_{out(i)} + \Sigma P^-_i \cdot c^-_{in(i)}} \right\}.$$

Hence, ΔV (= $\Phi F/RT$) can be determined. In particular, if the cations that permeate the membrane are Na^+ and K^+ and the only permeant anion is Cl^-, this expression becomes

$$\Delta V = V_o - V_i = \frac{RT}{F} \cdot \ln\left[\frac{P_K[K_{in}] + P_{Na}[Na_{in}] + P_{Cl}[Cl_{out}]}{P_K[K_{out}] + P_{Na}[Na_{out}] + P_{Cl}[Cl_{in}]}\right]. \tag{5.6a}$$

Or,

$$\Delta V = \frac{RT}{F} \cdot \ln\left[\frac{[K_{in}] + \alpha[Na_{in}] + \beta[Cl_{out}]}{[K_{out}] + \alpha[Na_{out}] + \beta[Cl_{in}]}\right] \tag{5.6b}$$

where $\alpha = P_{Na}/P_K$ and $\beta = P_{Cl}/P_K$.

Thus, it is often convenient to express permeabilities as ratios, to K^+ permeability, say, as indicated above.

Some texts refer to this equation as the Goldman–Hodgkin–Katz or GHK equation. See Chapter 7 of Weiss (1996) for further discussion of this derivation.

WORKED EXAMPLE

If a cell has internal and external concentrations as shown in Figure 5.9 below, (1) estimate the membrane potential from the Goldman equation and (2) calculate the net movement of K^+ and Na^+ across the membrane, given that the ratios of permeabilities P_K:P_{Na}:P_{Cl} is 1:0.05:0.45 and the actual K^+ permeability is 6×10^{-9} m/s.

1. From Equation 5.6

$$\Delta V = V_{oi} = 25\ln\left[\frac{1[150] + 0.04[15] + 0.45[125]}{1[5] + 0.04[145] + 0.45[9]}\right] = 25\ln(197/14.9) = 64.6\text{ mV}.$$

Note that this potential is very close to the Nernst potential for Cl^- ($V_{Cl} = -25\ln(9/125) = 65.8$ mV. In this case, the last factor in the numerator and denominator can be omitted (i.e., giving the Hodgkin–Katz equation), and a similar value for membrane potential would be obtained. Note also that the Nernst potential for Na^+ (–57 mV) is a long way from the membrane potential.

2. From Equation 5.4,

$$J_K = -P\Psi\left[\frac{c_{out} - c_{in}\exp(-\Psi)}{1 - \exp(-\Psi)}\right] = -6 \times 10 - 9(65/25) \cdot \left[\frac{5 - 150\exp(-65/25)}{1 - \exp(65/25)}\right]$$

$$= +103 \times 10^{-9}\text{mol/m}^2\text{/s} \quad \text{(the positive sign indicates flow from left to right)}$$

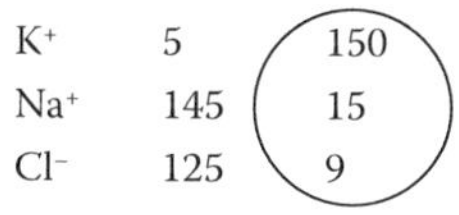

K^+	5	150
Na^+	145	15
Cl^-	125	9

FIGURE 5.9
A spherical cell. The values are of external and internal ion concentrations in mM.

$$J_{Na} = -6\times10^{-9}\times0.04\times(65/25)\left[\frac{145-15\cdot\exp(-65/25)}{1-\exp(-65/25)}\right] = -97\times10^{-9}\,\text{mol/m}^2/\text{s}.$$

The two fluxes are approximately the same and indicate the loss of K^+ and the gain of Na^+ suffered by a cell if the Na/K pump is rendered inoperative.

TUTORIAL QUESTIONS: Nos. 13–16.

5.4.5 The (Ussing) Flux Ratio Equation

This is a useful way of checking whether a particular ion is undergoing a mechanism of passive independent diffusion. It requires the influx and efflux to be estimated by a tracer method. Recalling equations used in the Worked Example in Section 5.4.2,

$$P_i = j_{in}(1 - e^{-\Psi})/(c_{out}\Psi) = j_{out}(1 - e^{-\Psi})/(c_{in}e^{-\Psi}\Psi).$$

Rearranging, we get

$$(j_{out}/j_{in}) = (c_{in}/c_{out})\exp(-\Psi) = (c_{in}/c_{out})\exp(-zFV_{oi}/(RT)) \tag{5.7}$$

or:

$$\ln(j_{out}/j_{in}) = \ln(c_{in}/c_{out}) - zFV_{oi}/(RT) = zF(V_N - V_{oi})/(RT),$$

where V_N is the Nernst potential $(RT/(zF))\ln(c_{in}/c_{out})$ for the ion in question.

The flux ratio equation can be extended to apply to transport in the general case:

$$RT\ln(j_{1\to2}/j_{2\to1}) = nzF(V_N - V_{21}) + \varphi \qquad \text{(J/mol)},$$

where $n > 1$ for self-interaction (diffusion along narrow channels) or <1 for facilitated diffusion (Cl^- exchange) and φ is a measure of active transport or other processes such as osmosis (i.e., driven by bulk flow of solvent).

WORKED EXAMPLE

In a sample of frog skin, bathed on both sides with Ringer's solution, Cl^- influx is measured to be 3×10^{-6} mol/m^2/s and the skin potential is 26 mV, inside positive. Estimate what the efflux will be (assume Cl^- is passively transported).

Answer

Referring to Equation 5.7, the ratio c_{in}/c_{out} is 1, because the solutions are identical, and V_{oi} is + 26 mV, with z = –1. The ratio j_{in}/j_{out} is thus exp(–(–1) × 26/26) = 2.72. Because j_{in} is 2×10^{-6} mol/m^2/s, j_{out} must be 1.1×10^{-6} mol/m^2/s.

TUTORIAL QUESTIONS: Nos. 17–21.

5.5 Electrical Measurements

5.5.1 Membrane Conductance

So far, the flux J of ions has been measured in molar terms, i.e., mol/m^2/s. To convert this to an electrical current density it is merely a matter of multiplying by a factor zF, giving units of amp/m^2. The total electrical current density across a membrane is a fairly complex function of voltage because the total current is made up from contributions from individual ions (see the Worked Example in Section 5.4.4.2 above). As a general rule, the relationship between the current of a particular ion i ($J'_i = zFJ_i$) and the applied potential difference V_{oi} is similar to that of a semiconductor junction, as shown in Figure 5.10.

The conductance, or inverse resistance, is usually expressed in units of siemens per unit area (S/m^2). There are two ways of defining this quantity, the slope conductance $g_i = \partial J'_i/\partial V = z_i F.\ \partial J_i/\partial V$ and the chord conductance $G_i = J'_i\ /(V_{oi} - V_N)$, where V_N is the Nernst potential (and also the potential at which $J'_i = 0$).

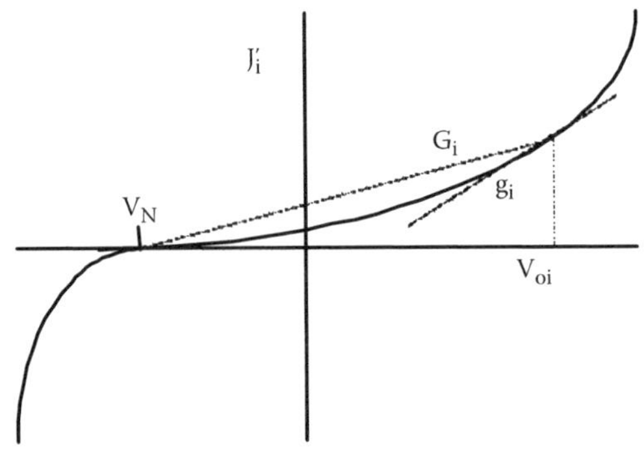

FIGURE 5.10
Electrical properties of membranes. A generalized current–voltage characteristic (actually, J'_i is usually a current *density*) showing the slope conductance gi and the chord conductance G_i. See text for an explanation of these quantities.

Chord conductance can be measured from tracer fluxes as follows:

$$J_i' = zFJ_i = zF(j_{out} - j_{in}); \qquad V_{oi} - V_N = \{RT/(zF)\} \ln(j_{out}/j_{in}) \qquad \text{(from Equation 5.7)}$$

So $G_i = z^2F^2(j_{out} - j_{in})/\{RT \ln(j_{out}/j_{in})\}$.

In general, the total membrane conductance g_m is given by the sum of the slope conductances, not the chord conductances, i.e., $g_m = \Sigma g_i$. When we study nerve membrane, we will discover that for some values of V_{oi}, g_{Na} is actually negative, whereas G_{Na} remains positive.

5.5.2 Interrelationships between Membrane Potential, Permeability, and Electrical Conductance

It is easy to show that for $j_{1\to2} \approx j_{2\to1}$, $G_i \approx (z^2F^2/RT)j_{2\to1}$.

Put $j_{1\to2}/j_{2\to1} = 1 + \delta$, where $\delta = (j_{1\to2} - j_{2\to1})/j_{2\to1}$ and is small: $\ln(1 + \delta) = \delta - \delta^2/2 + \delta^3/3 - \ldots\ldots$ If we neglect all but the first term, we get the required result.

The relationship between permeability and slope conductance can be examined by recalling the definition of differential conductance:

$$\begin{aligned} g_i &= \partial J_i'/\partial V = z_iF \cdot \partial J_i/\partial V \\ &= (\partial/\partial V) \cdot \left(\left(P_i z_i^2 F^2/RT\right) \cdot \left\{ \left(c_{i1} - c_{i2} \exp\left(z_iFV/RT\right)\right) \Big/ \left(1 - \exp\left(z_iFV/RT\right)\right) \right\} \right) \end{aligned}$$

putting

$$y = \exp(z_iF/RT)$$

we get

$$g_i = \{P_i z^3F^3V/(RT)^2\}(c_{i1} - c_{i2}y)/\{(1 - y)\ln(c_{i1}/(c_{i2}y)\}.$$

In other words, the conductance varies directly with permeability, but there is a complex dependence on membrane potential and concentration ratio.

TUTORIAL QUESTIONS: Nos. 26–30.

5.6 Active Ion Transport

5.6.1 Measuring Active Ion Transport—Short-Circuit Current Technique

A useful way of directly measuring active transport was developed by HH Ussing in the early 1950s. It was first applied to frog skin, and consists merely of measuring the external current required to reduce the skin potential to zero, when the skin is bathed with the same Ringer's solution on both the inside and outside surface. What is actually happening is that the external current supplies the companion negative charges for the sodium which is being actively transported from outside to inside of the skin. Because each Na^+ ion is thus neutralized, the PD goes to zero at a certain value of external current, this value being the same as the current due to active Na^+ transport. Ussing also showed, using radioactive tracers, that the net Na^+ movement (in electrical units) was the same as the short-circuit current (s.c.c.), i.e., $I_{sc}/A = zF(j_{Na,in} - j_{Na,out})$, where A is the cross-sectional area of the skin. In other epithelia, the active transport is more complicated, with other ions contributing to the s.c.c. See the Worked Example below for further information.

WORKED EXAMPLE

1. A sample of toad skin, 6.7 cm^2 in area, is short-circuited by a current of 0.137 mA. Before this current was applied, the (open circuit) membrane potential was 26 mV. What is the membrane conductance (in S/m^2) of this skin?
2. If the skin pumps only Na^+ actively, calculate this pumping rate in $mol/m^2/s$.
3. If the efflux of Na is 0.2 $mol/m^2/s$, what is the influx of Na?
4. If the efflux of Cl is the same as Na, what is the value of Cl conductance?

Answers

1. The conductance is given by the equation $I_{sc}/(AV_{oc}) = 0.137 \times 10^{-3}/(6.7 \times 10^{-4} \times 26 \times 10^{-3}) = 7.9\ S/m^2$.
2. The s.c.c. can be converted to molar quantities by the equation $I_{sc}/(AF) = 0.137 \times 10^{-3}/(6.7 \times 10^{-4} \times 9.65 \times 10^4) = 2.1 \times 10^{-6}\ mol/m^2/s$.
3. The s.c.c. is the difference between Na influx and efflux, so the influx value must be $2.1 + 0.2 = 2.3\ \mu mol/m^2/s$.
4. Under short-circuit conditions, the influx and efflux of Cl will be the same, so the conductance can be estimated from the equation $G_{Cl} = (z^2F^2/RT)j_{Cl} = \{(9.65 \times 10^4)^2/(8.31 \times 300)\} \times 0.2 \times 10^{-6} = 0.74\ S/m^2$ (around 10% of total conductance).

Ussing estimated the equivalent voltage of the "sodium battery" in the frog skin was around 160 mV, but that due to leakage of ions between the epithelial cells (the "shunt" pathway) the actual voltage across the skin is $190 \times R_{shunt}/(R_{series} + R_{shunt}) = 60$ mV. where R_{series} and R_{shunt} are the equivalent resistances of the series (within cells) and shunt (between cells) pathways (Figure 5.11).

5.6.2 Energetics of Ion Transport

Ion pumps are categorized into electrogenic (where they contribute to the membrane potential) and electroneutral (where they do not). Ussing argued that in the frog skin the pump was actually a one-to-one exchange of Na^+ for K^+ and that the different sides of the cell were selectively permeable to K^+ and Na^+ (giving rise to diffusion potentials), as shown in Figure 5.11.

$$V_{oi} = (RT/F)\{\ln(Na_o/Na_{cell}) + \ln(K_{cell}/K_i)\} = 25(\ln(120/5) + \ln(120/3)) \approx 160\ mV.$$

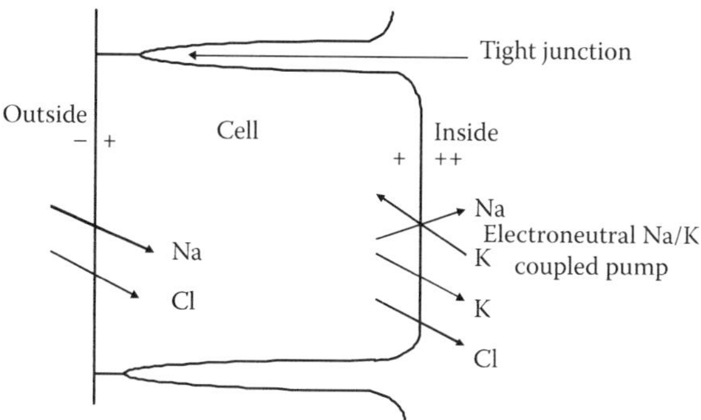

FIGURE 5.11
The Ussing model of the frog skin. The PD across the skin can be as much as 160 mV under some treatments to make the tight junctions really tight. The outward-facing cell membrane is Na^+-selective whereas the inward one is K^+-selective. The PD is the sum of the Na^+ and K^+ diffusion potentials, and the pump does not contribute to the PD.

In red blood cells, the situation is different: the membrane potential is only 10 mV and the outside concentrations of Na^+, K^+, and Cl^- are 155, 5, and 112 mM, whereas the internal concentrations are 19, 136 and 78 mM, respectively.

Suppose δm moles of Na^+ are pumped from the inside (i) to the outside (o) of the cell. The work done in doing this (to overcome electrical and concentration gradient forces) is

$$\Delta\mu = \delta m\{RT\ln[Na]_o + FV_o - (RT\ln[Na]_i + FV_i)\}.$$

Because the pump exchanges Na^+ for K^+, there will also be work done in pumping K^+ from o to i. Let δn be the number of moles of K^+ transferred when dm moles of Na^+ are transferred.

$$\Delta\mu' = \delta n\{RT\ln[K]_i + FV_i - (RT\ln[K]_o + FV_o)\}.$$

Let $V_{oi} = V_o - V_i$, as previously

$$\text{Work done} = \Delta\mu + \Delta\mu' = \delta mRT\ln\{[Na]_o/[Na]_i\} + \delta nRT\ln\{[K]_i/[K]_o\} + (\delta m - \delta n)FV_{oi}.$$

There is good evidence to suggest that for every three Na^+ pumped out, two K^+ are pumped into the cell. The total work done in doing this is (using V_{oi} in Volts)

$$3 \times 8.31 \times 310 \times \ln\{155/19\} + 2 \times 8.31 \times 310 \times \ln\{136/5\}$$
$$+ (3 - 2)96{,}500 \times 0.01 = 34{,}200\ \text{J} = 34.2\ \text{kJ}.$$

Compare this to the release of 29 kJ/mol ATP in the hydrolysis reaction

$$\text{ATP} = \text{ADP} + P_i.$$

5.7 Summary of Equations

Standard values

$$R = 8.31\ \text{J/mol/K};\ F = 96{,}500\ \text{C/mol};\ RT/F = 25\ \text{mV at } 18°\text{C}.$$

Equations:

1. **Diffusive Processes**

Fick's laws

$$J = -D\,dc/dx.$$

$$\partial c/\partial t = D\partial^2 c/\partial x^2$$

Stokes–Einstein

$$D = RT/(6\pi\eta aN) \text{ or } D = kT/(6\pi\eta a) \qquad \text{(because } R/N = k)$$

Einstein–Sutherland relation

$$D = uRT/(|z|F),$$

$$D = 2u^+u^-RT/((u^+ + u^-)|z|F).$$

2. **Electrochemical Equilibrium**

$$\begin{aligned} \underline{\mu} &= \mu^0 + RT\cdot\ln(\gamma c) + zFV \\ &= \mu^0 + RT\cdot\ln(c) + zFV \text{ for dilute solutions.} \end{aligned}$$

Nernst formula

$$V_2 - V_1 = V_{21} = \frac{RT}{zF}\cdot\ln\left(\frac{c_1}{c_2}\right) = -V_{12}$$

$$V_{21} = 58\log_{10}\left(\frac{c_1}{c_2}\right)(\text{mV})$$

Donnan condition

$$V_{21} = \frac{RT}{F}\cdot\ln\left(\frac{K_1}{K_2}\right) = \frac{RT}{F}\cdot\ln\left(\frac{Cl_2}{Cl_1}\right).$$

So

$$\left(\frac{K_1}{K_2}\right) = \left(\frac{Cl_2}{Cl_1}\right) = \left(\frac{Na_1}{Na_2}\right) \text{etc.}$$

In general, $([A_1^z]/[A_2^z])^{1/z} = r$, where r is the Donnan ratio. (thus inversion as shown above if z is negative)

Osmosis

$$\Pi = RT\,\Sigma\, C_I \qquad \text{(van't Hoff relation)}.$$

3. **Electrodiffusion**

Electrodiffusion equation

$$J = -(URT/F)(dc/dx) - zUc(dV/dx)$$

Diffusion potential

$$V_{21} = [(u^- - u^+)/(u^- + u^+)](RT/F)\ln(c_2/c_1) = [(u^+ - u^-)/(u^+ + u^-)](RT/F)\ln(c_1/c_2).$$

4. **Permeability and Goldman Equation**

Nonelectrolytes

$$P = J/(c_{in} - c_{out}) = J/\Delta c$$

$$P = D_m k/t$$

(D_m refers within the membrane.)

$$P = uRTk/(|z|Ft)$$

(Mobility u is within the membrane.)

Integration of electrodiffusion equation

$$\begin{aligned} J &= -\frac{P\Psi}{k}\left[\frac{c'_{out} - c'_{in}\cdot\exp(-\Psi)}{1-\exp(-\psi)}\right] \\ &= -P\Psi\left[\frac{c_{out} - c_{in}\cdot\exp(-\Psi)}{1-\exp(-\psi)}\right] \end{aligned}$$

where $\Psi = zVF/(RT)$

Electrolyte permeability (from above)

$$P = -J(1 - e^{-\Psi})/\{(c_{out} - c_{in}e^{-\Psi})\Psi\}.$$

For unidirectional tracers

$$P = j_{in}(1 - e^{-\Psi})/(c_{out}\Psi).$$

Can be written as $P = j_{in}/(c_{out}f)$, where f is a "field factor" $\Psi/(1 - e^{-\Psi})$.

Goldman equation

$$V_{oi} = \frac{RT}{F} \cdot \ln\left[\frac{P_K[K_{in}] + P_{Na}[Na_{in}] + P_{Cl}[Cl_{out}]}{P_K[K_{out}] + P_{Na}[Na_{out}] + P_{Cl}[Cl_{in}]}\right]$$

or

$$\Delta V = \frac{RT}{F} \cdot \ln\left[\frac{[K_{in}] + \alpha[Na_{in}] + \beta[Cl_{out}]}{[K_{out}] + \alpha[Na_{out}] + \beta[Cl_{in}]}\right],$$

where $\alpha = P_{Na}/P_K$; $\beta = P_{Cl}/P_K$.

5. **Flux Ratio and Conductance**

Flux ratio

$$(j_{out}/j_{in}) = (c_{in}/c_{out})\exp(-\Psi) = (c_{in}/c_{out})\exp(-zFV_{oi}/(RT))$$

or

$$\ln(j_{out}/j_{in}) = \ln(c_{in}/c_{out}) - zFV_{oi}/(RT) = zF(V_N - V_{oi})/(RT),$$

where V_N is the Nernst potential $(RT/(zF)\ln(c_{in}/c_{out})$ for the ion in question.

Conductance

$$g_i = \partial J_i'/\partial V$$

$$G_i = J_i'/(V_{oi} - V_N)$$

$$G_i = z^2F^2(j_{out} - j_{in})/\{RT\ln(j_{out}/j_{in})\}$$

$$g_m = \Sigma g_i.$$

(For some values of V_{oi}, g_{Na} is actually negative, whereas G_{Na} remains positive.)

Tutorial Questions

1. (a) State Fick's laws of diffusion, giving the units of the quantities involved.
   (b) Estimate the diffusion coefficient for NaCl in bulk solution given that the hydrated ion radii for Na^+ and Cl^- are 164 and 110 pm, respectively. Take water viscosity to be 10^{-3} kg m/s.
   (c) Estimate electrical mobility of these ions.
   (d) Make an estimate of the thickness of a membrane whose urea permeability is 2×10^{-5} m/s, given that the diffusion coefficient for urea in lipid is 1×10^{-9} m^2/s and the lipid/water partition coefficient is 2×10^{-4}. What value would be expected for a cell (plasma) membrane?

2. The concentration of a solute in a given solution only varies along the x direction. At time t = 0 the concentration is given by the expression

$$C_o = 0.01 + 0.05x^3$$

units of x in meterand C in mol/m^3

(a) What is the concentration gradient at x = 1?

(b) What is the diffusion coefficient if the flux at x = 1 and t = 0, is 1.5×10^{-4} mol/s/m^2?

3. Show that the diffusion coefficient for the chloride ion in water is 2×10^{-5} cm^2/s, given that the ionic mobility is 7.91×10^{-4} cm^2/s/V. Calculate the effective hydrodynamic radius of the ion given that the viscosity of dilute NaCl is 10^{-3} kg/m/s, and that charge separation can occur. (Take Avogadro's number as 6×10^{23}/mol.)

4. A nonselective membrane separates two 0.1 M solutions of LiCl.

(a) What will the PD be between these solutions?

(b) What will the PD be if one of these solutions is replaced by a 0.02 M solution of LiCl? (Give polarity of solutions).

(c) If the membrane were to become permeable to Li + only, what would the PD become? (Concentrations are as those in (b).)

(d) If the membrane were now permeable only to water and the concentrations were as in ii), what pressure difference would have to be applied, at 25°C to prevent bulk flow of water, and to what side?

(Mobilities: Li^+ 4.01×10^{-8}; Cl^- 7.91×10^{-8} m^2/s/V; use R = 0.08 L atm/mol/K.)

5. Where do liquid junction potentials occur, and what are they due to?

A 3M NaBr/Agar bridge dips into a 0.5 mol/L and a 0.005 mol/L solution of NaBr. What PD would you expect between two Ag/AgCl electrodes dipping into these solutions? (u_{Na}, u_{Br} are 5.19×10^{-4}, 8.13×10^{-4} m^2/s/V, respectively.)

Suppose a semipermeable membrane now separates these two solutions. Calculate the pressure (at 25°C) which would have to be applied to prevent the movement of water. Which side would it have to be applied to?

6. (a) A thin membrane separates two large well stirred solutions of NaCl. The mobilities of sodium and chloride ions in the membrane are u_{Na} and u_{Cl}, where $0 < u_{Na} < u_{Cl}$ and the concentrations of NaCl on sides 1 and 2 of the membrane are C_1 and C_2.

(i) Is this system in an equilibrium or steady state? Discuss the difference between the terms "steady" and "equilibrium" state when applied to the above situation.

(ii) If the membrane were changed so that $u_{Na} = 0$ and $u_{Cl} > 0$, is the system described above in a "steady state" or an equilibrium?

(b) J_{Cl} is the particle current density (mol/m^2/s) for chloride ions moving through the membrane when $u_{Cl} > 0$ and also $u_{Na} > 0$.

(i) Give a mathematical expression for J_{Cl} in terms of electric and concentration gradients within the membrane.

(ii) What additional assumptions are made in deriving the liquid junction equation from the expression you have given in answer to (b)(i).

7. (a) Define the following: electrical mobility, diffusion coefficient, membrane permeability, and partition coefficient (if giving a formula, explain the quantities involved and give units where appropriate).

(b) Two NaCl solutions are separated by a porous plug, assumed to act as a perfect liquid junction. What PD would be measured between the two Ag/AgCl electrodes, given that the hydrated ion radii of Na^+ and Cl^- are 164 and 110 pm, respectively (see Figure 5.12)?

8. (a) A simple physiological solution consists of the following: NaCl 120 mmol/L; KC1 2 mmol/L; $CaCl_2$ 1 mmol/L; $NaHCO_3$ 1 mmol/L. Assuming a membrane to be impermeable to these ions, but permeable to water, calculate the osmotic pressure of this solution in mOsmol/L and the hydrostatic pressure in Pa which would be needed to prevent water movement across this membrane if half-strength solution were placed on the other side. To which solution would the pressure be applied? (Assume that T = 27°C.)

9. (a) The permeability of the cell membrane to many substances is related to the oil/water partition coefficients of these substances. Comment on why this should be so, and why it is **not** so for small ions and water.

(b) In the system shown in Figure 5.13, the concentration of the impermeable anion R^{4-} is 10 mM.

(i) What will the concentrations of K^+ and Cl^- be on side 1?

(ii) What PD will exist across the membrane?

(c) The osmotic pressure of plasma is 7.479×10^5 Pa at 300 K. How many Kg of NaCl must be added to 1 m^3 of water to make it isotonic (equal osmotic pressure) with plasma.

10. A membrane separating two infinite solutions of $CaCl_2$ has a potential difference of +100 mV across it (and is at a temperature of 300 K). If the concentration of $CaCl_2$ on side 1 is 10^{-3} mol/m^3 and on side 2 is 20.086×10^{-3} mol/m^3 how much work is done in transferring 1 mol of Ca^{2+} from side 1 to side 2. (Assume that $CaCl_2$ is completely dissociated in solution and the potential difference is defined by $V_2 - V_1$.)

11. A 10 mM solution of protein is placed in a dialysis bag, which is permeable to all molecules but protein. The bag is in turn placed in a large 300 mM KCl solution. Assuming the volume of the dialysis bag to remain constant and the Donnan ratio (at equilibrium) to be 1.138.

(a) What is the final concentration of potassium ions within the dialysis bag at equilibrium?

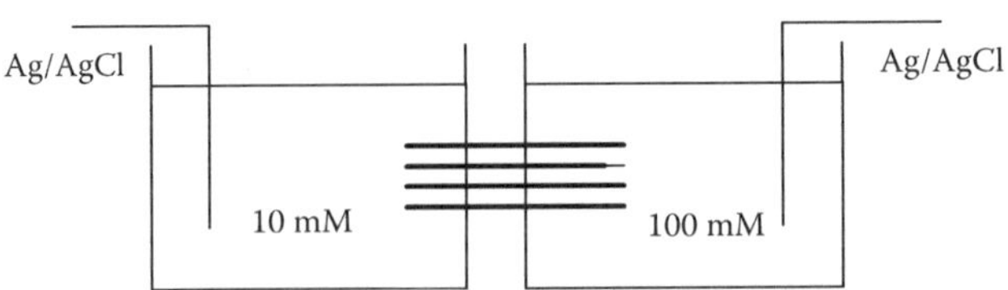

FIGURE 5.12
See Tutorial Question 7.

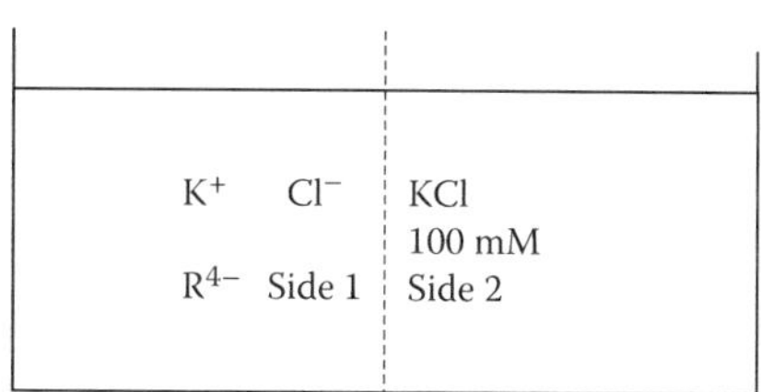

FIGURE 5.13
See Tutorial Question 9.

(b) What is the osmotic pressure within the dialysis bag at equilibrium if the protein molecules and ions behave ideally?

(c) What is the potential difference across the dialysis bag wall at 300 K?

(d) What **is** the charge on the protein molecule?

12. (a) In single muscle fibers the resting potential is 90 mV, inside negative; the internal concentrations of Na^+ and K^+ are 10.4 and 124 mmol/L and the external concentrations 110 and 2.3 mmol/L, respectively. Use the Hodgkin–Katz equation to estimate the ratio of Na to K permeabilities.

    (b) Using radioactive tracers, K^+ influx was found to be 55 nmol/m^2/s. Estimate the expected values of K and Na permeabilities.

13. From tracer flux experiments, the influx of K^+ into a nerve cell was estimated to be 9.0 pmol/cm^2/s. Use the equation

$$j_{(1\rightarrow 2)} = PC_2(VF/(RT))/(1 - \exp(-VF/(RT))$$

to estimate P_K, given that $[K_o] = 2$ mmol/L, $[K_i] = 130$ mmol/L and that $V_{oi} = 90$ mV (inside negative).

What is the equilibrium potential for K^+?

14. (a) Use the Goldman equation to show that the transmembrane potential of the unicellular organism, as shown in Figure 5.14, is approximately 90 mV. Which side is positive? Which ions are in equilibrium? If they are not, suggest why not. (Other experiments suggest that the ratio of Na^+ to K^+ permeability is 0.01 and the ratio of Cl^- to K^+ permeability is 0.09.)

    (b) K^+ influx is measured to be 80 nmol/m^2/s. Estimate K^+ permeability.

15. In single muscle fibers, the influx of Na, K, and Cl was found, using radioactive tracers, to be 40, 55, and 2 nmol/m^2/s, respectively. The concentrations of these ions

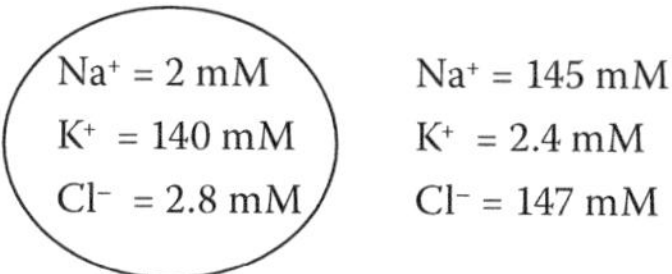

FIGURE 5.14
See Tutorial Question 14.

in the bathing solution were, respectively, 110, 2.3 and 77 mmol/L, and those in the intracellular fluid were 10.4, 124 and 1.5 mmol/L.

Use the measured PD of 100 mV (inside negative) to estimate permeability coefficients.

Estimate membrane potential using the Goldman equation.

Show that a Donnan equilibrium distribution exists for two of these ions.

16. (a) The membrane potential for a sample of red cells was found to be 25 mV, inside negative. The influx of Na, K and Cl was determined using tracers and the concentrations of these ions in the bathing solution and cell interior were also determined as follows:

	Na	K	Cl	
Influx	0.008	0.06	0.3	mol/m^2/s
Intracellular	15	30	60	mM
Bathing solution	150	10	160	mM

(i) Calculate the permeability coefficients for Na, K, and Cl.

(ii) What membrane potential is predicted by the Goldman equation?

(iii) What are the Nernst potentials for these ions?

(iv) Which ion would you suspect as being actively transported and why?

(b) Briefly discuss the assumptions made in the derivation of the Goldman equation.

17. In nonshort-circuited frog skin bathed by Ringer's solution on both sides, the PD is 50 mV inside positive. The Cl efflux is estimated to be 30 pmol/cm^2/s. Use the equation

$$j_{i(1\to2)}/j_{i(2\to1)} = (c_{i1}/c_{i2})\exp(-ziFV21/RT)$$

to estimate the Cl^- influx.

Suggest reasons why the Cl^- influx might be different from this estimate.

18. (a) A red blood cell is impaled by a microelectrode and the transmembrane potential clamped at 30 mV (inside positive). The internal Cl concentration is found to be 78 mmol/L and the external concentration 112 mmol/L. Chloride efflux is measured to be 6.1 nmol/m^2/s and the Cl influx 29 nmol/m^2/s. Show that these flux values are compatible with a passive independent mechanism of transport.

(b) From these data, estimate chloride conductance.

19. (a) In a certain experiment, the area of a skin sample, bathed on both sides by Ringer's solution, is 7 cm^2, and the average s.c.c. is 506 μA. Na^+ influx, by a radioisotope method, is 822 pmol/cm^2/s and Na^+ efflux and, by a second radioisotope, 83 pmol/cm^2/s. How much of the s.c.c. can be accounted for by Na^+ active transport?

(b) When the s.c.c. is removed, the (open circuit) skin potential is 25 mV (inside positive). If the Cl^- efflux is 92 pmol/cm^2/s, what would the Cl^- influx be if Cl^- movement is entirely passive?

(c) Estimate skin conductance and the proportion of this made up by the Cl^- conductance (assumptions as in (b)).

20. (a) A sample of frog skin, area 6.7×10^{-4} m^2, was short-circuited, and the s.c.c. was found to be steady at 253 μA. The Na^+ and Cl^- influxes were measured to be 4.2 μmol/m^2/s and 0.28 μmol/m^2/s, respectively.

    Assuming that the s.c.c. measures net Na influx, estimate

    (i) Cl^- efflux; (ii) Na^+ efflux; (iii) Cl^- conductance; and (iv) skin conductance, if the PD rises to 58 mV on removal of s.c.c.

    (b) The (open circuit) voltage of 58 mV is maintained during a period in which the Cl^- influx was measured at 1 μmol/m^2/s. Calculate

    (i) Cl^- efflux and (ii) Cl^- conductance during this period.

    (N.B., PD is inside positive; identical Ringer's solution on both sides of skin throughout experiment; flux values are unidirectional).

21. (a) Briefly describe the Ussing short-circuit current technique. Frog skin, bathed on both surfaces by Ringer's solution, is short-circuited. The current is found to be 220 μA for an area of 7×10^{-4} m^2, whereas the Na and Cl efflux was found to be 370 nmol/m^2/s and 520 nmol/m^2/s, respectively. Assuming Na to be the only actively transported ion (inward), estimate (i) Na influx, (ii) Cl influx, (iii) Cl conductance, and (iv) membrane conductance, if the open circuit voltage is 54 mV.

    (b) A nonshort-circuited frog skin was bathed on the inside by Ringer's solution and on the outside by Ringer's solution diluted five times. The skin potential (E_{io}) is 40 mV and the Cl efflux was found to be the same as that previously presented, 520 nmol/m^2/s. What is the Cl influx?

Bibliography

Aidley DJ. 1998. *The Physiology of Excitable Cells*, 4th Edition. Cambridge University Press, Cambridge, UK.

Enderle JD, Blanchard SM, Bronzino JD. 2000, 2005. *Introduction to Biomedical Engineering*. Elsevier-Academic Press, Burlington, MA, USA.

Guyton AC, Hall JE. 2006. *Textbook of Medical Physiology*. Elsevier Saunders, Philadelphia, PA, USA.

Hobbie RF, Roth BJ. 2007. *Intermedicate Physics for Medicine and Biology*. 4th Edition. Springer, New York.

Katz B. 1966. *Nerve, Muscle and Synapse*. McGraw-Hill, New York.

Malmivuo J, Plonsey RL. 1995. *Bioelectromagnetism* (online). http://www.bem.fi/book/index.htm. Viewed October 10, 2011.

Plonsey RL, Barr R. 1988, 2007. *Bioelectric Phenomena*. Plenum, New York.

Weiss TF. 1996. *Cellular Biophysics*, Volume 2, Electric Properties. MIT Press, Cambridge, MA, USA.

6

Excitability and Synapses

Andrew W. Wood

CONTENTS

This chapter will cover the phenomenon of *excitability* or the ability of certain cells to respond to electrical stimuli by undergoing rapid changes in membrane potential so that the polarity reverses; in other words, a cell which is normally positive on the outside becomes momentarily negative (for a particular region of the cell). This reversal of polarity is a basis of communication within different parts of a nerve or a muscle cell. The basis for communication from nerve to nerve or from nerve to muscle cell, involves information being transferred across a specialized gap or *synapse* between adjacent cells. This is concerned with the conversion of electrical signals into a chemical signal and then back to an electrical signal (although in some specialized synapses, there is no chemical step).

6.1 Nerve Cell or Neuron

The *neuron* (see Figure 6.1a) consists of a cell body (*soma*), with branches or *dendrites* and a long cylindrical *axon* (diameter: 1–15 μm; length: up to 0.5 m) extending from the *axon*

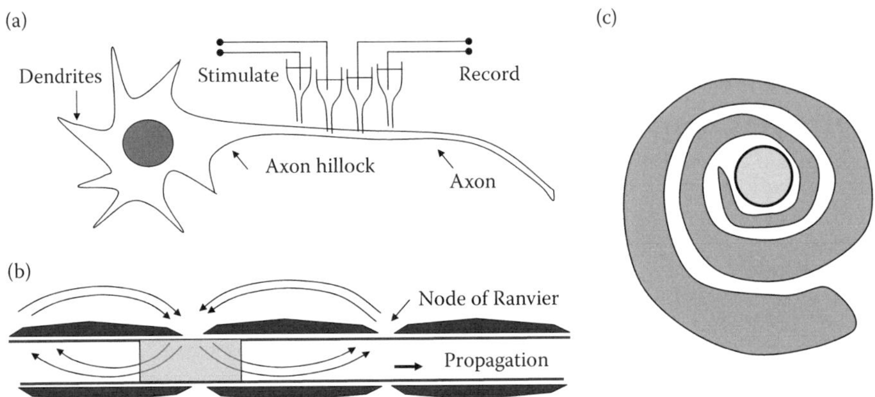

FIGURE 6.1
A typical neuron is shown in (a). In this case, the axon is unmyelinated. Microelectrodes are inserted at particular points to electrically stimulate and record from the axon. A myelinated axon is shown in (b), with the direction of current flow when the axon has been excited. Note that the current enters and leaves via the "nodes of Ranvier," which are the gaps between the enveloping Schwann cells shown in (c).

hillock. Axons can be *myelinated* or *unmyelinated*. Myelinated axons have an insulating sheath of *Schwann cells* (whose membranes consist chiefly of lipid-rich myelin) wrapped around in several layers (Figure 6.1b). The conduction velocity increases with increasing diameter and is typically in the range 20 m/s–50 m/s.

Unmyelinated axons lack this layer; in this case, conduction velocity is much slower (typically 5 m/s) and varies with the square root of the diameter. Information is carried along the nerves as a series of impulses consisting of reversals in electrical potential difference (PD) across the axon membrane; these impulses travel at constant speed, and the magnitude does not diminish during travel.

6.1.1 Resting Potential

When the axon is not carrying nervous impulses, the membrane PD remains constant at between 60 mV–100 mV, inside negative. An electrogenic sodium/potassium pump (as described in the previous chapter) is responsible for the maintenance of this PD. The membrane is 20 times more permeable to K^+ than Na^+, and the PD approaches the equilibrium or Nernst potential for K^+. In large axons, it is possible to extract and analyze the internal solution ("axoplasm") from the axon. The concentrations of K^+, Na^+, Cl^-, and an anion isethionate$^-$, which does not readily permeate the axon membrane, are given in Table 6.1, together with experimentally determined relative values of permeability of the first three ions. Note that the membrane permeability to potassium is 20 times that of sodium. Substituting these data into the Goldman equation (see previous chapter) gives an estimate of –61 mV for the resting potential (for 4°C), which is close to that observed (see also Worked Example in Section 5.4.4.2). Note these data give the following Nernst potentials: K^+: –85 mV; Na^+: 53 mV; Cl^-: –62 mV (negative indicates that the inside of the axon is negative). The resting PD is thus almost exactly the same as the equilibrium (Nernst) potential for Cl^-. For technical reasons, the Nernst potential for K^+ is estimated to be nearer –72 mV than –85 mV. Isethionate$^-$ acts to produce a Donnan system (see previous chapter). However, the ions are not in Donnan equilibrium.

TABLE 6.1

Experimentally Determined Values of (a) Concentration (mM) and (b) Permeability (Relative to K^+) of Major Ionic Constituents of Axoplasm (of a Giant Marine Squid) and Artificial Sea Water

(a)	**Sea Water (mM)**	**Axoplasm (mM)**	
K^+	10	340	
Na^+	460	50	
Cl^-	540	40	
Isethionate$^-$	0	400	
(b)	$\mathbf{P_K}$	$\mathbf{P_{Na}}$	$\mathbf{P_{Cl}}$
Rest	1	0.04	0.45
AP max	1	20	0.45

6.1.2 Action Potential

The nervous impulse or *action potential* (AP) consists of a sudden change in membrane potential to around 30 mV–50 mV, inside positive (a swing of around 100 mV). This approaches the Nernst potential for Na^+. This lasts for less than 1 ms (Figure 6.2).

The Goldman equation gives an estimate of +42 mV for the height of the action potential (using the "AP max" data from Table 6.1). Note that now, the permeability ratios for Na/K have reversed: the membrane is now 20 times more permeable to sodium than to potassium, a change of 200 times compared to the resting case. The resulting PD can be

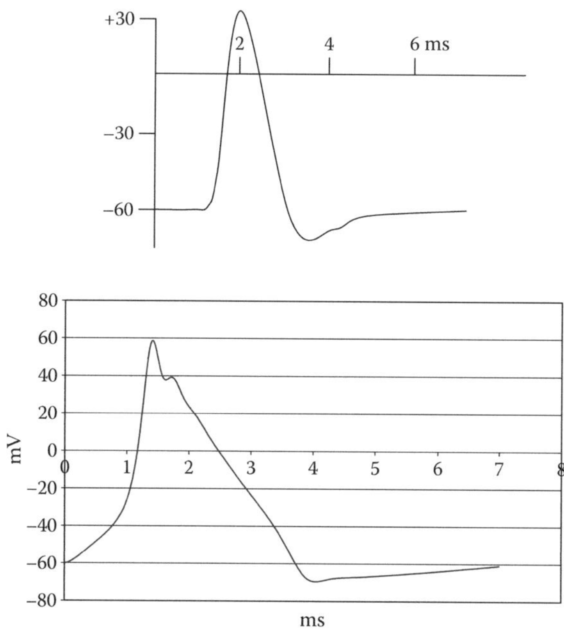

FIGURE 6.2
Upper: The action potential, measured across an axon membrane, as shown in Figure 6.1a. This particular example is of the squid giant axon, which has a resting potential of −60 mV and an action potential of +30 mV. Note the "overshoot" that occurs at around 4 ms. Lower: The action potential as calculated mathematically using the Hodgkin–Huxley (HH) model. The MATLAB code for doing this is shown in Figure 6.16.

recorded using the setup shown in Figure 6.1, with two pairs of microelectrodes, one for applying the transmembrane stimulus, the other to record the response. Figure 6.3 shows the idealized recordings from an axon: initially all microelectrodes are in the surrounding medium. The sequence shows firstly the change in PD when one of the recording electrodes penetrates the axon (PD = –90 mV) When the stimulating electrode penetrates the axon, the recorded PD becomes more negative (hyperpolarization) with negative stimulating pulses and less negative (depolarization) with pulses of the opposite polarity. For moderate pulses, these recorded PDs return to resting PD, approximately exponentially.

There is some *threshold* level of depolarization, which when exceeded gives rise to an action potential. The AP has the capacity to be propagated undiminished along the axon. The threshold shown here is at –60 mV, thus a depolarization of at least 30 mV is necessary to elicit an action potential. Note that stronger stimulating pulses still give rise to action potentials of the same height (around +30 mV). So, we get either a "full-blown" propagated action potential or nothing at all (except for a local transient PD change): this is known as the "all-or-nothing" law. Note, too, that the duration of the last, stronger, pulse is shorter, but it is still effective in generating an AP (the threshold is exceeded earlier). Normally the propagated AP proceeds from the axon hillock to the terminal end of the axon, but it can go backward up the axon if the stimulus is applied at the terminal end. If elicited in the middle of the axon, separate APs propagate in both directions from that point. The velocity of propagation is approximately constant as it courses along the axon. In an unmyelinated fiber, with a velocity of 5 m/s, the depolarized region (which lasts for 0.7 ms) will be around 3 mm in length. For short fibers, this represents a substantial proportion, if not all, of the axon. It should be understood that this region represents a reversal in the relative Na/K permeability ratio, permitting an inrush of Na^+, because of the high concentration of Na^+ in the outside solution. In fact, soon after the reversal of permeability ratio, this ratio

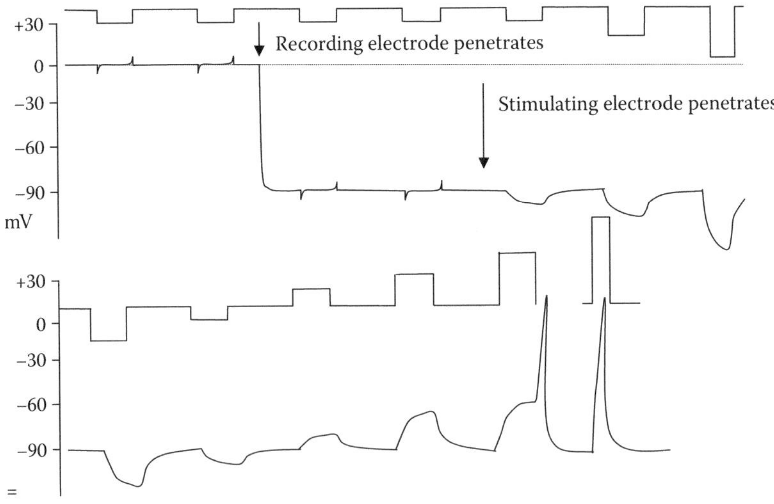

FIGURE 6.3
Diagram of experimental results from the arrangement shown in Figure 6.1a. Initially, hyperpolarizing stimuli are applied, and also initially, all four microelectrodes are in the surrounding bathing solution. When the recording electrode penetrates, a resting potential of –90 mV is obtained. Excursions from this value occur for the duration of applied current, once it is applied across the membrane. Depolarizing stimuli generate action potential "spikes" once the stimulus is greater than a certain threshold. (From Katz, B., *Nerve, Muscle and Synapse*, McGraw Hill, New York, 1966. With permission.)

returns to its normal ratio of favoring K^+ over Na^+ and allowing a subsequent (transient) efflux of K^+.

6.2 Membrane Excitability

6.2.1 Electrotonic Spread

If the voltage at a point on the membrane is changed in a manner that does *not* cause the membrane to reach threshold, there will be an exponential fall-off in voltage on either side of this point (this is termed *electrotonic spread*). This decrement is termed the *space constant* of the cell, and is of the order of a millimeter for a nerve axon. If the membrane voltage is allowed to return to its original value, it will do so exponentially in time, with a *time constant* of a few milliseconds: Section 6.3 derives the so-called *cable equation* for the nerve axon, which describes this behavior. (See Tutorial Questions 1–3.)

6.2.2 Concept of "Just Sufficient Stimulus" to Threshold

In order to reach threshold (and for the axon to fire) the stimulus has to be sufficient to change the membrane potential by a certain amount, typically a depolarization from the resting potential of 15 mV–30 mV or so. If the membrane potential is changed very slowly, the membrane adapts, so that a larger depolarization then becomes necessary for threshold to be reached. In fact, if the membrane potential is changed too slowly, the axon will never fire. For rectangular pulses, there is a minimum stimulus *strength* for an infinitely long *duration* of pulse required for the axon to fire. As the duration gets shorter, the strength needs to be increased in order for the axon still to fire. A plot of the strength versus duration for just sufficient stimuli follows the form of a rectangular hyperbola (approximately), indicating that the amount of positive *charge* withdrawn from the membrane (i.e., current strength × duration) is the relevant determinant of whether a stimulus is effective or not. The minimum stimulus of infinite duration referred to above is known as the *rheobase*, and the minimum stimulus duration for effective excitation at twice the rheobase is known as the *chronaxie*. Chronaxie values range as follows: nerves: 0.3 ms–31 ms; muscles 0.1 ms–1 ms, cardiac and smooth muscles have higher values. Note that values can be measured for both nerve and muscle fibers (since muscle fibers can be made to contract via direct electrical stimulation. Plots of minimum effective stimulus strength versus respective stimulus duration are known as *strength–duration* (SD) curves (Figure 6.4c).

The general features of the shape of the SD curve can be predicted from a very simple membrane model: a parallel resistor (R) and capacitor (C) combination (Figure 6.4b). If the stimulator is modeled as a current source I_s, then this must equal the sum of the individual currents through R and C, thus

$$I_S = I_C + I_R = C d\Delta V/dt + \Delta V/R$$

where ΔV is the *change* in membrane potential due to I_S. We can solve for ΔV as a function of t

$$\Delta V = I_S R(1 - \exp(-t/(RC))$$

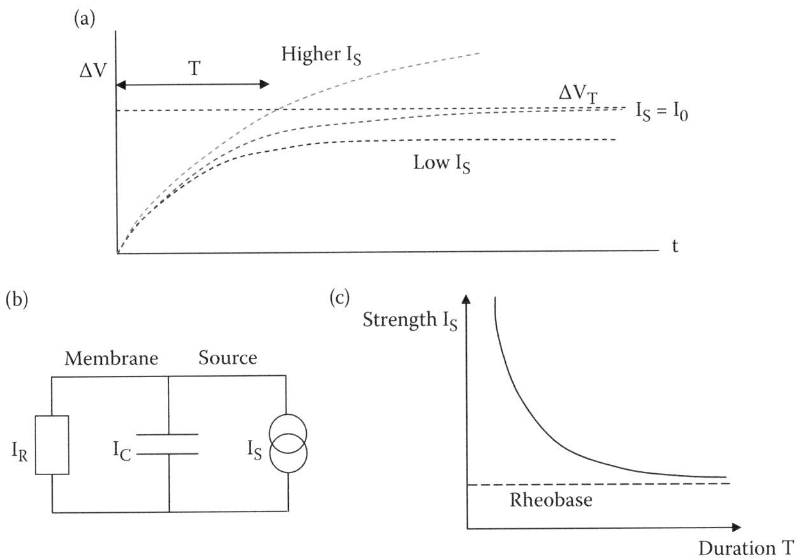

FIGURE 6.4
(a) Approach to membrane threshold. ΔV represents the difference from the resting potential caused by the stimulus, strength I_S. The pulse duration to get to the threshold depolarization ΔV_T is infinitely long for $I_S = I_0$, the rheobase, and T for the higher I_S value shown. A simple circuit representation of the membrane is shown in (b). The stimulating current from the source is represented by a capacitive and a resistive component.

If ΔV is equal to the change required to reach the threshold (ΔV_T) then the duration (t = T) that I_s has to flow for the threshold to be reached can be found. We can write

$$I_S = I_0/(1 - \exp(-T/(RC))$$

where $I_0 = \Delta V_T/R$, the rheobase, since this would be the value of I_S for an infinitely long pulse ($T \to \infty$). For $I_S = 2I_0$, the value of T is RC ln2, which is the chronaxie (you can easily verify this). Figure 6.4 illustrates that as the current is increased above I_0, the time before the threshold is reached diminishes. Note that the plot of T versus critical I_S will not be exactly a rectangular hyperbola, but will be close, especially for that part of the range where the duration is longer than the chronaxie.

After a nerve has "fired" (that is, an AP has been elicited) the effective threshold increases, and a stronger stimulus is be needed to get a second AP within about 2.5 ms–5 ms of eliciting the first one. This is known as the relative refractory period, because the nerve is resistant (hence refractory) to further stimulation. In fact, for times less than 2.5 ms, the threshold is effectively infinite, and no second AP is possible (this is known as the absolute refractory period). This sets an upper limit on number of impulses (or APs per second) which an axon can carry at around 400 Hz. This is illustrated in Figure 6.5, where the dashed line denotes where the threshold is at various times after the first stimulation.

How does this apply to a bundle of nerve fibers, like a toad sciatic, which has typically 10^5 individual fibers contained within the nerve trunk? Each fiber will fire at its own threshold and the stimulating current will vary according to where in the nerve trunk a fiber lies. With electrodes on the surface of the sciatic, it is obviously easier to stimulate the fibers closer to the electrodes than further away. Each fiber obeys the "all-or-nothing" law, but as the stimulus strength is increased, progressively more and more fibers are activated,

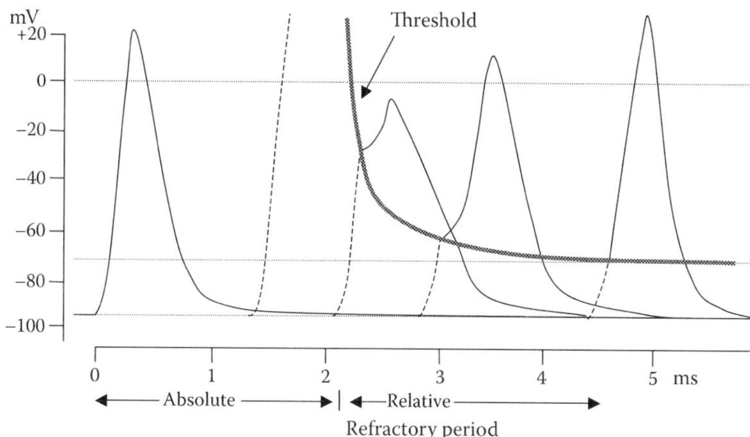

FIGURE 6.5
Absolute and relative refractory periods in a nerve. The right-hand side of the diagram shows the action potentials elicited by a second stimulus applied at various times after the first (at $t = 0$). The effective threshold is shown for various times after the first stimulus has been applied. It is effectively infinite for $t < 2$ ms. (With kind permission from Springer Science+Business Media, Berlin: *Human Physiology* 2nd ed., 1989, Schmidt, R. and Thews, G.)

until they are all activated. This accounts for the fact that when an AP is recorded from a sciatic nerve, as the stimulus is increased the amplitude of the response grows until a plateau is reached. A typical recorded extracellular AP is shown in Figure 6.6. The response is typically biphasic (that is, there is a positive going and a negative going phase) which arises because the AP passes first over electrode A and then electrode B, so the PD between them reverses as this happens. Recorded APs also show a "stimulus artifact" due to the passage of current directly along the nonexcitable parts of the nerve trunk. This can be minimized by the insertion of an earth electrode between the stimulating and recording electrodes. If, on the other hand, the nerve is in a conducting medium (as it would be, for example, if placed on filter paper soaked in a physiological solution), the currents arising

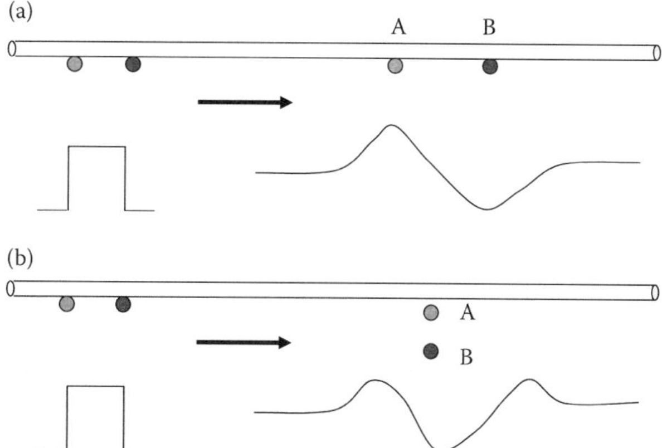

FIGURE 6.6
Biphasic and triphasic responses from amphibian sciatic nerve preparations. In (a), the recording electrodes contact the external surface of the nerve bundle surrounded by air, and in (b), the nerve is on moist filter paper and the recording electrodes are at right angles to the nerve bundle (recording volume conductor currents). Note the characteristic shapes of the responses.

from the AP will not be constrained to flow within the nerve trunk. If two electrodes are in the physiological solution and near to, but not actually touching the nerve, a triphasic action potential will typically be recorded. This reflects the fact that as the AP approaches the electrodes the current is flowing away from the nerve (see Figure 6.1), then when the AP is next to the electrodes it flows toward the nerve and then finally away from the nerve again as it passes by.

6.2.3 Propagation

The action potential is self-propagating: the inrush of Na^+ at the point of potential reversal withdraws charge from areas of the membrane ahead of this region. The threshold is reached at these regions ahead, so the potential reversal moves along to that point. The magnitude of action potential does not diminish with distance; it is thus also self-reinforcing. As the action potential passes, the membrane is left refractory for a few milliseconds or so as mentioned in the previous section.

In myelinated nerve, the depolarizations occur only at the *nodes of Ranvier* (the inrush of Na^+ at one node causing withdrawal of charge form the next node) and so the impulse goes via a series of jumps from one node to the next (this is called *saltatory conduction,* where "saltatory" indicates the jumping nature of this process). This is shown in Figure 6.1b.

6.2.4 Role of the Sodium Pump

If a number of impulses pass along an axon, the Na^+ concentration inside the axon rises and the K^+ concentration falls, to such an extent in some cases that the nerve fatigues and ceases to fire. The sodium/potassium pump will gradually restore the initial conditions if the nerve is allowed to rest. Each time an action potential occurs, the axon loses around 43 nmol/m^3 of K^+ and gains around the same amount of Na^+. This equates to approximately 10^{-11} mol/m^2 of area, per impulse. Compare this to the flux of K^+ or Na^+ during an action potential of around 2×10^{-4} mol/m^2/s during 1 ms (2×10^{-7} mol/m^2 per impulse). The Na/K pump is thus working at least three orders of magnitude more slowly than the magnitude of Na^+ inrush or subsequent transient K^+ efflux during the action potential. The axon thus relies on periods of nonstimulation to allow conditions to return to the original state. In fact, the nerve will become refractory if stimulated repeatedly, due to raised internal [Na^+].

6.3 Cable Theory Applied to Nerve Membrane: Membrane Properties

The axon can be modeled as a cable: in essence, it can be divided into segments, and the change in voltage and current in each section can be modeled using a simple electrical circuit, as shown in Figure 6.7.

Consider a section of axon membrane. With reference to Figure 6.7, the PD between points A and B is

$$V_o(A) - V_o(B) = -i_o R_o \delta x,$$

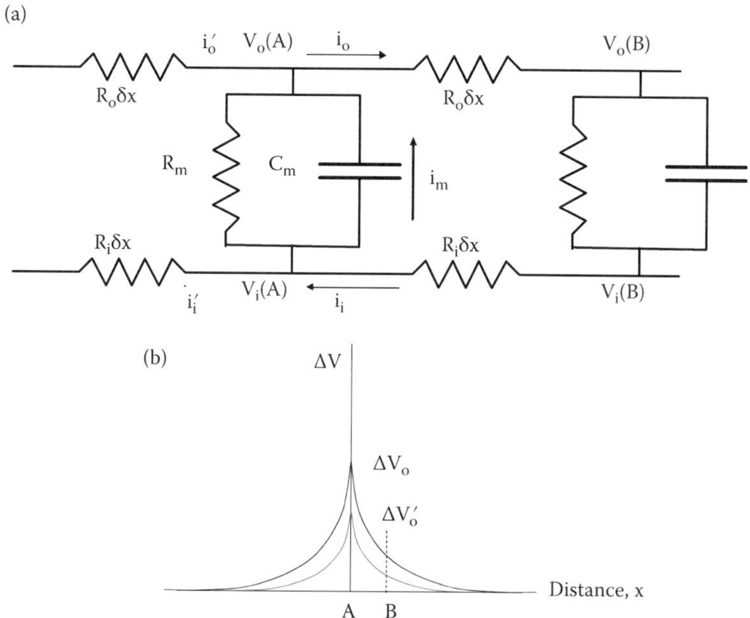

FIGURE 6.7
(a) Circuit representation of axon membrane giving rise to the cable equation. Each segment of axon is represented by a parallel resistor and capacitor combination (as in Figure 6.4b). i_o and i_i represent the currents flowing in the external and internal solutions, respectively. (b) Electrotonic spread. A small change in membrane potential is produced by the electrode at A. The change in PD on either side of A is shown. The lighter curve shows the PD change at a later time. At point B, the initial PD change and the change at subsequent times are given by the cable equation, as discussed in the text.

where i_o is the longitudinal current (A) and R_o is the external resistance (Ω m^{-1}) thus,

$$\delta V_o = -i_o R_o \delta x$$

or

$$\partial V_o/\partial x = -i_o R_0 \tag{6.1}$$

similarly,

$$\partial V_i/\partial x = -i_i R_i \tag{6.2}$$

and the change in internal current must be the same as the current through the membrane (note that i_m is in A m^{-1}, the current per unit length of axon).

$$\delta i_i = i_i' - i_i = -i_m \delta x = -\delta i_o \tag{6.3}$$

By definition

$$V_m = V_i - V_o$$

and differentiating this gives

$$\partial V_m/\partial x = \partial V_i/\partial x - \partial V_o/\partial x$$
$$= i_i R_i + i_o R_o$$

from Equations 6.1 and 6.2.

Differentiating again,

$$\begin{aligned}\partial^2 V_m/\partial x^2 &= -\partial i_i R_i/\partial x + \partial i_o R_o/\partial x \\ &= +\, i_m (R_i + R_o)\end{aligned} \tag{6.4}$$

from Equation 6.3.

R_o can often be ignored (because the external solution can often be regarded as infinitely large), so that

$$\partial^2 V_m/\partial x^2 \approx i_m R_i.$$

Membrane current in electrophysiology is usually measured in A/m^2 (i.e., a current *density*); this is related to the current per unit length i_m by dividing the latter by the circumference of the axon.

$$I_m = i_m/(2\pi a).$$

If we express the internal resistance (in Ω per meter in length) in terms of the resistivity of axoplasm ρ_i (in Ω m), we get

$$R_i = \rho_i/(\pi a^2);$$

thus,

$$I_m = \partial^2 V_m/\partial x^2 \cdot (\pi a^2/(\rho_i 2\pi a)),$$

i.e.,

$$I_m = \partial^2 V_m/\partial x^2 (a/(2\rho_i)).$$

Consider now the current through R_m and C_m, and putting $\Delta V_m = V_m - V_r$, where V_r is the resting potential:

$$i_m = \Delta V_m/R_m + C_m\, \partial V_m/\partial t$$

$$\partial^2 V_m/\partial x^2 = (R_i + R_o)[\Delta V_m/R_m + C_m\, \partial V_m/\partial t].$$

From Equation 6.4, putting R_o back into the expression, rearranging gives

$$\partial^2 V_m/\partial x^2 = (R_i + R_o) C_m\, \partial V_m/\partial t + (R_i + R_o)\Delta V_m/R_m$$

and then

$$R_m/(R_i + R_o)\partial^2V_m/\partial x^2 - R_mC_m\,\partial V_m/\partial t - \Delta V_m = 0$$

or

$$\lambda^2\partial^2V_m/\partial x^2 - \tau\,\partial V_m/\partial t - \Delta V_m = 0 \quad (6.5)$$

where $\lambda = \sqrt{[R_m/(R_i + R_o)]}$ is the space (or length) constant and $\tau = R_mC_m$ is the time constant. Typical values of space constant are around 2 mm–5 mm for large unmyelinated nerves.

Let us now consider some physical properties of biological membranes (Table 6.2), compared to artificial lipid bilayers, which can be fabricated over small apertures from thin films of the phospholipids found in cells.

Note in Table 6.2 how much larger the membrane resistance of a lipid bilayer is compared to an axon membrane. This is due to the presence of ion channels in the latter. Capacitance values, on the other hand are very similar. The various measures are interrelated as follows:

$C = C_m/(2\pi a) = \kappa\varepsilon_0/d$; $R = (1/G)$, where G is membrane conductance (S/m^2);
$R = \rho_m d$; $R_m = \rho_m d/(2\pi a)$;
Time constant $\tau = R_mC_m = \rho_m d/(2\pi a)\kappa\varepsilon_0(2\pi a)/d = \rho_m\kappa\varepsilon_0$
$R_i = \rho_i/(\pi a^2)$;
$R_o = \rho_o/A_e$, where A_e is effective cross-section of external solution through which membrane-associated currents pass (usually large), making R_o small.

Let us now apply the cable equation (6.5) to specific examples of electrotonic spread. Suppose a stimulus is applied by electrode A in Figure 6.7b which is not strong enough to elicit an AP. The membrane potential will change at point A by an amount ΔV_0, and on

TABLE 6.2

Lipid Bilayer Membrane Properties

Property	Biological Membrane	Bilayer Membrane
Thickness (d: nm)	6–10	6.7–7.5
Axon radius (a: m)	$5\text{–}500 \times 10^{-6}$	NA
Capacitance/unit membrane area (C: F/m^2)	$0.5\text{–}1.3 \times 10^{-2}$	$0.38\text{–}1.0 \times 10^{-2}$
Capacitance/unit length of axon (C_m: F/m)	$0.16\text{–}0.3 \times 10^{-6}$	NA
Dielectric constant (κ)	7	2–3
Membrane resistance (R: $\Omega\ m^2$)	$10^{-1}\text{–}10^{5}$	$10^{6}\text{–}10^{9}$
Membrane resistivity (ρ_m: Ω m)	1.6×10^{7}	
Resistance *in* unit length of axon (R_m: Ω m)	50–5000	
Resistivity of internal solution (axoplasm) (ρ_i: Ω m)	0.003–0.5	
Internal resistance of unit length of axon (R_i: $\Omega\ m^{-1}$)	$10^{8}\text{–}10^{12}$	
Resistivity of external solution (ρ_o: Ω m)	0.0022–0.0087	
External resistance of unit length of axon (R_o: $\Omega\ m^{-1}$)	Depends on experimental setup, but relatively low	

either side of this point, the recorded potential will die away exponentially with distance from point A. From the cable equation (putting $\partial V_m/\partial t = 0$), we get

$$\Delta V = \Delta V_0 \exp(-|x|/\lambda)$$

where λ is the space constant referred to above. If we now consider a specific location (x′, say, or point B) and assume that the $\partial^2 V_m/\partial x^2$ can be ignored (which may be a gross oversimplification), we can study how the voltage $\Delta V'$ varies with time. We get

$$\Delta V' = \Delta V_0' \exp(-t/\tau)$$

where $\tau = R_m C_m$. Note from Table 6.2 that the product $R_m C_m$ is the same as RC. This is related to the chronaxie, as we saw earlier.

WORKED EXAMPLE

(a) A subthreshold depolarization of 5 mV at a point x = 0 on an axon leads to smaller depolarizations on either side of this point, decaying away to zero. At a point x = ±2 mm the magnitude of depolarization is just 1.85 mV. Estimate the space (or length) constant for this axon.

Answer: Using the formula $\Delta V = \Delta V_0 \exp(-|x|/\lambda)$ above, we can write $-|x|/\lambda = \ln(\Delta V/\Delta V_0) = \ln(1.85/5) = -0.99$, and since x is 2 mm, λ must be $-2/(-0.99) = 2$ mm.

(b) At x = 0, the depolarization has fallen this same value, 1.85 mV, 3 ms after the moment the value was 5 mV. Estimate the time constant for this axon.

Answer, using the formula $\Delta V' = \Delta V_0' \exp(-t/\tau)$, but following the same method, we get $-(t/\tau) = \ln(1.85/5) = -0.99$ and since t is 3 ms, τ must be $-3/(-0.37) = 3$ ms.

(c) Given that the membrane capacitance is 1μF/cm² (0.01 F/m²), show that the membrane resistance is 0.3 Ω m².

Answer: $\tau = RC$, so $R = 3 \times 10^{-3}/0.01 = 0.3\ \Omega\ m^2$

(d) Estimate the resistance of the axoplasm (the solution inside the axon) per unit length (Ω/m), assuming that the resistance of the external solution per unit length is negligible in comparison to this and that the axon radius is 4×10^{-6} m.

Answer: Since $\lambda = \sqrt{(R_m/(R_o + R_i)}$ and R_o can be ignored, we can put $\lambda = \sqrt{(R_m/R_i)}$. From the equations given above, we see that $R = R_m 2\pi a$. Thus $\lambda = \sqrt{(R/(R_i 2\pi a))}$. Rearranging, we get $R_i = R/(\lambda^2 2\pi a) = 0.3/((2 \times 10^{-3})^2 2\pi 4 \times 10^{-6}) = 3 \times 10^9$ Ω/m. This fits with the range given above.

TUTORIAL QUESTIONS: Nos. 1 and 2.

6.4 Membrane Currents: Experimental Methods for Measuring

Earlier we saw that permeability data for the axon during the passage of an AP supported the notion of a transient change in permeability from K^+ to Na^+-favoring then finally back to K^+-favoring again. Using a *voltage clamp* (a rapid feedback amplifier), the axon membrane potential can be held at a desired potential (the resting potential, say) and then changed to a second potential virtually instantaneously (a voltage step). The earliest type of voltage clamp in nerves effectively clamped the entire axon membrane at the same potential; in more recent developments of this technique, only a very small area of membrane (or patch) is clamped at a particular potential. In either case, the currents that flow following this step in voltage can be monitored and decomposed into the various contributory ionic and displacement currents. In Section 6.5, there will be more description on how this can be accomplished. For the moment, it can be stated that the main currents of interest are K^+ and Na^+, and these can be analyzed separately. Knowing the appropriate electrical driving forces, the ion conductances can be estimated, from

$$G_K = J'_K/(V_m - V_K),$$

where G_K is the K^+ conductance (in S/m^2, or similar units), V_m is the membrane potential (inside relative to outside) and V_K is the Nernst (or equilibrium or reversal) potential for K^+ (which is around –72 mV for the squid axon). This is the same as the chord conductance, described in the previous chapter. In fact, the texts usually use the symbol g_K for K^+ conductance (even though it is not strictly speaking a slope conductance) and I_K for K^+ current *density* (even though the symbol I is usually reserved for *current*). The essential purpose of the voltage clamp apparatus is to discover how G_K and G_{Na} vary with time and with the value of the clamped potential. As we will see in Section 6.5, knowledge of this allows estimates to be made of these quantities during the normal propagated action potential. This, together with knowledge of the way electrical charge accumulates or is discharged from the membrane, allows a complete synthesis of the way that voltage changes during the action potential, which can be compared against the actual values shown in Figure 6.2.

6.4.1 How the Voltage Clamp Operates

Figure 6.8 shows a section of nerve axon in which fine electrode wires have been inserted. One of these wires represents the output from the operational amplifier OA, whose input is the *difference* between the membrane potential V_m measured between the middle two wires and a "command voltage" V_c, which is set by the experimenter, and can be changed in a stepwise manner between selected voltages. The output current is fed back across the axon membrane down to an earth wire via a measuring device for current (A). The direction of the current is such that the input voltage "seen" by the OA is driven to a low value, almost zero. In this situation, $V_m - V_c = \delta V$, where $\delta V \ll V_m$. To all intents and purposes, we can say that $V_m = V_c$. The actual value of δV is determined by the gain of OA, since a nonzero input is required to produce a nonzero output: a very large gain of OA (typically 10^5) guarantees that δV will be negligible. The adjustment of current I_m to produce $V_m = V_c$ takes place almost instantaneously and represents the ion currents or charge/discharge of the membrane to make this happen. Since the ion conductances are, in general, time-dependent, the feedback current (which is the same as the membrane current I_m) will change with time, as shown in the diagram. The early "blip" is current associated with the displacement of charge from the membrane, due to the sudden step change in voltage, whereas the inward, followed by the outward current is due to specific ion channels responding over time to the change in voltage (Figure 6.9a). The important feature of the voltage clamp is that it removes any voltage variations along the axon (apart from sudden step changes). This

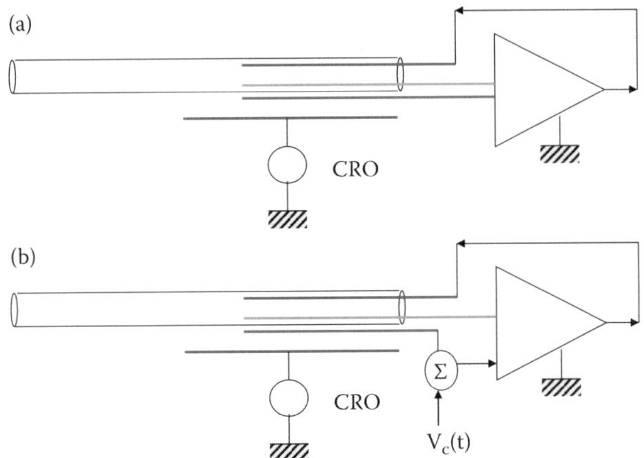

FIGURE 6.8
The voltage clamp apparatus. In (a), the membrane potential is monitored by the two fine wire electrodes on either side of the axon membrane and connected to the input of the operational amplifier. The output current is fed back across the membrane, and the current is monitored in the CRO (as the voltage across a standard resistor). This current is sufficient to maintain the input potential at a few microvolts, effectively zero. In (b), a "command potential" V_C is introduced, which will maintain the membrane at the same magnitude (since now, $V_m - V_C$ will effectively be zero). V_C is usually the resting potential $-V_r$ to start with, and then it changes to another selected value in a stepwise manner (see Figure 6.9).

means that variations with *time as the only variable* can be studied for each fixed V_m value (typical curves are shown in Figure 6.9).

Since the ion currents were known to be mainly Na^+ and K^+, the challenge, in the original experiments, was how to separate them. Initially this was done by substituting some of the Na^+ in the bathing solution, so that a Na^+ current would not occur. Subsequently, it was discovered that the currents could be blocked by specific agents: the most well known are tetrodotoxin (TTX), which blocks Na^+ channels, and tetraethylammonium (TEA), which

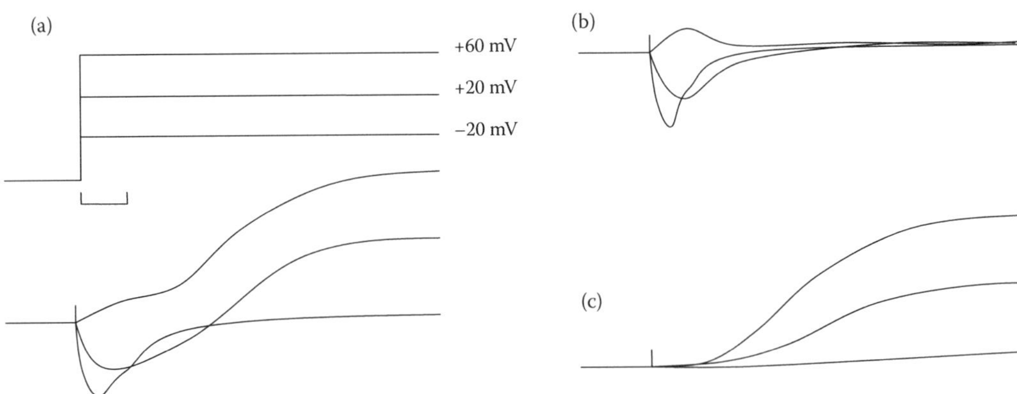

FIGURE 6.9
Recordings obtained from the voltage clamp circuit shown in Figure 6.8b. In (a), responses to voltage steps to −20, +20 and +60 mV are shown. The "blip" at the start is due to membrane capacitance. Inward current is represented by downward excursions. Note that for +60 mV, the current remains outward for the duration of the observation. The length of the horizontal bar denotes 1 ms. In (b), the nerve blocker TEA has been added, leaving just the Na^+ current and, in (c), TTX, leaving the K^+ current.

blocks K^+ channels. Effects on the voltage clamp currents by adding these agents are shown in Figure 6.9b. This reveals that the Na^+ current density I_{Na} peaks after less than 1 ms, then returns rapidly to zero, whereas I_K rises more slowly, reaching a plateau after around 5 ms. Details on how these phenomena can be described mathematically are given in Section 6.5.

6.4.2 Patch Clamping

More recent work has allowed tiny patches of membrane (a few microns in diameter) to be isolated. The method of isolating these relies on the production of micropipettes whose tips are smoothed, so that rather than penetrating the cell (as shown in Figure 6.1a) they form a tight seal with the membrane surface (called a "gigohm seal" because of the high electrical resistance between the inside and outside of the pipette when sealed against the membrane. The various ways of isolating a patch of membrane are summarized in Figure 6.10. This method has allowed the characteristics of individual channels to be studied, by voltage clamping, as above. The circuit is slightly modified, as shown in Figure 6.11 (compare with Figure 6.8b). Essentially, the current I_m is monitored for a small number of channels (ideally a single channel) whilst the

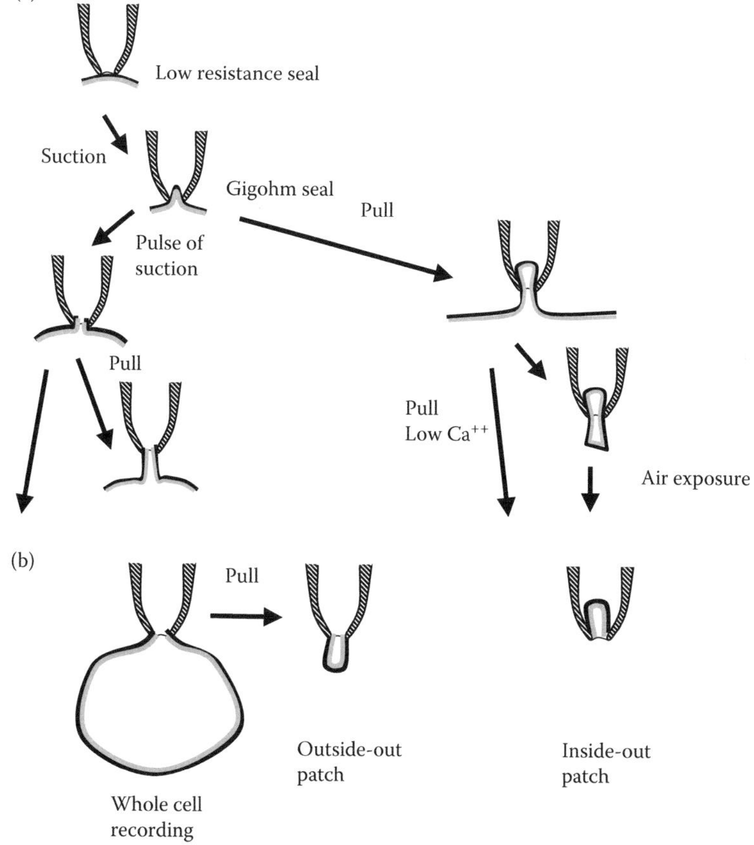

FIGURE 6.10
The various configurations of membrane patches in preparation for patch clamping, showing (L to R) whole cell recording, outside-out patches and inside-out patches. The two lipid layers are shown in black and gray to emphasize the way the membrane is oriented. (With kind permission from Springer Science+Business Media, New York: *Bioelectricity: A Quantitative Approach* 3rd ed., 2007, Plonsey, R. and Barr, R., altered.)

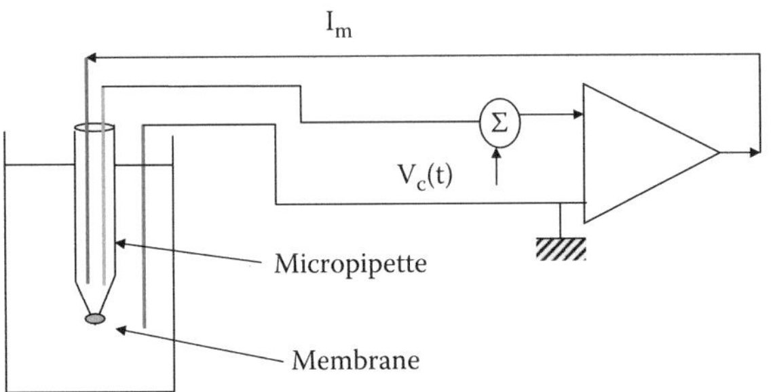

FIGURE 6.11
Patch clamp (compare with Figure 6.8b). The patch is maintained at a potential determined by V_C, and the current I_m to do this is monitored. Under certain conditions, it is possible to have only a single channel within the patch, and I_m will reflect the current through this one channel. In actual circuits, there is only one electrode within the micropipette, serving both to record the membrane potential and to deliver I_m.

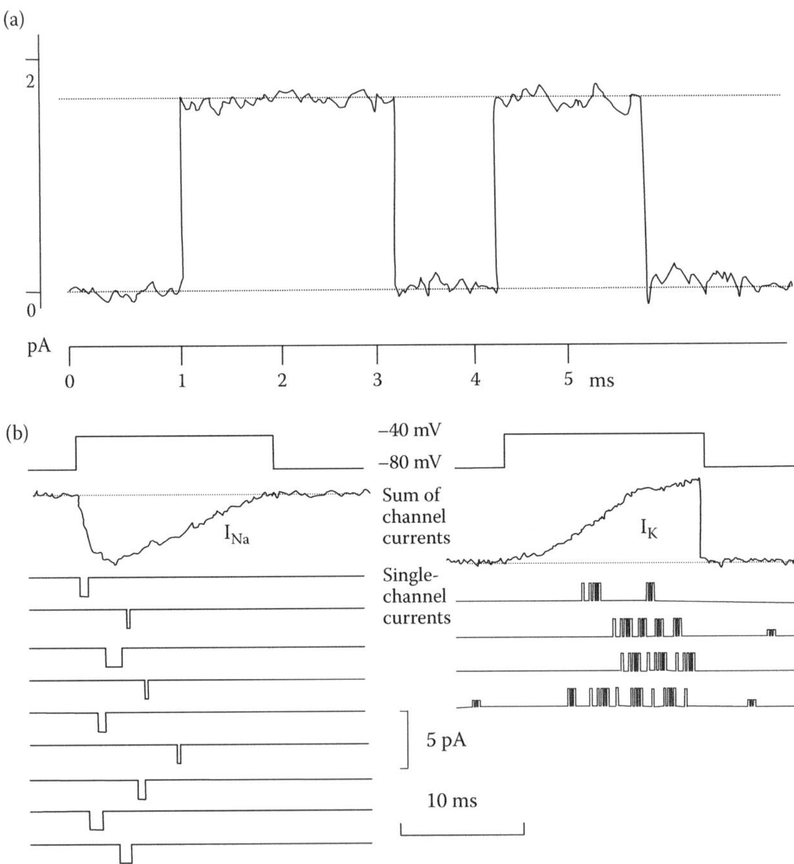

FIGURE 6.12
Top: Current from a single K^+ channel is shown for V_m = +10 mV. Note that the channel is either closed or open. Bottom: Currents from several single channels shown (Na^+ on left and K^+ on right) and compared with the combined current. Note that the combined currents resemble those shown in Figure 6.9b and 6.9c. (With kind permission from Springer Science+Business Media, Berlin: *Human Physiology* 2nd ed., 1989, Schmidt, R. and Thews, G., altered.)

membrane patch is held at specified values of V_m. These channels are found to switch in an unpredictable fashion from a closed to open state and then back again. The *probabilities* of individual channels being open follow a similar time course as for $|I_{Na}|$ and I_K shown in Figure 6.9b. An increase in current is thus equivalent to an increasing probability of individual channels being open. Figure 6.12b represents currents in several individual channels studied over several milliseconds. Note that each channel either carries the same few pA of current or nothing: if the voltage step were to be made larger, the channel currents would be larger.

6.5 Hodgkin–Huxley Theory

This theory was developed by Sir Alan Hodgkin and Sir Andrew Huxley, working at Cambridge University in the 1940s and 1950s. Most of the experimental work was on the squid giant axon as described above. The challenge was twofold: (1) to derive a mathematical theory which could predict the time course of I_{Na} and I_K during the application of various values of voltage clamp and (2) to see whether it would be possible to predict what would happen to membrane potential when the membrane was not clamped, but stimulated in the normal way. In the latter case the charging or discharging of the capacitor C_m would also contribute a current and therefore would affect the voltage. It was also assumed that each ion permeates the membrane by a separate route (ion-specific channels) and that this could be represented by the circuit shown in Figure 6.13. The batteries V_{Na} and V_K represent the Nernst potentials of these two ions (note that when $V_m = V_{Na}$ or V_K, the current of that particular ion is zero). The flow of other ions is lumped together as a "leak" channel, with its specific Nernst potential V_L. The conductances (or reciprocal resistances) of these channels are denoted G_{Na}, G_K and G_L, respectively, with the first two varying with respect to time (and also with respect to V_m). Further, it was assumed that a kinetic theory could be developed to account for the time course of these conductance changes (the Hodgkin–Huxley (HH) equations). The time course of $G_{Na}(t)$ and $G_K(t)$ is shown in Figure 6.14. These are derived simply by taking the data shown in Figure 6.9b,c and dividing the $I_{Na,K}$ values

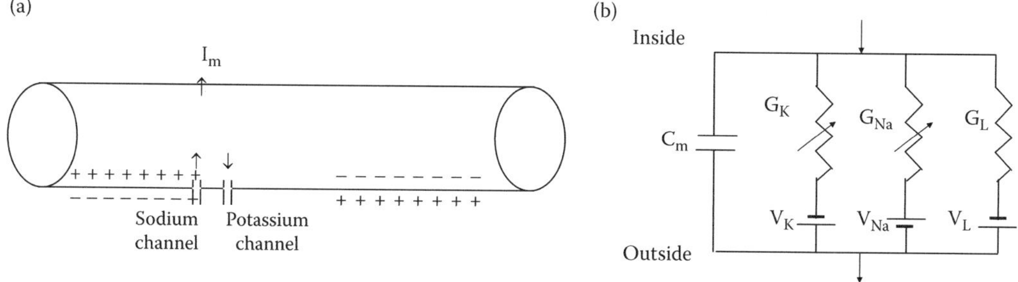

FIGURE 6.13
(a) A representation of the nerve axon, showing separate Na^+ and K^+ channels (for simplicity, only one of each shown). The depolarized portion of the axon is shown on the left (negative outside). (b) The HH electrical circuit representation of the axon membrane, with the Nernst potentials of the Na^+, K^+ and "leak" ions shown as electrical cells. The resistances are the inverse ion conductances, and the capacitor is the membrane capacitance.

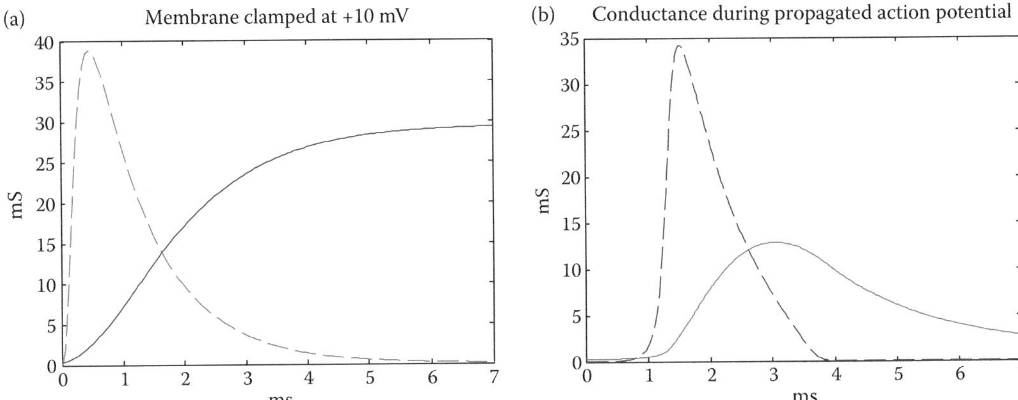

FIGURE 6.14
Ion conductance (a) during clamping and (b) during passage of action potential. Na^+ conductance shown dotted; K^+ conductance shown with full line.

at each time point by $V_m - V_{Na,K}$. For example, for a clamp potential of +10 mV (V_m), each value of I_{Na} will be divided by +10 – 55 = –45 mV and each value of I_K by +10 – (–72) = 82 mV. Since I_{Na} values are negative, G_{Na} will be positive. Since conductance is related to the intrinsic properties of ion channels (as the patch clamp data subsequently showed), any theory would need to make some assumptions about how the ion channels are controlled: the essential idea was that of a voltage-sensitive *gate* and, moreover, a gate whose properties would also vary with time. The time variance was assumed to arise from two competing forces (both voltage dependent), one tending to favor the gate opening and the other opposing it. This is common to many theories of receptor kinetics, with the overall rate of change of receptor occupancy being the difference between association and dissociation rates.

6.5.1 Modeling Conductance Changes

From the circuit diagram shown in Figure 6.13 we can write

$$I_m = C_m dV_m/dt + I_{Na} + I_K + I_L.$$

Here I_m = total membrane current density, I_{Na},I_K = sodium, potassium current density, I_L = "leak" pathway current density. Conventionally, the units are I_i in mA/cm². V_m in mV; C_m = in μF/cm² (the usual value conveniently taken as 1 μF/cm²). In voltage clamping experiments, $I_m = I_K + I_{Na}$ (that is, $dV_m/dt = 0$ after the initial step change and I_L can be ignored).

To model I_K, a concept is introduced of an entity or subunit within the channel that controls whether or not the channel would be open (or the degree to which the channel would be open). The probability of this subunit being in the correct position is given by a dimensionless factor *n* which varies between 0 and 1, which then determines the K^+ current density I_K by the following two equations:

$$I_K = G_K^0 n^4 (V_m - V_K);$$

$$dn/dt = \alpha_n(1 - n) - \beta_n n.$$

Here V_K is the Nernst potential for K^+ (–72 mV) and G_K^0 is the maximum K^+ conductance (which would occur if all channels were open simultaneously and whose value was determined experimentally to be 36 mS/cm^2). α_n and β_n represent rate constants for the channel opening and closing processes (or more precisely the gating process) and are equivalent to the association and dissociation rate constants referred to earlier. A large value of α_n gives a greater chance of the channel opening. The rate constants were found from experimental data to depend on V_m only and to have the following form:

$$\alpha_n = 0.01(-50-V_m)/\{\exp[(-50-V_m)/10] -1\};$$

$$\beta_n = 0.125 \exp[(-V_m -60)/80].$$

Note that the shape of the plot of $G_K(V_m(\text{constant}), t)$ in Figure 6.14 is sigmoidal, rather than of the form $A(1 - \exp(-t/\tau))$, which would be expected from the above formulae if $G_K \propto n$. If multiple n-type subunits are needed to be in place simultaneously for the channel to open, then the chance of a channel being open is substantially reduced. In fact, if n is the probability of a single n-type subunit being in place, then the probability of two such subunits being simultaneously in their proper locations is n^2, and if three, then the overall probability is n^3 and so on. The best fit to experimental data was obtained by setting this number to four, hence the dependence of the K^+ current on n^4.

For Na^+, the situation is slightly more complicated, because the channel apparently turns itself off after around 1 ms. Thus, in addition to a probability of certain subunits causing the Na^+ channel to open (the probability of each of these subunits being in place being *m*) there is also a "turn-off" (or inactivation) subunit, whose probability of being *absent* is given by *h*. There needs to be several m-type subunits present and one h-type subunit absent in order for the channel to open. The best fit was obtained for three "turn-on" subunits; thus, I_{Na} is proportional to m^3h.

$$I_{Na} = G_{Na}^0 m^3 h(V - V_{Na}),$$

where

$$V_{Na} = 55 \text{ mV and } G_{Na}^0 \text{ (the maximum Na}^+\text{ conductance)} = 120 \text{ mS/cm}^2.$$

The m-type subunits behave according to

$$dm/dt = \alpha_m(1 - m) - \beta_m m,$$

where

$$\alpha_m = 0.1(-35 - V_m)/\{\exp[(-35 - V_m)/10] -1\};$$
$$\beta_m = 4 \exp[(-V_m - 60)/18]$$

and the h-type subunit as

$$dh/dt = \alpha_h(1 - h) - \beta_h h,$$

where

$$\alpha_h = 0.07\exp[(-60 - V_m)/20];$$
$$\beta_h = 1/\{\exp[(-V_m - 30)/10] + 1\}.$$

It is assumed there is no dependence of the "leak" channel on time and it can be modeled as

$$I_l = G_l(V_m - V_l),$$

where

$$V_l = -50 \text{ mV and } G_l = 0.3 \text{ mS/cm}^2.$$

Note that at the resting potential m, n and h have the values (determined experimentally) of 0.05, 0.3 and 0.6. A few milliseconds after each step change in V_m, when a steady state is obtained (effectively $t = \infty$) the values of these parameters are given by $m_\infty = \alpha_m/(\alpha_m + \beta_m)$, and similarly for n_∞ and h_∞. The time constants for the exponential changes in m, n, and h following step changes are given by $\tau_m = 1/(\alpha_m + \beta_m)$ and so on.

Thus,

$$m(t) = m_\infty - (m_\infty - m_0)\exp(-t/\tau_m),$$

where m_0 = resting value of 0.05. Similar equations can be written for n(t) and h(t) with n_0 = 0.3 and h_0 = 0.6, respectively. Remember that $m_\infty n_\infty$ and h_∞ are functions of V_m alone (for a given constant temperature and Ca^{2+} concentration).

Note that as V_m goes more positive (or less negative), α_m, α_n and β_h *rise*, and β_m, β_n, and α_h *fall*.

Given the empirically determined values of these parameters as a function of V_m, these can be used in an inverse sense to calculate I_{Na} and I_K via the m(t), n(t) and h(t) equations for a particular value of clamped potential V_m, and these values should (and do) match those experimental values shown in Figure 6.9. This can be done by a simple MATLAB ® script file (or m-file), such as that shown in Figure 6.15.

WORKED EXAMPLE

Estimate the K^+ current (density) in mA/cm² at 2 ms after changing the clamped potential from –60 mV to +10 mV.

Answer

Firstly, we estimate $G_K(t = 2)$. Note that all the quantities are scaled in mV and milliseconds. Using the equation $\alpha_n = 0.01(-50-V_m)/\{\exp[(-50-V_m)/10] -1\}$ with $V_m = +10$, gives 0.601 ms^{-1} and $\beta_n = 0.125\exp[(-V_m - 60)/80]$ gives 0.052 ms^{-1}. This gives $\tau_n = 1/(\alpha_n + \beta_n) = 1.53$ ms and $n_\infty = \tau_n\alpha_n$ as 0.92. Thus, the value of *n* at t = 2 from $n(t) = n_\infty - (n_\infty - n_0)\exp(-t/\tau_n)$ and $n_0 = 0.3$ is $0.92 - 0.62\exp(-2/1.53) = 0.75$. Now, since $G(t) = G_K^0 n(t)^4$ we get $36.(0.75)^4 = 11.5$ mS/cm². Multiply this by the electrical driving force $V_m - V_K$ to get the current density. Take V_K to be –72 mV (see above), so $I_K = 11.5(+10 - (-72)) = 945$ μA/cm² = 0.945 mA/cm².

I_{Na} can be estimated in a similar manner. Note that the leak current is a mere 0.02 mA/cm².

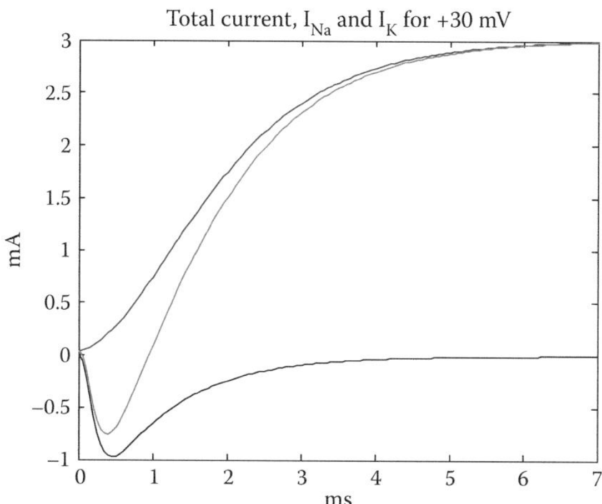

FIGURE 6.15
MATLAB® clamp currents:

```
function HHCLAMP2
%does the analytic solution of the voltage clamped axon

t = 0:.05:7;

v=30;
a1=0.01*(-50-eps-v)/(exp((-50-eps-v)/10)-1);
b1=0.125*exp((-v-60)/80);
a2=0.1*(-35-eps-v)/(exp((-35-eps-v)/10)-1);
b2=4*exp((-v-60)/18);
a3=0.07*exp((-v-60)/20);
b3=1/(exp((-30-v)/10)+1);
ninf=a1/(a1+b1);
minf=a2/(a2+b2);
hinf=a3/(a3+b3);
n0=0.3;
m0=0.05;
h0=0.6;

n=ninf-(ninf-n0)*exp(-t(:)*a1/ninf);
m=minf-(minf-m0)*exp(-t(:)*a2/minf);
h=hinf-(hinf-h0)*exp(-t(:)*a3/hinf);

y(:,1)=120.*m.^3.*h*(v-55)/1000;
y(:,2)=36.*n.^4*(v+72)/1000;
y(:,3)=y(:,1)+y(:,2);

plot (t,y)
```

6.5.2 Modeling the Propagated Action Potential

When an extra equation was included, allowing for simultaneous voltage changes, the H–H equations were found to accurately predict the time course of the *propagated* action potential. This equation equates the total membrane current density I_m to the sum of the ionic currents I_K and I_{Na}, and the charge displacement current ($C_m dV_m/dt$) needs to be included, as well as the leak current I_l; that is,

$$I_m = CdV_m/dt + I_{Na} + I_K + I_l = CdV_m/dt + G_{Na}(V_m - V_{Na}) + G_K(V_m - V_K) + G_l(V_m - V_l).$$

From Section 6.3 (cable theory), we saw that membrane current density depends on the second gradient of the surface potential with distance; that is,

$$I_m = \partial^2 V_m/\partial x^2 (a/(2\rho_i),$$

where a is the axon radius (0.25 mm approximately) and ρ_i is the resistivity of the axoplasm (with a value of 30 Ω cm). Now, since $\partial^2 V_m/\partial t^2 = (\partial x/\partial t)^2 (\partial^2 V_m/\partial x^2) = \theta^2 (\partial^2 V_m/\partial x^2)$, where θ is a velocity, we can write

$$I_m = \{a/(2\rho_i\theta^2)\} \partial^2 V_m/\partial t^2,$$

where θ is the (constant) propagation velocity (around 18 m/s).

So, in addition to the three m, n and h equations, written as $dm/dt = \alpha_m(1 - m) - \beta_m m$, etc., there is also a fourth differential equation:

$$\{a/(2\rho_i\theta^2)\} d^2V_m/dt^2 = CdV_m/dt + G_{Na}(V_m - V_{Na}) + G_K(V_m - V_K) + G_l(V_m - V_l).$$

Although this is second order (because of the d^2V_m/dt^2 term), it is possible to split this into two first-order equations by introducing a variable $dV_m/dt = y$. The equation now becomes

$$a/(2\rho_i\theta^2)\} dy/dt = Cy + G_{Na}(V_m - V_{Na}) + G_K(V_m - V_K) + G_l(V_m - V_l).$$

These five simultaneous first-order differential equations can be solved by the MATLAB function ODE23 (which is described more fully in Chapter 25). A full MATLAB script file is shown in Figure 6.16.

6.5.3 Insights from the Model

So, how would we describe, in words, what is happening? The following sequence of events occurs following the initial stimulatory event, which consists of the removal of positive charge from a particular spot on the surface of the axon, leading to a depolarization at this spot. This opens the Na channels and allows Na to rush in, further depolarizing the membrane (G_{na} increasing). Once the threshold is exceeded, this becomes a *positive feedback process* such that the change leads to a bigger change, and more Na rushes in. Soon the inactivation process starts, and the Na channels begin to close (controlled by the "h" parameter). The depolarization also causes the K^+ channels to open, but this process is slower than for Na^+. As these channels open, K^+ rushes out and the membrane returns to resting (after a brief period of overshoot, where the membrane potential moves toward the K^+ Nernst potential). The propagation of the AP can be described in relation to the theory as follows: The membrane currents have to flow in a circuit—the inrush of Na is accompanied by a withdrawal of charge further down the axon. This withdrawal of charge brings the membrane potential of the next region up to and beyond the threshold, so effectively, this is the next region where the AP occurs. The region upstream of these regions is refractory, so the threshold is effectively infinite there.

The theory also gives insights as to why the so-called "extracellular AP" is triphasic (see Figure 6.6), that is it has three excursions from the baseline, two in one direction, with the middle phase in the opposite direction. If electrodes were to be placed at right angles to the axon (or nerve bundle, for that matter) in the external medium, with both electrodes close to, but not touching the axon, the membrane current I_m changes direction twice as the AP passes the electrodes. Initially outward (as charge is withdrawn before the AP reaches that

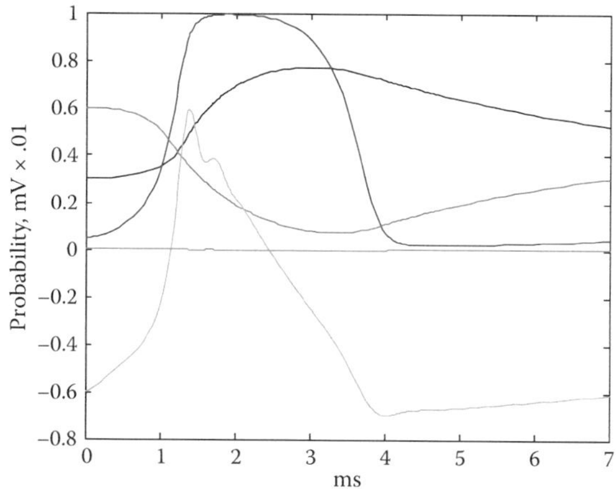

FIGURE 6.16
MATLAB propagated action potential: compare with Figure 6.2. Traces from top: h(t), n(t), m(t), V_m(t).

```
[t,y]=ode45(@hhprop2,[0 7],[0.3,0.05,0.6,-60,0]);
 y(:,4)=0.01*y(:,4);
 y(:,5)=0.002*y(:,5);
plot(t,y)

function yp =hhprop(t,y)
%does the runge kutta solution of the propagated action potential

y1 = y(1);
y2 = y(2);
y3 = y(3);
y4 = y(4);
y5 = y(5);

dy1dt=(1-y1)*0.01*(-50-eps-y4)/(exp((-50-eps-y4)/10)-1)-
y(1)*0.125*exp((-y4-60)/80);
dy2dt=(1-y2)*0.1*(-35-eps-y4)/(exp((-35-eps-y4)/10)-1)-y2*4*exp((-y4-60)/18);
dy3dt=(1-y3)*0.07*exp((-y4-60)/20)-y3/(exp((-30-y4)/10)+1);
dy4dt=y5;
dy5dt=-(36*(y4+72)*y1^4+120*(y4-55)*y2^3*y3+0.3*(y4-50)+1*y5)*8.483;

yp=[dy1dt
    dy2dt
    dy3dt
    dy4dt
    dy5dt];
```

point), the direction becomes inward when the AP is immediately below the position of the electrodes. Finally, there is a weaker current in the outward direction once the AP has passed by. From the cable equation, I_m is proportional to the second derivative of V_m, which would again imply a triphasic response.

Although the theory refers to unmyelinated nerves, it has been extended to predict the membrane potential changes in myelinated nerves as well, including modeling the current entering or leaving at the nodes of Ranvier. Since the propagation velocity is proportional to the distance between nodes, and since this distance depends on axon diameter (d), the dependency on d, rather than $\sqrt{d}$ (which it is for unmyelinated nerves) is explained. The latter dependency arises out of the variation of axon radius (a) with the square of velocity (θ) in the cable equation above. In addition to the extension of the HH theory to myelinated nerves, there has also been extension to describe the voltage changes during APs in skeletal and cardiac muscle cells. In the case of the latter, a computational model, with 11 separate ion channels, has been developed by Y Rudy and coworkers (see http://rudylab.wustl.edu/research/cell/publications/single.htm for further information, and for the MATLAB code).

So, to summarize,

- The *"voltage clamp"* allows I_{Na} and I_K and hence channel conductances to be characterized.
- The HH model provides a way of integrating these currents with capacitance and volume conductor currents.
- The HH equations provide a mathematical description of how the Na^+ and K^+ channels respond during an AP.
- This model also predicts the method of AP *propagation* in both unmyelinated and myelinated nerves, since the circulating currents tend to stimulate the region immediately ahead of the region undergoing reversal of potential.

6.6 Synapses

Nerve-to-nerve and nerve-to-muscle transmission is accomplished usually via a chemical step: The *synapse* (a "clasp") is the region where the nerve axon terminal or synaptic knob forms a close association with the dendrite of another nerve or "end-plate" region on a muscle fiber. In chemical synapses, there is a cleft of around 15 nm–20 nm in nerve–nerve and 50 nm–100 nm in neuromuscular synapses between the presynaptic and postsynaptic membranes. Minute quantities of specific chemical neurotransmitters cross this cleft in order to couple the excitation from one electrically excitable cell to the next. There are also synapses where the coupling is via direct electrical current: these are relatively rare but in specific locations such as the retina and hippocampal region of the brain.

6.6.1 Conversion of Electrical to Chemical Signals

The most common chemical neurotransmitter substance is acetyl choline (ACh), and such synapses are known as cholinergic. All neuromuscular synapses in skeletal muscles, many synapses in the central nervous system (CNS), and all nerve–nerve synapses in the peripheral nervous system (PNS) are cholinergic, together with those in the parasympathetic division of the autonomic nervous system (ANS) and in the ganglia of sympathetic division of the ANS.

The process of the transmission of the action potential from the first excitable cell (the presynaptic cell) and the second (the postsynaptic cell) can be summarized as several stages, which are shown in Figure 6.17, and summarized as follows (the numbers correspond to the locations shown in the diagram):

(1) The action potential's arrival at the synaptic knob of the presynaptic fiber leads to an influx of calcium.

(2) This triggers exocytosis of acetyl choline (ACh), which is contained in vesicles (with approximately 3000 ACh molecules per vesicle).

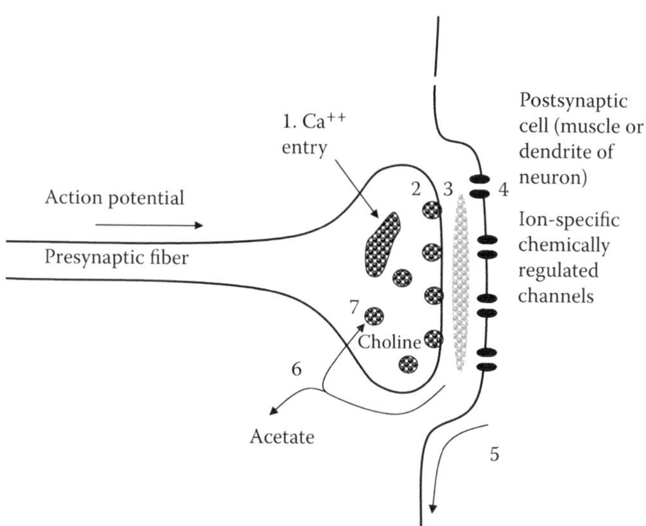

FIGURE 6.17
Synapses. See text for explanation of numbers.

(3) ACh crosses synaptic cleft and binds to ACh receptors on chemically regulated ion channels (selective for Na^+) on the postsynaptic cell membrane.

(4) As ACh binds, ion conductance goes up momentarily. Each "packet" or *quantum* due to the release of ACh from one vesicle gives rise to around 0.5 mV or depolarization, lasting 10 ms–20 ms. This is called a "miniature end-plate potential," or MEPP.

(5) If around 100 vesicles discharge within a few milliseconds, a depolarization of around 40 mV is obtained, enough to elicit an action potential in the postsynaptic membrane. This is called an "excitatory post synaptic potential," or EPSP.

(6) The ACh in the cleft (and on the receptors) is converted rapidly to choline and acetate, by an enzyme acetylcholinesterase (AChE).

(7) Choline (but not acetate) is taken up again by the synaptic knob (choline reuptake) and is formed again into ACh by coenzyme A and (as ACh) is then collected back into vesicles.

This process is characterized by a delay of 0.2 ms–0.5 ms at each synapse (due to the delay in ACh release and diffusion of ACh across the cleft). For neuromuscular synapses, the EPSP is generated by the temporal summation of sufficient MEPPs. (Figure 6.18). For nerve–nerve synapses, the axon hillock spatially and temporarily integrates EPSPs from typically hundreds of individual synapses over the dendrites (Figure 6.19a). The incoming neurons can also be inhibitory, in which case they produce a hyperpolarization or "inhibitory postsynaptic potentials" (IPSPs). Whether or not the axon fires depends on whether the EPSPs or IPSPs are stronger in the temperospatial summation. Inhibitory synapses near the axon hillock control how many EPSPs are required to cause the axon to "fire" (postsynaptic inhibition) (Figure 6.19b).

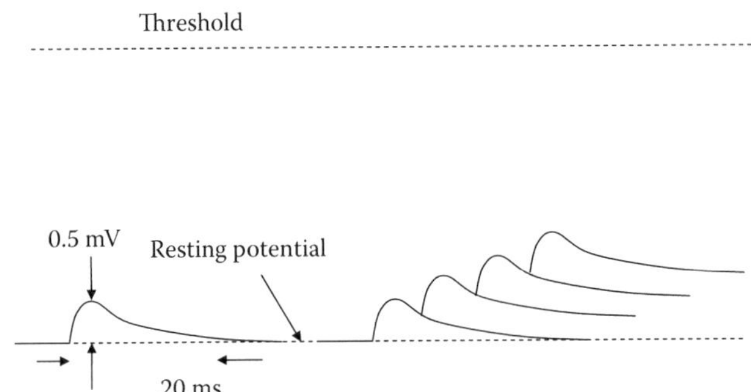

FIGURE 6.18
Summation of miniature end-plate potentials (MEPPs): single MEPP shown at left; at right, several MEPPs arriving within a few milliseconds of each other, giving rise to summation. Approximately 100 MEPPs are required to displace to postsynaptic membrane potential from its resting value (around 90 mV) to the threshold (–75 mV). The MEPP is a depolarization caused by the transient inflow of Na^+ (with some outflow of K^+).

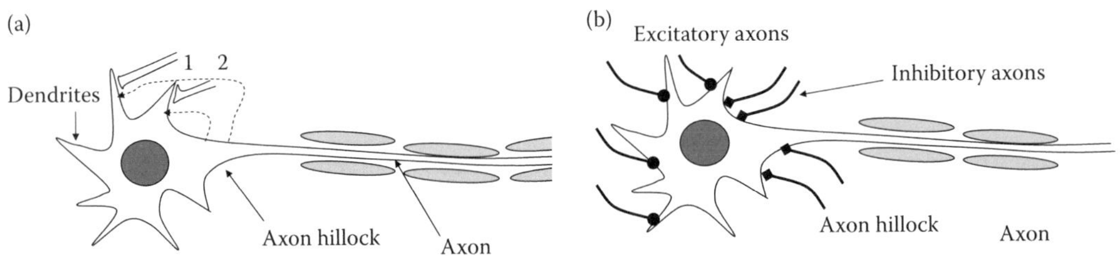

FIGURE 6.19
Excitatory postsynaptic potentials (EPSPs) and inhibitory postsynaptic potentials (IPSPs). The effects of two EPSPs arriving within milliseconds of each other form axons, shown as 1 and 2 in (a). The dotted lines indicate the direction of current flow, removing charge from the membrane surface at the axon hillock (where the threshold is nearest to the resting potential). Thus, there is spatial summation from individual EPSPs, and if there are sufficient numbers, the axon hillock will "fire," giving rise to a propagated action potential along the axon and involving the soma and dendrites, inhibiting further integration of EPSPs until the end of the refractory period. Postsynaptic inhibition is shown in (b) with several inhibitory neurons (producing IPSPs) modulating the number of EPSPs required to cause the axon to "fire."

An "interneuron" can dampen the secretion activity in a presynaptic Ia fiber, if activated a few milliseconds before it, inhibiting the motoneuron for 500 ms or so (presynaptic inhibition). The neurotransmitter from the interneuron is gamma-amino butyric acid (GABA), which inactivates presynaptic Na^+ channels (via "h" mechanism) (Figure 6.20). A chemical synapse acts as a "rectifier," since stimulating a postsynaptic axon cannot activate presynaptic fibers. Characteristic flows of current through channels in postsynaptic membrane have been studied by *patch clamping* (Figure 6.11).

Neurotransmitters can be categorized as "classical" and "neuromodulators." Table 6.3 summarizes typical members of each category and whether it is excitatory (E) or inhibitory (I). Their actions are compared in Figure 6.21.

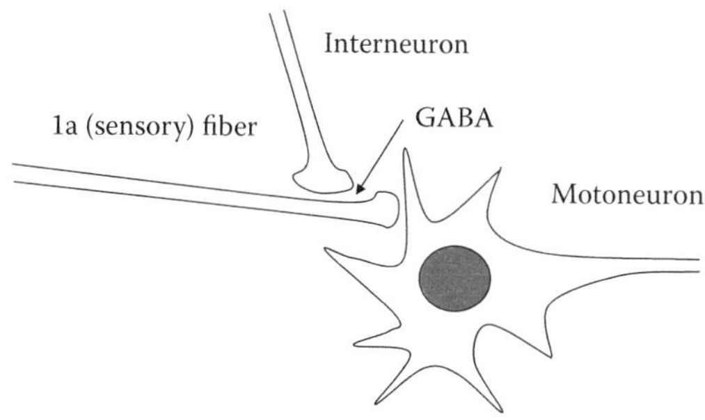

FIGURE 6.20
Presynaptic inhibition. The interneuron releases GABA at the axon bulb, which modulates the ability of the 1a fiber to excite the motoneuron.

TABLE 6.3
Categories of Neurotransmitters

"Classical" Neurotransmitters			
ACh		E	
Amino acids	GABA (from glutamate)	I	
	Glycine	I	
	Aspartate	E	
	Glutamate	E	
Monoamines	Noradrenaline (NAd)	E, I	(Ad and NAd also hormonal, secreted by adrenal medulla)
	Adrenaline (Ad)	E, I	(Ad and NAd also called catecholamines)
	Dopamine (DOPA)	I	
	Serotonin (5 HT)	I	(Converted in pineal gland to melatonin)
Neuromodulators			
Neuropeptides	Substance P		(Slow release and long action: axonal transport)
	Enkephalins		
Hormones	ADH		(As above)
	Oxytocin		
	Prostaglandins		
	(And several others)		
High energy	ATP		
	GTP		
Dissolved gas	Nitric oxide (NO)		
	Carbon dioxide (CO)		
Second messengers	Adenylate cyclase		
	Cyclic AMP		
	Calcium		

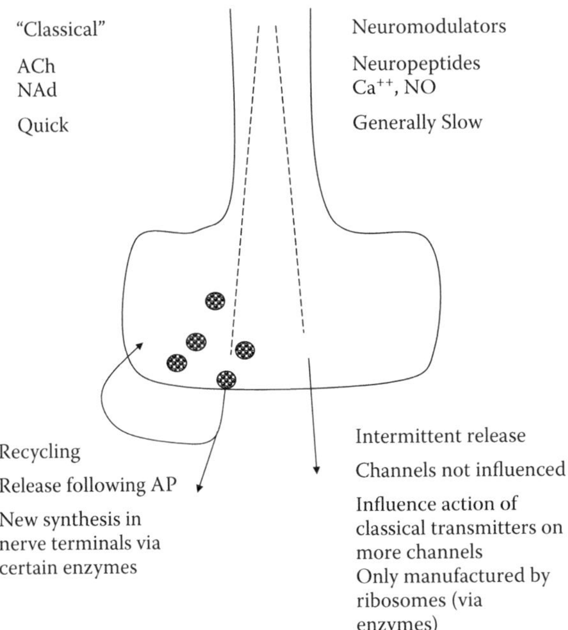

FIGURE 6.21
Comparison of actions of "classical" neurotransmitters with the neuromodulators.

Tutorial Questions

1. A spherical cell has a resting membrane potential of −65 mV, membrane resistance of 10^6 Ω, and membrane capacitance of 10^{-8} Farad. A brief current pulse passed through a centrally placed microelectrode is found to produce a depolarization of 10 mV, which decays exponentially:
   (a) Determine the time constant of this cell.
   (b) What time must elapse for the depolarization to decay to 5 mV?
   (c) When a pair of identical stimuli is delivered less than 6.93 ms apart, the cell undergoes an action potential. Assuming that temporal summation occurs, determine the threshold voltage of this cell.
2. The transmembrane potential at a particular point of a long cylindrical unmyelinated axon is increased by 10 mV. 12 mm from this point, this transmembrane potential increase has decayed to 0.498 mV. Assume the following conditions:
   (a) The transmembrane potential changes are static.
   (b) The resistivity of the intracellular medium is 5×10^{-1} Ω m.
   (c) The resistance of unit area of membrane is 1.0 Ω.
   (d) The extracellular resistance per unit length is zero.
   (e) Capacitance of unit area of membrane is 1.5 μF.

   Then, answer the following questions:
   (i) How is the length constant for such an axon defined?
   (ii) What is the length constant for this cell?

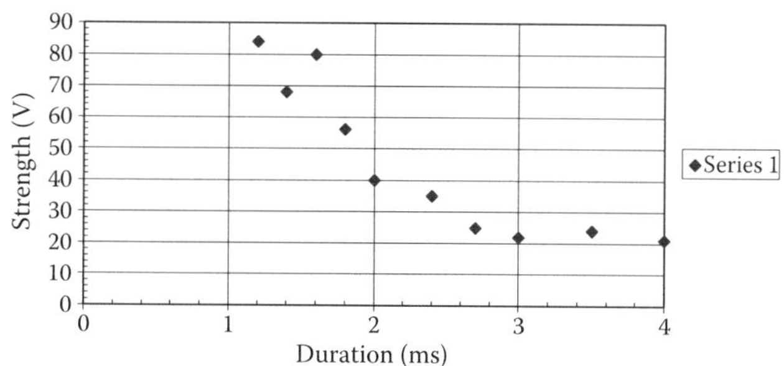

FIGURE 6.22
See Tutorial Question 3.

(iii) What is the time constant for this cell?

(iv) Derive an expression for the intracellular resistance per unit length.

(v) Using the length constant and the expression derived in (iv) determine the internal radius of the axon.

3. From the information given, estimate rheobase and chronaxie for this group of fibers. Explain exactly what is meant by "strength" and "duration" (Figure 6.22).

4. (a) The membrane current in squid axon, clamped at $t > 0$ at 0 mV (previously at –60 mV), both before and after tetrodotoxin (in high dosage) is shown in Figure 6.23.

Assuming that the equilibrium potentials for Na and K are +55mV and –72 mV, respectively, calculate the following:

(i) The magnitude and direction of peak I_{Na} and I_K.

(ii) The values of g_{Na} and g_K 2 ms after application of the clamp.

(b) Given that potassium conductance is given by the following:

$$g_K = g'_K n^4 = 36[(\alpha_n/(\alpha_n+\beta_n)) - ((\alpha_n/(\alpha_n+\beta_n)) - 0.3)\exp(-t(\alpha_n+\beta_n))]^4$$

where $\alpha_n = 0.01(-50-V)/\{\exp[(-50-V)/10] - 1\}$ and $\beta_n = 0.125\exp[(-V-60)/80]$, verify the result obtained for g_K in (b)-(ii) above (i.e., for $t = 2$ ms).

The time course of n versus t is shown (Figure 6.24). What are the values of n′ and t′ in this example?

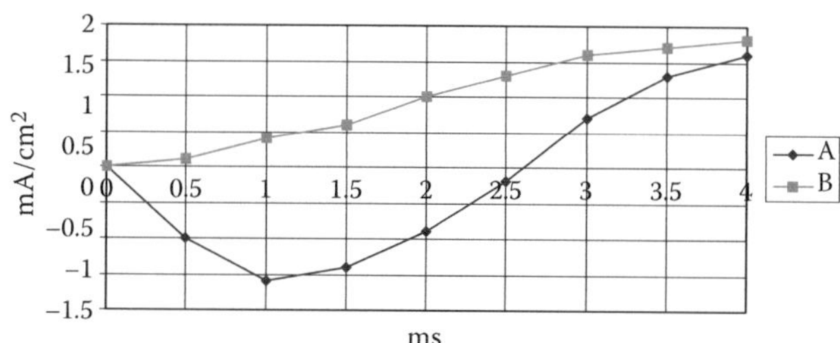

FIGURE 6.23
See Tutorial Question 4a.

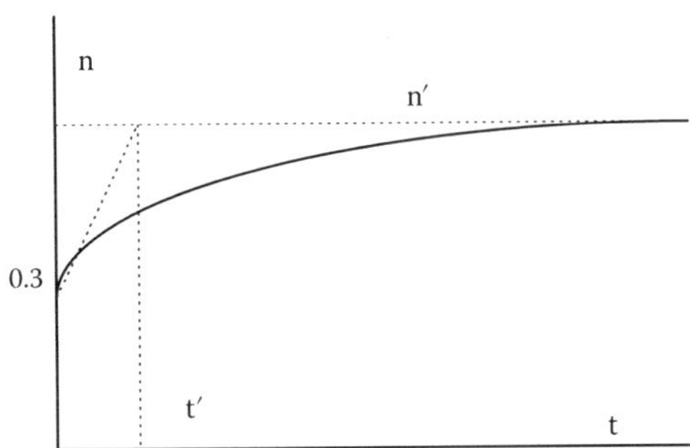

FIGURE 6.24
See Tutorial Question 4b.

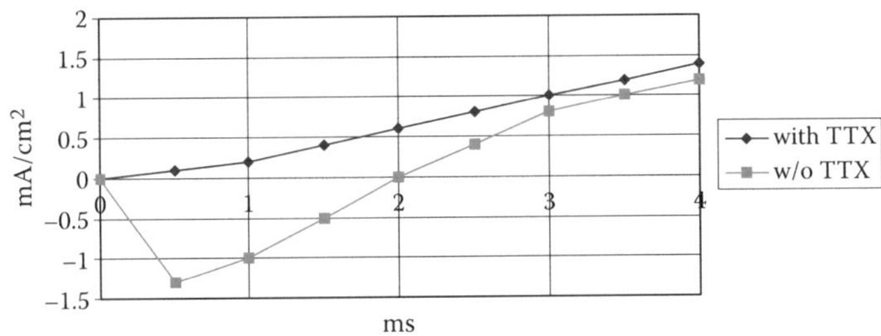

FIGURE 6.25
See Tutorial Question 5.

5. In the giant fiber of the marine worm *Myxicola*, the axoplasm is found to contain 45 mmol/L of Na^+. In a voltage clamp experiment, the axon is bathed in artificial sea water ($[Na^+] = 450$ mmol/L) and clamped so that the outside is 10 mV more positive than the inside. The total current is given by curve A (see Figure 6.25). Later, the PD is returned to resting potential (−65 mV: similar to squid axon), and then an excess of tetrodotoxin is added to the bathing solution. The axon is then clamped at the same voltage as before, and the total current is shown in curve B (Figure 6.25).
   (a) Calculate the reversal potential for Na^+ at 25°C.
   (b) Calculate the magnitude of the maximum Na conductance at this voltage.
   (c) Sketch the time course of the Na and K conductances at this voltage. (Assume that $E_K = -70$ mV.)

Bibliography

Aidley DJ. 1998. *The Physiology of Excitable Cells*, 4th Edition. Cambridge University Press, Cambridge, UK.

Enderle JD, Blanchard SM, Bronzino JD. 2000, 2005. *Introduction to Biomedical Engineering*. Elsevier Academic Press, Burlington, MA, USA.

Guyton AC, Hall JE. 2006. *Textbook of Medical Physiology*. Elsevier Saunders, Philadelphia, PA, USA.

Katz B. 1996. *Nerve, Muscle and Synapse*. McGraw Hill, New York.

Malmivuo J, Plonsey RL. 1995. *Bioelectromagnetism* (online). http://www.bem.fi/book/index.htm. Viewed October 10, 2011.

Plonsey RL, Barr R. 1988, 2007. *Bioelectric Phenomena*. Plenum, New York.

7

Skeletal Muscle Biophysics

Per Line

CONTENTS

7.1 Introduction

There are approximately 700 skeletal muscles in the body, and together they comprise about 40% of the body mass in a noncorpulent individual. When at work, muscles can use up to 1000 times more energy (as indexed by rate of ATP consumption) than when at rest, generating also copious amounts of heat as a by-product. This chapter will examine various biophysical aspects of muscle contraction but will begin with an overview of muscle structure.

7.2 Skeletal Muscle Structure

Skeletal muscles (also termed voluntary or striated muscle) are organized in a hierarchical manner (see Figure 7.1) and, at the gross level, are organs. Both ends of muscles merge with tough connective tissues known as tendons, and it is via the tendons that muscles are anchored to bone (or other connective tissue). The attachment end where the muscle tends to be fixed in terms of movement is known as the origin, whereas the end that attaches to a more movable structure is known as the insertion. For example, the gastrocnemius muscle has its origin at the femur and its insertion at the calcaneus (heel bone). However, determining the origin and insertion is not always as straight forward. Filling the gaps between different muscles are fibrous connective tissue structures known as fasciae. Surrounding the individual whole muscle is a fibrous elastic connective tissue called the epimysium.

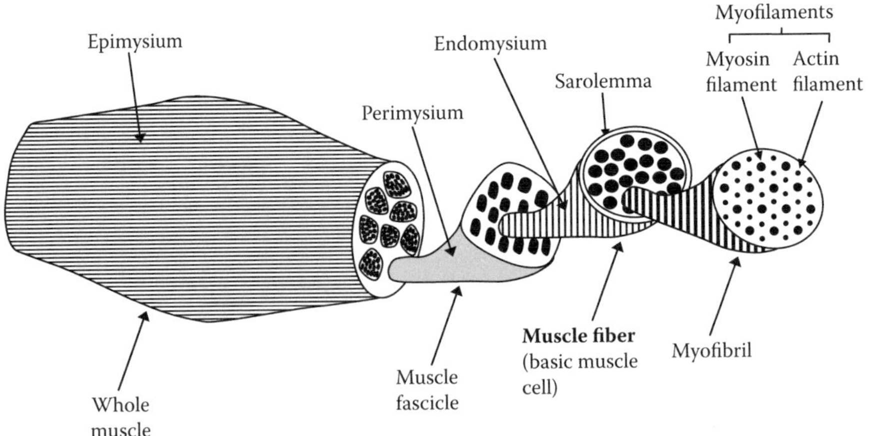

FIGURE 7.1
Muscle structure overview—from whole muscle to myofilaments.

At the next level down, the muscle is organized into tissue bundles known as fascicles. Fascicles are surrounded by connective tissue, the perimysium, and can vary greatly in size. In a human, a fascicle typically contains about 1000 muscle fibers, although this number can vary considerably. Depending on the form, a muscle cell can span the entire length of the muscle, and because of its enormous length (up to 30 cm)-to-diameter (typically ranges from about 10 μm to 100 μm) ratio, a muscle cell is often called a muscle fiber, or less frequently a myofiber. The endomysium is the connective tissue that surrounds individual muscle fibers. Underneath the endomysium is the muscle cell membrane known as the sarcolemma. The bulk volume (about 80%) of muscle fibers are made up of hundreds (or sometimes thousands) of structures called myofibrils. Myofibrils are on average a tiny 1 μm in cross-section (maximum diameter of 2 μm) but span the entire length of the muscle fiber (from a few millimeters up to 30 cm). The myofibril is divided into thousands of repeating units called sarcomeres (depending on the muscle fiber length)—the basic structural and functional units of contraction in the muscle.

7.2.1 Muscle Types

At the gross level, muscles are categorized into various types based on their general shape and the arrangement of the fascicles with respect to the tendons. The five general categories that muscles are classified into are fusiform, parallel, convergent, pennate, and circular muscles (see Figure 7.2). As an example, fusiform muscles, such as the *biceps brachii,* have a bulging central body (also known as belly), but tapered ends. Fusiform muscles are actually a type of parallel muscles, but because of their spindle shape, they are often put in a different category. Parallel muscles are uniform in width, with the fascicles running parallel to each other. As typified by the *rectus abdominis,* some parallel muscles are quite

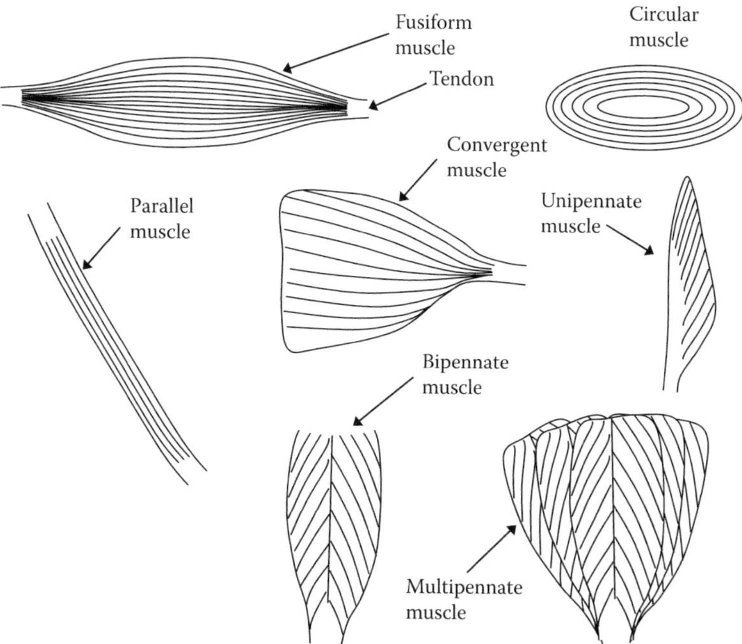

FIGURE 7.2
Muscle types.

long with an appearance of multiple bellies in series—separated by tendinous intersections (also called tendinous inscriptions). A muscle fiber can shorten by approximately 30% of its total length. Hence, parallel muscles can shorten by about the same amount because their muscle fibers run parallel to the muscles long axis. Pennate muscles, which are feather shaped, are divided into three subtypes: unipennate, bipennate, and multipennate. Circular muscles (also called sphincters), such as the *orbicularis oris* of the mouth, are located around body openings. Convergent muscles, such as the *pectoralis major* in the chest, are spread out in a fan-like manner.

7.3 Muscle Ultrastructure

Ultrastructure essentially refers to a biological structure that can only be revealed in fine detail by use of an electron microscope. In skeletal muscle cells, there are three important ultrastructures concerned with the contractile process: the sarcolemma and transverse tubules, the sarcoplasmic reticulum, and the myofibril.

7.3.1 Sarcolemma and Transverse Tubules

The outer covering of the muscle fiber consists of the sarcolemma. Making up the sarcolemma are a thin (20 nm to 30 nm) outer layer of connective tissue called the basal lamina (also known as basement membrane when combined with a thin layer of collagen fibrils called the reticular lamina) and an inner plasma membrane (also called plasmalemma or cell membrane) that constitutes the true boundary of the muscle fiber. The basal lamina is freely permeable and is composed of collagen and glycoproteins, whereas the plasma membrane contains specialized receptors and other protein channels that regulate input and output of molecules and ions—embedded in a typical proteolipid bilayer characteristic of most cell membranes. The basal lamina and plasma membrane are separated by a gap of about 20 nm. The sarcolemma is the outer covering of the muscle cell, whereas the transverse tubules (abbreviated to T-tubules), although distinct in composition, are inward perpendicular extensions of the plasma membrane into the sarcoplasm (a term used for the cytoplasm of a muscle cell). Filled with extracellular fluid, the interior of the T-tubule system is continuous with the extracellular space outside the muscle cell. T-tubules may connect directly to the plasma membrane or indirectly via caveolae (tiny vesicles). When initiated, an action potential travels across the sarcolemma in all directions. The action potential also travels along the extensive network of T-tubules inside the muscle fiber, triggering muscle contraction.

7.3.2 Sarcoplasmic Reticulum

In a muscle cell, the sarcoplasmic reticulum is a very specialized type of smooth endoplasmic reticulum and plays a pivotal role in the transmission of the nerve impulse to the contractile machinery of the muscle fiber. The sarcoplasmic reticulum consists of a network of tubules that enclose the myofibrils inside the muscle cell, as well as coming into close contact with the T-tubules. In the region of contact with the T-tubules the tubules of the sarcoplasmic reticulum are considerably enlarged, forming two separate cavities called terminal cisternae—one on either side of each T-tubule (illustrated in Figure 7.3).

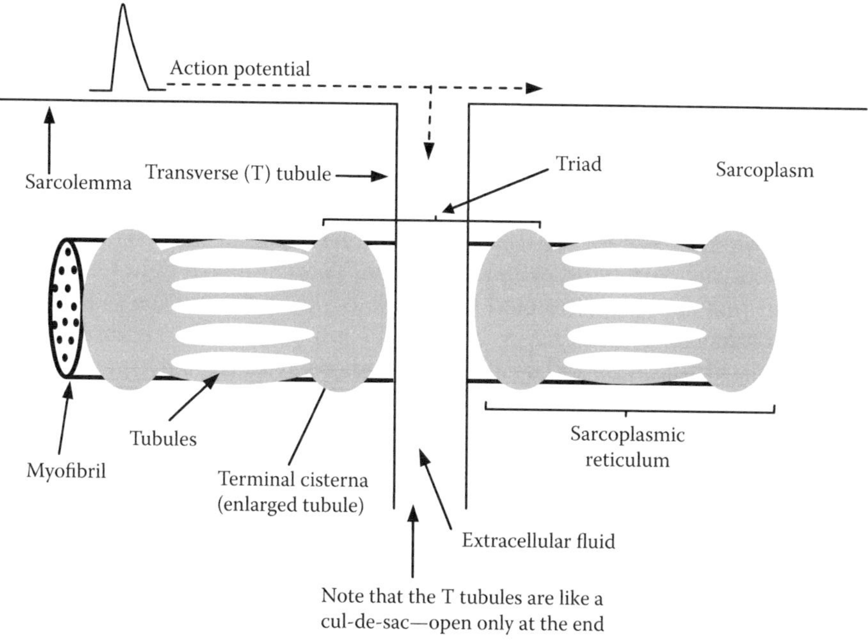

FIGURE 7.3
Sarcolemma, transverse tubules and sarcoplasmic reticulum.

Together, the one T-tubule and two terminal cisternae are referred to as a triad. It is at this intersection that calcium is released into the sarcoplasm to enable muscle contraction. A ryanodine receptor (RyR) calcium release channel, embedded in the sarcoplasmic reticulum membrane, releases calcium ions from the sarcoplasmic reticulum into the sarcoplasm. The RyR includes a large component in the sarcoplasmic domain that forms most of the so-called foot region that bridges the gap (11 nm to 14 nm) between the T-tubule and terminal cisternae membranes.

Apart from the triad region, the network of tubules in the sarcoplasmic reticulum contains a membrane rich with specialized calcium pumps (Ca^{2+} ATPase). These pumps reaccumulate the sarcoplasmic calcium ions back into the sarcoplasmic reticulum via active transport. After being absorbed through the tubule membranes in the central region of the sarcoplasmic reticulum, the calcium then moves to the terminal cisternae for storage. A protein called calsequestrin, which binds calcium and is largely confined to the terminal cisternae, where it is attached to the inner side of the membrane, ensures that the calcium is concentrated in these cavities. The calcium pump reaccumulates calcium until the sarcoplasmic calcium concentration gets below 10^{-8} M. Calcium reuptake is energized by ATP splitting. Since the rate of ATP splitting is very low below 10^{-8} M sarcoplasmic concentration of calcium ions, the reaccumulation ceases when it fills to this level. Inside the sarcoplasmic reticulum, the calcium concentration is between 0.5 mM and 1 mM, indicating the pump can work against a 10,000-fold or more concentration gradient.

7.3.3 Myofibril and Sarcomere

As mentioned earlier the myofibril is divided longitudinally into thousands of repeating units called sarcomeres. Sarcomeres are about 2.5 μm long in resting muscle and about

1 μm to 2 μm in diameter (the same diameter as the myofibril) and are the basic structural and functional units of contraction in the muscle. The primary structure of the sarcomere is organized around two longitudinally overlapping threadlike filaments (or myofilaments) called myosin filaments (or thick filaments) and actin filaments (or thin filaments), as indicated in Figure 7.4. Each sarcomere contains about 1000 myosin filaments, and about 2000 actin filaments attached to the Z line at each end. The dark regions or lines in the myofibril and sarcomere represent their appearance in a micrograph, with most of the A band appearing dark, except for the light H zone in the center of the A band. Within the H zone, there is a narrow dark band called the M line. There is also a light I band consisting of only the actin filament type, bisected by a dark Z line, such that the I band spans portions of two sarcomeres. The Z line (or Z disk) is composed of filamentous proteins and constitutes the longitudinal boundary of the sarcomere, with the sarcomere consisting of the myofibril between two adjacent Z lines. Actin filaments from both adjacent sarcomeres, as well as titin filaments, are connected at their ends to the Z line. The Z line also links actin filaments laterally to the longitudinal axis, forming a lattice arrangement. Titin is an elastic protein that runs from the Z line to the M line and assists in keeping the myofilaments aligned with each other, as well as helping prevent the muscle from overstretching. At a molecular weight of approximately 3,000,000 Da, titin (also known as connectin) is the largest known protein. In the A band, titin runs through the core of the myosin filament.

Running down the center of the H zone is the M line—bisecting the sarcomere. The M line is composed of several nonmyosin proteins that presumably link adjacent myosin filaments crosswise to the long axis, thus helping to stabilize the A band lattice. The A band is composed of myosin filaments and any portion of actin filaments that overlap the myosin filament—the latter known as the zone of overlap. The region at the center of the A band containing no actin filaments is the H zone. This region shortens upon muscle contraction, as does the I band, whereas the A band remains unchanged during contraction. The bare zone (also called pseudo H zone) within the H zone occurs in the middle of every myosin filament and is a region containing no cross-bridges (projections of the myosin heads),

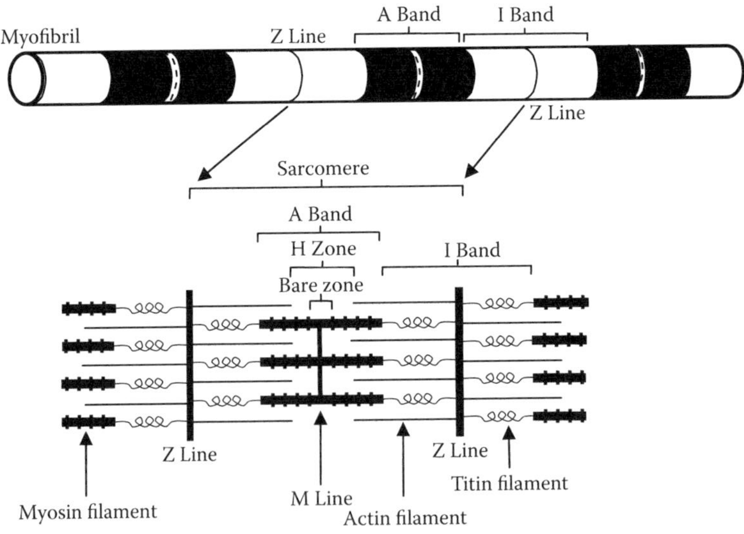

FIGURE 7.4
Sarcomere.

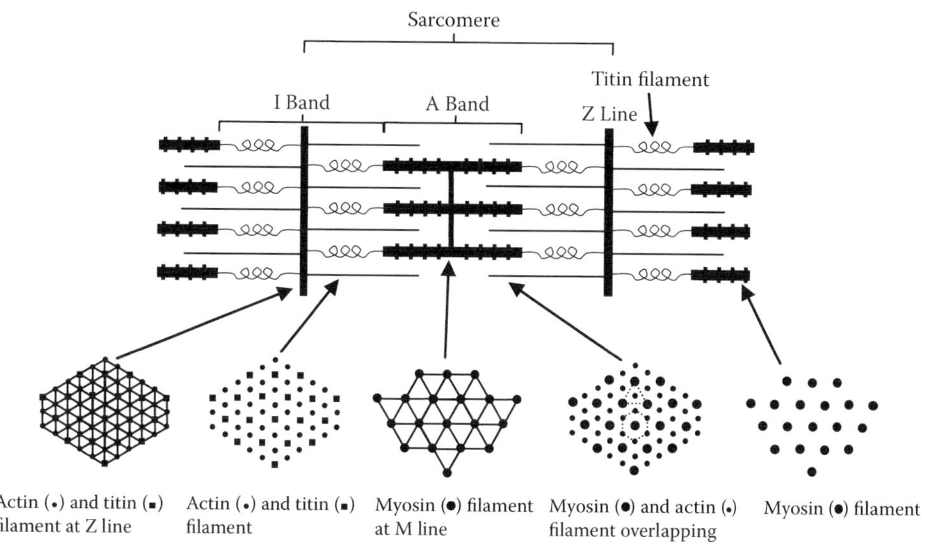

FIGURE 7.5
Lattice arrangement in sarcomere.

only the tail portion of the myosin molecule. Figure 7.5 illustrates the cross-section lattice arrangements of the filaments at various longitudinal positions along the sarcomere. Note that in the region of overlap, three myosin filaments surround each actin filament, and six actin filaments surround each myosin filament.

7.3.3.1 Myosin and Myosin (Thick) Filaments

Myosin (thick) filaments are composed of about 200 to 500 myosin molecules, as well as a titin filament (about 1 nm in diameter) running through its core, and are located within the A band of the sarcomere. The myosin filaments are between 1.5 μm and 1.6 μm long and from 10 nm to 15 nm in diameter. Each sarcomere contains about 1000 myosin filaments. As shown in Figure 7.6 the filament has paired myosin heads of the myosin molecules

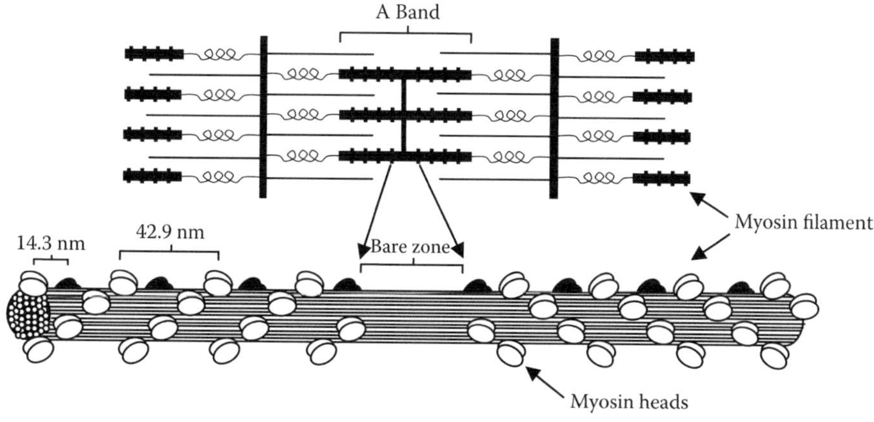

FIGURE 7.6
Myosin filament.

projecting outward in a spiral array—forming a helix type configuration along the length of the filament. There are actually three concurrent helixes about the same axis, each with nine myosin head pairs per turn. Note that there are thought to be three paired myosin heads (called a "crown") projecting every 14.3 nm along the longitudinal axis of the filament, with successive within crown pairs in cross-section being rotated 120° about the shaft of the filament axis. Successive crowns are rotated at 40° with respect to the shaft of the filament axis. Projections oriented in the same direction occur every 42.9 nm—this is the repeat distance of the structure—although the pitch of each helix is 128.7 nm (9 nm × 14.3 nm). The myosin molecules are packed such that the myosin head to tail directions are reversed near the midpoint of the filament, forming a bipolar structure.

The myosin molecule is composed of six separate subunits (protein chains)—yielding a combined molecular weight (mol. wt) of about 520,000 Da. It consists of two large subunits (mol. wt of ~220,000 Da each), commonly referred to as heavy chains, and four small subunits (mol. wt of ~20,000 Da each), known as light chains. The two heavy chains coil around each other, forming a double helix, but with one end separated and containing two free heads. The myosin molecule is about 150 nm long and 2 nm in diameter.

The myosin molecule can be split into several subfragments using certain enzymes. The enzyme trypsin, attacking the hinge region between the arm and the tail, splits the myosin molecule into the two components—heavy meromyosin (HMM) and light meromyosin (LMM)—as indicated in Figure 7.7. The LMM region is also known as the tail or body of the myosin molecule and consists of a large portion of the two heavy chains coiled around each other in a helical manner. The HMM fragment, using the enzyme papain, can be split further into more subfragments—two S1 globular head regions and one S2 arm region. The S1 head region is also referred to as a cross-bridge during muscle contractions. Two different light chains are attached to each S1 head, these being known as regulatory light chains (RLCs—also called 5,5′-Dithio-bis-2-nitrobenzoic acid [DTNB] light chains) and essential light chains (ELCs). The RLCs are thought to play an important modulatory role in the contraction of skeletal muscle. The ELCs are also known as alkali light chains

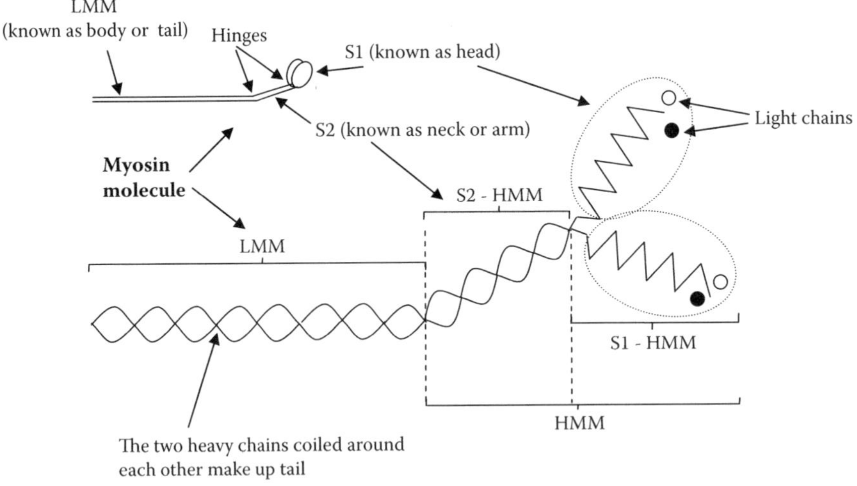

FIGURE 7.7
Myosin molecule.

and are necessary for the hydrolysis of ATP (adenosine triphosphate) to ADP (adenosine diphosphate). The S1 heads are the site of both the actin and ATP binding sites—there are no such binding sites on the S2 and tail regions of the myosin molecules.

7.3.3.2 Actin and Thin (Actin) Filaments

Thin (actin) filaments are about 7 nm–8 nm thick in diameter and about 1 μm long, and are composed of the proteins F actin, troponin, tropomyosin, and nebulin. The basic unit of the thin filament is a single protein molecule called G actin with a molecular weight of about 45,000 Da, comprising two similar globular domains partially separated by a cleft. G actin is actually more of an oblong (ellipsoidal) shape, with dimensions of about $6.7 \times 4.0 \times 3.7$ nm. When polymerized G actin forms F actin, the latter a filament with the appearance of a double stranded helix—as indicated in Figure 7.8. Hence, F actin filaments are essentially helical assemblies of G actin molecules. In passing from G actin to F actin one of the domains rotates with respect to the other by 20°, giving the G actin molecule in the filament a flatter appearance. The F actin filament comes in repeat units of 13 G actin molecules. As each molecule rotates close to 180° (more precisely 167°), an appearance is given of a double helix slowly winding. G Actin molecules occur along the long axis of the thin filament every 2.75 nm, in a staggered fashion.

Figure 7.8 illustrates the generally accepted model of the thin filament. Note that in addition to a string of G actin molecules (F actin), the thin filament consists also of the protein molecules tropomyosin and troponin (known also as the "regulatory proteins"). Tropomyosin molecules have a molecular weight of about 66,000 Da each and are composed of two coiled rod-like structures (two α-helical subunits) about 2 nm in diameter and 40 nm long. In the thin filament, they are assembled end-to-end, as illustrated by the tropomyosin contact points in the above diagram, forming a double helix. The tropomyosin molecules occupy the two grooves in the F actin structure, forming two strands that run the length of the thin filament. The rod-like tropomyosin molecules are each attached to seven G actin molecules on the F actin filament and, during the resting state, block the active sites on these G actin monomers.

Troponin molecules have a molecular weight of approximately 80,000 Da each, attaching to the tropomyosin strands at regular intervals of 38.5 nm along the thin filaments. Consisting of three separate subunits, the troponin molecule is often referred to as the troponin complex. The three subunits are Ca^{2+} binding component (troponin C—abbreviated as TnC), inhibitory component (troponin I—abbreviated as TnI) and tropomyosin-binding component (troponin T—abbreviated as TnT). Each TnC is composed of two domains, each with two Ca^{2+} binding sites, and thus can bind four calcium ions—numbered I to IV. Sites

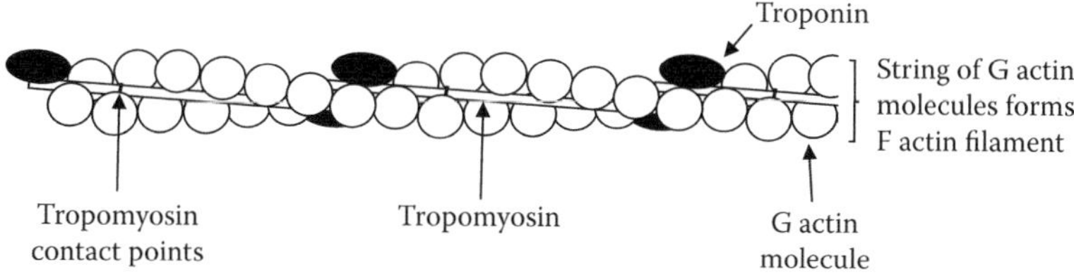

FIGURE 7.8
Actin (thin) filament. (After Ebashi, S., *Nature*, 240, 218, 1972. With permission.)

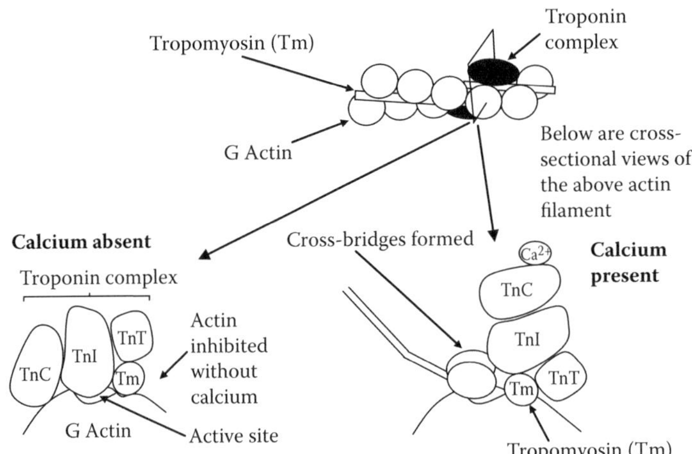

FIGURE 7.9
Schematic illustration of hypothetical interactions of the troponin components, G actin and tropomyosin in the absence and presence of calcium.

I and II bind calcium specifically, but with low affinity, whereas sites III and IV have high affinity Ca^{2+} binding sites that can also bind Mg^{2+} competitively. Although many details are not known, it is believed that calcium-induced conformational changes to TnC initiate other structural changes in the thin filament that allows myosin heads to bind to the active sites on the G actin molecules, as the troponin–tropomyosin complex is translocated—exposing the G actin active sites. When Ca^{2+} binds to TnC, the molecule is thought to interact with TnI such that the main inhibitory region of TnI is dissociated from G actin. The TnT subunit anchors troponin to the F actin filaments by binding to both TnI and tropomyosin. The conformational change in troponin somehow causes the tropomyosin molecule occupying the groove to be dragged away from the myosin binding sites on the G actin molecules, thus allowing myosin heads to attach to the active sites. A schematic illustration of hypothetical interactions of the troponin components, tropomyosin, and G actin is shown in Figure 7.9.

Another molecule associated with the actin filament is called nebulin—a large molecule (~800,000 Da) whose length correlates with the length of the thin filament in skeletal muscle. Its location on the thin filament is uncertain, as is its exact role, but it is thought to possibly be involved in regulating F actin assembly, in particular F actin filament length.

Before looking at the mechanism linking muscle action potentials to the exposure of the active sites on the G actin molecules of the thin filaments (excitation–contraction coupling), an overview of the neuromuscular junction will be considered next.

7.4 Neuromuscular Junction (NMJ)

Large, myelinated motor neurons (or motoneurons) innervate skeletal muscle fibers. Motor neurons emanate either from the anterior gray horn of the spinal cord or from brainstem nuclei. With very few exceptions (e.g., extraocular muscles), a muscle fiber is innervated by only one motor neuron. However, a motor neuron can innervate from about two to three muscle fibers (e.g., laryngeal muscles—where fine control is needed) to well over a thousand muscles fibers (e.g., the gastrocnemius, medial calf muscle—where brute strength is needed). Most motor units, which is the name given to a motor neuron and associated muscle fibers (the muscle fibers innervated by the motor neuron), usually consist of between 100 and 1000 muscle fibers. To quantify this value, the term innervation ratio (IR) is used; defined as the number of muscle fibers innervated by one motor neuron.

Where the motor neuron enters the muscle is referred to as the motor point. From here, the axon of the motor neuron branches out in order to supply axon terminals to the muscle fibers it innervates—at regions called neuromuscular junctions that are located approximately half way along the length of the muscle fibers. The zone of innervation is a term used to refer to the localization of these neuromuscular junctions at specific regions along the muscle fiber. Most muscle fibers have only one such neuromuscular junction. The neuromuscular junction is a cholinergic synapse, as it is a chemical synapse that releases the neurotransmitter acetylcholine (ACh).

As mentioned, the impulse from a motor neuron is transmitted to individual muscle fibers at specialized sites of connection between the sarcolemma and axonal ending called the neuromuscular junction (also known as the myoneural junction). As indicated in Figure 7.10, the region of sarcolemma lying directly under the axonal ending is called the motor end plate. Near the motor end plate, the motor neuron is no longer covered by a myelin sheath, and the axon is divided into short processes that are partly sunken into synaptic

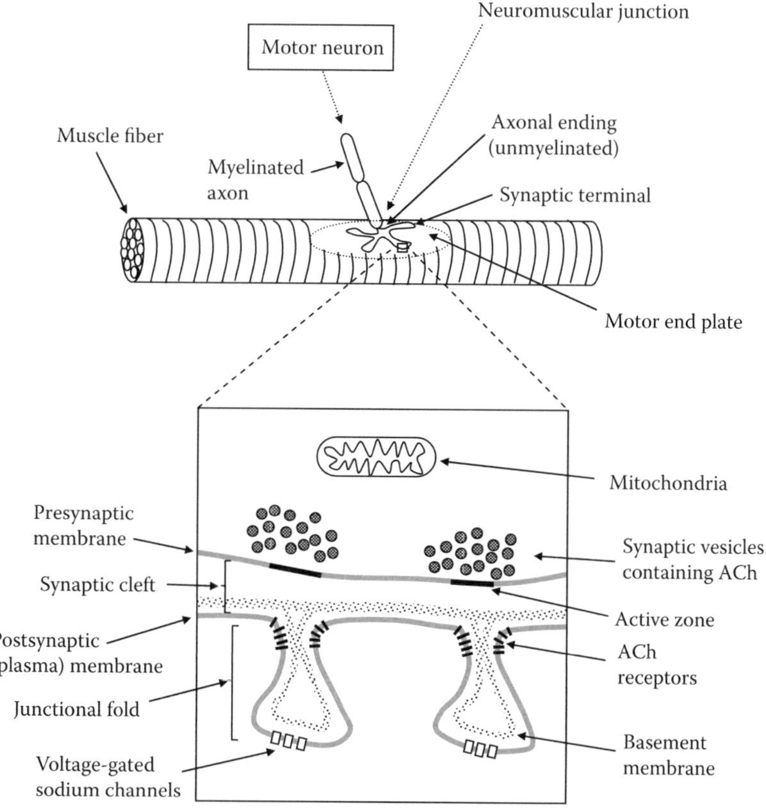

FIGURE 7.10
Neuromuscular junction (NMJ).

gutters (or troughs), whereby the motor end-plate region is depressed. The processes sunken in the synaptic gutters contain many swellings (varicosities), commonly referred to as synaptic terminals or boutons, that are covered by a thin layer of Schwann cells (Schwann cell sheath). These synaptic boutons contain vesicles filled with the neurotransmitter ACh. The space of about 50 nm between the synaptic terminals and motor end plate is filled with extracellular fluid and is called the synaptic cleft. The motor end plate also has junctional folds consisting of invaginations of the membrane. The crest of these junctional folds are particularly rich in chemically gated nicotinic acetylcholine receptor (nAChR) channels that transmit the incoming nerve stimulus (in the form of packets of the neurotransmitter acetylcholine) into the muscle fiber as an end-plate potential (local depolarization). This occurs because, when opened, the nAChR channels allow Na^+ and K^+ to pass through.

The mechanism by which an action potential along the axon of the motor neuron stimulates an action potential across the sarcolemma of the muscle fiber is as follows: Depolarization of the presynaptic axon terminal membrane leads to voltage-gated calcium channels opening up and, subsequently, the influx of Ca^{2+} into the axon terminal. The presence of Ca^{2+} somehow induces the release of ACh into the synaptic cleft from storage vesicles in the presynaptic terminal boutons. This occurs by a process known as exocytosis, where the vesicles containing ACh fuse with the synaptic terminal membrane at regions called active zones next to the synaptic cleft—resulting in the release of the ACh into this region. Once in the synaptic cleft the ACh quickly diffuses to the motor end-plate region, where it binds to nAChR channels embedded in the end-plate membrane. Only about a half of the ACh molecules released into the synaptic cleft make it across to the nAChR channels on the end plate. The reason for this high attrition rate is that the synaptic cleft has a high concentration of the enzyme acetylcholinesterase (AChE) which is very effective in breaking down ACh molecules—into the products acetate and choline. The basement membrane surrounding the junctional folds is particularly rich in AChE. Even ACh molecules that end up binding to the receptors are broken down by AChE within about 20 ms.

There are in the order of about 50 million nAChR channels in the end-plate region of the muscle fiber. When open a single channel causes about a 0.3 μV depolarization at the end plate. Usually two molecules of ACh are needed for a single channel opening. The ACh is released in quanta or packets of approximately 10,000 molecule per vesicle, resulting in an end-plate potential of about 0.5 mV—referred to as a miniature end-plate potential (MEPP). Although variable, roughly 125 vesicles (or quanta) of ACh are released into the synaptic cleft during a motor neuron action potential at the presynaptic terminal. When stimulated by a motor neuron action potential the resulting end-plate potential is from about 50 mV to 75 mV in the positive direction, resulting from the summation of the MEPPs; this from the opening of over 200,000 channels. This depolarization is well in excess of what is needed to trigger an action potential across the sarcolemma. The normal resting potential of a muscle fiber is about −90 mV, with the threshold to trigger an action potential in the sarcolemma being approximately −55 mV, as indicated in Figure 7.11.

At the neuromuscular junction, all end-plate potentials under normal circumstances are excitatory (depolarizing). The area of sarcolemma beside the motor end plate contains a high density of voltage-gated ion channels (both Na^+ and K^+ specific), similar to those in the nerve axon, which turn the end-plate potential into a muscle fiber action potential that is subsequently conducted across the rest of the sarcolemma. Membranes in structures such as the sarcolemma and axon are referred to as being excitable because they have a sufficient density of voltage-gated ion channels to propagate an action potential.

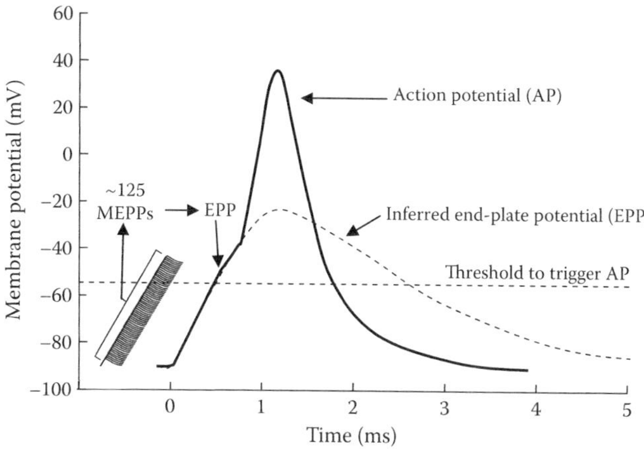

FIGURE 7.11
MEPP, EPP, and action potential.

The motor end plate is considered as supporting only graded potentials, which are critical as they depolarize the adjacent excitable membrane to threshold—thus triggering the action potential. However, voltage-gated sodium channels are known to be highly concentrated in the motor end-plate region, especially in the troughs of the junctional folds. A likely explanation for this is that it ensures there is sufficient current to trigger an action potential.

The muscle action potentials in muscle fibers are initiated and conducted similarly to action potentials traveling in unmyelinated axons of neurons. In muscle fibers, the velocity of conduction is about 3 m/s–5 m/s, and the action potential duration is between about 1 m/s and 5 m/s. This velocity of conduction is comparable to the speed of conduction in unmyelinated axons, but considerably slower than the conduction speeds of up to 120 m/s in myelinated large-diameter axons.

7.4.1 Nicotinic Acetylcholine Receptor (nAChR) at Motor End Plate

The nAChR is a ligand-gated channel, with a molecular weight of about 275,000 Da and a channel diameter of around 0.65 nm. Ligand-gated channels are often referred to as ionotropic receptors as their activity directly affects ion movement across the postsynaptic membrane, usually resulting in fast and direct neurotransmission. At the neuromuscular junction, the postsynaptic membrane is the motor end plate. As the receptors can be selectively activated by nicotine (the classical competitive agonist) they have been called nicotinic acetylcholine receptors (nAChRs). In the densest regions (at the crest of the junctional folds) a typical motor end plate has a density of nAChRs in the order of 10,000/μm^2. As shown in Figure 7.12, the nAChR at the motor end plate consists of five subunits ($2 \times \alpha$, $1 \times \beta$, $1 \times \delta$, $1 \times \gamma$) arranged symmetrically around an axis perpendicular to the membrane. There are approximately 500 amino acids in each of the five protein subunits, with each subunit containing the following: four segments (M1, M2, M3, M4) crossing the membrane, one large extracellular domain, and a large cytoplasmic domain. The two ACh binding sites are located between subunits α/δ and α/γ. The channel pore and gate are formed by the five M2 segments aligning together in the middle of the membrane. In the resting state, the M2 segments are kinked and form a hydrophobic ring that keeps the channel closed.

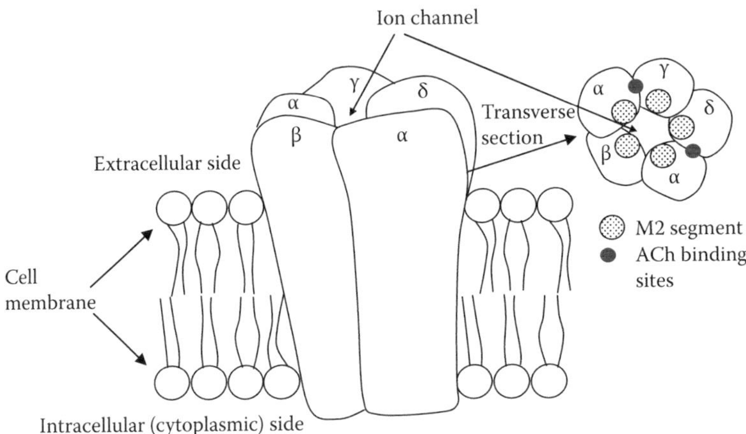

FIGURE 7.12
Nicotinic acetylcholine receptor (nAChR) channel.

Binding of ACh molecules is believed to remove the kink by causing a conformational change in the M2 segment, leading to hydrophilic amino acid residues lining the pore instead of hydrophobic ones. In this state, the channel is open, with small ions such as Na^+ (moving inward) and K^+ (moving outward) able to pass through. However, the sodium ions are driven by a stronger electrochemical gradient than are the potassium ions, resulting in more Na^+ moving into the muscle fiber than K^+ moving out. Therefore, the net effect is a depolarization of the motor end plate, and this is usually of sufficient magnitude to depolarize the adjacent excitable sarcolemma to below threshold, resulting in the opening of voltage-gated sodium channels and the generation of a muscle action potential.

7.5 Excitation–Contraction Coupling

Excitation–contraction coupling is a term used to describe the events linking the action potentials traveling across the sarcolemma to the making available of the active sites on the actin (thin) filaments for myosin heads to bind. After this, the muscle, as it contracts, goes into a repetitive cycle of events referred to as the cross-bridge cycle (see Section 7.6). The events prior to excitation–contraction coupling are described in Section 7.4.

The basic sequence of events in excitation–contraction coupling is illustrated in Figure 7.13. The action potential traveling across the sarcolemma also travels along the extensive network of T-tubules inside the muscle fiber, triggering muscle contraction. Although not fully understood, it is thought that the action potential traveling down the T-tubules causes a conformational change in a voltage-gated calcium channel in the T-tubule membrane known as the dihydropyridine receptor (DHPR). Hence, the DHPR acts as a voltage sensor for excitation–contraction coupling. It is thought that the change in the DHPR channel somehow opens up the ryanodine receptor (RyR) calcium release channel, embedded in the sarcoplasmic reticulum membrane, causing the release of calcium ions from the sarcoplasmic reticulum into the sarcoplasm. The calcium ions then diffuse into the sarcomere and bind to the troponin C subunits on the actin filaments, resulting in conformational

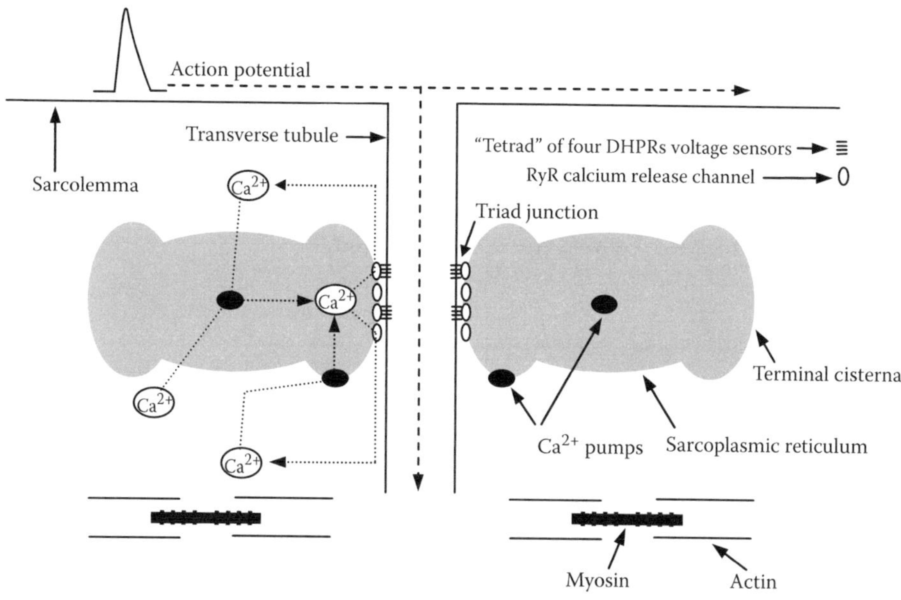

FIGURE 7.13
Excitation–contraction coupling. (Reprinted from *Prog. Biophys. Mol. Biol.*, Vol. 79, Dulhunty, A.F., et al., p. 47, Copyright (2002), with permission from Elsevier.)

changes and movement within the troponin–tropomyosin complex—that leads to the exposure of actin active sites for the myosin heads to bind, and hence, contraction of the muscle occurs.

In more recent times, the definition of excitation–contraction coupling has narrowed a bit, to only encapsulate the process coupling depolarization of the sarcolemma and transverse tubules to the release of calcium from the sarcoplasmic reticulum. This reflects the ongoing, yet still elusive, efforts to determine the coupling mechanism between the DHPR and RyR that leads to the release of calcium from the RyR channel embedded in the sarcoplasmic reticulum. It is thought to be a mechanical coupling involving conformational changes in the DHPR and RyR, but this has yet to be confirmed. In the junctions between the terminal cisternae membrane of the sarcoplasmic reticulum and the T-tubule membrane, "tetrads" of four DHPRs are coupled to every second RyR. Why only every second RyR is involved is also an unresolved question. The RyRs are the largest known ion channel proteins and consist of four subunits, each with a molecular weight of about 560,000 Da. Whilst the RyR and DHPR are the core components in the excitation–contraction coupling complex—forming the so-called foot structures in the junctional gaps, it has become evident that these components are the hub of a larger macromolecular complex called the calcium release unit. As such, the calcium release unit contains interactions between many proteins, each interaction impinging in some way on the excitation–contraction coupling process.

For example, some of the other proteins reportedly involved include the FK506 binding protein 12 (FKBP12), thought to be essential in terms of coordinating the opening of the RyR channels so that full conductance is achieved. Calmodulin is believed to act as a type of switch for a particular RyR and DHPR protein/protein interaction. Also, RyR activity is strongly inhibited when the calcium binding protein calsequestrin, located in the lumen of the sarcoplasmic reticulum, binds to a couple of proteins (triadin and junctin) in the sarcoplasmic reticulum membrane that also bind to the RyR. Although not fully understood, the interaction between these proteins may be involved in regulating calcium release from the RyR channel by detecting calcium levels and subsequently reducing channel activity when the calcium stores within the sarcoplasmic reticulum become depleted. There are many more protein/protein interactions in the calcium release unit, most of whose functions, like the ones discussed above, are not fully understood.

7.6 Cross-Bridge Theory

7.6.1 Sliding Filament Hypothesis

The sliding filament hypothesis was first discovered in 1954, by two different groups working independently. Andrew Huxley and Ralph Niedergerke were one group, with Hugh Huxley and Jean Hanson being the other group (the Huxleys were not related). However, on finding out that they were coming to similar conclusions, they later agreed to coordinate simultaneous publication in *Nature* of their respective papers. In essence, the sliding filament model established that during muscle contraction, the actin and myosin filaments stayed essentially constant, and that change in sarcomere length was caused by overlap of the two filaments. This explained why the A band, the only region where myosin filaments were found, stayed constant during contraction and why the I band, which had actin filaments attached to the Z lines, shortened during contraction. It was thought that the muscle contraction occurred by a sliding filament mechanism, mediated by cross-bridges seen in electron micrograph cross-sections. The sliding filament hypothesis is pretty much universally accepted now and has been for some time. Since then, work has focused on elucidating the mechanism whereby the cross-bridges produce a sliding force.

7.6.2 Cross-Bridge Theory

Cross-bridge theory is concerned with the molecular mechanism of muscular contraction. Essentially, it is concerned with explaining how the sliding filament hypothesis works, that is, explaining the mechanism whereby the cross-bridges produce a sliding force. Cross-bridges are the small outward projections from the myosin filaments. Generally, most definitions of a cross-bridge considers only the S1 myosin heads as cross-bridges (adopted here), although there are also rods (subfragment S2) that protrude from the myosin filaments—which connect the myosin heads to the myosin filament. The cross-bridge theory is the paradigm of muscular contraction and force production that has grabbed most of the attention since first proposed by Andrew Huxley back in 1957. The model has been updated over the years, and it does not account for all observed phenomena, but it is still the dominant model. The basic postulates of the cross-bridge theory are that the myosin heads (cross-bridge) are able to attach to the actin filament close by, given the right conditions. When attached, the cross-bridge undergoes some sort of a conformational change that ends up pulling the actin filament toward the M line, such that it slides past the myosin filament with the effect of shortening the sarcomere. This conformational change and/or movement of the cross-bridge is known as the power stroke (or working stroke). After this, the cross-bridge detaches and is available to reattach again provided conditions are right. Although thought to operate independently, numerous cross-bridges are activated at any one particular time, allowing a summation of forces produced by the many individual power strokes. As some cross-bridges detach from the actin filament, other cross-bridges attach, ensuring that continuous tension is applied during a contraction.

It is still unclear as to the proportion of cross-bridges that are active at any one time, and part of the uncertainty is whether both myosin heads in the cross-bridge contribute to the tension. For isometric contractions, estimates of cross-bridges that are active at any one time usually range from 15% to 50%. Also, in revised versions of the cross-bridge model, the rotation of the myosin head at the attachment point to actin was important, but subsequent evidence indicates that such a rotation may not be possible, as the attachment point appears to be fixed. Hence, instead of a swinging myosin cross-bridge model, as was outlined in 1969 (H.E. Huxley) and 1971 (A.F. Huxley and Simmons) and illustrated in Figure 7.14, a swinging lever-arm model is now more widely accepted. In the latter model, a version of which is illustrated in Figure 7.15, the stepwise change during the power stroke is believed to be caused by a conformational change in which the lever arm (consisting of a stretch of the myosin heavy chain to which two myosin light chains are attached), also called the regulatory domain, is thought to rotate about a hinge or fulcrum near the middle of the

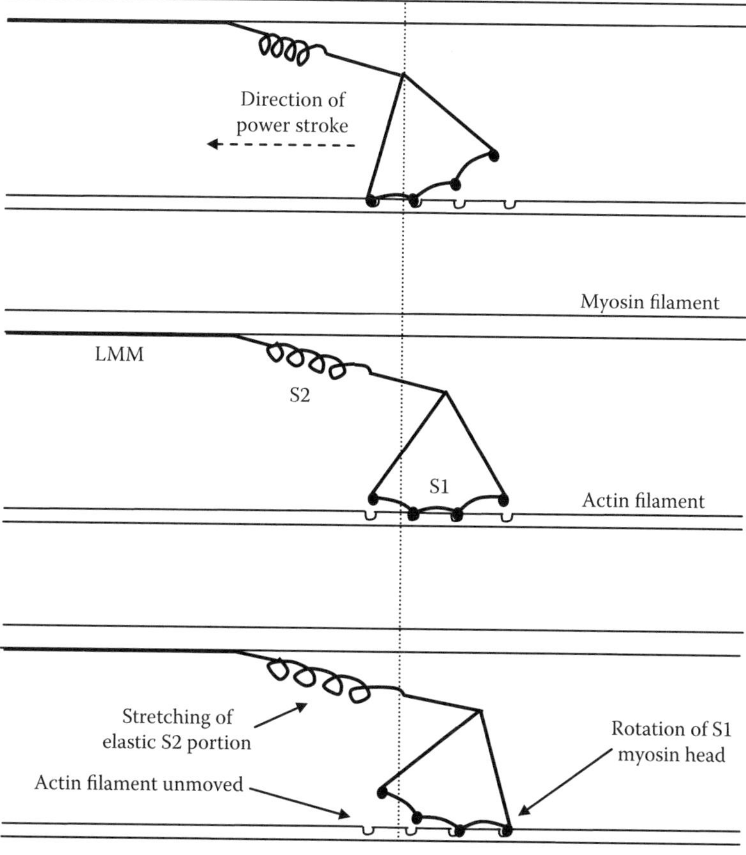

FIGURE 7.14
A.F. Huxley's 1971 swinging myosin cross-bridge model: illustration of an isometric contraction. (Huxley, A.F. *J. Physiol.* 1974. Vol. 243. p. 30. Copyright Wiley-VCH Verlag GmbH & Co. KGaA. Reproduced with permission.)

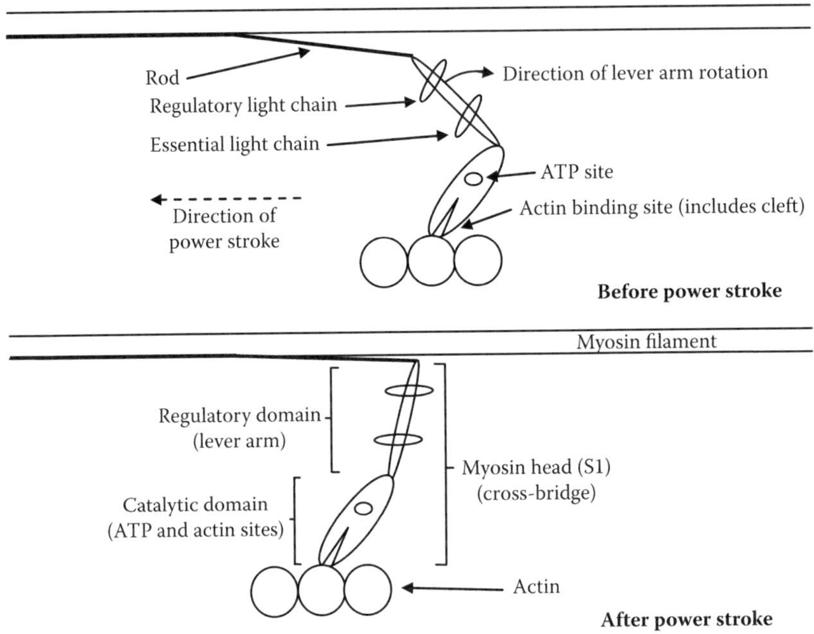

FIGURE 7.15
Swinging lever arm cross-bridge model: illustration of an isotonic contraction. (Reprinted with permission from Highsmith, S., *Biochemistry*, 38, 9791–9797. Copyright 1999 American Chemical Society.)

myosin head—where it flexibly joins the catalytic domain. As such, the anterior part (catalytic domain) of the myosin head remains fixed in its orientation with the actin filament, with the conformational change being mostly restricted to the posterior part (regulatory domain). This rotation of the lever arm is believed to slide the actin filament past the myosin filament, toward the M line and center of the sarcomere—this action essentially constituting the power stroke. The cross-bridge consists of two myosin heads, both connected to the same rod domain, but only one myosin head is shown in Figure 7.15, with the catalytic and regulatory domain constituting the myosin head (S1). The lever arm is also attached flexibly to a fibrous rod domain—the latter forming the link to the myosin filament.

Measurements of step sizes for the power stroke of a single cross-bridge vary between 5 nm and 15 nm, and estimates of the force from a single head vary between 2 pN and 10 pN. The average force of a single two-headed cross-bridge has yielded values of 1.7 pN and 3.4 pN in separate studies. Despite the above variations in measurements, the values give a rough guide as to the forces and distances involved.

7.6.3 Cross-Bridge Cycle

Although the cross-bridge theory has been modified over the years, the "older" swinging myosin cross-bridge model is still useful as a framework for outlining the cyclical events occurring during a muscle contraction, even though the conformational changes and mechanisms during the power stroke are still unresolved, as are some specifics of the reactions of the myosin head with actin and ATP (adenosine triphosphate). This recurring activity during muscle contraction is known as the cross-bridge cycle; this is illustrated in Figure 7.16 and described as follows:

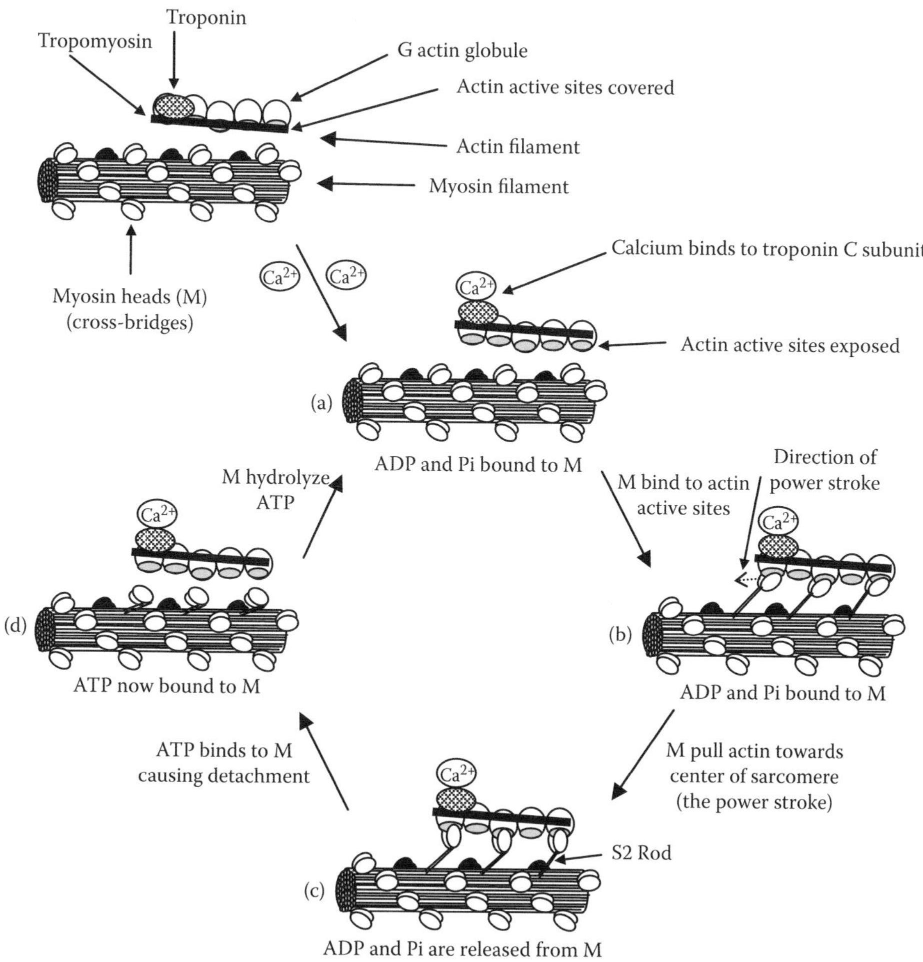

FIGURE 7.16
Cross-bridge cycle.

The energy that drives the power stroke during muscle contraction is chemical energy from ATP, which is split into the products ADP and inorganic phosphate (often designated as P_i) by hydrolysis. This ATPase activity occurs on the myosin head, with the cleavage products fastened there. During this process, the myosin head changes its position so that it extends more perpendicular to the actin filament. This movement or conformational change by the myosin head is often likened to energy being stored in a cocked spring, the energy being supplied by the hydrolysis of ATP. In the cross-bridge theory, a cocked myosin head will bind to active sites on the actin filament if calcium is present in the sarcoplasm surrounding the myofilaments. Calcium ions bind to the troponin C (TnC) subunits on the actin filaments, causing conformational changes and movement within the troponin–tropomyosin complex, leading to the exposure of the actin active sites. Once the actin active sites are exposed, the myosin head attaches to it, and this somehow triggers a conformational change in the myosin head that is known as the power stroke (see Figure 7.16b).

With the power stroke, the myosin heads pull the actin filaments toward the M line in the center of the sarcomere. The myosin molecules are arranged so that the myosin head to tail directions are reversed near the midpoint of the myosin filament, such that on each half of the myosin filament, there are myosin heads pulling actin filaments toward the M line. Once the power stroke has occurred, the myosin heads are "uncocked" and become free of the ADP and P_i, these products now having been released into the sarcoplasm (see Figure 7.16c). In this theory, the cross-bridges are still bound to the actin until ATP binds to the myosin heads, at which point the cross-bridges detach (see Figure 7.16d). Being detached, and with ATP bound to it, the myosin heads will now be "recocked" by hydrolyzing the bound ATP (see Figure 7.16a). Providing calcium is still present, this contraction cycle will repeat itself, again and again. During a contraction, a myosin head performs about five power strokes every second. When calcium ion concentration in the sarcoplasm is less than about 10^{-7} M, contraction ceases. The threshold for contraction is just under 10^{-6} M, with peak contraction reached by 10^{-5} M calcium ion concentration.

7.7 Muscle Spindle

Extrafusal fibers and intrafusal fibers are the two basic types of contracting skeletal muscle fibers. The extrafusal fibers are innervated by alpha motor neurons and are the fibers responsible for essentially all mechanical work performed by the muscle during a contraction, such as contractions producing movement, as well as resting muscle tone. Whilst also contracting, intrafusal muscle fibers do not contribute in any significant way to the work performed by the muscle because of its unique structure. Nuclear bag fibers and nuclear chain fibers are the two different types of intrafusal muscle fibers, with both types usually present in a muscle spindle. Muscle spindles (see Figure 7.17) are sense organs that are responsive to stretch, and are classified as proprioceptors (a group of mechanoreceptors). Intrafusal fibers are much shorter in length longitudinally then the extrafusal fibers, varying from about 4 mm to 10 mm in length, as well as smaller in diameter. A muscle spindle consists of a connective tissue capsule that encapsulates the 5 to 12 intrafusal muscle fibers within it. Muscles may contain anything from about a dozen to hundreds of muscle spindles, with spindles tending to be denser in muscles where fine control is required. For example, the lateral gastrocnemius muscle (a large and powerful muscle) may contain about five spindle capsules per gram of muscle, compared with about 119 spindle capsules per gram of muscle for the intrinsic Vth interossei muscle of the hand—the latter a small muscle requiring delicate control. The muscle spindle gets its name from its spindle-like shape, which has tapered ends and a round bulging central body. There are usually two to three times as many nuclear chain fibers than nuclear bag fibers in a muscle spindle. Nuclear bag fibers are further categorized as either dynamic nuclear bag fibers or static nuclear bag fibers, because there are histochemical differences between the two types of nuclear bag fibers, and the dynamic nuclear bag fibers are thinner and shorter than the static nuclear bag fibers.

Primary afferent type Ia fibers and secondary afferent type II fibers are the two types of sensory nerve fibers that innervate muscle spindles. The primary type Ia fibers have annulospiral endings that innervate the central portion of both nuclear chain fibers and nuclear bag fibers, whereas the secondary type II fibers, which can have either annulospiral endings or flower-spray endings (where the fiber spreads out like branches on a bush), also

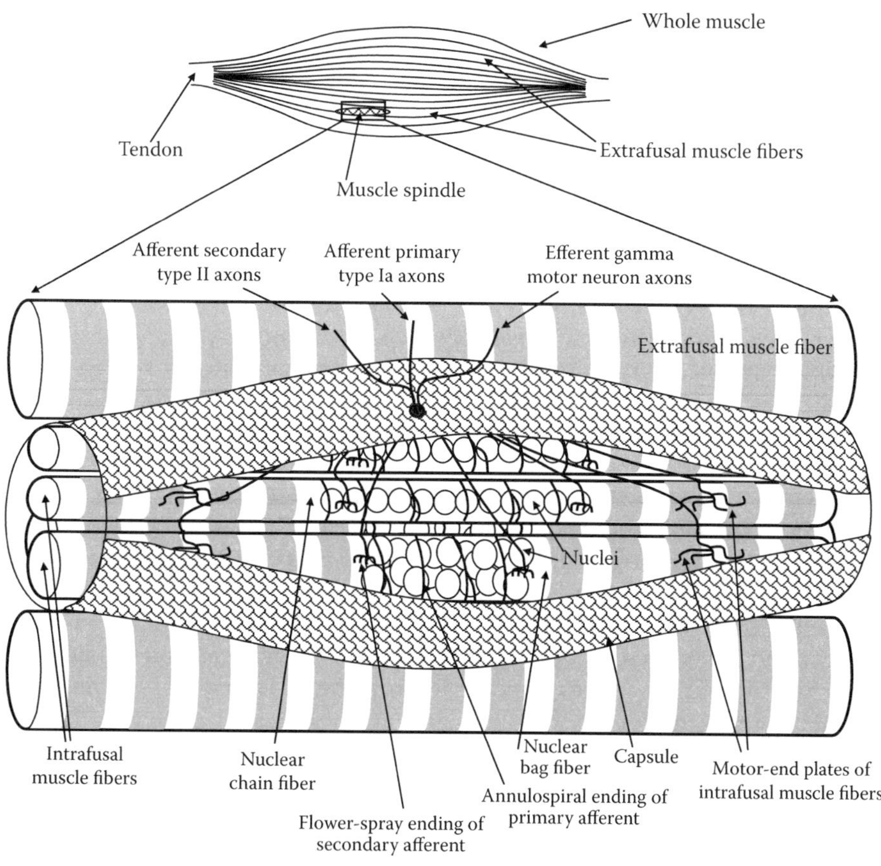

FIGURE 7.17
Muscle spindle.

innervate both types of intrafusal fibers, but more on the ends of the primary annulospiral endings. However, dynamic nuclear bag fibers lack secondary afferent type II innervation. Type Ia fibers are greater in diameter than type II fibers (average of 17 μm versus 8 μm), and transmit sensory signals faster (up to about 120 m/s). The motor neurons innervating intrafusal muscle fibers are known as gamma motor neurons, and two types are distinguishable. Dynamic gamma motor neurons innervate dynamic nuclear bag fibers, and static gamma motor neurons innervate nuclear chain fibers and static nuclear bag fibers. When dynamic gamma motor neurons are activated they tend to increase the sensitivity of the muscle spindles to velocity (rate of change of muscle spindle receptor length), whereas activation of the static gamma motor neurons increases the static sensitivity associated with a given amount of stretch. In the case of the latter both primary and secondary endings (from the afferent sensory neurons) will transmit signals continually in excess of a minute after being stretched. With regards to the former (the dynamic response), primary endings of the type Ia large-diameter fibers at the center of the dynamic nuclear bag fibers respond intensely in numbers, but briefly in duration (limited to the time period of the stretching).

As muscle spindles are stretched, for example, picking up a heavy weight, this causes stretching of the central portion of the intrafusal muscle fibers. Ion channels embedded in the surface of the sensory nerve endings innervating the central region of these intrafusal fibers are opened as the nerve endings are stretched. This causes depolarization of the sensory nerve terminals, and if threshold is reached, an action potential will be initiated at the initial segment of the axon. These nerve signals then send information about muscle length and the rate of change of muscle length to the spinal cord, cerebellum, and other brain regions. The sensory afferents may also synapse on an alpha motor neuron in the spinal cord, which in turn innervates the extrafusal fibers of the same muscle, causing contraction of the muscle and shortening of the muscle spindle and central receptor region. Antagonist muscles may also be inhibited by the spindle sensory afferents synapsing on inhibitory interneurons that inhibit the alpha motor neurons of the antagonist muscles from firing.

Stretching of the central sensory receptor region of the muscle spindle can occur both actively and passively. Passive stretching occurs when a muscle is passively lengthened by an event such as a knee jerk, elicited by striking the patellar ligament. In this instance, as the quadriceps muscles are stretched, so is the muscle spindles embedded within these muscles, resulting in the firing of the stretch sensitive sensory afferent nerve fibers within the central region of the spindles. These afferent axons synapse on alpha motor neurons in the spinal cord that innervate the extrafusal muscle fibers of the quadriceps muscles, leading to contraction of these fibers and shortening of the muscle (and causing the lower leg to move forward). This passive reflex is known as the knee jerk, and occurs without the involvement of gamma motor neurons.

As indicated in Figure 7.18, active stretch is brought about by the firing of gamma motor neurons, resulting in the stretching of the central portion of the intrafusal muscle fiber (Figure 7.18b). Stretching of the central receptor region leads to activity in the sensory neurons whose nerve endings innervate the central region, which in turn leads to contraction of the muscle as the afferent sensory neurons synapse on alpha motor neurons that innervate the muscle. The activation of alpha motor neurons results in the contraction of

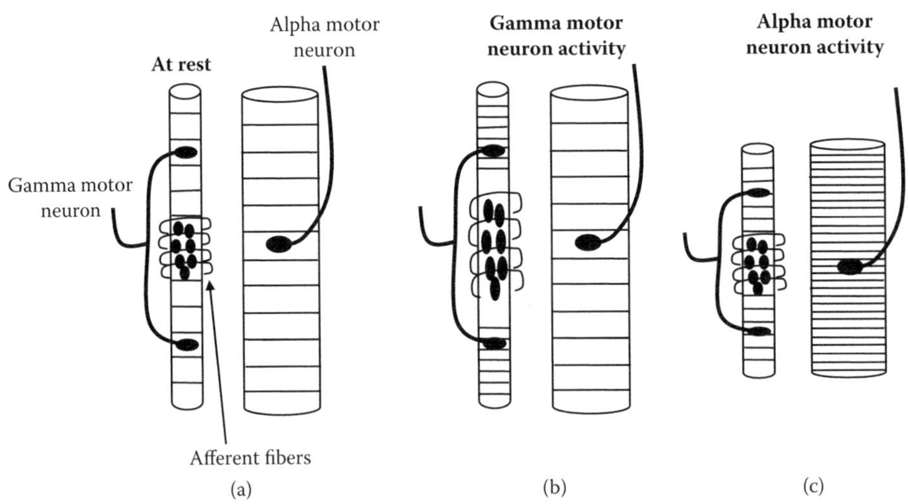

FIGURE 7.18
Muscle spindle activity.

the extrafusal muscle fibers, which in turn causes a shortening of the muscle spindle and a reduction in afferent spindle activity (Figure 7.18c). When carrying out voluntary motor activity that is transmitted from the motor cortex, the gamma motor neurons are usually coactivated, along with the alpha motor neurons, so that both extrafusal and intrafusal muscle fibers contract around the same time. Overall, the muscle spindle is shortened by this coactivation, due solely to the effects of the extrafusal muscle fibers contracting. However, the coactivation has a neutralizing effect on the central sensory receptor region, which has also been shortened by the whole muscle contracting. The stretching effect on the central portion of the fiber by the gamma motor neuron–induced contraction of the intrafusal fibers effectively counters the shortening of this region due to the contraction of the extrafusal fibers; hence keeping the length of the central part effectively constant during the muscle contraction.

The muscle spindle and its intrafusal fibers are oriented parallel to the extrafusal fibers, with the intrafusal fibers anchored at both ends to either the muscle spindle capsule or muscle connective tissue, with the muscle spindle itself being attached at both ends to muscle connective tissue. As such, the muscle spindle is stretched passively when the muscle is stretched, and shortened passively when the muscle contracts. When gamma motor neurons are activated, they do not change the length of the muscle spindle per se. This is because the central regions of the intrafusal muscle fibers are essentially free of contractile elements (i.e., myosin and actin filaments), being filled with nuclei and primary and secondary afferent nerve endings instead. Therefore, when the contractile parts of the intrafusal muscle fibers (situated off-center) are activated the centers (including receptor nerve endings) of the fibers are stretched in both directions, as the contractile elements at each end of the intrafusal fibers contract toward each respective end (the poles) (Figure 7.18b). The intrafusal fibers, although anchored at both ends, are not anchored to the central part of the fiber.

7.7.1 Golgi Tendon Organ (GTO)

In the context of discussing the muscle spindle, a brief mention of the Golgi tendon organ (GTO) is relevant. Whilst the muscle spindle gives feedback to the spinal cord and brain on the length and rate of change of length of a muscle, the GTO gives feedback on the tension and rate of change of tension in the muscle. The GTO is in series with the muscle fibers, at the point where they connect to the tendon, as opposed to the muscle spindles, which are in parallel with the extrafusal muscle fibers. GTOs consist of dendrites that branch and wrap around collagen fibers enclosed by a connective tissue capsule within the tendon. Connecting at one end of this GTO capsule are about 10 to 15 muscle fibers. When the muscle contracts, and hence tension is increased, the nerve endings of the GTO are stimulated leading to signals being transmitted to the spinal cord and brain. Whereas muscle spindles stimulate the muscle they are embedded in to contract, and inhibit antagonist muscles from contracting, GTOs do the opposite, that is, they inhibit the muscle they are embedded in from contracting and activate antagonist muscles to contract (via stimulating the alpha motor neuron innervating it). Rather than directly connecting with a motor neuron, GTOs usually synapse on interneurons in the spinal cord that then send either an inhibitory or an excitatory signal to the motor neuron. The afferent nerve fibers innervating the GTO are myelinated type Ib fibers, about 10 μm to 20 μm in diameter. However, as they branch out within the GTO capsule they eventually become unmyelinated. GTOs are usually between 1 mm and 1.5 mm long and about 0.5 mm in diameter, and in most muscles, there are roughly as many GTOs as there are muscle spindles.

7.8 Skeletal Muscle Mechanics

All sarcomeres in a myofibril, as well as all myofibrils in a muscle fiber, shorten essentially in unison in response to a muscle action potential traveling across the sarcolemma and transverse tubule system. The muscle fibers in a motor unit are considered to be activated simultaneously. However, the axon of a motor neuron branches out as it forms its many neuromuscular junctions on different muscle fibers. Hence, there are bound to be differences in conduction times of the action potential traveling from the motor neuron to the various presynaptic terminals, resulting in a slight asynchrony in the firing of muscle fibers within the same motor unit. When a muscle is used to do work there are usually several motor units activated asynchronously, but overlapping, so that there is a summation of forces. The more force required, the more motor units will be recruited in order to increase tension. Before looking at the mechanics of whole muscles, it is useful to first consider the mechanics of individual muscle fibers, as the mechanics of whole muscles are essentially the summation of the activity of individual muscle fibers.

7.8.1 Mechanics of Muscle Fibers

Usually sarcomere length is in the range of 1.6 μm to 2.6 μm, although this range will be exceeded if the muscle is shortened or stretched considerably. When a muscle fiber shortens during a contraction the amount of shortening possible depends on the number of sarcomeres in series along the myofibril. During a contraction the shortening of all sarcomeres are added together. The shortening of an individual sarcomere will usually be no more than about 2 μm or so, depending on the resting sarcomere length at the start of contraction. However, when thousands or even tens of thousands of sarcomeres are shortened at the same time, then added together this can result in the whole muscle shortening significantly—in some muscles by up to about 30% of its resting length.

The amount of force generated by a muscle fiber during contraction is proportional to the number of power strokes (pivoting cross-bridges), and the number of power strokes possible is proportional to the overlap between the myosin and actin filaments. As indicated in Figure 7.19c and 7.19d, maximal contractile force is achieved at sarcomere lengths between about 2 μm and 2.2 μm. At these lengths, all the cross-bridges of the myosin

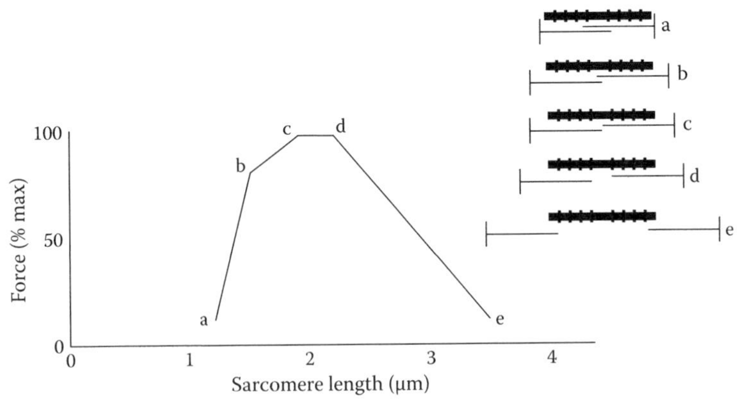

FIGURE 7.19
Length–tension relationship in sarcomere.

filaments are overlapped by the actin filaments, but not to the point where the actin filaments themselves overlap. In the case of the latter, this will disrupt the 3-D lattice of the filaments, causing interference in the binding of the heads of the cross-bridges to the active sites of the actin filaments—and subsequently a reduction in force generated (see Figure 7.19b). If the shortening continues, then the myosin filaments will be in contact with the Z lines, making further sarcomere shortening no longer possible, and the force generated will approach zero (see Figure 7.19a). Where the sarcomere has been stretched to the point that no myosin cross-bridges overlap the actin filaments, the force generated by the muscle also approaches zero (see Figure 7.19e).

Figure 7.20 shows how the force (or tension) develops over time during a single twitch contraction of a muscle fiber. A twitch includes the stimulus (depolarization of the motor end plate), a latent period, a contraction phase, and a relaxation phase. The duration of twitches vary greatly, from as brief as about 7 ms in some ocular muscle fibers, to as long as about 100 ms, as in muscle fibers of the *soleus* muscle of the calf. The latent period lasts about 2 ms, irrespective of the total twitch duration. During this period the muscle action potential travels across the surface (sarcolemma), as well as to the inside (the transverse tubules), of the muscle fiber, eventually resulting in the release of calcium from the sarcoplasmic reticulum into the sarcoplasm. The presence of calcium in the vicinity of the actin filaments results in the contraction of the sarcomeres. When the resultant contractile force has an external effect, in which the measurable tension starts to increase, then the contraction phase of the muscle twitch has begun. After peak twitch tension is achieved the relaxation phase begins, which corresponds with declining calcium levels due to the net reabsorption of calcium back into the sarcoplasmic reticulum. A reduction in calcium means fewer active sites will be available for the myosin head to bind to, and hence less force will be produced by the muscle fiber. The relaxation phase usually lasts longer than the contraction phase.

When muscle fibers are activated, they will usually be stimulated repetitively in order to generate the necessary force to carry out the required work, rather than as stand-alone individual twitches. Figure 7.21 illustrates how the tension summates when the muscle fibers are stimulated repetitively at different rates. If individual twitches are separated long enough to allow the internal environment of the muscle fiber to return to a normal resting state, then no summation will occur. If the muscle fiber is stimulated at the end of the relaxation phase, or soon thereafter, then a staircase type effect (called treppe) will occur whereby each subsequent twitch gets stronger until a maximum strength is reached. Stimulation during the relaxation phase itself produces a phenomenon known as unfused

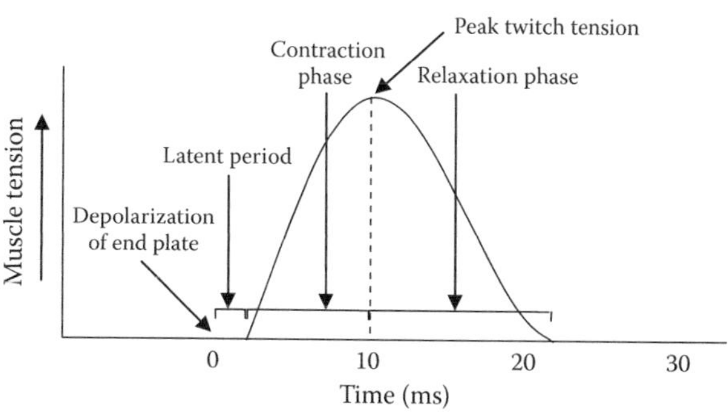

FIGURE 7.20
Tension during a single twitch.

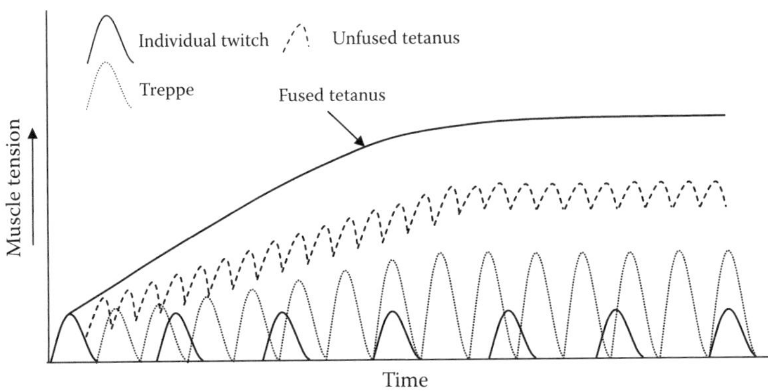

FIGURE 7.21
Tension summation.

tetanus (or incomplete tetanus), where the tension keeps steadily building up until a peak level is attained. Here, the tension of the twitch is added to the tension of the previous twitch, producing a summation of tension in a manner often referred to as wave summation. There are only partially decreases in tension between successive stimuli during this type of wave summation, producing a ripple-like effect in the observed graph. Finally, when the muscle fiber is restimulated before the previous relaxation phase begins, then the ripples in the unfused tetanus disappear, producing a smooth tension curve that rises until a plateau is reached. The frequency of stimulation at which the bumpiness disappears is known as the tetanic frequency (or fusion frequency), and the force is the tetanic force. The maximal tetanic force is the tension at the plateau of the curve. Individual twitch tension is usually about 10% to 25% of the maximum tetanic tension in the muscle of mammals.

Essentially, the changes in tension observed during the wave summation of the twitches is correlated positively with calcium availability in the cytoplasm, although some increases in muscle tension may also be caused by certain enzymes working more efficiently after the heating effects of preceding twitches. Stimulating the muscle much beyond the tetanic frequency will not see any significant further rises in tension, as there is already enough calcium to keep maximal possible active sites available for the power strokes that generate the tension. Fused tetanus is a phenomenon that occurs at a frequency between about 20 Hz and 60 Hz in most skeletal muscle, although this figure can vary considerably (up to 300 Hz—depending on the type of muscle fiber), with muscle fibers of briefer twitch duration having larger tetanic frequencies. However, the lower limit of the tetanic frequency will usually still be larger than the maximum likely frequency of stimulation of a muscle fiber by a motor neuron, which is about 25 Hz. Hence, most normal muscle fiber contractions will usually not involve fused tetanic contractions, but a mixture of the other types discussed.

7.8.2 Summation Activity of Whole Muscles

In an in vivo setting individual motor fibers do not usually contract by themselves, but collectively as members of a motor unit, and so when dealing with muscle as a whole it is more useful to consider tension summation at the motor unit level and above. A motor unit is made up of a single motor neuron and the total number of muscle fibers it innervates. In normal muscle, all muscle fibers in a motor unit will fire in unison in response to an action

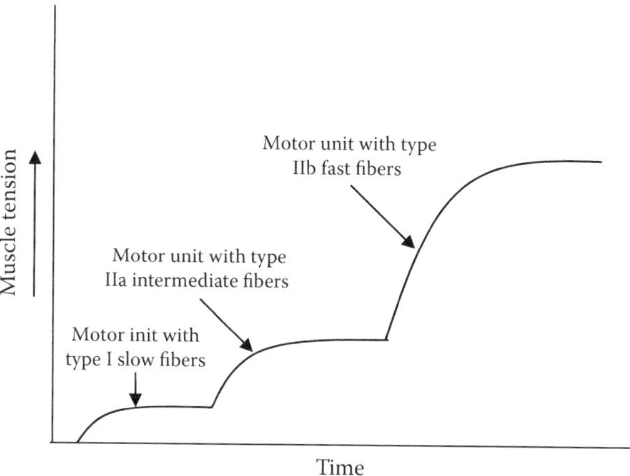

FIGURE 7.22
Recruitment of motor units.

potential stimulus from the motor neuron, and, depending on the strength of nerve stimulation, usually more than one motor unit will be activated. Hence, not only will the tension produced by all muscle fibers in an individual motor unit summate, but also the tension produced by all active motor units. Rarely will all motor units be activated at the same time during a sustained contraction; rather they will operate in a rotating fashion, allowing nonactive motor units time to recover somewhat between contractions. As there is overlap in contraction between different motor units during this rotational activity, the tension produced by different individual motor units will summate.

Another type of tension summation in muscle is called recruitment, as illustrated in Figure 7.22. In this process, motor units are only activated on a per need basis so that, for example, if minimum force is required, only the motor units containing small-diameter type I slow fibers will be recruited to contract. Note that all muscle fibers in a single motor unit will be of the same type, but that usually most muscles contain all three different types of motor units intermingled. As more force is required motor units containing type IIa intermediate fibers will be recruited, followed by the motor units containing large-diameter type IIb fast fibers—the latter being the most powerful of the three motor unit types.

7.8.3 Isometric Muscle Contractions

Muscle contractions are generally classified as either isometric or isotonic contractions, depending on whether the muscle changes length or not. Note that the tension developed by a muscle is a force, and is measured either as newtons (N) or as kilograms (kg) weight, where 9.8 N is equal to 1 kg weight. In isometric contractions, the tension generated by the muscle is less than the tension of the load on the muscle, and so the muscle cannot shorten. One way to measure the isometric tension characteristics of a muscle experimentally, such as the gastrocnemius muscle of a toad, is to fix both ends of the muscle so it cannot move, with one of the ends attached to a force transducer. The sciatic nerve is then stimulated causing the muscle to contract, with the tension generated measured via the transducer. The relationship between muscle length and tension in isometric contractions is illustrated in Figure 7.23. It shows that there is no change in the length of the muscle regardless of whether the muscle was stimulated to contract or not.

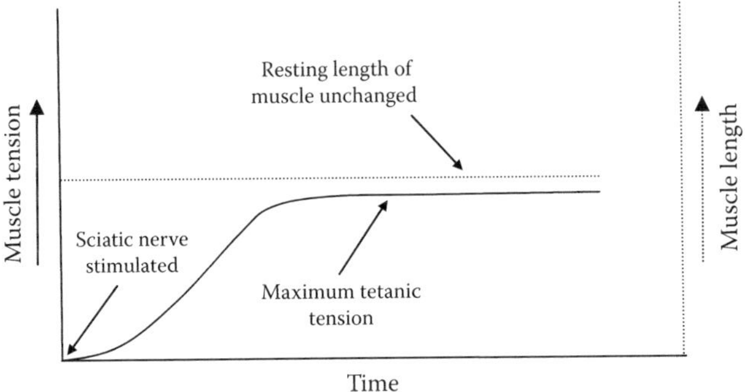

FIGURE 7.23
Isometric contraction.

By noting down the passive (resting) tension before beginning tetanic stimulation of the muscle, and then measuring the maximum tetanic tension (equivalent to the total tension in the muscle), a length–tension curve of the muscle can be obtained. In order to record a length–tension curve, the passive and total tension values need to be recorded for a range of muscle lengths, as indicated in Figure 7.24. Subtracting the passive tension from the total tension gives the active tension, which corresponds to the force that the muscle itself adds when contracting. As the muscle lengthens beyond the normal resting length, the passive tension increases until the total tension component is made up almost exclusively of the passive tension only. The passive tension is due to the resistance of the connective tissues in the muscle, and, although elastic initially, the connective tissues have great resistance when stretched considerably.

From Figure 7.24 it can be observed that there is a normal or optimal resting length of the muscle that will generate maximum isometric active tension when the muscle contracts. As the resting muscle length is shortened or elongated beyond this optimal range, the

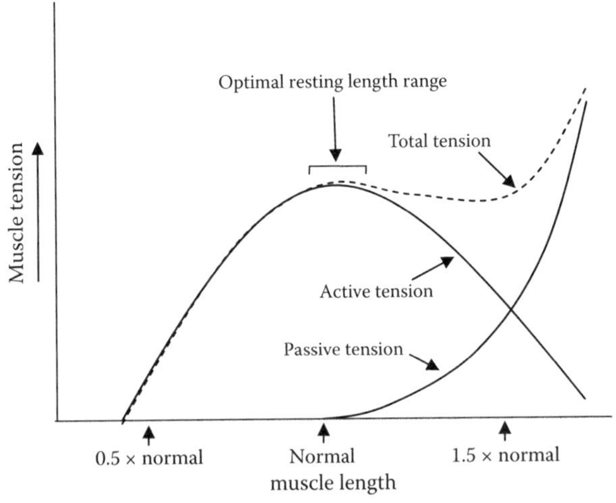

FIGURE 7.24
Length–tension curve in whole muscle.

maximum amount of force the muscle can produce when it contracts is diminished, for the same reasons discussed in relation to Figure 7.19. Joints usually prevent muscles from being stretched or compressed too far.

7.8.4 Isotonic Concentric Muscle Contractions

In isotonic muscle contractions, the lengths of the muscles change. When the muscle shortens, it is called a concentric contraction. Eccentric contractions involve the muscle actually lengthening, and will be discussed later. Experimentally, concentric contractions involve, for example, stimulating the sciatic nerve that causes the gastrocnemius muscle of a toad to contract. Tying one end of the muscle to a lever system that allows weights to be attached, the movement of the lever is then recorded when the muscle is stimulated to lift the load. From this, a relationship between load on the muscle and its corresponding shortening can be obtained.

Figure 7.25 shows a single (twitch) concentric isotonic contraction of a muscle. After the nerve is stimulated, the muscle soon begins to contract, but the muscle itself does not begin to shorten until the tension it generates slightly exceeds the resistance or load on the muscle. After this, the tension in the muscle remains constant until the contraction phase is over and relaxation of the muscle begins. Some important characteristics of a concentric contraction are highlighted by comparing twitch concentric contractions at different loads, as illustrated in Figure 7.26. The steepest slope of the shortening phase of the contraction can be taken as representing the constant velocity of shortening (V) of the muscle. As the load on the muscle increases, V decreases, as does the duration of the twitch and the total amount of shortening possible. The latter is because maximum isometric active tension for a muscle is only achieved within an optimal muscle length range, and so at shorter muscle lengths, the maximum force the muscle can produce is less. Hence, as the muscle is loaded up (with weights) more and more, it becomes more likely that at shorter lengths the isometric tension of the muscle will not be strong enough to overcome the resistance of the load. Also illustrated in the diagram is that the time between nerve stimulation (occurring at time zero) and shortening onset increases as the load on the muscle increases. This is because the heavier the load, the longer it takes for the muscle to overcome the resistance of the load.

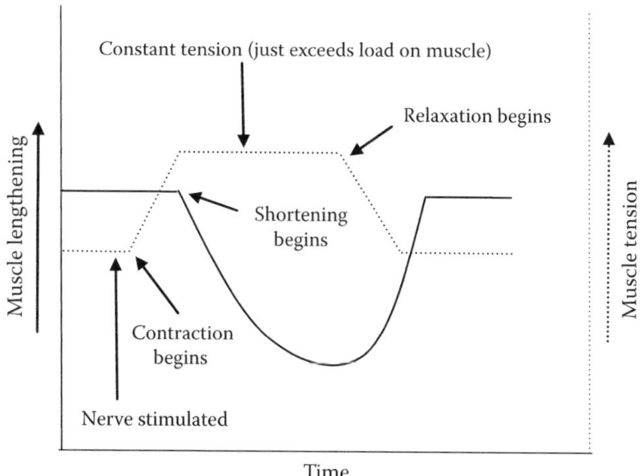

FIGURE 7.25
Concentric isotonic contraction.

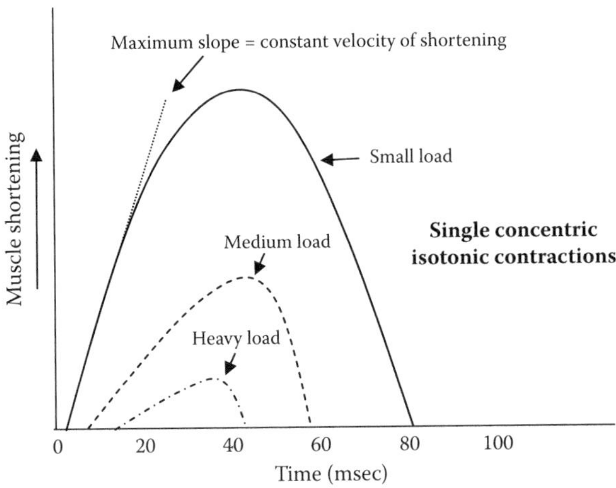

FIGURE 7.26
Relationship between load and shortening velocity.

7.8.5 The Hill Equation and Force–Velocity Curve

By keeping the muscle at the same initial resting length and recording V for different loads on the muscle, A. V. Hill's classic force–velocity curve is obtained, as indicated in Figure 7.27. The force–velocity curve, fitted to the experimental data, was used by Hill to obtain the force–velocity equation (Hill equation):

$$(P + a)(V + b) = b(P_0 + a) = \text{constant}$$

where P is the load on the muscle, V is the constant velocity of shortening, P_0 is the isometric tension, a is a constant with the dimensions of force (e.g., gram), and b is a constant with the dimensions of velocity (e.g., cm s^{-1}). If one experimentally records a series of values for

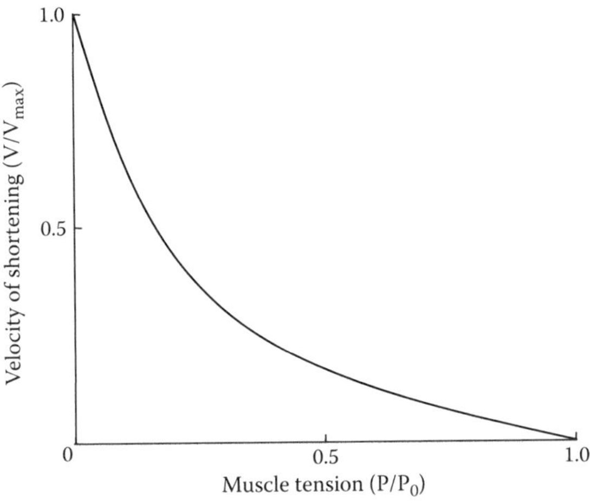

FIGURE 7.27
Force–velocity curve.

V and P, one can determine values for P_0 and V_{max} (see below) by extrapolating the plotted curve to where it intersects with the axes, as explained below. P_0 can also be measured directly in an experiment. With a series of values for V, P, and having determined P_0 and V_{max}, one can fit this data with Hill's force–velocity equation to determine the constants a and b in the equation.

The Hill equation can also be written in terms of velocity of shortening:

$$V = b(P_0 - P)/(P + a).$$

The Hill equation is not theoretically based, but derived empirically, although theoretical approaches have also yielded force–velocity equations. When P = 0, $V = V_{max} = bP_0/a$, as the maximum velocity of shortening (V_{max}) in a muscle occurs when no load is attached to the muscle. Hence, V_{max}, which can be difficult to directly measure, can be found on the force–velocity curve by noting where it intersects with the vertical axis (i.e., when P = 0). Note that the constant a typically varies between about $0.1P_0$ and $0.4P_0$, but is usually taken as $0.25P_0$. Hence, it follows from the equation $V_{max} = bP_0/a$ that V_{max} is about 2.5 to 10 times larger than b (depending on what value the constant a is). The constant b increases significantly with temperature, approximately doubling after a 10 degree Celsius (°C) heating near 0°C. When V = 0, $P = P_0$, as the isometric tension in a muscle is equivalent to the load on the muscle when it can no longer shorten. Hence, P_0 can be found on the force–velocity curve where it intersects with the horizontal axis (i.e., when V = 0).

The Hill equation only applies to concentric (shortening) contractions, not eccentric (lengthening) contractions, and is only valid during constant velocity of shortening (approximated as the steepest gradient of the shortening twitch contraction). It also only applies to sarcomere lengths between about 1.7 μm and 2.5 μm (the optimal resting length region), as the isometric tension (P_0) generated by the muscle at these lengths do not differ appreciably. The reason the Hill equation only applies to these muscle lengths is that it makes the assumption that P_0 does not vary when the muscle shortens during the experiment. From the length–tension curve (see Figure 7.24), it can be observed that P_0 does vary when muscle length is changed, as beyond the optimal resting length, P_0 decreases. However, an unchanging P_0 may be a reasonable approximation for muscle lengths in a realistic physiological environment, as normally muscles tend to operate in the plateau (optimal or peak) region of the length–tension curve.

7.8.6 Power Output

Work (measured in joules (J)) is force × distance. For example, when a force of 1 N is applied through a distance of 1 m, the work equals 1 J or 1 N-m. Power (measured in watts) is the rate at which work is done (or energy is expended), and is equal to force × velocity (or work ÷ time). From the Hill equation,

$$W = PV = Pb(P_0 - P)/(P + a)$$

where W is the power output, P is the load on the muscle, V is the velocity of shortening, P_0 is the isometric tension, and a and b are constants. If $P = P_0$ or P = 0 then the work done is zero, and hence the power output (W) becomes zero.

W reaches a maximum when $dW/dP = 0 = P^2 + 2aP - aP_0$.

Calculating the roots of the above quadratic equation gives

$$P = a[(1 + P_0/a)^{1/2} - 1].$$

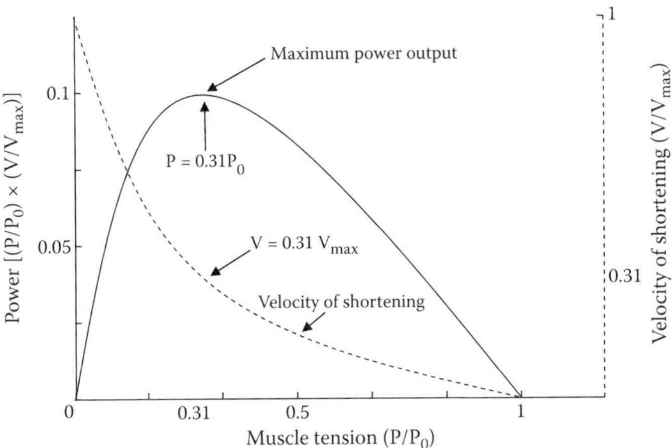

FIGURE 7.28
Power–force curve.

If $a = 0.25P_0$ (a representative value, although muscle fibers do vary), then W reaches a maximum at $P = 0.31P_0$.

If instead $a = 0.40P_0$, then W reaches a maximum when $P = 0.35P_0$.

Also, from the Hill equation,

$$V/V_{max} = a(P_0 - P)/P_0\,(P + a).$$

If $a = 0.25P_0$, then

$$V/V_{max} = 0.25(1 - P/P_0)/(P/P_0 + 0.25).$$

It was earlier calculated that at $a = 0.25P_0$, W reaches a maximum when $P = 0.31P_0$. Hence, using $P/P_0 = 0.31$ in the above equation, $V/V_{max} = 0.31$. Therefore, at $a = 0.25P_0$ maximum W occurs when $V = 0.31V_{max}$ and $P = 0.31P_0$, as also indicated in Figure 7.28. Hence,

$$\text{Maximum Power Output } (W_{max}) = 0.31P_0 \times 0.31V_{max} = 0.1(P_0 \times V_{max}) \text{ (watts)}.$$

The maximum power output usually corresponds to the greatest efficient in terms of getting work done in response to energy input, and this has practical consequences when, for example, setting up the gears for a cyclist. It is important for the cyclist to adjust gear ratios so that the legs are contracting at optimum speed against optimum load whatever the incline. Hence, as the incline steepens, the load increases, and the velocity of shortening decreases. Therefore, to get back to the optimum condition one has to change to lower gears to decrease the load on the muscle and increase the speed of shortening.

7.8.7 Isotonic Eccentric Muscle Contractions

The Hill equation does not apply to isotonic eccentric muscle contractions—these being contractions where the muscle lengthens. If the force applied to a muscle is greater than the isometric tension of the muscle, then the muscle will lengthen. At first, the muscle lengthens slowly, but when the applied forces proceed beyond about twice the isometric tension (P_0), the lengthening is quite rapid. Although the muscle is lengthening, this is due to external forces overcoming the resistance put up by the muscle, and so is not indicative of any different active process produced by the muscle fibers per se, as there is no contractile process for muscle fiber elongation. Only through opposing muscle contractions, gravity, and elastic forces, or a combination of these processes, will muscle fibers return to their original length after a contraction. A lot of physical activities involve both concentric and eccentric isotonic contractions. For example, keeping your elbow on a table and lifting a weight with your hand involves a concentric contraction—as the force produced by the muscle during the flexion overcomes the resistance posed by the weight. However, when you lower the weight during extension, the *biceps brachii* muscles may still contract to some degree to help control the extension movement, the extension brought about mostly by gravity in this instance.

7.9 Muscle Energetics

Energetics essentially deals with energy. It involves studying the transformation and flow of energy. Hence, muscle energetics has to do with the transformation and flow of energy resulting from the contraction of muscles. In muscles, chemical energy is converted into mechanical energy (the work done by the muscle) and heat. In this section, this heat production will briefly be examined, as will be the chemical energy source (see energy metabolism). Chemical energy is the energy liberated in a chemical reaction or energy absorbed in forming a chemical compound. It is the energy stored in the chemical bonds of a substance.

7.9.1 Heat Production

Skeletal muscles perform many important and vital bodily functions, including producing movement and maintaining posture, as well as supporting and protecting internal tissues and organs. Another critical function of skeletal muscles is to help maintain body temperature via the heat dissipated during muscular contractions. At the onset of and during muscle contraction, energy is supplied by the hydrolysis of ATP, with one mole of ATP generating approximately 48 kJ of energy. For every ATP molecule split, this amounts to about 19 kT (where k is Boltzmann's constant of 1.381×10^{-23} JK^{-1} and T is absolute temperature expressed in units of Kelvin [K]). However, only about 40%–50% of this chemical energy is converted to mechanical energy (work done), with the remainder dissipated as heat, the latter specifically referred to as initial heat. This relationship can be summed up in the following equation:

$$-\Delta E = h + w$$

where ΔE is the chemical energy expended (the minus sign indicates an exothermic chemical reaction—one where heat is generated), h is the heat produced, and w is the work done by the muscle. All three forms of energy are measured in joules.

The heat produced as the result of a muscular contraction can be divided into initial heat and recovery heat, the latter relating to the heat produced after the relaxation of the muscle. Initial heat is the heat produced at the onset of and during muscle contraction, and can be subdivided into several categories. Activation heat represents the large amount of energy expended by the muscle immediately after stimulation, associated with the processes whereby the contractile elements within the muscle fibers are turned on. The energy produced by maintaining the muscle in an active contracting state is commonly referred to as maintenance heat, a state that immediately follows the activation heat state. Also, within the initial heat category is the heat of relaxation, which relates to the heat produced by the muscle during relaxation.

The above describes the initial heat of an isometric contraction. An isotonic shortening contraction also produces the above heat producing reactions, but there is also an additional heat produced, known as the heat of shortening. As the muscle shortens, extra heat is produced, in proportion to the distance shortened. This release of extra heat in proportion to muscle shortening, associated with the heat of shortening, is known as the Fenn effect. When relaxation is over there is a prolonged period where there is resynthesis of ATP to precontraction levels, resulting in heat production that is above normal resting levels, which is known as recovery heat.

7.9.2 Energy Metabolism

As already mentioned energy for muscle contraction is supplied by the hydrolysis of ATP, as represented by the following reaction:

$$\text{ATP} + \text{H}_2\text{O} \rightarrow \text{ADP} + \text{P}_i.$$

The above reaction ensures that the myosin head is cocked and ready to do a power stroke whenever the active sites on the actin filaments are exposed. The hydrolysis of ATP also provides the energy whereby calcium ions are transferred back into the sarcoplasmic reticulum lumen via the calcium pumps. Whilst difficult to quantify precisely, the energy demands of contracting muscles are huge. About 2500 molecules of ATP are split per second by every myosin filament during a contraction. Each sarcomere contains about 1000 myosin filaments, and a myofibril contains about 10,000 sarcomeres end to end (depending on the length of the muscle fiber), with an adult muscle fiber containing about 2000 myofibrils. Using the above approximate estimates indicates that there may be in the order of 20 billion myosin filaments in a muscle fiber. Hence, if every second of a skeletal muscle contraction consumes about 50×10^{13} ATP molecules for every muscle fiber that is active then energy demands of even small muscles, consisting of just a few thousand muscle fibers, are large.

Not all muscle fibers are activated during a contraction, and many motor units are activated on a rotating basis. However, even if only a small percentage of fibers are activated at any instant in, for example, a quadriceps muscle (which contains at least about a million muscle fibers), then the demand for ATP will be huge for any exercise that involves continuous use of this muscle. The problem is that muscle fibers store at best only a few seconds worth of ATP, and so muscle activity will soon cease and enter a state analogous to rigor mortis unless an alternative supply of ATP is accessed rapidly, as ATP is needed not only to provide the energy for the power strokes, but is also required for the actin and myosin filaments to detach immediately after the power stroke. There are fortunately other sources of ATP in the cells, in particular three different ways that the muscle fiber can synthesize ATP: phosphorylation of ADP by phosphocreatine, aerobic metabolism, and anaerobic metabolism.

7.9.2.1 Phosphorylation of ADP by Phosphocreatine (PC)

The process of obtaining ATP by phosphorylation of ADP by phosphocreatine (also known as creatine phosphate) involves tapping into the energy reserves that are stored in the molecule PC. This is achieved by phosphorylation of ADP, which essentially involves adding a phosphate group to ADP to yield ATP, as indicated in the following reaction:

$$\text{PC} + \text{ADP} \rightarrow \text{ATP} + \text{C}$$

where C is creatine. Creatine is a nitrogenous acid found in muscle tissue either free, or combined with phosphorous to form PC. PC is an organic molecule capable of storing energy, but in order for the muscle fiber to use this energy, the above chemical reaction first needs to take place. However, it is very much a reversal of the reaction whereby PC was formed in the first place:

$$\text{ATP} + \text{C} \rightarrow \text{ADP} + \text{PC}.$$

Usually in the resting condition, excess ATP is produced by aerobic metabolism, and some of this ATP is then used to create rapidly available energy reserves by the above chemical reaction, with the energy in the ATP transferred to creatine for storage. The mechanism of obtaining ATP from the reverse reaction involves the reaction being catalyzed by the enzyme creatine kinase (also known as creatine phosphokinase), whereby the high-energy phosphate group from the PC is transferred to ADP to produce ATP. Obtaining ATP via phosphorylation of ADP by phosphocreatine is a very rapid process, and allows the muscle to carry out another 10 s to 15 s of sustained contractions before this energy source is depleted also. If sustained activity is to be continued beyond the time that the stored PC and ATP are used, then ATP will need to be supplied mainly by anaerobic metabolism.

7.9.2.2 Anaerobic Metabolism

Anaerobic metabolism is also known as glycolysis, and is a process whereby ATP is derived from either glycogen reserves in the muscle fiber or glucose from the blood. In glycolysis glucose, either derived directly from the blood or by the breaking down of glycogen (a polysaccharide chain of glucose molecules), is converted to pyruvic acid by a chemical process that results in a gain of ATP. In this reaction, one glucose molecule yields two molecules of pyruvic acid. The actual process costs the cell two ATP molecules, but it later gains four ATP molecules, and so there is a net gain of two ATP molecules. If only moderate activity is required of the muscle, then the pyruvic acid becomes one of the energy sources (fuel substrates) used by the mitochondria to produce ATP via aerobic metabolism. Each pyruvic acid molecule yields an additional 17 molecules of ATP if used as an energy source in aerobic metabolism. Hence, a glucose molecule has the potential to yield 34 additional ATP molecules if the products of the glycolysis, the two pyruvic acid molecules, are also used in aerobic metabolism.

Anaerobic glycolysis comes into play when the available PC and ATP stores in the muscle fiber are exhausted, but intense activity (usually 70% or more of maximal possible rate) is still required of the muscle. Under these conditions the mitochondria is producing ATP at the maximum rate it can, but requiring oxygen for the process, and with the oxygen required now being far greater than that which can be supplied, aerobic metabolism can only supply about a third or so if the required ATP for maximum levels of activity. Under these conditions, glycolysis can supply the extra required ATP needed, for about another 60 s or so after the PC and ATP stores are exhausted, but after this time period, the muscle becomes fatigued because of the effects that lactic acid has on the muscle. When there is accumulation of pyruvic acid in the muscle fiber as a result of the mitochondria not being able to utilize it at the rate that it is supplied by glycolysis, then the pyruvic acid gets converted to lactic acid instead. Lactic acid is, along with pyruvic acid, a three carbon molecule, but contains an additional two hydrogen molecules. There are several explanations for the buildup of lactic acid. The most common explanation is that peak exercise leads to an excess of hydrogen, as there is insufficient oxygen to oxidize it all, and it is the acceptance of these extra hydrogen molecules by pyruvic acid that converts it to lactic acid.

The build of lactic acid in the muscle increases the acidity of the muscle, as evidenced by the accompanying lower pH levels, eventually resulting in the muscle suffering fatigue. Muscle fatigue relates to the inability of a muscle to contract, even though it may still be receiving stimuli in the form of action potentials. Although many specifics of muscle fatigue are uncertain, raising the concentration of hydrogen ions may cause conformational changes in certain muscle proteins, which may alter their activity. Proteins thought to be adversely affected as a result include the calcium release pump, as well as actin and myosin. Also, intense exercise of short duration may result in failure of the excitation–coupling mechanism by, for example, the buildup of potassium in the extracellular fluid of the T-tubules to the extent that there is a disturbance in the membrane potential. It may also be possible that the ADP and P_i accumulation in the sarcoplasm during peak exercise may slow the speed of cross-bridge cycling.

Once the intense exercise is over the muscle then has to get rid of the lactic acid that has accumulated, as well as replenish the stores of PC and glycogen. The lactic acid diffuses out of the muscle fibers and into the blood, and subsequently ends up in the liver. In the liver, the lactic acid is converted back to pyruvic acid. Part of this pyruvic acid is then metabolized by the mitochondria to yield ATP, with this ATP then being used as the energy source to convert the remaining pyruvic acid back into glucose. After this, the glucose ends up back in the bloodstream and subsequently in the muscle fibers, where it provides the building blocks for replenishing the glycogen stores. ATP and PC levels are restored by aerobic metabolism in the muscle fibers after the intense exercise has ceased. Aerobic metabolism also supplies the ATP needed to reconstitute the glycogen reserves from glucose. Hence, oxygen consumption in the immediate postexercise period will be greater than normal resting levels until the muscle is restored to its preexercise state. Although useful in the short term, anaerobic metabolism is an inefficient way of generating ATP from glucose, with only two molecules of ATP the net gain from one glucose molecule, as opposed to 36 molecules of ATP that can be generated via aerobic metabolism of glucose. In both instances, glucose is broken down into two molecules of pyruvic acid, but with anaerobic metabolism, the pyruvic acid gets converted to the inconvenient by-product lactic acid, whereas in aerobic metabolism, it yields an additional 34 molecules of ATP.

7.9.2.3 Aerobic Metabolism

When at rest or during moderate levels of exercise, most of the ATP in the muscle fiber is provided through aerobic metabolism. In the resting condition, fatty acids absorbed from the circulation are broken down in the mitochondria in order to generate the ATP. However, when moderate exercise is undertaken the demand for ATP is greater and so, in addition to fatty acids, pyruvic acid (obtained from glycolysis) is now also destructively metabolized in order to produce ATP. At this stage, oxygen is still abundant enough to prevent any buildup of lactic acid. Inside the mitochondria, the fuel substrates (whether fatty acids, pyruvic acid, or even amino acids) are acted upon by enzymes so products of the reaction can enter the tricarboxylic acid (TCA) cycle (also known as the Krebs cycle or the citric acid cycle). The TCA cycle involves a series of enzymatic reactions of these fuel substrates, that when coupled with a system known as the electron transport system, serves to generate ATP. In moderate, long-duration exercise depletion of fuel substrates, such as glycogen stores, are correlated with the onset of fatigue. Also, dehydration and low blood glucose are known to exacerbate fatigue. At peak exercise, aerobic metabolism cannot keep up with the demand for ATP, mainly due to insufficient oxygen supply, and so, anaerobic metabolism tends to predominate during periods of intense activity.

7.10 Types of Muscle Fibers

Classifying skeletal muscle fiber types is not necessarily a straight forward issue, as there are several different classification techniques, and agreement is not always found between these techniques. Histochemical, morphological, biochemical or physiological characteristics can be used to describe and classify muscles. In this overview of the subject, muscle fibers will be categorized as slow, intermediate, and fast muscle fiber types; classification into these three types being generally accepted.

7.10.1 Slow versus Fast Muscle Fibers

Other names for slow fibers include slow oxidative fibers, slow-twitch fibers, slow-twitch oxidative fibers, red fibers and type I fibers, whereas alternative names for fast fibers include fast glycolytic fibers, fast-twitch fibers, fast-twitch glycolytic fibers, white fibers and type IIb fibers. In regards to intermediate fibers, these are also referred to as fast oxidative fibers, fast-twitch oxidative fibers and type IIa fibers. As many of the characteristics of intermediate fibers lie somewhere between fast and slow fibers, it is more instructive to compare fast and slow fibers.

The speed of contraction is greater in fast fibers compared to slow fibers. In an individual twitch, peak force is achieved earlier in fast fibers, and the rate of relaxation is more rapid in the twitch of fast fibers. The differences in the speed of contraction are the result of differences in various proteins of the muscle fibers, such as contractile, regulatory, and calcium binding proteins. These proteins are known as isoforms, being similar but not identical in amino acid sequence. In regard to tetanic stimulation of muscle fibers, the fusion frequency is known to be higher in fast fibers. Whilst the average density of myofilaments (actin and myosin filaments) in cross-sections of muscle fibers does not differ significantly, fast fibers have more of these myofilaments as they have larger diameters. The size of fiber diameter correlates with tension that the muscle can develop, with fast fibers capable of generating greater tension. Capillaries surrounding slow fibers are significantly more in number than those that surround fast fibers, as are the number of mitochondria. The red color of the slow muscle fibers is due to it being surrounded by numerous capillaries, as well as the slow fibers having high myoglobin content. Myoglobin is a red iron-containing protein pigment, similar in function to hemoglobin, which stores oxygen in muscle fibers that will be used in the mitochondria to produce energy. Being low in both of these characteristics, fast fibers appear white or pale.

In terms of cross-bridge cycling, which is essentially determined by the rate the myosin head of the muscle fiber splits ATP (ATPase activity), fast fibers cycle at about four times the rate of slow fibers. The rate of ATPase activity of the predominant muscle fiber type in a muscle correlates positively with the maximum velocity of shortening of the whole muscle. Cross-bridge cycling also correlates with the amount of ATP used, with the fast fibers, which cycle much quicker, having greater demand for ATP when they contract compared to slow fibers. Slow fibers tend to be predominant in muscles such as, for example, the soleus muscle of the calf. The soleus muscle is heavily used in endurance activities, such as running, and slow fibers are much better suited to this. A muscle in which fast fibers predominate is the gastrocnemius, also of the calf. The gastrocnemius muscle is involved in more explosive activities, such as jumping, that fast fibers are much better at. However, nearly all muscles are composed of both fast fibers and slow fibers, although the ratio of these fibers differ in different muscles, and based on heredity the ratio can also differ in the same muscle between different people.

7.10.2 Motor Unit Types and Muscle Fiber Plasticity

Recall that a motor neuron, along with all the muscle fibers it innervates, is referred to as a motor unit. Muscle fibers within a motor unit tend to always be of the same type such that slow-twitch (S) motor units consists of type I slow fibers, fast fatigue resistant (FR) motor units consist of type IIa intermediate fibers, and fast fatiguable (FF) motor units consist of type IIb fast fibers. However, a muscle usually consists of a mixture of different types of motor units that are intermingled. Hence, a fast fiber of a FF motor unit may be positioned alongside slow fibers from an S motor unit. Motor units are usually recruited in order, with the smaller motor units first (usually type S), followed by type FR, and lastly the larger type FF. This order of recruitment also generally reflects the order of power in the muscles, with the type FF motor units producing the greatest tension.

The growth in size of a muscle (hypertrophy) is not caused by an increase in the number of muscle fibers, but an increase in the size of existing muscle fibers, as muscle fiber numbers remain pretty much unchanged in adults during their lifetime. Endurance training produces a slight decrease in diameter. Hypertrophy affects primarily fast fibers, as a result of high intensity exercise, such as weight lifting. More specifically, hypertrophy is caused by an increase in the number of actin and myosin filaments in the muscle fiber. This causes the myofibrils within the muscle fibers to enlarge, as the extra myofilaments are incorporated into the existing myofibrils, leading to larger-diameter muscle fibers. It is also thought that the myofibrils may split as a result of high intensity exercise, caused by disruption of the Z lines, with the resultant two myofibril fragments growing back to their original size. Hence, hypertrophy of a muscle that leads to increased force of contraction is thought to be caused by a combination of more myofilaments in existing myofibrils and more myofibrils per se.

Not only can a muscle fiber change in size in response to, for example, exercise, but there is also evidence of conversion from one type to another type. This is most common between type IIb and type IIa fibers. During endurance training, type IIb fibers can transform into type IIa fibers. During severe deconditioning or injury to the spinal cord, type I to type II conversions are possible, although evidence for the reverse conversion (type II to type I) is lacking, except in chronic electrical stimulation experiments, which show that such conversions are possible. Table 7.1 compares characteristics of slow and fast skeletal muscle fibers.

TABLE 7.1

Comparing Characteristics of Slow and Fast Skeletal Muscle Fibers

Characteristic	Slow	Fast
Primary ATP synthesis	Aerobic (oxidative phosphorylation)	Anaerobic glycolysis
Fusion frequency	Low	High
Contraction speed	Slow	Fast
Fiber diameter	Smaller	Larger
Density of myofilaments in cross-sectional area	Similar to fast	Similar to slow
Tension developed	Low	High
Capillaries	Many	Few
Mitochondria	Many	Few
Color	Red	White
Myoglobin	High	Low
Fatigue rate	Resistant to fatigue	Fatigues quickly
Cross-bridge cycling	Slow	Fast
ATPase activity	Slow	Fast
Glycogen stores	Low	High
Sarcoplasmic reticulum calcium pumping	Moderate	High

7.11 Development of Skeletal Muscle

The formation of muscle cells (fibers) is referred to as myogenesis. Undifferentiated cells called myoblasts, derived from the mesoderm, are the precursor of muscle cells. There are three primary embryonic germ layers, the middle of which is the mesoderm. All embryos initially form these three cellular germ layers, the other two layers being the ectoderm and the endoderm. After this, the organs and tissues of the body then form from these three germ layers—a process that involves cells becoming distinct or specialized. Myoblast cells arise from premyoblastic cells in the dermamyotome. The dermamyotome is a term used to describe cells in the dorsolateral region of a segmental mass of mesoderm called the somite. Somites occur in pairs, and begin development at the cranial end of the neural groove (a midline indentation that gradually deepens until the dorsal end fuses to form the neural tube—the latter a hollow structure that gives rise to the central nervous system), and add sequentially in the caudal direction on either side of the neural groove and notochord (a structure important in the organizing of the neural tube). The premyoblastic cells begin to differentiate into myoblasts at 4 to 5 weeks of gestation.

Soon after they are formed, myoblast cells migrate to form a tissue mass called the myotome in the center of the somite. The myoblast cells located in the myotome then differentiate within to form the axial musculature. Other myoblast cells migrate away from the dermamyotome to the lateral regions of the embryo, where they form the musculature of the limbs and body wall (hypaxial musculature). Whether in the myotome or in the lateral regions, myogenesis proceeds in a similar way. The single nucleus myoblasts become spindle-shaped and then assemble together and fuse to form multinucleate myotubes (see Figure 7.29), which are developing muscle fibers that take on a tubular shape and attach at both ends to tendons and the developing skeleton. Unfused myoblasts form the satellite cells in muscle tissue, and are sandwiched between the basement membrane and the

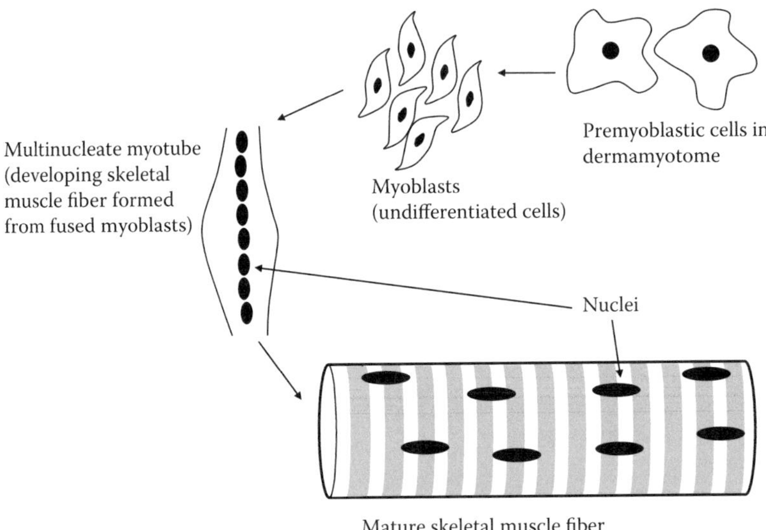

FIGURE 7.29
Some key events during skeletal muscle development.

plasma membrane of the sarcolemma. Satellite cells are relatively inactive until stimulated into action by injury to the muscle, whereby its actions assist in repairing damage to the muscle tissue. The initial formation of most muscle groups have occurred by 9 weeks of gestation, with contractile proteins being synthesized, as well as the beginnings of neuromuscular junctions observed. Differentiation of muscle fibers is completed by the end of gestation, or thereabouts, with no new muscle fibers formed from myoblasts after this, although in the case of injury they can later be generated with the assistance of satellite cells. However, existing fibers will grow in size from birth to adulthood as the body develops.

7.12 Electromyography

Electromyography (EMG) is the technique associated with measuring the electrical activity associated with skeletal muscle. Electric potentials measure the work that an electric field has to perform in order to move electric charges, and are expressed as volts. As such, the EMG is the summation of electric potentials from many activated muscle fibers, with the contribution to the potential from each muscle fiber being related to its distance from the recording electrodes. The EMG amplitude most strongly reflects the muscle fibers closest to the electrodes. More specifically, the EMG measures the effect that changing concentrations of ions around the muscle fibers has at where the electrodes are located. The concentration of ions around a muscle fiber membrane changes considerably when an action potential travels across it, and it is the summation of action potentials from all firing muscle fibers that yield the interference type pattern of the EMG signal. That is, waves of depolarization spreading out over many muscle fibers (this being the movement of the action potentials) are recorded at a distance as a difference in electrical potential between two electrodes placed over the muscle (if a bipolar recording), resulting in the EMG signal.

Monopolar recordings measure the difference in potential of an electrode placed over the muscle with respect to an electrode connected to ground.

However, muscle fibers generally do not fire in isolation, but in conjunction with all other members of the same motor unit. Instead of action potential, the name given to the electrical potential measured as a result of the summed activity of all fibers in a motor unit firing is motor unit action potential (MUAP). Hence, usually the EMG signal is composed of the sum of all the MUAPs activated, although EMG signals from individual muscle fibers also occur (for example, myotonic discharges and end-plate spikes). Two general types of EMG are recognized. When the EMG is obtained by invasively inserting a needle electrode into the muscle, it is most commonly known as needle EMG. On the other hand, if electrodes are placed noninvasively on the skin over the muscle (or muscles), then it is referred to as surface EMG. Both types of EMG have detailed procedures for performing recordings, and analyzing the signal, but describing these methods is beyond the scope of this chapter.

Most of the time spent carrying out needle EMG examinations involve assessing MUAPs from contracting muscles, usually in order to assist in the clinical evaluation of patients with possible neuromuscular disorders. Although the presence of single MUAPs can be detected using surface electrodes, to see the fine details requires the insertion of needle electrodes. Although similar in shape, the duration of the MUAP, typically between about 5 ms and 15 ms, is usually longer than the action potential of a single muscle fiber (usually between about 1 ms and 5 ms). As shown in Figure 7.30, the MUAP has certain characteristics. The duration is the time period from the initial deflection of the MUAP away from baseline up until it finally returns to baseline, and is the parameter that most accurately reflects the number of muscle fibers and area of the motor unit. Hence, a longer duration MUAP usually reflects a motor unit with more muscle fibers and area than a shorter MUAP. The amplitude typically only reflects fibers close to the electrode, and is a measure of the amplitude (peak-to-peak) of the main spike. As such, increased amplitude is associated with larger-diameter muscle fibers, a greater density of muscle fibers in the region, and greater synchrony of firing of muscle fibers in the vicinity. MUAPs usually have amplitude between 100 μV and 2 mV. Phases are the number of times the MUAP crosses the baseline plus one, whereas turns are changes in the direction of the MUAP waveform

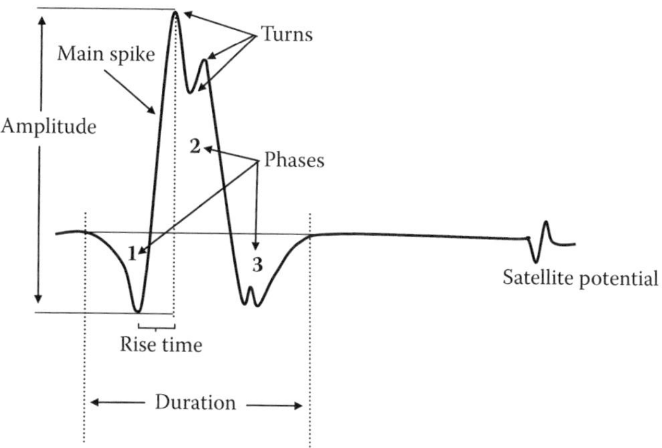

FIGURE 7.30
Motor unit action potential (MUAP). (Daube, J.R. and Devon, I.R. *Muscle and Nerve*. 2009. Vol. 39. p. 255. Copyright Wiley-VCH Verlag GmbH & Co. KGaA. Reproduced with permission.)

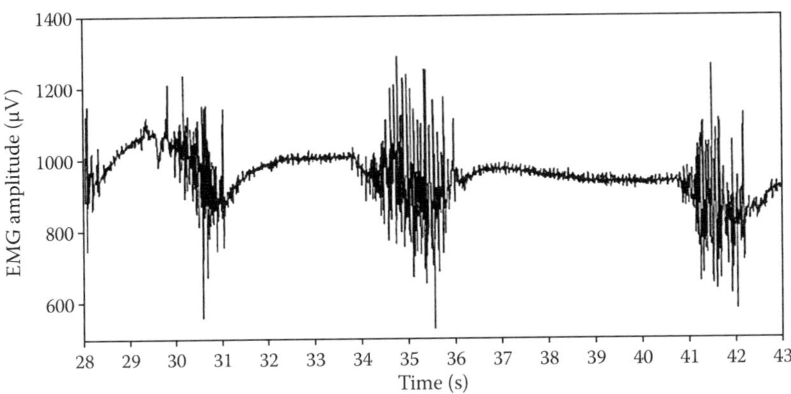

FIGURE 7.31
Bursts of recorded surface EMG.

where the baseline is not crossed. Both phases and turns indicate the same thing, with an increase in the number of phases and/or turns reflecting less synchronous firing of the muscle fibers within the motor unit associated with the MUAP. Most MUAPs normally consist of two to four phases. The duration of the main spike positive-to-negative MUAP component is known as the rise time, and it indicates the proximity of the needle electrode to muscle fibers in the motor unit. An increasing rise time indicates the electrode is more distant from any fibers in the motor unit. Only MUAPs with a rise time of less than 500 μs are thought to indicate suitable enough electrode placement for proper analysis of the MUAP waveform. Satellite potentials represent early-reinnervated muscle fibers that are time-locked potentials lagging the MUAP.

Although easier to perform than needle EMG, surface EMG (see Figure 7.31) lacks the fine resolution of needle EMG, being limited more to investigating the general activity of the entire muscle. The size of the surface EMG varies considerably, from about 1 μV to over 1 mV. The former amplitude involves small signals, such as single motor unit responses from small muscles in a relaxed state, whereas much larger EMG signals are recorded during isometric contractions or movement involving a large muscle or muscle group. There is a close positive correlation between the tension in an isometric contraction and the size of the surface EMG. To quantify this correlation the EMG signal is usually rectified and integrated. The frequency range of the EMG signal is very broad, from about 1 Hz to 1000 Hz, or even higher. Hence, the requirement of the EMG recording system can differ considerably; depending on what part of the EMG signal is to be measured. The EMG signal can be examined throughout movement. By using high gain amplifiers, and sampling data at high speed, the resultant EMG signal can also be used to indicate the tension generated in a muscle before the actual beginning of movement.

7.13 Comparison of Skeletal, Cardiac, and Smooth Muscle

One of the distinct characteristics of skeletal muscle fibers (cells) is the length that some of them attain (up to 30 cm), whereas cardiac muscle cells (also called cardiocytes) and smooth muscle cells are usually no longer than 100 μm and 200 μm, respectively. The diameter/width of skeletal muscle fibers (10 μm–100 μm) tends also to be greater than that of cardiocytes (10 μm–20 μm) and smooth muscle cells (2 μm–10 μm). Another distinct feature of skeletal muscle fibers is that they have multiple nuclei, whereas cardiocytes and smooth muscle cells usually contain only one centrally placed nucleus. Occasionally, cardiocytes may have more than one nucleus. In skeletal muscle fibers the nuclei are located just underneath the plasma membrane, and in a small-diameter fiber there is estimated to be about 200–300 nuclei/mm of muscle fiber length. As for external appearance (as viewed through a light microscope), skeletal and cardiac muscle fibers appear striated, due to the structural organization of the myofilaments within the sarcomeres, whereas smooth muscle is nonstriated. Smooth muscles are not organized into myofibrils and sarcomeres but instead form an irregular network throughout the cell, with the actin filaments attached to structures called dense bodies. In smooth muscles, intermediate filaments and dense bodies form a type of cytoskeleton in the cytoplasm. There are also no T-tubules in smooth muscle (only caveolae), whereas T-tubules are present in both skeletal and cardiac muscle fibers. The shape of smooth muscle is spindle like—tapering off at the ends (see Figure 7.32). Skeletal muscle and cardiocytes have cylindrical shapes. However, the smaller in size cardiocytes connect to other cells longitudinally and may also be branched, whereas skeletal muscle fibers (excluding intrafusal muscle fibers in muscle spindles which are anchored at both ends to either the muscle spindle capsule or muscle connective tissue) attach to tendons or aponeurosis (a wide tendinous sheet). As such, skeletal muscles are closely associated with the skeletal system. Cardiac muscle is located in the walls of the heart, whereas smooth muscle are located within walls of visceral organs such as digestive, respiratory, urinary, and reproductive tracts, as well as blood vessels, iris of eye, and piloerector muscles of hair follicles.

Gap junctions, which are membrane protein linkages between two adjacent cells that form channels uniting the cytoplasm of the two cells, are present in single-unit smooth muscle fibers, as well as at the intercalated discs (where adjacent cardiac muscle cells connect together) of cardiocytes, but absent from skeletal muscle fibers. Smooth muscle cells are categorized into two major types. Single-unit smooth muscle fibers (also known as visceral smooth muscle) are connected via gap junctions, and so an action potential in one, or only a few fibers, is enough for the electrical activity to spread so as to cause contraction in the entire sheet of tissue. Multiunit smooth muscle cells are innervated more similarly to skeletal muscle fibers, with neurotransmitters released from axon terminals

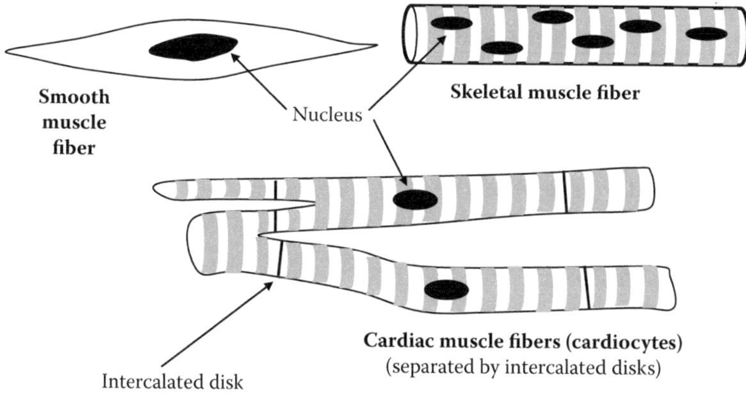

FIGURE 7.32
Comparison of skeletal, cardiac, and smooth muscle.

(or varicosities), except that a smooth muscle fiber may be closely associated with more than one motor neuron. Also, several multiunit smooth muscle cells may be influenced by the neurotransmitters released by one motor neuron. Gap junctions in multiunit smooth muscle are rare. A feature present in cardiocytes and smooth muscle cells (in single-unit muscle), but not in skeletal muscle, is autorhythmicity. Essentially, autorhythmicity is the ability to contract independently and rhythmically. In the case of the heart, an in-built pacemaker triggers a wave of depolarization that leads to contraction of the cardiocytes in the walls of the heart chambers. Cardiac muscle is, however, innervated by motor neurons from the autonomic nervous system, which can alter (increase or decrease) the rate of pacing (heart rate) by the pacemaker cells, as well as the force of contraction. Smooth muscles are also innervated by the autonomic nervous system. Both cardiac muscle and smooth muscle can be influenced by hormones. In terms of energy metabolism, skeletal muscle fibers rely primarily on aerobic metabolism at rest and during moderate levels of activity,

TABLE 7.2

Comparing Characteristics of Skeletal, Cardiac, and Smooth Muscles

Characteristic	Skeletal Muscle	Cardiac Muscle	Smooth Muscle
Nuclei	Multiple	Usually single	Single
Cell size, length	100 μm–30 cm	50 μm–100 μm	30 μm–200 μm
Cell size, width	10 μm–100 μm	10 μm–20 μm	2 μm–10 μm
Gap junctions	Absent	Present at intercalated discs	Present in single-unit smooth muscle, but rare in multiunit
Contraction speed	Fastest	Intermediate	Slowest
Calcium source	Sarcoplasmic reticulum (SR)	Extracellular fluid and SR	Extracellular fluid and SR
Calcium regulation	Troponin on actin filaments	Troponin on actin filaments	Calmodulin on myosin heads
Hormonal influence	None	Adrenaline	Multiple
Neural control	Somatic motor division	Autonomic nervous system	Autonomic nervous system
Contraction initiated by	Motor neuron releasing acetylcholine (ACh)	Automaticity (pacemaker cells)	Automaticity, chemicals, stretch
Autorhythmicity	No	Yes	Yes—single-unit smooth muscle
Source of energy	Aerobic and anaerobic	Aerobic	Mostly aerobic
External appearance	Striated	Striated	Nonstriated
T-tubules	Present	Present	Absent
Effect of motor neuron stimulation	Excitatory	Excitatory or inhibitory	Excitatory or inhibitory
Cell shape	Long and cylindrical	Branching	Spindle-shaped
Location	Associated with skeletal system	Walls of the heart	Walls of viscera, blood vessels, iris of eye, etc.
Attachment of actin filaments	To Z line	To Z line	To dense bodies
Can be tetanized?	Yes	No	Yes
Relative fatigability	Variable, but relatively fast	Resistant	Resistant
Organization of myofilaments	Sarcomere structure	Sarcomere structure	No sarcomere structure
Terminal cisternae	Present	Absent	Absent

but are predominantly anaerobic (glycolysis) at high levels of exertion. Cardiocytes rely almost exclusively on aerobic metabolism, whereas with smooth muscle the energy is supplied mostly by aerobic metabolism. Cross-bridge cycling occurs much more slowly in smooth muscle compared with skeletal muscle.

Whereas skeletal and smooth muscle can be stimulated at high frequencies to tetanus, where the tension increasingly builds until a maximum level is reached, no such wave summation and tetanic contractions occurs in cardiac muscle. In cardiocytes the refractory period ends around the time the muscle has nearly completely relaxed, with the refractory period (about 150 ms to 300 ms) about as long as the contraction period (200 ms or more), and so wave summation and tetanus does not occur. If it did, the heart would not pump blood effectively, if at all. The shorter refractory period is for atrial muscle, and the longer period is for ventricular muscle fibers. In skeletal muscle fibers, the refractory period is about 1 ms to 2 ms in duration, with a much longer relative contraction period (total time of twitch contraction) of about 7 ms to 100 ms. With skeletal muscle fibers, the action potential duration is about 1 ms to 5 ms, compared to about 200 ms to 300 ms or more for cardiocytes. In smooth muscle, the action potentials can vary from 10 ms to 50 ms in spike action potentials, and up to 1000 ms in plateau action potentials. In terms of contraction speed, skeletal muscle contracts the fastest of the three muscle different muscle types, followed by cardiac muscle, which takes roughly 10 times as long to contract. The contraction period of smooth muscle is about 30 times as long when compared to skeletal muscle. Another feature of difference is that in skeletal and cardiac muscle, calcium regulation that leads to contraction is by troponin and its effect on actin filaments, whereas in smooth muscle, it is by calmodulin and its effect on myosin heads. In skeletal muscle, the source of calcium for contraction is the sarcoplasmic reticulum, whereas in cardiocytes and smooth muscle, it is from both the extracellular fluid and sarcoplasmic reticulum. Table 7.2 compares characteristics of skeletal, cardiac, and smooth muscles.

7.14 Muscle Diseases

Muscle diseases caused by defects in the muscle itself, particularly if it is skeletal muscle, are known as myopathies. As such, these myopathies are primary, as the disease originates in the muscle itself. Factors that affect the function and health of muscles may also be caused by diseases or abnormalities of the nervous system (e.g., motor neuron disease), known as neuropathies, or it may be the immune system which is the primary cause of the disorder (e.g., myasthenia gravis). In such instances, the myopathy is secondary to the main cause of the disorder. There are two general categories of myopathies, those where the muscle fibers shrink in size but are not destroyed (atrophic myopathies), and those where the muscle fibers are destroyed (destructive myopathies). The causes of muscle diseases can be acquired, such as with infection by viruses or parasitic worms (as in the case of trichinosis), or be genetic in origin, with the muscular dystrophies being examples of the latter. Myopathies may be indicated by problems such as fatigue, muscle weakness, cramps, atrophy, pain, stiffness, and involuntary movements, although it should also be considered that some of these symptoms may ultimately result from a disease of the nervous or immune system, rather than the muscle tissue per se. Here, a few of the better-known muscle diseases will be briefly considered, regardless of whether the disorders are intrinsic to the muscle itself, or secondary to nervous or immune system disorders.

Muscular dystrophies are myopathies caused by genetic errors in genes that are important in muscle function, and are progressive disorders where muscle fibers are destroyed and replaced with fat and fibrous tissue. Of the muscular dystrophies, Duchenne muscular dystrophy (DMD) is probably the best known, as well as the most serious type. DMD is X-linked recessive in its method of inheritance, with two years the mean age of onset, and usually only occurring in males. The gene affected codes for the protein dystrophin, with the mutation causing an absence of the protein. Dystrophin is an important structural protein that helps stabilize the sarcolemma by linking thin filaments to the anchoring proteins of the sarcolemma. Symptoms include progressive muscle weakness, with death often occurring between the age of 15 years and 30 years caused by cardiac problems and respiratory paralysis. In the less severe Becker muscular dystrophy, the dystrophin protein is deficient but still produced. Hence, whilst similar features are observed as in DMD, Becker muscular dystrophy progresses more slowly, allowing patients to live considerably longer (mean age at death being 42 years).

Autoimmune diseases occur when the immune system creates antibodies to attack the body's own tissues. One common autoimmune disease is *myasthenia gravis*, where antibodies are produced that attack and destroy the nicotinic acetylcholine (ACh) receptors at the motor end plate of the neuromuscular junction in skeletal muscle fibers. Although the release of acetylcholine neurotransmitters may be normal, in myasthenia gravis there are fewer receptors for the ACh to bind to. Hence, adequate end-plate potentials may not be generated at the motor end plate when it is stimulated by a motor neuron, and so no action potential is produced. The end result is that the muscle does not contract, or extraordinary effort is needed to perform relatively simple movements. Drooping of the eyelids, difficulty in talking and swallowing, general muscle weakness, and abnormally high fatigue after moderate exercise are characteristics of myasthenia gravis. There is no known cure, although symptoms can be temporarily controlled to some degree by drugs. For example, acetylcholinesterase inhibitory drugs, such as neostigmine, allow the functional ACh receptors remaining to be activated longer, allowing for larger end-plate potentials to be produced. If severe enough, the patient may die of paralysis of the respiratory muscles.

Trichinosis is a myopathy caused by infection of the muscle tissue by the larvae of the parasitic worm *Trichinella spiralis*. Human infection usually takes place through eating undercooked parasitized pork, or through eating the meat of more exotic animals, such as walrus or bear. After ingestion, the larvae develop rapidly into adult worms in the intestinal tract. Subsequently, a second generation of larvae are produced which then invade many body tissues (including muscles), using the systemic circulation to assist in the migration. In muscle tissue, the larvae grow to reach a size of up to 1000 μm in length and about 30 μm in width. Once established in the muscle the larvae become enclosed in a cyst, remaining viable for several years, with muscle fibers penetrated by the larvae, as well as some adjacent ones, ending up being destroyed. Muscle pain, weakness, swelling of infected muscle tissue, as well as diarrhea, are among the symptoms of this disorder. However, muscular symptoms may not be evident unless the larvae count reaches about 100 per gram of tissue. Most affected are the muscles of the eyes, tongue, diaphragm, chest, and legs. Treatment is carried out using the drugs thiabendazole and corticosteroids concurrently.

Motor neuron diseases (MND) are generally progressive and degenerative neuropathies of unknown origin. There are several different forms of MND, with the most common variant known as amyotrophic lateral sclerosis (ALS). Symptoms of the disease include muscle weakness, atrophy, fasciculations, spasticity, and cramps. Being a neurodegenerative disease, involving the progressive destruction of fibers in the pyramidal tract and motor neurons in the ventral horn of the spinal cord, the prognosis of patients with ALS is bleak. ALS, also known as Lou Gehrig's disease (after the famous US baseball player), occurs in the 55- to 60-year age group most frequently, and most patients die within 3 to 5 years of diagnosis. Death commonly occurs as a result of respiratory difficulties.

Inflammatory myopathies include acquired skeletal muscle diseases of manifold etiology and development. The most commonly seen types are classified as idiopathic (a disease with unknown cause), and include inclusion body myositis (IBM), polymyositis (PM) and dermatomyositis (DM), with the latter two considered variants of the same disorder by many. The main symptoms of these inflammatory myopathies include loss of muscle tissue and muscle weakness, particularly in the limb-girdle and proximal limb muscles. In DM there is usually also skin rashes associated with the disorder, particularly of the eyelids, face, and hands. DM tends to be remitting in terms of progression after several years, but the availability of treatment with steroids and other antiinflammatory drugs means that it is rare for the full history of the disease to unfold. Being somewhat acutely acquired, patients with DM are likely to seek treatment early, although the age of onset varies considerably (20 years to 80 years). With PM, the disease declines in severity after many years, but if left untreated during this period, the patient may be left with a considerable residual muscle weakness. PM has a more chronic course compared with DM, with patients only likely to seek medical assistance a year or more after symptoms appear, as opposed to weeks with DM. The onset of PM peaks between 30 and 60 years of age, and drug treatment is similar to that of DM. In IBM peak offset is between 40 and 60 years of age, and because the muscle weakness develops very gradually it may take several years for a patient to seek medical opinion about their symptoms. IBM progresses continually, albeit very slowly, with the patient unable to walk within about 20 years after onset. Unlike PM and DM, there is no effective treatment for IBM, and there is even doubt as to the disease representing a primary inflammatory myopathy.

Tetanus is a preventable disease, through immunization, where a protein neurotoxin released by bacteria, which has invaded human tissue through a wound, causes painful skeletal muscle spasms. The bacterium is known as *Clostridium tetani*, and the toxin it releases is called *tetanospasmin*. The toxin manages to be absorbed into somatic motor neurons at the neuromuscular junctions, and is subsequently transported by retrograde transport to the central nervous system (CNS). In the CNS, the neurotoxin is taken up by inhibitory interneurons and once inside causes the release of inhibitory neurotransmitters to be blocked, so that the inhibitory control mechanism of motor neurons is no longer functional, resulting in sustained and uncontrolled contractions of the body. Also known as lockjaw, the muscle spasms usually start with the jaw and then spread throughout the body within 2 to 3 days of the first symptoms appearing. The muscle spasms last for about a week, and the mortality rate is about 50% if left untreated. Treatment is available for unimmunized patients, involving tetanus antitoxin to deactivate any remaining toxin and antibiotic therapy to kill the bacteria. Although in some patients muscle relaxant drugs and sedatives are enough, most patients have to be paralyzed and ventilated. Efficient treatment can reduce the mortality rate to about 10%. Those that survive the disease usually have no long-term aftereffects.

In many ways, this brief excursion into muscle diseases has only touched the tip of the iceberg, so to speak. It is a very large area, and to cover it adequately would require a separate dedicated volume entirely. For a much more comprehensive treatment of the subject, the book edited by Walton, Karpati, and Hilton-Jones (*Disorders of Voluntary Muscle*, 6th edition, Churchill Livingstone, New York, 1994) is recommended.

Study Questions

1. The contractile filaments in an average length muscle fiber consume in the order of about 50×10^{12} ATP molecules for every second they are activated. Given that the hydrolysis of one molecule of ATP on the myosin head generates approximately 19 kT, calculate how much chemical energy is converted to mechanical energy in 3 s of sustained contraction by the filaments if 40% of the chemical energy is dissipated as heat. The temperature in the sarcoplasm is 310 K.
2. (a) Sarcomeres at rest are about 2.5 μm long. A resting muscle fiber is measured to be 20 cm long. How many sarcomeres are there in each myofibril of this muscle fiber?
   (b) Each sarcomere contains about 1000 myosin filaments, with each myosin filament consisting of about 350 myosin molecules. If the above muscle fiber contains about 2000 myofibrils, roughly how many myosin molecules are there in the fiber?
3. The mean innervation ratio of the *gastrocnemius, medial* muscle of the calf is 1800. Given that the number of muscle fibers it has is 1,170,000 and that there are about 10,000 sarcomeres in each of the 100 myofibrils of each of these muscle fibers, how many motor units are there in this muscle?
4. The cat soleus muscle is composed of 28,000 muscle fibers. A section of the nerve supplying the soleus gives a total count of 450 myelinated axons of which 60% are motor. Of the motor axons, 30% supply muscle spindles and therefore do not contribute to the force developed by the contracting muscle. What is the mean innervation ratio for this soleus muscle?
5. A muscle is composed of 28,000 muscle fibers that are all of similar size and type. Assuming that the force developed by a motor unit is proportional to the number of muscle fibers it contains, calculate the number of muscle fibers in the smallest and largest motor units (30 mN and 390 mN) if the maximum tetanic force developed by the muscle is 26.5 N.
6. The force a muscle can develop is usually expressed not simply as so many newtons, but as the force per unit area. An estimate of the cross-sectional area of the muscle can be obtained if muscle weight and length are known. Assuming a density of 1 g/cm^3 for muscle tissue, the volume is calculated; from this and muscle length the cross-sectional area is calculated. It is assumed here that the area is uniform along the whole length of the muscle. For a soleus muscle 0.075 m long and weighing 0.003 kg, calculate the force per unit area, given that the maximum tetanic force developed by the muscle is 26.5 N, and that in a muscle such as soleus the fibers run only 50% of the whole length.
7. (a) A muscle is 50 mm in diameter. Each muscle fiber is 50 μm in diameter. Approximately how many muscle fibers are there in the muscle?
   (b) There are 2000 motor axons innervating the muscle. What is the average innervation density?
   (c) The muscle exerts a maximum tetanic tension of 80 kg. What is the maximum tension per cm^2 muscle?
   (d) What is the average maximum tension per muscle fiber?
8. Calculate the vertical power output of a person of mass 72 kg running at top speed (2.4 m/s) up a single flight of stairs (height, 3.2 m).

9. If a 20-cm-long parallel muscle contracts fully, what will be its postcontraction length?
10. The resting potential of a muscle fiber is –90 mV, with the threshold to trigger an action potential being –55 mV. What would be the minimum number of simultaneous MEPPs expected to generate a muscle action potential?
11. A patient is complaining about muscle weakness, with intense effort required to perform even simple tasks. Upon examination, the alpha motor neurons in the patient's muscle function normally, with the axon terminals releasing normal amounts of acetylcholine into the synaptic clefts. However, no contraction is observed. With the administration of an acetylcholinesterase inhibitory drug, the muscles are now observed to contract when stimulated. What disorder is the patient likely suffering from and why?
12. A motor unit action potential (MUAP) has an amplitude of 1.8 mV, two phases, a rise time of 300 μs, and a duration of 14 ms. What do these parameters indicate about the muscle fibers within the motor unit associated with the MUAP?

Skeletal Muscle Mechanics with Emphasis on Hill Equation

13. Given that $P_0 = 2500$ N and $V_{max} = 0.11$ m/s, calculate the maximum power output of this muscle (in watts).
14. If $P_0 = 775$ N and $a = 0.30P_0$, what load (P) on the muscle will yield the maximum power output? Give the final answer in kilograms.
15. If $V_{max} = 0.14$ m/s and $a = 0.25P_0$, what speed (V) will yield the maximum power output? Note also that the power output reaches a maximum when $P = 0.31P_0$.
16. In a series of isotonic contractions, a frog muscle lifted seven loads at the velocities indicated:

    Load (g): 2.8, 8, 16, 27, 40, 51, 67

    Velocity (cm/s): 4, 2.9, 1.8, 1.1, 0.6, 0.3, 0.04

    When plotted, the data gives the following Hill's force–velocity curve:

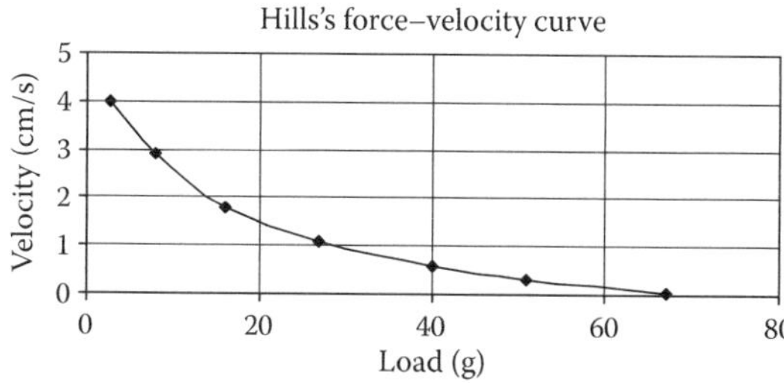

    From the plot, estimate values of V_{max} and P_0. Then fit the above data with Hill's force–velocity equation to determine values for the constants a and b.
17. Given the alternative form of the Hill relationship:

$$V/V_{max} = a/P_0(1 - P/P_0)/\{(P/P_0) + (a/P_0)\}$$

    and the knowledge that

$$a/P_0 = b/V_{max} = 1/4$$

(a) Calculate the speed at which $P = (1/3)P_0$.

(b) What is the power output at this point?

Answers to Study Questions

1. k = Boltzmann's constant = 1.381×10^{-23} JK^{-1} and T = absolute temperature = 310 K.

$$50 \times 10^{12} \times 19 \times 1.381 \times 10^{-23} \times 310 \times 3 \times 60/100 = 7.32 \times 10^{-6} \text{ J.}$$

2. (a) $0.2/2.5 \times 10^{-6}$ = 80,000 sarcomeres.

   (b) $350 \times 1000 \times 80,000 \times 2000 = 26 \times 10^{12}$ myosin molecules.

3. The innervation ration (IR) is the number of muscle fibers innervated by one motor neuron. Hence, number of motor units (M_u) = N/IR, where N = the total number of muscle fibers in muscle. Therefore, M_u = 1,170,000/1800 = 650 motor units. Other information is irrelevant.

4. 450×0.6 = 270 motor axons.

$$270 \times (1\text{–}0.3) = 189 \text{ innervating fibers.}$$

   So mean innervation ratio is 28,000/189 = 148.

5. Force per fiber = 26.5/28,000 = 0.95 mN.

   So the numbers in the smallest and largest motor units are 30/0.95 = 32 and 390/0.95 = 411, respectively.

6. Volume must be 3 ml, so cross-sectional area is 3/7.5 = 0.4 cm^2 or 4×10^{-5} m^2.

   Force per unit area is thus $26.5/4 \times 10^{-5}$ = 66 MNm^{-2}. Fiber % length is irrelevant.

7. (a) Ratio of areas is ratio of diameters squared = $(10^3)^2 = 10^6$ (assuming complete packing).

   (b) Innervation ratio = $10^6/2000$ = 500.

   (c) $F_{max} = 80 \times 9.81$ N; tension per $cm^2 = 80 \times 9.81/(2.5^2 \times \pi) = 40$ N (cm^{-2}).

   (d) Tension per fiber is $80 \times 9.81/10^6$ = 0.78 mN.

8. Power = F × v_{vert} = mg v = $72 \times 9.81 \times 2.4$ = 1.7 kW.

9. 20×0.7 = 14 cm (because of their structure parallel muscles can shorten by approximately the same amount as the muscle fibers within it, which is about 30% of their total length).

10. 90 – 55 = 35 mV. One MEPP results in an end-plate potential of about 0.5 mV. Thus, 35/0.5 = 70 MEPPs.

11. Myasthenia gravis. In this autoimmune disease, antibodies are produced that attack and destroy nicotinic acetylcholine receptors at the motor end plate in skeletal muscle fibers. The use of anticholinesterase drugs allows the remaining functional acetylcholine receptors to be activated longer, allowing for larger end-plate potentials to be produced.

12. Amplitude of 1.8 mV indicates large-diameter fibers. Two phases indicates synchrony of firing in fibers surrounding the electrode. A rise time of 300 μs indicates

electrodes placed close enough to fibers for proper analysis. Duration of 14 ms indicates a motor unit with many muscle fibers.

13. Maximum power output (W_{max}) = $0.1(P_0 \times V_{max}) = 0.1 \times 2500 \times 0.11 = 27.5$ watts.
14. Maximum power output occurs when $P = a[(1 + P_0/a)^{1/2} - 1]$. Substituting $a = 0.30P_0$ into the equation yields $P = 0.324P_0$. Given $P_0 = 775$ N, then $P = 0.324 \times 775 = 251$ N. As 1 kg = 9.8 N, then the load (P) on the muscle in kg = 251/9.8 = 25.6 kg.
15. Maximum power output occurs when $V/V_{max} = a(P_0 - P)/P_0\ (P + a)$. If $a = 0.25P_0$, then $V/V_{max} = 0.25(1 - P/P_0)/(P/P_0 + 0.25)$. It is given that power output reaches a maximum when $P = 0.31P_0$. Hence, using $P/P_0 = 0.31$ in the above equation, $V/V_{max} = 0.31$. Therefore, $V = V_{max} \times 0.31 = 0.14 \times 0.31 = 0.043$ m/s.
16. Extrapolating the curve of the plotted data to the x and y axes, respectively gives a value of ~70 g for P_0 and ~5 cm/s for V_{max}. If we use the value P = 16 g, the corresponding V value is 1.8 m/s. Thus (using the Hill equation $(P + a)(V + b) = b(P_0 + a)$, as well as $V_{max} = bP_0/a$),

$$(16 + a)(1.8 + b) = (70 + a)b = (5 + b)a \text{ (two equations with two unknowns)}$$

From the second equation, $70b + ab = 5a + ab$; so $b = 5a/70$

In the first equation, $(16 + a)(1.8 + 5a/70) = (70 + a)5a/70$

$$28.8 + 80a/70 + 1.8a + 5a^2/70 = 5a + 5a^2/70$$

$$28.8 = 5a - 1.8a - 80a/70 = 5a - 1.8a - 1.143a = 2.057a$$

Thus, a = 14 g, b = 5a/70 = 1.0 cm/s (hence, $a = 0.2P_0$).

Using other P and V data from above will give roughly similar results (a varies between about 14.0 g and 15.6 g; therefore b will vary between 1.0 cm/s and 1.1 cm/s).

17. (a) $V/V_{max} = ¼(1 - 1/3)/(1/3 + ¼) = 1/6/(7/12) = 2/7$.
    (b) $W = (P/P_0 V/V_{max}) = 2/7.1/3 = 2/21$.

Bibliography

Aidley DJ. 1998. *The Physiology of Excitable Cells*, 4th Edition. Cambridge University Press, Cambridge.

Bagshaw CR. 1993. *Muscle Contraction*, 2nd Edition. Chapman & Hall, London.

Beck WS. 1971. *Human Design: Molecular, Cellular, and Systematic Physiology*. Harcourt Brace Jovanovich, Inc., New York.

Caldwell JH. 2000. Clustering of sodium channels at the neuromuscular junction, *Microsc. Res. Tech.*, 49:84–89.

Carlson FD, Wilkie DR. 1974. *Muscle Physiology*. Prentice Hall, Inc., Englewood Cliffs, New Jersey.

Cullen MJ, Landon DN. 1994. The normal ultrastructure of skeletal muscle. In: Walton J, Karpati G and Hilton-Jones D. (Editors), *Disorders of Voluntary Muscle*, 6th Edition. Churchill Livingstone, New York, pp. 87–137.

Daube JR, Rubin DI. 2009. Needle electromyography. *Muscle Nerve* 39:244–270.

Dulhunty AF, Pouliquin P. 2003. What we don't know about the structure of ryanodine receptor calcium release channels, *Clin. Exp. Pharmacol. Physiol.* 30:713–723.

Dulhunty AF. 2006. Excitation–contraction coupling from the 1950s into the new millennium. *Clin. Exp. Pharmacol. Physiol.* 33:763–772.

Dulhunty AF, Haarmann CS, Green D, Laver DR, Board PG, Casarotto MG. 2002. Interactions between dihydropyridine receptors and ryanodine receptors in striated muscle. *Prog. Biophys. Mol. Biol.* 79:45–75.

Ebashi S. 1972. Calcium ions and muscle contraction. *Nature* 240:217–218.

Gomes AV, Potter JD, Szczesna-Cordary D. 2002. The role of troponins in muscle contraction. *IUBMB Life* 54:323–333.

Finer JT, Simmons RM, Spudich JA. 1994. Single myosin molecule mechanics: Piconewton forces and nanometer steps. *Nature* 368:113–119.

Fitzgerald MJT, Folan-Curran J. 2002. *Clinical Neuroanatomy and Related Neuro*science, 4th Edition. W.B. Saunders, Edinburgh.

Greenstein B, Greenstein A. 2000. *Color Atlas of Neuroscience: Neuroanatomy and Neurophysiology.* Thieme, Stuttgart.

Guyton AC, Hall JE. 2006. *Textbook of Medical Physiology*, 11th Edition. Elsevier Saunders, Pennsylvania.

Harford J, Squire J. 1997. Time-resolved diffraction studies of muscle using synchrotron radiation. *Rep. Prog. Phys.*, 60:1723–1787.

Henneman E. 1974. Peripheral mechanisms involved in the control of muscle, Chapter 22. In: Mountcastle VB (Editor), *Volume One: Medical Physiology*, 13th Edition. The C. V. Mosby Company, Saint Louis, pp. 617–635.

Herzog W. 2000. Cellular and molecular muscle mechanics, Chapter 3. In: Herzog W (Editor), *Skeletal Muscle Mechanics: From Mechanisms to Function.* John Wiley & Sons, Ltd, West Sussex, UK, pp. 33–52.

Herzog W. Muscle. 2007. In: Nigg BM, Herzog W (Editors), *Biomechanics of the Musculo-skeletal System*, 3rd Edition. John Wiley & Sons, Ltd, West Sussex, England, pp. 169–225, 2007.

Herzog W, Ait-Haddou R. 2002. Considerations on muscle contraction. *J. Electromyogr. Kinesiol.* 12:425–433.

Highsmith S. 1999. Lever arm model of force generation by actin–myosin–ATP. *Biochemistry* 38:9791–9797.

Hill AV. 1938. The heat of shortening and the dynamic constants of muscle. *Proc. R. Soc. Lond. (Biol.)* 126:136–195.

Hilton-Jones D. 1994. The clinical features of some miscellaneous neuromuscular disorders. In: Walton J, Karpati G and Hilton-Jones D. (Editors), *Disorders of Voluntary Muscle*, 6th Edition. Churchill Livingstone, New York, pp. 967–987.

Holmes KC. 1997. The swinging lever-arm hypothesis of muscle contraction. *Curr. Biol.* 7:R112–R118.

Holmes KC. 2009. Actin in a twist. *Nature* 457:389–390.

Huxley AF. 1957. Muscle structure and theories of contraction. *Prog. Biophys. biophys. Chem.* 7:255–318.

Huxley AF. 1974. Muscular contraction (Review Lecture). *J. Physiol.* 243:1–43.

Huxley AF. 2000. Mechanics and models of the myosin motor. *Phil. Trans. R. Soc. Lond. B* 355:433–440.

Huxley AF. 2000. Cross-bridge action: Present views, prospects, and unknowns, Chapter 2. In: Herzog W (Editor), *Skeletal Muscle Mechanics: From Mechanisms to Function.* John Wiley & Sons, Ltd, West Sussex, UK, pp. 7–31.

Huxley AF, Niedergerke R. 1954. Interference microscopy of living muscle fibres. *Nature* 173:971–973.

Huxley AF, Simmons RM. 1971. Proposed mechanism of force generation in striated muscle. *Nature* 233:533–538.

Huxley HE. 1969. The mechanism of muscular contraction. *Science* 164:1356–1366.

Huxley HE. 2000. Past, present and future experiments on muscle. *Phil. Trans. R. Soc. Lond. B* 355:539–543.

Huxley HE. 2004. Fifty years of muscle and the sliding filament hypothesis. *Eur. J. Biochem.* 271: 1403–1415.

Huxley HE, Hanson J. 1954. Changes in the cross-striations of muscle during contraction and stretch and their structural interpretation. *Nature* 173:973–976.

Jones DA, Round JM. 1990. *Skeletal Muscle in Health and Disease: A Textbook of Muscle Physiology.* Manchester University Press, Manchester.

Kandel ER, Siegelbaum SA. 1991. Directly gated transmission at the nerve–muscle synapse, Chapter 10. In: Kandel ER, Schwartz JH and Jessell TM (Editors), *Principles of Neural Science*, 3rd Edition. Appleton & Lange, Norwalk, Connecticut, pp. 135–152.

Karpati G, Currie GS. 1994. The inflammatory myopathies. In: Walton J, Karpati G and Hilton-Jones D (Editors), *Disorders of Voluntary Muscle*, 6th Edition. Churchill Livingstone, New York, pp. 619–646.

Keynes RD, Aidley DJ. 1991. *Nerve and Muscle*, 2nd Edition. Cambridge University Press, Cambridge.

Klein MG, Schneider MF. 2006. Ca^{2+} sparks in skeletal muscle. *Prog. Biophys. Mol. Biol.* 92:308–332.

Marieb EN, Hoehn K. 2010. *Human Anatomy & Physiology*, Eighth Edition. Pearson Benjamin Cummings, San Francisco.

Martini FH, Nath JL. 2009. *Fundamentals of Anatomy & Physiology*, 8th Edition. Pearson Education, Inc. San Francisco.

Martini FH, Welch KW. 2005. *A & P Applications Manual*. Pearson Education, Inc. San Francisco.

McArdle WD, Kath FI, Katch VL. 2001. *Exercise Physiology: Energy, Nutrition, and Human Performance*, 5th Edition. Lippincott Williams & Wilkins, Baltimore.

Molloy JE, Burns JE, Kendrick-Jones J, Tregear RT, White DCS. 1995. Movement and force produced by a single myosin head. *Nature* 378:209–212.

Oda T, Iwasa M, Aihara T, Maédia Y, Narita A. 2009. The nature of the globular-to fibrous-actin transition. *Nature* 457:441–445.

Ohtsuki I. 1999. Calcium ion regulation of muscle contraction: The regulatory role of troponin T. *Mol. Cell. Biochem.* 190:33–38.

Pallis CA, Lewis PD. 1994. Involvement of human muscle by parasites. In: Walton J, Karpati G, Hilton-Jones D. (Editors), *Disorders of Voluntary Muscle*, 6th Edition. Churchill Livingstone, New York, pp. 743–759.

Preston DC, Shapiro BE. 1998. *Electromyography and Neuromuscular Disorders: Clinical–Electrophysiologic Correlations*. Butterworth-Heinemann, Boston.

Rhoades R, Pflanzer R. 2003. *Human Physiology*, 4th Edition. Brooks/Cole, Pacific Grove.

Rohkamm R. 2004. *Colo Atlas of Neurology*. Georg Thieme Verlag, Stuttgart.

Saladin KS. 2004. *Anatomy and Physiology: The Unity of Form and Function*, 3rd Edition. McGraw-Hill, New York.

Salmons S (Editor). 1995. Muscle. In: Williams PL (Chairman of Editors), *Gray's Anatomy*, 38th Edition. Churchill Livingstone, New York, pp. 737–900.

Schmidt RF. 1983. Motor systems, Chapter 5. In: Schmidt RF, Thews G (Editors), *Human Physiology*. Springer-Verlag, Heidelberg, pp. 81–110.

Scott W, Stevens J. 2001 Human skeletal muscle fiber type classifications. *Phys. Ther.* 81:1810–1816.

Silverthorn DU. 2010. *Human Physiology: An Integrated Approach*, 5th Edition. Pearson Benjamin Cummings, San Francisco.

Slater CR, Harris JB. 1994. The anatomy and physiology of the motor unit. In: Walton J, Karpati G, Hilton-Jones D (Editors). *Disorders of Voluntary Muscle*, 6th Edition. Churchill Livingstone, New York, pp. 3–32.

Squire JM. 1997. Architecture and function in the muscle sarcomere. *Curr. Opin. Struct. Biol.* 7:247–257.

Stern RM, Ray WJ, Quigley KS. 2001. *Psychophysiological Recording*, 2nd Edition. Oxford University Press, Oxford.

Stone TW. 1995. *Neuropharmacology*. W.H. Freeman and Company Limited, New York.

Szczesna D, Zhao J, Potter JD. 1996. The regulatory light chains of myosin modulate cross-bridge cycling in skeletal muscle. *The Journal of Biological Chemistry* 271:5246–5250.

Trinick J. 1992. Understanding the functions of titin and nebulin. *FEBS Lett*. 307:44–48.

Volkenshtein MV. 1983. *Biophysics*. Mir Publishers, Moscow.

Von Tscharner V, Herzog W. 2007. EMGe. In: Nigg BM and Herzog W (Editors), *Biomechanics of the Musculo-skeletal System*, 3rd Edition. John Wiley & Sons, Ltd, West Sussex, England, pp. 409–452.

Vrbová G, Gordon T, Jones R. 1995. *Nerve–Muscle Interaction*, 2nd Edition. Chapman & Hall, London.

Walsh FS, Doherty P. 1994. Cell biology of muscle. In: Walton J, Karpati G and Hilton-Jones D (Editors), *Disorders of Voluntary Muscle*, 6th Edition. Churchill Livingstone, New York, pp. 33–61.

Widmaier EP, Raff H, Strang KT. 2004. *Vander, Sherman, Luciano's Human Physiology: The Mechanisms of Body Function*, 9th Edition. McGraw-Hill, New York.

Woledge RC, Curtin NA, Homsher E. 1985. *Energy Aspects of Muscle Contraction*. Academic Press, London.

Wray D. 1994. Neuromuscular transmission. In: Walton J, Karpati G, Hilton-Jones D (Editors), *Disorders of Voluntary Muscle*, 6th Edition. Churchill Livingstone, New York, pp. 139–178.

Yamada K. 1999. Thermodynamic analyses of calcium binding to troponin C, calmodulin and parvalbumins by using microcalorimetry. *Mol. Cell. Biochem*. 190:39–45.

Zierler KL. 1974. Mechanism of muscle contraction and its energetics, Chapter 3. In: Mountcastle VB (Editor), *Volume One: Medical Physiology*, 13th Edition. The C. V. Mosby Company, Saint Louis, pp. 77–120.

Part III

Heart and Circulation

8

Cardiac Biophysics

Andrew W. Wood

CONTENTS

8.1 Introduction

The constant need of cells for a supply of oxygen for metabolism and for the removal of metabolic products (such as CO_2) requires a mass transport system. The network of capillaries in the human body is such that no cell is more than 20 μm–30 μm from blood vessel. In simpler organisms, as soon as the number of cells exceeds around a few thousand, there is a rudimentary circulation in place, which can be either open or closed. The earthworm, for example, has a closed circulation, consisting of two vessels, one of which contains a contractile region and a series of valves. Fish have a single circuit with a two-chambered heart. Crocodiles and amphibia, on the other hand, have two circuits, with a three-chambered heart (blood from the two circuits mix to a certain extent). Birds and mammals have four-chambered hearts, with two completely separate circuits. Hence, essentially, the human heart consists of two pumps side by side, each consisting of two chambers, an *atrium* (or *auricle*) and a *ventricle*. The atrium (which is now a common architectural term) refers to a receiving space, and the ventricle comes from a word meaning "a small belly." Although the heart is two independent pumps in parallel, they share a common electrical activation system to control muscular contraction. There are thus two major approaches to studying the heart: mechanical aspects (heart as a pump) and electrical aspects (heart as a collection of excitable muscle fibers). Each of these is dealt with in the sections below.

8.2 Mechanical Events: The Heart as a Pump

As just mentioned, the heart can be thought as being *two* pumps: the right side pumping into the lungs (pulmonary circulation) and the left into the rest of the body (systemic circulation). The output from the right side must equal that of the left (except for short periods of adjustment). The simplest analysis of the relationship of the heart to the circulation is to make an analogy with Ohm's law. In Figure 8.1, the heart is represented by two current sources (circles). The cardiac output Q (as volume flow per unit time, such as m^3/s) is analogous to an electrical current and the pressure p (measured in Pa) like a voltage. The total peripheral resistance (TPR) is thus $(p_1 - p_2)/Q \approx p_1/Q$, since $p_2 \approx 0$. Note that the same value for Q occurs in both the left and right side (if this were not so, then one of the circuits would have to take up the extra volume, depleting the other: there are brief moments of adjustment between the two sides, however, lasting a few beats). Because the mean pressure in the systemic arteries is around five times that in the pulmonary, the TPR of the systemic circuit must also be around five times that in the pulmonary (actually, the TPR determines the arterial pressure, not vice versa).

Because the left side of the heart generates much higher pressures, the mass of cardiac muscle (*myocardium*) surrounding the left ventricle is much larger than for the right. In fact, the right ventricle is more of a crescent shape in cross-section (Figure 8.2), with the outlet to the pulmonary arteries actually further to the left than the outlet to the main systemic artery, the *aorta* (note that left and right refer to the organ in the chest, and not as it appears on a printed page). The other noteworthy feature is that the myocardial fibers run mainly circumferentially in the inner layers and from the apex to the base in the outer layers surrounding the ventricles. The atria are similar but with far less muscle mass. Between the atria and the ventricles is a *fibrous skeleton* constructed of tough proteins such as keratin and elastin, which support the four heart valves (Figure 8.3).

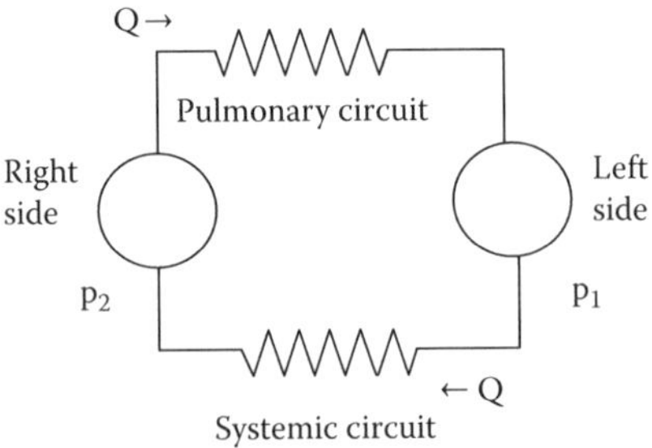

FIGURE 8.1
An electrical circuit analog for the human circulation. Q is flow rate (equivalent to electrical current) and p_1 and p_2 are pressures (voltages). The two circles represent pumps (or current sources).

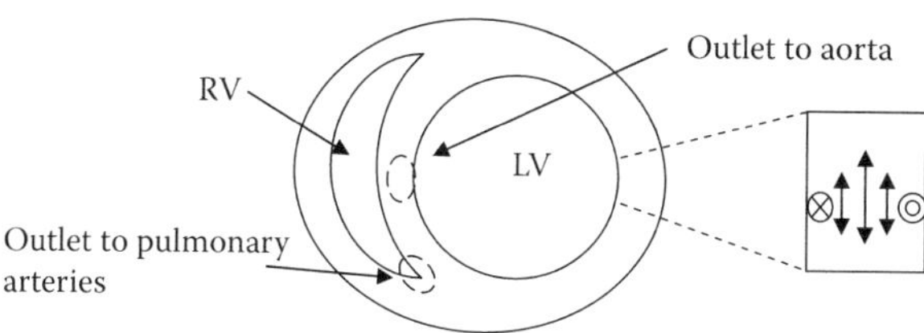

FIGURE 8.2
Cross-sectional view of the ventricles: note the greater mass of muscle surrounding the left ventricle (LV) and the arrangement of muscle fibers in the wall (⊗ and ◎ indicate fibers running perpendicular to the plane of the diagram.

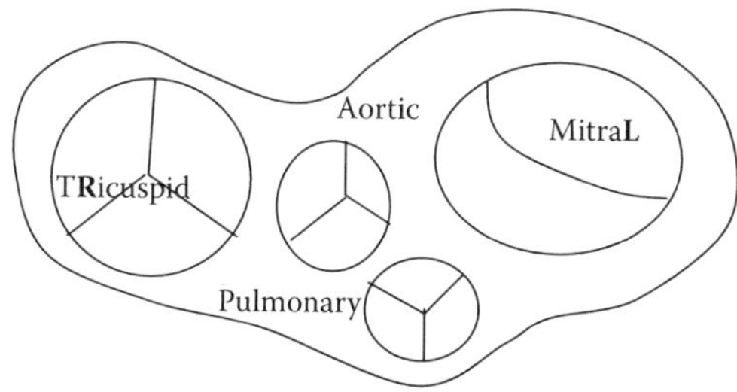

FIGURE 8.3
Diagram of the arrangement of heart valves in the fibrous skeleton. Note that the "R" in "TRicuspid" and the "L" in "MitraL" helps to remember which is on which side of the heart (however, note that mitral also has an "R"!).

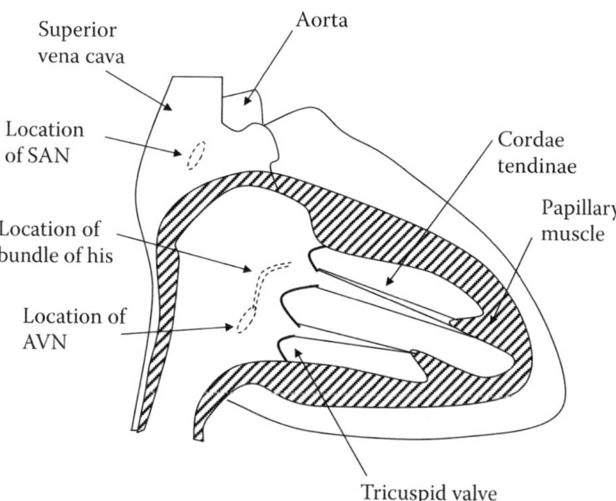

FIGURE 8.4
View of heart sliced open to show cordae tendinae and papillary muscles in the right ventricle. Location of components of the electrical conduction system also shown.

The valves are remarkable structures: the leaflets are 0.1 mm thick and will undergo 3×10^9 opening and closing cycles during a lifetime. Closing is passive: when the pressure on the ventricular side exceeds that on the atrial side, the leaflets come together, quite forcibly and aided by the anchoring ligaments, the *cordae tendinae*. These attach to muscular processes on the walls of the ventricle (the *papillary* muscles), which further ensure a close fit: the mitral valve in particular has to withstand a difference in pressure of around 15 kPa (0.15 atm) (Figure 8.4).

The valves, of course, prevent backflow between ventricles and atria and between major outflow arteries (aorta and pulmonary) and the ventricles, but note that there are no valves at the entry into the atria, on either side. When the atria undergo contraction (impelling blood into the ventricles), some backflow occurs, but this is minimal, in view of the greater pressure and momentum of inflowing blood. The pumping action of the heart is *pulsatile*, that is, the flow into the aorta and pulmonary artery is zero for part of the cycle, depending on whether the A and P valves are open or shut. Because of the way the electrical activation system causes both ventricles to contract together, the M and T valves close within milliseconds of each other, followed around 0.3 s later by the A and P valves closing. These events mark the major divisions of the cardiac cycle, which will now be described.

8.2.1 Cardiac Cycle

The major divisions are *systole* ("coming together") and *diastole* ("moving apart"). These divisions refer to events in the ventricles (there is also an atrial systole, which occurs late in ventricular diastole). The start of systole is when M and T valves *shut*, and the start of diastole happens when A and P valves *shut*. These moments correspond to the familiar *heart sounds* (known as S1 and S2), which are due to a sudden cessation of blood flow, causing reverberations through the heart and surrounding tissues. These sounds are commonly referred to as "Lubb" and "Dupp" as indicated in Figure 8.5 below. S2 is higher in pitch (median frequency 50 Hz) than S1 (median frequency 20 Hz). These are the sounds heard in a *stethoscope* or in any microphone with good low-frequency characteristics placed in contact with the chest wall. Particular valves are best heard in specific locations, as indicated in the second part of Figure 8.5.

The cardiac cycle can be thought of as the *fibrous skeleton* moving up and down—as the atria expand, the ventricles contract, and vice versa.

This is not quite accurate, but is a good approximation. It helps in remembering which valves are open at a particular moment (see Figure 8.6).

The phases of the cardiac cycle are given different names in different books, but Figure 8.7 (sometimes referred to as a "Wiggers diagram") has the numbers 1–6 in circles. These refer to the following:

During systole:

1. Isovolumetric contraction: All valves are shut; there is a rapid pressure rise but no outflow. The ventricular volume is the same as the end of the previous diastole (end-diastolic volume [EDV]).
2. Ejection: Aortic (and pulmonary) valves are open, and ventricular volume falls until these valves close again (this volume is called the endsystolic volume [ESV]; the stroke volume [SV] is the difference between EDV and ESV). Note that the ventricular pressure increases to a maximum (as the muscle continues to contract) and then begins to fall. The aortic valve closes when the ventricular pressure falls below the pressure in the aorta.

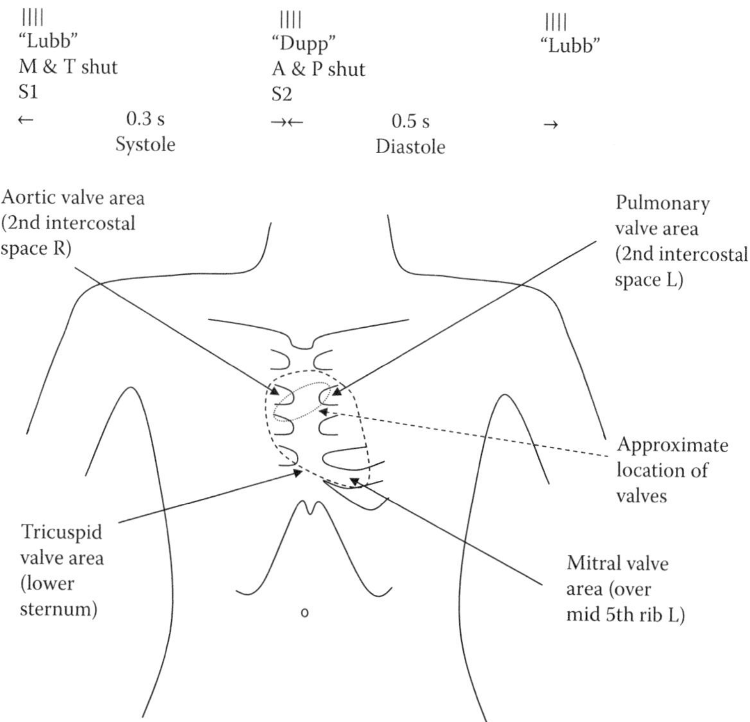

FIGURE 8.5
Heart sounds. Top: Timing of the major heart sounds S1 and S2 ("Lubb" and "Dupp"), signaling the start of ventricular systole and diastole. Lower: Positions over the chest where the sounds from particular valves are best heard. Note that these do not coincide with the positions directly above the valves concerned.

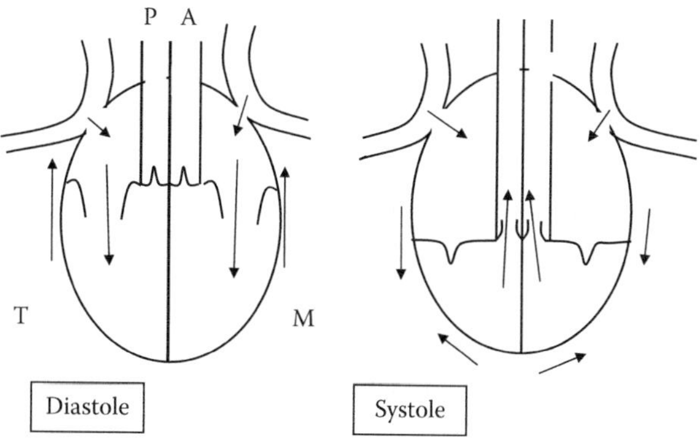

FIGURE 8.6
The cardiac cycle can be envisaged as the fibrous skeleton (supporting the four valves) moving up and down under the alternating action of the ventricular and atrial muscles. Note that the total volume of the four chambers remains approximately constant and that the valves close passively as a result of buildup of pressure.

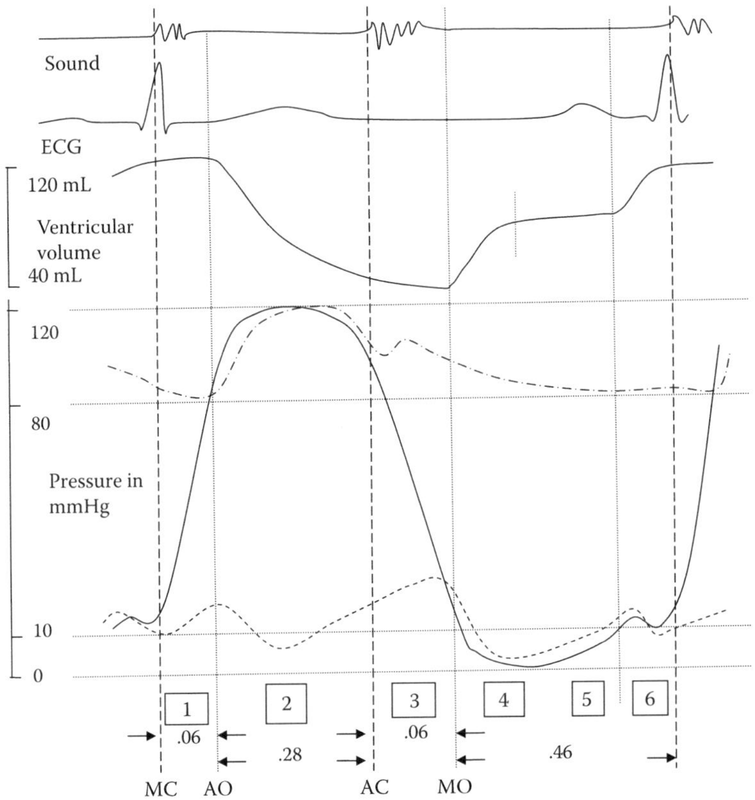

FIGURE 8.7
Wiggers Diagram for left side of heart. From top: Heart sounds, showing S1 and S2; ECG, showing R wave timing; ventricular volume, showing end-diastolic and end-systolic volumes of 120 and 40 mL, respectively; pressures in mmHg in left ventricle (full line), aorta (dot dash), and left atrium (dashed). Numbers in boxes refer to phases of cardiac cycle (see text), and the approximate timing for some of these phases in seconds is shown below them. Bottom: Mitral valve closure and opening are denoted MC and MO, respectively; aortic opening and closure are denoted AO and AC, respectively.

During diastole:

3. Isovolumetric relaxation: All valves are shut; there is a rapid pressure fall but no change in volume from ESV.
4. Rapid filling: Mitral (and tricuspid) valves are open; blood flows through from atria; pressure in ventricles continues to fall as muscle relaxes.
5. Slow filling: Continuation of item 4, but at a slower rate. Note that the ventricular volume does not change much during this time.
6. Atrial systole: Here the atria themselves contract in response to the pacemaker (sinoatrial node [SAN]—see below) being activated. An extra 15 mL or so of blood is transferred from atria to ventricles.

Note from the diagram that the pressure in the aorta ranges from 80 mmHg to 120 mmHg and that the pressures in most large arteries are similar. These minimum and maximum pressures are known as *diastolic* and *systolic arterial pressures*, respectively (for obvious

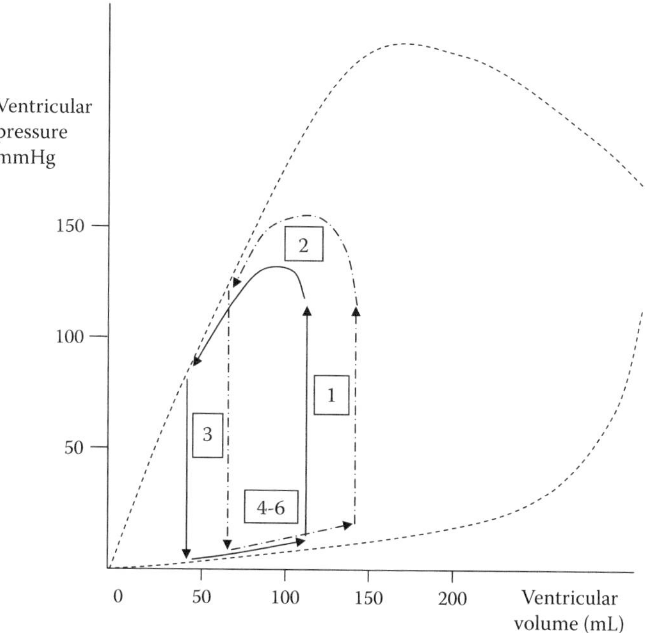

FIGURE 8.8
Left ventricular "stroke work" curves. Two loops are shown. The numbers correspond to the phases referred to in the previous diagram. The loop on the left represents a normal situation; that on the right represents where end-diastolic volume (EDV) has been raised, for whatever reason. The area in the loop represents the amount of work done by the ventricular muscles (stroke work). In the case of raised EDV, the amount of stroke work is higher than normal, illustrating Starling's law of the heart. The upper dotted curve represents the potential pressure that the ventricle could develop if inflow and outflow were prevented and the cardiac muscle was maximally stimulated. The lower dotted curve represents the pressure in an unstimulated ventricle, again with inflow and outflow prevented (reflecting the passive tension of the muscle).

reasons). Note that the rather sharp notch (dicrotic notch) in aortic pressure occurs just after the A valve closes (Figure 8.7). Figure 8.8 is a plot of the way pressure varies with volume during the various phases of the cardiac cycle. The enclosed area is the 'work done', as further discussed in Section 8.6.5.

The pressure in the atria is largely determined by movements of the fibrous skeleton, rising as the skeleton moves up and falling as it moves down. The flow in the aorta is maximum during the initial rapid ejection and actually flows in reverse for a brief moment just after the aortic valve closes; this is because of the elastic recoil of the aortic wall.

8.2.2 Coronary Circulation

The myocardium requires constant blood perfusion: around 10% of the cardiac output (Q in Figure 8.1) is immediately diverted into the L and R *coronary arteries*. The opening to these are just downstream of the A valve leaflets and are at the bottom of specialized *pouches of Valsalva*. The fluid dynamics is such that when the valves are open, there is an eddy current that diverts blood into the coronary arteries, and when the A valve is shut, the elastic recoil of the aorta continues to force blood into them.

The coronary arteries are shown in Figure 8.9: the major veins run in parallel with the arteries, and blood from these veins flows back into the R atrium via an opening known as

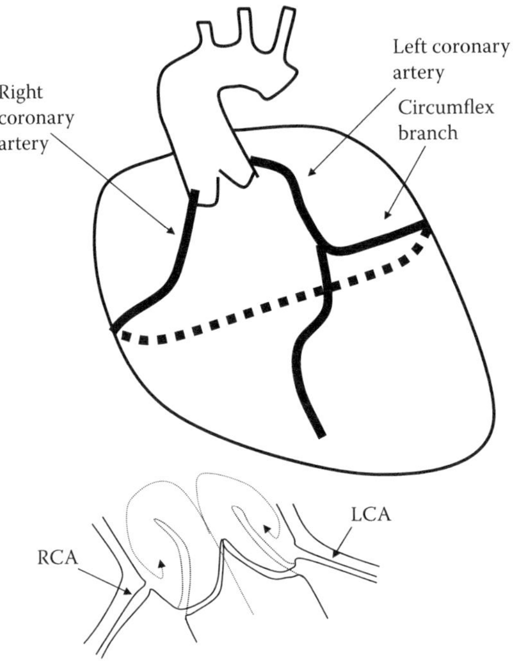

FIGURE 8.9
Top diagram: Coronary arteries, showing their origin behind the aortic valve leaflets. The main branches are shown; the smaller branches are omitted for clarity. Lower diagram: Aortic valve leaflets in the opened and closed position, showing the sinuses of Valsalva from which the openings to the right and left coronary arteries (RCA and LCA) lead. Even when fully open, eddy flow patterns ensure a steady flow into these arteries.

the *coronary sinus*. Disease of the coronary arteries is a leading cause of death, and methods for diagnosing and treating this disease will be discussed in Chapter 10.

8.2.3 Comparison between Cardiac Muscle and Other Muscle Types

Cardiac muscle is intermediate between the skeletal and smooth muscles, as Figure 8.10 and Table 8.1 show. Given certain treatment, skeletal muscle will begin to resemble cardiac, and similarly, cardiac muscle can resemble smooth muscle.

Chapter 6 discussed some of the experimental methods used to investigate skeletal muscle. Similar methods can be used to study cardiac muscle (especially the papillary muscle, which is a convenient shape for these types of investigation). Note that the passive tension is quite considerable at the peak of the active tension, making characterization of the "active state" (defined as shortening velocity at zero load) quite difficult. Nevertheless, the work of Sonnenblick et al. (1965) achieved this through extrapolation (as shown in Figure 8.11).

It can be shown that the maximal rate of change in ventricular pressure during isometric contraction can be correlated to maximal shortening velocity and therefore gives an indication of the state of health of ventricular muscle.

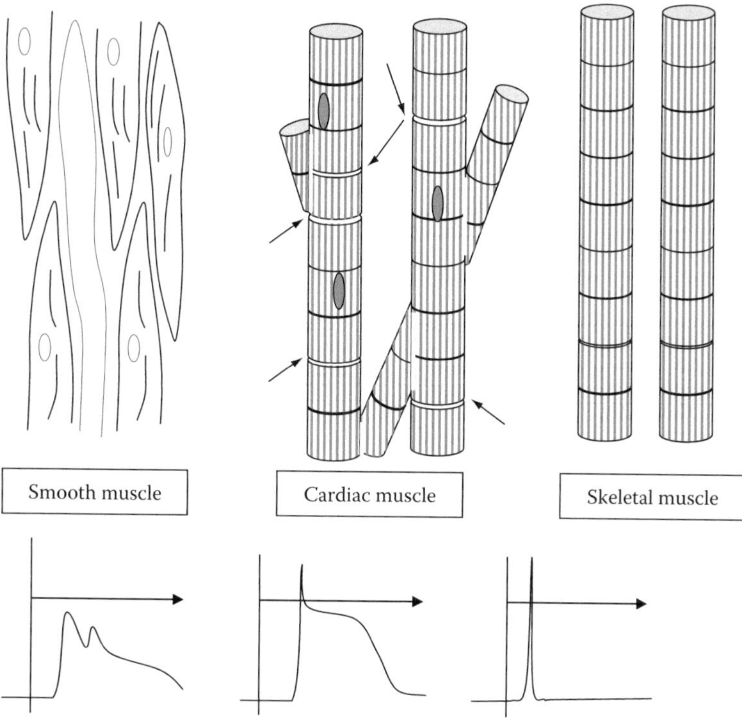

FIGURE 8.10
Comparison of smooth, cardiac, and skeletal muscle fibers. Top diagram: Representation of anatomical arrangement of cells. Note the intercalated disks (arrows) in the case of cardiac muscle. Lower diagram: Action potentials of the three muscle types (voltage and time on vertical and horizontal axes, respectively). See Table 8.1 for intercomparisons between cell types.

TABLE 8.1
Comparisons between Muscle Cell Types

Features Common between Smooth and Cardiac Muscles	Features Common between Cardiac and Skeletal Muscles
Anatomical	Anatomical
Central nucleus	Cross-striations
Syncytial arrangement of cells	Cylindrical shape
	Color
Functional	Functional
Autogenic excitation	Length–tension relationships
Inherent rhythmicity	Contraction time
Intercellular transmission of electrical excitation	
Autonomic nervous system control	
	After denervation, skeletal muscle can resemble cardiac muscle by exhibiting spontaneous excitation.
	After particular treatment, cardiac muscle can exhibit summation and tetanus (features of skeletal muscle), which it would not normally do.

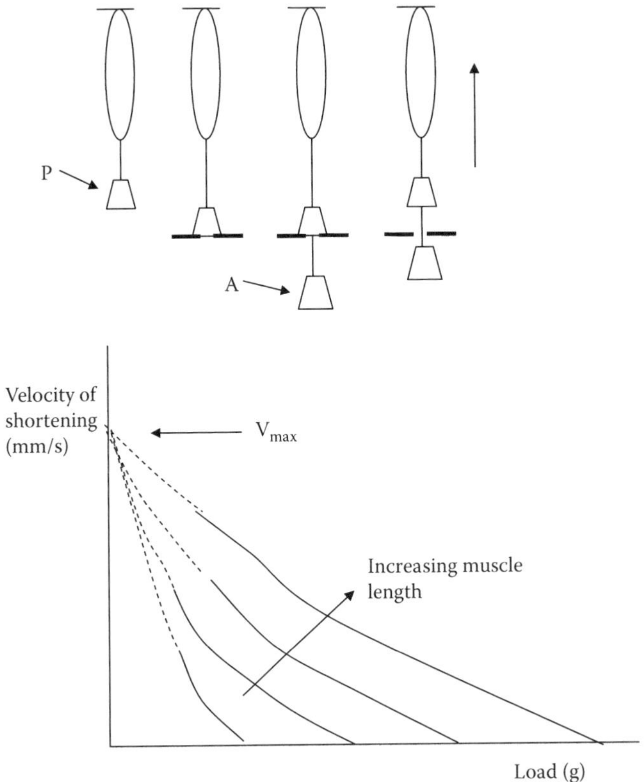

FIGURE 8.11
Experiments on papillary muscle. Top: A preload (P) is added to stretch the muscle, then an afterload (A) is added to produce a range of total loads for different initial muscle lengths (horizontal axis of lower graph). The muscle is then electrically stimulated, and the shortening velocity is recorded (vertical axis of lower graph). The experimental points can be extrapolated to a single point on the velocity axis (V_{max}).

8.3 Electrical Events in the Cardiac Cycle

This brings us now to the other aspect of heart function: the myocardium as an assembly of electrically activated muscle fibers. Two points must be emphasized: the heart rhythm is autogenic (that is, it is generated by the heart tissue itself), and both sides of the heart (right and left) are activated more or less in synchrony. The activation is in two distinct phases: from a region in the top of the atria down through the atria as far as the fibrous skeleton; and then after a delay of around 1/7 second, the ventricles activated from the interventricular septum, then from the apex (bottom) up through the outer walls, and back to the fibrous skeleton.

Each cell in the myocardium has the property of autogenic rhythmicity; that is, even disaggregated and isolated, they will continue to exhibit regular membrane depolarizations followed by contractions, as shown in Figure 8.12. The pacemaker region or SAN is in the atrial wall near the superior vena cava and has the fastest inherent rhythm of all cardiac tissues (even isolated strips of ventricle will contract spontaneously, but the inherent rhythm will be slower). The SAN thus acts as a basic "drumbeat" to which all other

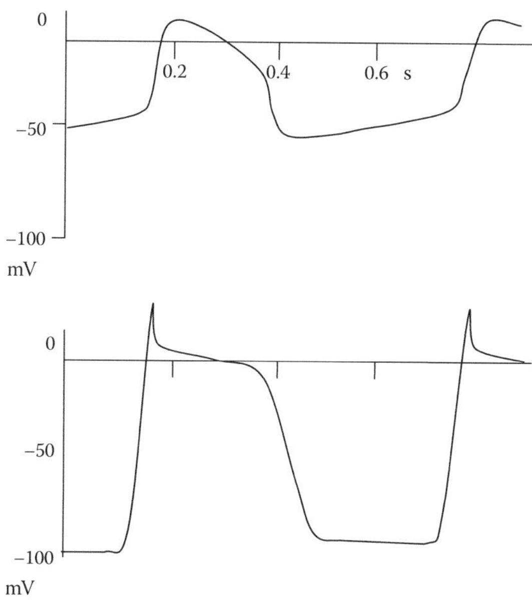

FIGURE 8.12
Membrane potentials recorded from SAN cells (upper) and ventricular cells (lower).

cardiac tissues respond. The regular depolarizations can be measured using a microelectrode inserted into SAN cells. These show a characteristic unstable membrane potential that oscillates between around −60 mV and +10 mV. A ventricular cell action potential is shown in comparison (Figure 8.12).

A "wave of depolarization" spreads out (at a speed of 1 m/s) from the SAN, with direct electrical stimulation of one muscle fiber by another until the fibrous skeleton is reached. The wave cannot go any further than this because of the insulating nature of the fibrous skeleton.

This wave is also sensed by the atrioventricular node (AVN), which is embedded in the wall between the atria. There are actually some internodal pathways between the SAN and AVN that facilitate this. Fibers in the AVN conduct the depolarization wave at a slow rate (0.2 m/s) down into something resembling an insulated cable. This cable (AV bundle or bundle of His) is composed of rapidly conducting fibers (5 m/s) called *Purkinje fibers*. These fibers pass through the fibrous skeleton where they branch to form a left and right bundle, running down the wall separating the right and left ventricles. These branch further to deliver the wave of depolarization to all regions of the ventricle within a few milliseconds. Having received this wave of depolarization, the ventricles then contract, with the apex (bottom) contracting a little ahead of the base (the region near the fibrous skeleton).

The depolarized cells are negative on the outside, as opposed to the nondepolarized regions, which are positive. Because some parts of the ventricles (for example) depolarize ahead of the rest, the heart behaves like a *dipole*, which is a region of positive charge separated from a region of equal negative charge. The field surrounding a dipole and the associated equipotential lines are shown in Figure 8.13 below. The potential difference between two arbitrary points can clearly be estimated from the values of the equipotentials they lie on. For a given radial distance from the center of the dipole, this difference is

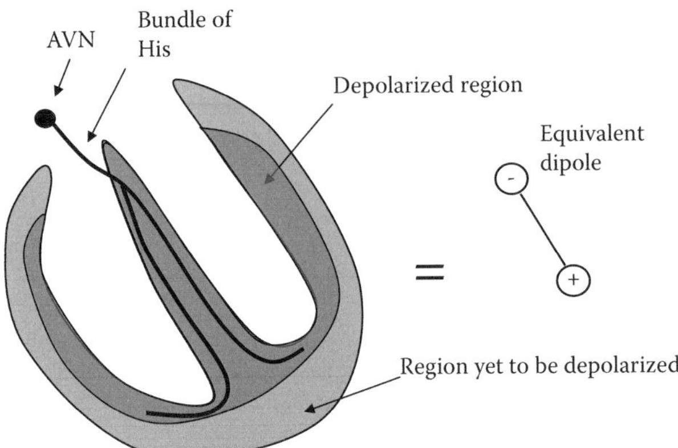

FIGURE 8.13
The conduction system of the heart. The atrioventricular node (AVN) is in the wall of the right atrium, which depolarizes in response to a wave of depolarization moving across the atrial walls from the initial depolarization of the SAN (see Figure 8.4). The depolarization in the AVN is conducted after a slight delay via the bundle of His to the two branches to the two ventricles. As the wave of depolarization arrives in the ventricular muscle mass, there are two distinct regions, a depolarized and a nondepolarized region, with the former moving into the latter. At any given instant, the two regions can be represented by a positive and a negative charge, with varying separation and direction. This equivalent dipole is shown to the right of the diagram and is the basis for the "precordial" (or "before the heart") field, which can be measured on the surface of the body (see also Figure 11.1).

maximal along the line joining the position of the two charges and zero at a point at right angles to it (Figure 8.13).

The effects of this dipole can thus be detected some distance from the heart, specifically, on the skin. In fact, if the two wrists are connected to a sensitive voltmeter, a potential difference can be detected at some times during the cardiac cycle. This is the origin of the ECG, and a typical trace is shown in Figure 8.14.

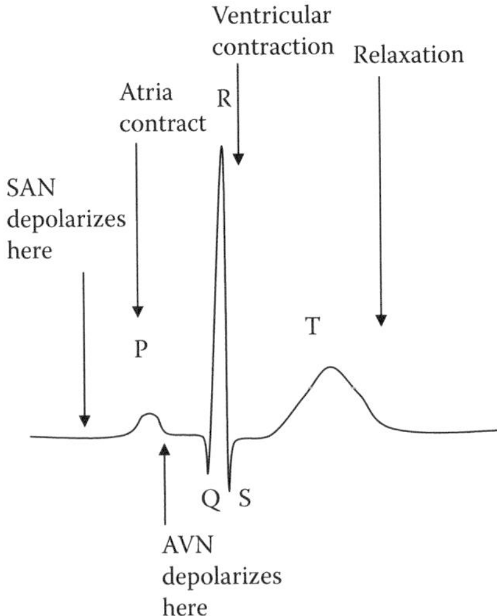

FIGURE 8.14
Features of the electrocardiogram (ECG), showing how these relate to the events in the conduction system shown in Figure 8.13.

The P wave corresponds to atrial depolarization, the QRS complex to ventricular depolarization (note the 1/7 s delay), and the T wave is when the ventricles return to their resting state (repolarization)—remember that ventricular action potentials have long plateau regions, lasting for about 1/3 s. The "equivalent dipole" representing the state of depolarization of the heart is thus varying in magnitude and direction during the cardiac cycle. With the use of an array of several skin electrodes, it is possible to trace the dipole position and strength: it actually completes three revolutions each beat. In Chapter 11, the process of recording and interpreting the ECG is described in greater detail.

8.4 Basic Heart Equation

Cardiac output (CO), or Q in Figure 8.1, is the volume flow of blood in either the pulmonary or the systemic circuit. It is normally measured in liters per minute, but for calculations involving SI units, it is best converted to cubic meters per second. A normal resting value is 6 L/min, or 100 mL/s, which converts to 10^{-4} m^3/s. There are two main factors determining CO: the amount of blood ejected from either ventricle per beat (the SV) and the heart rate (HR), or the number of heart beats per unit time. The relationship is as follows:

$$\text{Cardiac output} = \text{heart rate} \times \text{stroke volume} \quad \text{or} \quad Q = HR \times SV.$$

Heart rate varies between 45–220 beats per minute (bpm; average = 60 bpm). It relates to *membrane* properties of muscle fibers (particularly of the pacemaker region). Agents that modify the rate of depolarization of the pacemaker fibers (adrenaline, acetyl choline [ACh], etc.) will determine the value of HR; these are known as *chronotropic* (time-altering) effects.

Stroke volume varies between 70 mL and 105 mL per beat (average = 80 mL per beat). The value relates to the ability of muscle fibers to *contract*. Again, agents such as adrenaline alter this ability (these are known as *inotropic* [strength-altering] effects).

Cardiac output ranges from 3 to 25 L/min (average = 6 L/min). More correctly, this is measured in the pulmonary artery, because coronary circulation draws off blood just downstream of the aortic valve leaflets, so its measurement in the aorta will be an underestimate.

8.5 Control of the Heartbeat

8.5.1 Chronotropic Effects

The SAN and other regions of the heart receive a nervous supply from the vagus nerve (transmitter: ACh) and from the sympathetic nervous system (transmitter: noradrenaline). The vagus is normally active, keeping the heart rate down to around 70 bpm. If the cardio-inhibitory center is activated, ACh release from the vagus slows the heart rate by slowing the rate of depolarization of the SAN cells to the threshold. Alternatively, if the cardio-acceleratory center is activated, the release of NAd has the reverse effect: a quicker depolarization to the threshold. Increased temperature (i.e., fever) has a similar effect.

8.5.2 Starling Effect

This is named the Starling law and the Frank–Starling principle in UK and US texts, respectively. This refers to the increased cardiac output (hence, SV) in isolated heart preparations in which the muscle is stretched, that is, by inflating the ventricle with extra blood during diastole (EDV increase). This is similar to the effect seen in skeletal muscle fibers where the strength of contraction increases as the initial length of the muscle increases. It is equivalent to stating that the strength of heart contraction adjusts to meet the demands placed on it. For example, if the heart is made to pump against an increased TPR (termed *afterload*), the CO will be maintained via more forceful beats. The mechanism for this is that for the first few beats after the TPR value is increased, the heart will expel less than it receives, and the ventricular fibers will increase their length, through the increase in EDV. This effect was demonstrated by Starling in animal preparations in which the heart of an experimental animal was perfused by blood from an external circulation; in the intact animal, the effect is difficult to demonstrate. This is explained in the next paragraph. Starling in fact demonstrated this "law" by plotting atrial pressure (which would determine EDV) and CO (see Davson and Eggleton 1968). In fact, because CO is the same from the right and left ventricles, he measured the pressure in the *right* atrium, which is very similar to venous pressure. In large-scale modeling of the circulation and understanding of the adjustments made in the circulation due to a variety of effects, Guyton has made much of this basic relationship between CO and venous pressure (Hall 2011).

8.5.3 Inotropic Effects

Cardiac muscle is sensitive to hormones such as adrenaline, noradrenaline, and thyroxine in the way that the individual fibers are able to generate tension. For a given initial fiber length (see Starling above), the amount of tension developed can be increased by the action of adrenaline, for example. In exercise, the heart tends to be emptied more efficiently by this mechanism; that is, ESV decreases as SV increases. This is opposite to the Starling effect, with increased SV from increased EDV. This effect of adrenaline on muscle fiber contraction characteristics is termed an effect on *myocardial contractility* and the effect is described as a *positive inotropic effect*, as compared to a positive *chronotropic* effect if heart rate is increased. These effects normally cloak the Starling effect, but this is not to say that it does not occur.

8.6 Pulsatile Pressure

8.6.1 Nature of "Blood Pressure"

Arterial pressure is often stated as 120/80, or similar. This refers to 120 mmHg systolic, 80 mmHg diastolic, where 1 mmHg is an old unit of pressure and is equal to 130 Pa. This is the pressure inside an artery *above atmospheric*. Since atmospheric pressure is 760 mmHg approximately, arterial systolic pressure is 120/760 or 0.16 atmospheres in *excess* of atmospheric, or 1.16 atm in absolute terms. In some vessels (the heart chambers for example), the pressures become negative for part of the cardiac cycle.

Mean blood pressure is really the time-weighted average excess pressure in an artery at the same vertical height as the heart. Since the variation of pressure with time has an asymmetrical profile (see below), the mean pressure is usually taken as follows:

$$p_m = P_d + (1/3)\,(p_s - p_d).$$

In other words, this is equal to diastolic plus one-third pulse pressure.

The pressure inside arteries and veins varies with vertical height above or below the heart. Consider an imaginary cylinder drawn in the fluid (in this case, blood), with the axis of the cylinder vertical. If p_1 and p_2 are the pressures acting at the bottom and top surfaces of the cylinder (both of area = A) and the cylinder is Δz thick, the forces are in equilibrium if

$$p_1A = p_2A + g\rho\Delta zA$$

that is, $\Delta p = g\rho\Delta z$.

Since the density of blood, ρ, is around 1070 kg/m^3, with each centimeter we move vertically down in the vascular system, the pressure increases by $9.81 \times 1070 \times 0.01 = 105$ Pa, or $105 \times 100/\ 13300 = 0.8$ mmHg, approximately.

In large animals, such as the giraffe, which can be 6 m tall, the change in blood pressure in the head when it moves from feeding from trees to drinking from a river is 480 mmHg. It certainly needs very strong artery walls to prevent overdistension as the head goes down and collapse of the vessels as the head goes up. In the case of the human lying down and standing up, Figure 8.15 gives the mean pressures (in mmHg) in these two circumstances in both arteries and veins.

Note that the pressure inside the veins of the head is subatmospheric when standing. Tissue pressure usually causes them to collapse. The reader may consider why, if the pressure in the arteries in the feet is greater than that in the heart, the blood does not flow backward.

Lying					
Arterial	95	←	100	→	95
Venous	5	←	2	→	5
	Head		Heart		Feet

Standing Arterial		Venous
51	Head	−40
↑		↑
100	Heart	2
↓		↓
183	Feet	88

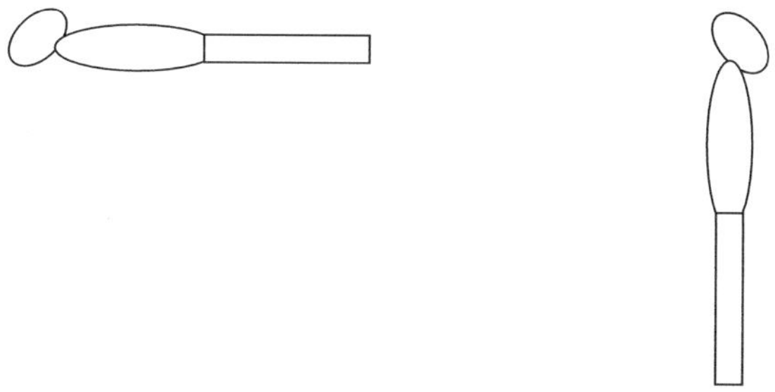

FIGURE 8.15
Comparison of pressures within arteries and veins in mmhg (above atmospheric) for a person lying down and standing.

WORKED EXAMPLE

Question: Show by calculation why, in the standing human, if the mean arterial blood pressure measured in the upper arm is 95 mmHg, the same quantity measured in the leg will be 175 mmHg (assuming the points of measurement to be 1 m apart).

Answer: the difference between the two measurements is 80 mmHg, which is 80/100 × 13.3 kPa, or 10.64 kPa. Recalling that $\Delta p = g\rho\Delta z$, with $g = 9.81$ m/s^2 and $\rho =$ 1070 kg/m^3, the right-hand side is thus 10.5 kPa for $z = 1.0$, which is within experimental variation.

8.6.2 Why Is the Pressure Pulsatile? Is It Pulsatile in Capillaries and Veins Also?

This is because the heart is a pulsatile pump, and pulsatile flow leads to pulsatile pressure. Rather like a bicycle tire's inner tube, the vascular system behaves like a large, pressurized, elastic reservoir. It also resembles an electric circuit, with pressure represented by voltage. Each part of the vasculature can be represented by parallel resistances. The major part of the resistance is made up by the arterioles, which control the arterial blood pressure and maintain a mean value of around 95 mmHg. The bulk of the pressure drop is across these vessels. Although the capillaries are much smaller in diameter, many of them are in parallel, so the overall resistance is less. The pulsations in pressure are just detectable in the capillaries if the flow is artificially stopped by occlusion.

In most small veins, there are no pulsations in pressure, but in the larger ones, pulsations in nearby arteries, muscular contractions, respiration, or contraction of the heart itself cause pressure fluctuations.

8.6.3 How Is the Tension in the Vessel Walls Related to Pressure?

This is given by the so-called law of Laplace, which can be represented by the following simple formula:

$$T = pr.$$

Here, T is wall tension in pascal-meters, p the pressure in pascals, and r the radius in meters. During the left ventricular isovolumetric contraction phase (see Figure 8.7), the radius is constant, so the pressure is proportional to wall tension. The rate of change of this tension with time is also a measure of the rate of shortening of muscle fibers, if they were allowed to do so ("V_{max}"). In Chapter 11, the importance of measuring the rate of pressure rise during this phase as an index of the health or otherwise of the heart muscle ("myocardial contractility") is discussed.

8.6.4 How the Pressure Varies throughout the Circulation

Imagine a very small organism or "microbot" moving through the circulation but carrying a pressure transducer. The x-axis is the distance that the organism has moved. Note that the pulses are present in the arterioles, where the fall-off of pressure is greatest (Figure 8.16).

8.6.5 Energy Expended by Heart per Beat

Each time the ventricles contract, the muscles do work against the pressure in the aorta (on the left side) and the pulmonary arteries (on the right). Similar to a piston in a cylinder, the work done is $\int p.dV$, or $p'.\Delta V$, where p' is the mean aortic or pulmonary artery pressure (95 and

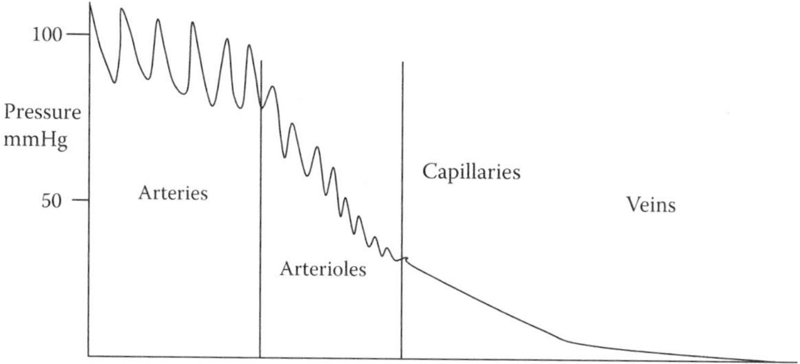

FIGURE 8.16
Pressure variation through the systemic vasculature. Note that the pressure is pulsatile in the arteries and arterioles and that the fall in pressure is greatest across the arterioles.

16 mmHg, respectively) and ΔV is the SV. These values have to be converted to pascals and cubic meters, respectively, in order to obtain work done in joules. The ventricles also impart momentum to the blood, reflected in the kinetic energy (KE) density ($1/2\,\rho <u^2>$, where $<u^2>$ is the average velocity squared for the entire cardiac cycle). For the moment, we will assume that the KE component can be ignored (the next chapter will discuss this further), but taking ΔV as 80 mL ($= 80 \times 10^{-6}$ m^3) and p′ as 95/100 × 13,300 Pa, we obtain a value of just over 1 J per beat for the left ventricle (and 0.17 J for the right ventricle). Since the average heart beat is 70/60 s, or 1.17 times per second, the rate of doing work (power) is around 1.2 W. As a comparison, the power of a compact fluorescent lamp bulb is around 15 W. This is only approximate: compare with Figure 8.8, where the *Work Done* was represented by the area of the loop. However, remember that the SV ΔV is the difference between the two vertical lines in the loop and p′ is the average of the upper part of the loop. The lower part (which is very small anyway) is ignored.

Tutorial Questions

1. In the horse, the resting heart rate is around 30 bpm, and the SV is 900 mL. Show that the cardiac output is around 27 L/min. Convert this to cubic meters per second.
2. A giraffe is 6 m tall. Calculate the change in pressure in arteries in the head as the animal changes from reaching for high leaves to drinking from a pool at its feet. Express this in atmospheres, given that 1 atm = 1.01×10^5 Pa.
3. Given that the pressure in the ventricles can reach 120 mmHg and that the volume at end-diastole is 120 mL, estimate the tension in the wall in pascal-meters at the beginning of the ejection phase, assuming the ventricle to be spherical ($V = 4/3\pi r^3$).

Bibliography

Burton AC. 1972. *Physiology and Biophysics of the Circulation*, 2nd Edition. Year Book Medical Publishers, Chicago.

Davson H, Eggleton MG. 1968. *Starling and Lovatt Evans Principles of Human Physiology*, 14th Edition. Churchill, London.

Hall JE. 2011. Guyton and Hall. *Textbook of Medical Physiology*, 12th Edititon. Elsevier, Philadelphia, PA, USA.

9

Rheology of Blood

Andrew W. Wood

CONTENTS

9.1 Introduction

Blood is a remarkable fluid. In the 1817 "New Medical Dictionary" by Robert Hooper, the writer notes that "some . . . have considered it as alive and have formed many ingenious hypotheses in support of its vitality." The previous chapter described that the purpose of having a circulation was to transport materials to where they are needed. Blood is a complex mixture of cells, proteins, and electrolytes. Among other requirements, blood needs to be able to transport, for example, oxygen (O_2) to cells in such a way that it is stored within the blood until it reaches the capillaries, where it is delivered to the cells that need it. It is possible to estimate the rate at which O_2 needs to be transported by considering the rate at which it is absorbed within the lungs. Normally, this is around 250 mL/min or 4 mL/s or 4/22.4 = 0.18 mmol/s. Assuming that the body weighs 70 kg and thus occupies around 7×10^4 mL, the requirement is approximately $4/(22 \times 7 \times 10^4) = 2.6 \times 10^{-6}$ mmol/mL/s. This may seem a small figure, but the solubility of O_2 in saline is approximately 6 mg/L or 0.2×10^{-3} mmol/mL, and the diffusion coefficient is 2×10^{-9} m^2/s. If the lungs were to rely on dissolving in plasma and then diffusion alone, the best rate of delivery would be around 1.5×10^{-5} mmol/s (assuming a lung exchange area of 75 m^2 and a diffusion path length of 2 μm). This is a factor approximately 104 less than the actual figure given above. Because O_2 uptake increases by a factor of around four times during exercise, the need for enhanced delivery becomes apparent. Several important hormones, such as insulin and adrenalin, also are delivered via the blood stream, and these have to be delivered such that an appropriate level of concentration is reached in target organs within a certain time frame.

To fully understand the behavior of blood as a fluid, it is important to have a clear understanding of the way that fluids, in general, behave and then blood in particular. Rheology is the study of the way fluids flow; there are a number of basic principles, which will now be described.

9.2 Properties of Fluids

There is a fundamental relationship between the velocity of fluid (u) and the flow rate (which will be denoted by Q). The flow rate is given in units of volume per unit time, such as mL/min or (in SI units) m^3/s. The flow rate usually refers to an enclosed space, such as a tube, and the two are related by the following:

$$Q = uA = u\pi r^2$$

where A is the cross-sectional area, and r is the tube radius, for a circular cross section. Note that 1 m^3/s = 10^3 L/s = 10^6 mL/s.

If the fluid has a velocity which varies over the cross section (which it normally does), then u represents a mean velocity averaged over the cross section ($\bar{u}$) rather than the value at a particular location.

WORKED EXAMPLE

Question

Estimate the mean velocity of blood during systole in the aorta, given an aortic diameter of 20 mm and a resting cardiac output (CO) of 6 L/min. Assume that systole lasts for 0.3 s and diastole 0.5 s.

Answer

First, express CO in SI units: 6 L/min is 100 mL/s or 100×10^{-6} m^3/s. However, blood is ejected into the aorta only during systole, so the flow rate during this time (Q′) must be ((0.3 + 0.5)/0.3) 100×10^{-6} m^3/s. The radius r is 0.01 m, so putting $u = Q'/(\pi r^2)$, we get $(8/3)\, 100 \times 10^{-6}/(\pi\, 10^{-4}) = 0.84$ m/s or 84 cm/s.

9.2.1 Viscosity

As mentioned, an important aspect of rheology is the way that fluid is able to deform: a viscous fluid such as lubricating grease or margarine does not deform easily (when, e.g., a block of one of these is placed between two flat plates and, then, the plates moved in different directions), whereas other fluids, such as milk or water, are much easier to shear.

A pile of smooth A4 sheets of paper will display some of the aspects of viscous flow of fluids. If the top sheet of a fresh ream of printer paper is gently stroked repeatedly in a particular direction, the top sheet will move, carrying the sheets below to a lesser extent. One way to "fan" a pile of smooth papers is to use the knuckle to execute a circular motion on the top sheet; the others will slowly follow to produce a fan. In the first case, the force is being applied tangentially to the top sheet; it amounts to a stress (force/area) where the area concerned is the area of the sheet (see Figure 9.1). The sight adhesion forces between sheets cause the effects on the top sheet to be transferred to the sheets below. Essentially, what prevents all the sheets moving at the same speed is the friction between them. Viscosity is fluid friction. Because each of the sheets is moving (slowly!), we can consider how the speeds vary with distance down the pile (or in this case, from the bottom, see diagram). If the pile of papers were replaced by a block of margarine, a similar thing would happen, but in this case, the layers would be infinitesimally small.

Suppose for the moment that the speed (velocity) increases linearly from the bottom, then there is a constant velocity gradient du/dz all the way through the pile (or block). The basic law of viscosity is $S = \mu du/dz$, where S is tangential stress (F/A), and μ is a constant coefficient of viscosity (SI unit: Pa s). The gradient du/dz generally is not constant, and thus, the stress (in the fluid, rather than the applied stress) will vary with z. Fluids for which stress is proportional to the rate of shear (du/dz) are known as Newtonian (μ is constant).

There are fluids for which μ is not a constant. These are known as non-Newtonian and include familiar examples such as cream (which exhibits a yield stress so that it will not shear until a certain stress is exceeded, which is why it is difficult to pour) and ceiling paint (which shows greater viscosity with low shearing rate, which makes it less likely to drip). "Silly putty," which can be drawn out if the rate of shear is low but resists deformation if stressed rapidly (when dropped on the floor, it will bounce), is an example of the reverse of ceiling paint.

The viscosity coefficient of water varies with temperature but is around 10^{-3} Pa s (1 mPa s). There is an older unit, the poise. One centipoise (cP) is equal to 1 mPa s. Whole blood has a viscosity coefficient of around 5 mPa s, but blood is a non-Newtonian fluid, that

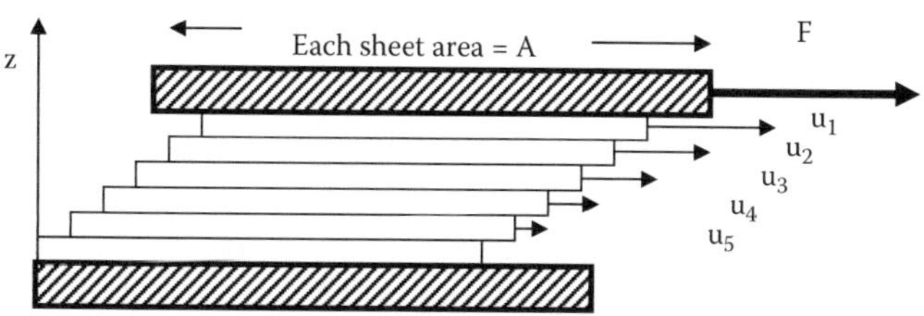

FIGURE 9.1
A fluid between two plates, one is moving, and the other is stationary, is imagined as a series of sheets (lamellae) sliding over each other. The shaded plate shown at the top is being moved with a force F to produce the velocities u_1, u_2, ... in each of the lamellae. The velocity of the highest and lowest lamellae, which are in contact with solid boundaries, are u_1 and zero, respectively.

is, μ depends on du/dz. This is discussed further below and complicates the accurate measurement of μ for blood. In many arteries and arterioles, the behavior is approximately Newtonian, and the relation between blood pressure and flow is given by Poiseuille's law (see below), which is derived by considering viscous forces.

9.2.2 Conservation of Mass

In a tube whose cross section varies, if the tube is rigid, the amount of fluid entering a section is the same as the amount leaving (unless there are tributaries: i.e., other tubes connecting within the section). In other words,

$$u_1A_1 = u_2A_2 = Q$$

where the u's are velocities in *m/s*, the A's are cross-sectional areas in m^2, and Q is flow rate in m^3/s. This is a very important fundamental relationship; if the tube tapers, the velocity of flow will increase—this is the principle used in the nozzles of fire and garden hoses to produce high-velocity jets. If an artery narrows because of disease, then the velocity of blood flow will increase along that section, with increased turbulence.

If the tube is elastic and V is the volume of a particular section,

$$dV/dt = Q_1 - Q_2 = u_1A_1 - u_2A_2.$$

Here, there is a difference between influx and efflux, and this is represented by the rate of change of volume of the section itself. In an artery, for example, the wall is elastic and expands momentarily as the pressure rises with each heartbeat. This will be discussed further in the next chapter.

WORKED EXAMPLE

Question

Water enters a flexible tube at 10 mm/s and emerges at 15 mm/s. The diameters of the inlet and outlet are 10 mm and 6 mm, respectively. What will the rate of expansion of this tube be?

Answer

From the above equation, the rate of change of volume dV/dt will be $10^{-2} \times \pi \times (0.005)^2 - 1.5 \times 10^{-2} \times \pi \times (0.003)^2 = (7.85 - 4.24) \times 10^{-7}$ m^3/s = 0.36 mL/s.

9.2.3 Conservation of Energy

In large arteries and veins, the viscous dissipation of energy is unimportant (see Reynolds number [Re], below), and energy is conserved along imaginary streamlines* drawn in the fluid. Because there are no velocity components at right angles to the streamline, no kinetic energy crosses it (see also Figure 9.2). The energy at any point on the streamline is a sum of three terms: kinetic energy, potential energy, and pressure energy. When this is expressed per unit volume, we get

$$\tfrac{1}{2}\rho u^2 + \rho gz + p = H \qquad \text{(Bernoulli's equation)*.}$$

* A streamline is a line in the fluid such that the direction of flow is always tangential to this line.

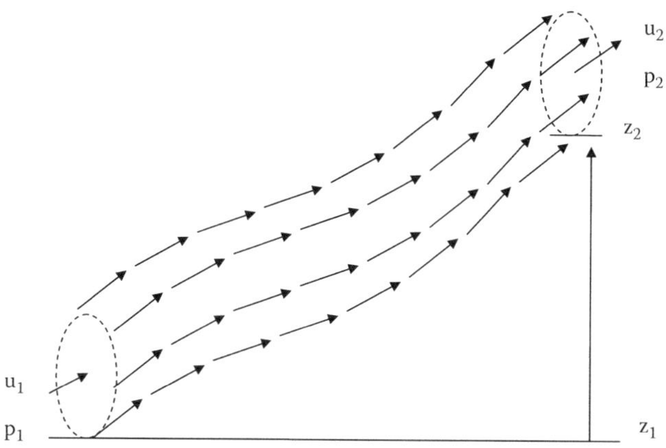

FIGURE 9.2
An imaginary tube constructed of streamlines within the flowing fluid. Because the velocity vectors are all pointing along the walls of the tube and there are no velocity components at right angles to these, there is no loss of total energy in going from one end of the tube to the other (as long as frictional, or viscous, forces can be ignored).

Here, ρ is blood density, u is velocity, g is gravitational acceleration, z is vertical height, and p is pressure. In unsteady flow (such as arterial flow), streamlines are not the same as pathlines.* H is a constant and is sometimes known as the "head" (as in "head" of pressure).

For a fluid at a constant vertical height, regions where the velocity is higher will be where the pressure is less, compared with other regions along the same streamline. This is used, for example, in the Venturi principle, where fluid is forced through a constriction (a "throat"). The low pressure in the constriction can be used to generate a vacuum (see Figure 9.3).

This also is the reason why the pressure in a fluid decreases as vertical height increases. In the previous chapter, we saw that pressures in the arteries of the head are around one third of those in the feet (we also saw the variation with height $\Delta p = g\Delta z$).

9.2.4 Laminar and Turbulent Flow Patterns

Flow patterns, which are regular, are known as laminar, and those, which involve swirling motion, are known as turbulent. As the speed of flow increases, there is a transition from the first to the second. This can be observed in the patterns of smoke emanating from a stack; as the speed of emission increases, vortices and eddies start to be observed in the plume, and the flow pattern becomes less stable. These swirling motions greatly enhance the mixing of fluids; rapid stirring of milk into tea with a spoon will rapidly intersperse the two fluids, whereas slower stirring preserves the milk in streaks.

* A pathline is the path taken over time of an individual particle in the fluid. Leaves dropped into a river from a bridge will illustrate pathlines. On the other hand, water passing a certain point (such as a protruding branch) follows a "streakline." Streamlines are different again, but for steady flow, streamlines follow pathlines.

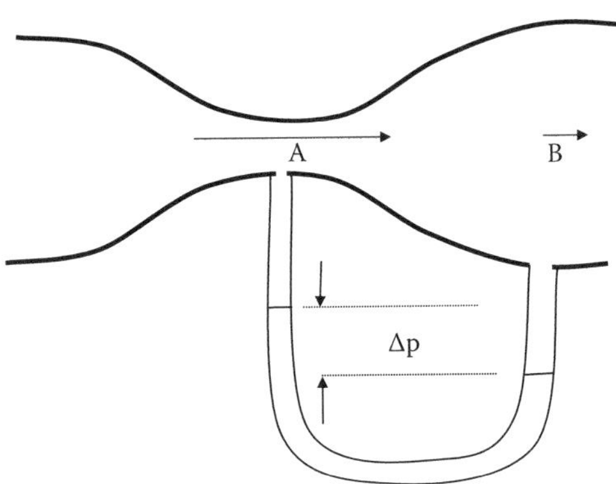

FIGURE 9.3
The Venturi principle. Assuming Bernoulli's equation can be applied to the region between points A and B, the difference in velocity implies that there will be a difference in pressure, with a lower pressure at point A.

9.2.5 Steady and Unsteady Flow

Flow can be laminar but unsteady, that is, the pattern varies with time. In this case, pathlines do not follow (instantaneous) streamlines. Pathlines in rivers can be observed easily by dropping leaves from a bridge and watching their progress. Two such leaves dropped in quick succession in the same region will probably follow similar paths. If the river is influenced by the movement of waves downstream, then the two leaves will follow different paths. Because blood flow in the arteries and elsewhere is pulsatile, the flow is unsteady, and this adds to the complexity of the analysis of blood flow in the body.

9.2.6 Reynolds Number

This is a dimensionless number, which represents the ratio of inertial to viscous forces, that is, whether the fluid velocity is sufficient to overcome frictional forces. The precise definition depends on the environment, but for flow in a circular tube, this is usually given as follows:

$$\mathrm{Re} = \mathrm{ud}/\nu, \text{ where the kinematic viscosity } \nu = \mu/\rho.$$

Reynolds number is useful for two main reasons: it is a way of assessing whether viscosity can be ignored (and Bernoulli's equation can be safely applied) and secondly for determining whether the flow will be laminar or turbulent. If Re» 1, viscosity can be ignored, and if Re is greater than approximately 2000, the flow is turbulent. There is in fact a transitional region of Re values between 1500 and 2500 where the flow will alternate unpredictably between laminar and turbulent. As an example, the flow velocity in the aorta is roughly 0.5 m/s, the diameter is 0.02 m, and the density is just over 10^3 kg/m^3. Using the value of μ given above, this gives a value for Re of 2000. It should be noted that occasionally, the equation is given with radius rather than diameter, in which case, the relevant Re values are half of those just discussed.

9.2.7 Poiseuille's Law

The derivation from first principles is given in Section 9.4. In a long tube, once a flow pattern has been established in which the pressure falls linearly with distance (see next chapter on "developed flow patterns"), the rate of shear (du/dr) is proportional to r (the radial distance from the centerline), and the velocity profile is parabolic, the ratio of pressure drop Δp to flow rate Q over a distance ℓ is given by the following:

$$\Delta p/Q = 8\mu\ell/(\pi r4) \qquad \text{Poiseuille's law.}$$

Note that this formula is a fluid analog of Ohm's law: Δp is analogous to potential difference and Q to electrical current. The analogy becomes even more striking if u rather than Q is used in the formula; it becomes the following:

$$\Delta p/\bar{u} = 8\,\mu\ell/(r^2) \qquad \text{Poiseuille's law: another form.}$$

The right-hand side has more than a passing resemblance to the resistor formula $R = \rho\,\ell/w^2$ for a resistor ℓ long and w wide in both directions and resistivity ρ. More details will be given in the next chapter on the influence of the vessel walls on the flow patterns within the blood vessels.

9.2.8 Navier–Stokes Equations

This can be used to solve general fluid dynamics problems. A form of these equations suitable for fluid flow in tubes is derived at the end of this chapter. In the tube, the radial distance is denoted by r and the axial distance by x; the relevant equation is as follows:

$$\frac{d^2u}{dr^2} + \frac{1}{r}\frac{du}{dr} - \frac{1}{\mu}\frac{dp}{dx} = \frac{1}{\nu} \times \frac{du}{dt}.$$

If we can ignore the first two terms of this equation, we get (remembering that $\nu = \mu/\rho$):

$$-\frac{\partial p}{\partial x} = \rho\frac{\partial u}{\partial t}.$$

9.3 Properties of Blood

9.3.1 Blood Composition

Blood is a complex mixture containing cells, proteins, and electrolytes. The components can be listed as follows:

9.3.1.1 Formed Elements

1. Red cells (erythrocytes—from "erythros," red). There are approximately 5×10^6 per mm^3. They are formed in the bone marrow from pluripotent stem cells, but once they differentiate and reach maturity, they lose their nuclei. At that stage,

they are essentially bags of hemoglobin (Hb), the globular protein involved in O_2 transport. In fact, Hb makes up approximately 35% of the red cell by weight. Each red cell is a biconcave disk 2 μm thick and 8 μm in diameter, which, because of this "half-deflated beach ball" structure, is highly deformable. Red cells are able to flow along the smallest capillaries, which are 3 μm in diameter.

2. White cells (leucocytes: various subtypes—named from "leukos," white). There are 8×10^3 per mm^3, so they are around 1000 times less numerous than red cells. They perform many functions, mainly to do with immune responses, which involves the digestion of bacteria and dead tissue.
3. Platelets or thrombocytes are small cell fragments 2 μm–3 μm in diameter (thus, smaller than erythrocytes), which are involved in wound healing by releasing several important factors at the appropriate moment.

9.3.1.2 Liquid Particles

Chylomicrons are large lipoproteins, which are involved in fat (lipid) transport. These are created in the small intestine, where they enter first the lymphatic system and then the bloodstream (via the thoracic duct of the lymph system). They are then delivered to the liver, where they are then metabolized. They consist mainly of triglycerides and form a spherical structure with a diameter of a few hundred nanometers.

9.3.1.3 Plasma

The composition of plasma can be divided into two main components: inorganic ions and proteins.

Inorganic ions have the following approximate concentration in mM: Na^+ 151; Cl^- 106; K^+ 5; HCO_3^- 36; Ca^{2+} 2.5; Mg^{2+} trace; phosphate 0.4. Note that the major cations add to around 158 mM and anions to 142 mM, the difference being made up by negatively charged macromolecules. It is customary to represent the osmotic pressure (OP) of plasma as the sum of these molar amounts (158 + 142 = 300 mOsmol/L). Saline solutions having the same OP as plasma ("isotonic") have a concentration of 150 mM or 9 g/L. Converted to SI units, 300 mOsm/L = 300 Osm/m^3. This, multiplied by RT (8.31 J mol^{-1} K^{-1} × 310 K), gives the OP in Pa (800 kPa or 6000 mmHg).

The protein content of plasma amounts to approximately 68 g/L, consisting of albumin (mainly), globulin, and fibrinogen. This latter is involved in clotting, in which it is converted to fibrin. Anticoagulants *prevent this happening* and have to be added to blood before centrifugation if *plasma* is to be collected. If clotting occurs in a blood sample, the clear liquid collected is called *serum*. Anticoagulants include the following: heparin, warfarin (coumadin), oxalate, citrate, and Ca^{2+} chelators (see Martini and Ober 2005). These proteins contribute to OP of 25 mm Hg (3.3 kPa), which, although small in comparison to the isotonic saline figure above, represent the main determinant of water movement across the walls of the capillaries, because the small clefts between cells are too small to allow plasma proteins through to the extracellular fluid space but are big enough to allow Na^+ and other small ions through. The 300 mOsm/L figure is relevant for individual cell membranes (such as the red cells themselves) because Na^+ and Cl^- do not readily permeate. If a dilute saline solution were to be infused intravenously, there is a danger of red and other cells swelling and bursting.

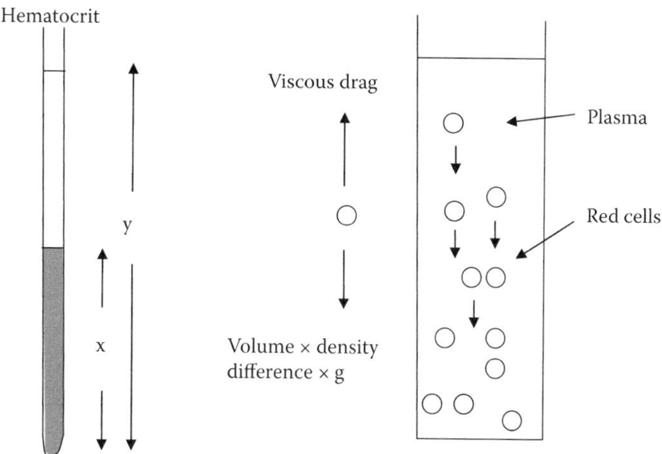

FIGURE 9.4
Left: the definition of Hct. Whole blood is centrifuged for 5 min or so in a parallel-sided tube, and then, the ratio of the packed cell volume to the total volume (represented as x/y) is taken. This converted to a percentage is the Hct. Right: ESR. Red cells (erythrocytes) can be considered as spheres, which will attain a terminal velocity u very quickly within the blood plasma. This steady velocity (which is the ESR) represents a balance between downward gravitational and upward viscous forces. This can easily be measured by noting the time taken for the top of the tube to become clear. A number of diseases give rise to abnormal ESR values.

9.3.2 Hematocrit

Hematocrit (Hct) is the volume % represented by red blood cells (normally 40%–47%). It is usually measured by spinning a small blood sample in a centrifuge for 5 min or so until the red cells form a packed mass at the bottom of a sample tube. If the blood sample is put into a thin tube with parallel sides (as shown below), the Hct is simply the ratio of the column of red blood cells to the total column (x/y × 100). Hct is low in anemia (where the control system for red cell production is impaired) and high in polycythemia, which can either be due to an overproduction of red cells in the bone marrow or a decrease in plasma volume (Figure 9.4).

9.3.3 Erythrocyte Sedimentation Rate

A blood sample left to itself will slowly clear from the top—this is what happens much more quickly in a centrifuge. The rate of sedimentation (or velocity in m/s) is given by the formula

$$u = 2(\rho_{rbc} - \rho_{plasma})a^2g/(9\mu),$$

where a is the effective radius of the red cells, and μ is the plasma viscosity. The term $\rho_{rbc} - \rho_{plasma}$ represents the difference between red cell and plasma density, usually around 50 kg/m^3. The effective radius can be bigger than 4 microns because the red cells tend to stack to form aggregates or *rouleaux*.

Values usually range around a few centimeters per hour. Erythrocyte sedimentation rate (ESR) is elevated in a range of conditions including tuberculosis, cholera, and some cancers. It is low in sickle-cell anemia and where there is low plasma protein. The derivation of this is straightforward and illustrates some useful concepts. The red cells descend because of gravitational force: this force is given by gravitational acceleration g times the effective mass ($\rho_{rbc} - \rho_{plasma}$) × volume or ($\rho_{rbc} - \rho_{plasma}$) $4\pi a^3/3$. This force is opposed by the viscous frictional force, which is given by Stoke's law $F = 6\pi\mu au$. When the red blood cell attains a constant velocity, the two forces are equal, and the formula at the beginning of this section is obtained. It also is instructive to use this formula to check how long blood should be spun in a centrifuge to get effective separation. The effective acceleration is given, not by g this time but by $r\omega^2$ or $4\pi^2 rf^2$, where r is the radius of the centrifuge rotor, and f is the number of rotations per second. For example, in a centrifuge with radius of 0.1 m and speed of 600 rpm (10 rotations per second), the acceleration is 40 × g approximately, and velocity of sedimentation is 0.3 mm/s. The red blood cells will thus take 170 s to travel 50 mm in a blood tube.

9.3.4 Plasma Viscosity

The viscosity of plasma is fairly easy to measure (in contrast to whole blood, see below). Values are around 1.2 mPa s at 37°C, but rather higher values are obtained at room temperatures.

9.3.5 Effective or Apparent Viscosity of Whole Blood

Presence of red blood cells modifies viscosity because of extra shearing involved. Figure 9.5 below shows how the velocity gradient becomes steeper (which is the shearing rate) because of the need to flow around the red cell. There is a "no slip" condition at the surface of the cell, which means that the adhering fluid has to flow at the same speed as the red cell itself (and hence, the speed is the same on all surfaces of the cell). The viscosity appears larger because greater stress has to be applied to overcome the extra shearing, although the actual plasma viscosity does not change. The effective (or apparent) viscosity of the whole blood is nearer 5 mPa s, but the precise value depends on where in the cardiovascular

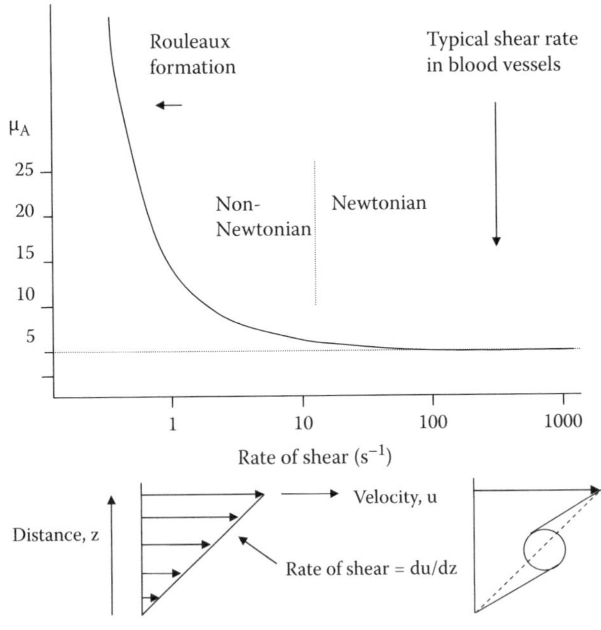

FIGURE 9.5
Variation of apparent viscosity of the whole blood with rate of shear. At low shear rates, blood exhibits yield stress. (Adapted from Caro CG et al., *The Mechanics of the Circulation*, Oxford University Press, Oxford, 1978.)

system is being considered and under what circumstances. Some of the viscous behavior of the whole blood can be predicted from theories of deformable spheres in fluid. For example, some of the equations derived to predict the viscosity of an emulsion, such as milk (containing minute fat droplets), can be adapted for blood. Some of the features of sickle-cell anemia (and the related disease, thalassemia) can be mimicked using a simple formula derived by Einstein for rigid spheres in fluid:

$$\mu_{(blood)} = \mu_{(plasma)}(1 + 2.5\ \text{Hct}/100).$$

There are three effects worthy of note (Figure 9.6):

1. Effective viscosity rises very steeply as *rate of shear* falls below 1 S^{-1} (due to rouleaux formation).
2. Effective viscosity rises as hematocrit rises (can rise to around 10 for Hct = 80%).
3. Paradoxically, effective viscosity falls as tube diameter falls below 0.2 mm (but rises again as diameter falls below 0.01 mm). This is due to axial accumulation of red cells, giving rise to a relatively cell-free layer close to the vessel wall. This is known as the Fåhraeus–Lindqvist effect.

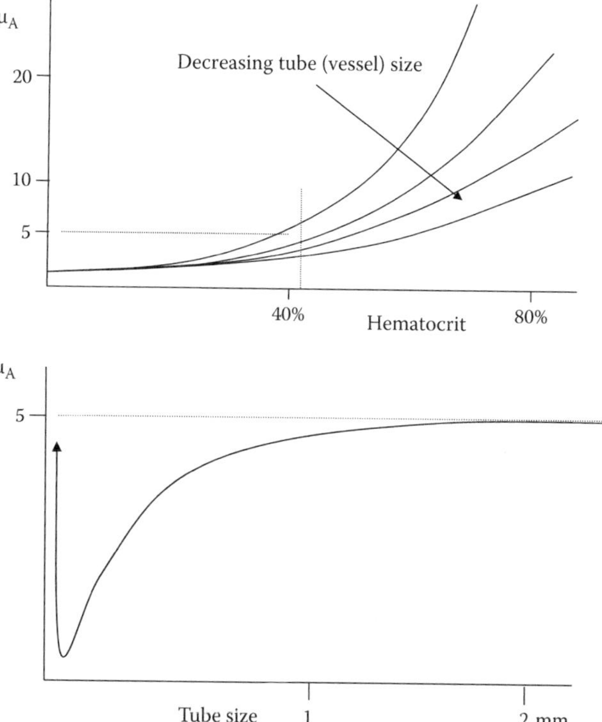

FIGURE 9.6
Variation in apparent viscosity (μ_A) of the whole blood. Top: variation of μ_A with hematocrit determined in tubes of various diameters. Lower: the Fåhraeus–Lindqvist effect—the reduction in μ_A as blood vessel diameter reduces. Below 0.01 mm, μ_A rises due to restricted movement within the vessels. This effect can also be seen in the top diagram by noting the fall in μ_A for smaller vessels at constant Hct (dotted vertical line).

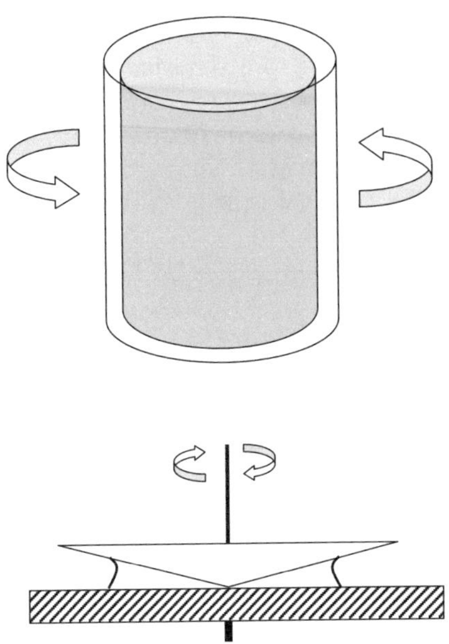

FIGURE 9.7
Diagrams of viscometers. Top: the concentric cylinder type—fluid under test is placed between the cylinders. The torque on the inner cylinder as the outer cylinder is rotated is measured. Lower: the cone and plate type. The fluid is introduced between the rotating cone and the stationary plate. A sensitive strain gauge measures the torque on the plate.

9.3.6 Measurement of Viscosity of Whole Blood

There are three broad types of viscometer, which have been used to estimate the effective (apparent) viscosity of blood under various conditions. These are described below (see also Figure 9.7).

9.3.6.1 Capillary Type

This measures the flow produced through a series of capillary tubes for a particular pressure between the ends of the tubes. This type assumes flow to be fully developed (i.e., a parabolic velocity profile exists within the tube), and the Poiseuille formula is applied. Tubes have to be <100 μm in diameter, and it is rather difficult to produce tubes with uniform diameter throughout their length. Some workers have used animal limbs (in vivo) as a viscometer; obviously, many assumptions have to be made, but the Poiseuille equation can be applied if indwelling pressure and flow transducers can be made to operate in the main arteries and veins to the limb. The chief assumptions relate to assuming that the arterioles mainly determine viscosity; the average radius and length have to be known as does the number.

9.3.6.2 Rotational

1. Concentric Cylinder

   This is basically two cylinders, one stationary, and the other is rotating at angular velocity ω. There needs to be some means for determining the torsion on the stationary cylinder. Because the shear rate in fluid is proportional to r^2 (see derivation below), there is a need for the thickness of the fluid layer to be very much less than r for constant shear rate in gap. If the fluid exhibits a "yield stress," the shearing is underestimated (see Figure 9.7, upper).

2. Cone and Plate

   This is considered to be the best method for measuring blood viscosity, and most commercial viscometers are now of this type. The cone rotates, and the torsion on the plate around an axis is measured. The cone has a very shallow shape, such that the angle in the space where the fluid is placed is around 3°. This angle seems to ensure that the shearing rate is independent of r and that this is almost as true for non-Newtonian as Newtonian fluids (Figure 9.7, lower).

9.4 Derivations

9.4.1 Derivation of Poiseuille's Equation

Consider the flow along a tube to be like a series of concentric shells sliding over each other (Figure 9.8).

Consider points *A* and *B*, distance *L* apart, on a shell of radius *r*. Suppose pressure at A is *p* and that at *B* is $p + \Delta p$, the shell is Δr thick.

$$\text{Net pressure force acting on shell} = -\Delta p \times \textit{shell} \text{ cross-sectional area}$$

$$= -\Delta p\pi\left((r+\Delta r)^2 - r^2\right)$$

$$\cong -\Delta p\pi \times 2\pi r\Delta r$$

Net viscous force = difference between stresses (in Pa) at inside and outside surfaces of shell × area of shell surface

$$= S_2 \times 2\pi(r+\Delta r)L - S_1 \times 2\pi rL$$

$$= \mu\left(\frac{du}{dr}\right)_{r+\Delta r} \times 2\pi(r+\Delta r)L - \mu\left(\frac{du}{dr}\right)_r \times 2\pi rL$$

$$= \mu \times 2\pi L\frac{d}{dr}\left[\left(\frac{du}{dr}\right)\times r\right]\delta r \text{ as } \Delta r \to \delta r \to 0$$

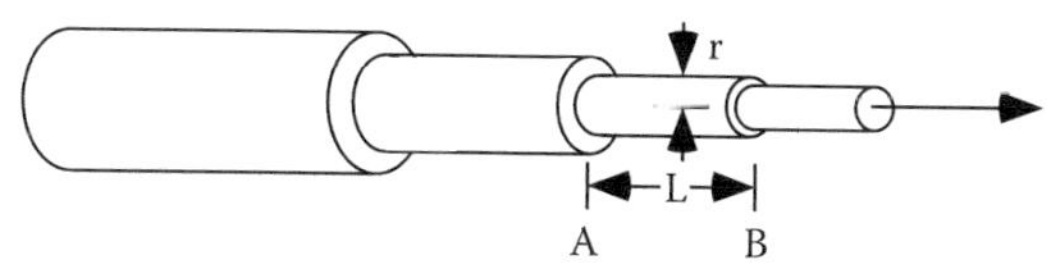

FIGURE 9.8
Derivation of Poiseuille's law. Flow in a circular cross section tube is imagined as a series of concentric cylinders sliding over each other. The derivation considers a short length of a particular cylindrical shell AB.

These two forces must be equal:

$$\mu \times 2\pi L \frac{d}{dr}\left[\left(\frac{du}{dr}\right)\times r\right]\delta r = -\Delta p \times 2\pi r \delta r \quad *.$$

First integration gives the following:

$$\left(\frac{du}{dr}\right)\times r = -\frac{\Delta p \times r^2}{2\mu L} + c \quad **,$$

now when $r = o$ (center of tube),

$$\frac{du}{dr} = 0, \text{ so } c = 0$$

second integration gives

$$u = \frac{-\Delta p}{4\mu L}\times r^2 + k$$

because of "no slip" conditions, when $r = a$, $u = o$, (a is tube radius), thus

$$k = \frac{a^2 \Delta p}{4\mu L}.$$

We then have

$$u = \frac{\Delta p}{4\mu L}(a^2 - r^2),$$

but we usually measure volume flow rate Q, where

$$Q = \bar{u}A = \bar{u}\pi a^2.$$

Hence, we have to find the mean velocity, that is, from

$$\bar{u} = \frac{\int_0^{\pi a^2} u\,dA}{\int_0^{\pi a^2} dA} = \frac{\int_0^a \frac{\Delta p}{4\mu L}(a^2 - r^2)\times 2\pi r\,dr}{\pi a^2}$$

$$Q = \bar{u}\pi a^2 = \int_0^a \frac{2\pi \Delta p}{4\mu L}(a^2 r - r^3)\,dr$$

$$= \frac{\pi \Delta p}{2\mu L}\left|\frac{a^2 r^2}{2} - \frac{r^4}{4}\right|_0^a = \frac{\pi \Delta p}{2\mu L}\left(\frac{a^4}{2} - \frac{a^4}{4}\right),$$

thus

$$Q = \frac{\pi a^4}{8\mu L}\Delta p \qquad \text{[Poiseuille's law]}.$$

9.4.2 Equation of Motion for Unsteady Flow

If the flow in the cylindrical tube is varying with time, a mass x acceleration term must be included in the force equation. The equation on the previous page marked * becomes

$$\mu 2\pi L \frac{d}{dr}\left(\frac{du}{dr} r\right)\delta r + \Delta p 2\pi r \delta r = \rho 2\pi r L \delta r \frac{du}{dt}$$

$$\rho 2\pi r L \delta r = \text{density} \times \text{volume} = \text{mass}$$

$$\frac{du}{dt} = \text{acceleration}.$$

Divide by $\mu 2\pi L \delta r$

$$\frac{d}{dr}\left(\frac{du}{dr} \times r\right) + \frac{\Delta p r}{\mu L} = \frac{\rho}{\mu} \times r \frac{du}{dt}.$$

If L is made small $\frac{\Delta p}{L} \rightarrow -\frac{dp}{dx}$, where x is the distance along the tube (the minus sign because we have tacitly assumed $p_A > p_B$), thus,

$$\frac{1}{r}\frac{d}{dr}\left(\frac{du}{dr} \times r\right) - \frac{1}{\mu}\frac{dp}{dx} = \frac{\rho}{\mu} \times \frac{du}{dt}$$

$$\frac{d}{dr}\left(\frac{du}{dr} \times r\right) = \frac{d^2u}{dr^2} + \frac{du}{dr}; \text{ NB } \frac{\rho}{\mu} = \frac{1}{\nu}; \nu = \text{Kinematic viscosity}$$

that is,

$$\frac{d^2u}{dr^2} + \frac{1}{r}\frac{du}{dr} - \frac{1}{\mu}\frac{dp}{dx} = \frac{1}{\nu} \times \frac{du}{dt}.$$

This is the equation we saw earlier. If external forces are acting, these should be included in the equation. In general, for a three-dimensional case,

$$\nabla^2 \underline{u} - \frac{1}{\mu} grad\, p + \frac{\underline{F}}{\mu} = \frac{1}{\nu} \times \frac{du}{dt} \qquad NB\ \nabla^2 \equiv \frac{\partial^2}{\partial x^2} + \frac{\partial^2}{\partial y^2} + \frac{\partial^2}{\partial z^2},$$

where $\underline{u}$ and $\underline{F}$ are vector quantities. This equation assumes the fluid incompressible. For a compressible fluid, more complicated but general sets of equations apply, which are alternative forms of the Navier–Stokes equations.

Tutorial Questions

1. A sample of heparinized blood has been allowed to stand for 1 hr, and a clear layer 5 cm deep has formed at the top. Estimate the effective radius of the red cells (as if they were spheres), given that the viscous force is $6\pi\mu av$, where v is the terminal velocity, and a is the radius, and also given that the densities of plasma and red cells are 1030 kg m^{-3} and 1100 kg m^{-3}, respectively. How does this compare with the normal red cell radius?

   Why is ESR elevated in disease?
2. In an attempt to measure plasma viscosity, 0.8-mm-diameter silicone balls $\rho = 1300$ kg m^{-3}) are allowed to fall through a 10-mL sample of plasma contained in a measuring cylinder, where they attain a terminal velocity (u) of 6 cm s^{-1}. From these data, given that the frictional force is $6\pi\mu au$ and that the 10-mL sample weighed 10.3 g, calculate plasma viscosity μ (a is ball radius).
3. A flow meter placed on the dog aorta upstream from the arterial branches registers a (time and radial position) mean velocity flow of 18 cm s^{-1}. From X-ray digital movie file, the stroke volume is estimated at 35 mL, and the aortic diameter at the point of velocity measurement is estimated at 18 mm. The heart rate is 98 bpm. Estimate volume flow rate in the aorta (CO) and, hence, the volume flow rate in the coronary circulation.

   Taking blood viscosity to be 5×10^{-3} Pa s show why turbulent flow might be expected, bearing in mind that peak flow rate may exceed mean flow by four times.
4. The average pressure drop across the arterioles in a bat's wing is 5 kPa. Given that each is 30 μm in diameter, 200 μm long, and that 5 mL (= 5 cm^3) flows into the wing per minute, estimate the number of arterioles in the wing. (Resistance per unit length = $8\mu/\pi r^2$.) If the diameter falls by 10%, calculate the change in pressure drop.
5. Poiseuille's law states that the fluid resistance per unit length in circular cross-section tubes is = $8\mu/\pi r^2$. Given that the average flow velocity in the arterioles is 0.05 m/s, the average diameter is 50 μm, and the average length is 1 mm, what pressure difference exists between the ends of these vessels? (Take blood viscosity to be 5×10^{-3} Pa s).
6. (a) In the inferior vena cava just above the confluence of the iliac veins, the pressure was measured at 22 mmHg, with the subject standing upright. At this point, the vessel diameter was 1.2 cm, and the approximate steady velocity of flow was 30 cm s^{-1}. The pressure probe was then moved 21 cm vertically up the vena cava, where it measured 5 mmHg. Use Bernoulli's equation to estimate the flow velocity at this point.

   If the vessel diameter at this point is 1.4 cm, calculate the total volume flow rate from the tributary veins joining the vena cava between the two points in question. (Assume the veins retain a circular cross section.)

   Discuss the applicability of Bernoulli's equation to this problem.

(b) Outline the difference between laminar and turbulent flow. Will the flow at the point in the vena cava where the velocity is 30 cm s^{-1} be turbulent?

7. Given a CO of 6 L min^{-1}, calculate the velocity of blood along arterioles, assuming there are 20,000 of these vessels in parallel, each with a diameter of 0.1 mm. If each is 1 mm long, calculate the pressure drop across this length (resistance per unit length is given by $8\mu/\pi r^2$).
8. In a particular artery, the flow profile is observed to be parabolic along a 10-cm length. A pressure/flow probe estimates the time mean pressure at the upstream end to be 10 kPa and the mean whole flow velocity to be 10 cm s^{-1}. At this point, the artery is 2 mm in diameter. What would you expect: (a) the pressure and (b) the flow rate to be at the point 10 cm downstream. If you were to use Bernoulli's equation, you would get an incorrect answer. Why?

References

Caro CG, Pedley TJ, Schroter RC, Seed WA. 1978. *The Mechanics of the Circulation*. Oxford University Press, Oxford.

Martini FH, Ober WC. 2005. *Fundamentals of Anatomy and Physiology*, 7th Edition. Prentice Hall, Upper Saddle River, NJ, USA.

Bibliography

McDonald DA. 1974. *Blood Flow in Arteries*. E. Arnold, London.

Mountcastle VB (ed). 1974. *Medical Physiology*. Vol 2, Chapt. 32–45. C V Mosby, St. Louis.

Pedley TJ. 1980. *The Fluid Mechanics of Large Blood Vessels*. Cambridge University Press, Cambridge, UK.

10

The Vascular System: Blood Flow Patterns in Various Parts of the Circulation

Andrew W. Wood

CONTENTS

So far, we have considered the basic mechanics and electrical activation of the heart and the remarkable properties of blood. We have also seen that the circulation consists of various classes of blood vessel, arteries, arterioles, capillaries, venules, and veins. In this chapter, we will consider the characteristic blood flow patterns in each of these, together with an examination of the special arrangements in the brain, the lungs, and the heart itself. Finally, the mechanisms of control of blood pressure (BP) and heart rate will be explained. Lastly, major diseases of the cardiovascular system and their treatment will be discussed.

10.1 Arteries

10.1.1 Structure of Wall

The endothelium has the power of regeneration when damaged. Smooth muscles are arranged longitudinally near the heart, then helically further away. Consequently, nerve stimulation produces only a 5% decrease in aortic radius but a 20% decrease in femoral artery radius. Walls display an increase in Young's modulus on distension above 25%. Walls also have their own blood supply, named the vasa vasorum (Figure 10.1).

10.1.2 Atherosclerosis

This is literally "porridge hardening." The condition refers to "fatty streaks" or accumulations under endothelium (maybe due to a defect in lipid (cholesterol) transport mechanism). These become roughened patches if the endothelium breaks down then regrows. Further degeneration leads to hard scar tissue, which narrows the vessels, encouraging blood stasis and, hence, clot formation. Parts of the clot can break off (embolus) and lodge downstream (embolism). Most are absorbed (if small), but others pose a danger if they block a vital artery, especially the coronary arteries, giving rise to ischemia (no blood perfusion) and infarction (tissue degeneration). In many developed countries, approximately 40%–50% of all deaths result from arterial diseases of this type.

10.1.3 Entrance Region versus Fully Developed Flow Region

In Section 9.2.7 theory behind Poiseuille's Law was introduced: Poiseuille noted in his original experiments that in a long pipe fed by a tank, the pressure fell quite rapidly in the region close to the tank (entrance region), then fell steadily (with a linear gradient) further away (fully developed flow region). The velocity profile in the entrance region is rather flat, converting to a parabolic shape in the fully developed flow region (Figure 10.2). The length of the entrance region can be estimated from an empirical formula and an estimate of the Reynold's number (Chapter 9). The reason the fully developed flow is parabolic can be seen from the derivation of Poisueille's law in the previous chapter. Because of the extra

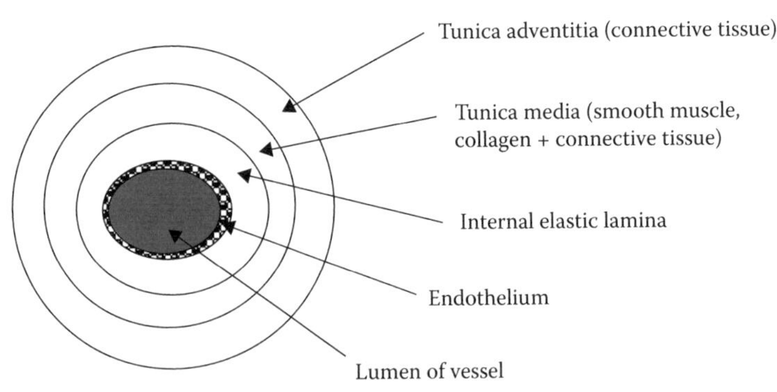

FIGURE 10.1
A diagrammatic representation of the wall of an artery (in reality, the thickness is around 5% of the diameter) showing the endothelial layer in relation to the surrounding elastic layers.

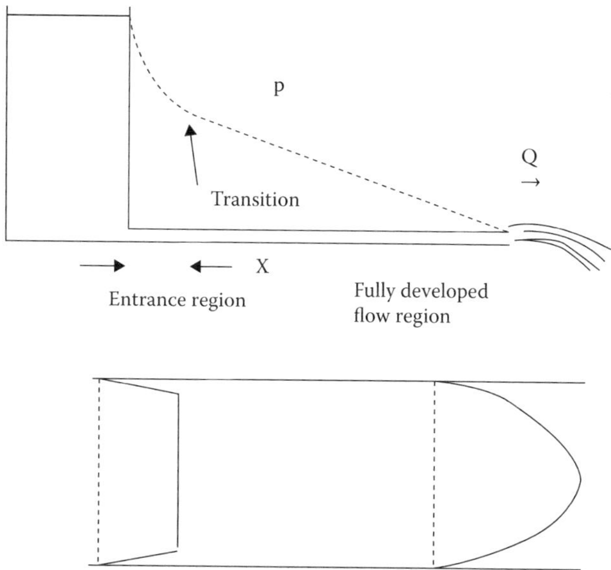

FIGURE 10.2
A fluid flowing from a large reservoir via a long horizontal tube. The pressure inside this tube falls in a non-linear fashion at first, then transitions to a region (the "fully developed flow" region) where the fall in pressure is linear with distance. The entrance region length X is related to the Reynolds number and the type of flow occurring (laminar or turbulent).

shearing occurring in the entrance region, the effective fluid resistance (given by dp/Q) is greater than that for the fully developed flow region (where it is constant).

In the aorta, there is a combination of entrance region pattern and fully developed flow pattern (entrance region length varies a bit with the phase of cardiac cycle but is around 20 cm–30 cm). In Tutorial Question 1 at the end of this chapter, some empirical equations are given to estimate this length. There are a number of features of aortic and large artery flow patterns, which will now be described. First, the flow is higher on the *inside* of bends in the entrance region (because the pressure is higher on the outside of the bend and, hence, the velocity is less, because of the Bernoulli effect). However, when the flow is fully developed, the faster velocity in the middle of the vessel gives rise to a different pattern, called secondary motions. These arise in the abdominal aorta and consist of a helical motion of blood associated with bends due to the greater inertia of blood in the center of the vessel. Axial streaming of red blood cells as they tend to move away from vessel walls is probably related to Bernoulli (because high velocity implies low pressure); this implies that the margins of the vessels are relatively cell-free.

10.1.4 The Windkessel Model of the Arteries

The arterial system is assumed to behave like a single compliant reservoir or chamber (Windkessel means "air chamber"; see Figure 10.3). $Q_H(t)$ represents the inflow from the heart (this is zero during diastole but is like a half cosine during systole). Q(t) is the "runoff" into the arterioles and capillaries (assumed constant), and V is the volume of arterial chamber (around 1 L). $P_a(t)$ is the arterial pressure, which is assumed to be proportional to V; that is, $p_a(t) = V(t)/C$, where C is the *compliance* of the artery, which is related to wall

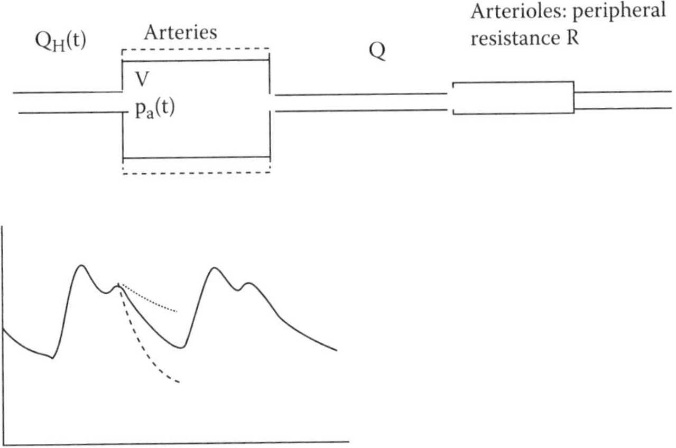

FIGURE 10.3
The Windkessel model of the arterial system. Top: the basic model: $Q_H(t)$ is the time-varying flow rate from the heart into the aorta, and Q is the runoff (assumed constant) into the arterioles and capillaries. The arteries are assumed to have a varying volume (V) and pressure $p_a(t)$. Bottom: the arterial pressure waveform showing the approximately exponential falloff during diastole. The dotted and dashed lines correspond to low and high values of the product of peripheral resistance and arterial compliance.

elasticity (formally, the increase in volume per unit increase in pressure: a stiff artery has a low compliance and vice versa). Because venous pressure is approximately zero, we can put $p_a = Q.R$ (Ohm's law) and from the continuity equation

$$dV/dt = Q_H - Q \text{ or}$$

$$dV/dt = -Q \text{ during diastole.}$$

Thus, $Cdp_a/dt = -p_a/R$, giving

$$p_a(t) = p_a(0)(1 - \exp(-t/(RC)).$$

Thus, the rate of descent of pressure during diastole increases for low values of arterial compliance ("hardened" arteries) or low peripheral resistance: the larger pulse pressure $(p_s - p_d)$ is a feature of certain types of atherosclerosis.

This is a useful prediction of the Windkessel model, but more accurate arterial models need to take the propagation of pressure waves along the artery walls into consideration, thus an interest in pulse wave velocity (PWV).

10.1.5 Pulse Wave Velocity

Most people are familiar with "feeling for the pulse" by placing (usually) the third and fourth fingers in the groove between the tendons to the thumb at the wrist. This pulsation is due to the elastic distension of the artery as the extra amount of blood associated with each left ventricular contraction passes through the artery (Figure 10.4). The pulse can also be felt ("palpated") by inserting the fingers between the tendons at the end of the biceps in the armpit. If it were possible to measure both simultaneously (which is by the deft use of both hands or by using electronic equipment), it will be noted that the pulses are not simultaneous because the pulse is propagating with a certain speed along the artery wall.

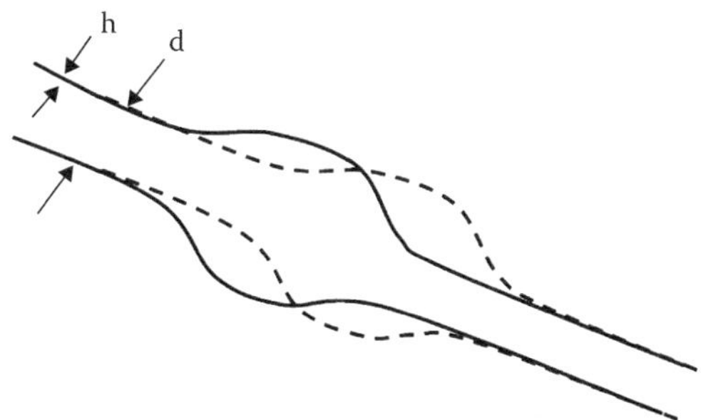

FIGURE 10.4
An exaggerated representation of a pulse wave in an artery (diameter d, wall thickness h). The expulsion of blood leads to a "bulge" in a region of the artery. The elastic recoil of the wall tends to restore the original diameter, but like a wave on a string, the bulge tends to move along the artery with a velocity determined by the combined properties of the artery wall and the blood density. This velocity is more or less independent of the flow rate (or flow velocity) of blood in the artery.

Note the speed (or velocity) is a property of the vessel wall elasticity rather than the fluid flow rate. The Moens–Korteweg formula (below) is a good predictor of PWV, and values in the range 5 m/s–10 m/s are usually obtained.

$$c^2 = Eh/(\rho d)$$

where c is the PWV, E is the wall Young's modulus, h is the thickness, d is the diameter, and ρ is the density of blood. This equation is derived in full at the end of this chapter. The average speed of blood in the artery itself is usually much lower than this, typically 0.3 m/s. This is analogous to the mechanical pulsation felt at the nozzle end of a garden hose as the water tap is turned full on, which will be felt long before the actual water leaving the tap reaches the nozzle.

WORKED EXAMPLE

In a particular arterial segment, the PWV is determined to be 8 m/s. Estimate the (Young's) modulus of elasticity (stating units) of the arterial wall, given the parameters at the start of this paper and a wall-to-diameter ratio of 0.04. Assume blood density to be 1080 kg/m³.

Answer

Using the Moens–Korteweg equation, we can discover Young's modulus E:

$$E = c^2\rho d/h = 8^2 \times 1080/0.04 = 1.73 \text{ MPa}.$$

In arterial disease, the radius decreases, the wall thickness increases, and the walls become "stiffer" (hence, E increases). These combine to increase PWV. Note that the Young's modulus for the artery wall is around 0.5 MPa at physiological values of the arterial pressure. If the internal pressure is made to rise abnormally (so that the diameter is cause to increase by 30% or more), the value for E rises steeply (this is caused by the elastic fibers acting to "stiffen" the wall as the artery is distended).

10.1.6 Kinetic Energy of Blood Flow in Arteries

When blood is ejected from the heart, in addition to having overcome the back pressure due to the peripheral resistance, the ventricles also have to impart momentum to the blood, which may attain speeds of up to 1 m/s in the aorta. During isovolumetric contraction, the blood is stationary within the ventricles. The average speed in the aorta or pulmonary artery can be estimated as follows: the average volume flow rate during systole will be given by $\Delta V/T$, where ΔV is the stroke volume, and T is the ejection time (remember that the flow rate is essentially zero during diastole). Using the formula $Q = uA$, we can estimate the average velocity as follows:

$$u = (\Delta V/T)/((1/4)\pi d^2).$$

The mass of blood involved is thus the stroke volume ΔV multiplied by blood density ρ. Thus, the amount of kinetic energy (KE) imparted to the blood in the aorta is $1/2\ \rho \Delta V u^2$.

WORKED EXAMPLE

Estimate the % KE in terms of total energy per beat in the left and right ventricle, respectively, assuming a stroke volume of 80 mL, an ejection time of 0.28 s, and diameter of 24 mm in both aorta and common pulmonary artery.

Answer

First convert stroke volume to SI units—80 mL = 80×10^{-6} m^3. Now, estimate u = $(80 \times 10^{-6}/0.28)/(1/4\ \pi\ (24 \times 10^{-3})^2) = 0.63$ m/s (which is a reasonable value). The KE is thus $\frac{1}{2}.1080(0.63)^2 = 0.017$ J/beat. This compares with 1 J/beat obtained in Section 8.6.3 for the left ventricle (thus, the KE% is less than 2%). However, for the right ventricle, the $\int p.dV$ work was only 0.17 J, so in this case the %KE is 10%. In exercise, although the SV increase may only be 25%, because the formula has essentially ΔV^3, this amounts to a doubling of KE. This, coupled with a slight shortening of T, leads to the KE% being around 25% for the right ventricle in exercise.

10.2 Blood Flow in the Heart Muscle (Coronary Flow)

The two main branches of the *coronary circulation* split off from the aorta just behind the aortic valve leaflets (in the pouches of Valsava). When the valve is closed (diastole), blood is forced into the coronary arteries by back pressure. When the valve is open (systole), eddy currents ensure a constant perfusion also during this period. After perfusing the heart muscle (myocardium), the blood returns to the right atrium via the coronary sinus. Elsewhere in the body, noradrenaline tends to constrict blood vessels, but in the coronary circulation, the opposite happens—vessels are dilated (they are also in the skeletal muscle). Oxygen tension is very important: if this is low, the coronary vessels dilate. One very common heart ailment is *angina pectoris* or chest pain associated with spasm or other disease of the coronary vessels. It commonly afflicts sufferers following exertion or emotions affecting cardiac output (CO). This can be treated with nitroglycerin, amyl nitrite, or related compounds. Blood clots (usually as a result of arterial disease or *atherosclerosis*—see above) can lodge in the coronary vessels, forming an embolus. This can lead to large portions of the myocardium being starved of oxygen (due to lack of blood flow or *ischemia*), leading to degeneration of the muscle or *myocardial infarction* (MI), commonly known as heart attack.

10.3 Blood Flow in the Brain

The *circle of Willis* (see Figure 10.5) ensures adequate perfusion of the brain tissue despite occlusion in a particular artery. The body control systems attempt to maintain flow through the brain (which accounts for 13% of CO) at all costs. *Fainting (syncope)* is actually a strategy the body uses to restore flow to the brain, despite a reflex slowing of the heart (*bradycardia*: see below): as a person collapses to the ground, pressure in the arteries of the head increases. In *hemorrhage*, blood is shunted from the skin to the brain (hence, pallid appearance).

Arterioles are not sensitive to noradrenaline but to dissolved gas concentrations. As CO_2 increases and O_2 levels fall, the arterioles dilate due to raised pH.

Cerebral capillaries have a tight endothelium, providing the basis of the *blood–brain barrier.*

Cerebral venules drain into sinuses (which are large, flattened cavernous spaces).

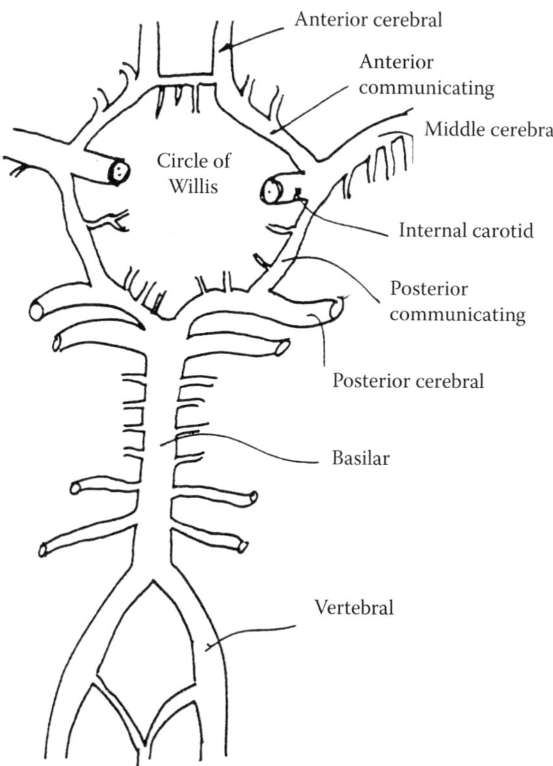

FIGURE 10.5
Diagram of the arteries making up the circle of Willis. These are as they would appear if looking down on to the base of the skull. The vertebral arteries come into the skull via the openings in the cervical vertebrae and thence the skull and the internal carotids via canals in the temporal bones of the skull.

10.4 Blood Flow in the Lungs

The range of pressures in the pulmonary artery (carrying deoxygenated blood) is 8–25 mmHg (compared with 80–120 mmHg in the aorta), and the pressure in the pulmonary veins is 5 mmHg. Despite the CO increasing six times in exercise, the pressure gradient does not alter appreciably. This implies a sixfold drop in pulmonary capillary resistance. How this resistance regulates itself is not entirely understood. Alveolar pulmonary capillaries are very short and flattened. The diffusion path length between the gas phase of air in the alveoli and the liquid phase in the capillaries is quite short (around 2 μm). The four lipid bilayers do not present a large diffusional barrier because CO_2 and O_2 are both lipid soluble. Emboli in the circulation are trapped in the lung tissue, where they break up without much danger. Three important hormones affecting the vasomotor tone are modified in the lung: inactive angiotensin I is converted to the active angiotensin II; the vasodilator bradykinin is inactivated and 5HT (serotonin), which is a vasoconstrictor with a role in wound healing, is removed from circulation.

10.5 The Control of Arterial Pressure

Control of blood pressure (BP): specialized pressure sensors in the major arteries (principally, the carotid) called *baroreceptors* (like *baro*meter) monitor the arterial BP; if it is too high, the vasomotor center is inhibited, and fewer impulses travel along the sympathetic nerves to the arterioles (also, the adrenal medulla puts out less noradrenaline). This allows the arterioles to get wider, lowering arterial pressure.

Contrast this with the control of heart rate: the parasympathetic part of the autonomic nervous system (ANS; vagus nerve) slows the rate (*bradycardia*), whereas the sympathetic part (noradrenaline as the neurotransmitter) causes acceleration (*tachycardia*). These nerves originate from the cardiac center, which has an inhibitor and accelerator part, respectively. These together are known as the *chronotropic* (as in *chrono*meter) effects on the heart. The ANS also affects the strength of contraction of the heart (see Chapter 8). These effects are known as *inotropic* effects on the heart. Because peripheral resistance p_a/Q (p_a is the arterial pressure, and Q is the CO) is determined by $1/r^4$, where r is the effective arteriolar radius, the arterial pressure is only partially controlled this way. Because Q is determined by both heart rate and stroke volume (the latter being under inotropic control), the interactions are quite complex. Nevertheless, when, in exercise, the value of Q rises five to six times, the rise in p_a is really quite minor.

10.6 Arterial Disease: Diagnosis and Treatment

Although arterial BP commonly increases in exercise and in response to psychological stress, in a significant subset of the population, the value while resting increases over a period of months or years to abnormally high values. This state is known as *hypertension* and is defined as a sustained elevated diastolic pressure (usually 95 mmHg taken as the dividing line).

Hypertension is a strong risk factor in both *cerebral vascular accident* (stroke) and MI (heart attack). It occurs in approximately 71% of males and 85% of females in the population over 75 years and 12% of males in the age range of 20–34 years, with an upward trend for this younger age group. See also the discussion of *angina pectoris* in Section 10.2. Primary or *essential* hypertension (which accounts for 90% of cases) has no other apparent cause. In secondary hypertension, a cause, such as a hormone-secreting tumor, can be identified. The causes of primary hypertension are not fully understood, but hypotheses include the following: a hyperreactive vasomotor center (or abnormal stress response); an abnormal NaCl retention or hyperreactivity of arterioles to high NaCl levels (with a strong genetic component); and a positive feedback due to hyperreactive arterioles giving rise to further damage in these vessels because of high BP, especially in renal arterioles. Risk factors include age; family history, being overweight and inactive; smoking (perhaps via nicotine action on the blood vessels); and diet with a high salt intake, high alcohol intake, low potassium, or low vitamin D. Psychological stress can lead to elevated BP levels, but this can be ameliorated by stress management techniques.

10.6.1 Treatment of Hypertension

Lifestyle changes are the most useful line of primary treatment (increase of physical activity, changing of diet to low fat and low salt, cessation of smoking, and limiting alcohol intake), but in more severe cases, this has to be supplemented by drugs, either of a single type or in combination. There are a range of options, each with a different mode of action to reduce arterial BP. Drugs have a pharmacological name but also go under a brand name, with the same or similar compounds having several different brand names. Common brand names are given in italics in Table 10.1.

10.6.2 Biochemical Tests

Low-density lipoprotein ("bad" lipoprotein) is associated with cell uptake of cholesterol and hence atherogenesis (the formation of fatty streaks). Its normal range is 1.6 mM–4.7 mM and should be low. On the other hand, high-density lipoprotein ("good" lipoprotein) promotes cholesterol removal (normal range 0.8 mM–2.0 mM) and should be high. If low-density

TABLE 10.1

Drugs for the Treatment of Hypertension

Type	Action	Example(s)
Adrenergic beta-blockers	Reducing CO and hence BP	Examples: propranolol (*Inderal*), oxprenolol (*Trasicor*)
Calcium blockers	Reduced force of contraction of heart, hence BP (also vasodilator action)	Verapamil (*Verpamil*)
Vasodilators	Reduce peripheral resistance, hence BP	Hydralazine (*Apresoline*)
Angiotensin-converting enzyme inhibitors	Similar to vasodilators	Captopril (*Capoten*)
Sympatholytic	Reduction of sympathetic activity	Guanethidine (*Antipres*), reserpine (*Serpasil*), methyldopa (*Aldomet*)
Diuretics	Reduction of blood volume	Frusemide (*Lasix*)

lipoprotein/high-density lipoprotein ratio is greater than 6, this is a strong indicator of *coronary artery disease.*

Diagnosis or monitoring of recovery from MI can be tested by measurement of the following: lactate dehydrogenase isoenzymes (LDH; heart/brain): (LDH_1/LDH_2 >1 for MI), and creatinine phosphokinase (CPK): CPK_1/CPK_{TOTAL} >5% (for MI).

10.7 The Microcirculation

10.7.1 Description

The purpose of having a circulation is to transport gases, nutrients, and hormones to each cell in the body and to clear away products of metabolism (CO_2 for example). The microcirculation consists of a fine mesh of capillaries with associated arterioles and venules. There also is a parallel system of fine vessels called the lymphatic system, which aids in the recycling of tissue fluid (which is mainly water). The true (blood) capillaries are around 200 μm and around 40 μm apart, so no individual cell is more than a few neighbors away from a capillary.

The microcirculatory unit is shown in the stylized Figure 10.6. Because capillaries are of the order of 6 μm in diameter, they individually provide a high resistance to flow, but because there are so many of them providing parallel pathways, the overall resistance of the capillary bed is in fact less than that of the arterioles, which provide the main element of control of blood flow into the microcirculation. The smooth muscle surrounding the arterioles and forming the precapillary sphincter is influenced by innervation from both the sympathetic and parasympathetic system, the former releasing noradrenaline, which causes most smooth muscles to constrict, except in the heart and skeletal muscles where the opposite effect, dilatation, occurs.

There is considerable variation in the size of gaps between the endothelial cells forming the capillaries; the largest gaps occur in the liver and in the intestines, and large regions where the opposite sides of the cell have coalesced to form "fenestrations" allow lipid-soluble substances (such as dissolved gases) to quickly permeate through. The whole of the inner surface area is thus available for such substances to diffuse through. On the other hand, lipid-insoluble substances (such as ions, water, and urea) can only diffuse through the "gaps," so the effective area for diffusion is much, much less (in fact, these gaps are only 8 nm wide: 1/1000 of the cell radius).

10.7.2 Bulk Flow of Fluid across the Capillary

The flow depends on the balance of forces across the capillary layer and the fluid resistance of the layer. The balance of forces involves both the difference in hydrostatic pressure between the capillary fluid and the surrounding tissue (or interstitial) fluid and the difference between the osmotic pressure in those two places. The main contribution to blood plasma osmotic pressure in this case is the plasma protein, especially albumin, which amounts to some 25 mmHg. In this case, the NaCl does not contribute because it is able to permeate the layer easily, whereas protein cannot. This balance is expressed as the

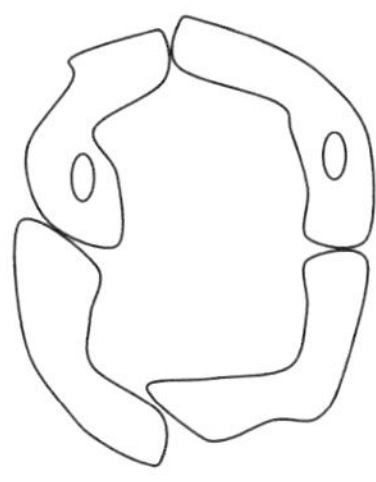

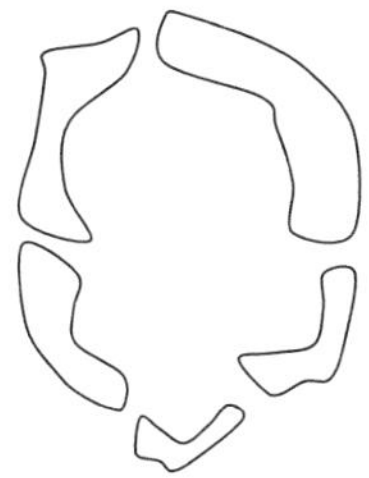

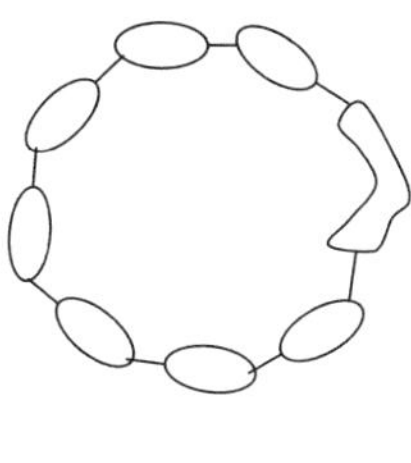

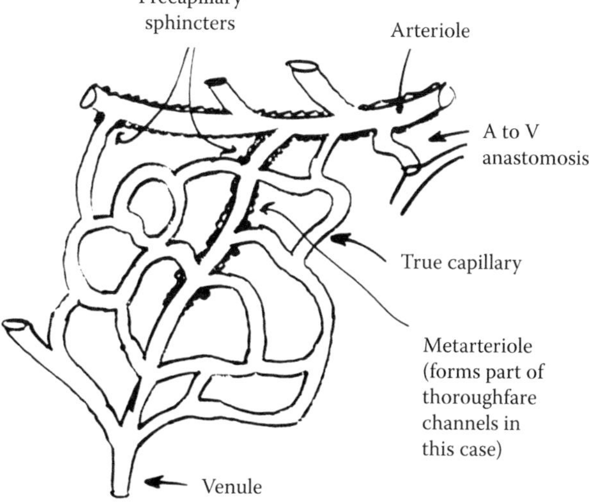

FIGURE 10.6
Microcirculation. Top: endothelial arrangements in three types of capillary. Bottom: the "microcirculatory unit."

following equation (referred to in British textbooks as the Starling hypothesis, after E. H. Starling, who also has a law of the heart named after him):

$$\Delta p_F = (p_c - p_t) - (\Pi_p - \Pi_t);$$

where Δp_F is the net filtration force, p_c is the capillary pressure, p_t is the tissue pressure, Π_p is the plasma (also known as oncotic or colloid) osmotic pressure, and Π_t is the tissue osmotic pressure.

Normally, Π_t would be expected to be zero (because very little protein finds its way into the interstitial or extracellular fluid), but values between 3 mmHg and 15 mmHg are reported. Similarly, p_c would not be expected to be much different from atmospheric, in fact, experimental values vary between –7 mmHg and +3 mmHg. However, it is customary to make the assumption that p_t and Π_t cancel.

10.8 The Veins

In considering blood flow patterns in the veins, the flow rate of blood increases as downstream pressure (e.g., right atrial pressure, P_{RA}) falls, but eventually, flow reaches a plateau because the vessel collapses because the internal pressure becomes less than the surrounding tissue pressure. When this happens, the vascular resistance rises because the effective radius falls. Figure 10.6 illustrates this, with a sharp inflection point where the tissue pressure and P_{RA} are the same. Note that when P_{RA} exceeds a certain value, all venous return (VR) ceases. The upstream pressure (which is not exactly the pressure in the major veins but is rather the pressure in the entire system if the heart were suddenly to stop beating and blood were rapidly moved from arteries to veins) is unable to overcome P_{RA} in this situation. This pressure has been termed "mean systemic pressure" or P_{MS} by Guyton (Hall 2011). The lower part of Figure 10.7 shows how a simple analog of the venous system constructed of flexible tubing inside a rigid cylinder will exhibit similar behavior to that shown in the upper part.

In normal cardiovascular function, there are various factors, which aid VR. First, skeletal muscles, on contraction, squeeze the blood in the veins past the internal semilunar valves. In the calf, for example, the act of walking aids the VR by this "muscle pump" mechanism. Second, on breathing in, the negative pressure in the chest causes a greater pressure gradient

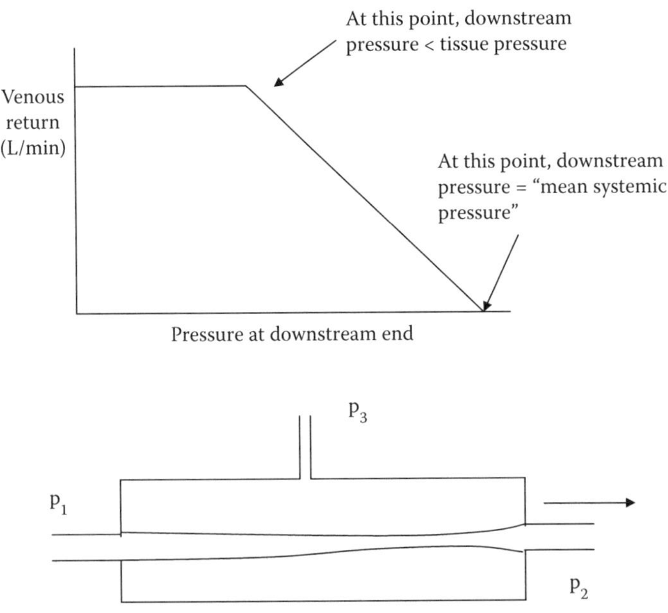

FIGURE 10.7
Venous return curves. This behavior can be observed by studying the flow rates through a flexible tube inside a rigid inflatable chamber as shown in the lower part of the diagram. As pressure p_2 falls below p_3, part of the flexible tube collapses, which limits the flow rate.

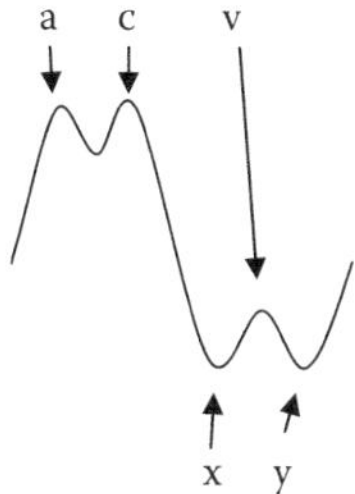

FIGURE 10.8
A typical venous pressure waveform. The origin of the features a, c, and v are as follows: "a" is due to the rise in back pressure during atrial systole; "c" corresponds to the bulging of the mitral and tricuspid valves into the atria during ventricular systole (coinciding with the carotid pulse, from which it is named) and "v" corresponds to the fibrous skeleton moving upward (toward the atria) toward the end of systole. The other features (x and y) have no particular significance.

by lowering the thoracic pressure and, hence, P_{RA}. Third, ventricular filling: as the atria empty, this lowers the P_{RA} and, hence, pressure in the vena cavae. If a pressure transducer is introduced into a major vein, a waveform similar to that shown in Figure 10.8 is obtained.

10.9 Guyton's Model of the Circulation

In the above section, we saw that the flow of blood back to the heart is limited by the tendency of the veins to collapse if the internal pressure is too low. The pressure gradient, in the normal situation where VR is unimpeded, is $P_{MS} - P_{RA}$, where P_{MS} is the "mean systemic" pressure and P_{RA} is the right atrial pressure. The mean systemic pressure is a theoretical concept: if the heart were suddenly stopped and blood allowed to distribute very rapidly through the circulation such that the pressure was everywhere the same, this would be P_{MS}. Guyton estimated that, in humans, this would be about 7 mmHg, normally (Hall 2011). Increased blood volume, sympathetic stimuli, and skeletal muscle contraction would tend to increase P_{MS}. On the other hand, alterations to the resistance to VR (for whatever reason) would tend to alter the (negative) slope of the VR curve. Some of these effects are shown in Figure 10.9.

Because VR is the same as CO if averaged over several beats of the heart, the point at which the VR and CO (or Starling) curves cross can be thought of as representing the equilibrium or operating point for the cardiovascular system. Many effects on the CV system can be understood in terms of effects on the VR/CO curves.

From the work of Sarnoff and others, the CO curve has been shown to be very sensitive to myocardial contractility or the strength of contraction of the heart. The effects of sympathetic and parasympathetic stimulation are shown below. If the heart begins to fail, contractility falls. Initially, the system is able to compensate for this by additional activity of the sympathetic system or alternatively increased extracellular fluid (which shifts the VR curve to the right and the CO curve to the left). The price paid is the high PRA values and consequent distension of the heart and associated pulmonary edema. In the absence of sympathetic compensation, the heart failure would rapidly lead to a fatality.

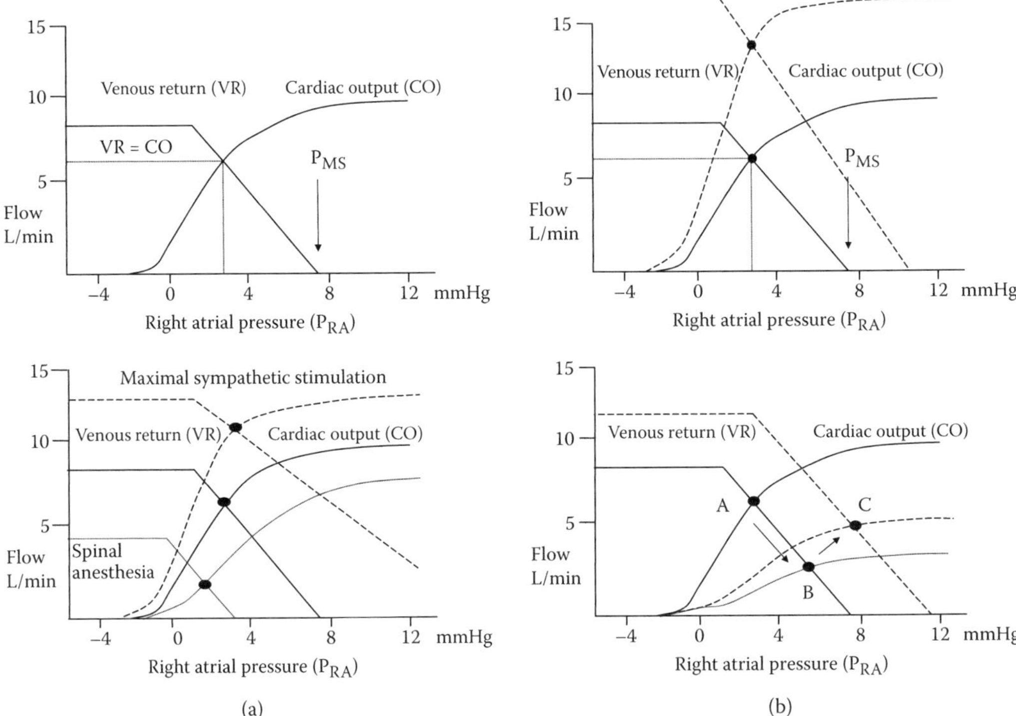

FIGURE 10.9
(a) Guyton diagrams. Top: normal VR and CO curves. Bottom: sympathetic effects on myocardial contractility and on vascular tone. Dashed: maximal; dotted: minimal, induced by spinal anesthesia. (b) Top: effects of exercise (dashed) versus normal (full). Note that in exercise, there is an overall reduction in peripheral resistance (compare maximum sympathetic stimulation above). Bottom: cardiac failure: A, normal; B, following heart failure; C, sympathetic reflexes operating. Note increased P_{RA}, which represents danger of increasing fluid retention.

On the other hand, in exercise, both CO and VR curves move as before (VR to the right and CO to the left) compared to normal. In this way, a CO of some six times normal can be obtained with modest increase in filling or right atrial pressure (P_{RA}).

10.10 Derivation of Moens–Korteweg Equation

By analogy with familiar expressions for wavespeeds in strings, springs and in compressible media, we get the following for wavespeed for propagation along an elastic tube (Lighthill 1986)

$$c = \frac{1}{\sqrt{\rho D}} \tag{10.1}$$

where c is the wavespeed, ρ is the density, D is the wall distensibility ≡ $\Delta A/A_o/\Delta p$ where ΔA is the ↑ in cross-sectional area for an ↑ in internal pressure of Δp.

From the elasticity equation:

$$S = E \times \frac{\pi \Delta d}{\pi d_o} \text{ (i.e., Hooke's law assumed to hold)}$$

where S is wall stress, E is Young's modulus, Δd is diameter increase, d_o is initial diameter.

$$S = \frac{\Delta p d_o}{2h}$$

where h is wall thickness, from the "Laplace" law.

Now:

$$\frac{\Delta A}{A_o} = \frac{\pi\left(\frac{d}{2}\right)^2 - \pi\left(\frac{d_o}{2}\right)^2}{\pi\left(\frac{d_o}{2}\right)^2} = \frac{d^2 - d_o^2}{d_o^2}$$

$$= \frac{(d + d_o)(d - d_o)}{d_o^2} \cong \frac{2d_o(d - d_o)}{d_o^2} \tag{10.2}$$

$$\cong \frac{2\Delta d}{d_o}.$$

Thus, $\frac{\Delta p d_o}{2h} = E\frac{\Delta A}{2A_o}; \frac{d_o}{h} = E\frac{\frac{dA}{A_o}}{\Delta p} = ED$ (from Equation 10.2)

Thus, $D = \frac{d_o}{Eh}$

and $c = \sqrt{\frac{Eh}{\rho d_o}}$ (from Equation 10.1)

since $\frac{h}{d_o} \cong 0.05$ and $E \cong 4 \times 10^5$ Pa

$c \cong 4.3$ m/s.

Tutorial Questions

1. A catheter is inserted into the pulmonary artery (diameter, d = 2.5 cm) to measure pressure and flow simultaneously. The waveforms are as shown in Figure 10.10:
   (a) Show that stroke volume ΔV is approximately 60 mL and, hence, calculate CO.
   (b) Estimate whether flow is turbulent or not at this point. What are the features of turbulent and laminar flow and of entrance region flow?
   (c) Use the formula $\bar{p}\Delta V + \rho\Delta V\left(\frac{\Delta V}{0.25 \times \pi d^2 \times T}\right)^2$ to estimate the total work done by the right ventricle per beat and the percentage of this which appears as KE.

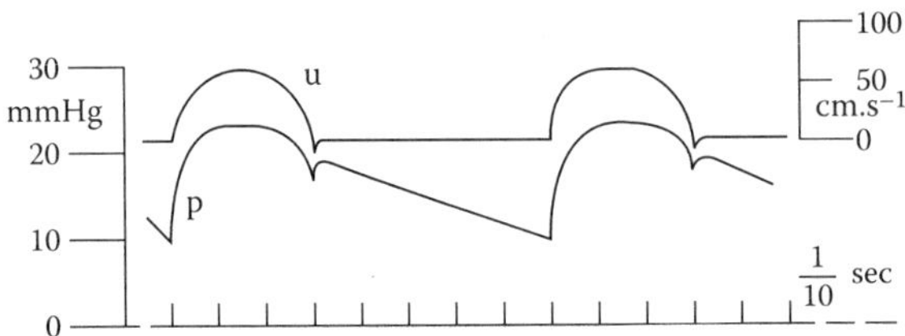

FIGURE 10.10
See Tutorial Question 1.

2. In clinical practice, the pressure in the pulmonary artery often is recorded at the same time that CO is determined (from the right heart). In a typical investigation, the CO was determined to be 5.4 L/min, and the pulmonary arterial pressure waveform was as shown in Figure 10.11:

   At the entrance to this vessel, the diameter was estimated to be 18 mm.

   (a) Calculate the average velocity of flow during systole at this point.

   (b) Estimate whether the flow at this point is laminar or turbulent.

   (c) Estimate the distance along the pulmonary artery before the flow becomes fully developed. Laminar flow: $X = 0.03\ d\ (Re)$; turbulent flow: $X = 0.69\ d\ (Re)^{0.25}$.

   (Neglect taper).

3. The diagram (Figure 10.12) represents a branch in the long arteries.

   Pulse wave monitors are placed at A, B, and C. The pulse arrives at B 20 ms after A and at C 15 ms after B.

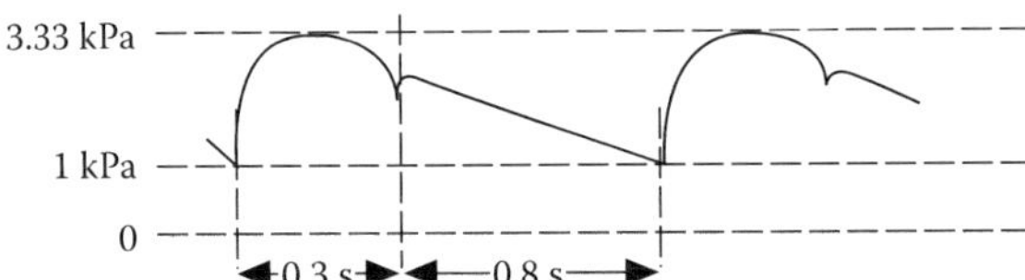

FIGURE 10.11
See Tutorial Question 2.

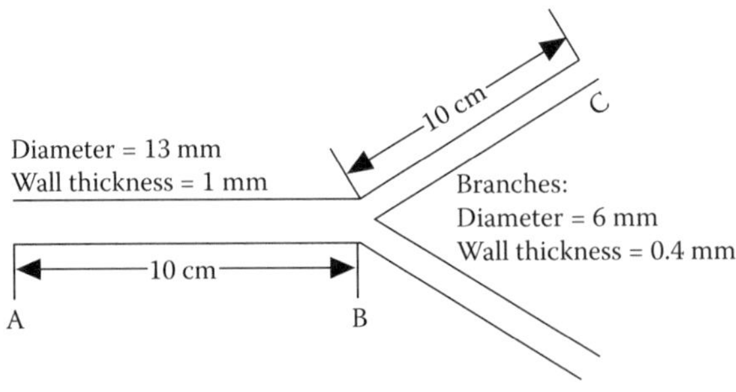

FIGURE 10.12
See Tutorial Question 3.

Calculate:

(a) The wave speed

(b) The Young's modulus of the segments AB and BC

4. A pulse wave detector is placed on the dog aorta at the iliac bifurcation, another is placed 10 cm upstream on the aorta, and a third is placed 10 cm downstream on one of the iliac arteries. The pulse wave delay between the first two is 17 ms, and that between the first and the third is 10 ms.

   Calculate:

   (a) The distensibility, and

   (b) The Young's modulus of the aorta and iliac artery, given that the ratio of wall thickness to vessel diameter is 0.05.

Bibliography

Hall JE. 2011. *Guyton and Hall Textbook of Medical Physiology*, 12th Edition. Elsevier-Saunders, Philadelphia, PA, USA.

Lighthill MJ. 1986. *An Informal Introduction to Fluid Mechanics*. Oxford University Press, Oxford.

11

Cardiovascular System Monitoring

Andrew W. Wood

CONTENTS

11.1 Introduction

The electrical events within the heart, together with the pressures and flows throughout the circulation, offer a great deal of useful data upon which accurate diagnosis of abnormality can be obtained. Much of this monitoring is of continuous data, and analysis can be carried out in real time, permitting prompt intervention when necessary.

This chapter will deal first of all with electrical events (the electrocardiogram, or ECG) and then will discuss methods for measuring the total blood flow from the heart (cardiac output, or CO), followed by methods of measuring pressure and flow within the cardiovascular system. Finally, methods for measuring the oxygen saturation of the blood (SaO2%) and the characteristic heart sounds will be discussed.

11.2 ECG Monitoring

In Chapter 8 we saw how, as the "wave of depolarization" moved first through the atria and then, after a certain delay in the bundle of His, through the ventricles, this was equivalent to an electrical dipole of varying strength and direction. At the height of the R wave, the dipole has maximal strength, and the direction is usually around 20° to the left of the midline of the body and pointing slightly forward (although this varies a great deal between individuals). This situation can be visualized with reference to Figure 11.1 where the lines of electric force and the orthogonal isopotentials are shown. Clearly, if two electrodes are on a line on the surface of the body, are in the same direction as the dipole within the chest (A and B in the diagram), then a strong R wave signal will be obtained. Conversely, if electrodes on the chest are at right angles to the dipole direction (C and D), then zero R wave will be obtained (although nonzero signals may be obtained for other times during the cardiac cycle). The voltages at the positions of the R and L arms and L foot can be estimated from the diagram. What this shows is that the left upper arm is 0.15 units more positive than the corresponding

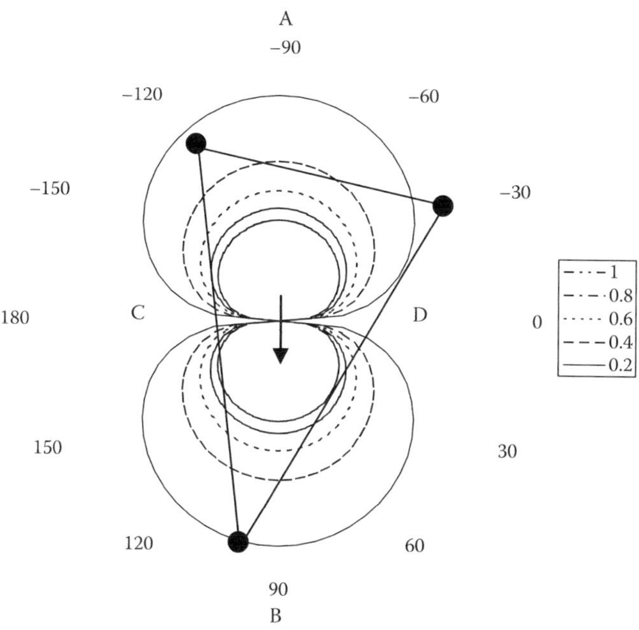

FIGURE 11.1
The dipole is shown in a vertical orientation. Normally the tip of the arrow would be pointing to the left, at approximately 60° to the vertical (30° to the horizontal). The triangle represents the effective position of the R and L arms and the L foot. The top of the triangle would normally be horizontal.

position on the right arm (–1.1 – (–0.95) = –0.15). Similarly, the left abdominal regions are 1.95 and 2.1 units more positive than the left and right arm, respectively (1 – (–0.95) = +1.95 and 1 – (–1.1) = +2.1). Note that the sum of the first two is equal to the third (1.95 + 0.15 = 2.1).

When these potentials are measured at the wrists and the ankles, the values will be smaller, but the ratios would be similar to those just stated. The three standard leads of the ECG, measured from these positions, are I, left minus right arm (LA – RA); II, left leg minus right arm (LL – RA); and III, left leg minus left arm (LL – LA). In practice, these are measured using differential amplifiers, with the right leg (RL) as Earth (or virtual Earth, as will be explained). They are referred to as *bipolar* electrodes. Since these positions (LA, RA, LL) are approximately 60° from each other (this is not quite the case in Figure 11.1, but the error is small), it is conventional to construct an equilateral triangle (called the Einthoven triangle, after the work of the Dutch Noble laureate Willem Einthoven, 1860–1927) to represent the situation. Conventionally, the zero of the angle is taken as the direction from RA to LA, as shown in Figure 11.2.

From Figure 11.2, the dipole (representing the heart during the QRS complex) in this set of axes is directed approximately 45° below the horizontal and to the right. This is known as the axis deviation. Normal values are between –30° and +90°. From the diagram, it is obvious that the magnitude of lead I (V_I) will be M cos θ. It is easy to show (by rotation) that the other two leads are as follows:

Lead II: $V_{II} = M\cos(\theta + 60) = \frac{1}{2}M(\cos\theta - \sqrt{3}\sin\theta)$;

Lead III: $V_{III} = M\cos(\theta + 120) = -M\sin(\theta + 30) = \frac{1}{2}M(-\cos\theta - \sqrt{3}\sin\theta)$

Note again that $V_I + V_{III} = V_{II}$.

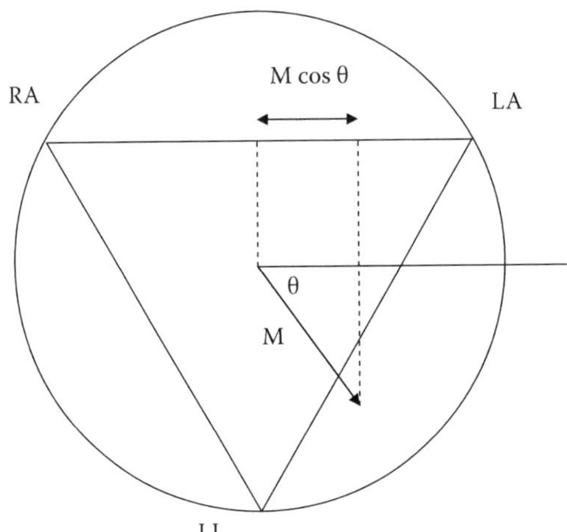

FIGURE 11.2
The Einthoven triangle, showing the axis deviation of the heart, θ. RA and LA indicate right and left arms, and LL indicates left leg.

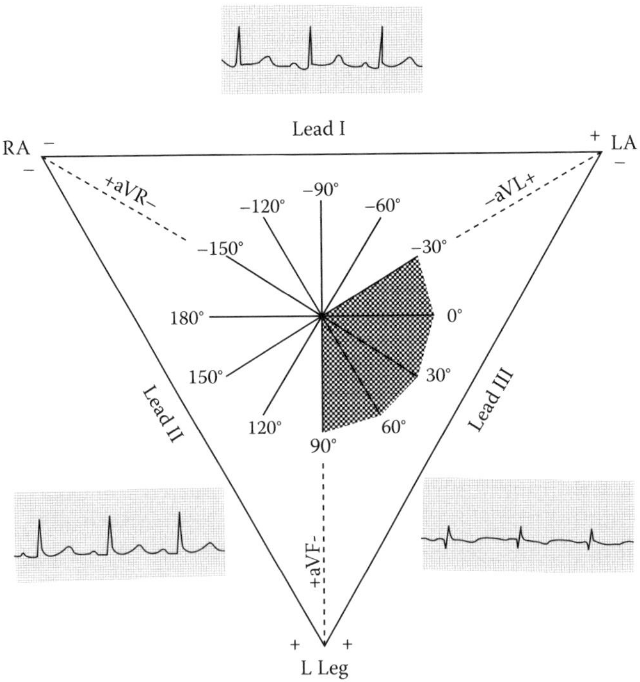

FIGURE 11.3
The three standard leads (I, II, and III, as indicated). Note that with the axis deviation at 30°, lead III is showing a very small QRS complex, because the axis is directed at right angles (orthogonal) to the line joining LA and LL.

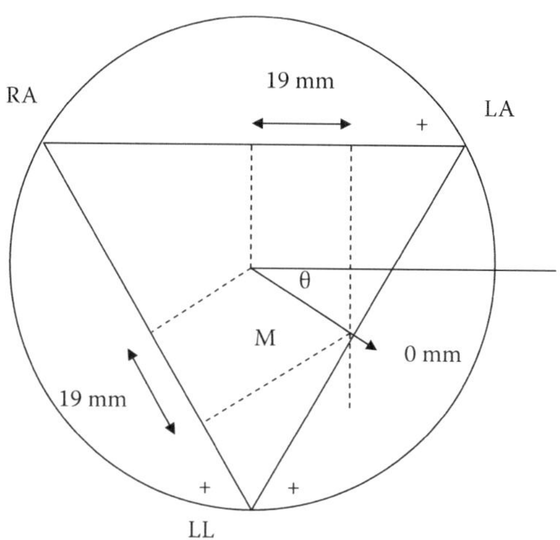

FIGURE 11.4
This illustrates the magnitudes of leads I and II read from the tracings in the previous diagram. See text for procedure for determining θ.

If we now take an actual ECG trace of leads I, II, and III (Figure 11.3), then working back the other way, three lengths representing the magnitudes of the R waves (actually the R minus the Q or S wave), can be used to "discover" the equivalent dipole direction and strength. In the traces shown in Figure 11.4, V_I (which is R – S/Q) is 19 mm, V_{II} is 19 mm and V_{III} is 2.5 – 2.5 = 0 mm.

The axis deviation (which in this case is +30°) can be found by the construction in Figure 11.4, where each V_I, V_{II}, and V_{III} are represented by a distance from the midpoint of each side of the triangle and perpendiculars are dropped from each of the three points.

In addition to the standard leads, there are two other sets of leads: first, the augmented leads, formed by measuring between the LA, RA, and LL, and the average of the other two leads in each case (achieved in practical ECG machines by connecting the other two leads together in each case). These are denoted aVR, aVL, and aVF (F for foot), respectively. These three directions (+30, –30, and +90) form a second equilateral triangle at 30° to the first, so the magnitudes of the signals can be derived by simple trigonometry. The third set of leads, the six "chest leads," are formed by measuring between each of the six positions over or near to the heart, as shown in Figure 11.5, and a second electrode formed by connecting V_I, V_{II}, and V_{III} together. Since this second electrode represents an average voltage over the chest region, the chest leads are referred to as "unipolar." As might be expected, the R wave is normally very prominent at V_4.

Modern ECG recorders produce on a single A4 sheet 4-s segments of the standard leads, the augmented leads and the six chest leads obtained in parallel. The normal sensitivity of the recording is 1 cm per mV, with a time base of 50 mm/s, although these can be varied. The contamination by 50/60 Hz "hum" from nearby power cords and other instrumentation can be minimized by the use of differential recording (the 50/60 Hz signal is a "common-mode" signal which can be rejected by a factor of 10^5 or more) together with "notch" filtering at the appropriate frequency. Early ECG instruments connected the RL directly to Earth, but because of the possibility of "microshock" due to inadvertent contact with other charged surfaces, the RL is nowadays driven by a current to actively reduce the common-mode voltage to a low value. If the patient is exposed to an accidental high voltage, the driver circuitry saturates, effectively ungrounding the patient and preventing large current flows through the body.

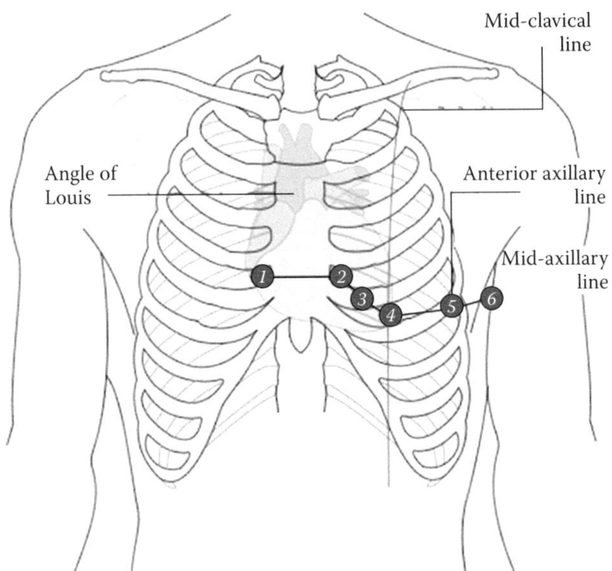

FIGURE 11.5
Standard locations for the six chest leads. (http://www.nottingham.ac.uk/nursing/practice/resources/cardiology/function/chest_leads.php).

11.3 Cardiac Output Monitoring

Cardiac output (CO; also denoted by Q) is the volume of blood ejected from *either* the left or the right side of the heart in unit time. Since the coronary circulation leaves the aorta just downstream of the aortic valve, the best site for measuring CO is in the pulmonary artery. It is equal to stroke volume (SV) × heart rate (HR) and is a very useful parameter to measure both in intensive care and in sport and fitness assessment.

Most cardiac output determinations are based on the *Fick principle*. This principle is basically a statement of conservation of mass: the total amount going into a compartment of fixed volume must equal the amount coming out at the effluent points combined. More succinctly, "what goes in must come out." Consider the following diagram (Figure 11.6) in which a substance is flowing along the tube at flow rate Q and more of the same substance is being injected at rate R. The concentration, initially C_1, rises to C_2 after the extra substance has had time to mix.

The amount leaving in time δt is the volume × concentration, and since volume is flow rate × δt, this will be $C_2Q\delta t$. This will equal the amount of substance going in from the left plus that injected, i.e., $C_1Q\delta t + R\delta t$. Thus,

$$Q = R\delta t/((C_2 - C_1)\delta t) = R/(C_2 - C_1) = \text{rate of addition/concentration difference.}$$

Many methods for cardiac output determination are based on this principle. The earliest used is the *direct Fick method*. In this method the substance whose concentration is measured is most commonly blood oxygen, and the rate of addition is obtained from the rate at which oxygen is added to the blood, which is the same as the rate oxygen is consumed if a subject inhales air from a bell jar. Here the exhaled air is returned to the bell jar via a

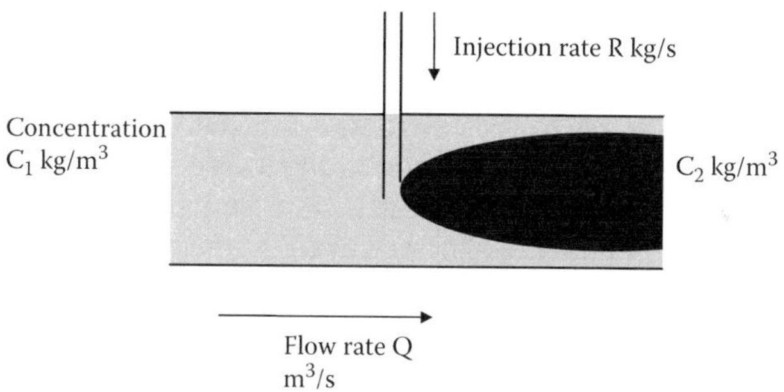

FIGURE 11.6
The principle of dye dilution for the determination of blood flow rate Q. The dye spreads out to give a concentration of C_2 across the vessel at some point downstream.

carbon dioxide absorber, so the rate of O_2 uptake shows itself as a general decline in the level of the bell jar.

The blood O_2 concentrations are measured by withdrawing a sample of arterial blood (using a needle), and the blood going toward the lungs is sampled in the pulmonary artery using a catheter ("mixed venous" sample). In words,

$$\text{Cardiac Output} = O_2 \text{ Consumption Rate/Arteriovenous } O_2 \text{ Difference.}$$

This equation at first sight is counterintuitive. If O_2 is being consumed at a constant rate, the AV O_2 difference would get bigger as CO falls. Since low CO is often associated with low arterial O_2 (in recovery from operations, for example), this would require venous O_2 to fall even more. Low venous O_2 would normally be associated with greater tissue O_2 usage, such as would occur in exercise. This confuses several phenomena. The best way to think of the situation is to consider the control of cardiac output, which is related directly to tissue O_2 demand, which in high in exercise and low during recovery from operations. Thus, the main determinant of CO in the above equation is the numerator. In the denominator, the arterial O_2 normally does not alter much, and the venous O_2 will increase slightly in exercise and decrease in sleep or under anesthesia, where there is greater efficiency of O_2 extraction.

A similar determination can be done by measuring CO_2 evolution and blood CO_2 levels. In this case, the equation is

$$\text{Cardiac Output} = CO_2 \text{ Evolution Rate/Arteriovenous } CO_2 \text{ Difference.}$$

The *indirect Fick method* does away with measuring gas concentrations in blood samples: instead it assumes that, in expired air, end-tidal CO_2 (measured in convenient concentration units, such as a percentage*) closely approximates arterial partial pressure. It assumes also that if the subject rebreathes from a bag, the concentration in the bag after 10 or so breaths is close to mixed venous.

* Typically, end-tidal CO_2 is given as 45%: this means that in 100 mL of blood, if all of the CO_2 can be liberated and measured in a volume meter as a gas at a standard temperature and pressure, it would occupy 45 mL.

11.3.1 Dilution Methods for Cardiac Output Determination

Dilution methods rely on the addition of a substance foreign to the body (such as a dye, a radioactive substance, or a gas such as acetylene) such that $C_1 = 0$. Some dilution methods rely on temperature rather than concentration, and these will be dealt with separately. Rather than a steady infusion of this foreign substance (which could be dangerous) the addition is as a *bolus* or *slug* and in as short a duration as possible. The equation becomes (integrating top and bottom and remembering that integrating a rate of infusion over time gives *amount added*):

$$Q = \int R dt / \int C_2 dt = m / \int C_2 dt = \text{mass added/area under concentration vs. time curve}$$

Some write this as $Q = m/(\hat{C}T)$; where $\hat{C}$ is the average concentration measured during the time the dye is washed out of the system and T is the time between when the dye appears then disappears in samples. Although this is a simpler equation to remember, there are several difficulties in actually applying it to data. Effectively, it represents the area $\int C_2 dt$ as an equivalent rectangle, area $\hat{C}T$.

If the cardiovascular system were very large the curve of C_2 versus t would look something like the graph shown below, for dye injected quickly into a large vein leading directly to the heart and then samples withdrawn from a major artery, such as in the forearm (Figure 11.7).

At point x, the dye was injected, and then at some later time, y, it began to be detected in blood samples. The blood concentration reaches a peak and then falls off more or less exponentially, as the concentration in the central blood pool (the heart chambers and lungs) gets progressively less as fresh blood arrives there. The falling exponential represents the *washout* of the dye from the central blood pool compartment. Of course, in a 6-L volume of circulation, the initial bolus will eventually recirculate to the central pool and then to the collection point, albeit greatly dispersed by this time. The concentration will begin to rise again, as shown by the dotted line.

Dye dilution has several disadvantages, not the least of which is the recirculation problem, and it is now no longer used in clinical practice (although the principle is still used in research applications and helps in the understanding of a range of perfusion situations involving foreign agents). Common dyes used in the past are Evans' blue and indocyanine green.

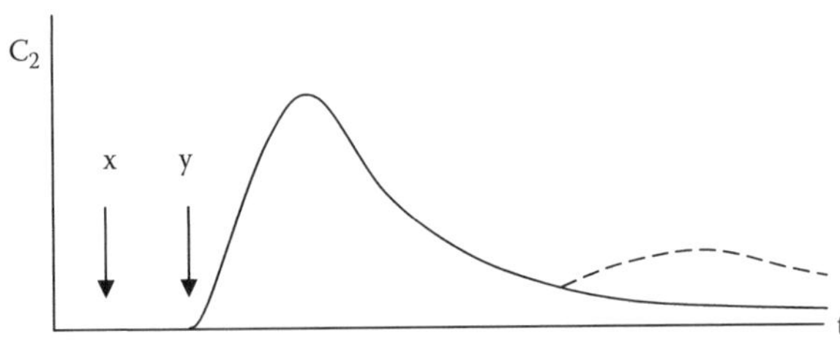

FIGURE 11.7
A dye dilution curve. Dye is injected at point x, and the dye begins to appear in the samples withdrawn downstream at time y. The dotted line shows the effect of recirculating dye, which needs to be eliminated when estimating the total area under the curve.

11.3.2 Thermal Dilution

A very popular method for measuring cardiac output in a clinical situation uses, instead of a colored dye, a few milliliters of saline injected at around 20°C temperature via a catheter directly into the right atrium. In this case, rather than measuring concentration, the slight drop in temperature in the blood due to the addition of this bolus of cold saline is measured, usually in the pulmonary artery, using a thermistor. In a single passage through the heart the saline is dispersed throughout the blood and is assumed to attain blood temperature (37°C). The equation for blood flow rate, which in the pulmonary artery is the CO, is easy to derive:

$$\text{Heat Gained by Bolus} = \text{Heat Lost by Blood.}$$

Or more precisely,

Mass of Saline × Specific Heat of Saline (s_i) × Temperature Rise in Bolus = Integrated Mass of Blood Affected × Specific Heat of Blood (s_b) × Weighted Average Drop in Blood Temperature,

where

Mass of Saline = Volume Injected × Density of Saline = $V\rho_i$

Integrated Mass of Blood Affected = Flow Rate in Pulmonary Artery × Increment in Time × Density of Blood = $\int Q dt \, \rho_b$.

Hence, we have an expression involving CO (Q). The weighted average fall in temperature is given by $\int \Delta T dt / \int dt$, where ΔT is the instantaneous fall from the unperturbed blood temperature, normally 37°C. Remember that Q itself is an average flow rate (in L/min) taken over several heart cycles. The $\int dt$ is just the time over which the fall in blood temperature is detectable, and appears both in numerator and denominator of the right hand side of the first equation.

Thus,

$$V\rho_i s_i (T_b - T_i) = Q\rho_b s_b \int \Delta T dt, \qquad \text{or} \qquad Q = V\rho_i s_i (T_b - T_i) / \{\rho_b s_b \int \Delta T(t) dt\}.$$

Here T_b is initial blood temperature, T_i the injectate temperature and $\Delta T(t)$ the instantaneous drop in temperature in the blood as the bolus passes through the pulmonary artery. A typical thermal dilution curve might look like that shown in Figure 11.8.

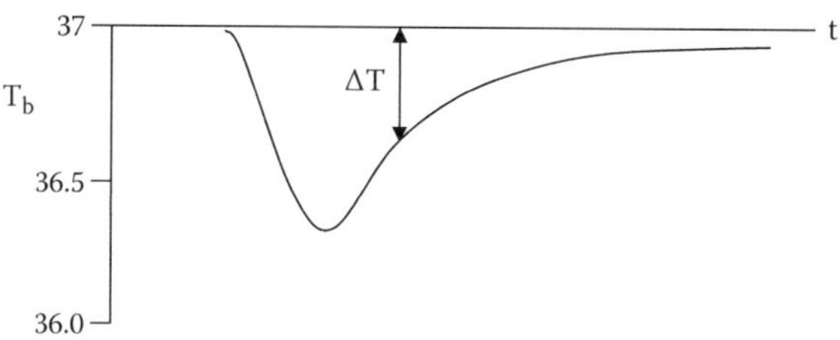

FIGURE 11.8
Thermal dilution curve. The graph is of blood temperature downstream of the point of injection of a bolus of cool saline. The curve is an inverted version of Figure 11.7, and there is no recirculation, thus estimating area enclosed by the curve is much easier.

Typical values are V around 5 mL; ρ_i = 1018 kg/m^3; s_i = 0.964; T_b = 37°C; T_i = 20°C; ρ_b = 1057 kg/m^3; s_b = 0.87. The factor $\rho_i s_i/(\rho_b s_b)$ is quite close to unity. If, in the above example, the weighted average fall in blood temperature is 0.3°C over a 5-s period, the CO (Q) is given by 5 × 1 × (37 – 20)/(0.3 × 5) = 57 mL/s = 3.4 L/min.

As with the dye dilution example, a high CO would correspond to a faster washout, so the area represented by $\int \Delta T(t)dt$ would be less.

The thermal dilution measurement is usually carried out via a Swan–Ganz cathether, which has an inflatable balloon at its tip to assist in placement, and which can also be used to measure pressures (including pulmonary capillary wedge pressure, or PCWP) (see Figure 11.9a).

The Swan–Ganz catheter is inserted into the neck vein and then on into the SVC, the RA, the RV, and then the PA.

The catheter has four lumens: (1) the distal pressure (tip), (2) the proximal pressure/saline injection (20 cm from the end), (3) the inflatable balloon (1 cm from the end), and (4) the wires to the thermistor (4 cm from the end). For CO measurement, saline is injected into the RA, and then the temperature is measured in the PA at the points shown. For PCWP, the balloon is inflated momentarily while the distal pressure is being measured.

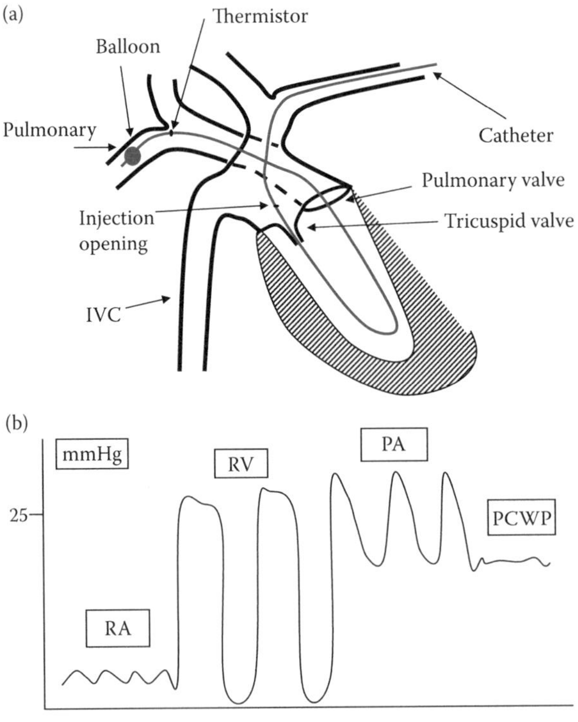

FIGURE 11.9
(a and b) The Swan–Ganz catheter, shown as the light-colored line in the upper diagram. (a) The points of injection and temperature measurement are shown, together with the inflatable balloon used to guide the catheter into the appropriate regions. (b) The lower diagram is a pressure trace, measured at the catheter tip, showing the characteristic waveforms obtained as the catheter is advanced. RA, RV, PA, PCWP are right atrium, right ventricle, pulmonary artery, and pulmonary capillary wedge pressure, respectively.

The catheter is guided into place through the partial inflation of the balloon—the blood flow carries the balloon along the same direction. The actual position of the tip can be gauged by distal pressure waveform, as shown in Figure 11.9b.

11.3.3 Derived Indices

Since larger people tend to have larger values of CO, when comparisons are made, the CO values are usually adjusted. This is called the cardiac index, and the body surface area (BSA), rather than the body mass, is used to derive it. BSA can be estimated from an empirical formula (du Bois formula – see Chapter 1 Section 5) that combines height and weight to come up with values in square meters.

$$CI = CO/BSA = SV \times HR/BSA.$$

Comparing the action of the ventricle to a piston in a cylinder, in which the work done each stroke is the average pressure times the volume of fluid displaced:

Left Ventricular Stroke Work (LVSW) = SV × (Mean Arterial Pressure (MAP) – Pulmonary Capillary Wedge Pressure (PCWP) + Kinetic Energy (KE) Term.

(KE is usually ignored; PCWP is sometimes ignored)

$$\text{LVSW Index} = \text{LVSW/BSA} = \{(\text{MAP} - \text{PCWP})\text{CI/HR}\}/\text{BSA}.$$

KE is sometimes estimated from $\frac{1}{2}\rho SV2u^2$, where u is the mean blood flow in the aorta (which is equivalent to the more familiar $\frac{1}{2}$ mu^2). Since SV is ejected in time T, where T is the ejection time (0.28 s) and the aortic radius is a, we can use $u = SV/(\pi a^2 T)$ in the formula above. ρ is of course blood density. In exercise and in the pulmonary circulation, KE cannot be ignored (because the arterial pressure is so much less), and the KE component can account for up to 25% of the stroke work (see also Chapter 10 Section 1.6).

11.3.4 Other Methods of Measuring Cardiac Output

11.3.4.1 Nuclear Cardiology

This involves the injection of a radioisotope that tags either the blood or the myocardium. A *gamma camera* to form radioactivity images at particular instants during the cardiac cycle is used (see chapter on medical imaging for further information). The ECG can be used to "gate" image acquisition, so that data from several cardiac cycles can be accumulated. Each R wave is used to begin timing typically 25 epochs, each 30 ms long. Thus, if the end of systole and diastole occur (say) in epochs number 2 and 23, the SV can be estimated from the difference between the end diastolic volume and the end systolic volume (EDV – ESV). These volumes can be estimated from 2-D images (from the gamma camera) by assuming they can be revolved around the long axis to produce a spheroid.

11.3.4.2 X-Ray Fluoroscopy

The X-ray machine is below a couch on which patient lies. Radio-opaque substance injected into a vein which shows up the edges of the heart chamber in a video acquired from a digital X-ray imager. As above, software can calculate ventricular volume from area by assuming the chamber to be an oblate spheroid. From EDV – ESV = SV and knowledge of HR, CO can be calculated. Computed tomography (CT) is also used following a similar approach as above with ECG gating.

11.3.4.3 Ultrasound

There are a number of methods: the most common one is to use a transesophageal Doppler probe. This gives blood velocity u in the aorta. If the radius of the aorta can be measured using ultrasonic echo methods, CO is given by $Q = u\pi r^2$.

11.3.4.4 Magnetic Resonance Imaging

This is similar to nuclear cardiography; images are acquired using the ECG to "gate" corresponding phases from different cardiac cycles.

11.3.4.5 Chest Impedance

This is known as *impedance plethysmography* and is a noninvasive method which can be used in nonclinical monitoring (such as in fitness assessment). However, the absolute accuracy is poor, although it is able to identify percentage changes quite well. It involves passing a high-frequency (50 MHz) current through the chest (from the neck to waist regions). As the blood is ejected into the aorta, this expands and reduces the chest impedance slightly. The volume change can be estimated from the following formula (the Kubicek formula).

$$\Delta V = -\rho(L/Z_0)^2\Delta Z$$

where ΔV is the volume of blood ejected into the aorta, ρ is blood resistivity (around 1.5 ohm-m), L_0 the distance between the neck and waist electrodes, Z_0 the average impedance (resistance) of the chest (around 28 Ω) and ΔZ the change in impedance between systole and diastole. However, in order to compensate for "run-off" into the abdominal regions of the aorta, it is customary to replace the ΔZ term by $(dZ/dt)_{initial}T$, where T is the ejection time (around 27 ms). ΔV is then taken as an estimate of SV, which when multiplied by HR gives CO.

11.4 Pressure Measurement

We saw in Chapter 7 the importance of measuring blood pressure, especially arterial pressure. Pressure within blood vessels can either be measured invasively (using a narrow saline-filled catheter introduced into the vessel via an incision in the skin) or noninvasively (by, for example, measuring the tendency for the tissue overlying an artery to deform due to the pulsatile nature of the blood flow). The latter is obviously more convenient for the patient, but the former is more accurate. Noninvasive methods will be described first.

11.4.1 Noninvasive Methods for Measuring Arterial Blood Pressure

11.4.1.1 Riva–Rocci–Korotkov (or Korotkoff) Method

This is the common method for measuring "blood pressure" in routine medical examination. It consists of a cuff around the upper arm and some method for detecting sounds of turbulent flow (Korotkoff sounds) within the brachial artery. Originally these sounds were detected by listening via a stethoscope (auscultation), but more recent designs have incorporated low-frequency microphones or other detectors within the cuff to analyze these sounds. The method consists of raising the cuff pressure above the systolic arterial pressure and then allowing the air to escape slowly from the cuff. The cuff pressure is monitored using a mercury column (called a plethysmograph) or in modern instruments and electronic pressure sensor. The pressure is measured in mmHg (the height of the mercury column: 100 mmHg = 13.3 kPa), even in electronic sensors. When the cuff pressure falls below arterial systolic pressure, small amounts of blood are able to "spurt" past the region of the arm occluded by the cuff, which can be plainly heard in the stethoscope or picked up by the microphone (Korotkoff sounds). The pressure where these sounds are first heard is thus the systolic pressure in the artery. As the cuff pressure continues to fall, the turbulent blood flow is able to be maintained for more and more of the cardiac cycle until eventually the sounds become muffled and then disappear. This point represents the diastolic pressure (there is a tendency now to use the disappearance rather than the muffling point as indicative of diastolic pressure). See Figure 11.10 for further details. Careful observation of the top of the mercury column will reveal very slight oscillations in height which begin as the systolic pressure is reached (as cuff pressure falls) and stop once diastolic pressure has been passed. Some electronic methods detect these oscillations in pressure (the so-called oscillometric method) to determine the systolic and diastolic pressure points.

11.4.1.2 Constant Pressure Monitoring in the Finger (Finapres)

This consists of a rigid tube which fits over a finger, with a small airtight seal around the lower part of the finger. As the arterial blood enters the finger, normally the volume of

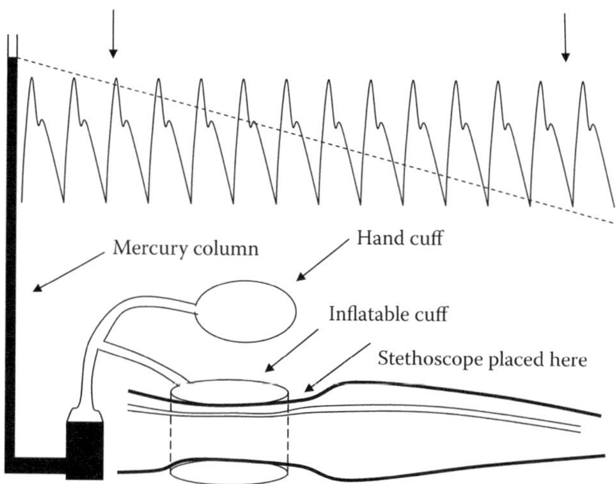

FIGURE 11.10
Riva–Rocci–Korotkoff method for measuring arterial blood pressure. Vertical arrows: commencement and disappearance of Korotkoff sounds.

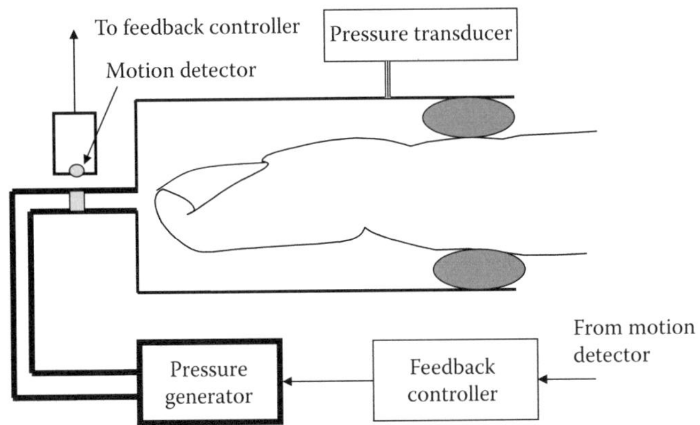

FIGURE 11.11
The Finapres method for obtaining arterial pressure waveform.

the finger expands to accommodate this extra blood volume. In this device a pressure is applied within the tube to maintain the volume of the finger constant (a "volume clamp"). The pressure required to do this equals the pressure in the artery (see Figure 11.11). Since the pressure is continuously adjusted, the output is likewise a continuous signal, and the details of the arterial waveform can be recorded, calibrated in mmHg.

11.4.1.3 Finger Plethysmograph (Finger Pulse Meter or Pulse Oximeter)

This finger-clip device is routinely used to detect a pulse in emergency situations and can be used to measure oxygen saturation (SaO2)% (see later section).

Light from a red light emitting diode (LED) is reflected from bone in finger. The path length (and hence the intensity of light detected by the photoresistor) varies with the volume of blood in finger varies in time with blood pulses. The trace recorded has the same features as the arterial waveform (with a dicrotic notch, etc.) but the signal can only be calibrated in mmHg if there is an independent way of doing this. However, this method has been used to estimate ventricular ejection time (see Figure 11.12).

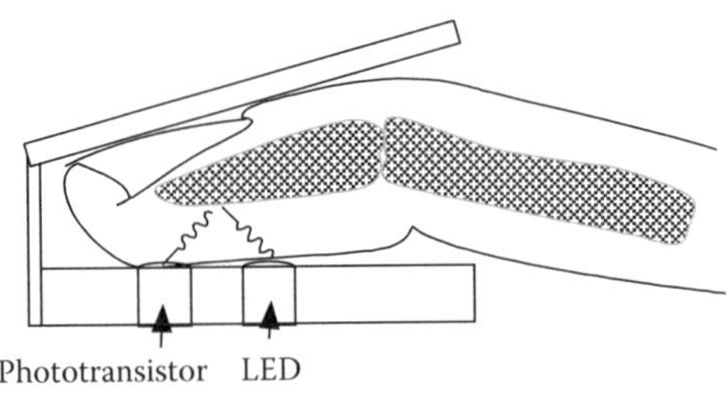

FIGURE 11.12
The finger plethysmograph. Light from the LED is reflected from the bone and the intensity arriving at the phototransistor varies with the volume of blood in the region traversed by the light.

11.4.1.4 Pulse Tonometry

This recognizes the fact that the difference between internal and external pressure (Δp) in a vessel is related to the wall tension T and radius r by the so-called Laplace formula $\Delta p = T/r$. If a sufficiently heavy object is placed over the artery such that one portion becomes flat, r is effectively infinite and Δp will be zero. In other words, the force divided by the area of the "rider" will equal the arterial pressure. This method for measuring arterial pressure waveforms is not widespread, but on the other hand, this method (or a modification of it) is commonly used for estimating intraocular pressure.

There are no practical, reliable, methods for measuring *venous* pressure waveforms noninvasively.

11.4.2 Invasive Methods for Measuring Arterial Pressure

In order to introduce a catheter into an artery, some method has to be used to avoid blood loss. The Seldinger technique uses a sealed rigid tube (or trochar) which is introduced into the artery. The seal prevents blood loss. A guidewire is then fed into the trochar and then the trochar removed. The hollow catheter is then fed over the guidewire into the artery and then the guidewire removed (Figure 11.13).

The catheter is then advanced to the site of measurement. In the case of coronary investigations, the catheter can be advanced from the groin, up the aorta into the pouches of Valsalva, where the tip can enter the right or left coronary artery (special spring-loaded

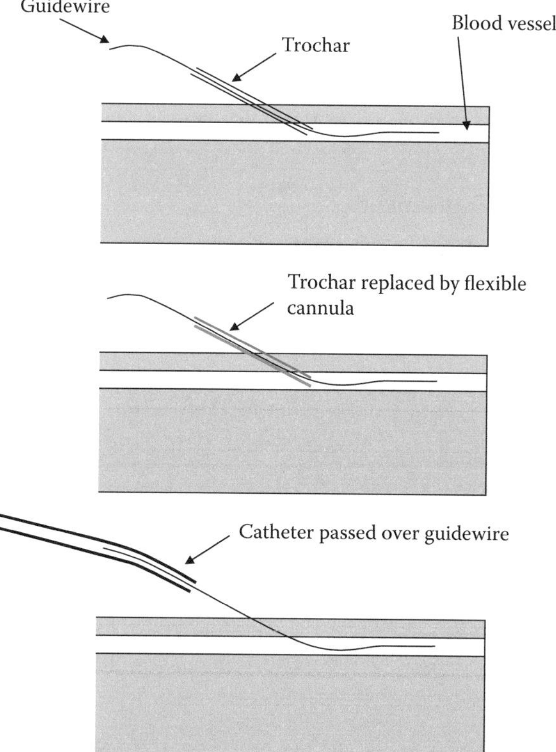

FIGURE 11.13
The Seldinger technique for inserting catheters into blood vessels.

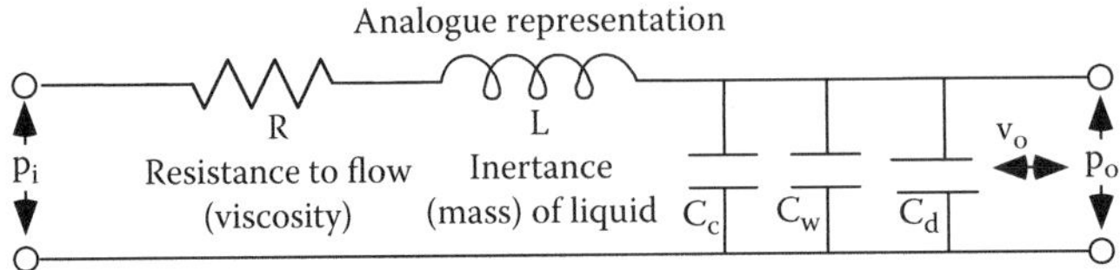

FIGURE 11.14
The equivalent electrical circuit for a fluid-filled catheter connected to a pressure transducer. p_i is the actual pressure and p_0 is the pressure recorded by the transducer. C_C, C_W, and C_d are the compliances of the catheter, the catheter fluid and the transducer diaphragm, respectively.

tips are used to select one or the other). This placement is used mainly for the injection of radio-opaque dye in fluoroscopic investigations rather than measurement of pressure. In order to measure pressure, a transducer has to be connected to the proximal end of the catheter. These transducers will be described below. The catheter–transducer system can be represented by the circuit diagram shown in Figure 11.14. Note that there is a potential for signal distortion because of the mass of saline in the catheter and the compliance of the measuring system (particularly the catheter itself). This imposes some restrictions on the design of the catheter and the transducer, as we will see.

11.4.3 Invasive Methods for Measuring Venous or Right-Heart Pressure

Catheterizing a vein is a less risky procedure than an artery in terms of potential for blood loss. The Swan–Ganz catheter already described in connection with the thermal dilution method of measuring cardiac output. Because it is a multilumen device (see Figure 11.9), the Seldinger technique is impractical, and a simple incision is made instead. The inflated balloon is used to guide the catheter into the correct locations in the right heart.

11.4.4 Types of Pressure Transducers

Basically, all pressure transducers convert the displacement or flexure of a diaphragm into an electric signal. Most (but not all) use the principle of the unbalanced Wheatstone Bridge to do this. This principle can be understood by referring to Figure 11.15a.

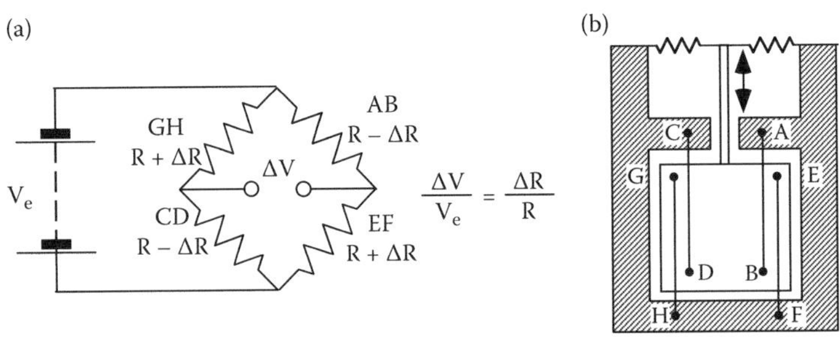

FIGURE 11.15
(a and b) The unbalanced Wheatstone Bridge and the arrangement of resistive wires in the unbonded strain gauge-type pressure transducer. The letters in (b) correspond to those in (a). ΔV is the signal which is recorded and V_e is the excitation voltage. ΔR is the change in resistance produced by the diaphragm being displaced upward in this case.

The voltage labeled ΔV is proportional to the small fractional change in resistance, if the fraction is less than around 5%. Pressure transducers arrange for a change in pressure to produce a small change in resistance of some element within the transducer.

$$\Delta V = \frac{R_2 V_e}{R_1 + R_2} - \frac{R_4 V_e}{R_3 + R_4}.$$

If R_1, $R_4 = R + \delta R$ and R_2, $R_3 = R - \delta R$, then $\Delta V = \frac{\delta R}{R} V_e$.

This forms the basis of strain gauges, which are often formed by printing miniature copper conducting strips on to a plastic substrate. This can then be bonded to metal bars such that when the bar is flexed, the copper changes length slightly and hence resistance. This arrangement is known as a bonded strain gauge.

11.4.4.1 Unbonded Strain Gauge Wire (Statham P23 and Related Models)

This was popular for clinical work in the 1970s and is shown in Figure 11.15b. As the metal foil diaphragm moves downward the wire resistance elements AB and CD get longer (so resistance ↑) and resistance elements GH and EF shorten (so R ↓); these wires are connected into a Wheatstone Bridge arrangement as shown. A typical sensitivity is 10 mV/100 mmHg, with a volume displacement of 04 mm^3/100 mmHg. This latter figure represents the transducer compliance and is important in predicting the dynamic response of the system (see below). For a wire, the resistance formula is $R = \rho L/(\pi r^2)$, where ρ is resistivity (Ω m) L is length and r radius of wire. Thus, $\delta R/R = \delta\rho/\rho + \delta L/L - 2\delta r/r$, and since $\delta L/\delta r = -\sigma$ (Poisson's ratio), $\delta R/R = \delta\rho/\rho + (1 + 2\sigma)\delta L/L$. The gauge factor $G = (\delta R/R)/(\delta L/L)$ is a measure of how much change in resistance the added pressure is likely to produce. For metals, G ~ 1.6, which is fairly modest, however for semiconductors, G ~ 150, thus a very small change in dimension will give a very much larger change in resistance.

11.4.4.2 Silicon Diaphragm

These devices are used in the majority of clinical pressure transducers in use today. The metal foil diaphragm has been replaced by a small silicon wafer, with insulated p-type semiconducting regions etched into a silicon dioxide layer with aluminum contacts deposited at the end as shown in Figure 11.16a.

These have the advantage of being very robust, but on the other hand they are very temperature dependent. This is compensated for by having some of the p-type regions acting as temperature sensors, so that the results can be electronically compensated. The sensitivity is around 5 mV/100 mmHg (for 10 V excitation) with a displacement of 0.001 mm^3/100 mmHg.

11.4.4.3 Linear Variable Differential Transformer

Formerly this was a common form in clinical pressure transducer, although problems with having to use ac activation and phase detection to distinguish between positive and negative pressures has now meant that other designs are now favored. Thus, it is now rarely found in clinical applications, although the method is important to understand in relation to a general method for measuring displacement or pressure (Figure 11.17).

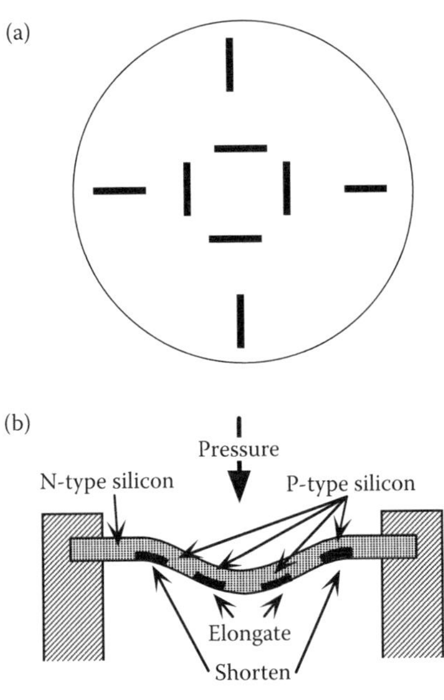

FIGURE 11.16
(a and b) Solid-state pressure transducer, with a silicon wafer as a diaphragm. Flexing the diaphragm leads to elongation of some of the insulated p-type regions and shortening of others. The measurement arrangement is similar to Figure 11.15a.

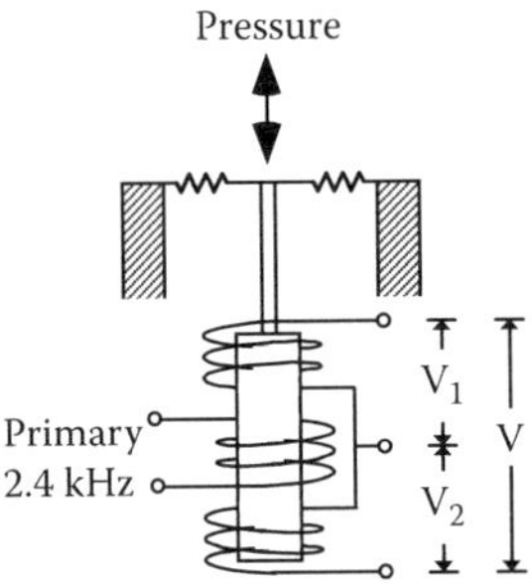

FIGURE 11.17
The linear variable differential transformer type of pressure transducer, with AC excitation.

The primary is activated by typically 2.4 kHz current. When the core is symmetrically placed between the two secondary coils the induced voltages $V_1 = -V_2$, so the total is zero. If the pressure presses the diaphragm down (or up) the voltages do not cancel and $V_1 - V_2 = V$, which is proportional to pressure. Typical sensitivity is 20 mV/100 mmHg.

11.4.4.4 Fiber Optic Catheter Tip

A number of manufacturers supply tip-mounted pressure transducers that rely on the flexing of a membrane to alter the amount of returning light delivered by fiber optics (see Figure 11.18). These can be as small as 0.25 mm (outside diameter). They are particularly

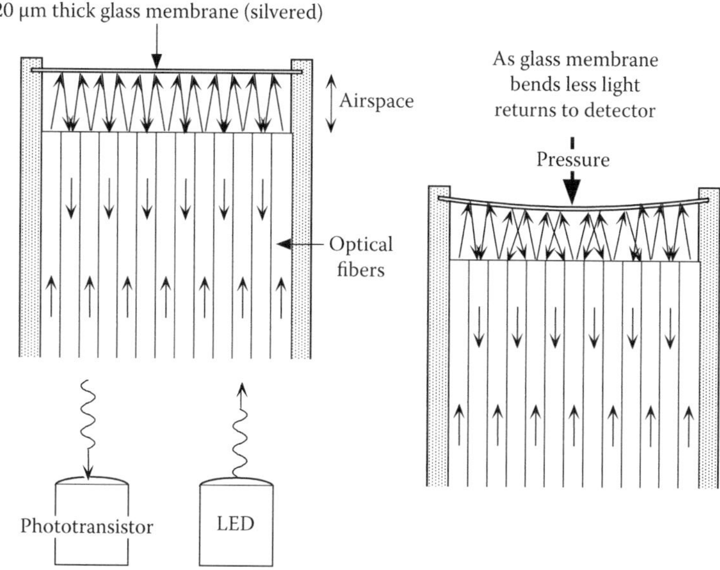

FIGURE 11.18
Fiber optic catheter tip type of pressure transducer. Flexing of the membrane leads to varying amounts of light returning to the phototransistor.

immune from electromagnetic interference and are very stable. The signal conditioning unit can compensate for temperature variations, and the overall accuracy is around ±1 mmHg. The added advantage is that the dynamic response is not determined by a fluid-filled catheter, as it is in other types.

11.4.5 Dynamics of Fluid-Filled Catheters

As just mentioned, most clinical pressure measurements are made at the proximal end of a saline-filled catheter, typically around 1.5 mm in diameter. Since there is often around 1 m between the distal end of the catheter and the point at which the transducer is attached, the overall system has to be modeled to ensure that the measured pressure waveforms are accurately represented. In addition to their use in cardiology, fluid-filled catheters or tubes (where the fluid could be a gas) are used in other branches of clinical monitoring, lung physiology, for example. The catheter–transducer system is represented by an electrical circuit as shown in Figure 11.15. Elements R, L, and C represent fluid viscosity, fluid mass, and the combined compliance (that is, compressibility or expandability) of the fluid, the diaphragm in the transducer and the catheter itself. Simple expressions can be derived for R, L and C (in terms of fluid properties) as below.

Resistance to Fluid Flow (see also Chapter 9):

$$R = \frac{8\mu l}{\pi r^4} \text{: Poisieulle's law.}$$

(NB, μ is the viscosity of the saline, and l and r are the length and radius of the catheter.)

Inductance or inertance of liquid in vessel:

$$L = \frac{\rho l}{\pi r^2}.$$

This arises because the fluid mechanical equivalent of the definition of inductance, $L = v/(di/dt)$ is $L = \Delta p/(dQ/dt)$ and there is a fundamental hydrodynamic relation $\pi r^2 \rho (dQ/dt) = \Delta p/l$: the equation for L follows—here Q refers to flow rate (m^3/s), ρ the density of fluid and Δp the change in pressure over a distance l).

The compliance of the system is the equivalent of capacitance (dq/dv in electrical and dV/dp in mechanical terms); the sum of compliances of component parts are

C_c = catheter material compliance = Typically 10^{-13} m^3 Pa^{-1}
C_w = catheter fluid compliance = (~ 0.53×10^{-15} m^3 Pa^{-1} mL^{-1})
C_d = transducer diaphragm displacement (m^3 Pa^{-1}) typically 0.1 mm^3 per 100 mmHg $\cong 7.5 \times 10^{-15}$ m^3 per Pa.

The total compliance C is the sum of C_c, C_w, and C_d, although the first of these is usually at least an order of magnitude greater than the other two.

11.4.5.1 Evaluating Pressure Transducer/Catheter Performance: "Step Response" Test

The frequency response of a catheter can be inferred by applying a sudden increase (or decrease) in pressure (a "step") to the distal end of the catheter (the one usually placed within the artery). A typical response would look like the waveform in Figure 11.19.

This is an exponentially decaying sinusoid, with a decay time constant of τ and a period T_d between adjacent maxima. Ideally the signal should go straight to the final dotted value without either any wiggles or alternatively a slow exponential fall-off.

Essentially,

$$\tau = \frac{2L}{R} \leftrightarrow \frac{2\rho l}{\pi r^2}\frac{\pi r^4}{8\mu l} = \frac{\rho r^2}{4\mu}$$

and l/τ is called the *damping factor* (denoted β)

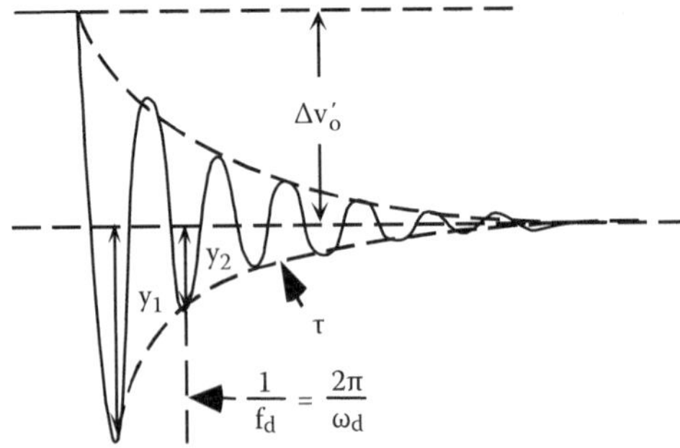

FIGURE 11.19
The step test for catheter response testing. The measured pressure may exhibit a "log decrement" type of response, where the natural frequency ω_n and the relative damping factor ζ can be estimated from the features of this waveform (see text).

ω_d (radians s^{-1}) is the *damped angular frequency* = $2\pi f_d$, where f_d is the frequency in Hz (= $1/T$) in the diagram.

As the damping (the value of β) becomes less, ω_d gradually increases until the system is undamped and $\omega = \omega_n$ is the *undamped (or natural) angular frequency*:

$$\omega_n \frac{1}{\sqrt{LC}} = \sqrt{\frac{\pi r^2}{\rho l C}}.$$

Dividing β by the natural frequency ($\zeta = \beta/\omega_n$) is called the *relative damping factor*, which when substituting the equations above gives

$$\varsigma = \frac{4\mu}{r^3}\sqrt{\frac{lC}{\pi\rho}} = \frac{\Lambda}{\sqrt{4\pi^2 + \Lambda^2}}; \Lambda = \ln\left(\frac{y_1}{y_2}\right).$$

Here, Λ is called the *log decrement*, with amplitudes $y_{1,2}$ shown in Figure 11.16.

The catheter–transducer system is analogous to a weight hanging from a spring contained in a cylinder of oil. The viscosity of the oil and the spring constant will determine the damping factor and the mass of the suspended object and the spring constant will determine the natural frequency of vibrations. For high fidelity recordings of pressure, the objective is to get a relative damping factor of around 0.6 (called critical damping). Very small values of ζ give rise to sustained oscillations ("ringing") and large values to a slow response, in which high frequency components are absent. The natural frequency of the system f_n should also be high enough so as not to interfere with the highest frequency components we wish to measure (normally around 25 Hz). Note from the equations above that any air bubbles in the saline will cause C_w to rise and hence ω_n will fall and β will rise, degrading the response of the measuring system. Similarly, blood clots in the end of the catheter will cause r to fall and hence similar changes to ω_n and β as given above (Figure 11.20).

WORKED EXAMPLE

The compliance of a catheter-transducer system is estimated as 2×10^{-13} m^3/Pa. Given that the catheter is 1.5 m long and has an internal diameter of 0.8 mm (radius 0.4 mm) determine

i) the natural frequency (in Hz) and
ii) the relative damping factor for this measurement system.

Assume density of fluid to be 1000 kg/m^3; viscosity 0.001 Pa s.

Answer

Using the equation for natural frequency, $\omega_n^2 = \pi \times (0.4 \times 10^{-3})2/(10^3 \times 1.5 \times 2 \times 10^{-13})$ = 1,676. Thus ω_n = 41 rad/sec or 6.5 Hz. The damping factor = $((4 \times 10^{-3})/(0.4 \times 10^{-3})^3) \times \sqrt{[(1.5 \times 2 \times 10^{-13})/(\pi \times 10^3)]} = 0.61$. Thus although the damping factor is close to optimal, the natural frequency is far too low to be able to record up to 25 Hz accurately.

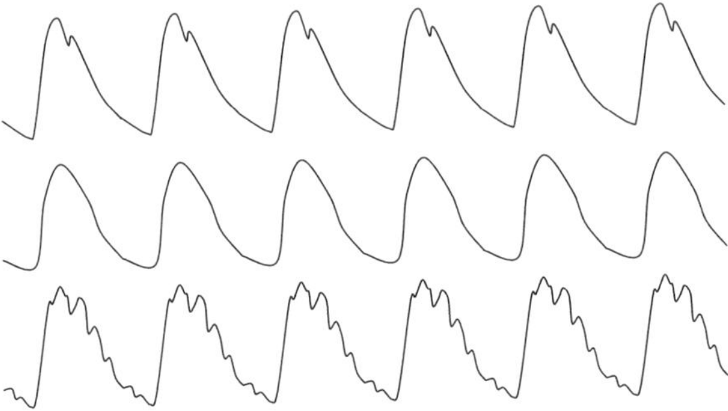

FIGURE 11.20
Distortion of the arterial waveform (top) due to the presence of an air bubble (middle) or uncontrolled movement within the vessel (bottom). The bottom two traces represent overdamped and underdamped systems, respectively.

11.4.5.2 Using Catheters to Estimate Heart Valve Abnormalities: Gorlin Formula

It is possible in disease for the edges of the valve leaflets to adhere to each other, thus reducing the effective area through which blood can flow when the valve opens (called stenosis). One way of estimating the severity of this is to measure the drop in pressure Δp as a catheter is pushed through (or pulled back, against the direction of flow) while the orifice is open. Normally, Δp should be very small, but as the area of the orifice gets smaller Δp will grow larger. If we apply Bernoulli's principle (see Chapter 9) to the streamline through the orifice (and neglecting gravitational effects), we get (see Figure 11.21)

$$\frac{1}{2}\rho u_1^2 + p = \frac{1}{2}\rho u_2^2 + p_2.$$

If we assume $u_2 \gg u_1$, we can neglect u_1 and, of course, $u_2 = Q/A$, where Q is flow rate through the orifice and A is the area of the orifice (or more accurately the smallest area of flow constriction, as shown in Figure 11.21). We can thus write

$$A = Q\sqrt{\rho}/(2C_d\sqrt{\Delta p})$$

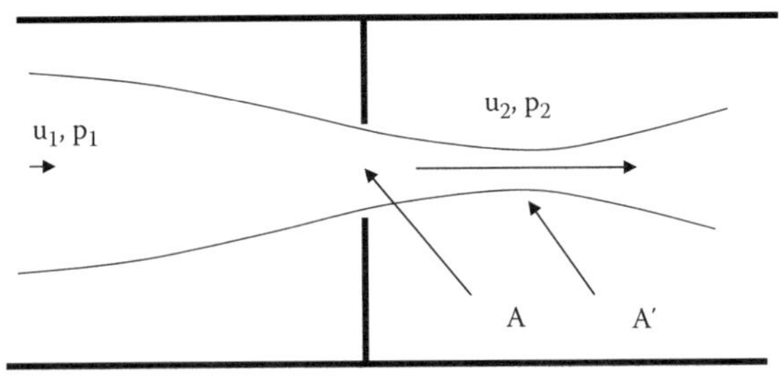

FIGURE 11.21
The principle of estimating the orifice area using the Gorlin formula. Valve stenosis is represented by the narrowed region shown. The streamlines (see the discussion of the Bernoulli equation in Chapter 9) reach the narrowest area A′ some distance downstream from the actual orifice, giving rise to the need for applying a correction factor.

where Δp is $p_1 - p_2$ and C_d is a factor to adjust for the orifice area A being somewhat larger than the smallest area of flow constriction A′. For the mitral valve C_d is 0.6, but for the other three valves it is 0.85. If Δp is measured in mmHg and Q in mL/min, then a conversion factor has to be included (this is usually given as 44.3, without explanation). Q can be estimated from cardiac output by adjusting for the fraction of the cardiac cycle the flow actually occurs (e.g., CO × 8/3, if the flow occurs during systole and systole represents 3/8 of the cardiac cycle). As an example, if Δp is 50 mmHg and CO is 5 L/min, then the formula gives 0.75 cm^2, approximately.

11.5 Flow Measurement

This refers to measuring the velocity of blood, u (in m/s), rather than the volume flow rate, Q (in m^3/s), although, of course, if the cross-sectional area is known (πr^2) the latter can be estimated from the former by $Q = u\pi r^2$. Although in the past a variety of principles were used to estimate flow velocity (electromagnetic flowmeters, hot-wire anemometers etc.) the principle of ultrasound Doppler is now the main basis of this measurement.

11.5.1 Ultrasonic Methods for Measuring Flow (Particularly of Blood)

11.5.1.1 Basic Properties of Ultrasound

Ultrasound is sound whose frequency is above the audible range of humans (i.e., $f > 20$ kHz). The usual range of frequencies used in medical diagnosis and therapy is 1 MHz–10 MHz. The penetration of ultrasound into tissue decreases as the frequency increases, so the high frequencies are used for superficial blood vessels and the lower frequencies for the deeper ones. At 10 MHz the wavelength of sound is 0.15 mm in tissue (in which sound velocity is 1500 m/s) so the trade-off is much better spatial resolution at these higher frequencies. At any of these frequencies air absorbs the ultrasound very strongly, so transducers have to be coupled to the skin surface via a gel of some description. Ultrasound has been widely used to produce images of tissue, since any boundary between two layers (for example muscle and blood) will produce an echo which will give clues as to its position. Initially though, we will concentrate on the exploitation of the Doppler principle for the measurement of fluid flow.

11.5.1.2 Production and Detection of Ultrasound

A number of materials exhibit the property of piezoelectricity, that is, the production of a mechanical strain in a material subjected to an electric field, or conversely the production of a tiny electric field across the material when compressed. Materials with good piezoelectric properties include quartz and certain ceramics, such as lead zirconate/titanate or PZT. Small crystal discs less than 1 mm thick can thus act both as transmitters (like loudspeakers) and receivers (like microphones) (see Figure 11.22b). The two faces have a metallic film applied which are then soldered to wires. If a pulse of a few volts with a duration of less than a microsecond is applied the crystal it will "ring" at its natural mechanical frequency and ultrasound will be propagated into surrounding regions. In Doppler applications, however, the electric field is applied as a continuous sinusoidal voltage at the resonant frequency of the crystal.

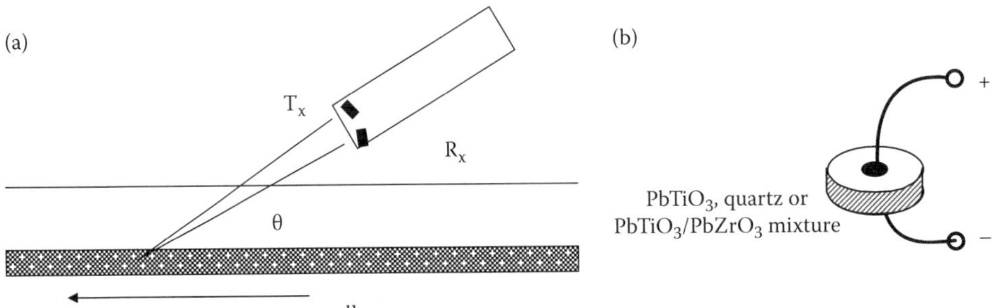

FIGURE 11.22
(a) Principle of ultrasonic Doppler flow measurement. There are two piezoelectric crystals side by side. In this case, the upper one (T_x) emits a continuous ultrasonic wave, and the lower crystal (R_x) picks up the reflected waves from the moving blood cells. There are two Doppler shifts in frequency due to the relative motion of source and receiver. (b) A piezoelectric crystal with metallic electrodes on the two faces to either impose a time varying voltage (T_x) or to pick up small voltages induced by mechanical movements due to returning ultrasonic echoes (R_x).

11.5.1.3 Doppler Principle

A full description of this can be found in any elementary tertiary physics text. A fine pencil of sound waves enters the skin and encounters the moving blood corpuscles in the artery below the skin (see Figure 11.22). An observer on the blood cells would detect a frequency of $f_s' = f_s(1 - u \cos \theta/c)$ if traveling away from the source, which is the transducer, held at an angle θ to the skin. The echo from the cell acts as a secondary source which is then detected by the transducer as a frequency of $f_s'(1 + u \cos \theta/c)^{-1}$. The final frequency is thus

$$f_s[(1 - u \cos \theta/c)/(1 + u \cos \theta/c)] \approx f_s\,(1 - 2\,u \cos \theta/c).$$

Or in terms of Doppler shift

$$\Delta f = f_s 2\,u \cos \theta/c.$$

Thus, for a transducer with transmitter frequency f_s of 10 MHz, an angle θ of 60° and a blood velocity of 0.5 m/s the Doppler shift is 3.33 kHz. This is within the audible range of frequencies and most ultrasonic Doppler instruments employ some way of directly converting this to an audible monitoring signal.

If the flow were toward the transducer rather than away from it the frequency received by the transducer will be greater then the transmitter frequency f_s by an amount $\Delta f = f_s 2\,u \cos \theta/c$. In other words, flow toward and flow away from the transducer of the same magnitude (and at the same angle) both give rise to identical Doppler shifts. If a vein and an artery are running side by side there needs to be some method of distinguishing the two flows. Indeed, in major arteries there may be a reversal of flow for part of the cycle, so again, this needs to be identified.

11.5.1.4 Transit Time

This uses pulses of ultrasound. It is often used to measure nonclinical fluid velocity in tubes, but has fallen out of favor for the clinical setting. The velocity of blood either adds to

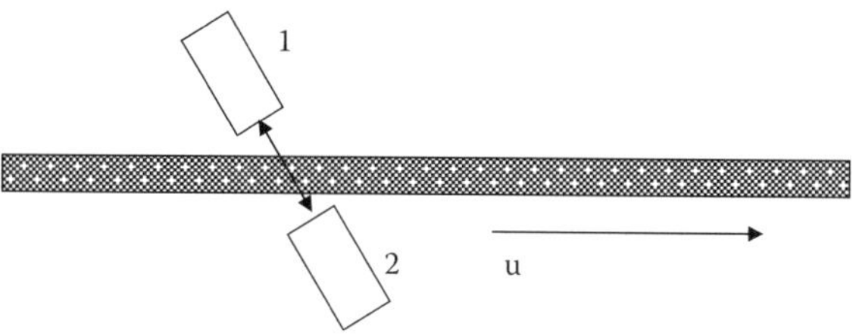

FIGURE 11.23
A "time-of-flight" method for measuring blood flow using two transducers, one to emit a pulse, the other to measure the transit time for the received pulse.

or subtracts from the transit time from one transducer to the other (Figure 11.23). D is the distance between the two transducers.

$$T_{1\to 2} = D/(c + u \cos \theta);\ T_{2\to 1} = D/(c - u \cos \theta),$$

so

$$\Delta T \approx 2D\,u \cos \theta/c^2.$$

11.5.2 Ultrasound Doppler Flow Probe: Signal Processing

The following description is of a basic ultrasound Doppler blood flow measuring instrument. These can be purchased relatively inexpensively for basic assessment of blood flow, often in a veterinary environment. Referring to Figure 11.24, the block in the top left of the diagram represents the ultrasonic generator, with is a continuous sine wave of frequency

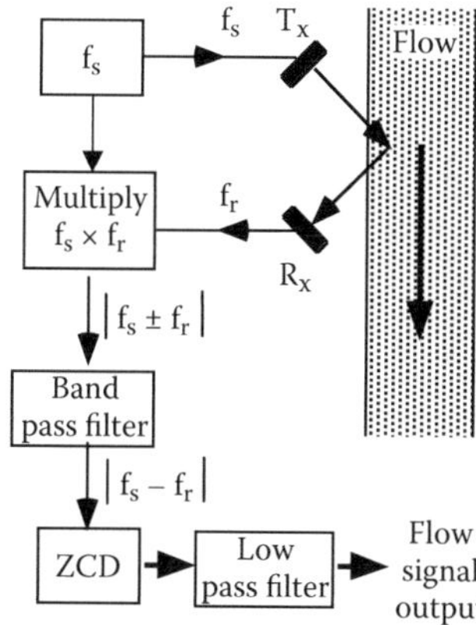

FIGURE 11.24
Signal processing for the ultrasonic Doppler flowmeter. The received frequency f_r will be less than the transmitted frequency f_s. The two waveforms are multiplied electronically (the transmitted waveform will have been suitably attenuated). This will result in a waveform containing frequencies at both $2f_s$ (approximately) and $f_s - f_r$. The second of these frequencies (in the kHz range) is then selected and the value of the frequency $(f_s - f_r)$ estimated using a zero crossing detector (ZCD). Since this gives rise to a series of pulses corresponding to the zero crossings a low pass filter is needed to convert this pulse sequence into a smooth waveform.

f_s fed to a piezoelectric crystal T_x. The generated ultrasound enters the blood vessel and the red blood cells scatter the ultrasound which is detected by crystal R_x (in reality, these crystals are usually side by side), whose frequency content has changed to reflect the Doppler shifts produced by the blood flow. If the flow was uniform, there would only be a single Doppler shift, 10 MHz (f_s) down to 9.997 MHz (f_r), say. The two signals f_s and f_r, which are sinusoidal, are multiplied electronically. If we assume for simplicity that both signals have unity amplitude and are in phase we can express this multiplication as

$$\cos(2\pi f_s t) \times \cos(2\pi f_r t) = \tfrac{1}{2}\,[\cos(2\pi(f_s + f_r)t) + \cos(2\pi(f_s - f_r)t)].$$

In other words, a sum of two sinusoids, one oscillating at 19.997 MHz and the other at 3 kHz, the latter being the Doppler shift signal, proportional to blood velocity. A low pass filter is used to isolate the latter signal, which is shown diagrammatically in Figure 11.24.

Note as the velocity varies the Doppler frequency varies. The amplitude represents the strength of the echoes, which is to do with a number of factors, principally the volume of blood insonated. There are a number of ways of performing a frequency to voltage conversion. The simplest is a zero crossing detector which gives a pulse output for every positive going signal (in other words, it counts the number of sinusoids per unit time, Figure 11.25).

This sequence of pulses for each positive going variation is then low pass filtered, giving a "saw tooth" pattern, which rises and falls depending on how many pulses there are in a given time interval (Figure 11.26). Normally, the saw teeth are too small to be seen.

This instrument is only able to detect the major velocity component. In order to detect a range of velocities (represented by a frequency spectrum, say, from 0 kHz to 3.5 kHz), a means of analysis of the frequency content of the Doppler signal has to be employed. One method that is easy to understand is to have a serries of narrow band filters, so that the 0- to 3.5-kHz spectrum is divided into 35 × 100-Hz wide bands, with the strength of the signal in each band representing the amount of blood flowing at the speed represented by that band. A further complication is that in arterial flow, there is a reverse component for a short interval within each cycle, so some means has to be found to distinguish the case where $f_r > f_s$ from that where $f_s > f_r$. This means is called the *directional Doppler* technique.

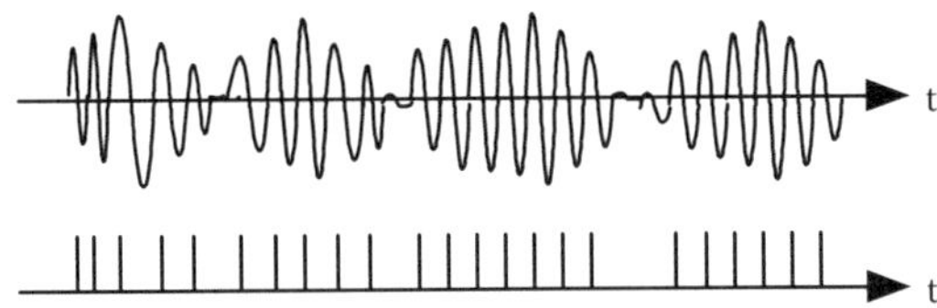

FIGURE 11.25
See Figure 11.24—the zero crossings from negative to positive are shown. Each indicates the start of a new cycle, and this indicates the instantaneous frequency.

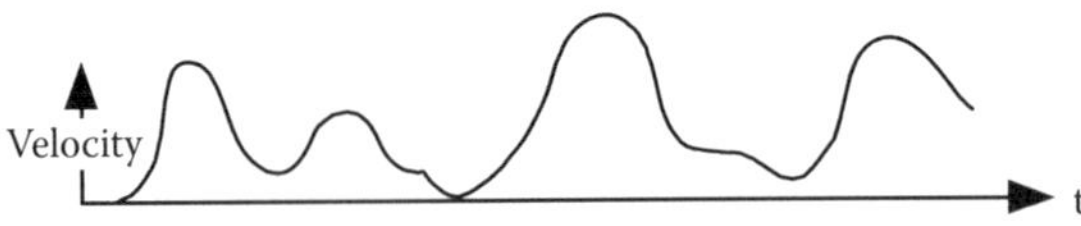

FIGURE 11.26
The smoothed version of the ZCD output from Figure 11.25.

11.5.2.1 Directional Doppler

One easy wayto achieve this is to multiply the received signal ($A_1\cos[2\pi(f_s + \Delta f)t + \phi]$) with an attenuated version of the transmitted signal ($A_0\cos(2\pi t)$). If we make $A_1 \approx A_0$, remembering that $\cos \alpha \cos \beta = ½[\cos (\alpha - \beta) + \cos (\alpha + \beta)]$, we get two components, one oscillating at a frequency of Δf (the Doppler shift) and the other at approximately $2f_s$. If a similar operation is carried out with a 90° phase shifted version of the transmitted signal ($A_0\sin(2\pi t)$), the oscillating component at the Doppler shift frequency Δf is also 90° phase shifted. If Δf is positive (flow toward), this 90° phase-shifted signal is *retarded* with respect to the unshifted signal, and if the flow is away, the signal is phase *advanced*. Logic circuits (triggered from threshold crossing) can distinguish these two situations (Figure 11.26).

11.5.3 Ultrasound Imaging in Cardiology

Cardiac ultrasound monitoring usually involves an integrated instrument which is primarily to produce real-time images of the heart and blood flow in individual vessels, but also has the capability of performing Doppler measurements at the same time. There is also a mode that allows the movement of heart valves to be followed, which uses a different technique from Doppler, called "M" mode. Color coding is often used to display the flow information on a 2-D image.

There is a fuller description of ultrasonic imaging in Chapter 23. However, for completeness, a brief overview will be presented here. Historically the ultrasonic modes have been classified as A, B, scanned, and M (originally TM).

11.5.3.1 A (Amplitude) Mode

This is the earliest and is in fact a display of the actual echo signal envelope (Figure 11.27). Echoes occur from interfaces (for example blood/muscle at the heart chamber wall) due to acoustic impedance mismatch. Unlike in the Doppler case, the transducer acts as both a transmitter (loudspeaker) and a receiver (microphone), since the data is the response to a single ultrasound pulse.

11.5.3.2 B (Brightness) Mode

Here the intensity of the "spot" on the display screen is modulated by the intensity of the echo. The more intense the echo, the brighter the spot. The position of the spot is determined by the beam direction (the on-screen vector represents the direction in relation to defined coordinates). See Figure 11.28.

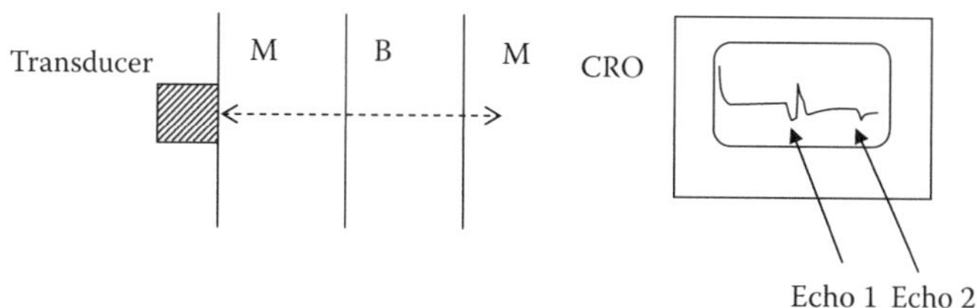

FIGURE 11.27
"A" mode ultrasound: the raw signal resulting from echoes from the MB and BM interfaces (it is assumed here that the ultrasound beam passes through the ventricle). M is (cardiac) muscle; B is blood.

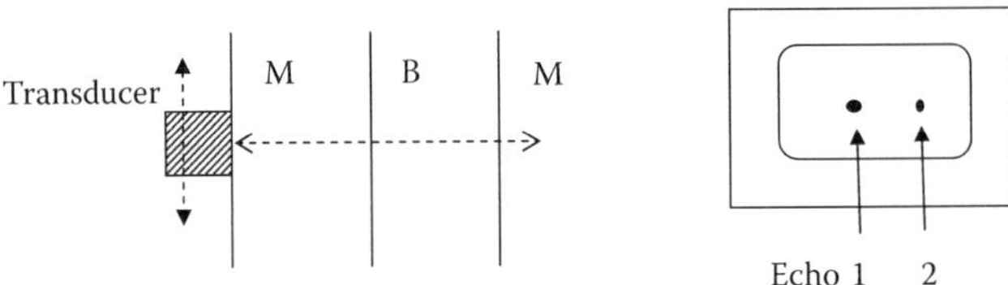

FIGURE 11.28
"B" mode ultrasonic imaging. This time the MB and BM interfaces produce echoes whose positions are represented by the horizontal axis on the CRO screen and their magnitudes by the brightness of the spot.

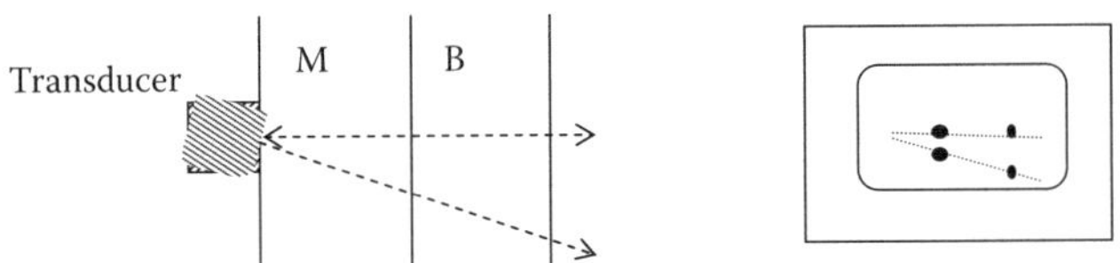

FIGURE 11.29
Scanned "B" mode imaging. The direction of the ultrasound beam is represented on the screen, with the echoes in their proper positions. If the pulse repetition rate is high, there will be an almost continuous line of dots for the MB and BM interfaces.

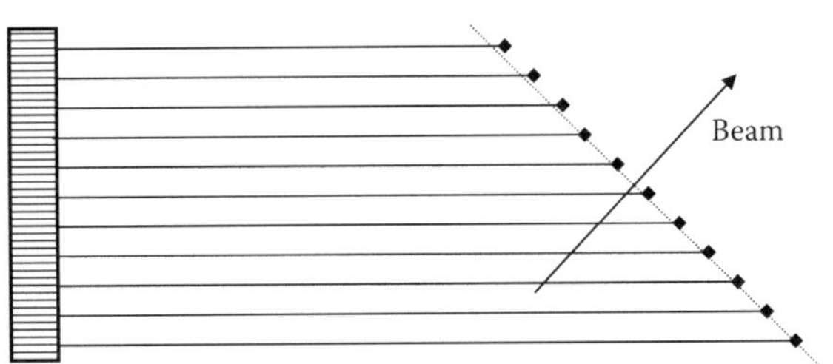

FIGURE 11.30
An electronically steered ultrasonic beam. Each element in the transducer (on the left) produces a pulse which can be delayed to various degrees from those on either side. Following the Huyghens principle, the advancing set of individual wavefronts will sum to give a synthetic wavefront. In this case the wavefront is equivalent to a plane wavefront advancing in the direction shown.

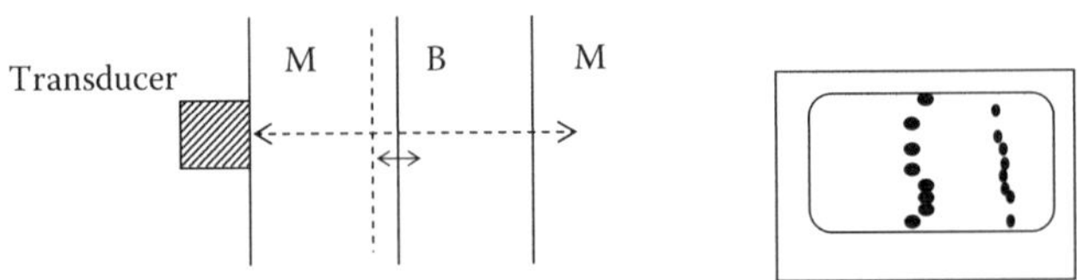

FIGURE 11.31
"TM" or "M" mode of ultrasonic imaging. The beam is held in a fixed position. In this case the line of dots follows the movement of the MB and BM interfaces (in this case due to the movement of the ventricle wall during the cardiac cycle).

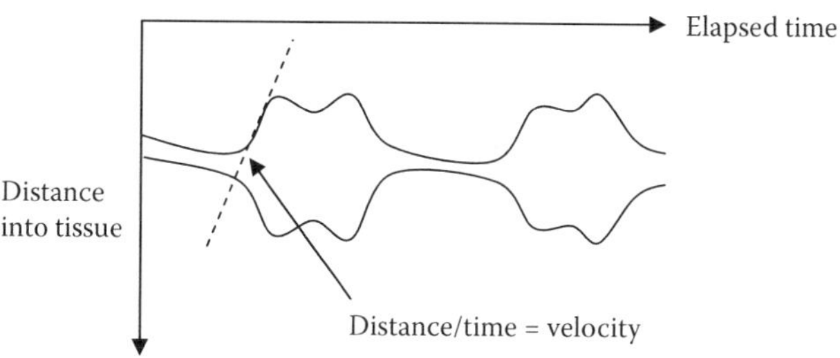

FIGURE 11.32
Representation of mitral valve movement. The two traces represent the anterior and posterior valve leaflets as the open and close. The slope indicated by the dotted line can be used to estimate valve opening or closing velocity.

As the transducer is moved in the direction shown (Figure 11.29), the position of the spots in the y direction on the display can move to correspond. We can also scan over the region in a "fan" of individual beam directions each direction corresponding to a single pulse sent into the tissue. The "spots" on the screen are made to persist and the position on screen follows that of the transducer.

Next, a "fan" of spots is obtained as the beam is progressively scanned over a (2-D) region. The apex of the "fan" represents the entry point into the body and the plane of the scan is approximately at right angles to the surface of the body. Scanning can be done manually, mechanically or (as is more usual in cardiology) electronically. Electronic scanning uses a phased array of tiny transducers to artificially construct a wavefront (Figure 11.30).

By changing the beam direction really quickly, it is possible to do a fan-shaped scan over a particular area in around 50 ms. Images can be updated faster than the "flicker fusion frequency," so they appear as a movie in real time. By controlling the movement into the page, a 3-D image can be obtained from "slices" (slightly delayed).

Doppler information can also be obtained, either "continuous" (giving information about flows at all distances from the transducer) or "pulsed," where the returning echoes can be "gated" to select only those returning within a small time interval (and therefore from a particular depth in the tissue).

11.5.3.3 M (or TM; Time–Motion) Mode

This is specifically for following heart valve motion. For this, the beam direction is fixed, but the display scrolls to give a "history" of where the interface structures are. If, for example, the tissue–blood interface oscillates to and fro (as it would at the surface of the mitral valve), the trace in Figure 11.31 would be obtained.

Displays are usually rotated through 90°, so that downward represents further into the tissue and L–R indicates the passage of time. Valves have a characteristic "M" (or "W") shaped signature, and the maximum slope gives the closing or opening velocity of the valve leaflet (which will be less in disease) (Figure 11.32).

11.6 Analysis of Heart Sounds

The two principal heart sounds, SI and SII, corresponding to the sudden closure and "water hammer effect" of the two atrio-ventricular valves (M and T), and the other two valves (A and P), respectively, have characteristic frequencies centered around 35 Hz and 50 Hz, respectively. In order to record these sound electronically (rather than auscultation using a stethoscope) a microphone with low frequency sensitivity is needed. Many of these couple chest wall movement to a piezoelectric sensor, similar to that shown in Figure 11.22b. The advantage of electronic recording is that the timing and character of the sounds can be analyzed in relation to other signals, such as the ECG and oximeter signals (see below).

11.7 Oximetry: Measurement of Blood Oxygen Saturation

Oxygen saturation percentage (SaO2) can be estimated from measuring the amount of light reflected from the bloodstream (or transmitted through it) at two specified wavelengths. The measurement is based on the Beer–Lambert law (Figure 11.33). This states that the intensity of light is attenuated exponentially in solution due to absorption of light energy by particular molecules in solution.

$$I = I_0 \exp(-A(\lambda)),$$

where I and I_0 are emerging and incident intensity, respectively (in W/m^2), and $A(\lambda)$ is the absorbance of the sample (which varies with wavelength of light λ).

For a sample of blood, path length Δx, in which we assume that the hemoglobin is the main absorber, we can write

$$A(\lambda) = [Hb_{total}]\, \Delta x \,\{a_0(\lambda)C_0 + a_r(\lambda)C_r\},$$

where the first term on the right hand side is the total hemoglobin concentration (145 g/L, or thereabouts), and C_0 and C_r are the fractions of hemoglobin in the oxidized (HbO) and reduced (Hb) forms, respectively. $C_0 + C_r = 1$. $a_0(\lambda)$ and $a_r(\lambda)$ are the absorptivities of HbO and Hb at particular wavelengths (these are intrinsic properties of the hemoglobin molecule). The variation with wavelength is indicated below (Figure 11.34). At 805 nm, $a_0(\lambda') = a_r(\lambda')$; this is called an isobestic point. At this wavelength, we can write

$$A(\lambda') = [Hb_{total}]\, \Delta x \,\{a_0(\lambda')C_0 + a_r(\lambda')C_r\} = [Hb_{total}]\, \Delta x \,\{a'\},$$

since $C_0 + C_r = 1$. If we denote absorbance at 605 nm as $A(\lambda'')$, we can put

$$A(\lambda'')/A(\lambda') = (a_0(\lambda'')C_0 + a_r(\lambda'')C_r)/a', \text{ or more simply, } (a_0C_0 + a_rC_r)/a'.$$

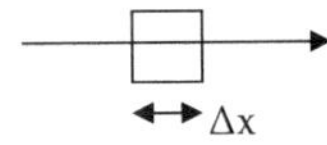

FIGURE 11.33
The Beer–Lambert law. A beam of light traverses the block of material such that the light path within this material is Δx. The ratio of light intensity leaving the block relative to that entering is given by $\exp(-A(\lambda))$ where $A(\lambda)$, the absorbance, depends both on the concentration of material within the block and Δx, the path length. The constant of proportionality, $a(\lambda)$, is known as the absorptivity. If there is more than one substance within the block, then each will contribute to $A(\lambda)$.

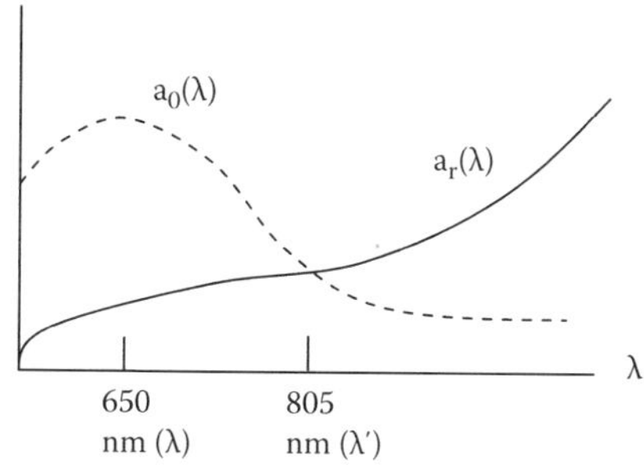

FIGURE 11.34
Absorptivities for oxygenated (dotted) and deoxygenated, or reduced, (full line) forms of hemoglobin. Note they are equal at 805 nm (isosbestic point) and have the maximum difference at 650 nm.

Now since $C_0 \times 100$ is the SaO2(%), and since $C_r = 1 - C_0$ we put this quantity on the left hand side:

$$\begin{aligned} SaO2 = C_0 \times 100 &= \{A(\lambda'')/A(\lambda'))\ (a'/(a_0 - a_r)) - a_r/(a_0 - a_r)\} \times 100 \\ &= \text{slope } A(\lambda'')/A(\lambda') - \text{intercept.} \end{aligned}$$

For reflectance systems, the effects are reciprocal; hence,

$$SaO2 = k_1\ R(\lambda')/R(\lambda'') - k_2,$$

where k_1 and k_2 are the slope and intercept for this particular relationship and the R values are the reflectances at the wavelengths given above.

11.8 Diagnosing Congenital Defects of the Cardiovascular System

In the newborn, congenital defects need prompt corrective surgery and early diagnosis is important. Oxygen saturation readings (via catheters) and characteristic heart sounds can be combined (see Figures 11.35 and 11.36 for some examples).

11.8.1 Ventricular Septal Defect (VSD)

This consists of an opening between ventricles with a L to R shunt of freshly oxygenated blood, making the SaO2 unusually high in the RV. There is a murmur following SI, due to the turbulent flow during systole through the defect. Note that SII splits because of the delayed end of systole in the RV.

11.8.2 Atrial Septal Defect (ASD)

This is an opening between atria which is normal during foetal life, but normally seals up just after birth. If it persists, there is a L to R shunt of oxygenated blood into the RA, which is especially apparent during early systole, where the bulging of the M valve into the LA propels blood through the defect.

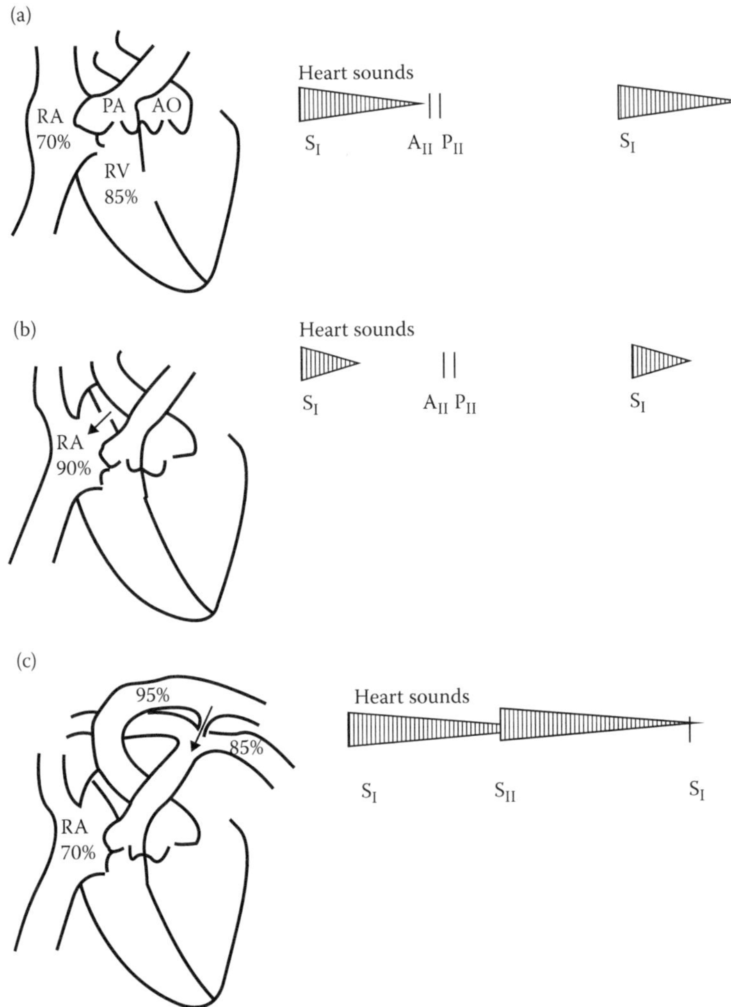

FIGURE 11.35
Congenital defects. Values of %SaO2 and heart sound representations for (a) ventricular septal defect, (b) atrial septal defect, and (c) patent ductus arteriosus.

11.8.3 Patent Ductus Arteriosus

Again this is normal during foetal life and seals just after birth. Where it persists, very high SaO2 readings are obtained in the pumonary artery (PA), where values of 70% might normally be expected. Since the aortic pressure is so much greater than that in the PA, a turbulent flow occurs throughout the cardiac cycle, and can be picked up by the heart sound detector.

11.8.4 Tetralogy of Fallot

As the name implies, there are four components to it: pulmonary stenosis (PA narrowing); VSD; aortic valve overlying RV (giving rise to a R to L shunt); RV enlargement. Here the characteristic pressures, obtained from catheterization, are diagnostic.

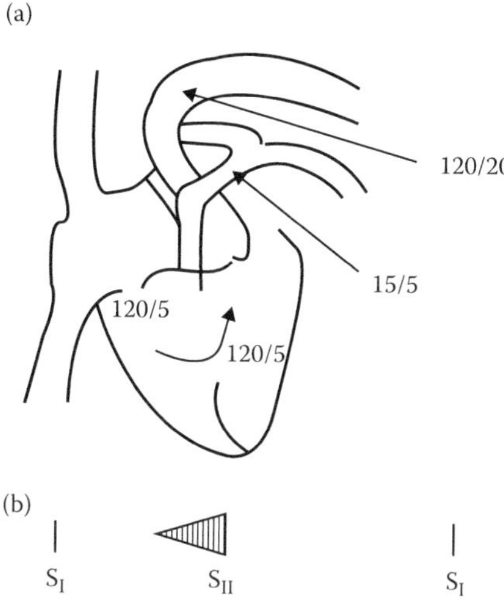

FIGURE 11.36
Congenital defects. Tetralogy of Fallot.

11.8.5 Aortico-Pulmonary Window

This is an opening between AO and PA for large proportion of where they are in contact.

11.8.6 Transposition of Great Vessels

Fortunately, it is rare, but here, the aorta and PA are interchanged, so RV leads to AO and LV to PA, and the two circuits do not communicate.

11.8.7 Coarctation of Aorta

This is a narrowing of the aorta in the thoracic region, giving rise to increased flow to the head and arms and reduced flow to the legs. The raised pressure leads to an incompetent aortic valve (i.e., aortic regurgitation), which has a characteristic sound.

Tutorial Questions

1. Indocyanine green (3 mg) is injected quickly into a large vessel, and samples are taken at 1-s intervals from an artery. Dye concentrations in these whole blood samples are as follows (in mg L^{-1}):

   0, 0, 0, 0.2, 1.0, 12.1, 19.5, 19.8, 12.4, 7.6, 4.7, 3.0, 4.6, 5.1, 5.2, 7.5.

   (a) Use a quick method to estimate cardiac output in L min^{-1}. Show working.

   (NB, for a falling exponential $\frac{c_n}{c_{n+1}}$ is constant.)

   (b) If, after 10 min the blood concentration has fallen to 1.2 mg L^{-1}, estimate total blood volume, stating assumptions made.

2. 8 mg of Evans' blue is injected quickly into a vein, and blood samples were taken at 1-s intervals from an artery. Dye concentration in plasma of these samples is as follows (in mg L^{-1}):

   0, 0, 0, 0.3, 1.1, 12.0, 19.4, 19.5, 12.2, 7.6, 4.7, 3.0, 4.6, 5.1, 5.2, 7.5. (Ht = 45%).

   (a) Make an estimate of the cardiac output. (Hint: for a falling exponential, $\frac{c_n}{c_{n+1}}$ remains constant.)

   (b) Estimate the volume of blood between the sites of injection and sampling.

3. 10 mg of indocyanine green is injected as a 1 mL bolus into the inferior vena cava via a venous catheter. Samples are withdrawn via an arterial catheter in the brachial artery, the catheter leading to an optical absorption flow through cell and then to a motor-driven syringe. The photo detector output together with the output when standard solutions are drawn through the cell are given in Figure 11.37.

   Estimate cardiac output from these data, showing all working.

4. The curve shown was obtained from thermal dilution by injecting 10.0 mL of dextrose at 20.2°C. Note the calibration pulse: assume linearity (Figure 11.38).

   (a) Briefly describe how the equation below is derived;

   (b) Estimate cardiac output.

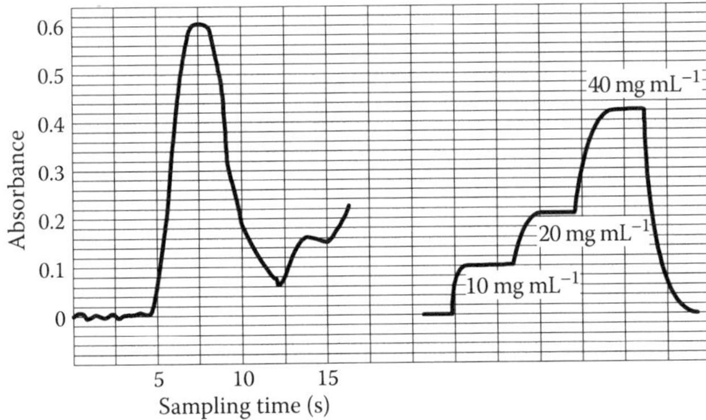

FIGURE 11.37
See Tutorial Question 3.

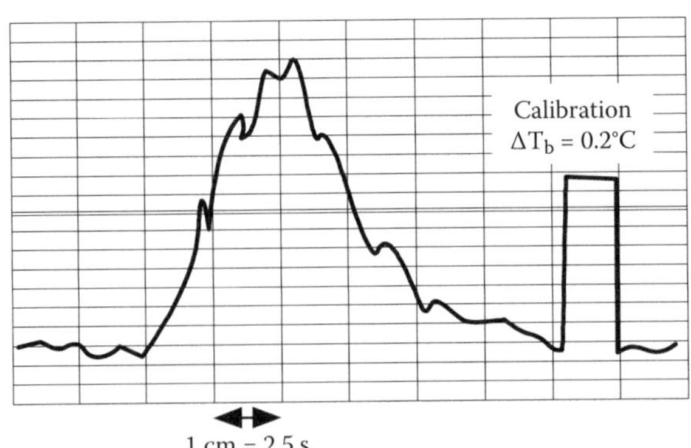

FIGURE 11.38
See Tutorial Question 4.

$$(\text{CO}) \times \rho_b s_b \int \Delta\, T_b\, dt = V\rho_i s_i\, (T_b - T_i)$$

Densities (kg/m^3): $\rho_i = 1018$; $\rho_c = 1057$. Specific heats: $s_i = 0.964$; $s_b = 0.87$; $T_b = 36.9°C$.

5. Describe the method of positioning, and the design of, a catheter for the measurement of CO using a thermal dilution technique. Figure 11.39 represents the temperature changes in the right heart in response to the injection of 10 mL of saline at 20°C.

   Calculate CO, given that

$$Q \times s_b\rho_b \int \Delta T\, dt = s_i\rho_i V(T_b - T_i)$$

   and

$$\frac{s_i\rho_i}{s_b\rho_b} = 1.06.$$

6. Explain how total blood flow to the leg can be estimated using electrical impedance measurements (Figure 11.40). Hence, estimate limb blood flow from the diagram given. The voltage is measured between electrodes 10 cm apart, and the

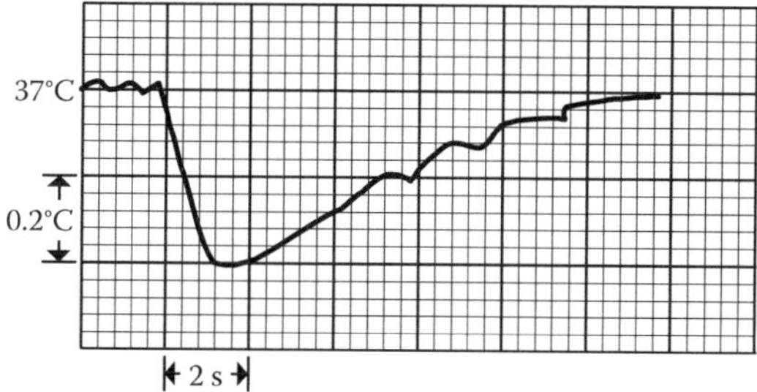

FIGURE 11.39
See Tutorial Question 5.

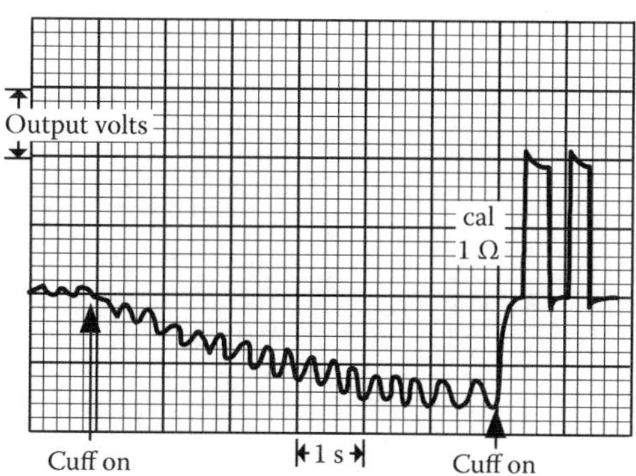

FIGURE 11.40
See Tutorial Question 6.

resting impedance between these is 20 Ω. Take blood resistivity to be 150 Ω cm. (NB, $\Delta V = -\rho\left(\frac{L}{Z_o}\right)^2 \Delta Z$.)

7. Venous outflow from the leg is occluded intermittently and the rectified AC voltage between two foil band electrodes 15 cm apart is shown in Figure 11.41, together with calibration pulses. The DC resistance between these electrodes is 11.7 Ω, and the resistivity of blood is 150 Ω cm.

   Estimate limb blood flow, given that the ratio of volume to impedance change is given by the factor $-\frac{\rho L^2}{Z_o^2}$.

8. From the data given below (see also Figure 11.42), calculate the following:
   (a) Resting cardiac output;
   (b) Cardiac output in exercise;
   (c) The % change in total peripheral resistance due to exercise.

$$\Delta V = \rho\left(\frac{L}{Z_o}\right)^2 \left(\frac{dZ}{dT}\right)_{max} \times T;\ \rho = 150\ \Omega\ \text{cm};\ L = 23\ \text{cm})$$

Resting: $Z_o = 25\ \Omega$, BP = 119/78

Exercise: $Z_o = 22\ \Omega$, BP = 135/89

9. Figure 11.43 represents the movement of the anterior mitral leaflet. What is its maximum speed of motion given that the velocity of ultrasound in tissue is 1.25 m ms^{-1}.

   Explain your method.

10. On catheter "pull-back" across a particular mitral valve, a pressure gradient of 2 kPa (15 mmHg) was observed in diastole. Calculate the area of the orifice using the Gorlin formula $A = \frac{Q}{0.6}\sqrt{\frac{\rho}{2\Delta p}}$ assuming a cardiac output of 6 L min^{-1} (100 mL s^{-1}) and a blood density of 1030 kg m^{-3}.

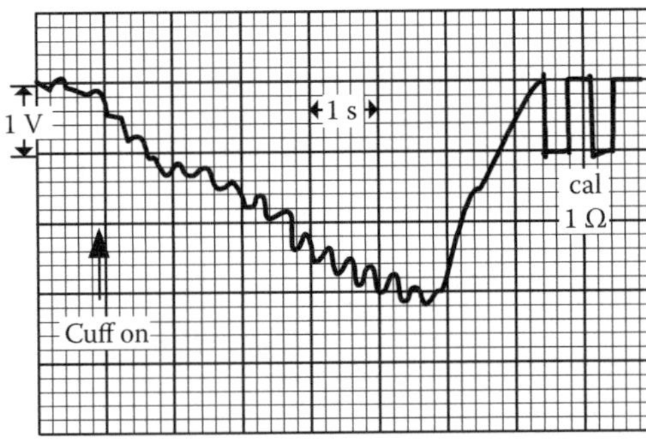

FIGURE 11.41
See Tutorial Question 7.

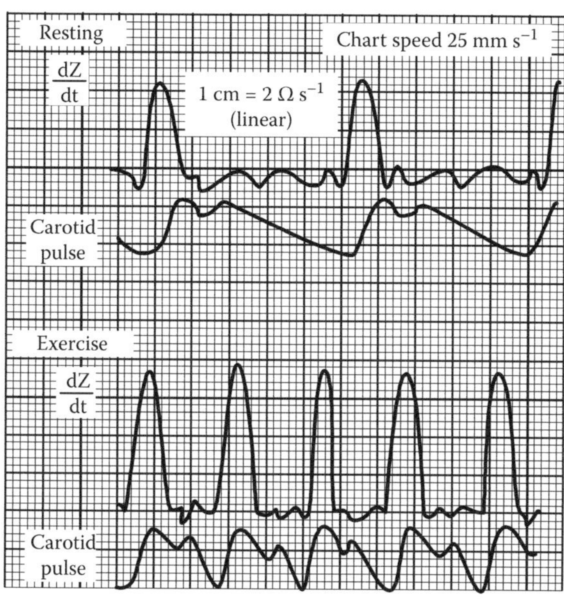

FIGURE 11.42
See Tutorial Question 8.

11. (a) The compliance of a catheter–transducer system is estimated by injecting 10 μL (= 10^{-8} m^3) fluid into the open end of the catheter. The corresponding rise in pressure recorded is 150 mmHg (= 20 kPa). Given that the catheter is 1 m long and has an internal diameter of 1.1 mm (radius 0.55 mm), determine (i) the natural frequency (in Hz) and (ii) the relative damping factor for this measurement system. Assume density of fluid to be 1000 kg/m^3; viscosity, 0.001 Pa s.

    (b) If it were possible to alter the diameter, but not the length or the compliance, determine whether it is possible, by selecting a suitable diameter, to record faithfully up to the 20th harmonic in the arterial pressure waveform (assume an HR of 1 beat/s).

    (c) Explain how a typical pressure transducer, for use in a cardiology department, works.

12. (a) An arterial catheter is 1 m in length, has an internal radius of 0.3 mm and a compliance of 10^{-14} m^3/Pa. Determine its natural frequency when filled with

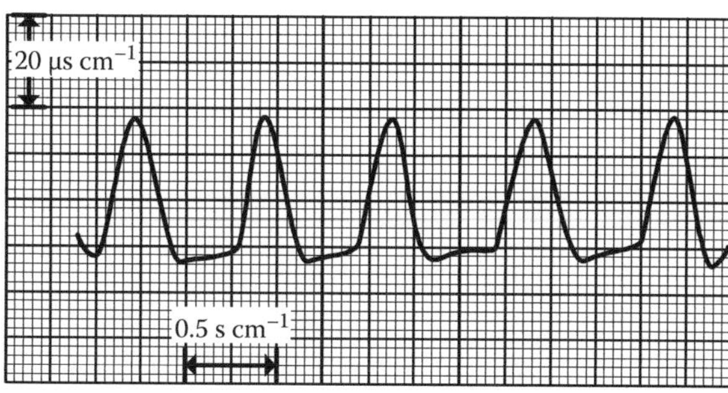

FIGURE 11.43
See Tutorial Question 9.

saline (density 10^3 kg/m^3; viscosity 10^{-3} Pa s), assuming that the transducer compliance and the saline compressibility can be neglected.

(b) Would this catheter be suitable for arterial blood pressure monitoring? How might the recording be improved if it were suitable?

(c) Could the same information be gained noninvasively? Explain your answer.

13. (a) An arterial catheter is 1 m in length, has an internal radius of 0.4 mm and a compliance of 2×10^{-14} m^3/Pa. Determine its natural frequency (in Hz) when filled with saline (density 10^3 kg/m^3; viscosity 10^{-3} Pa s), assuming that the transducer compliance and the saline compressibility can be neglected.

(b) What relative damping factor would be expected from this system?

(c) How would the step response look (approximately)?

(d) Would this catheter be suitable for arterial blood pressure monitoring? What factors would adversely affect the ability of this system to accurately record arterial blood pressure?

(e) Describe the limitations for noninvasive continuous recording of arterial blood pressure and attempts to overcome these.

Part IV

Lungs, Kidneys, and Special Monitoring

12

Respiratory Biophysics

Bruce R. Thompson and Joseph Ciorciari

CONTENTS

12.1 The Respiratory System

The primary function of the respiratory system is to transfer oxygen into the blood stream and carbon dioxide out. Oxygen is needed for cellular aerobic metabolism, and the removal of carbon dioxide is required to maintain acid–base balance. This is coordinated to maintain electrolyte and water balance in the face of changing metabolic requirements—sleep, activity, exercise, and food consumption. The respiratory System behaves like an *ideal gas exchanger*. It does this by providing the following functions:

- Provide a large contact area between air and blood (membrane allows diffusion and minimal resistance to gas flow).
- The inspired air must be saturated with water vapor and heated to tissue temperature to protect delicate tissues from damage.
- The distribution of gas and blood in the many exchange units should be closely matched.
- The gas exchange must be proportional to the uptake of oxygen and production of carbon dioxide by the cells.

12.2 Structure and Function

12.2.1 The Airway Tree

12.2.1.1 Upper Airway

The anatomical structure of the respiratory systems starts with the upper airway, which includes the nose, mouth, pharynx, and the larynx. The primary function of the upper airway is to assist in ventilation, humidification, and protection of the airways by allowing coughing, swallowing, and finally speech. There are many sets of muscles that are involved to perform these functions, which are controlled and coordinated by complex neuromuscular interactions.

During inspiration, the phrenic nerve discharges causing the diaphragm to contract and lower. Muscles in the upper airway and also the tongue contract, which abduct the vocal cords and pushes the tongue forward, which decreases the resistance of the upper airway.

12.2.1.2 Respiratory Airways

The airways within the lung are a series of branching tubes, which become shorter and narrower and, by the nature of continuously branching structure, become more frequent as they penetrate further to the periphery of the lung (Figure 12.1). Immediately below the upper airway is the trachea, which branches into the left and right main bronchi, which in turn divide into smaller bronchi. This branching continues until approximately 16 generations of bifurcations normally defined as the respiratory bronchioles. These are the smallest airways that do not accompany the alveoli and normally approximately 1 mm–2 mm in

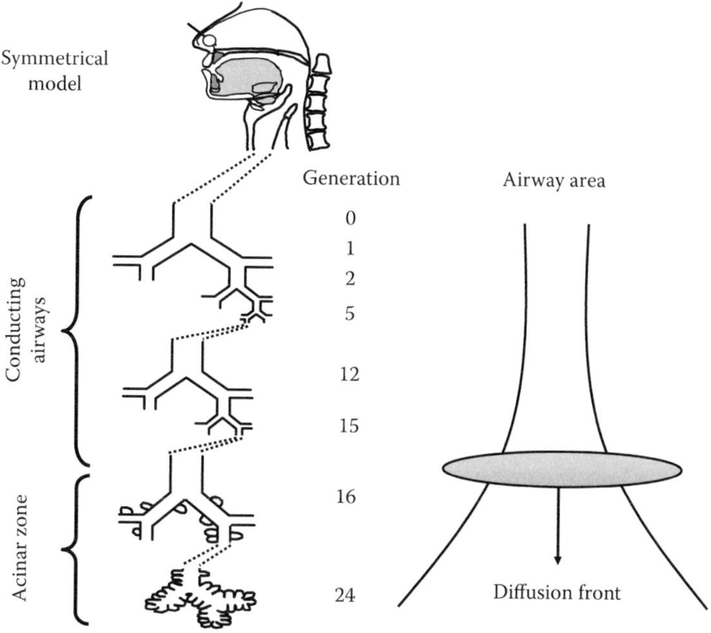

FIGURE 12.1
The branching or generations of the airways. Gas exchange occurs in the acinar zone.

size. These generations of airway comprise the conducting airways whose primary function is the bulk transport of gas. Gas flow down these airways is by convection. Another important anatomical feature of the conducting airways is the presence of cartilage rings that support these airways. Cartilage is present in airways down to the bronchioles (generation 4–5) and plays a significant part during expiration in terms for maintaining airway patency.

Another important feature of the conducting airways is that no gas exchange occurs, and as such, these airways are what comprise the anatomical dead space, that is, airways that are involved in ventilation, but no gas exchange occurs. The total volume of these airways is approximately 150 mL and can be measured as described later in the chapter.

From the terminal bronchioles, the airways further divide and start forming the respiratory bronchioles and alveolar ducts, which are completely lined with alveoli. These airways are what make up the acinar zone of the lung and whose primary function is gas exchange. The divisions of this zone are only a few millimetres, however the majority of the lung is made up of this gas exchange region. Depending on gender, height, and age, to name a few, the total volume of this zone is around 3–5 L.

12.2.1.3 Ventilation during Inspiration

During inspiration, the diaphragm drops, creating a negative pressure inside the pleural cavity. Gas will enter the lungs by convective flow and travel down as far as the terminal bronchioles (generation 16). From this point, as the total cross-sectional area of all airways combined is so large; bulk gas flow becomes zero and gas transport is by a process of diffusion to the alveoli. This process is very rapid as the distance traveled is very small (<2 mm), and the diffusion front is very large due to the large total area of the alveolar ducts.

12.2.1.4 Lung Volumes and Dead Space

Static lung volumes are a number of volumes and capacities, which describe the makeup of the lung and are usually measured with a spirometer and devices such as a body plethysmograph. Volumes and capacities can be seen on Figure 12.2.

When a subject is breathing normally, the volume of air that they breathe in or out is called the tidal volume (VT). When the subject breathes in maximally, the volume breathed in is called the inspiratory capacity (IC). At this point, the subject is at their total lung capacity (TLC). The subject then breathes out as far as they can voluntarily; the volume

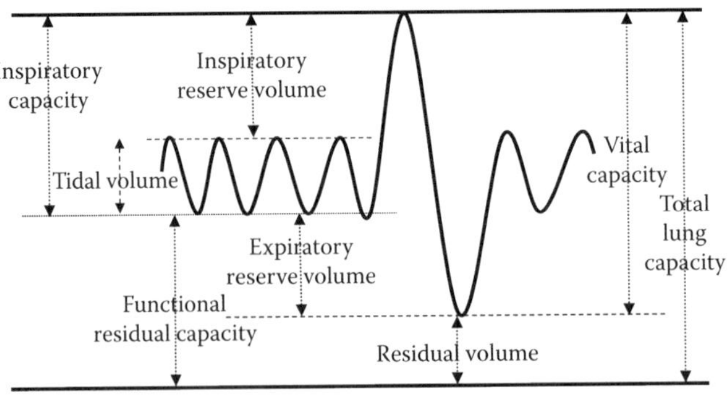

FIGURE 12.2
Lung volumes and capacities.

exhaled is the vital capacity (VC). At this point, gas still remains in the lung, and this volume is called the residual volume (RV). Understandably, the RV, functional residual capacity (FRC), and TLC cannot be measured by a spirometer as the subject would need to blow out the complete contents of the lung. To measure such volumes, devices, such as the body plethysmograph or a gas dilution technique, need to be employed.

Dead space, as described earlier, is the volume of the lung that does not contribute to the gas exchange process. There are two types of dead space: anatomical and physiological. Anatomical dead space is the volume of the conducting airways typically, 2 mL/kg body weight or ~150 mL. Physiological includes the anatomical dead space but also includes lung units with increased inequality between ventilation and blood flow ($\dot{V}/\dot{Q}$) within the lung.

To measure physiological dead space, the Bohr method is used. The Bohr method relies on the assumption that all CO_2 in the expired air comes for the gas exchange region of the lung. Therefore, the partial pressure of exhaled CO_2 ($PECO_2$) multiplied by the tidal volume (VT) will equal the alveolar P CO_2 ($PACO_2$) multiplied by the difference between the VT and dead space (VD).

If

$$PECO_2 \text{ X } VT = PACO_2\,(VT - VD),$$

therefore,

$$VD/VT = (PACO_2 - PE\,CO_2)/PACO_2.$$

Because the alveolar ($PACO_2$) and arterial ($PaCO_2$) partial pressure for CO_2 are virtually identical in normal subjects, the equation can be rewritten as follows:

$$VD/VT = (PaCO_2 - PE\,CO_2)/PaCO_2,$$

where $PaCO_2$ = arterial partial pressure of CO_2.

In normal subjects, VD/VT is <0.3. However, in patients with lung disease leading to increased heterogeneity between ventilation and blood flow, VD/VT can increase.

12.2.1.5 Various Volumes and Capacities of the Lung

To understand various aspects of pulmonary physiology, a number of volumes and capacities need to be defined (with reference to Figure 12.2). These can be measured during passive and forced exhalation spirometry procedures (see Section 12.10 in this chapter).

- *Tidal volumes (VT):* With normal breathing, the volume of air that they breathe in or out is called the tidal volume.
- *Inspiratory capacity (IC):* When the subject breathes in maximally, the volume breathed in is called the inspiratory capacity
- *Total lung capacity (TLC):* While fully inspired, the subject is at their total lung capacity.
- *Vital capacity (VC):* The volume of air that a subject voluntarily exhales after maximum inhalation.
- *Forced vital capacity (FVC):* The volume of air that a subject forcibly exhales after maximum inhalation.

- *Residual volume (RV):* The remaining volume of air after a maximum exhalation.
- *Expiratory reserve volume (ERV):* The reserve volume remaining after a VT exhalation and not including the RV.
- *Inspiratory reserve volume (IRV):* The reserve volume available after a VT inhalation and not including the RV.
- *Functional residual capacity (FRC):* The volume of air remaining in the lung after VT exhalation.
- *Forced expiratory volume ($FEV_{1.0}$):* The volume of air that a subject forcibly exhales after maximum inhalation after 1 s.
- *Forced expiratory ratio (FER):* The solution of $FEV_{1.0}/FVC$ expressed as a percentage.

Their relationships can be summarized as

$$TLC = VC + RV$$

$$TLC = IC + FRC$$

$$TLC = RV + ERV + VT + IRV$$

$$VC = IRV + VT + ERV$$

$$VC = IC + ERV$$

$$FRC = ERV + RV.$$

These volumes and capacities are affected by age, height, and gender. Figure 12.3 illustrates the relationship of VC and RV with respect to age. Other respiratory function tests, such as diffusion capacity, are also affected by age, but these will be discussed later in the chapter.

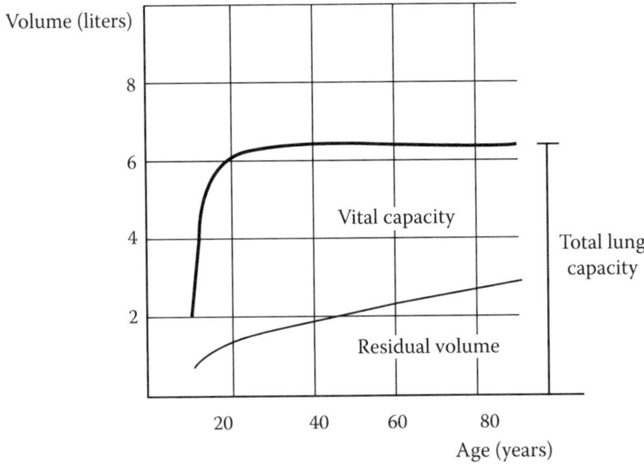

FIGURE 12.3
A graphical illustration of how lung volumes change with age. Note that average height is used. Note the rise in RV suggestive of loss of elasticity, compliance, and capacity loss.

WORKED EXAMPLE

A young male is referred to a lung function unit for respiratory assessment.

1. What patient history and physical characteristics would be recorded?
2. Briefly describe the test to determine the patient's pulmonary VC.
3. The results of such a test were FVC = 2.6 L, FEV = 1.5 L. Calculate his FER.
4. In which category of disease could this patient be included at this stage? What disease is the patient probably suffering from? Explain how you arrived at this conclusion. What further clinical procedures and or lung function tests could be carried out to confirm the diagnosis? Describe these tests and the probable results, assuming your diagnosis is correct.

Solution

1. Age, height, weight, gender, history of illness, smoker?
2. Standard lung function testing to measure lung volumes with spirometry techniques (bellows, wet and dry spirometry, transducer [pneumatograph]). Subject with mouthpiece and connected to equipment does quiet breathing (VT) and then is asked to take a slow deep breath and then slowly exhale as far as they can go, with the technician coaching them on. Normal TV breathing resumes. Repeat test three times.
3. FER = $FEV_{1.0}/FVC = 1.5/2.6 = 58\%$ severe obstructive.
4. Chronic bronchitis and chronic obstructive lung disease (COLD). Flow volume tests and typical obstructive curve.

12.3 Neurological Control of Respiration

There are a number of mechanisms, which control ventilation. These include neural, chemical, and other reflex mechanisms.

12.3.1 Neural Mechanisms

Inspiration and expiration cells are found in the brain stem, which help regulate the frequency of breathing as well as the VT. Specifically, these are found in the medulla and pons.

In the medulla, there are cells associated with inspiration and expiration. These cells inhibit each other when active. Within the medulla, there are two groups of cells: *the dorsal respiratory group and the ventral respiratory group.* The *dorsal respiratory group* is primarily composed of inspiration cells with some expiratory cells. This group is also the initial relay and integration site for carotid bodies, aortic bodies, and vagal pulmonary mechanoreceptors. Inspiratory neurons project to the inspiratory spinal motoneurons.

The *ventral respiratory group* is composed of both inspiration and expiration cells. These project to the phrenic, intercostals, and abdominal motor neurons. This group is regarded as the final integrating site of the brain stem respiratory complex. Within the pons are the *apneustic and pneumotaxic centers.* The *apneustic center* (lower) affects inspiratory neurons. Any damage here causes large inspratory grasps and transient expiratory efforts. It also affects inspiratory neurons of the medulla. The *pneumotaxic center* (upper) accelerates breathing and inhibits inspiration; it inhibits inspiratory medullar centers both directly and via apneustic centers. The cerebral cortex also is responsible for affecting inspiratory and expiratory medulla. This is why we can voluntarily hyperventilate and hypoventilate.

12.3.2 Chemical Mechanisms

The regulation of $PaCO_2$ is of prime importance. There are chemoreceptors for both oxygen and carbon dioxide. The CO_2 chemoreceptors are located centrally and peripherally. They are located on the ventral side of the medulla. This region reacts in seconds to local application of H^+ or dissolved CO_2. In the periphery, chemoreceptors are found in the carotid and aortic bodies. These respond to an increase in PCO_2 via reflex mechanisms, which stimulates ventilation (glosso-vagus). This system is more sensitive to sudden changes in CO_2.

12.3.3 Other Reflex Mechanisms

Other mechanisms also influence ventilation. For example, the *inflation reflex* promotes expiration. At maximum inflation, the inspiratory center is inhibited. The *deflation reflex* promotes inspiration; during deflation, the expiratory center is inhibited. *Local muscle stretch receptors* also are involved; for example, the muscle spindle stretch receptors (in diaphragm and intercostals muscles of rib cage) monitor the length of muscle activated

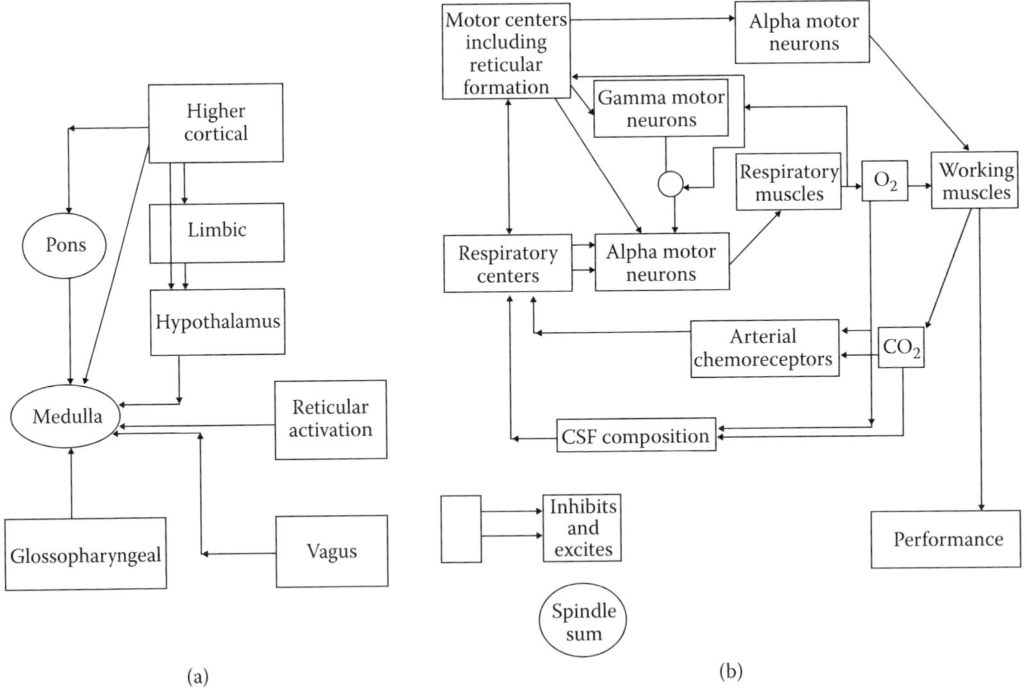

FIGURE 12.4
(a) Control of ventilation: control and feedback systems associated with the unconscious and conscious control of breathing. (b) Detailed control system of respiration (CSF).

by the gamma motor neuron. Complex interaction between muscle receptors occurs for the control of ventilation (see Figure 12.4a and b). Finally, *irritant receptors* detect noxious gases/dust.

12.4 Muscles Associated with Breathing

Many muscles and muscle systems work together to assist with the process of respiration. During inspiration, the *diaphragm* muscle contracts. This contraction forces the abdominal contents down and forward. This leads to an increase in vertical dimension of the chest cavity and causes the rib cage to lift and move forward. During normal tidal breathing, the diaphragm moves about 1 cm vertically. However, with forced inspiration/expiration, it may move by 10 cm. The diaphragm is innervated by phrenic nerve via C3, C4, and C5.

The *external intercostal muscles* (which are innervated by the spinal cord at the same level) connect the adjacent ribs, sloping downward and forward. During contraction, these muscles pull the ribs up and out. The accessory muscles also assist in inspiration. The scalene group of muscles elevates ribs 1 and 2, the sternomastoids raise the sternum, and the alae nasi cause the flaring of the nostrils.

During the process of expiration, most muscles return to equilibrium (relax), and the lung and chest wall return to a preinspired state by elastic forces. This process is referred to as a *passive* process. However, during the *active* expiration process, air can be forced out during exercise. The muscles involved in this active process are the abdominal wall muscles: rectus abdominus, internal and external oblique, and transverse abdominus. These muscles force the diaphragm upward. The internal intercostals also pull the ribs down and inward.

12.5 Pulmonary Circulation and Blood Flow

The main pulmonary artery receives mixed venous blood that is pumped by the right ventricle. The artery is similar to the airway branches as far as the second-degree bronchioles. Moving from the top to the periphery of the lung, the arteries, veins, and bronchi run in proximity to one another. However, in the periphery of the lung, the veins separate to run between the lung lobules, whereas the arteries and bronchi run down to the center of the lobules. The arteries separate into capillaries and form a dense network or mesh, which forms the walls of the alveoli.

The capillaries have a diameter of 10 μm, just large enough for a red blood cell, and the capillary segments are very short, forming an almost continuous sheet of blood around the alveoli wall. Oxygenated blood returns via the pulmonary veins and eventually unite to form large veins, which drain into the left atrium.

12.5.1 Pressure in the Pulmonary System

Overall, the pressures in the luminary circulation are significantly lower than that of the systemic circulation (see Table 12.1).

TABLE 12.1

Comparison of Blood Pressures in the Pulmonary and Systemic Circulatory System

Pressure (mmHg)	Pulmonary Circulation	Systemic Circulation
Systolic	25	120
Diastolic	8	80
Mean	15	100

As the pressures in the pulmonary circulation are significantly less than that in the systemic circulation, the pulmonary artery wall is significantly thinner than that seen in the systemic arteries and has less smooth muscle.

12.6 Respiratory Mechanics

The lungs are a very dynamic organ. As previously discussed, there are many muscles involved in the breathing process. In addition, there are many muscles within the airways themselves that are very dynamic during respiration. Moreover, the lungs are also very compliant and distensible (elastic). These elastic properties can critically affect the efficiency of gas exchange. Therefore, there are many opposing forces on the lung and chest wall during the process of respiration to overcome various resistances.

12.6.1 Pressure Relationships in the Thoracic Cavity and the Lung

The flow of gases in and out of the lung is, in part, a result of negative or positive pressures within the lung that are generated by the movement of the diaphragm and intercostal muscles of the chest wall. All pressures that are described are all in relation to atmospheric pressure. There are a number of pressures in the lung and chest wall that need to be understood (Figure 12.5).

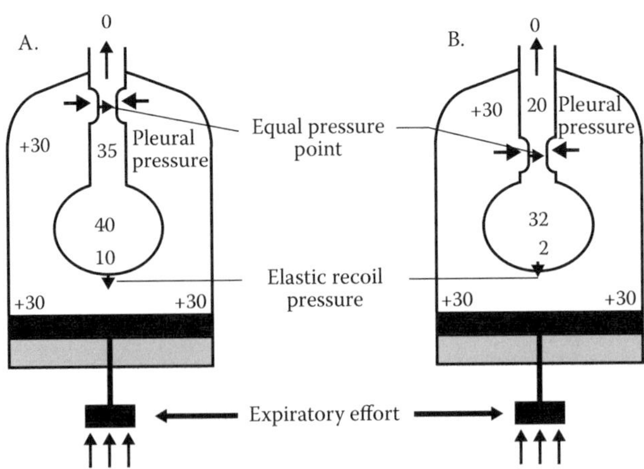

FIGURE 12.5
Pressures in the lung and chest wall during exhalation in the normal and diseased lung.

12.6.1.1 Intrapulmonary Pressure

The intrapulmonary or intraalveolar pressure, Palv, is the pressure within the alveolus. This pressure does fluctuate with respiration; however, at either end-expiration or end-inspiration, this pressure potentially equalizes with atmospheric pressure in people with no lung disease because ultimately the alveoli are directly connected to the outside environment via the airways.

12.6.1.2 Intrapleural Pressure

The intrapleural pressure is the pressure within the pleural cavity, that is, the space between the outside of the lung and the inside of the chest wall. This fluid space is only a couple of millimeters in size. Importantly, the intrapleural pressure is always negative relative to the alveolar pressure and also atmospheric pressure. If the intrapleural pressure was not negative compared with the alveolar pressure or atmosphere, then the lung would collapse.

The negative intrapleural pressure is generated by a couple of opposing forces. Because of the elastic nature of the lungs, the lungs' natural tendency is to recoil. In addition, the surface tension of the alveoli and the fluid surrounding the alveoli (surfactant—discussed later) also leads to contraction. However, both these forces have opposing forces from the chest wall. The natural tendency of the chest wall is for expansion and, therefore, pulls the thorax outward. As there is a thin fluid layer between the lung and chest wall, this virtually adheres the lung to the chest wall. This is analogous to microscope slides held together with a thin film of fluid. The two slides can slide against each other easily; however, significant force needs to be applied to pull them apart. Therefore, with the lungs recoiling inward against an outward recoil of the thorax, an equilibrium is achieved between these two opposing forces, leaving a slight negative pressure of the intraalveolar pressure relative to the intrapleural pressure. The difference between these two pressures is termed the transpulmonary pressure.

The negative transpulmonary pressure and fluid coupling of the lung to the chest wall is fundamentally important, as homeostasis of these in part allow air to move in and out of the lungs. Alterations in the elastic recoil of the lung or disruption of the fluid layer of the lung can lead to the collapse of the lung.

12.6.2 Dynamic Collapse of Airways

Imagine a subject taking the biggest breath that they can up to TLC, at which point they maximally exhale to RV and at which point they inhale again up to TLC maximally. If we plotted flow against volume, a characteristic shape of this curve is achieved and is termed the flow volume curve (see Figure 12.6). There are a number of aspects of this curve that are described later in the section on pulmonary function testing. However, one aspect of the curve that is worth discussing at this point is that this curve is the envelope of what can be achieved for an individual. That is, it is virtually impossible to penetrate the maximum flow volume loop that each individual can achieve. Therefore, no matter at what lung volume or what degree effort is exerted, it is not possible to breach the outer boundaries of the flow volume loop.

The reason for this phenomenon is the dynamic collapse of airways during exhalation. If you refer to Figure 12.5, during the exhalation, 30 cm H_2O in this example is exerted by the diaphragm and chest wall onto the lung. In addition, the elastic recoil of the lung is 10 cm H_2O, leading to 40 cm H_2O at the alveolar level. As this is happening during

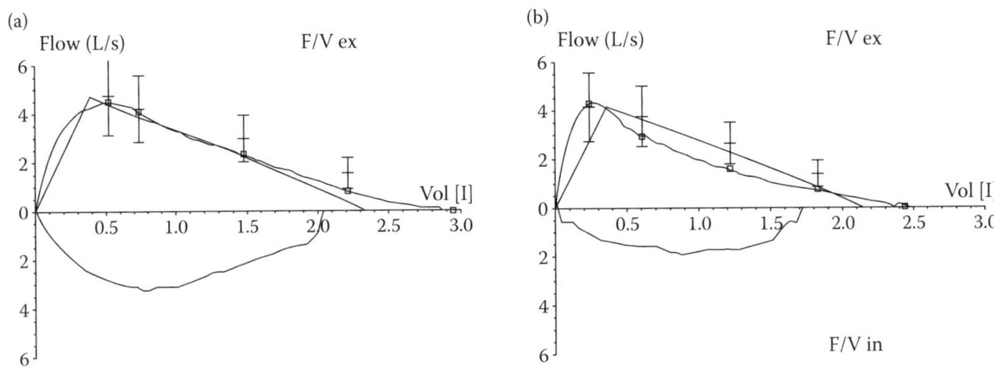

FIGURE 12.6
Flow volume curves normal (a) and mild obstruction (b) versus predicted flow volume.

exhalation with an open glottis, there is a direct connection between the alveolus and the air. Therefore, there is a progressive decline in pressure from the alveolus, which is at 40 cm H_2O, and the mouth is open to atmosphere. The point where the pressure inside the airway is equal to the intrapleural pressure is called the equal pressure point. Moving proximally (i.e., toward the mouth), the pressure inside the airway will be less than the intrapleural pressure leading to the collapse of the airway. Under normal conditions with the subject breathing tidally, airway collapse does not occur because the equal pressure point occurs in airways, which have cartilage to support them from collapsing. However, the equal pressure point is moveable with lung volume. As lung volume decreases, the equal pressure point moves more distally mainly because the elastic recoil of the lung decreases, hence a lower pressure differential between the distal and proximal airways and, finally, the mouth. As the equal pressure point moves more distally, it moves to airway generations, which are not surrounded and supported by cartilage. These airways are significantly more compliant and, therefore, prone to collapse under these conditions.

12.6.3 Pressure Volume Curve

The lung in the chest cavity is similar to a balloon in a glass jar. If we could measure the pressure between either inside the balloon relative to the atmosphere or the pressure between the balloon and the glass jar and plot this against lung volume, we are to generate what is known a pressure volume curve (Figure 12.7). Applied to the human, it is possible to measure the intrapleural pressure via an esophageal balloon placed above the diaphragm. Also, it is possible to obtain the intrapleural pressure by having the subject breathe on a mouth piece in a body plethysmograph and closing the shutter on the mouth piece. Measuring the pressure at the mouth when there is no flow will give the pressure inside the lungs. If either of these pressure are plotted over a range of lung volumes, then the pressure volume curves generated for either of these technique will an indication of the compliance of the lung or the compliance of the lung and chest wall combined.

From Figure 12.7, the lung/chest wall pressure volume curve is nonlinear and plateaus out at higher lung volumes. That is, it requires greater pressure to inflate the lung, and

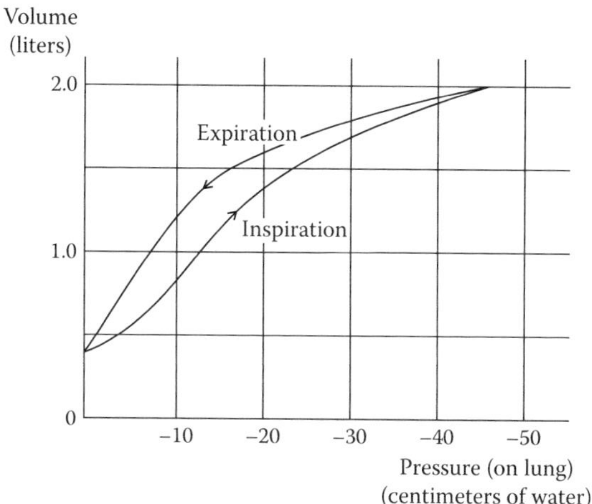

FIGURE 12.7
Pressure volume curves for expiration and inspiration.

therefore, the lung is less compliant at higher lung volumes. This concept is important especially in the context of understanding the physiological consequences of various lung diseases. Some lung diseases, such as pulmonary fibrosis, decrease lung compliance, whereas others, such as emphysema, decrease lung compliance.

12.6.4 Compliance of Airways

Compliance is a measure of "stiffness" of the lung and is quantified by the gradient of the slope of the pressure/volume curve. For an adult male with normal lung capacity, the compliance is measured as 200 mL/cm H_2O. With high transpulmonary pressure, compliance decreases (P/V curve is flatter). A decrease in compliance can occur due to a number of factors; pulmonary venous pressure may increase, causing the lung to engorge with blood, and with alveolar edema, the alveoli cannot expand. There may also be a decrease in compliance with fibrosis of the lung or if the lung has been unventilated for a long period. Age, emphysema, and alterations in the elasticity of the pleura of the lung can increase compliance.

It is not only possible to measure compliance of the whole lung and chest wall as one unit. It is also important to acknowledge that all airways have mechanical properties to stretch and behave differently at different lung volumes. Importantly, airway diameter also is important because, in general, smaller airways are more distensible than larger airways. Finally, factors, such as the tone of the smooth muscle that surrounds the airways, have an impact on the overall stiffness of the airways. The measurement of individual airway compliance has largely been confined to animal studies and ex vivo studies. However, recently, it has been possible to measure airway compliance noninvasively.

High-resolution computed tomographic scans are taken at three different lung volumes, TLC, FRC, and a volume midway between TLC and FRC. From the high-resolution computed tomographic scans, individual lumen diameter is able to be measured at each lung

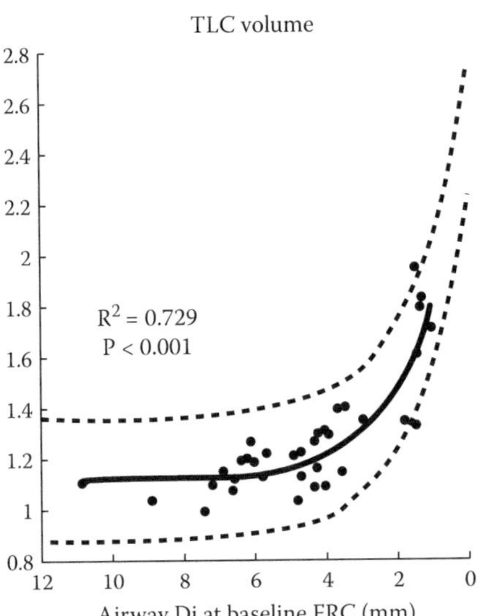

FIGURE 12.8
Change in internal diameter of airway from FRC to TLC against airway diameter at FRC. (Kelly VJ, Brown NJ, King GG, Thompson BR (2010). *Med and Biol Eng and Computing*, 48, 489–496.)

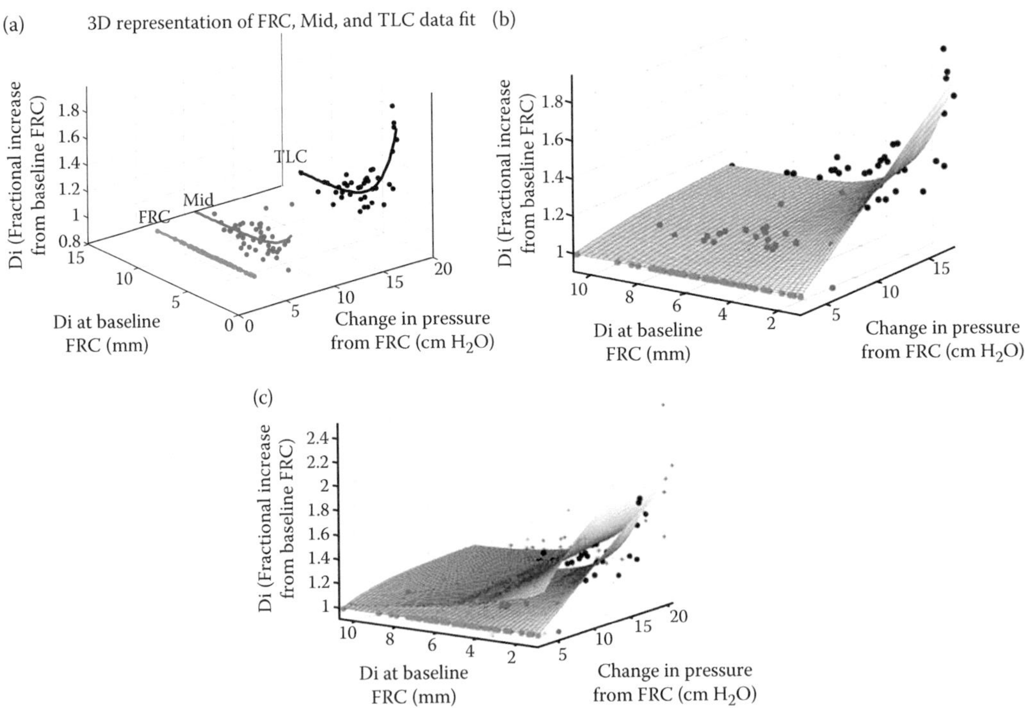

FIGURE 12.9
Three-dimensional plots of change in internal airway from FRC, against internal airway diameter at FRC and change in airway pressure: (a) individual data, (b) interpolated surface, (c) two surfaces, with the bottom surface prebronchodilator and the top surface postbronchodilator.

volume. Figure 12.8 is a plot of fractional change in internal airway diameter from FRC to TLC compared with airway diameter at FRC. From this figure, the airways that changed the most, that is, increased, and were therefore more distensible are the small airways. If we also measure pressure volume curves, we also are able to measure the change in esophageal pressure from FRC to the midvolume and also to TLC. This gives us the ability to display the data in three dimensions, that is, internal diameter of individual airways at FRC against fraction change in internal diameter from FRC and change in pressure from FRC (Figure 12.9a). From this analysis, we can create a three-dimensional surface (Figure 12.9b). From this analysis, it can be seen that it is the small airways that open the most and are therefore more compliant. Importantly, the greatest change in airway diameter of the small airways occurred between the midvolume and TLC, demonstrating that significant pressure was required before the airway would open and, therefore, becomes less distensible.

Using the above analysis, it is possible to repeat the above measurement postbronchodilator to observe the effect of smooth muscle tone (Figure 12.9c). From the postbronchodilator surfaces, there is a significant increase in the fractional inner diameter of the airways from FRC to the midvolume postbronchodilator compared with the same volume change prebronchodilator. That is, the airways have become significantly more compliant between FRC and midvolume postbronchodilator compared with the prebronchodilator surface.

12.6.5 Surface Tension

The surface tension of the alveoli is very important in maintaining open alveoli during respiration. The reason for this is that alveoli not surprisingly vary in size, and therefore, the surface tension will also vary. The surface tension of sphere is inversely proportion to its radius as can be seen in the following equation: pressure = 4T/R (Laplace's law). If we perform the simple experiment where we connect two balloons of different radii together with a stop cock in between (Figure 12.10) when the tap is opened, the smaller balloon will empty into the larger balloon. The reason for this can be described by the equation above. The balloon with the smaller radius will have a higher pressure inside

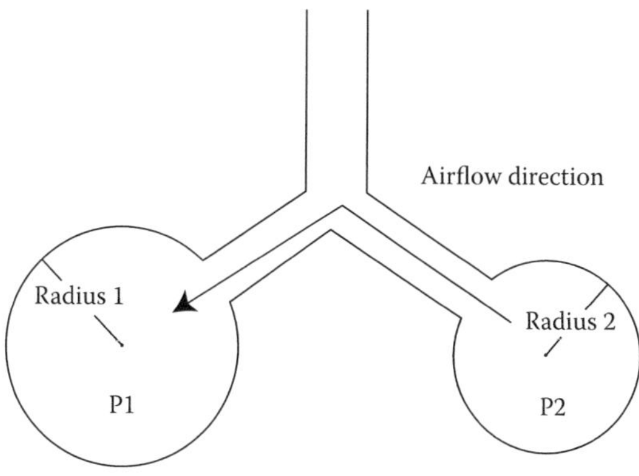

FIGURE 12.10
Surface tension: pressure in bubble 2 (P2) > pressure in bubble 1 (P1); therefore, air moves from smallest to largest bubble. Pressure = 4T/R.

it than the larger balloon due to the inverse relationship between surface tension and radius.

The reason why this does not happen in a real lung is that the alveoli have a thin film of liquid surrounding them called surfactant, which equilibrates the surface tension across the millions of alveoli. Surfactant is a phospholipid that is produced by specific cells lining the alveoli. Although the exact properties are not clear, surfactant has a number of properties that ensure stability of alveoli and prevent alveolar collapse or atelectasis. First, surfactant lowers the surface tension of the alveoli and, therefore, makes the lung more compliant. This is clearly seen when ventilating very premature infants that, after birth, require ventilatory support. These very small babies have very underdeveloped lungs with limited surfactant. When ventilating these babies, it is common to ventilate at much higher pressures than that required by an adult because the lung is not compliant due to the lack of surfactant in the infant. Second, surfactant ensures stability between alveoli. As mentioned before, the varying size of alveoli will tend to remain intact regardless of the size or pressures within. With loss of surfactant leading to difficulty in inflating the lungs, some areas collapse (atelectasis), followed by the filling of alveoli with fluid (transudate).

12.7 Concentration of Gases

The concentration of a gas is the amount of gas per unit volume, e.g., Air = 77% N2 + 20% O_2 + 3%H_2O. In arterial blood, the oxygen concentration (O_2 content) for Hb = 15, CaO_2 ~ 20 mL O_2/100 mL blood.

12.7.1 Partial Pressure of Gases

We can find the partial pressure of a gas by multiplying its concentration by the total pressure, for example, 20.93% of air is O_2. If the barometric pressure is 760 mmHg, the partial pressure of O_2 (PO_2) is as follows:

$$(20.93/100) * 760 = 159 \text{ mmHg}.$$

Now, the total pressure of any gas is equal to the sum of the partial pressure of its constituent gases. For example, air pressure = $PO_2 + PN_2 + PCO_2 + PH_2O$.

This is not entirely true as water vapor acts independently of the whole gas, so in the lung, PH_2O = 47 mmHg

Therefore,

$$PO_2 + PCO_2 + PN_2 = 760 - 47 = 713 \text{ mmHg}.$$

Also,

$$PO_2 = 20.93/100 * 713 = 149 \text{ mmHg}.$$

WORKED EXAMPLE

At TLC, the intrapleural pressure is determined to be –7 cm H_2O in a subject who is able to develop an intraalveolar pressure of +35 cm H_2O during forced expiration. Draw a schematic diagram of the situation indicating intrapleural, mouth, intraalveolar, mean airway, and transpulmonary pressures.

Answer

See diagram below.

At TLC, IP = –7 cm H_2O and IAP = 35 cm H_2O. Then, transpulmonary pressure is IAP – IP = 35 – (–7) = 42 cm H_2O.

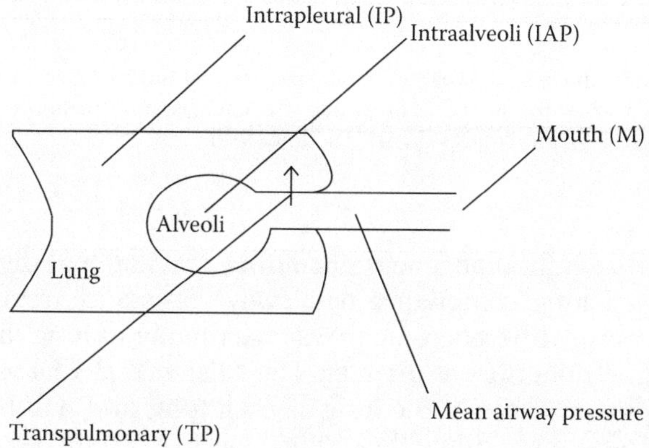

12.8 Gas Exchange within the Lung

12.8.1 Gas Mixing within the Lung

Mixing efficiency describes the adequacy with which the inspired gas mixes with gas already in the lung. It has long been known in patients with obstructive lung disease that it can take many breaths for gases to mix and equilibrate evenly throughout the lungs, which may be attributed to gas being trapped in some areas of the lung, ventilatory inhomogeneity, sequential emptying, and dead space. Various tests have been developed to study gas mixing efficiency within the lung, and most of these use a mechanism where a gas that is insoluble, and not normally present in the atmosphere, is continuously sampled while the subject is breathing normally. Washing curves of the tracer gas can then be studied to see how long it takes for the gas to equilibrate within the lung. The nitrogen washout technique uses the opposite mechanism (Figure 12.11). In the nitrogen washout test, the subject breathes a gas mixture that has nitrogen replaced with another gas. Nitrogen is then measured while the subject is breathing the new gas mixture, allowing the construction of a washout curve.

Convection-dependent inhomogeneity (CDI) results from different lung units having different specific ventilation (inspired volume to initial alveolar volume ratio). There are

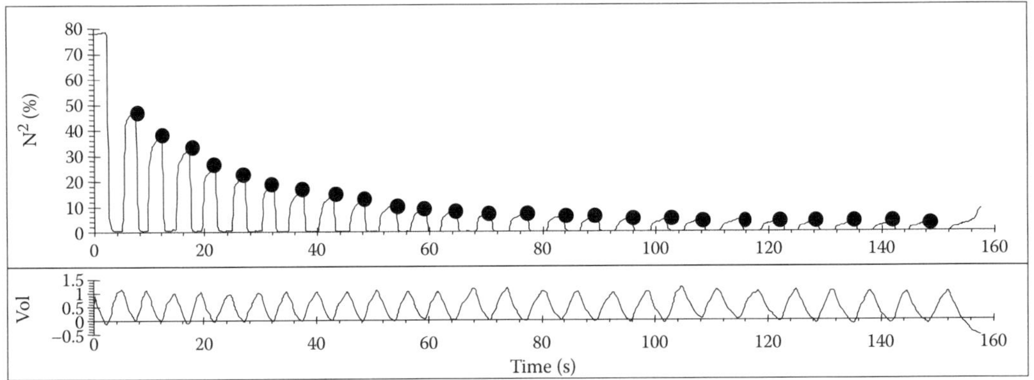

FIGURE 12.11
In the nitrogen washout test the subject breathes a gas mixture that has nitrogen replaced with another gas. Nitrogen is then measured while the subject is breathing the new gas mixture, allowing the construction of a wash-out curve (Nitrogen [N^2 %] and Volume [liters] vs. time [sec]).

numerous causes for such inhomogeneity including gravitational distortion of the lung, differences in regional lung compliance or airways resistance, regional differences in transpulmonary pressure, differences in airway narrowing in lung disease, and destruction of lung tissue in a nonuniform manner. The existence of CDI will always result in alterations to the washout of gas in the lung as each lung unit will have a characteristic number of breaths it requires to dilute the resident gas to a certain degree. Numerous analyses of such multiple breath washouts have been performed. CDI may be accompanied by sequencing of flow from different lung units, which have been shown to produce a slope of expired gas concentration in single breath washouts.

However, more recent modeling studies have shown that much of the slope of the alveolar plateau in single breath washouts and, by implication, much of the inefficiency of gas mixing are due to diffusion CDI (DCDI). DCDI is a complex interaction between convective and diffusive transport that inhibits mixing in any asymmetric lung structure, even in the absence of differences in specific ventilation between units. As the mechanism depends on the interaction between convection and diffusion, it is only important in those regions of the lung where transport due to these two mechanisms is of comparable magnitude. Convective transport decreases as the periphery of the lung is approached due to the rapid branching structure of the airways. In humans, at about the level of the entry of the acinus, the necessary conditions for DCDI exist. The fact that DCDI is responsible for much of the phase III slope of the single breath nitrogen washout explains why this test is sensitive to apparently minor degrees of small airways disease, as it is directly affected by changes in the structure of the lung in the periphery.

12.8.2 Gas Exchange across the Alveolar–Capillary Membrane

Gases move through conducting airways via convection. At the alveolar level, the cross-sectional area is very large, so the velocity of conduction approaches zero. Therefore gases must travel via diffusion. For gases to move from the alveolus to the blood stream, the gas needs to go from a ventilated environment across the alveolar capillary membrane into the blood via a network of capillaries that surround the alveolus. The rate of diffusion of the gas across the alveolar wall is dependent upon the equation given by Fick's law of diffusion:

$$V_{gas} = AD/T\ (PA - P\overline{v}) \tag{12.1}$$

where V_{gas} is the rate of uptake of gas, A is the area of the membrane, D is the diffusivity of the gas, PA is the partial pressure of the gas in the alveolus, $P\overline{v}$ is the partial pressure of the gas in the blood entering the lung, and T is the thickness of membrane

The diffusivity of a gas is proportional to the solubility of the gas and inversely proportional to the square root of its molecular weight. The quantities A, D, and T in the Fick equation are usually combined to give one diffusion term due to the difficulty in measuring each separately.

Although all gases obey Fick's law, the rate of transfer of gas into the blood stream is dependent upon the rate of blood flow because $P\overline{v}$ is relatively small compared with PA. The time for which gas exchange can occur, known as the capillary transit time, is typically 0.75 s at rest and decreases to 0.25 s in exercise and high cardiac output states. Capillary transit time is greater than the time needed for equilibration of the partial pressures of most gases.

12.8.3 Diffusion-Limited/Perfusion-Limited Gases

Gases can be broadly categorized as either perfusion limited or diffusion limited. Perfusion-limited gases are rapidly taken up into the blood stream, and the rate of uptake is dependent upon the rate of blood flow through the lungs. As the blood quickly becomes saturated (within a third of the way along the capillary), a back pressure develops, limiting the rate of gas uptake for the rest of the bloods transit. However, diffusion-limited gases are gases, which are poorly soluble in blood, not chemically inert, and are almost able to solely combine with hemoglobin and, therefore, not develop a back pressure in the plasma. Uptake of diffusion-limited gases is independent of the rate of blood flow through the alveolar capillaries. Gases, such as carbon monoxide and nitrous oxide, have an affinity with hemoglobin that is over 250 times that of oxygen, and therefore, the blood has a large capacity to hold these gases. The rate of the gas uptake into the blood stream will therefore be dependent upon the diffusion properties of the alveolar capillary membrane.

Oxygen uptake lies in between these two situations as it is fairly insoluble in blood plasma but combines with hemoglobin quite rapidly, and under normal conditions, its uptake is perfusion limited. In diseased lungs and possibly very fit subjects at extreme exercise, oxygen uptake may become diffusion limited.

12.8.4 Transfer Rate of Diffusion-Limited Gases

The transfer rate of a diffusion-limited gas can be divided into components of resistances given by the following equation:

$$\frac{1}{T_L} = \frac{1}{D_m} + \frac{1}{\vartheta Vc} \tag{12.2}$$

where T_L is the is the transfer factor of the gas, D_m is the membrane diffusion capacity, Vc is the pulmonary capillary blood volume, and ϑ is the reaction rate of the gas with hemoglobin.

12.8.5 Gas Transport in Blood-Dissolved/Chemically Bound

Once a gas has entered the blood, it can be carried in two forms. Gases can either be dissolved in plasma and therefore contribute to the total partial pressure of gases dissolved or by combining with other substances, such as hemoglobin, thereby increasing the total volume of gas in the blood.

Gases dissolved in plasma, which are not in chemical combination with other substances, such as hemoglobin, follow Henry's law, which states that the volume of gas that can be dissolved in a liquid is proportional to the partial pressure of the gas in the gaseous phase.

12.8.6 Gas Transport of Nitrogen in the Blood

Respiratory gases O_2, CO_2, and N_2 are transported by different mechanisms in the blood. Nitrogen forms the bulk of the composition of atmospheric air and is carried by the blood in simple solution in the plasma. As nitrogen also has a low solubility, it can only be carried in small quantities. Under normal conditions, nitrogen has little importance in the overall gas exchange process. However, under hyperbaric conditions (e.g., while diving) where the partial pressure of the gas is significantly higher, an increase in the amount of nitrogen that is dissolved in solution will occur.

12.8.7 Gas Transport of Oxygen in the Blood

Oxygen is transported predominantly in combination with hemoglobin, although a small percentage of the gas is dissolved in solution in the plasma. Of the total amount of oxygen carried by the blood, 97% of the gas is in combination with hemoglobin, and the amount is determined by the oxygen–hemoglobin dissociation curve (Figure 12.12). The curve is of curvilinear shape in which the steep part of the curve demonstrates that large amounts of oxygen can be unloaded for small changes in the alveolar partial pressure of oxygen. The flat part of the curve shows that even for large changes in PaO_2, there are only small changes in the arterial oxygen saturation. The uptake of oxygen and, therefore, the position of the oxygen–hemoglobin dissociation curve can be affected by temperature, acidity of the blood, and 2,3-diphosphoglycerate.

12.8.8 Gas Transport of Carbon Dioxide in the Blood

Carbon dioxide is carried by the blood in three main forms. These include being dissolved in the plasma, in combination with carbamino-hemoglobin, and as bicarbonate. Carbon dioxide is twenty times more soluble than oxygen in blood plasma, and as a result, approximately 10% of carbon dioxide that is off loaded into the lung from the blood is from plasma solution (see Figure 12.13). The bulk of carbon dioxide transport is in the form of bicarbonate (~80%), resulting from the reaction of carbon dioxide with water in the presence of the enzyme carbonic anhydrase (CA). The chemical equation for this process is as follows:

$$CO_2 + H_2O \overset{CA}{=} H_2CO_3 = H + HCO_3^-.$$

The first reaction is slow within the plasma but is much faster within the red blood cell because of the abundance of the enzyme CA. The second reaction of the ionic dissociation

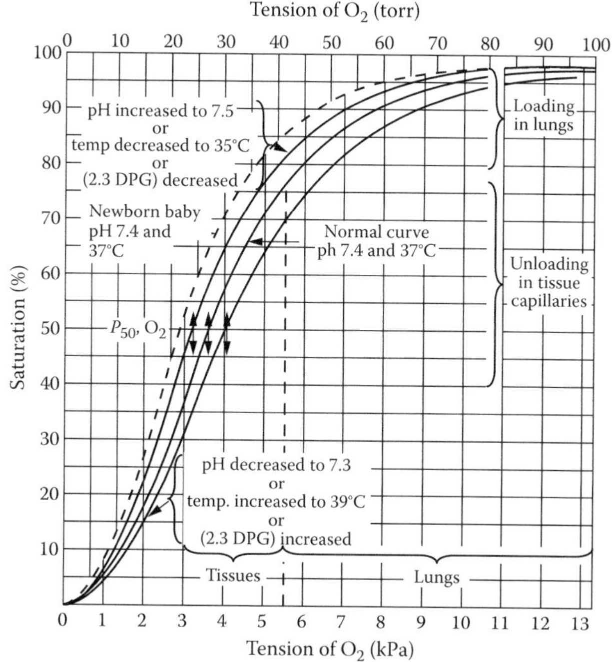

FIGURE 12.12
Oxygen–hemoglobin dissociation curves. (From Cotes JE, *Lung Function: Assessment and Application in Medicine,* 5th Edition, Oxford UK: Blackwell Scientific Publications, 1993; Kelly VJ, Brown NJ, King GG, Thompson BR (2010). *Med and Biol Eng and Computing*, 48, 489–496.)

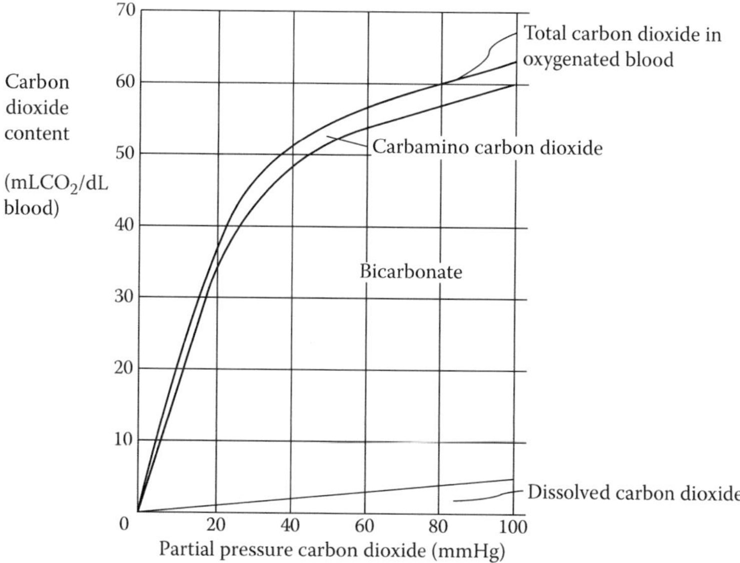

FIGURE 12.13
CO_2 dissociation curves.

of carbonic acid is fast and is not aided by any enzymes. The hydrogen ions from this second reaction react with hemoglobin as follows:

$$H + HbO_2 = H + Hb + O_2.$$

This reaction reduces the amount of acid; therefore, the presence of reduced deoxygenated hemoglobin in the peripheral blood helps in the loading of carbon dioxide from the peripheral capillaries. Conversely, the oxygenated hemoglobin assists with the unloading of carbon dioxide from the pulmonary capillaries. This process is known as the Haldane effect.

12.8.9 Ventilation–Perfusion Inequality

Optimal gas exchange requires a similar distribution of ventilation and perfusion throughout the lung. In normal subjects, the lungs are influenced by gravity, resulting in more blood being at the base of the lung. Similarly, the weight of the lung results in a pleural pressure gradient (more negative at the top) so that the upper lung units are on a stiffer part of the compliance relationship, and ventilation is therefore greatest at the lung bases. These mechanisms do not achieve perfect $\dot{V}/\dot{Q}$ balance so that $\dot{V}/\dot{Q}$ ratios remain somewhat reduced at the bases than the apices in normal lungs (Figure 12.14).

In patients with lung disease, the mismatch between ventilation and perfusion can be greatly increased relative to normal subjects, resulting in significant disturbances to gas exchange. In the normal upright lung, both ventilation and blood flow increase toward the base of the lung; however, blood flow increases more than ventilation, resulting in $\dot{V}/\dot{Q}$ decreasing from 3 at the top to 0.3 at the bottom of the lung.

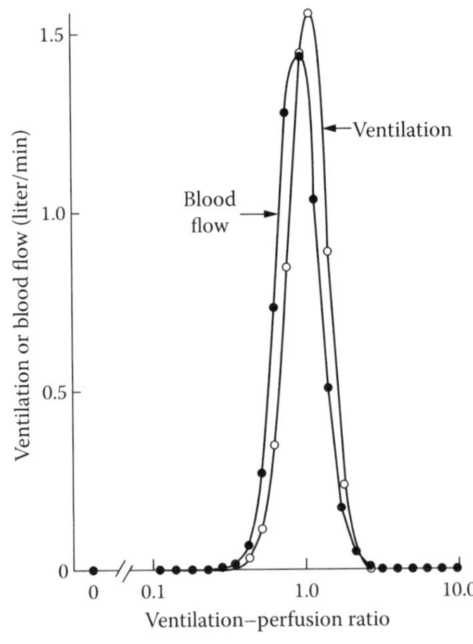

FIGURE 12.14
Ventilation or blood flow versus ventilation–perfusion ratio in a normal subject. (From West JB, *Respiratory Physiology: The Essentials*, 8th Edition, Baltimore USA: Lippincott, Williams & Wilkins, 2008.)

In the 1960s and 1970s, work was performed on the quantification of the relationship between ventilation and perfusion. This work was based on the graphical representation of ventilation perfusion because algebraic manipulations of the gas exchange equations were only manageable to a certain value. The final stage of the development was to display continuous distributions of ventilation perfusion ratios in healthy and diseased lungs, which was made possible by the use of complex equations more easily solved by the modern computer. From this work, the multiple inert gas elimination technique for creating continuous distributions of ventilation perfusion inequality was developed. A typical distribution is shown in Figure 12.14, which displays ventilation and perfusion against the natural logarithm of ventilation, divided by perfusion. Every point on the curve represents a compartment of the lung model used in the development of the equations. Therefore, if a single point is taken from the abscissa, thus giving a ventilation perfusion ratio, the independent values of both ventilation and perfusion can be obtained from the ordinate for that particular compartment.

Normal ranges for blood gases:

Arterial

$PaO_2 = 90 - 110$ mmHg

$PaCO_2 = 35 - 45$ mmHg

$pH = 7.35 - 7.45$

$HCO_3^- = 25 \pm 4$

Venous

$PvO_2 = 40$ mmHg

$PvCO_2 = 50$ mmHg

$pH = 7.30$

12.9 How Can Blood Gases Be Altered?

A number of factors and states can alter blood gas levels; some are discussed next.

12.9.1 Environment

The inspired partial pressure of oxygen is fundamental to the interpretation of PaO_2. It is obvious that under conditions of hyperoxia, for example, supplemental oxygen or in hyperbaric chamber, the PaO_2 will be increased. Conversely, under hypoxic environmental conditions, for example, at altitude, the PaO_2 will be decreased. Although measuring PaO_2 at altitude in the laboratory setting is rare, supplemental oxygen in the clinical setting, however, is relatively common, and before results are interpreted, it is essential that the FiO_2 is known.

12.9.2 Ventilation

Changing ventilation (e.g., by exercise, nasal ventilation) will have a direct effect on pH, PO_2, and PCO_2. Table 12.2 provides a brief overview of the relationship between ventilation and arterial blood gas values.

TABLE 12.2

Relationship between Ventilation and Arterial Blood Gases

	pH	$PaCO_2$	PaO_2
Normal	7.40	40	90
Hyperventilation	+	–	+
Hypoventilation	–	+	–

Note: + = increase, – = decrease.

12.9.3 Disease

Physiological causes of hypoxemia occurs by five primary mechanisms. These are as follows:

(1) Decrease in PiO_2 (i = inspired)
(2) Decrease in alveolar ventilation
(3) Low V/Q units
(4) Shunt (pulmonary/extra pulmonary)
(5) Diffusion impairment.

In the first three conditions, the A–a gradient is normal (i.e., the alveolar–arterial equation), and in the last two, the A–a gradient is abnormal. The alveolar air equation relates mean PAO_2 to PiO_2 and $PaCO_2$. This provides an assessment of the overall effectiveness of gas exchange in the lung.

$$PAO_2 = PiO_2 - (PaCO_2/0.8)$$

$$\text{A–a gradient} = PAO_2 - PaO_2 = PiO_2 - (PaCO_2/0.8) - PaO_2.$$

Normal range ~15–20 mmHg.

The alveolar arterial PO_2 difference gives an indication of V/Q inequality. The alveolar PO_2 is calculated as an ideal PAO_2, that is, the PAO_2 the lung would have if there was no ventilation perfusion inequality, and it was exchanging gas at the same respiratory exchange ratio as the real lung.

12.9.4 Hypercapnic Respiratory Drive

The hypercapnic drive is controlled by the central chemoreceptors located on the ventral surface of the medulla. This is a rapidly responding system that acts on blood CO_2 via H^+ ions of cerebral spinal fluid (CSF). The hypercapnic drive is much stronger than the hypoxic drive and is therefore the main controller for respiration. Prolonged changes in pH of CSF is normalized by HCO_3^- diffusion across the blood–brain barrier over 36–48 hours. Increasing blood CO_2 increases the production of H^+ ions, which stimulates the central chemoreceptors, leading to an increase in ventilation.

12.9.5 Hypoxic Respiratory Drive

The hypoxic drive is controlled by the peripheral chemoreceptors located in the carotid and aortic bodies. They respond to decreasing $PaO_2 < 500$ mmHg (also decreasing pH and increasing $PaCO_2$); however, the response increases exponentially when the PaO_2 decreases below 100 mmHg. Decreasing blood O_2 stimulates the peripheral chemoreceptors, leading to an increase in ventilation.

As stated previously in normal subjects, the hypercapnic drive is the main controller for respiration; however, patients under conditions of prolonged hypercapnia (e.g., patients with long-standing emphysema) may have nearly normal CSF pH and, therefore, abnormally low ventilation for the $PaCO_2$. This suggests that the CO_2 response has been blunted over time and that the dominant controller would be the hypoxic drive. The following gases give an indication of this effect. PaO_2 = 50 mmHg and $PaCO_2$ = 60 mmHg.

12.9.6 Administration of Oxygen

As stated previously, chronic hypercapnea may lead to a blunted CO_2 response; therefore, ventilation is controlled by the hypoxic drive. This can lead to a significant problem when patients in this situation are implemented with supplemental oxygen. Giving supplemental oxygen will increase PaO_2. If the PaO_2 is high enough where the hypoxic response is significantly diminished, this will effectively eliminate the hypoxic drive, leading to a significant rise in $PaCO_2$ and potentially making the patient respiratory compromised. This is a significant contraindication for the implementation of oxygen therapy.

12.9.7 Acid/Base Balance

The lung has a profound role in maintaining acid/base balance for the whole body. The lungs indeed excretes approximately 10 000 mEq of carbonic acid each day compared with the kidneys, which only excrete a 1% of that amount of fixed acids. CO_2 and, therefore, the lungs have a central role in the maintenance of pH by altering alveolar ventilation and, therefore, the removal of CO_2. The mechanism of control is by the following:

1. Plasma buffers, which react quickly (<1 ms)
2. Respiratory adjustment, which is slightly slower than plasma buffers (1 s–10 s), and
3. Renal adjustment, which reacts slowly (hours–days).

12.9.8 Buffer Systems

The major system involves HCO_3^-, Hb, and phosphate; however, the most important is HCO_3^-. The behavior of which follows the Henderson–Hasselbalch equation.

$$H+ + HCO_3^- = H_2CO_3 = CO_2 + H_2O$$

$$pH = 6.1 + \log[HCO_3^-]/0.3.PaCO_2.$$

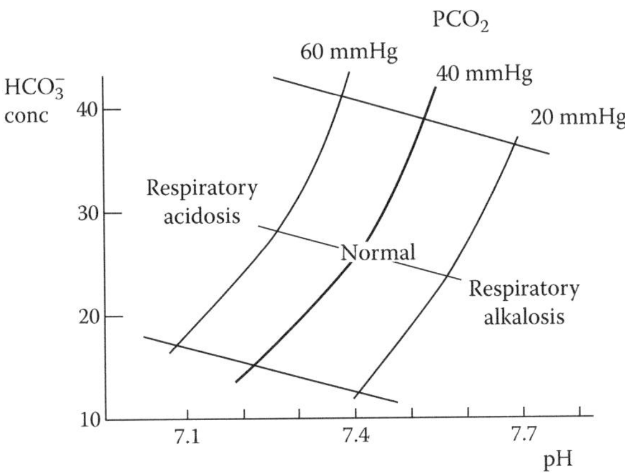

FIGURE 12.15
The Davenport diagram.

The relationship between pH, $PaCO_2$, and HCO_3^- can be demonstrated on the Davenport diagram (see Figure 12.15). The normal value for blood pH is 7.4 (middle point on the diagram). When pH changes due to metabolic or respiratory changes, the diagram illustrates the potential compensatory mechanisms, namely HCO_3^-, and carbonic acid to return pH back to normal. These adjustments are discussed next. Typical gas value associated with respiratory and metabolic acidosis and alkalosis are given.

12.9.9 Respiratory Adjustment

As $PaCO_2$ is inversely related to pH, altering ventilation and therefore $PaCO_2$, will have a significant impact in maintaining pH homeostasis. From the Henderson–Hasselbalch equation, an increase in H^+ ions will lead to an increase in CO_2. As described previously, respiratory control is very sensitive to changes in $PaCO_2$. Therefore, increases in $PaCO_2$ will stimulate the central chemoreceptors in the brain, leading to an increase in ventilation, which in turn will decrease $PaCO_2$. This mechanism is described as a respiratory acidosis. Conversely, a decrease in H^+ ions will lead to a decrease in CO_2 and therefore ventilation will decrease leading to a rise in pH. This is known as a respiratory alkalosis.

12.9.9.1 Examples

Respiratory Acidosis

pH = 7.25

$PaCO_2$ = 60 mmHg

PaO_2 = 53 mmHg

HCO_3^- = 25

Compensated Respiratory Acidosis

pH = 7.36

$PaCO_2$ = 65 mmHg

PaO_2 = 55 mmHg

HCO_3^- = 35

Respiratory Alkalosis

pH = 7.57

$PaCO_2$ = 28 mmHg

PaO_2 = 161 mmHg

HCO_3^- = 24

12.9.10 Compensated Respiratory Alkalosis

Not really physiological as the subject would have to maintain excessive ventilation for many hours to days. In this situation muscle fatigue normally happens before compensation would occurs.

12.9.10.1 Examples

Metabolic Acidosis

pH = 7.25

$PaCO_2$ = 40 mmHg

PaO_2 = 78 mmHg

HCO_3^- = 17

Compensated Metabolic Acidosis

pH = 7.36

$PaCO_2$ = 29 mmHg

PaO_2 = 97 mmHg

HCO_3^- = 17

Metabolic Alkalosis

pH = 7.49

$PaCO_2$ = 40 mmHg

PaO_2 = 61 mmHg

HCO_3^- = 32

12.9.11 Compensated Metabolic Alkalosis

For this physiological scenario, this form of compensation is again not really physiologically because the compensation process would require the human subject to hypoventilate. This is something the body will not ordinarily do especially if the subject is hypoxic.

12.10 A Summary of Lung Diseases

There are a large number of respiratory diseases, but this chapter cannot cover each in sufficient detail; however, they can be characterized based on their main actions on respiratory mechanics and overall gas exchange. These are listed here in the following eight categories: obstructive and restrictive, pulmonary vascular, occupational, malignant, infectious, genetic, and neurological diseases. Note that some diseases can be classified in more than one category as they progress.

1. *Obstructive Diseases.* These are diseases, which lead to airflow obstruction. These include; asthma, bronchitis (excessive mucous), emphysema (many forms), chronic bronchitis, and localized tracheal obstruction.
2. *Restrictive Diseases.* These are diseases, which impact on the mechanics or, more specifically, the elasticity or compliance of chest wall movement. These fall into the following four subcategories.
   (a) *Disease of the chest wall:* increased stiffness (scoliosis), decreased volume (thoracoplasty).
   (b) *Diseases of the pleura:* increased stiffness (fibrothorax), decreased volume (pneumothorax).
   (c) *Disorders of the respiratory muscles:* decreased strength (amyotrophic lateral sclerosis).
   (d) *Disorders of the lung:* increased stiffness (diffuse interstitial fibrosis), decreased volume (pneumonectomy), sarcoids (granulomatous tissue), collagen disease, lymphangitis carcinomatosa (cancer in the pulmonary lymph).
3. *Pulmonary Vascular Disease.* These are diseases associated with the lung arterial and venous blood supply systems. These include pulmonary edema (accumulation of fluid), pulmonary embolism (clots), pulmonary hypertension, and pulmonary arteriovenous fistula (shunts).
4. *Occupational Diseases.* These diseases tend to be associated with inhaling atmospheric pollutants such as particles, chemicals, or gases. These can lead to a condition such as *pneumoconiosis*. These pollutants are carbon monoxide, sulfur oxides, hydrocarbons, particulate matter (e.g., wood pulp), nitrogen oxides, photochemical oxidants, cigarette smoke, coal dust, silica (silicosis), asbestos (asbestosis), iron oxides (siderosis), beryllium (lesions), and chalk dust.
5. *Malignant Diseases.* These diseases are associated with tumor and/or cancerous growths in the lung tissue. In many cases, these are associated with occupational issues; however, a genetic component also may be a factor. These include bronchial carcinoma and alveolar carcinoma.
6. *Infectious Diseases.* These diseases are associated with bacterial, viral, or fungal pathogens, for example, pneumonia tuberculosis (bacillus). Most of these conditions are contagious with contact of infected humans or animals.
7. *Genetic.* A condition that is genetic and affects the growth and development of lung tissue is the disorder cystic fibrosis. This disorder affects other organs as well (pancreas).
8. *Neurological.* As mentioned previously in this chapter, the control of ventilation is complex, and any faults in the feedback mechanisms associated with the respiratory drive may lead to a variety of breathing-related difficulties such as gasping, fast breathing, or cessation of breathing during sleep (sleep apnea).

12.11 A Lung Function Testing Summary

During a respiratory function test, spirometry is the main tool for quantifying the effects of respiratory disease and response to treatment. However, other tests are available to identify subtle changes in respiratory function. These tests are briefly summarized here.

12.11.1 Spirometry

This technique is a very useful one to assess the volume and flow characteristics of normal and unhealthy lungs during inhalation and exhalation. The instrumentation measures these volumes and flow rates using a variety of instrumentation techniques such as the pneumotach flow sensors and displacement transducers. The participant basically inhales and exhales into a closed system, allowing the measurement of the volumes moving into and out of the system. In reference to Figures 12.2 and 12.6, all volumes, except RV and TLC, can be measured. Spirometry can be used to investigate whether an illness has obstructive or restrictive characteristics. This is done by calculating the relationship $FEV_{1.0}/FVC$. The forced expiratory volume ($FEV_{1.0}$) is the volume of air expired rapidly in 1 s after inhaling to a maximum. The forced VC is the total air exhaled rapidly after maximal inhalation. If this ratio is greater than approximately 80%, then the disease is considered restrictive; however, if it is under approximately 80%, then it is considered obstructive.

In summary,

$$\text{Forced expiratory ratio (FER)} = FEV_{1.0}/FVC.$$

12.11.2 Flow-Volume Loops

As illustrated in Figure 12.6, a flow-volume curve shape can be used to also determine the restrictive or obstructive nature of a respiratory disease. The telltale sign for an obstructive disease is the time it takes to completely exhale. This is usually longer and tapers for a few more seconds. This shows that capacity may be less than normal, but due to the obstruction, the exhalation takes longer.

12.11.3 Other Flow Measurements

As illustrated in Figure 12.6, the flow-volume curve shape can be used to also determine a number of patterns including airflow obstruction. The telltale sign for an obstructive disease is the concavity in the expiratory limb of the flow volume curve. In some situations the FVC will also be reduced.

Peak flow: peak flow rate (mL/s)

$FEF_{50\%}$: forced expiratory flow at 50% of VC

$FEF_{75\%}$: forced expiratory flow at 75% of VC

12.11.4 Body Plethysmography

This device enables the measurement of all volumes, flows, and procedures (FVC), but by pressure differences between chest wall, thoracic cavity, mouth pressure, and atmospheric pressure, then the RV and FRC can be estimated. This device is far more accurate at measuring these respiratory volumes and assisting in detecting subtle changes associated with respiratory illness. It operates on the principal of Boyle's law ($P \times V = k$); the product of pressure and volume for a gas is constant in a chamber and under isothermal conditions.

With the participant sitting in the body plethysmograph and panting against a closed airway, the FRC can be measured. The ratio of the change in volume with respect to change in pressure is measured. Total thoracic volume at FRC can be measured.

$$FRC = P_{atm}(\Delta V/\Delta P).$$

Total thoracic volume at FRC can be measured.

$$TLC = IC + FRC$$

$$RV + TLC - VC$$

(Note: P_{atm} = atmospheric pressure)

12.11.5 Gas Dilution Techniques

One technique, which also is useful in measuring the distribution of the RV, is the nitrogen washout method. This technique was discussed earlier in Section 12.7. Other techniques for measuring ventilation/perfusion inequalities also are described in Section 12.7 and illustrated in Figure 12.14.

12.11.6 Gas Transfer/Diffusion

This test measures the lungs' overall ability to transfer oxygen from within the alveoli to the blood. The gas transfer factor is equal to the volume of gas taken up over a defined period of time, divided by the pressure difference for the gas in the alveolar regions and the capillaries.

To estimate the transfer factor, we need to use a gas that is solely diffusion limited, for example, carbon monoxide (CO). Carbon monoxide has higher affinity to bind on to haemoglobin than oxygen.

The measurement for diffusing capacity is called transfer factor; diffusion for carbon monoxide (D_{CO}) or D_L is made with the following relationship.

$$V_{gas} \text{ is approx} = A/T^* D(P1\text{-}P2)$$

where V_{gas} is the rate of transfer, D is the diffusion coefficient, A is the area of gas transfer (blood–gas barrier) – (50–100 m^2 in human), and $T < 0.5$ μm.

Assumptions:
Ideal conditions for diffusion
D is proportional to solubility of gas
D is proportional to 1/molecular weight
and $D_L = AD/T$, then $D_L = V_{gas}/PA_{gas.}$

Then $D_{CO} = V_{CO}/PA_{CO}$ mL/min/mmHg (D_{CO} is the rate of CO transfer per mmHg partial pressure).

12.11.7 Imaging

The imaging techniques, such as X ray and magnetic resonance imaging, are useful in detecting structural changes to the lung tissue. The advantages of these technologies are described elsewhere in this book.

12.12 Summary of Respiratory Terms, Facts, and Formulae

GAS PRESSURE

Symbols/Nomenclature/Formulae

P_{gas} – Partial pressure of a gas

F_{gas} – Fraction of a gas

C_{gas} – Concentration of a gas- gas content

BP – Barometric pressure (atmospheric pressure)

$P_{gas} = F_{gas} \times PB$

$$C_{gas} = F_{gas} \times \text{Volume (for gaseous phase)}$$

For blood phase-associated with P_{gas} solubility and affinity for physicochemical carriers (Hb—hemoglobin). In respiratory, P_{gas} is most often used.

Units:

- mmHg = torr = 0.133 kPa
- mmHg = 1.36 cm H_2O (cm H_2O = 0.77 mmHg)

Gas laws:

- The following gas laws will be referred to at various times:
- Boyle's law (P and V)
- Charles' law (P and T)
- Avogadro's principle (No. of mols)
- Ideal gas law (PV = nRT)
- Dalton's law of partial pressure
- Henry's law of solubility

Normal blood gases values:

- Arterial
- PO_2 95–100 mmHg, PCO_2 36–44 mmHg, pH 7.36–7.44
- O_2 content 21 vol%, CO_2 content 48–50 vol%
- Venous
- PO_2 37–42 mmHg, PCO_2 42–48 mmHg, pH 7.34–7.42
- O_2 content 15–16 vol%, CO_2 content 52–54 vol%

Normal vital capacity:

Adults:

- Males: $-8.7818 + (0.0844 \times \text{Height(m)}) + (-0.0298 \times \text{Age})$
- Females: $-2.9001 + (0.0427 \times \text{Height(m)}) + (-0.0174 \times \text{Age})$

Note BSA = body surface area

Children:

- Males: 250 mL/year
- Females: 200 mL/year

Vital capacity is maximal at about the age of 20 years and then declines at the rate of about 1% per annum.

Questions

Question 1

(a) Draw and label a diagram detailing the organization of the human lung. With this drawing, also highlight the physiological zone (acinar zone) and the anatomical dead space (conducting zone).

(b) Draw and label a diagram detailing the mucociliary transport system within the human lung. Highlight and label the specialized cells found in this region.

(c) Draw and label a diagram detailing the complex interaction between the heart–lung system.

Question 2

Outline the neurological and humoral mechanisms that control respiration, indicating (e.g., by flow diagram) how their actions are integrated to achieve this control.

Question 3

Describe the oxygen dissociation curve. What does it tell us about compensatory mechanisms associated with blood?

Question 4 (**multiple choice questions: encircle the most appropriate answer**)

1. Most of the CO_2 in blood is transported as
   (a) CO_2 dissolved in plasma
   (b) Bicarbonate (HCO_3^-) ions
   (c) Part of acid hemoglobin
   (d) Carbonic acid
   (e) Serum albumin
2. Which of the following will increase the oxygen-carrying capacity of hemoglobin?
   (a) High temperature
   (b) pCO_2
   (c) Acidity

(d) 2,3 Bis-phosphoglycerate (2,3 BPG)
(e) None of the above

3. The vital centers regulating respiration, heart beat, and blood pressure are located in the
   (a) Cerebral cortex
   (b) Medulla oblongata
   (c) Hypothalamus
   (d) Spinal cord
   (e) Postcentral gyrus
4. Oxygen diffuses into the blood from the alveoli, and carbon dioxide diffuses into the alveoli from the blood until the respective pressures of these gases in the blood of the pulmonary vein
   (a) Exceeds those pressures within the alveoli
   (b) Equal those pressures within the alveoli
   (c) Equal the respective pressures of these gases in the blood of the pulmonary artery
   (d) Are less than those pressures within the alveoli
   (e) Equal the respective pressures of these gases in the blood of the right atrium of the heart
5. During respiration, one area of the respiratory system is always subatmospheric, and the pressure in that area is defined by the
   (a) Interalveolar pressure
   (b) Intrapulmonic pressure
   (c) Intraalveolar pressure
   (d) Interpulmonic pressure
   (e) Intrapleural pressure
6. The presence of surfactant in the lungs means that the
   (a) Lungs are moist
   (b) Surface tension of the alveolar fluid is lowered
   (c) Lungs are thinner
   (d) Heart is protected
   (e) pH of the pleural fluid is normal
7. For inspiration to occur,
   (a) The lungs must expand, which decreases lung volume and does not affect lung pressure
   (b) The lungs must expand, which decreases lung volume and increases pressure in the lungs
   (c) The lungs must expand, which increases lung volume and increases pressure in the lungs
   (d) The lungs must expand, which increases lung volume and decreases pressure in the lungs
   (e) None of these statements are correct

8. If the level of CO_2 in the blood drops too low,
   (a) Acid hemoglobin formation is favored
   (b) Ventilation increases
   (c) Breathing increases
   (d) HCO_3^- combines with H^+ to form carbonic acid, which then dissociates to form CO_2 and water
9. Which of the following would result in increased ventilation or an increased respiratory rate?
   (a) An increase in PCO_2 of the blood
   (b) A decrease in PO_2 of the blood
   (c) An increase in H^+ concentration of the blood
   (d) All of the above
10. Oxygen diffuses into the blood from the alveoli, and carbon dioxide diffuses into the alveoli from the blood until the respective pressures of these gases in the blood of the pulmonary vein
   (a) Equal those pressures within the alveoli
   (b) Exceeds those pressures within the alveoli
   (c) Are less than those pressures within the alveoli
   (d) Equal the respective pressures of these gases in blood of the pulmonary artery

Answers

MCQ 1–10, in order, 1.c, 2.b, 3.b, 4.b, 5.e, 6.b, 7.d, 8.d, 9.a, 10.a.

Bibliography

Cotes JE. 1993. *Lung Function: Assessment and Application in Medicine*, 5th Edition. Blackwell Scientific Publications, Oxford, UK.

West JB. 2008. *Respiratory Physiology: The Essentials*, 8th Edition. Lippincott, Williams & Wilkins, Baltimore, USA.

13

Renal Biophysics and Dialysis

Andrew W. Wood

CONTENTS

The kidneys are vital to the maintenance of equilibrium between the food and drink we consume and the excretion of fluid in the form of urine. We also lose fluid through perspiration and, to a lesser extent, respiration. The kidneys and associated control systems are able to monitor fluid balance and, by extension, circulatory blood volume. In more primitive animals, such as amphibia, the skin also is a major component in the regulation of fluid balance, so a portion of what is now known about the function of the human kidney has been inferred from the effects of hormones and other agents on the rate of ion transport across amphibian skin. In Chapter 5, the importance of measuring short-circuit current was highlighted.

This chapter also will briefly touch on diseases of the kidney and, in particular, some of the bioengineering aspects of kidney dialysis, the "artificial kidney."

13.1 Kidney Function

The mammalian kidney serves three main functions. The kidneys (1) regulate the constancy of interior environment, particularly in relation to mineral salts; (2) excrete waste, particularly metabolites such as urea, uric acid, creatine, and creatinine; and (3) conserve valuable foodstuff, such as glucose and amino acids.

Function (1) implies that no matter what the composition is of the food and drink we ingest (within bounds) or how much fluid we lose through sweating or excreting, the salt content of the blood plasma remains remarkably constant at around 150 mM. The amount of extracellular water, as well as the blood volume (and hence pressure), also is regulated in the long term by the kidneys. The second function (2) is what is normally associated with the kidneys, but the third function (3) is extremely important to the overall efficiency and conservation of metabolic fuel and complex molecules, which require a significant cellular effort in their production.

There are three processes, which go with the three functions (which do not correspond one-for-one): (1) filtration of blood, while retaining large molecules (proteins) and organelles (such as red blood cells); (2) selective reabsorption of valuable substances via active transport (e.g., NaCl via the Na^+/K^+ATPase pump or via a specific Cl^- pump); and (3) tubular secretion of certain substances, especially noxious ones, actively added to urine (particular reference will be made later on to para-amino hippurate or PAH).

Before discussing the details of kidney structure, some basic facts will be presented. First, blood flow through the kidneys is approximately 1500 mL/min or 25% of cardiac output. Excretion rate of urine ranges from 0.1 mL/min to 20 mL/min, with an average of 1 mL/min (or 1.5 L/day).

Maximum concentration of salt in urine is 750 mM NaCl (which is around five times that of blood plasma). On the other hand, on occasions, the urine can contain very little salt (there is a tendency for higher urine flow to be associated with lower salt content). This is indicative of the wide range of control the kidneys are able to exert, coping with situations where large quantities of fluids are imbibed to the opposite, where an individual may be prevented from drinking anything for long periods.

13.1.1 Kidney Structure

The two kidneys are embedded deep within the abdomen, close to the spine and in front of the lowest (12th) ribs. Note that the left kidney is slightly higher than the one on the right. Each kidney is supplied by a short (renal) artery leading directly from the abdominal aorta. Once inside the kidney, these branch to form several interlobar arteries, which form hundreds of arcuate arteries (running in arcs of circles, see Figure 13.1). These then further branch intro interlobular arteries, which then form the smallest branches to around 10^6 spherical structures known as *glomeruli* (literally, 'little balls'). At the far side of these, the venous supply more or less parallels the arteries, culminating in the two renal veins.

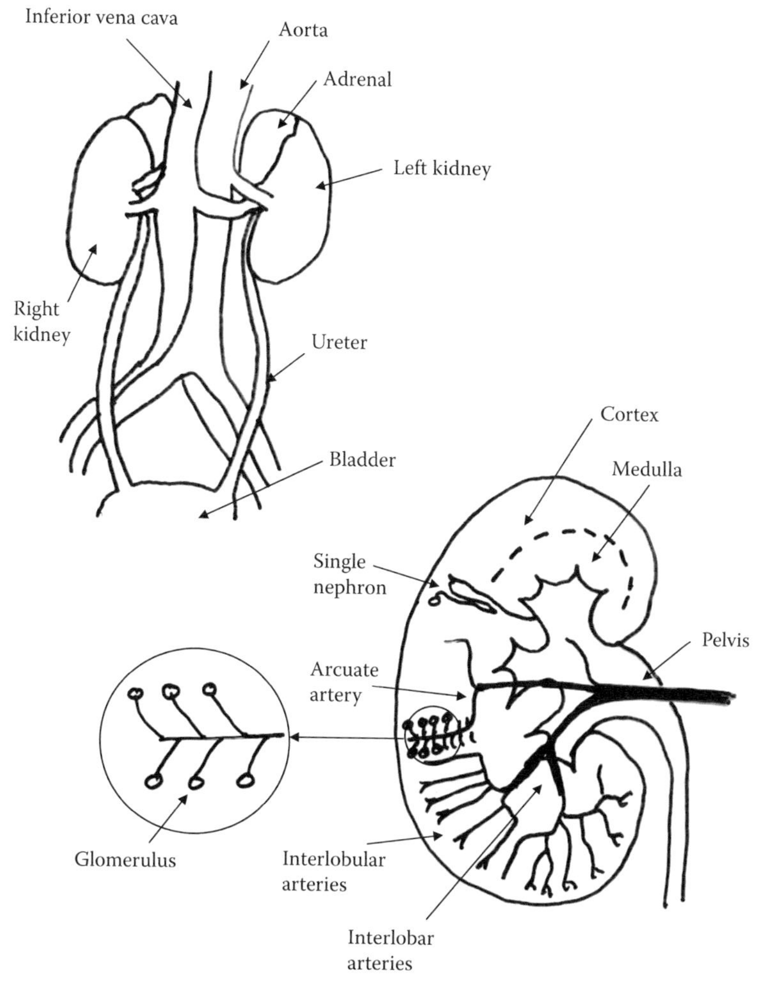

FIGURE 13.1
Gross anatomy of the renal systems, showing the renal arteries and the large veins. The terminal arterioles and renal capillaries contained in the glomeruli are also represented.

In the next section, we will see how the urine collects from each of the 10^6 glomeruli, draining into the renal pelvis, then on into the ureters to the bladder. The renal tissue itself is divided into two broad regions, the renal cortex (on the outside) and the medulla (toward the pelvis). On top of each kidney is a region of related tissue, the adrenal glands, which each have a cortical and medulla region (secreting *cortico*steroids and adrenaline and related catecholamines, respectively).

Note that the glomeruli each represent the interface between the blood circulation and the epithelial layer.

The glomerulus is where the filtration takes place: each one is connected to a tortuous tube called the *renal tubule*. This, together with the glomerulus and associated blood vessels, is called the *nephron*, the functional unit of the kidney.

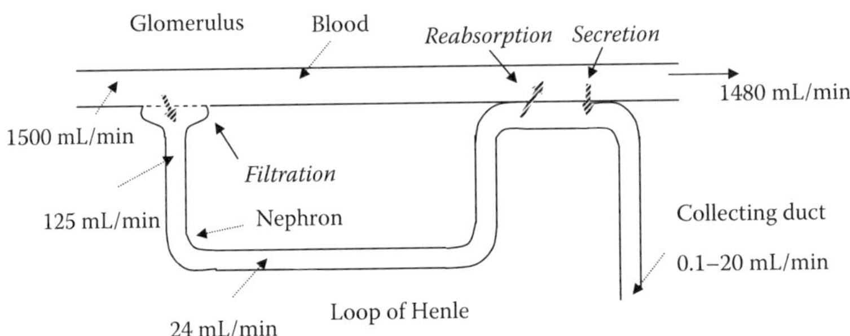

FIGURE 13.2
Highly diagrammatic representation of the nephrons lumped as a single nephron, showing sites of filtration, reabsorption, and secretion. Note fluid flow rates.

Since all nephrons are more or less identical, we can lump them all together to produce a diagram for kidney *function* (a similar thing is done for the alveoli of the lung) (Figure 13.2).

Note the relative lumped flow rates for all of the nephrons and blood vessels: most of the 20% of blood *filtrate* that enters the nephron in the glomerulus is returned to the blood supply via the *vasa recta* (VR) system (the blood capillaries associated with the nephrons). This is 20% of the flow rate of blood *plasma*, which is approximately 650 mL/min. The total blood flow (of which, plasma is around half) to the kidneys falls very slightly from 1500 mL/min to 1480 mL/min from the renal artery to the renal vein. The VR (meaning "straight vessels") are shown diagrammatically in Figure 13.7 and consist of a network of arteries, capillaries, and venules surrounding the hairpin-shaped sections of the nephrons (shown in Figure 13.3). The reabsorption and secretion of substances by the nephron occur initially into or out of the interstitial fluid but, from there, are transported very efficiently to or from the VR.

Glomerular filtration rate (GFR) is the volume flow rate into all the nephrons together (i.e., 125 mL/min). This represents the flow occurring across the glomerular capillary walls (across the endothelial layer), thence across the epithelial layer into the top end of the nephron (called the Bowman's capsule). The GFR is determined by the following: (1) the permeability of the capillaries in the glomerulus to water (K), and (2) the balance between hydrostatic and osmotic forces between the blood vessels and nephron (net filtration pressure Δp). The glomerulus acts as a sieve: anything larger than protein molecules (including, of course, blood cells) is unable to cross. The watery filtered plasma is known as "ultrafiltrate." The inorganic component is mainly NaCl at a concentration of 150 mM. Because NaCl completely dissociates and because the nephron epithelium is poorly permeated by both Na^+ and Cl^-, the osmotic pressure (OP) of the ultrafiltrate is given as 150 × 2 = 300 mOsm/L. The osmole was defined in Chapter 4: 300 mOsm/L is equivalent to 300 Osm/m^3 × RT = 0.77 MPa or 5800 mmHg, or 7.5 atm.

The net filtration pressure is given by the following equation:

$$\Delta p = (p_g - p_c) - (\pi_g - \pi_c).$$

Here, p_g and p_c are the hydrostatic pressures in the glomerulus and the Bowman's capsule, respectively, with indicative values of 60–80 and 15 mmHg, respectively. The

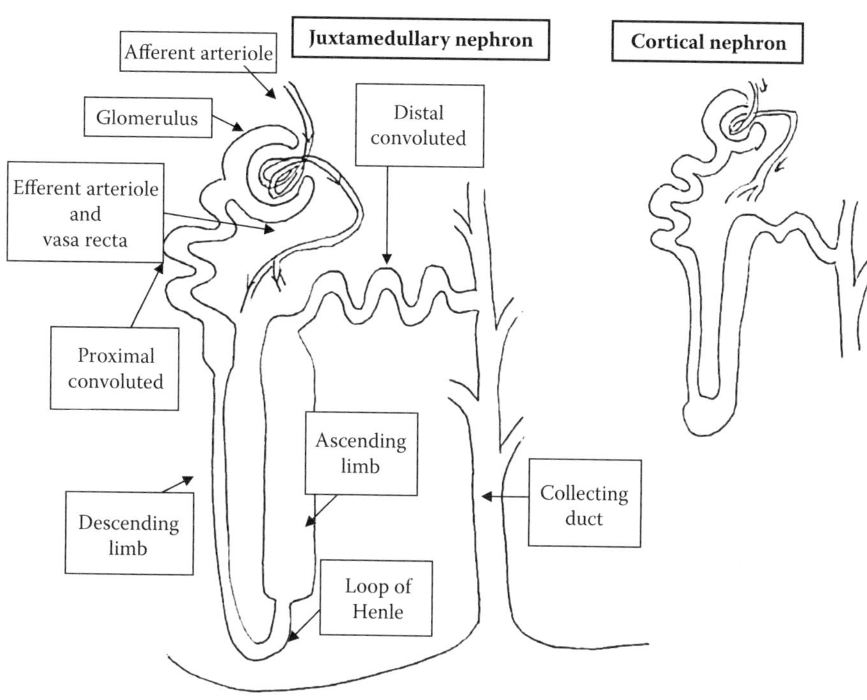

FIGURE 13.3
Shows representative nephrons of the two populations: juxtamedullary and cortical. Note also the thick portion of the ALLH.

terms π_g and π_c represent OPs in the glomerulus and capsule with values of 30 mmHg and 0 mmHg, respectively. These latter values require some explanation. First, π_c is nonzero in certain diseases (such as proteinuria, where protein appears in the urine), and second, the glomerular OP is determined by those molecules unable to pass easily through the gap junctions in the glomerular capillary endothelium, mainly albumin in the plasma. At 30 mmHg, this is far less than the 5800 mmHg noted above to be the osmotic equivalent of the NaCl in the ultrafiltrate. This is because Na^+ and Cl^- can easily permeate the glomerular capillary walls but not the tubular epithelium further down in the nephron. Thus, the plasma protein concentration is the main determinant of π_g. Finally, p_g varies according to the relative diameters of the afferent and efferent arterioles, which both have smooth muscle sphincters under the autonomic nervous system control.

$$\Delta p = (60 - 15) - (30 - 0) = 15 \text{ mmHg}.$$

Note that if arterial pressure (and hence p_g, which would normally be a little lower) falls below 55 mmHg, renal filtration and, hence, urine flow cease.

After the Bowman's capsule (see Figure 13.3), the filtered plasma (known as ultrafiltrate) passes in turn along the proximal convoluted tubule (PCT), the descending then the ascending limb (of the loop of Henle—the hairpin), thence via the distal convoluted tubule (DCT) to the collecting duct (CD). This last part (CD) receives fluid from many individual

DCTs, which then passes into the renal pelvis, then via the ureters to the bladder. The convolutions of the PCT and DCT may look random, but there is a close association between the afferent and efferent arterioles of the glomerulus and the DCT, which will be elaborated on later. The hairpin arrangements come in two types: juxtamedullary and cortical—the first extends down almost as far as the renal pelvis, but the second is much shorter. The latter also have a capillary network, which lacks the flow regulation of the arterioles in the VR. This regulation does allow the proportion of blood flowing to the two types of nephrons to vary.

The glomerular capillaries form a fine network with a large surface area in contact with the epithelial layer of the Bowman's capsule, allowing efficient fluid exchange (Figure 13.4). The endothelial cells lining these capillaries have fenestrations ("windows") where patches of membrane from the opposite walls of the cell coalesce, allowing high permeability to lipid-soluble substances. There also are gap junctions between adjacent cells, these being sufficiently "tight" to prevent small proteins, such as albumin, to cross into the Bowman's capsule. The epithelial layer, which is separated from the endothelial cells by a thin supporting basement membrane, is characterized by interdigitating processes from adjacent cells with narrow (around 80 nm) gaps between them. Note that the diffusion distance from the capillary wall to the Bowman's capsule is of the order of a few micrometers. There is an overall negative charge, which further inhibits proteins from leaving the capillaries. On the other hand, water plus ions can filter through at around 125 mL/min, making the value for K to be around 8 mL/min/mmHg. Note that most of this is reabsorbed during passage into the CD, where the final flow rate could be as low as 0.1 mL/min (but is normally nearer 1 mL/min) (Figure 13.2).

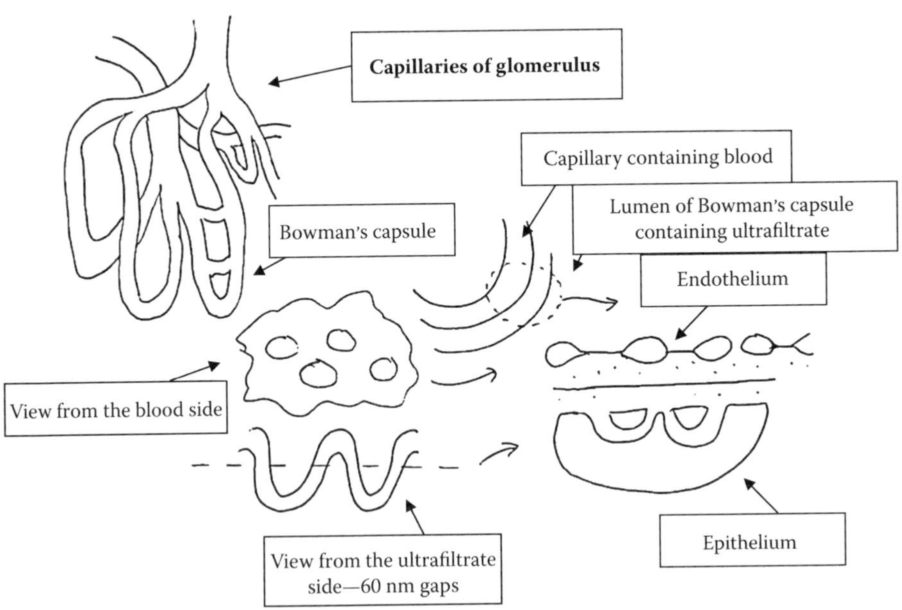

FIGURE 13.4
The junction between capsular endothelial lining of blood capillaries and the epithelial lining of the tubule. Lower left shows the view of the face of these layers, and lower right shows a cross-sectional view of the two layers.

13.2 Kidney Function Tests: Formulae

Because the ultrafiltrate collected in the renal pelvis is not modified on its passage to the bladder and then on micturition (voiding of the bladder), quite a lot can be inferred about kidney function by collecting urine over a particular time (e.g., 1 h) and then taking a blood sample. The volume voided in this time, together with the concentrations of test substance x in both the blood plasma and the urine, provides the basis for most kidney function tests. Fundamental to these is the concept of *clearance*, which will now be discussed.

13.2.1 Renal Clearance

For a test substance *x*, the renal clearance C_x is calculated as follows:

$$C_x = \frac{\dot{V}U_x}{P_x}$$

C_x has units of flow, such as mL/min, the same as "V dot," the urine flow rate.

P_x is the concentration of x in plasma, and U_x is the concentration of x in urine, in the same units (e.g., g/mL). This can be interpreted by studying Figure 13.5. In essence, the amount of x, which has appeared in the urine at the end of the collecting period (e.g., 1 min) will be $\dot{V}U_x$, and this originally will have been in the blood plasma. If we represent the flow rate (in mL/min) of plasma through the kidneys as Q, then the original amount of x in the plasma entering the kidney in 1 min would have been QP_x. The amount leaving will be $QP_x - \dot{V}U_x$. Another way to look at this is to imagine the volume leaving in 1 min to be made up of two compartments, one having been completely cleared of x and the other still having x at its original concentration, P_x. The volume per minute completely cleared is C_x (= $\dot{V}U_x/P_x$, as above), and the remaining portion is $Q - C_x$ (this can be obtained by

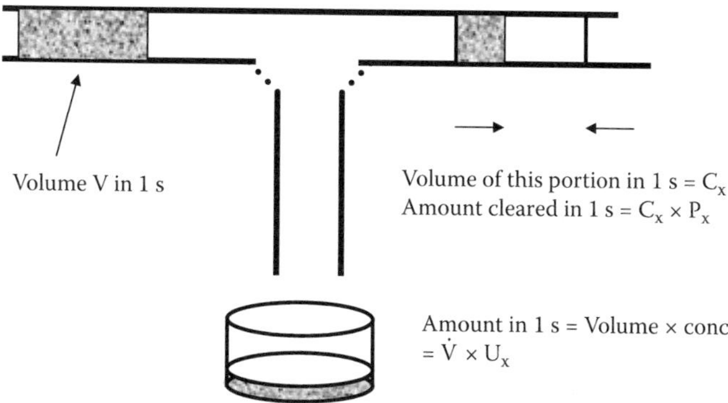

FIGURE 13.5
Renal clearance. The shaded volume V on the left (considered to be within the renal arteries) moves through the kidney (represented by the t-junction) in 1 s. Some of the material appears in the urine. Conceptually, the volume V on the side leaving the kidney (via the renal veins) has two portions: a volume C_x, which has been completely cleared of the substance in question, and a volume $V - C_x$, where the substance is still at its original concentration. Of course, in actuality, the substance in the renal veins has a concentration $P_x(V - C_x)/V$. Usually, clearance values C_x are given in units of mL/min because these give convenient values to remember.

dividing the amount-leaving equation above by P_x). Of course, the concept of a completely cleared portion is completely artificial, but it does provide a simple way of judging how a particular substance is eliminated by the kidney. Values range from <0.1 mL/min for amino acids to 600 mL/min for PAH (which was mentioned in the introduction). If C_x is close to Q, this means that most of the test substance x is eliminated; in other words, the cleared volume is the same as the volume of plasma flowing through per minute. If the value of C_x is close to zero, this means that none of the substance appears in the urine; either none was allowed across to the Bowman's capsule or, if it was, all of it was subsequently reabsorbed. Figure 13.2 reminds us that, although around 20% of the plasma was initially filtered into the tubule, most of that was returned to the blood supply.

So, in summary, this clearance volume per unit time is virtual rather than real because it represents *a portion of plasma volume from which the substance x has been cleared altogether* (the remaining portion having x in its original concentration).

13.2.2 Glomerular Filtration Rate

This refers to the total flow of ultrafiltrate out of the blood vessels of the glomerulus and into the PCT via the Bowman's capsule, that is, that part of the renal plasma flow separated into the renal tubules. A substance x, which is freely filtered but is not secreted or reabsorbed, will estimate GFR. There are two main candidates, *inulin* (which has to be infused) and *creatinine* (which is produced naturally). The latter is the more common method, although the former is more accurate. Thus,

$$\text{GFR} = C_{\text{Creatinine}}.$$

The normal value is around 120 mL/min–125 mL/min; it varies slightly with body mass or body surface area.

13.2.3 Effective Renal Plasma Flow

Measuring C_x for a substance, which is actively and efficiently secreted into the ultrafiltrate, such that the clearance is effectively 100%, can be used to estimate the plasma flow rate (Q). As we have seen, the most effectively cleared substance known is PAH, which again is naturally occurring. Thus,

$$\text{ERPF} = C_{\text{PAH}} = Q.$$

13.2.4 Tubular Reabsorption and Secretion

Once GFR is estimated, the contribution of filtration (as opposed to reabsorption of secretion) to a given clearance value can be calculated. If a substance is being secreted into the tubules, C_x – GFR will be positive and will represent the amount of clearance attributable to *secretion*.

$$P_x(C_x - \text{GFR}) = U_x\dot{V} - P_x\text{GFR}$$

represents, in g/min, the rate of secretion (e.g., x could be PAH).

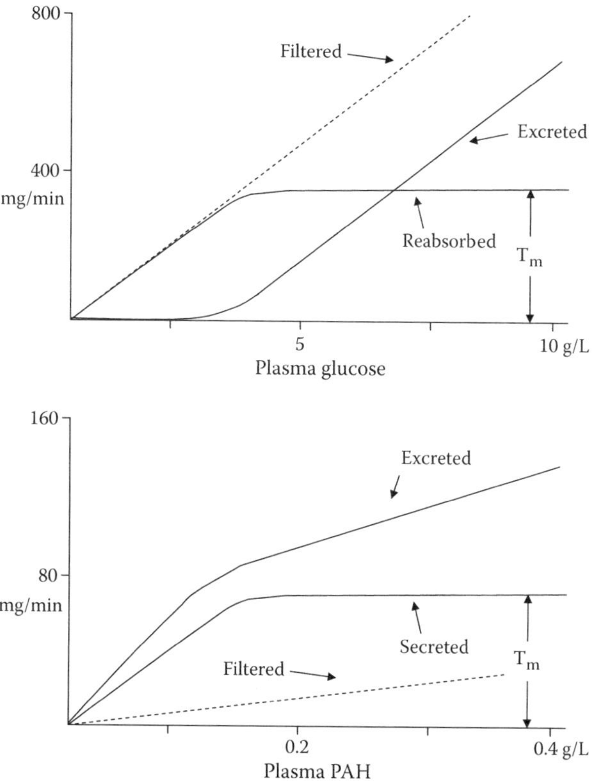

FIGURE 13.6
Top. Variation of tubular reabsorption rate with concentration of the material (in this case, glucose) in plasma. Note the rate excreted is the difference between filtered and reabsorbed. Lower: renal excretion of PAH is a combination of active secretion and filtration. Note that both the reabsorption and secretion reach a plateau above a particular level of plasma glucose or PAH.

Similarly, if reabsorption is occurring, GFR > C_x, or

$$P_x GFR - U_x \dot{V}$$

is the rate of *reabsorption* (Figure 13.6).

13.2.5 Maximal Tubular Reabsorption and Secretion

Raising the amount of substance x in the plasma will not necessarily lead to more efficient reabsorption and secretion compared with lower levels. Both reabsorption and secretion mechanisms show the phenomenon of *saturation*; that is, at high values of P_x, no further enhancements in either process are obtained. For reabsorption, increasing amounts are excreted (Figure 13.6, upper), and the C_x value rises toward the GFR value. For secretion, increasing amounts are retained (Figure 13.6, lower), and the C_x value falls toward the GFR value. Both mechanisms involve membrane carriers, the availability of which controls the overall rate. In Figure 13.6, T_m represents the maximal rate of reabsorption or secretion.

13.2.6 Renal Plasma and Blood Flow

As mentioned above, PAH clearance (C_{PAH}) is a convenient way of estimating effective renal plasma flow (ERPF). This gives a normal value of around 600 mL/min. To measure the whole blood flow (i.e., the flow rate of plasma plus red blood cells) the hematocrit (Hct) needs to be measured. This allows the *effective renal blood flow (ERBF)* to be estimated:

$$ERBF = C_{PAH}(100/(100 - Hct)).$$

Sometimes, a more accurate estimate of blood flow is required. In this case, the tiny PAH concentration in the renal vein $[PAH]_{venous}$ also is measured and the flow is estimated from the *Fick* principle.

$$\text{Total renal plasma flow} = [PAH]_{arterial}C_{PAH}/([PAH]_{arterial} - [PAH]_{venous}).$$

13.2.7 Filtration Fraction

This is the ratio of filtration to effective plasma flow; thus,

$$\text{Filtration fraction (FF)} = GRF/ERPF = C_{creatinine}/C_{PAH}.$$

The normal value is around 0.2. It represents the fraction of plasma volume, which enters the kidney tubules via the glomerulus/Bowman's capsule. In arterial hypertension, ERPF decreases (because of increased afferent arteriolar resistance), so FF increases; in glomerulonephritis (inflammation of the glomerular tissue), GFR decreases, so FF decreases.

13.2.8 Osmolar Clearance and Free Water Clearance

This applies clearance concepts applied to water itself (i.e., x = water). Osmolar clearance is a simple application of the clearance formula:

$$C_{osm} = U_{osm}\dot{V}/P_{osm}$$

where U_{osm} and P_{osm} refer to the OP of urine and plasma, respectively.

Free water clearance (FWC) is similar to the secretion and reabsorption formulae above:

$$FWC = \dot{V} - C_{osm} \text{ (mL/min)}.$$

If positive, this indicates diuresis or dilution; if negative, dehydration or concentration. If, we use GFR instead of C_{osm} in the above formula, we can estimate the rate of tubular reabsorption of water (GFR will always be greater than $\dot{V}$, which is in the range 0.1 mL/min–10 mL/min)

$$T_{H2O} = GFR - \dot{V}.$$

In fact, in most circumstances, over 99% of filtered water is reabsorbed.

WORKED EXAMPLE

The following data relates to plasma and urine samples obtained from a patient:

	Plasma	Urine
Creatinine (g/L)	0.011	0.89
Para-amino hippuric acid (g/L)	0.009	3.1
Glucose (g/L)	3.2	13.1
Osmolarity (mOsm/L)	295	385
Urine volume in 1 h (mL)		85

Estimate the following:

1. GFR
2. FF
3. Tubular reabsorption rate for glucose
4. FWC, explaining what these tell us about the state of this patient.

Answer

1. Using the formula $C_x = U_x\dot{V}/P_x$, we get, for creatinine, 0.89 × (95/60)/0.011 = 115 mL/min, which is an estimate of GFR and is within the normal range.
2. Using the same equation for PAH, we get 3.1 × (85/60)/0.009 = 488 mL/min, which is the ERPF; FF is GFR/ERPF = 0.24, which is rather high (and is a consequence of the low ERPF).
3. Tubular reabsorption is given by the equation $T_R = GFRP_x - U_x\dot{V}$ = 115 × 3.2 − 13.1 × (85/60) = 349 g/min (note units), which is close to the maximal reabsorption rate for glucose.
4. Finally, FWC is given by $C_W = \dot{V} - (U_o\dot{V}/P_o)$ = 85/60 − (385/295) × (85/60) = −0.43 mL/min. A negative value indicates that the kidney is concentrating urine.

13.3 Electrolyte Balance: Countercurrent Multiplication in the Nephron

The loop of Henle and the VR system form separate *countercurrent exchange* systems (see Figure 13.7). The essential feature is that the thick *ascending limb of the loop of Henle (ALLH)* causes reabsorption of NaCl from the tubule lumen *without* water accompanying. The reabsorption of NaCl is via a *Na-K-2Cl symporter,* which actively transports 1 each of Na^+ and K^+, together with 2 Cl^- ions, all from the lumen into the tubular epithelial cell (this is an electroneutral pump). Because of the extraction of NaCl, the OP (given in mOsm/L) progressively *falls* (as one goes up the ALLH) because the water is left in the tubule. The NaCl, as it gets reabsorbed, goes first into the surrounding interstitial fluid and then into the VR system (and hence back into the bloodstream). Some of this NaCl diffuses back into the *descending limb of the loop of Henle (DLLH)* where it joins the flow of ultrafiltrate down to the actual loop. The NaCl that is not convected away by the bloodstream is thus recycled

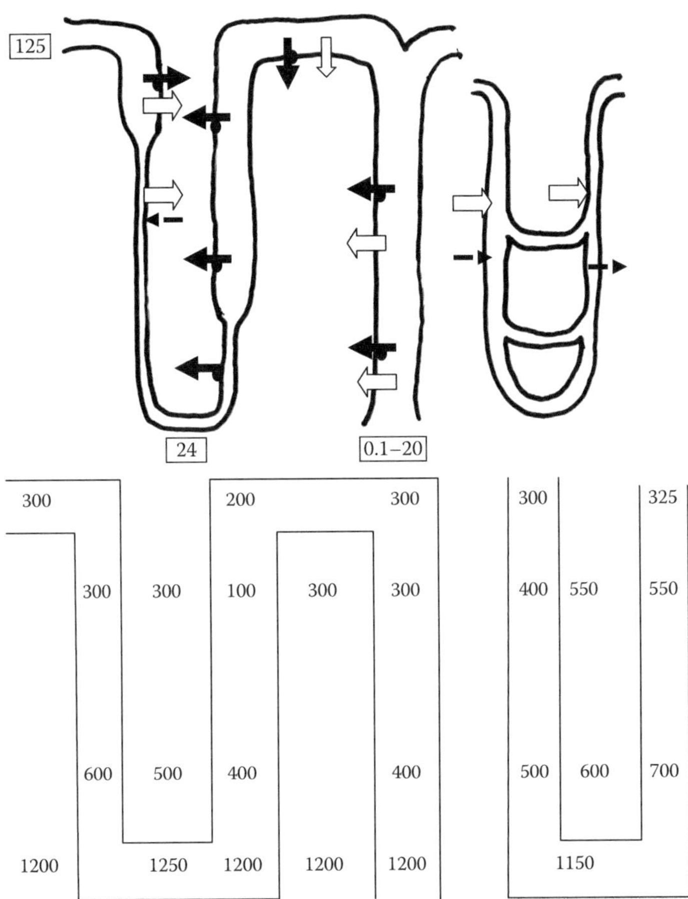

FIGURE 13.7
Countercurrent multiplication of NaCl in the nephron. Top: full dark arrows indicate active Na^+ and/or Cl^- pumping. Dotted arrows: passive Na^+ movement. Open arrows: water movement. Lower: indicative osmolarity values in mOsm/L. The nephron and the VR system shown diagrammatically on left and right, respectively. The numbers in boxes represent flow rate in mL/min.

back to the loop where the concentration rises to high levels (because the NaCl has nowhere else to go). The term countercurrent multiplication indicates that because of this recycling of NaCl, the concentration becomes multiplied, that is, there is a process of accumulation. The longer the limbs of the loops of Henle, the higher will be the concentration at the bend of the loop. In fact, certain desert rodents have exceptionally long limbs of the loop of Henle, and the loop OP can be as high as 2200 mOsm/L. This process can easily be modeled mathematically, as will be shown at the end of this chapter. The process also can be understood from the simple simulation shown in Figure 13.8. For simplicity, this allows the Na^+ pumping and flow into the hairpin to occur separately. After a few repeats of the "pump then flow" cycle, the concentration (and hence OP) is seen to accumulate in the bend. The interstitial fluid and the VR system will have similar OP values at similar positions in neighboring regions to the limbs of the loop. Note that, if anything, the OP of the ultrafiltrate is less when leaving the hairpin system than when it arrived. Diluted urine, with low OP and at a flow rate of 20 mL/min, can be seen as a default state: normally,

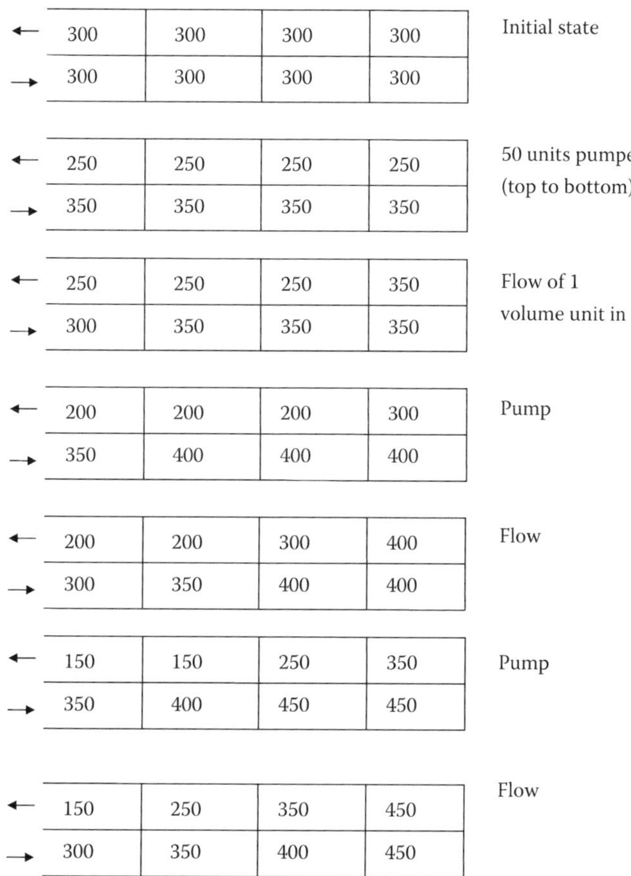

FIGURE 13.8
A simple simulation of countercurrent multiplication, assuming that volume flow and Na^+ pumping take place in separate steps.

the action of hormones, particularly the antidiuretic hormone (ADH) will cause the OP to be higher and the flow rate lower in the collecting duct (CD).

In the *proximal (convoluted) tubule*, the reabsorption of NaCl is *isotonic*, that is, water follows the NaCl, the OP stays the same, and the volume flow rate falls. In the distal convoluted tubule (*DCT*), the active transport of ions is important. In the first (or early) section, Na^+ is transported together with Cl^-, producing small potential differences, but later on, the Cl^- movement is passive, giving rise to much larger potential differences from the Na^+/K ATPase pump, which in turn drives the passive reabsorption of other ions. The *juxtaglomerular apparatus (JGA)* detects the ionic composition of the ultrafiltrate and feeds signals back to the resistance vessels (arterioles) via renin and, hence, angiotensin levels (see Figure 13.9). The *CD* has a varying permeability to water (the permeability is controlled by ADH). When the permeability is large, the CD fluid OP approaches the values in the interstitial fluid, that is, as high as 1200 mOsm/L. When the permeability is low, as mentioned above, it has a similar OP to the distal tubule or can, in fact, be lower because further absorption of NaCl can occur.

Figure 13.7 is a summary of the OP and of the Na^+ and H_2O movements in the kidney tubule (left) and VR (right). Active transport is shown with full arrows, and passive

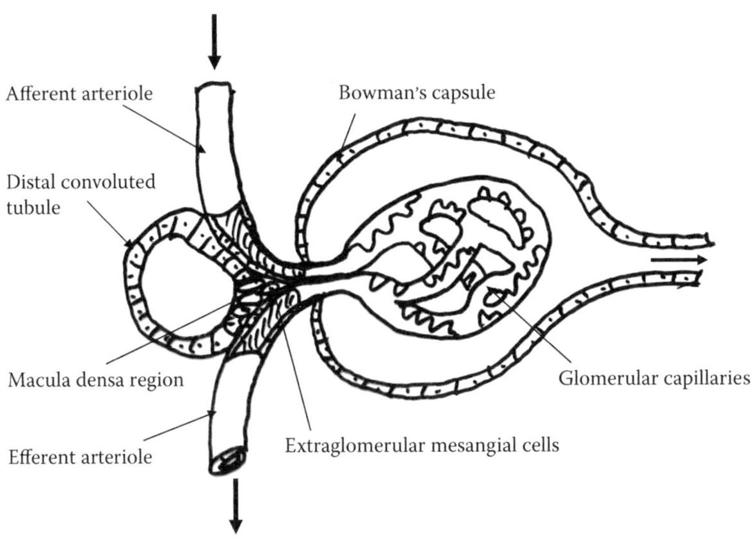

FIGURE 13.9
The JGA, showing the macula densa region between the DCT and the afferent and efferent arterioles.

movement of Na^+ and water is shown by dotted and open arrows, respectively. For simplicity, the active and passive movements of other ions and of urea are not shown. The numbers in the boxes indicate flow rate in mL/min (see also Figure 13.2).

13.3.1 Experimental Evidence for the Information Shown in Figure 13.7

It is instructive to briefly review some of the experimental evidence, which has allowed the understanding of kidney function at the tubular level. Some of the earlier evidence was inferred from large volumes of tissue, but more recent experiments have used microelectrodes and micropipettes to study mechanisms at the cellular level.

1. *Freezing Point Depression:* The amount by which the freezing point of pure water falls below 0°C can be related to salt content and, hence, to OP. The rat kidney is unusual in that it only has a single lobe, so when sliced into thin slices using a tissue slicer as shown in Figure 13.10a, the sequential slices, initially from the cortex, are deeper into the medulla. As the slices are taken further and further away from the cortex, the freezing point depression (and, hence, the osmolarity) gets larger.
2. *Microperfusion Techniques:* There are several versions of this technique. They use micropipettes into an individual tubule to introduce oil (to isolate a particular region), followed by test perfusion fluid. A particular arrangement is shown in Figure 13.10b. Certain amphibia (such as the newt) have tubules very close to the outer kidney surface, and thus, the tubules are accessible to the introduction of micropipettes. This fluid can then be recovered later and analyzed for solute content (to measure rates of secretion or reabsorption). Changes in the length of the aqueous test solution "bubble" indicates water movement across the epithelial wall.

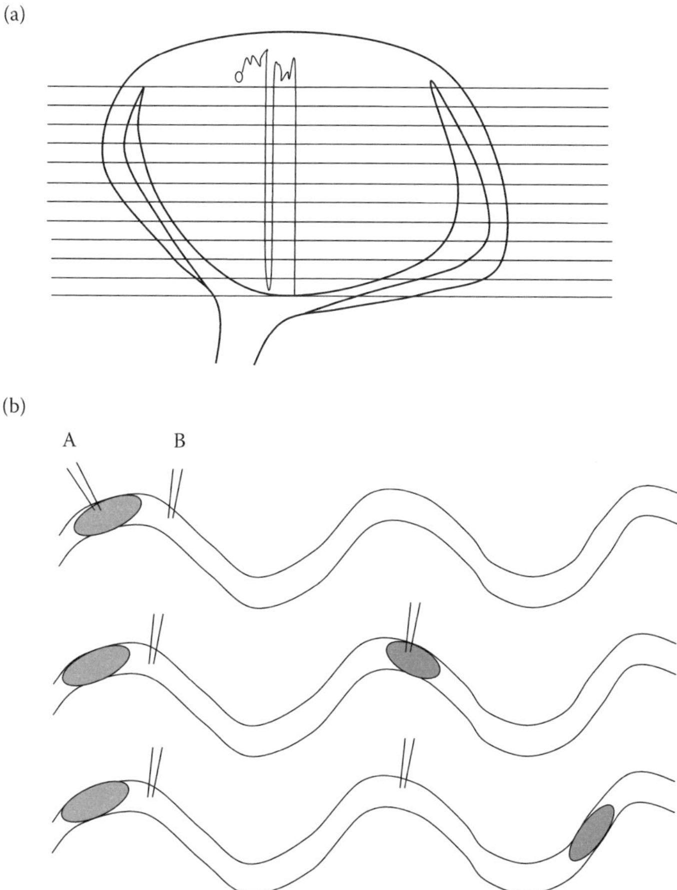

FIGURE 13.10
Analytical methods for identifying renal electrolyte dynamics. (a) Freezing point depression in rat kidney slices, identifying increasing [NaCl] the further away from the glomeruli (one of which is shown as a small circle) the slice is taken. (b) Stop flow technique, showing two micropipettes A and B inserted into the lumen of the nephron. An oil drop is introduced by A, and then, the pipettes move to produce an isolated perfused section of the tubule.

3. *Microelectrodes:* As described in Chapter 3, glass microelectrodes can be introduced into individual tubular cells to measure transmembrane potentials. It is possible to isolate individual nephrons, which can then be immobilized to allow impaling of individual cells. The effects of hormones and drugs, which act on the kidney can then be studied. As mentioned, sheets of epithelial tissue from amphibia (skin and urinary bladder) have active transport properties, which mimic, to a certain extent, the processes in the mammalian kidney tubules.
4. *Radioactive Tracers:* To measure the flux rate of individual components (Na^+, K^+, urea, etc.), radioactive forms of these components (such as ^{24}Na, ^{42}K, and ^{12}C-urea) have been added to test solutions within the tubules to "label" the movement of those components. The flux rate can be inferred from the clearance rate from the test solution or the rate of appearance in previously unlabeled tissue.

13.4 Control of Kidney Function

There are two major ways in which the kidney function is controlled: neural and hormonal. Neural control largely springs from the sympathetic nervous system acting to regulate the blood flow within the kidney and the distribution of blood flow to particular regions of the kidney. The afferent and efferent arterioles are the major target for this control. These also are affected by particular hormones, as described below.

13.4.1 Hormones Affecting the Kidney

1. Antidiuretic Hormone or Vasopressin

   Where does it come from?

   It is secreted from neurohypophyseal region of the pituitary (see Chapter 15).

   What does it do in the kidney?

   It acts on CDs to increase water reabsorption (hence, antidiuresis) by making the CD epithelial layer more permeable to H_2O.

   What causes it to be released?

   If plasma OP is too high (i.e., too salty), osmoreceptors in the hypothalamus of the brain (see Chapter 15) signal to the neurohypophysis to release ADH, hence, water is retained to return OP to normality.

   What is its structure?

   It is a nonapeptide (i.e., it is a polypeptide made from nine amino acids, as shown in Figure 13.11.

   Antidiuretic hormone is related to the hormone related to oxytocin, which causes uterine contractions and lactation during birth. The oxytocin molecule has Phe* replaced by Ileu and Arg* by Leu.

   The disease diabetes insipidus (which is not the same as diabetes mellitus) is due to an impairment of the pituitary function (perhaps due to tumor) causing reduced or no ADH to be released. Its main symptom is the excretion of 20 L/day (14 mL/min) of urine and, in consequence, a constant thirst experienced by the sufferer. It is treated with a synthetic form of ADH.

2. Aldosterone

   Where does it come from?

   It is secreted from the adrenal cortex (hence, *cortico*steroid and the name of the class that aldosterone belongs to— mineralo*corticoi*ds).

   What does it do in the kidney?

```
Glu(NH2) – Asp(NH2)– Cys – Pro – Arg
   |           |       |          |
  Phe    –    Tyr  –  Cys      Gly(NH2)
```

FIGURE 13.11
The structure of ADH. For the meaning of peptide abbreviations, see Chapter 4.

It acts on the distal tubule cells where it stimulates NaCl active transport (by stimulating specific RNA production) out of the tubular lumen and into the interstitial space. Water is then reabsorbed as a consequence (as the interstitial OP increases).

What causes it to be released?

The adrenal cortex, when stimulated by angiotensin II (see below), releases aldosterone.

What is its structure?

It has a typical steroidal structure (and is related to cholesterol and cortisol) as shown in Figure 13.12.

3. Angiotensin II

Where does it come from?

There are several stages to this process. The enzyme *renin* is secreted by granular cells in the macula densa region of the JGA (see Figure 13.9). These cells are next to the afferent arteriole and are affected by the fluid in the nearby DCT. Renin, which enters the bloodstream in the afferent arteriole, converts *angiotensinogen* into *angiotensin I*, which is physiologically inactive. Angiotensin I is converted to *Angiotensin II* in the lungs via another converting enzyme.

What does it do in the kidney?

It does two things: (1) it constricts blood vessels, raising the blood pressure (BP), and (2) it stimulates aldosterone production (and hence NaCl and water retention, which also raises BP via the increased blood volume caused as a consequence by water retention).

What causes it to be released?

Again, there are two things: (1) reduced blood flow in the afferent arterioles and (2) reduced OP in the DCT. This second mechanism is a direct negative feedback loop via the JGA. Reduced OP in the DCT indicates the ultrafiltrate to be possibly too diluted, so the macula cells are stimulated to secrete more renin, eventually leading to the stimulation of aldosterone via angiotensin II and, hence, to water retention.

What is its structure?

It is an octopeptide but without cross linkage; that is,

H–Asp–Arg–Val–Tyr–Ileu–His–Pro–Phe–OH.

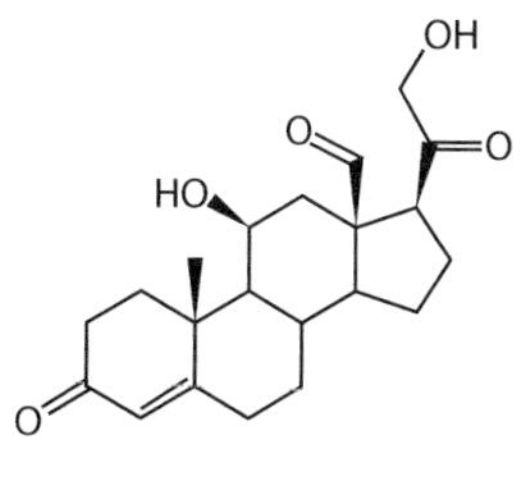

OH
O
O
HO
O

FIGURE 13.12
Structure of aldosterone, showing the characteristic steroid structure.

4. Atrial Natriuretic Peptide

   Where does it come from?

   It is secreted from the atrial endothelium of the heart.

   What does it do in the kidney?

   Dilates the afferent and constricts the efferent renal arterioles, hence causing GFR to increase via increased filtration pressure. It also causes greater blood flow through the VR and inhibition of sodium pumping in the DCT.

   What causes it to be released?

   It is released in response to increased stretching in the atria due to raised blood volume. Again, this is an example of a negative feedback loop in that the response to raised blood volume is increased GFR and reduced water reuptake, leading to greater excretion of water and a consequent lowering of the blood volume.

   What is its structure?

   It is a sequence of 28 amino acids with a disulphide bone between two cysteine residues at positions 7 and 23, giving rise to a ring-like structure.

13.5 Renal Pathophysiology and Therapeutic Drugs

13.5.1 Edema

This is the name given to the accumulation of excessive interstitial fluid. It can be localized or generalized. Cells lose their rounded appearance as the surrounding fluid volume increases (shown diagrammatically and somewhat exaggeratedly in Figure 13.13). Areas of the skin appear swollen as a consequence of this accumulation of tissue water. Localized edema does not involve the kidney and is due to inflammation (increasing capillary permeability) or venous blockage (as in glaucoma with raised ocular pressure). Generalized edema, on the other hand, often has strong links to the kidney. In particular, in kidney disease, generally described as *nephrotic syndrome* (characterized by proteinuria), there often is damage to the glomerular capillaries. This leads to a rise in tubular OP, so GFR falls, leading to water retention. Second, congestive heart failure gives rise to reduced blood

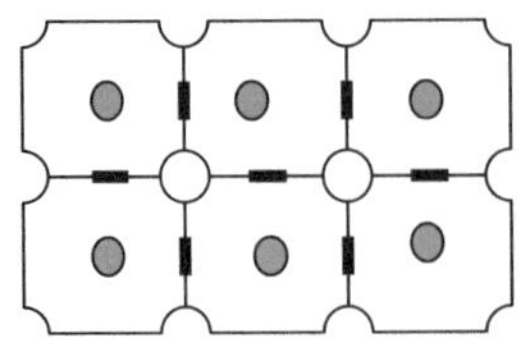

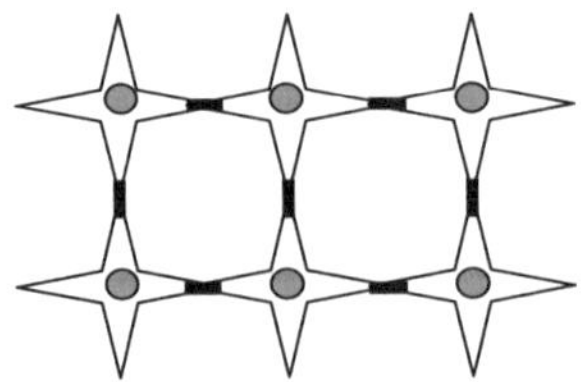

FIGURE 13.13
A depiction of a layer of healthy cells (left) compared with cells in the edematous tissue (right).

pressure due to falling cardiac ouput (due to myocardial insufficiency), so consequently, aldosterone is secreted in response to low juxtaglomerular pressure. Aldosterone then causes water retention via stimulation of NaCl reabsorption. The treatment of edema is via the administration of *diuretics*—the name given to drugs, which increase urine flow rate. These are of several different types.

1. *Pump inhibitors:* These slow the rate of NaCl reabsorption (and hence water uptake) from the tubules, so more water is excreted. They inhibit particular aspects of active ion transport. Some are Na^+ transport inhibitors (such as furosemide or frusemide, commonly sold under the name *Lasix*). This is a very common form of treatment for edema. It acts on the ALLH (and for this reason, often is called a loop diuretic) and specifically the Na-K-2Cl symporter. Other forms of loop diuretics include edacrynic or ethacrynic acid (*Edecril/Edecrin*) and "mercurials" (*Mersalyl*). The problem with loop diuretics is that they tend to cause K^+ loss because of the inhibition of the K^+ uptake by the symporter.
2. *Aldosterone antagonists:* These are equivalent to pump inhibitors in the slowing of NaCl uptake, in that aldosterone-stimulated Na^+ transport is inhibited, but these are more specific to Na^+ and are thus "K^+-sparing." There are two varieties: those which prevent Na^+ entry into the DCT epithelial cells, such as amiloride (*Amizide*), and those which displace aldosterone from binding sites in the same cells, such as spironolactone (*Aldactone*), which is an inactive aldosterone analog.
3. *Carbonic anhydrase inhibitors:* The enzyme carbonic anhydrase is present in tubular cells, and it catalyzes the formation of carbonic acid, which rapidly dissociates to form hydrogen and bicarbonate ions as follows:

$$CO_2 + H_2O \rightarrow H_2CO_3 \rightarrow H^+ + HCO_3^-.$$

   HCO_3^- acts as a companion ion to Na^+, so inhibiting its production inhibits the ability of the cell to carry out Na^+ pumping (from the cell into the interstitial fluid). As in 1 and 2, this leads to water loss.
4. *Osmotic diuretics:* The intravenous infusion of a complex sugar (mannitol) to a plasma level where the tubular reabsorptive capacity is exceeded (see Figure 13.6) causes raised tubular OP and a smaller than normal OP gradient between the tubule and the interstitial fluid. This inhibits the rate of water reabsorption, and hence, water is lost.

13.5.2 Kidney Disease

This can be divided into acute and chronic forms, the latter referring to conditions, which fail to respond to treatment over a period of weeks or months. The causes of acute renal failure can be as follows: (1) postrenal (i.e., between the kidney and the urethra), for example, blockage of the ureter by kidney stones; (2) prerenal (due to failure of other systems, such as heart failure or blood loss or where arterial pressure falls below 40 mmHg); or (3) intrarenal (i.e., within the kidney itself). Intrarenal disease can arise from several causes, including (1) ischemic injury (in which there is lack of regional or total blood flow for around 40 min); and (2) toxic injury (resulting from an overdose of certain antibiotics or organic compounds, such as CCl_4). A proportion of cases of intrarenal disease progresses to chronic renal failure. In these cases, the treatment options are as follows: (1) transplant, which is not always possible if a suitable donor is unavailable, or (2) kidney dialysis. The latter option involves bioengineering aspects, which will be considered in the next section.

13.6 Kidney Dialysis

The patient is required to visit the kidney dialysis center for three nights per week, with the dialysis carried out over several hours. An external arteriovenous shunt is created in the arm (in which an artery and a vein are surgically brought to the skin surface and a short tube is inserted to connect the two together when not undergoing dialysis). During dialysis, the tube is removed, and the artery and vein are connected to the "blood in" (b_0) and "blood out" (b_L) ports of the dialyzer (Figure 13.14). The other ports are "dialysis fluid in" (d_L) and "dialysis fluid out" (d_0). The blood and dialysis fluids are drawn through the dialyzer by rotary pumps.

The dialysis fluid contains normal amounts of things needed by the body and zero amounts of accumulated metabolites, such as urea and creatinine, which need to be eliminated. Table 13.1 shows a comparison of the major constituents of dialysis fluid compared with normal plasma and the plasma of an individual with chronic renal failure (uremic plasma). Dialysis fluid needs highly purified water as its solvent because any traces of material, such as aluminum, are able to be accumulated over the many hours and days of the dialysis. Typically, several hundred liters of dialysis fluid are used per session. The purified water is usually produced on-site by reverse osmosis with additional purification using ion exchange resins. The "dialysis fluid out" is discarded.

As Figure 13.14 shows, dialyzers are basically countercurrent exchangers with a cellophane-type membrane to facilitate exchange. Although modern dialysis units consist of hollow capillaries (containing the blood) surrounded by the dialysis fluid flowing in the opposite direction (a countercurrent system), it is easier to analyze the system mathematically by assuming the membrane to be a single rectangular sheet, L units long and w units wide. The detailed derivation is given at the end of this section, but to not break the continuity, the result is stated below. For a countercurrent exchange system, the concentration

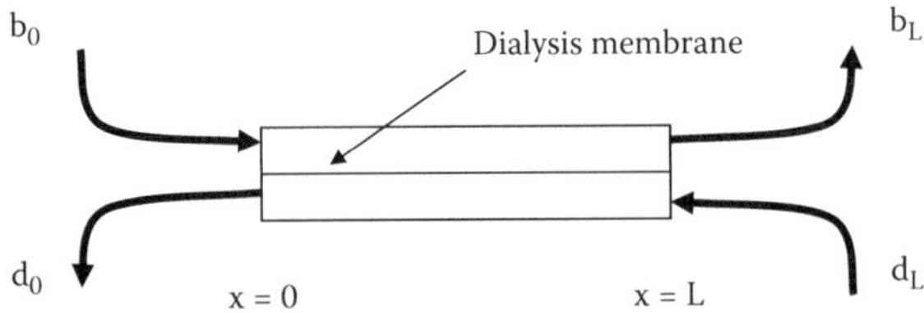

FIGURE 13.14
A diagrammatic representation of a kidney dialyzer. b_0 and b_L indicate blood concentrations on the inlet and outlet, respectively, and d_L and d_0 indicate the inlet and outlet dialysis fluid concentrations, respectively.

TABLE 13.1

Comparison of Constituent Concentrations between Normal Plasma, Kidney Dialysis Fluid, and the Plasma of a Patient Prior to Undergoing Dialysis

Constituent	Normal Plasma	Dialyzing Fluid	Uremic Plasma
Electrolytes (mM)			
Na^+	142	133	142
K^+	5	1	7
HCO_3^-	27	36	14
HPO_4^{2-}	1.5	0	4.5
$Urate^-$	0.3	0	2
SO_4^{2-}	0.3	0	1.5
Nonelectrolytes (mM)			
Glucose	5.6	6.9	5.6
Urea	4.3	0	33
Creatinine	0.09	0	0.53

Note: Concentrations of other constituents (Ca^{2+}, Mg^{2+}, Cl^-, and lactate) do not differ significantly between the three fluids.

of a particular substance at various points along the exchanger is given by the following (these are the starred equations in the derivation further on):

$$b(x) = b_0\,\{[\exp(-\alpha L)/Q_d - \exp(-\alpha x)/Q_b]/[\exp(-\alpha L)/Q_d - 1/Q_b]\}\text{ and}$$

$$d(x) = b_0\,\{[\exp(-\alpha L)/Q_d - \exp(-\alpha x)/Q_d]/[\exp(-\alpha L)/Q_d - 1/Q_b]\}$$

where $\alpha = Pw((1/Q_b) - (1/Q_d))$.

Here, P is the permeability of the membrane to the particular substance, w is the width of the membrane sheet (i.e., wL is its area), and Q_b and Q_d are the volume flow rates of blood and dialysate, respectively. In addition, b(x) and d(x) refer to the blood and dialysate concentrations at position x along the membrane sheet.

If $Q_b = Q_d$, we use different equations (because both numerator and denominator are tending to zero in the above equations):

$$b(x) = b_0((L - x) + Q/Pw)/[L + Q/Pw]$$

and

$$d(x) = b_0(L - x)/[L + Q/Pw].$$

13.6.1 Clearance

On analogy with the clearance formula we saw earlier, we can write the following:

$$C_x = Q_d d_0/b_0.$$

Because the quantity $Q_d d_0$ is the amount/unit time in the dialysis fluid, the amount lost/unit time by the blood will be

$$Q_b(b_0 - b_L).$$

So C_x also is given by the following:

$$Q_b(b_0 - b_L)/b_0.$$

Using the equations given above, we have

$$C_x = Q_b(1 - e^{-Z})/(1 - (Q_b/Q_d)e^{-Z}) \text{ if } Q_b \neq Q_d$$

where $Z = AP(1 - Q_b/Q_d)/Q_b$ and $A = wL$ the area of the membrane, or

$$C_x = Q/(1 + Q/(AP)), \text{ if } Q_b = Q_d = Q.$$

For a parallel flow (that is, not countercurrent) arrangement, this is as follows:

$$C_x = Q_b(1 - e^{-Y})/(1 + Q_b/Q_d),$$

where $Y = AP(1 + Q_b/Q_d)/Q_b$.

These expressions can be used to show that countercurrent arrangements are, on the whole, more efficient. If we take the following typical values, $Q_d/Q_b \cong 600$ mL/min ÷ 200 mL/min and $AP/Q_b \cong 3$ (dimensionless), we also can express clearance as a percentage; that is,

$$100(b_0 - b_L)/b_0 \text{ or } 100C_x/Q_b.$$

Substituting the typical values give efficiencies of 90.5% and 73.5% for countercurrent and parallel flow, respectively. Note that, in general, efficiency improves as Q_b is made slower and Q_d is made quicker.

13.6.2 Detailed Derivation

Consider the following pair of tubes, each of unit cross-sectional area and one unit wide, in steady state and in the absence of pressure gradients (Figure 13.15).

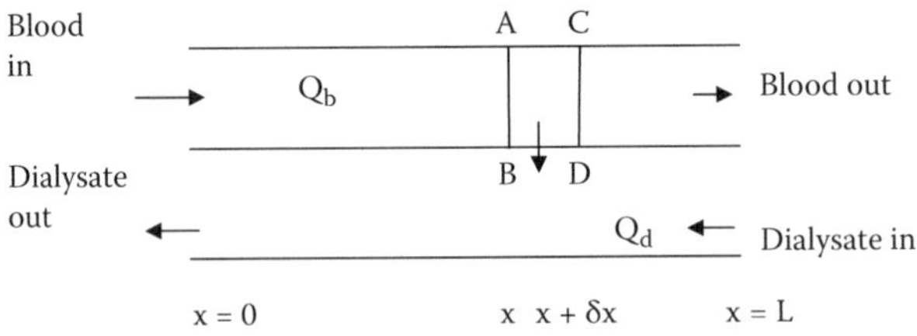

FIGURE 13.15
Derivation of equations to determine the concentration profiles within a kidney dialyzer. See text for explanation.

Consider concentration of a particular substance b(x) and d(x) at the point x in the blood and dialysate, respectively. From Fick's first law (see Chapter 5), the mass transferred across the common wall between x and x + δx in time δt is given by the following:

$$P[b(x) - d(x)]dx\delta t.$$

The mass entering across AB and leaving across CD in $\delta t = Q_b b(x)\delta t$ and $Q_b b(x + \delta x)dt$, respectively. In the steady state, increase in mass in element ABCD is zero, so

$$Q_b(x)\delta t - Q_b b(x + \delta x)\delta t - P(b(x) - d(x))dx\delta t = 0 \tag{13.1}$$

thus, $Q_b\ \delta b/\delta x = P(d - b)\delta x$ and, also, $-Q_d \delta d/\delta x = P(b - d)$.

Replacing incremental changes with infinitesimal changes (i.e., δ/δx by d/dx, etc.), dividing by Q_b, Q_d then adding the following:

$$d(b - d)/dx = -P((1/Q_b) - (1/Q_d))(b - d).$$

This is easily integrated, giving

$$b - d = k \exp(-\alpha x), \text{ where } \alpha = P((1/Q_b) - (1/Q_d)).$$

Now, substituting this into Equation 13.1,

$$db/dx = -Pk \exp(-\alpha x)/Q_b$$

and on integrating

$$b(x) = K + Pk \exp(-\alpha x)/(\alpha Q_b)$$

and from Equation 13.2

$$d(x) = K + Pk \exp(-\alpha x)/(\alpha Q_d).$$

The boundary conditions are as follows: at x = 0, $b(0) = b_0$, and at x = L, d(L) = 0 (i.e., there is zero concentration of the substance at the dialysate inlet; if this is not the case, then a different boundary condition would be applied). This gives the following:

$$b_0 = K + Pk/(\alpha Q_b) \text{ and } 0 = K + Pk \exp(-\alpha L)/(\alpha Q_d)$$

from these equations. The constants k and K can be eliminated to give *

$$b(x) = b_0 \{[\exp(-\alpha L)/Q_d - \exp(-\alpha x)/Q_b]/[\exp(-\alpha L)/Q_d - 1/Q_b]\}$$

and

$$d(x) = b_0 \{[\exp(-\alpha L)/Q_d - \exp(-\alpha x)/Q_d]/[\exp(-\alpha L)/Q_d - 1/Q_b]\}.$$

These equations are those we saw before, and they give the concentrations in the blood and dialysate as a function of distance x. If $Q_b = Q_d$, it can be shown that these equations represent a pair of parallel lines. Note that as $Q_b \to Q_d$, $\alpha \to 0$, but by differentiating the numerator and the denominator, it can be shown that

$$b = b_0((L - x) + Q/P)/[L + Q/P] \text{ and } d = b_0(L - x)/[L + Q/p],$$

and the difference $b - d$ is $b_0/(1 + LP/Q)$ (constant along the length of the dialyzer).

Note also, that if the substance is already present in the dialysate on entry, that is, at $x = L$, $d(L) = d'$ say, then the equations for b and d (marked * on the previous page) become

$$b = d' + (b_o - d')\{\text{as above}\} \text{ and } d = d' + (b_0 - d')\{\text{as above}\}.$$

13.6.3 Efficiency

The mass leaving the blood and entering the dialysate in unit time (and for unit width) is given by the following:

$$\int_0^L P\,(b - d)dx = -Q_b \int_0^L (db/dx)dx = Q_b(b_0 - b_L).$$

Clearance, which is mass removed per unit initial concentration (recall rationale behind formula $C = UV/P$), is thus $Q_b(b_o - b_L)/b_0$ mL/min (or other flow units).

An expression for b_L is given by the equations *, which, after rearranging, gives for $b_0 - b_L$

$$b_0\{[1 - \exp(-\alpha L)]/[1 - Q_b\exp(-\alpha L)/Q_d]\}$$

and for clearance,

$$Q_b\{[1 - \exp(-\alpha L)]/[1 - Q_b\exp(-\alpha L)/Q_d]\},$$

thus relating clearance (which can be measured experimentally) to α and L, which are, in turn, related to the area and permeability of the dialyzer material. In fact (and now generalizing to a membrane that is w units wide),

$$\alpha wL = (PwL/Q_b)\{1 - (Q_b/Q_d)\} = (PA/Q_b)\{1 - (Q_b/Q_d)\}.$$

For parallel flow (concurrent) arrangements, the equations for b and d become as follows:

$$b = d + (b_o - d')\exp(-\beta wx) \text{ where } \beta = P(1/Q_b + 1/Q_d)$$

$$d = d' + \{(b_o - d')\,Q_d\,(1 - \exp(-\beta wx))/(Q_b + Q_d)\};$$

this is for the more general case in which the dialysis fluid has a concentration d′ entry. These equations apply, of course, not only to the artificial kidney but also to any situation involving flows of material in the manner indicated and satisfying the initial assumptions. In particular, they can be applied to the maternal–fetal exchange system in the placenta and also, in modified form, to the kidney itself (see next section). If b and d are interpreted as temperatures and P is the thermal conductivity of the common wall of the tubes, the heat exchanger equations are obtained.

The following MATLAB code represents the concentrations (or temperatures) as a function of distance (x) for both countercurrent and parallel flow arrangements. The parameters can be altered to explore their effects. Enter 'dialysis' at the MATLAB® prompt.

```
function dialysis
%This code calculates the concentrations of a substance in the blood (b)
% and the dialisate(d) at various positions along a dialyser, assumed to
% be 20 cm long. The membrane area is 20 x 500 cm2 and the permeability
% 0.05 cm/min. The flow rates of blood and dialysate are 150 and 500 mL/min
% respectively. The first plot is for countercurrent, the second for
% parallel flow.

x = 0:1:20;

L = 20;
w = 500;
Qb = 150;
Qd = 500;
b0 = 100;
alpha = 0.05*(1/Qb - 1/Qd);
Z=exp(-alpha*w*L)/Qd - 1/Qb;
b = b0*(exp(-alpha*w*L)/Qd - exp(-alpha*w*x(:))/Qb)/Z;
d = b0*(exp(-alpha*w*L)/Qd - exp(-alpha*w*x(:))/Qd)/Z;

y(:,1)= b;
y(:,2)= d;

plot(x,y)

beta = 0.05*((1/Qb) + (1/Qd));

d1 = b0*Qd*(1 - exp(-beta*w*x(:)))/(Qb + Qd);
b1 = d1 + b0*exp(-beta*w*x(:));

y1(:,1) = b1;
y1(:,2) = d1;

figure
plot(x,y1)
```

WORKED EXAMPLE

How much cellophane (m^2) is required in a countercurrent dialyzer to produce a creatinine clearance of 130 mL/min if the blood flow is 100 mL/min and the dialysis fluid is 300 mL/min, given that the permeability of cellophane to creatinine is 0.022 cm/min.

Answer

We use the formula $C = (b_{in} - b_{out})Q_b/b_{in}$ to estimate clearance, which we need to be 130 mL/min. Using the following equation:

$$b_{in} - b_{out} = b_{in}\{(1 - \exp(-Z))/(1 - [Q_b\exp(-Z)/Q_d])\}, \text{ where } Z = AP\{1 - (Q_b/Q_d)\}/Q_b;$$

we can put $C = Q_b(1 - \exp(-Z))/(1 - (Q_b/Q_d)\exp(-Z))$. From the above, $Q_b/Q_d = 1/3$, and $Q_b = 100$ m/min, and we can write the following:

$$130 - 130(1/3)\exp(-Z) = 100 - 100\exp(-Z).$$

Collecting the following terms:

$$30 = (100 - 130(1/3))\exp(-Z)$$
$$\exp(-Z) = 0.692, \text{ so}$$
$$Z = 0.368 = (0.022) \times A \times (1 - (1/3))/100$$
$$A = 0.368 \times 100/((2/3) \times 0.022) = 2500 \text{ cm}^2 = 0.25 \text{ m}^2.$$

13.7 Countercurrent Flow of NaCl in the Nephron

13.7.1 Simple Theory

In the last section, the case where the flow is the same in both tubes ($Q_b = Q_d$) was considered. If concentration is plotted against distance x, we get the graph shown on the left in Figure 13.16, material diffusing from upper to lower curve. On the right, we see the situation in the kidney, where material (NaCl) is pumped from the ascending to the DLLH. The two tubes are of course joined at the bend of the loop. The NaCl is swept into the bend in the loop by the flow, and then, some of it is prevented from reaching the exit by being transported back into the descending limb. We saw this in the simple simulation shown in Figure 13.8. Unfortunately, the mathematical derivation given in the last section does not easily apply because of the following: (1) active transport rate is not proportional to the concentration difference, and (2) it does not deal with the situation at the bend in the loop.

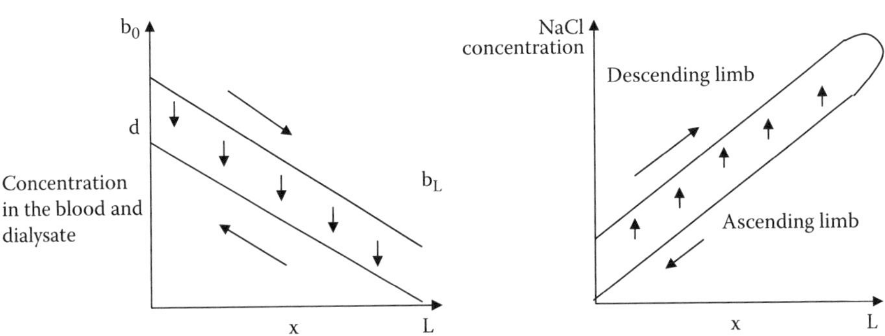

FIGURE 13.16
Left: the variation of blood and dialysate concentrations in a dialyzer in which the flow rates of blood and dialysate are the same. The arrows indicate constant transfer rate of material from blood to dialysate over the length of the dialyzer. Right: the arrangement in the loop of Henle, showing its similarity to the dialyzer system, but in this case, NaCl is pumped from a low to a higher concentration.

13.7.2 Theory Involving the VR System

This assumes that rather than the NaCl being pumped out of the ALLH and then immediately entering the DLLH, that the NaCl is taken up by the VR system, and it is this system that multiplies NaCl concentration at its loop bend. The interstitial fluid (and therefore, the DLLH) then equilibrates with the VR concentrations. Because a cross-sectional view of the arrangement of tubules and VR vasculature reveal a more or less random pattern (the ALLH and DLLH are not particularly close to each other—no more than the ascending and descending parts of the VR are—there is good reason to accept this simplification to be representative. Figure 13.17 is thus a representation of the VR.

As before, from conservation of mass consideration in element ABFG,

$$(C_1 + dC_1) = QC_1 + JSdx + P(C_2 - C_1)\ Ddx$$

$$dC_1/dx = JS/Q + PD(C_2 - C_1)/Q.........(1)$$

$$-dC_2/dx = JS/Q - PD(C_2 - C_1)/Q.........(2)$$

with the boundary conditions $C_1 = C_0$ at $x = 0$, and $C_1 = C_2$ at $x = L$.

If we add Equations (1) and (2), then integrate, we get the following:

$$d(C_1 - C_2)/dx = 2JS/Q;$$

$$C_1 - C_2 = 2JSx/Q + K = 2JS(x - L)/Q$$

Then, substituting for $C_1 - C_2$ in Equation (1) gives further integration:

$$C_1 = C_0 + JS\{1 + 2PDL/Q\}x/Q - JPSDx^2/Q^2$$

$$\text{and } C_2 = C_1 - 2JS(x - L)/Q.$$

Substituting realistic values for the parameters J, L, S, and Q give the behavior shown in Figure 13.18, which is not unlike the behavior shown in Figures 13.7 and 13.8.

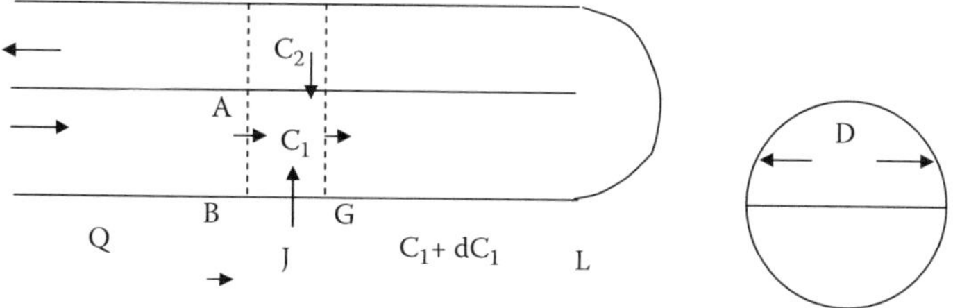

FIGURE 13.17
Representation of VR in a simple theory, which can be used to simulate the osmolarity profiles in the nephron. Here, S is the semicircumference of the tube, D is the diameter of the tube (wall width) C_1, and C_2 = concentrations of NaCl in tubes 1 and 2. J = influx of substance (assumed to be NaCl pumping rate out of the ascending limb of the loop of Henle and also assumed to be constant over the entirety of the outward facing surface, area 2LS). P = permeability coefficient for NaCl across the separating wall and Q = fluid flow rate (assumed constant).

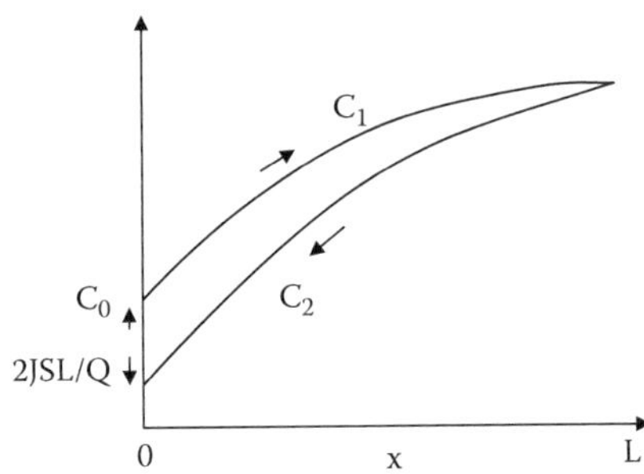

FIGURE 13.18
Values of [NaCl] in the loop of Henle predicted from the simple kidney model. Compare with Figures 13.8 and 13.16.

Tutorial Questions

1. In a clinical investigation, a standard creatinine and PAH clearance was performed. The results of analyses of plasma and urine samples gave the following data:

   1-h urine volume: 75 mL

   Plasma concentrations: creatinine 0.06 mg/mL; PAH 0.01 mg/mL; glucose 6 g/L

   Plasma osmolarity: 290 mOsm/L

   Urine concentrations: creatinine 6 mg/mL; PAH 5 mg/mL; glucose 320 g/L

   Urine osmolarity: 490 mOsm/L

   (a) Estimate GFR, ERPF, and FF from these data. Are they normal?

   (b) Given that TR = GFRPx – Ux $\dot{V}$, calculate the tubular reabsorption for glucose. How does tubular reabsorption vary with plasma glucose level?

   (c) Calculate FWC. Is the kidney concentrating or diluting?

2. (a) Explain, with the aid of diagrams, the principle of kidney dialysis.

   (b) A creatinine clearance of 125 mL/min is required for a countercurrent renal dialyzer. Calculate the area of cuprophan membrane required (permeability to creatinine: 2×10^{-2} cm/min) given the following conditions:

   (i) If the flow rate of blood and dialyzer fluid is both 200 mL/min

   (ii) If the dialyzer flow is 600 mL/min, and the blood flow remains at 200 mL/min

$$b_{in} - b_{out} = b_{in}/\{1 + Q/(AP)\} \text{ for } Q_b = Q_d;$$

$$\text{or } b_{in}\{1 - \exp(-Z)\}/\{1 - (Q_b/Q_d)\exp(-Z)\}, \text{ otherwise,}$$

$$Z = (AP/Q_b)\{1 - (Q_b/Q_d)\}.$$

   Hint: collect exponentials on one side of the equation.

3. In an artificial kidney dialyzer unit, creatinine clearance was determined to be 75 mL/min for a blood and a dialysate flow rate both at 100 mL/min. (Assume a countercurrent arrangement.)

(a) What would the creatinine concentration be in blood returning to the patient, if the initial level was 18 mg/L?

(b) If the permeability of the dialysis membrane to creatinine, P, is known to be 1.7×10^{-4} cm/s (1×10^{-2} cm/min), estimate A, the surface area of the membrane.

(c) What would the creatinine clearance become if the dialysis fluid flow rate were increased to 300 mL/min?

4. (a) Explain the principles of operation of a kidney dialyzer.

   (b) In a dialyzer, the membrane permeability to urea is 3×10^{-2} cm/min. Calculate urea clearance (as a percentage) if the membrane area is 1 m^2 (= 1×10^4 cm^2) and the flow rates of dialysis fluid and blood are 500 mL/min and 150 mL/min, respectively.

5. A simplified artificial kidney is shown in Figure 13.19.

   The semipermeable membrane is such that $PL/Q_b = 2.0$, where P is the permeability, L the length of the membrane, and Q_b is the blood flow (= 200 mL/min). Also, the dialysis fluid flow rate (Q_d) is 500 mL/min. Given that the difference in urea concentration between the blood entering and leaving the artificial kidney ($b_{in} - b_{out}$) is given by the following formulae:

$$\text{for parallel flow: } b_{in} - b_{out} = b_{in}\{1 - \exp(-\alpha L)\}/\{1 + (Q_b/Q_d)\},$$

where

$$\alpha = P[1 + (Q_b/Q_d)]/Q_b,$$

for countercurrent flow:

$$b_{in} - b_{out} = b_{in}\{1 - \exp(-\beta L)\}/\{1 - (Q_b/Q_d)\exp(-\beta L)\},$$

where

$$\beta = P(1 - (Q_b/(Q_d))/Q_b.$$

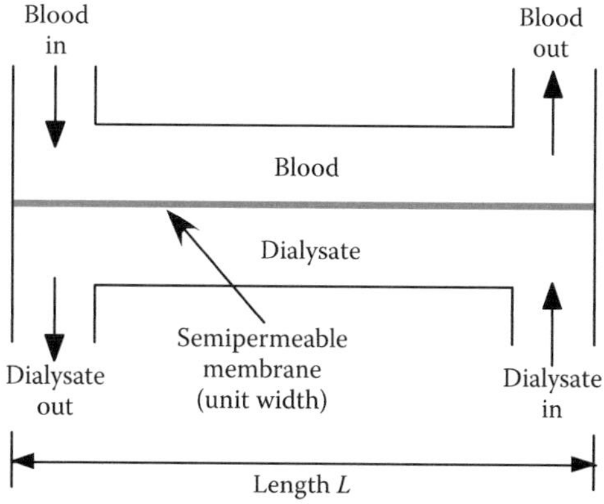

FIGURE 13.19
See Tutorial Question 5.

Calculate the efficiencies of the two flow arrangements, using the expression

$$100(b_{in} - b_{out})/b_{in} \text{ for efficiency.}$$

Sketch the way that concentration of urea in the blood and dialysate varies as a function of distance (x) along the membrane for the following: (1) parallel flow, and (2) countercurrent flow arrangements.

6. A renal dialyzer has an effective area of 1 m^2 and a urea permeability of 3.3×10^{-4} cm/s (0.02 cm/min). Blood passes through the dialyzer at 200 mL/min.
   (a) Show that a urea clearance of approximately 120 mL/min is achieved if dialysis fluid flows in a countercurrent fashion at 500 mL/min.
   (b) Calculate the urea concentrations in the blood returning to the patient and also in the dialyzer fluid, if the original blood urea (plasma urea) is 600 mg/L.
   (c) Calculate the clearance, under the same conditions, of another substance whose molecular weight is 100 times that of urea. Assume that permeability is inversely proportional to the square root of the molecular weight and that the constant of proportionality is the same for both substances.

   Hint: Assume that the standard countercurrent exchanger equations apply, that is, clearance

$$= \{1 - \exp(-X)\}/\{(1/Q_b) - (\exp(-X)/Q_d\},$$

   where

$$X = PA(1 - (Q_b/Q_d))/Q_b.$$

7. From the following data, calculate the following: (a) GFR, (b) ERPF, (c) FF, (d) total renal plasma flow, and (e) total renal blood flow.

   Plasma: inulin 0.21 mg/mL; PAH 0.018 mg/mL; (arterial sample) 0.002 mg/mL (venous sample); Hct 44%.

   Urine: inulin 17.3 mg/mL; PAH 7.6 mg/mL; 1-h volume 88 mL.
8. In a renal function test, the following data were obtained:

	60-min volume (mL)	Creatinine (mg/L)	PAH (mg/L)
Urine after 1 h	73	910	3800
Urine after second hour	48	670	3100
Plasma after 1 h		7.3	5.2

   (a) Estimate the following: (i) GFR, and (ii) ERPF, both in mL/min. Are they normal?
   (b) What factors affect the quantities in part (a)?
   (c) Explain what is meant by the term "maximal tubular reabsorption" for a substance such as glucose.

9. From the following data,

   Urine volume in 1 h: 55 mL

	Plasma	Urine
Creatinine (mg/mL)	0.012	1.4
PAH (mg/mL)	0.011	5.3
Osmolarity (mOsm/L)	295	830

   (a) Estimate GFR, ERPF, and FF from these data. Are they normal?
   (b) Calculate osmolar clearance and FWC. What is the significance of these quantities?

10. From the following data, calculate the following: (a) GFR, (b) ERPF, (c) FF, (d) total renal plasma flow, and (e) total renal blood flow.

    Plasma: inulin 0.21 mg/mL; PAH 0.018 mg/mL; (arterial sample) 0.002 mg/mL (venous sample); Hct 44%.

    Urine: inulin 17.3 mg/mL; PAH 7.6 mg/mL; 1-h volume 88 mL.

11. In a renal function test, the following data were obtained:

	60-min volume (mL)	Creatinine (mg/L)	PAH (mg/L)
Urine after 1 h	73	910	3800
Urine after second h	48	670	3100
Plasma after 1 h		7.3	5.2

    (a) Estimate the following: (i) GFR and (ii) ERPF, both in mL/min. Are they normal?
    (b) What factors affect the quantities in part (a)?
    (c) Explain what is meant by the term "maximal tubular reabsorption" for a substance such as glucose.

12. From the following data,

    Urine volume in 1 h: 55 mL

	Plasma	Urine
Creatinine (mg/mL)	0.012	1.4
PAH (mg/mL)	0.011	5.3
Osmolarity (mOsm/L)	295	830

    (a) Estimate GFR, ERPF, and FF from these data. Are they normal?
    (b) Calculate osmolar clearance and FWC. What is the significance of these quantities?

13. Determine the following: (a) GFR, (b) FF, and (c) FWC from the following data

    Plasma: creatinine 9.4 mg/L; PAH 8.7 mg/L; normal osmolarity

    Urine: Creatinine 0.68 g/L; PAH 3.2 g/L

    Volume 117 mL/h; specific gravity (SG) 1.007 (see Supplementary Data).

Comment on the physiological significance of these results.

Supplementary Data

Table of relationship between SG, osmolarity, and depression of the freezing point of urine.

SG	1.000	1.005	*1.007*	1.010	1.015	1.020	1.035
Osmol/L	0.0	0.2	*0.3*	0.4	0.6	0.8	1.4
Freezing point (°C)	0	–0.37	*–0.56*	–0.74	–1.1	–1.5	–2.6

Note: Values for glomerular filtrate shown in italics.

Bibliography

Hall JE. 2011. *Guyton and Hall Textbook of Medical Physiology,* 12th Edition. Elsevier-Saunders, Philadelphia, PA, USA.

Schmidt RF, Thews G. 1989. *Human Physiology,* 2nd Edition, Springer-Verlag, Berlin.

14

Cardiopulmonary Perfusion and Advanced Surgical Techniques

Andrew W. Wood

CONTENTS

This chapter will survey some of the major ways that biomedical engineering applications have impacted on surgery. The first of these is the use of rotary pumps and gas exchangers to take over the functions of the heart and lungs in open-chest surgery. The other two sections deal with the miniaturization of surgical techniques and the application of robotics to surgical practice. A brief mention will also be made to the development of implantable total artificial hearts.

14.1 Cardiopulmonary Bypass

Cardiopulmonary perfusion refers to situations in which the function of the heart and lungs is taken over by an artificial device: in popular parlance, this is referred to as the heart–lung machine. The basic idea is to immobilize the heart so that it can be operated upon; this means that the blood has to be diverted away from both the right and left sides of the heart and, since the lungs rely on being perfused by blood in order for gas exchange to take place, this function has to be done externally. This type of operation, pioneered in the early 1960s, is termed *open heart surgery*, which involves opening up the chest and pleural cavity. When the pleural cavity is opened, the lungs deflate and hence could not contribute to oxygenation even if circulation through them was maintained. More modern techniques try to use minimally invasive or "keyhole" surgery to reduce trauma. However, this is only effective in certain situations, and the open operations, particularly for heart and combined heart and lung transplants, continue to be the operation of choice.

In addition to transplants, the following procedures also are commonly performed using cardiopulmonary bypass (CPB): valve replacements (using prostheses or animal-derived

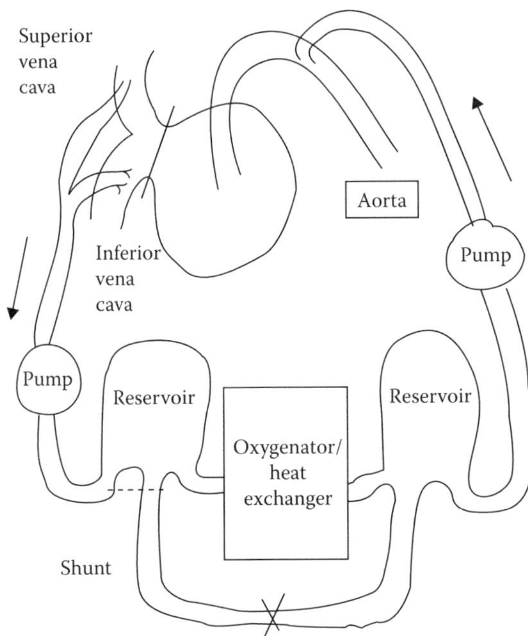

FIGURE 14.1
Connections from the vena cava to the CPB circuit, via the pumps, heat exchanger, and oxygenator, for return in this case to the aorta.

valves), coronary artery grafts (often referred to as triple or quadruple bypass surgery), and repair of septal defects and repair of aneurisms (using woven cylindrical prostheses). Successful operation of CBP requires multidisciplinary skills, including a good knowledge of human physiology with an understanding of biophysical and engineering principles. Hospitals carrying out this type of surgery employ teams of *perfusionists* whose job is to operate and maintain such equipment and to monitor the patient's condition (in relation to this equipment) during surgery.

Briefly, the procedure is as follows: the patient is maintained on a respirator as the sternum (breastbone) is divided, the pleural cavity is opened, and then the heart is exposed. A large flexible cannula is inserted into the vena cava (often a pair of cannulae, one each into the superior and inferior vena cavae). The patient's blood is then run into a circuit, as shown in Figure 14.1 (previously primed with blood obtained from transfusion), and then run back into the aorta as shown. Clamps are then applied to the vena cavae. The aortic valve prevents backflow from the aorta into the heart once the heart is opened (the site of incision depending on the operation). An additional sucker is inserted into the heart to empty the cavities, which are then filled with a cooled K^+-rich solution, which depolarizes the myocardial membranes and causes the heart to stop beating. The blood in the external circuit is also cooled, to reduce metabolic demands of sensitive tissue, particularly the brain. The principal components of the heart–lung machine are rotary pumps, the oxygenator (for CO_2 removal), and the heat exchanger. Each of these will now be described.

14.1.1 Rotary Pumps

The function of the heart is taken over by a bank of rotary (or roller) pumps. These need to satisfy a number of requirements, including pumping of blood at rates continuously

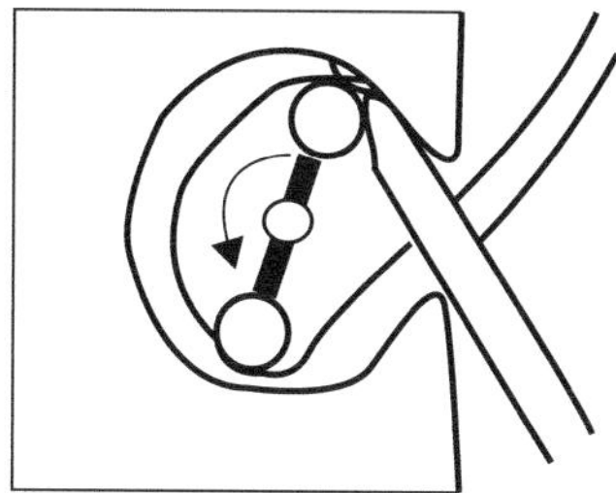

FIGURE 14.2
Diagram of a rotary pump for use in cardiopulmonary bypass (CPB) equipment.

variable in the range 0 L/min–5 L/min, a pumping action with minimal hemolysis (bursting of blood cells), absence of sharp edges to cause blood clotting, ease of sterilization (i.e., disposable parts), low priming volumes, and the ability to pump against 180 mmHg (extreme systolic pressure). The rotary pump is surprisingly simple in design, the squeezing action against the silicone rubber tube providing the pumping action. The tubing is self-feeding into the roller cavity at the start of operation, and the tubing itself is disposable (see Figure 14.2). The flow pattern, of course, is not as pulsatile as it would normally be, but many studies have shown that this is not a major problem. As shown in Figure 14.1, one pump each takes over the function of the right and left heart and other pumps in the bank can be used to remove blood from the operation site and to infuse drugs and so on. There are extra pumps (up to five altogether) in case of faults developing in the main two pumps. The pump speeds are controlled independently, continuously variable and can be operated manually in the event of major power failure (however, there is usually emergency back-up electrical power available). Regarding measures to prevent blood clotting in the system, materials such as silicone rubber, cellophane and Teflon™ (PTFE) are suitable, because of their low "wettability," smooth surface and their chemically inert nature.

14.1.2 Heat Exchanger

During bypass, it is usual to cool the patient's blood to 25°C or lower, to reduce O_2 requirement of sensitive tissue (e.g., brain). It is usual in an operation lasting longer than 2–3 h to warm the patient periodically. As mentioned, the heart is stopped by perfusing pericardium with cold solution rich in K^+ (cardioplegic solution). In order to rapidly cool and warm the patient, a countercurrent heat exchange system is used as shown in Figure 14.3. The wall between the blood and the cooling water is Teflon-coated stainless steel or other metal. The water is temperature-controlled by running it through a large thermostatically controlled water bath, with provision for both heating and cooling.

14.1.3 Oxygenator

Modern oxygenators are almost entirely formed from capillary bundles within a solid cylinder. These membrane-type oxygenators are similar to renal dialysis units, except that the membrane is polyethylene or silicone rather than cellulose, because the membrane needs to be O_2 and CO_2 permeable rather than to particular ions. The indicative calculation below shows that an equivalent of 1.7 m^2 of membrane area is needed: the problem is getting rid of CO_2 rather than getting O_2 into the blood. The capillary tubes are made from hydrophobic microporous materials with a pore size sufficiently small to

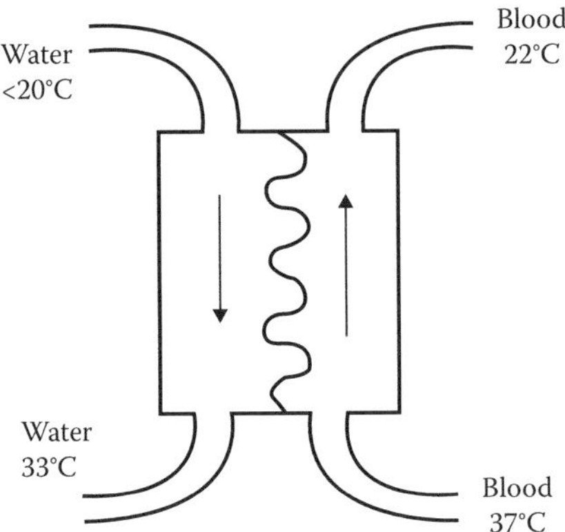

FIGURE 14.3
A heat exchanger for rapid cooling and rewarming the circulation while on CPB.

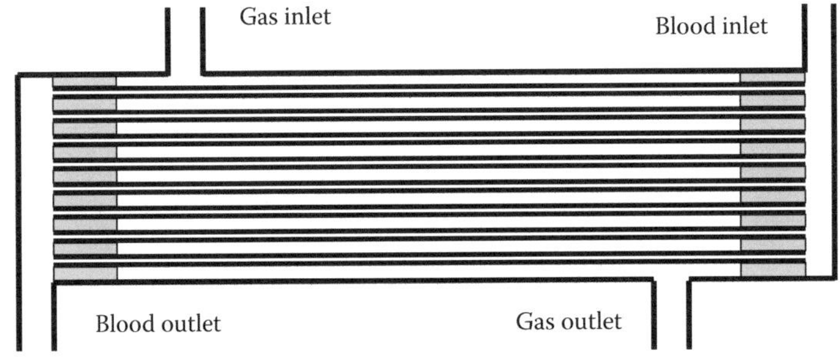

FIGURE 14.4
Capillary membrane oxygenator. Blood is conducted along many thousands of silicone capillaries, surrounded by humidified gas.

prevent plasma filtration and which are defect-free when formed into 0.3 mm diameter tubes (Figure 14.4).

Oxygenators are now almost entirely membrane types: earlier types included a) film type (involving rotating blood-coated disks in O_2 atmosphere) or b) bubble type, in which O_2 was bubbled directly through blood. These had problems with sterility and removal of residual bubbles.

In the derivation given below, the ratio of CO_2 to O_2 driving forces: $\Delta p'_{CO2}/\Delta p'_{O2} = 45/713 = 0.06$ and the ratio of transfer coefficients K_{CO2}/K_{O2} is approximately 5, so the ratio of gas transfer rates $Q_{CO2}/Q_{O2} \sim 5 \times 0.06 = 0.3$. In other words, CO_2 removal is around 1/3 as efficient as O_2 inflow. It is instructive to compare parameters associated with capillary membrane oxygenators with the real pulmonary circulation. This is shown in Table 14.1.

14.1.3.1 Oxygenator Performance Testing: Theory

In evaluating performance testing, the worst case would be a parallel flow arrangement—as mentioned, the flow arrangement is usually countercurrent and therefore more efficient.

The basic equation for gas to blood transfer (or vice versa) for a parallel flow arrangement (see Chapter 13 for derivation) is

$$Q = KA\{\Delta p_{inlet} - \Delta p_{outlet}\}/\ln(\Delta p_{inlet}/\Delta p_{outlet})$$
$$= KA\Delta p'$$

Here Q is in mL/min, A is the total membrane area (in cm^2), K is the mass transfer coefficient (in units dictated by the other parameters) and is a feature of the membrane material and Δp_{inlet}, Δp_{outlet} is the difference in partial pressure (in mmHg) between gas and blood for either O_2 or CO_2 at the inlet and the outlet of the oxygenator, respectively. The area of the membrane is what we want to estimate. The oxygenator can be represented by the diagram in Figure 14.5.

The desired partial pressures of O_2 and CO_2 in blood returning to the patient are 95 mmHg and 40 mmHg, respectively. In order to estimate Δp_{outlet} we need to know the corresponding gas pressures at the outlet. We can calculate this by knowing how much O_2 the body will require per liter of blood and the amount of CO_2 the body will produce.

For O_2 the content in blood is usually given as a volume percentage, 19.7% in the arteries and 14% in the veins. The amount going into the tissues is thus 5.7%, or 57 mL/L blood. If the perfusion rate is 4 L/min (note this is significantly below resting

TABLE 14.1

Comparison of Physical Characteristics between the Human Lung and a Cardiopulmonary Bypass Oxygenator Unit

Comparison	Lung	Oxygenator
Transfer area	70 m^2	2 m^2
Diffusion distance	2 μm	80 μm
Capillary length	1 mm	150 mm
Capillary diameter	5 μm	300 μm
Hematocrit	Normal	Blood diluted with plasma

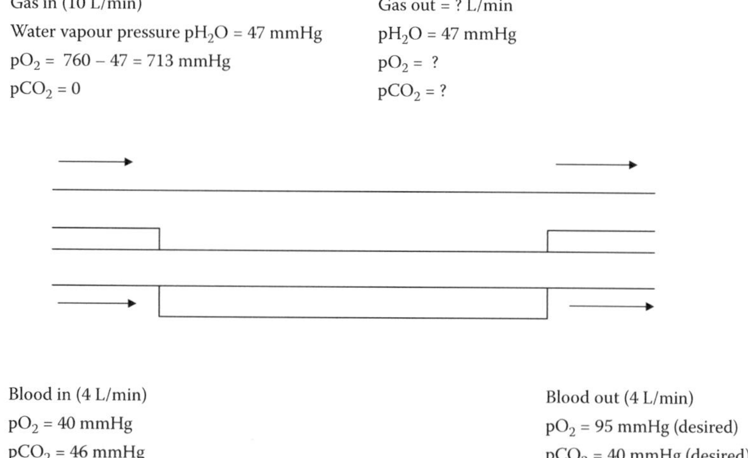

FIGURE 14.5
Representation of the membrane oxygenator, with "worst-case" scenario of gas and blood flowing in the same direction.

cardiac output, but this is normal during perfusion), this represents a demand of 57 × 4 = 228 mL/min or 0.228 L/min.

For CO_2: the venous content is 52.5% and the arterial 48%; thus, 4.5% is released from the tissues. For 4 L/min, this is 18% or 180 mL/min, or 0.18 L/min.

The volume flow rate at the outlet is thus 10 – 0.23 + 0.18 = 9.95 L/min. We can estimate pO_2 by estimating how much of this 9.95 L/min is represented by O_2. The proportion entering is 10 × 713/760 L/min, and 0.23 L/min enters the blood (and hence the tissues). So the *proportion* in the gas leaving will be {(10 × 713)/760 – 0.23}/9.95. We multiply this by 760 to get the pO_2 to be 699 mmHg. The proportion of outflowing gas made up of CO is given simply by 0.18/9.95. Again, multiplying this by 760 gives pCO_2 as 13.7 mmHg.

For O_2:

$$\Delta p' = \{(713 - 40) - (699 - 95)\}/\ln\{(713 - 40)/(699 - 95)\} = 638 \text{ mmHg}.$$

Volume transfer rate Q at 37°C will be the same as tissue consumption rate (230 mL/min). For typical membranes, K for O_2 is 0.5 mL/min/m^{2}mmHg,

$$\text{So } A = Q/(K\Delta p') = 230/(0.5 \times 638) = 0.7 \text{ m}^2.$$

For CO_2:

$$\Delta p' = \{(46 - 0) - (40 - 13.7)\}/\ln\{(46 - 0)/(40 - 13.7)\} = 35.3 \text{ mmHg}.$$

Q is CO_2 removal rate (180 mL/min). CO_2 permeability of membrane is 2.7 mL/min/m^{2}mmHg.

So

$$A = Q/(K\Delta p') = 180/(2.7 \times 35.3) = 1.9 \text{ m}^2$$

and is therefore a much larger requirement. From Table 14.1, each capillary has a length of 0.15 m and a diameter of 0.3 mm, so the surface area of each tube (π.d.l) is 1.4×10^{-4} m^2, so $1.9/1.4 \times 10^{-4}$ = 13,000 capillaries are needed. Since each of these occupies 7×10^{-6} m^2 in cross-section, a total area in excess of 0.1 m^2 is needed (there will be space between individual capillaries). This gives around 0.4 m diameter for the oxygenator. Commercially available oxygenators are around three times smaller in diameter, perhaps due to improvements in CO_2 permeability of the materials used and because of the higher efficiency of countercurrent flow arrangements.

14.1.4 Replacement Heart Valves and the Artificial Heart

One of the main uses of CPB is to allow the replacement of heart valves that have become diseased. These diseases include stenosis (inability to fully open due to partial fusing of the leaflets), regurgitation (improper closure allowing backflow during systole) or prolapse (another form of incomplete closure, where one of the leaflets bulges into the atrium during systole). These faulty valves can be replaced by a transplanted biological valve from a mammal, such as a pig (this is called a xenograft), from a human donor (called an allograft or homograft), or even, under some circumstances, from the patients themselves

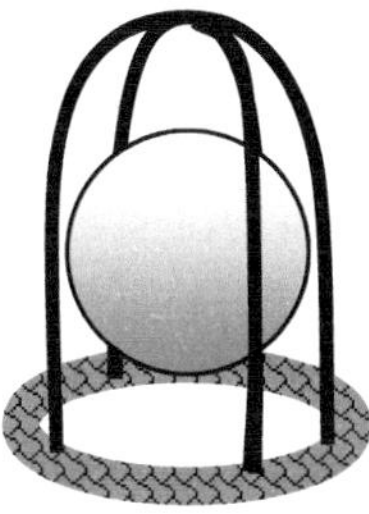

FIGURE 14.6
Artificial heart valves. Left: The Starr–Edwards ball and cage type. Right: The "tilting disk" type.

(autograft). In the latter case, the pulmonary valve is swapped for the patient's own aortic valve, with an allograft used in place of the pulmonary valve. However, these tend not to last as long as mechanical valves. Originally, these took the form of a silicone ball held in a stainless steel cage (Starr–Edwards type), with a seating ring of woven synthetic fabric, such as Dacron (which is sutured into the valve opening after the faulty valve has been excised). More recently, these have been superseded by "tilting disk" types, where the circular disk pivots around an alloy pin in the seating ring (Figure 14.6). Although there were instances of metal fatigue in earlier types, leading to the disk breaking free, more recent designs are very reliable. Most now consist of two tilting leaflets, which offer a greater opening area than the single disk type, but suffer from some added regurgitation. In addition to valves, prosthetic Dacron sleeves are used to replace sections of major arteries that have suffered aneurism, that is, a weakening of the wall allowing a dangerous bulging of the vessel at a particular point. We can recall that the "Law of Laplace" (see Chapter 10) states that the wall tension is the product of the pressure inside the artery and the radius. If the wall becomes weakened, allowing the radius to rise, then the wall tension needs to increase for a given pressure. If, because of the weakness, the wall is unable to produce this tension, then the radius continues to rise, and the vessel will eventually rupture, leading to hemorrhage into a body cavity. This highlights the need for rapid intervention.

This section would not be complete without some reference to advances in ventricular assist devices (VADs) (which supplement the output from a failing heart as a bridge to recovery) and total heart replacement devices (as a bridge to a transplanted organ). One of the difficulties has been to supply adequate power to the device without the power supply being an encumbrance for the recipient. Recent advances in battery technology have allowed both the device and the rechargeable power supply to be implanted, which means that no wires penetrate the skin. There are several designs for a total artificial hearts and VADs which rely on pneumatically driven membranes or magnetically driven impellers. At the present time, VADs are more widely used than total artificial hearts, since the survival time after implantation of the latter still needs to be extended to achieve acceptance. Good clinical and bioengineering perspectives can be found in Feldman (2010) and Bronzino (2006), respectively.

14.2 Minimally Invasive (Keyhole) Surgery

As indicated above, there is a tendency toward surgical techniques that avoid opening the thoracic or abdominal cavity (and indeed, surgical techniques in general involving large incisions). This has been made possible mainly by endoscopic techniques, which are described in greater detail in Chapter 24. Instruments are inserted through incisions of 1 cm or less and the endoscope is used to view the surgical area. The main advantages of these techniques are less tissue trauma, less chance of sepsis, and faster recovery time. The main disadvantage is the limited access, which may mean that the operation has to revert to a normal "open" operation in the event of an unforeseen complication.

14.3 Robotic Surgery

An extension of "keyhole" surgery is for the instruments to be under servomechanical control rather than direct control of the surgeon. In this way, there can be reverse amplification of movement (hand movements of the order of centimeters become fractions of a millimeter when translated to the instrument inside the patient). The range of movements can be increased, to allow access not normally possible to a surgeon using conventional instruments. The surgeon's view would normally be of a remote high-definition video display rather than direct viewing through an eyepiece. Recent innovations have included "haptic rendering" (in which the tissue mechanical resistance is fed back to provide the surgeon with perception of the force being applied to tissue, to prevent trauma) and 3-D displays, to give depth perception. The surgeon is therefore seated away from the patient at a console. This has the remote display and the two-handed controls for operating the instruments within the patient. Each instrument has more flexibility than the human wrist, having, in addition to the six translation and rotation degrees of freedom the opening and closing of the specialized instrument at the tip (miniature forceps, cutters, needle drivers, retractors, and so on). These instruments are attached to three or four robot arms forming the "patient-side cart" (Figure 14.7). Since the communication between console and cart is entirely electronic, there is little restriction on the distance between the two. In fact, in 2001, there was a famous "Lindbergh Operation" (named after the famous aviator who was the first to fly solo across the Atlantic), where surgeons in New York removed the gall bladder of a patient in Strasbourg, France. In addition to the minimally invasive robot cart, the operation was made possible by the use of high-speed optical fiber communication between the two centers. This example of "telesurgery" highlights the ability to deliver advanced surgical techniques to remote or rural areas lacking this degree of specialization. At a lower level of sophistication, a two-way video link between centers can provide expert input to the center with less specialization.

Tutorial Questions

1. A water saturated stream of pure oxygen at a temperature of 37°C is fed into the membrane (oxygenator) of a heart–lung machine. Using the information below determine the minimum area of the oxygenator membrane required to transfer 250 mL/min of O_2 into the blood stream.

   Permeability of the membrane at 37°C to O_2 is 0.5 mL m^{-2} min^{-1} $mmHg^{-1}$. Assume that the transfer of oxygen into the blood stream is determined primarily by the properties of the oxygenator membrane.

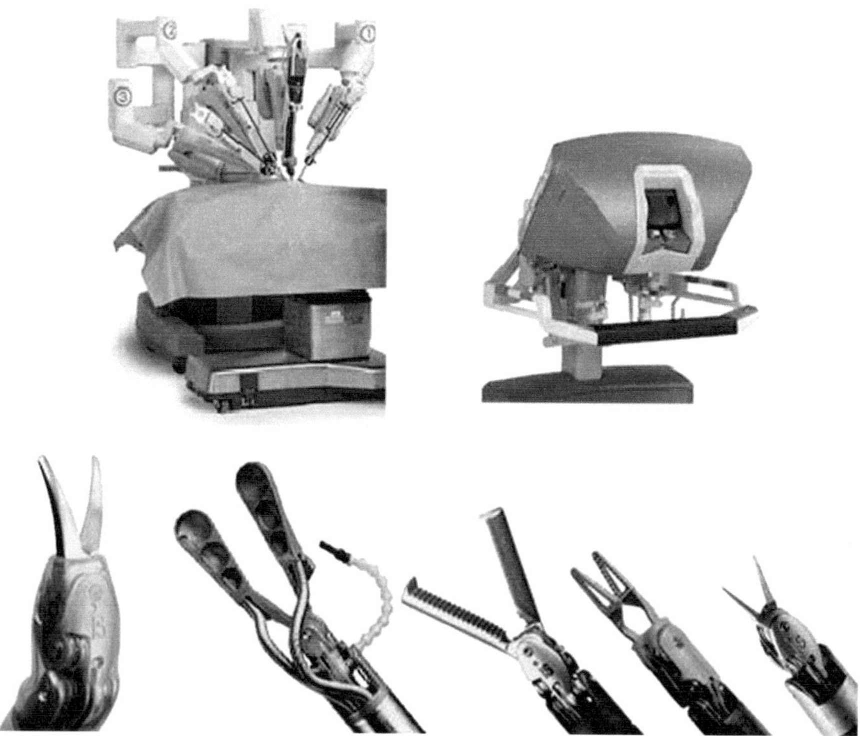

FIGURE 14.7
The da Vinci robotic surgery system. Top left: The patient-side cart showing three robot arms. Top right: The surgeon console showing viewing screen and binocular magnification. Bottom: A selection of surgical instruments placed at the end of robot arms. Diameter of shank: approximately 4 mm. (Courtesy of Intuitive Surgical Inc., Sunnyvale, CA, USA.)

Atmospheric pressure	760 mmHg
Partial pressure of H_2O at 37°C	47 mmHg
Partial pressure of O_2 in mixed venous blood	40 mmHg
Partial pressure of O_2 in gas leaving the oxygenator	700 mmHg
Partial pressure of O_2 in blood leaving oxygenator	95 mmHg

2. An oxygenator for a CPB circuit consists of a polyethylene membrane of area 2 m^2 and which allows CO_2 to pass through at a rate of 30 mL/min/m^2 for a 10 mmHg pCO_2 gradient at BTPS. Assuming that the CO_2 in a patient's blood has to be reduced from 52% to 40% (corresponding to a pCO_2 of 46 and 40 mmHg, respectively) before being returned to the patient, show that the oxygenator would be adequate for cardiac outputs ranging from 2 to 5 L/min.

   Volume transfer rate of gas (mL/min) = KAΔP′, where K is the mass transfer coefficient and A is the membrane area.

$$\Delta P' = \frac{\Delta P^{\text{inlet}} - \Delta P^{\text{outlet}}}{\ln(\Delta P^{\text{inlet}}/\Delta P^{\text{outlet}})},$$

where ΔP refers to the pCO_2 difference between blood and gas. Take pCO_2 at gas outlet to be 7 mmHg and 18 mmHg for cardiac outputs of 2 and 5 L/min, respectively. The inlet gas is humidified O_2.

References

Bronzino JD. 2006. *The Biomedical Engineering Handbook*, 3rd Edition. CRC Press/Taylor & Francis, Boca Raton, FL, USA.

Feldman A. 2010. *Heart Failure Management*. John Wiley & Sons, Chichester, UK.

Part V

The Central Nervous System

15

Organization of the Human Central Nervous System

Per Line

CONTENTS

15.1 Introduction

The central nervous system (CNS) consist of roughly 100 billion neurons, and about an order of magnitude more neuroglia. Neurons (nerve cells) and neuroglia (glial cells) are the two types of neural tissue comprising the CNS. Neurons are excitable cells that have the ability to transmit electrical signals without loss, allowing them to communicate with other neurons and cells by chemical and electrical signals at synapses. An average neuron has several thousand synapses (links) with other neurons. A neuron typically consists of a cell body, dendrites, and axon. Generally, the cell body contains organelles similar to other cells. The dendrites are typically short and highly branched processes that receive (via synapses) signals from other nerve cells. Axons are usually long and cylindrical, and typically only branch (forming collaterals) toward the end of their length. Neurons are classified into several structural categories, the main ones being bipolar, unipolar, and multipolar. Bipolar neurons, located in special sense organs such as the eye, are small with very distinct axon and dendritic processes. Unipolar neurons, where essentially the dendrites and axon are continuous (the cell body being off to one side), are the predominant type of neurons in the peripheral nervous system (PNS). Some unipolar neurons may have axons a meter or so long (such as sensory nerves from toes), as do some axons of multipolar neurons, specifically motor neurons (which control muscles in the toes). On the other hand, some neurons have axons as short as 100 μm. Most of the neurons in the

CNS are multipolar, with all motor neurons controlling skeletal muscles being multipolar. Multipolar neurons have multiple dendrites and one axon. Axons can vary in diameter from about 0.2 μm to 20 μm, with the smallest diameter nerve fibers usually unmyelinated.

Neuroglial cells are the supporting cells of the CNS, there being four types of neuroglia in the CNS: ependymal cells, astrocytes, oligodendrocytes, and microglia. The ependymal cells line the ventricles, and some specialized ependymal cells secrete cerebrospinal fluid (more later). Astrocytes are the most numerous CNS neuroglia and are involved in many functions, including maintaining the blood–brain barrier, creating structural support for neurons, and regulating interstitial fluid (the fluid occupying spaces between cells) composition. One function of the oligodendrocytes is to produce and maintain the fatty covering (known as myelin sheath) around axons in the CNS (Schwann cells perform this role in the PNS). Microglial cells are involved in the destruction and clean up of pathogens, removal of debris, and promotion of tissue repair.

As a whole the nervous system is divided into two main divisions (see Figure 15.1), the CNS and PNS. The PNS consists of all nerves external to the CNS, comprising 12 pairs of cranial nerves and 31 pairs of spinal nerves, plus their associated ganglia. A ganglion (plural ganglia) consists of groups of neuronal cell bodies located together outside the CNS, while the term *nucleus* (plural *nuclei*) is usually used when referring to neuronal cell bodies

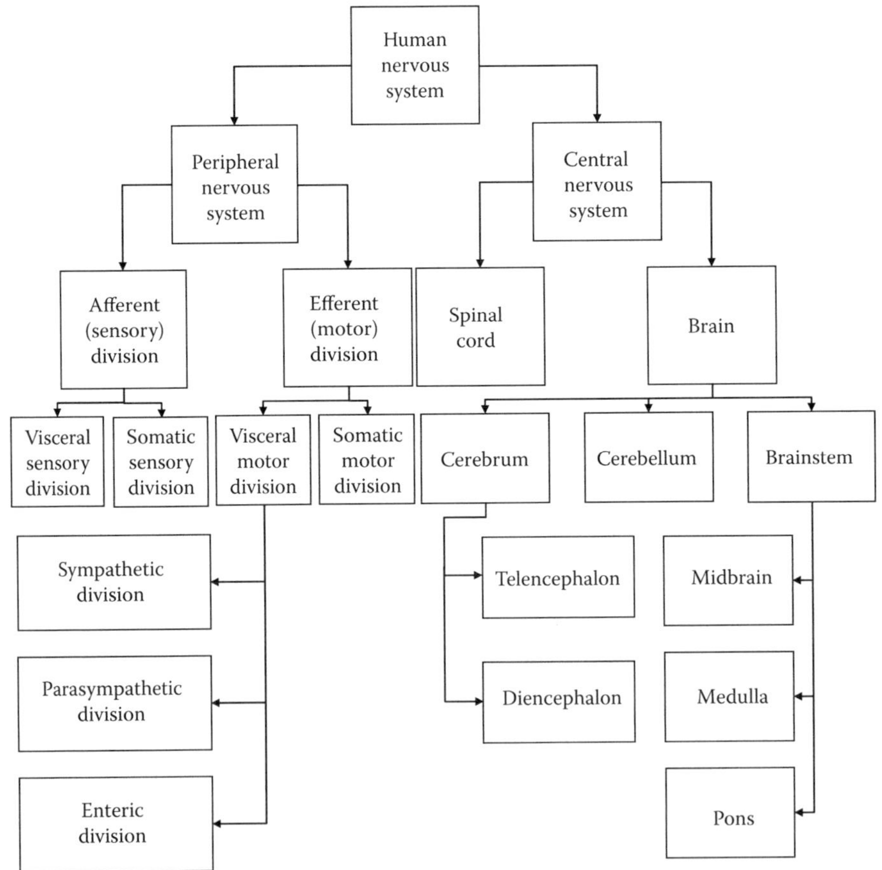

FIGURE 15.1
Nervous system divisions.

located together inside the CNS, but positioned subcortical (below the cerebral cortex) or in the spinal cord. The cerebral cortex refers to gray matter (CNS tissue containing predominantly neuronal cell bodies) on the surface of the cerebral hemispheres.

Essentially, the PNS is composed of two main parts: an afferent (sensory) division (transmits signals toward the CNS) and efferent (motor) division (transmits signals away from the CNS). The afferent division is further subdivided into a visceral sensory division and somatic sensory division, with the former transmitting signals from receptors in internal organs, whereas the latter sends information from receptors in peripheral tissue. In regards to the efferent division, it is also separated into two subdivisions; a somatic motor division that controls skeletal muscle, and a visceral motor division that is also known as the autonomic nervous system (ANS). Smooth muscle, cardiac muscle, adipocytes (fat cells), and glands are regulated by the sympathetic and parasympathetic subdivisions of the ANS. The sympathetic subdivision sends out motor commands through nerves originating in the thoracic and lumbar regions of the spinal cord, whereas the parasympathetic subdivision sends out motor commands through nerves originating in the brainstem (cranial nerves) and sacral region of the spinal cord. In general, the sympathetic system inhibits or opposes the physiological effects of the parasympathetic system, and vice versa. There is also a third less known subdivision of the ANS, called the enteric division, which innervates the gastrointestinal tract. In regards to cranial nerves, these are 12 pairs of PNS nerves that arise from the brainstem, diencephalon, and telencephalon. They contain the axons of sensory and motor nerves that, respectively, enter and exit the CNS.

The focus of this chapter is the CNS, which is divided into two parts: the brain (encephalon) and the spinal cord (medulla spinalis). Protecting the brain is the cranium, which encases it, whereas the vertebral column surrounds and protects the spinal cord. The brain is further partitioned, being made up of three major components; the cerebrum, cerebellum, and brainstem. The cerebrum consists of the telencephalon (the two cerebral hemispheres) and the diencephalon. The brainstem is composed of the midbrain, pons, and medulla. Hence, there are seven major divisions of the CNS: spinal cord, medulla, pons, cerebellum, midbrain, diencephalon, and telencephalon.

To navigate through the CNS one needs to be familiar with anatomical terms in relation to direction and planes of section. The longitudinal axis of the CNS runs from the inferior (situated beneath) end of the spinal cord to the anterior (situated to the front of) end of the frontal cortex, with a bend in the midbrain region. Running perpendicular to the longitudinal axis is the dorsoventral axis. As a result of the midbrain flexure, dorsal (pertaining to the back) is equivalent to posterior (located behind) in the brainstem and spinal cord, but becomes synonymous with superior (situated above) in the cerebrum. Similarly, ventral (pertaining to the abdomen) is equivalent to anterior in the brainstem and spinal cord, but becomes synonymous with inferior in the cerebrum. Rostral means situated toward the nasal region, whereas caudal pertains to the tail or hind parts. Lateral is toward the side, whereas medial is toward the middle of the body. Proximal is nearer to a point of reference, such as the body, whereas distal is farther away. Ipsilateral relates to the same side of the body, whereas contralateral relates to the opposite side.

There are three main anatomical planes of brain section, the horizontal, coronal, and sagittal planes. Sagittal sections are cut between the midline and lateral sides of the brain. Coronal (also known as frontal or transverse) brain sections are cut between the anterior and posterior sides. Horizontal brain sections are cut parallel to the longitudinal axis, being cut between the superior and inferior ends. A median (also known as midline or midsagittal) section divides the CNS into two similar halves. A section cut at right angles to the longitudinal axis in the spinal cord and brainstem is often referred to as a cross section. Also, when describing nerve fiber tracts, the term usually gives information about the origin and termination of the fibers. For example, a corticospinal tract originates in the cerebral cortex (the "cortico" part) and terminates in the spinal cord (the "spinal" part).

15.2 Development of the Central Nervous System

The CNS, including neuroglial cells and neurons, are derived from the neural plate (a thickened portion of the dorsal ectoderm—the external of the three primary germ layers that includes the ectoderm, mesoderm, and endoderm). Neural induction is the name given to the process, not fully understood, whereby the neural plate is induced to form the CNS by signaling molecules sent from the mesoderm (including notochord). The neural plate is formed on about day 18 (see Figure 15.2a), and early on in its development into the CNS it becomes a structure called the neural tube (see Figure 15.2c) by a process of infolding and fusion. The formation of the neural tube from the neural plate is called primary neurulation. As the infolding begins, the growing indentation is known as the neural groove (see Figure 15.2b). When the neural groove fuses at its dorsal end it then becomes the neural tube. The cavity within the neural tube is known as the neural canal, and gives rise to the ventricular system. During the infolding process cells at the dorsal end of the neural groove give rise to specialized migratory cells called neural crest cells. Neural crest cells give rise to most of the sensory neurons in the PNS, particularly

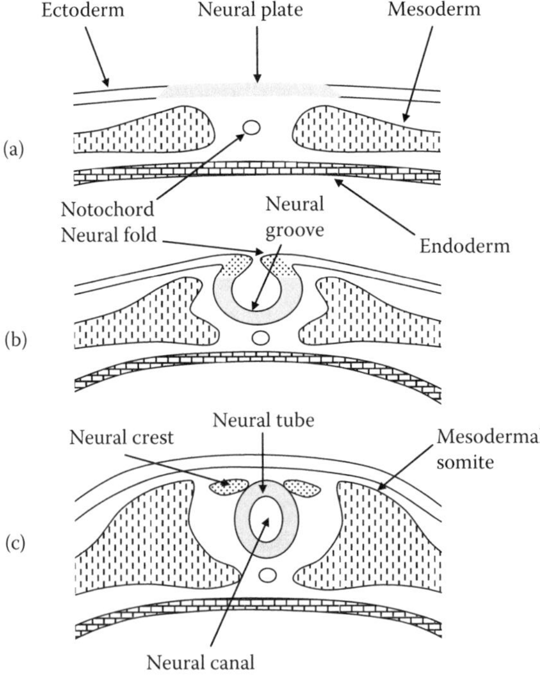

FIGURE 15.2
a, b, and c illustrate cross sections of the developing embryo at various stages of neurulation during the third and fourth week postconception.

those with cell bodies in the dorsal root ganglia (collection of cell bodies from sensory neurons located in the PNS, but close to the spinal cord) that form part of the spinal nerves, as well as neurons that make up the ganglia of the ANS. Epithelial cells lining the neural tube generate almost all the neurons and neuroglial cells that will form the CNS. This lining of the walls of the neural tube with epithelial cells is known as the ventricular zone. The neuroglial cells in the PNS are derived from the neural crest cells, as are the two innermost connective tissue membranes, the pia and arachnoid mater. The mesoderm gives rise to the outermost dura mater.

The fusion of the neural folds, along the sides of the neural groove, into the neural tube begins in the middle of the developing embryo, with ongoing fusion proceeding both in rostral and caudal directions until complete closure has occurred along the whole length of the neural groove. The opening at the caudal end closes last, at about four weeks or so after fertilization, with the rostral opening closing a day or two earlier. Prior to closure these openings at either end of the neural tube are called neuropores. Running bilaterally alongside the neural tube in the developing embryo are segmental masses of mesoderm, known as mesodermal somites, that later develop into vertebrae, muscle, and dermis. The rostral end of the neural tube will develop into the brain, while the caudal end will form the spinal cord. By the end of the first month of development the rostral end of the neural tube has formed an embryonic brain consisting of three divisions (vesicles), the forebrain (prosencephalon), midbrain (mesencephalon), and hindbrain (rhombencephalon). By the end of the fifth week the forebrain has differentiated into the telencephalon and diencephalon, and the metencephalon (pons and cerebellum) and myelencephalon (medulla) have emerged from the hindbrain. The telencephalon soon becomes the largest of the CNS major divisions, due mainly to the enormous expansion of the cerebral cortex. During peak cell proliferation, the CNS grows at a rate of approximately 250,000 neurons per minute. At birth the brain weighs an average of about 370 g, reaching an average maximum weight of around 1380 grams in an adult human. Almost all neurons of the adult brain are present before birth, and few are added after birth. Hence, the growth of the brain after birth is not due to new neurons being added. Proliferation of neuroglial cells, myelination (formation of myelin sheaths) of axons, and the formation of new nerve connections, are some of the main causes of brain growth after birth.

15.3 The Spinal Cord (or Medulla Spinalis)

The rostral two thirds of the adult vertebral canal are occupied by the spinal cord, the latter a cylindrical type neural structure roughly 45 cm long in an adult, and varying between 1 cm and 1.5 cm in width along its length (see Figure 15.3). Rostrally, the spinal cord is continuous with the medulla of the brainstem at the level of the foramen magnum, the latter being the hole in the skull that the spinal cord passes through. Caudally, the spinal cord terminates roughly at the level of the second lumbar vertebra. The termination point is known as the conus medullaris, and emerging from it is a connective tissue filament, called the filum terminale, that runs down and attaches to the coccyx (consists of three to five vertebrae at the caudal end of the vertebral column, which usually become fused together in adults), thus anchoring the caudal end of the spinal cord to the vertebral canal. Also, from the caudal end of the spinal cord the lumbar and sacral nerve roots (collectively called the cauda equina) descend variable distances before emerging out of the vertebral

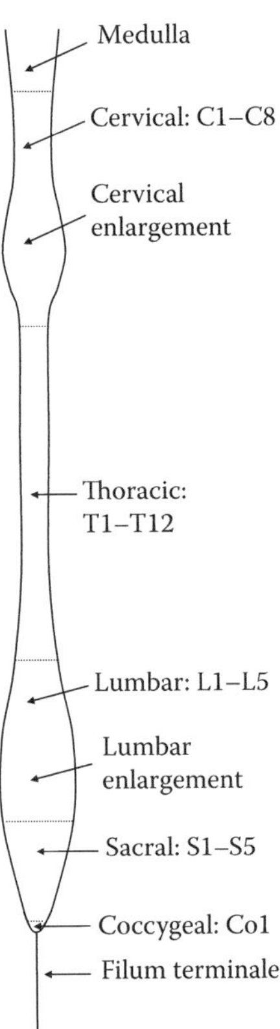

FIGURE 15.3
Illustration of spinal cord.

column, from their respective intervertebral foramina. The ventral (anterior) median fissure and dorsal (posterior) median sulcus (groove) are landmarks that divide the spinal cord into similar, left and right, halves.

Although showing no evidence of internal segmentation, for convenience the spinal cord is divided into 31 spinal segments based on where the 31 pairs of spinal nerves emerge from the spinal cord. In adults, the naming of the more caudal spinal cord segments are more reflective of which intervertebral foramina of the vertebral column the spinal nerves exit, rather than the actual location of the spinal cord segments in the vertebral canal. A letter and number is associated with each spinal cord segment, for example, L2 (lumbar 2), based on the name of the spinal nerve associated with that segment. The name of the spinal nerve, in turn, is related to the name of the vertebra adjacent to where the spinal nerve exits the vertebral column, with some rules. For example, spinal nerves caudal to the first thoracic vertebra are named from the vertebra immediately superior to the nerve; whereas the more rostral cervical spinal nerves are named from the vertebra immediately inferior to the nerve (spinal nerve C8 is in a transition zone, between vertebrae C7 and T1).

The 31 pairs of spinal nerves are subdivided, based on their location along the longitudinal axis of the CNS. Essentially, the 31 pairs of spinal nerves are grouped into five spinal cord regions; the cervical (C), thoracic (T), lumbar (L), sacral (S), and coccygeal (Co) regions. There are eight pairs of cervical nerves, 12 pairs of thoracic nerves, five pairs of lumbar nerves, five pairs of sacral nerves, and one pair of coccygeal nerves. The cross-sectional area of the spinal cord is largest around the caudal lumbar region (L4–L5; called the lumbar enlargement) and central cervical region (C4–C5; called the cervical enlargement), due to an increase in neurons (gray matter) supplying the lower and upper limbs, respectively.

Incoming (afferent) sensory information from the dorsal roots (posterior roots) is received by each spinal segment. Dorsal roots are the axons of sensory neurons that emerge from dorsal root ganglia (collection of nerve cell bodies outside the CNS), and both come in pairs for every spinal segment. The dorsal root fans out into many dorsal rootlets before entering the spinal cord (see Figure 15.4). Hence, sensory connections project into the dorsal (posterior) section of the spinal cord. The ventral roots (anterior roots) contain outgoing (efferent) information in the form of axons of motor neurons whose cell bodies are located in the ventral gray matter of the spinal cord. These axons control somatic or visceral effectors (a muscle cell or gland innervated by a motor neuron) and project out from the ventral (anterior) section of the spinal cord. The axons of the motor neurons emerging from the spinal cord makeup many ventral rootlets that subsequently fuse together to form the ventral root. When the dorsal and ventral roots are fused together, they form a single spinal nerve. Hence, spinal nerves are mixed nerves, consisting of both sensory and motor fibers.

The gray matter of the spinal cord, consisting mostly of neuronal cell bodies and their dendrites, resembles a butterfly configuration when viewed in cross section (see Figure 15.5). Surrounding the gray matter is white matter. Gray matter in the spinal cord is divided into regions called Rexed's laminae (10 layers), based on neural architecture and longitudinal organization—the laminae run parallel to the long axis of the spinal cord. The dorsal horn (posterior horn) is formed by laminae I through VI of the spinal gray matter—these laminae being arranged like flattened sheets. These nuclei generally receive incoming sensory information from the PNS, particularly in regards to pain, temperature, and touch. The ventral horn (anterior horn) is formed by Rexed's laminae VII to IX and contains motor nuclei. These laminae take on more of a rod like appearance. Laminae X

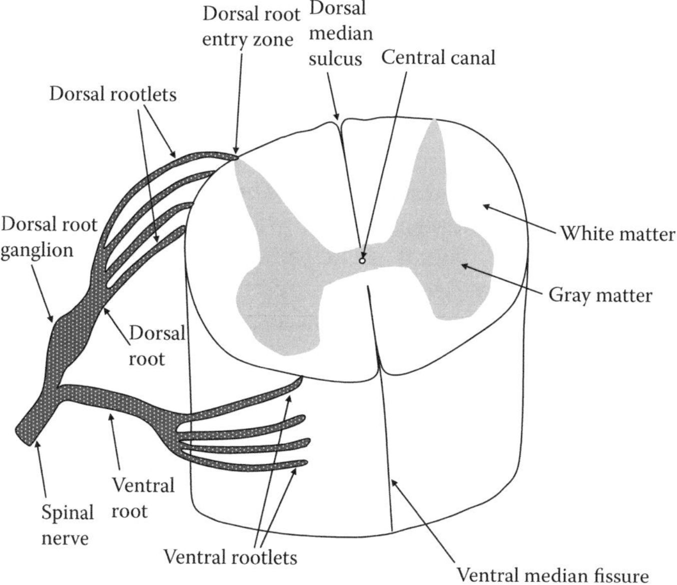

FIGURE 15.4
Diagram of a segment of the spinal cord showing the position of the dorsal and ventral roots and rootlets.

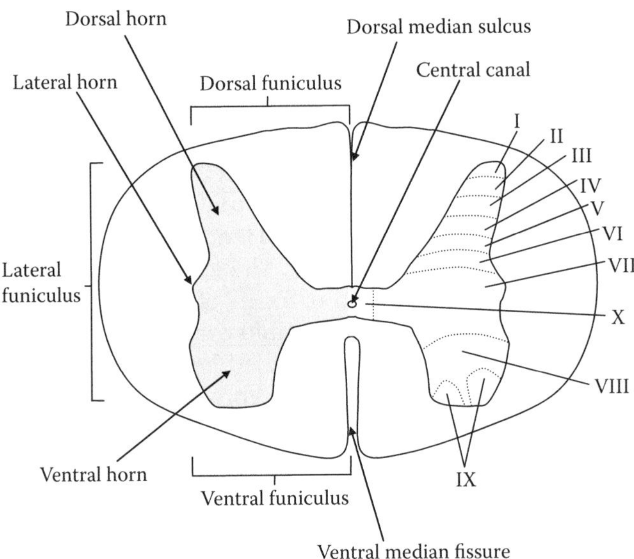

FIGURE 15.5
Figure shows a cross section through the spinal cord, including the approximate locations of Rexed's laminae (I to X).

is the gray matter surrounding the central canal, the latter being a narrow canal that is an extension of the ventricular system into the spinal cord. Lateral horns are present only in the lumbar and thoracic regions.

The white matter in the spinal cord, consisting of both myelinated and unmyelinated axons, is divided into three regions called funiculi (also known as columns). These are the dorsal (posterior) funiculus, ventral (anterior) funiculus, and lateral funiculus. In the spinal cord, the amount of white matter increases in the rostral direction as axons are added to ascending nerve tracts and axons leave descending tracts. Tracts are groups of axons within the CNS that have the same origin and destination and hence carry the same type of information. White matter within the funiculi of the spinal cord is commonly referred to in terms of tracts or fasciculi, the latter being distinct collection of nerve fibers that usually contain more than one tract. The dorsal funiculus consists of two prominent ascending fasciculi, the cuneate fasciculus and gracile fasciculus. The lateral funiculus contains both descending (e.g., lateral corticospinal tract) and ascending (e.g., dorsal spinocerebellar tract) nerve fibers. The ventral funiculus contains several tracts, primarily of descending fibers, including the ventral corticospinal tract, vestibulospinal tract, and reticulospinal tract.

15.4 The Brainstem

15.4.1 Medulla (or Medulla Oblongata)

The medulla is about 3 cm long and is the most caudal part of the brainstem, being continuous with the spinal cord below and the pons above (see Figure 15.6). On the anterior

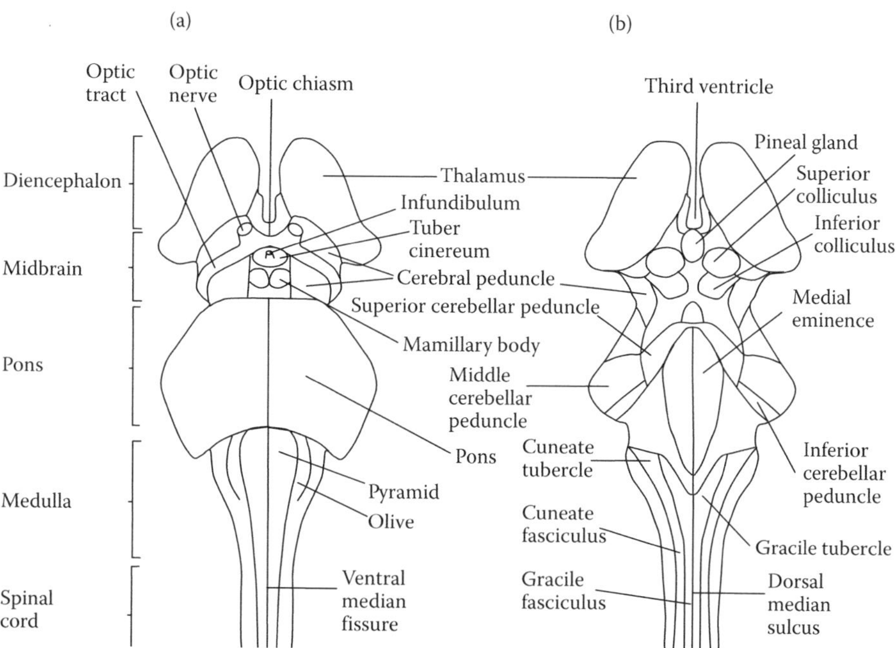

FIGURE 15.6
(a) Illustration of ventral (anterior) surface of the brainstem and diencephalon. (b) Illustration of dorsal (posterior) surface of the brainstem and diencephalon. Note that none of the cranial nerves in the brainstem are shown.

surface of the medulla are two prominent bulges referred to as pyramids, consisting of corticospinal tracts. The pyramids are part of the pyramidal system (involved in voluntary control over skeletal muscle) that begins at the primary motor cortex, consisting of both corticospinal and corticobulbar tracts, the latter being cortical projections that terminate on cranial nerve motor nuclei in the brainstem. Note that nuclei usually refers to clusters of neurons (i.e., gray matter) located beneath the surface of the CNS. They appear grayish, similar to the cortical areas on the surface. About 85% of these corticospinal axons decussate (cross the midline) in the caudal portion of the medulla, and then subsequently descend into the spinal cord as the lateral corticospinal tracts (one for each half of the spinal cord). The remaining fibers descend as the anterior corticospinal tracts and cross over further down the spinal cord instead, in the target area. On the posterior surface of the medulla the cuneate fasciculus and gracile fasciculus continue on from the spinal cord. These ascending axons synapse on the cuneate nucleus and gracile nucleus, respectively, in the dorsal part of the medulla, resulting in two swellings called the cuneate tubercle and gracile tubercle. Above the pyramidal decussation, the axons from the cuneate and gracile nuclei project anteriorly, being collectively referred to as the internal arcuate fibers. After they have crossed the midline the internal arcuate fibers become part of the medial lemniscus (bundle of nerve fibers) brainstem tract. The medial lemniscus tract consists of axons traveling rostrally through the brainstem on the way to their termination point in the thalamus. Lateral to the pyramids in the upper half of the medulla is an oval-like bulge corresponding to the location of the inferior olivary nucleus, which is the largest part of a complex known as the olives (or olivary bodies). Neurons in the inferior olivary nucleus have axons that project to the contralateral half of the cerebellum as the olivocerebellar tract, and as such constitute a significant part of the inferior cerebellar peduncle. Hence, the inferior olivary nucleus is an important relay nucleus to the cerebellum, being involved in the coordination and control of movements.

The medulla contains gray matter (motor and sensory nuclei) associated with the vestibulocochlear (VIII), glossopharyngeal (IX), vagus (X), spinal accessory (XI), and hypoglossal (XII) cranial nerves. Also situated in the medulla is part of the reticular formation, a loosely organized collection of nuclei that makes up a considerable part of the dorsal portion of the brainstem. Through its substantial interconnection with other regions of the CNS, the reticular formation is involved in the regulation of activities such as coordination and control of movements, sleep, pain perception, alertness, and visceral activity. Cardiovascular centers (controls, for example, heart rate and force of contractions) and respiratory rhythmicity centers (controls inspiratory and expiratory respiratory responses) are part of the reticular formation in the medulla. Any communication between the brain and spinal cord has to involve nerve fiber tracts that traverse through the medulla.

15.4.2 Pons (or Pons Varolii)

The pons (in Latin means bridge) is about 2.5 cm long and is situated between the midbrain above and the medulla below. Typically, the pons is divided into an anterior (or ventral) region, referred to as the basal (or basilar) pons, and a posterior (or dorsal) region, known as the tegmentum (this posterior region known as the tegmentum is not just present in the pons, but also extends the length of the medulla and midbrain). The anterior pons has a large bulging appearance and contains longitudinal and transverse running fiber bundles, as well as the pontine nuclei. A primary role of the pontine nuclei is to relay neural information from the ipsilateral cerebral cortex to the controlateral cerebellar hemisphere via the transverse running fibers of the middle cerebellar peduncle. The axons from the relay neurons in the pontine nuclei cross the midline within the pons before entering the cerebellum through the middle cerebellar peduncle. There are three cerebellar peduncles, the superior, middle, and inferior cerebellar peduncles, and the fibers that make up these peduncles (large collections of axons) connect the cerebellum with the rest of the CNS. While the middle cerebellar peduncle is an afferent pathway only, the inferior cerebellar peduncle, running through the more caudal part of the pons, contains both afferent and efferent pathways. The afferent fibers of the inferior cerebellar peduncle include axons from the vestibular nuclei (located in both the pons and medulla), as well as axons from other brainstem regions, and the spinal cord. Efferent fibers of the inferior cerebellar peduncle include axons that project from the deep cerebellar nuclei and vestibulocerebellar cortex of the cerebellum to the reticular formation and vestibular nuclei, respectively, of the brainstem. The superior cerebellar peduncle consists largely of efferent axons that emanate from neurons in the deep cerebellar nuclei (whose input in turn is from the cerebellar cortex), which project (after a relay in the thalamus—the main relay center of sensory information destined for the cortex) to the primary and premotor areas of the cerebral cortex. The fibers cross the midline in the caudal midbrain at the decussation of the superior cerebellar peduncle. Most of these fibers go on to the thalamus, although some end elsewhere, including the red nucleus (involved in limb movement control) of the midbrain. Although a minority, the superior cerebellar peduncle also contains afferent axons that synapse in the cerebellum.

The respiratory rhythmicity centers in the medulla are controlled by the pneumotaxic center and apneustic center located in the rostral and lower pons, respectively (both part of the reticular formation). Also considered part of the reticular formation is the nucleus locus ceruleus, located in the upper pons, which is thought to be involved in several functions, including responses to stress, as well as sleep and wakefulness. The dorsal surface of the pons forms part of the floor of the fourth ventricle. The pons contains nuclei associated with the trigeminal (V), abducens (VI), facial (VII), and vestibulocochlear (VIII) cranial nerves, with the nuclei of the latter being the vestibular and cochlear nuclei (these nuclei being located in both the pons and medulla). The major longitudinal running descending fiber bundles, in the anterior pons, include the corticospinal and corticobulbar tracts. Prominent ascending tracts include the continuation of the medial lemniscus and medial longitudinal fasciculus within the pontine tegmentum, although the medial longitudinal fasciculus also contains a small bundle of descending fibers.

15.4.3 Midbrain (or Mesencephalon)

The midbrain, at about 1.5 cm long, is the shortest and most rostral portion of the brainstem, being situated between the diencephalon above and the pons below. In cross section, the midbrain can be divided into three regions, the tectum (most posterior part), the tegmentum (the central part, which is a continuation of the pontine tegmentum), and the basis pedunculi (most anterior part; also referred to as the crus cerebri). As such, the substantia nigra (in Latin means black substance) constitutes a considerable part of the posterior basis pedunculi, although the substantia nigra is sometimes considered as another separate region. If this is the case, then the basis pedunculi (as defined above) minus the substantia nigra is sometimes known as the pes pedunculi. Alternatively, the midbrain is divided by some into just a tectum and cerebral peduncle; the latter defined as including all structures of the midbrain on each side except the tectum. The periaqueductal gray matter, which surrounds the cerebral aqueduct, and is located in the transitional region between the tectum and tegmentum (which is involved in diverse functions, including being part of a pain suppression circuit) is another structure that is sometimes indicated as a separate subdivision of the midbrain, although it is also sometimes included as either part of the tegmentum or tectum.

The midbrain contains nuclei associated with the oculomotor (III) and trochlear (IV) cranial nerves, the oculomotor nucleus and trochlear nucleus, respectively (both located in the periaqueductal gray matter), with both nuclei involved in motor functions to do with eye movement control. Also, the mesencephalic nucleus of the trigeminal nerve (V) extends from the pons into the midbrain. The inferior and superior colliculi nuclei and associated fiber tracts, are the main features of the tectum. The inferior colliculi, located in the caudal midbrain, is a relay center for auditory information, from the cochlear nuclei to the medial geniculate body of the thalamus, with the latter, in turn, projecting to the primary auditory cortex in the temporal lobe. Neurons in the superior colliculi, located in the rostral midbrain, play an important role in both eye movement control and visual reflexes. The ascending medial lemniscus tract is located within the tegmentum, as is the reticular formation and the prominent red nucleus, the latter playing a minor role in upper limb movement control. The red nucleus receives fibers from the cerebellum and motor areas of the cerebral cortex and gives rise to the descending rubrospinal tract (a minor tract terminating in the cervical spinal cord) and rubroolivary tract, the latter terminating in the inferior olivary nucleus of the medulla, which, in turn, projects to the cerebellum. The anterior portion of the basis pedunculi consists of massive bundles of descending nerve fibers that are either destined for the spinal cord (corticospinal fibers) or brainstem (corticobulbar fibers). The substantia nigra, a component of the basal ganglia, and part of the posterior basis pedunculi, is subdivided into two regions, the substantia nigra pars compacta (adjacent to the tegmentum) and substantia nigra pars reticulata (adjacent to the basis pedunculi). The substantia nigra pars compacta contain efferent dopaminergic projection neurons whose axons terminate in the striatum (the input nuclei of the basal ganglia are sometimes collectively referred to as the striatum); degeneration of these dopaminergic terminations being a feature of Parkinson's disease. The substantia nigra pars reticulata contains gamma-aminobutyric acid (GABA) projecting neurons that synapse primarily at the thalamus.

15.5 Cerebellum

Located posterior to the pons (see Figure 15.7) and connected to the pons via the superior, middle, and inferior cerebellar peduncles, is the cerebellum (in Latin means diminutive of cerebrum, which essentially means "little brain"). The cerebellum plays a major role in maintaining equilibrium (balance) and coordinating muscle contractions. Any communication between neurons inside and outside the cerebellum must be carried out through axons in the cerebellar peduncles. The cerebellum is made up of two hemispheres that are connected via a centrally placed vermis (midline region of cerebellar cortex). A flap of inner dura mater, the tentorium cerebelli, separates the cerebellum from the overlying occipital cortex. The cerebellar cortex is highly convoluted, similar to the cerebral cortex, with ridges or folds referred to as folia in the cerebellar cortex, rather than as gyri (the term

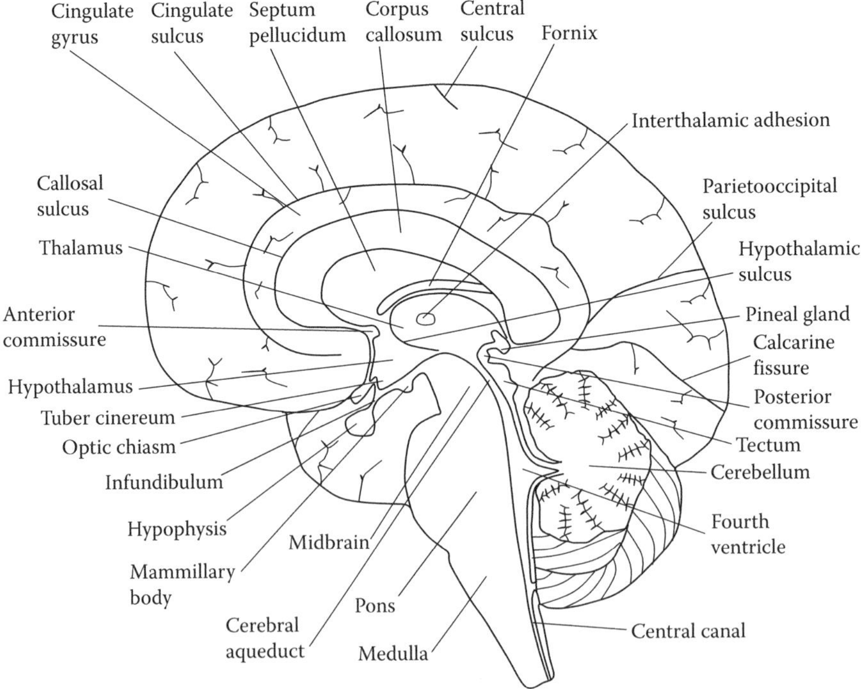

FIGURE 15.7
A midsagittal section illustrating some of the prominent features of the medial brain surface.

used in relation to the cerebral cortex). A number of these folia are organized into groups, called lobules, separated by fissures; there being ten recognized lobules in the cerebellar cortex. Two particularly deep fissures, the primary fissure and posterolateral fissure, divide the lobules into three lobes: the anterior lobe, posterior lobe, and flocculonodular lobe.

Similarly to the cerebral cortex, the cerebellum has a thin layer of gray matter, known as the cerebellar cortex, which contains neuronal cell bodies. While most of the cerebral cortex has six layers (see below), the cerebellar cortex has only three layers. The cerebellar cortex is rich in neurons, in particular Purkinje cells (large, branching neurons that can receive input from up to 200,000 synapses) and granule cells, with the Purkinje cells being the only neurons to have axons that emerge from the cerebellar cortex. The actions of Purkinje cells are entirely inhibitory, their axons terminating in the deep cerebellar nuclei of the cerebellum for relay into the brainstem and thalamus. One exception is with some of the efferent axons from the vestibulocerebellar cortex, which may project directly to the vestibular nuclei of the brainstem. Inhibitory cerebellar interneurons are the Golgi cells, basket cells, and stellate cells. Granule cells are the only excitatory interneurons within the cerebellum (i.e., with cell bodies in cerebellar cortex). Climbing fibers and mossy fibers are the two major inputs (both of them excitatory) to the cerebellum, including deep cerebellar nuclei. In the cerebellar cortex, the target of both the climbing fibers (direct synapse) and mossy fibers (indirect synapse—via the granule cells) are the Purkinje cells.

Internally, there is white matter in the cerebellum, consisting of afferent and efferent nerve fibers. Because the branching pattern of white matter in the cerebellum resembles a tree, particularly as seen in a sagittal section, it is known as the arbor vitae (tree of life). Within this white matter are four bilaterally paired nuclei known collectively as the deep cerebellar nuclei: the fastigial nucleus, emboliform nucleus, globose nucleus, and dentate nucleus. The globose and emboliform nuclei are together called the interposed nuclei. The Purkinje cells of the cerebellar cortex project their axons through the white matter to synapse at these deep cerebellar nuclei.

From a functional perspective, the cerebellum is somewhat imprecisely divided into three functional subdivisions, the spinocerebellum, vestibulocerebellum, and pontocerebellum (also called neocerebellum or cerebrocerebellum), each named from their primary neural information source. The vestibulocerebellum is essentially the flocculonodular lobe, and mostly receives axons from the vestibular nuclei of the brainstem, as well as axons directly from the vestibular ganglion. Efferent axons from the vestibulocerebellar cortex project to the fastigial nucleus of the cerebellum and vestibular nuclei of the brainstem. The vestibulocerebellum is important in eye movement control and in coordinating the actions of muscles involved in maintaining equilibrium. The spinocerebellum, consisting chiefly of the vermis of the anterior lobe and adjoining regions of both hemispheres, receives sensory information from the spinal cord, as well as other brain regions. Fibers from the spinocerebellar cortex project to deep cerebellar nuclei, the fastigial nucleus, and the interposed nuclei. The spinocerebellum is important in controlling posture and movement. The pontocerebellum, involved in the planning of movements, consists of the lateral hemispheres in both the anterior and posterior lobes, as well as the superior vermis in the posterior lobe. Afferents to the pontocerebellum originate primarily from the pontine nuclei (constituting most of the middle cerebellar peduncle), with these axons terminating in both the pontocerebellum and dentate nucleus. In turn, the dentate nucleus primarily projects to the primary motor area of the cerebral cortex, via the ventral lateral nucleus of the thalamus, although some fibers branch off to the red nucleus. The input to the pontine nuclei comes mainly from the cerebral cortex, so the pontocerebellum tracts largely convey information from the cerebral cortex to the cerebellum. Axons of Purkinje cells from the pontocerebellar cortex end in the dentate nucleus.

15.6 Diencephalon

The two major components of the diencephalon are the thalamus and hypothalamus. Other components include the epithalamus, subthalamus, optic chiasm, third ventricle, mamillary bodies, hypophysis (pituitary gland), and infundibulum (infundibular stalk or pituitary stalk). The mamillary bodies (containing mamillary nuclei) and optic chiasm (point of crossing over of the optic nerves) are often considered part of the hypothalamus, and the region they bind is known as the tuber cinereum. The epithalamus includes the habenular nuclei, habenular commissure (interconnects habenula nuclei), pineal gland (also known as pineal body or epiphysis cerebri), and posterior commissure (epithalamic commissure). The pineal gland, attached to the posterior aspect of the third ventricle, is a small conical-shaped structure that produces and secretes the hormone melatonin. The subthalamus contains extrapyramidal motor nuclei, including the globus pallidus, subthalamic nucleus, and zona incerta. The globus pallidus and subthalamic nucleus are considered part of the basal ganglia circuit.

The diencephalon is divided into halves by the third ventricle, with most of its components bilateral (i.e., paired structures). The diencephalon is the central core of the cerebrum, and is located above the midbrain, with the roof of the thalamus forming the superior border. The internal capsule, consisting of axons traveling to and from the cerebral cortex, forms the lateral borders of the diencephalon. The diencephalon's anterior border can be considered to be bound by the most rostral portion of the third ventricle and the anterior commissure (axons that interconnects olfactory and anterior temporal lobe structures of each cerebral hemisphere), with the posterior commissure (axons that interconnect midbrain structures in the separate brainstem halves) forming the posterior border. The floor (inferior surface) of the diencephalon is formed by hypothalamic structures.

The thalami are bilaterally paired egg-shaped gray matter structures that have been divided into numerous nuclei. A major role of the thalamus is to serve as a relay center for sensory information traveling from subcortical regions (e.g., spinal cord, brainstem) to the cerebral cortex. The thalamocortical projections are to the ipsilateral side of the cerebral cortex, and there is usually a reciprocal corticothalamic projection. In addition to receiving input from axons of spinal cord and brainstem nuclei, the thalamus also receives input from other brain structures, including the cerebellum and basal ganglia, as well as the retina of the eyes (via optic nerves [II] and tracts). Based on their locality, the thalamic nuclei can be divided into six groups. The six nuclear groups are the lateral nuclei, medial nuclei, anterior nuclei, intralaminar nuclei, midline nuclei, and reticular nucleus. The thalamic nuclei are also classed as being either relay nuclei or diffuse-projecting nuclei. The relay nuclei project to specific regions of the cortex (e.g., the lateral geniculate nucleus projecting to the primary visual cortex), whereas the diffuse-projecting nuclei project to wide regions of the cortex (e.g., the intralaminar nuclei are thought to be involved in regulating the level of cortical activity and arousal). The thalamic reticular nucleus only project axons to other thalamic nuclei, not to the cerebral cortex, and is involved in coordinating the activity of neurons inside the thalamus.

Despite its small size, the hypothalamus is critically involved in many body functions in regards to homeostasis and survival. Some of the activities the hypothalamus is involved in include regulating body temperature, blood circulation, the sleep–wake cycle, stress response, sexual behavior, appetite, and water balance. A major role of the hypothalamus is as the central region for the control of the two main divisions of the autonomic nervous system, the sympathetic and parasympathetic nervous systems, but not the third division, the enteric nervous system, which operates independent of the CNS. Another major role of the hypothalamus is in the regulation of endocrine functions through its control of the release of hormones from the hypophysis. The hypothalamus also serves as a center that coordinates the activities of the autonomic nervous system and endocrine system. In the diencephalon, the hypothalamus is located inferior to the thalamus. The hypothalami are bilaterally paired structures surrounding the part of the third ventricle that is inferior to the hypothalamic sulcus, the latter a shallow groove separating the hypothalamus from the thalamus on the medial surface. The hypothalamic nuclei can be divided into three mediolateral zones (from the perspective of a coronal section): the periventricular zone (most medial), middle zone, and lateral zone.

The lateral hypothalamic nucleus occupies the length of the lateral zone. This zone is important in integrating and transmitting information about emotions. The periventricular nucleus and arcuate nucleus comprise the periventricular zone. Parvocellular neurosecretory neurons are mostly located in the nuclei of the periventricular zone, these neurons being important in regulating endocrine hormones from the anterior lobe of the hypophysis (known as adenohypophysis). Parvocellular neurons release chemicals into the capillaries of the hypophyseal portal circulatory system in the median eminence region (part of proximal infundibulum) that either promote or inhibit the release of hormones from secretory cells further down in the adenohypophysis. The other part of the hypophysis, the posterior lobe (known as the neurohypophysis), receives input from magnocellular neurons that have their cell bodies located in the middle zone of the hypothalamus, primarily the supraoptic nucleus and paraventricular nucleus. The axons from these nuclei project into the neurohypophysis where they release oxytocin (causes contraction of the uterus during labor and stimulates milk flow from the mammary glands) and vasopressin (also known as antidiuretic hormone [ADH], a hormone that increases the reabsorption of water by the kidney). These two hormones are released from the terminals of the magnocellular neurosecretory neurons in the neurohypophysis directly onto capillaries of the systemic circulation.

The hypothalamus is thought to be involved in regulating the body's circadian rhythm through the indirect actions of the suprachiasmatic nucleus with the pineal gland. The tiny suprachiasmatic nucleus, embedded in the upper surface of the optic chiasm, receives a direct input from the retina, via the retinohypothalamic tract, thereby allowing visual stimuli to synchronize the 24-hour internal clock (or circadian rhythm) of the body. The suprachiasmatic nucleus controls circadian rhythms primarily through local connections with another hypothalamic nucleus, the paraventricular nucleus, which in turn projects to sympathetic preganglionic neurons in the spinal cord. After synapsing on postganglionic neurons in the superior cervical ganglion the signal is finally projected via postganglionic nerve fibers to the pineal gland, where the amount of melatonin secreted is regulated by this incoming stimulus. Melatonin is a hormone that is produced by the pineal gland in darkness, but not in bright light. It is released into the blood outside the blood–brain barrier, although direct release into the cerebrospinal fluid cannot be ruled out. Regardless, melatonin easily crosses the blood–brain barrier, and the level of this hormone in the brain is thought to influence the 24-hour day/night rhythm.

15.7 Telencephalon

The telencephalon is the cerebrum minus the diencephalon and consists of the two cerebral hemispheres. Each cerebral hemisphere has a heavily folded thin surface covering of gray matter, known as the cerebral cortex, and an inner core consisting of numerous nerve fibers, referred to as the white matter. Also, and located deep within each cerebral hemisphere, are masses of gray matter commonly referred to as the basal ganglia, as well as a lateral ventricle in each hemisphere. The folds or ridges on the surface of the cerebral cortex are called gyri, while the grooves or incisures are called sulci. Very deep sulci may instead be referred to as fissures, for example, the longitudinal fissure (or sagittal fissure) that separates the left and right hemispheres. Only about one third of the cerebral cortex is exposed on the brain surface, the remaining surface area confined to the sulcal spaces. The highly convoluted nature of the cerebral cortex allows the cortex to expand its volume by increasing surface area, rather than increasing its thickness. However, the cerebral cortex does vary in thickness, from about 2 mm to 4 mm, being thickest in motor and association areas, and thinnest in primary sensory areas. The cerebral cortex is divided into two basic types, neocortex and allocortex. The neocortex contains six cell layers and constitutes about 95% of the cerebral cortex, with the allocortex making up the other 5% of the cerebral cortex. The allocortex consist of varied types, although primarily cortex that is part of the hippocampal formation and olfactory system, but have in common that they are cortex that contain less than six cell layers. The thickness of the six layers (I–VI) of neocortex varies, as does the density of nerve cells within each layer. For example, the primary motor cortex has a thin layer IV, whereas the primary visual cortex has a thick layer IV. The surface or outermost layer (layer I) is known as the molecular layer, while the innermost layer (layer VI) is known as the fusiform or multiform layer. All output from the cerebral cortex comes from axons belonging to neurons known as pyramidal cells. Pyramidal cells are all excitatory and their cell bodies are shaped like a pyramid, ranging in height from about 10 μm to 100 μm, with the larger sized pyramid cells (also known as Betz cells) found primarily in layer V. There are several types of cortical interneurons, but the major types are classed as stellate cells. There are three general types of efferent projections of axons from pyramidal cells located in the output layers of the cerebral cortex: association, commissural, and projection. Association fibers connect parts of the same hemisphere, for example, the superior longitudinal fasciculus links the frontal and occipital lobes. Commissural fibers connect corresponding regions of the two hemispheres, with the major cerebral commissure being the corpus callosum. Projection fibers descend to the basal ganglia, thalamus, brainstem, or spinal cord.

The surface of the cerebral cortex is divided into the frontal, parietal, occipital, and temporal lobes (see Figure 15.8), with a fifth lobe, called the limbic lobe, surrounding the medial margin of the hemispheres. The cingulate gyrus (see Figure 15.7) and parahippocampal gyrus are the main surface features of the limbic lobe. The limbic lobe is part of a larger brain system, called the limbic system, which also includes structures such as the hippocampal formation, amygdala, septal area, and some thalamic nuclei. Learning, memory, and emotions are some of the main functions the limbic system is thought to be involved in. The paired cerebral hemispheres are divided by the longitudinal fissure, with the corpus callosum located at the floor of the longitudinal fissure. The blunt tips of the occipital, frontal, and temporal lobes are the respective poles. The central sulcus separates the somatic sensory and motor areas, this division serving also as separator of the frontal and parietal lobes.

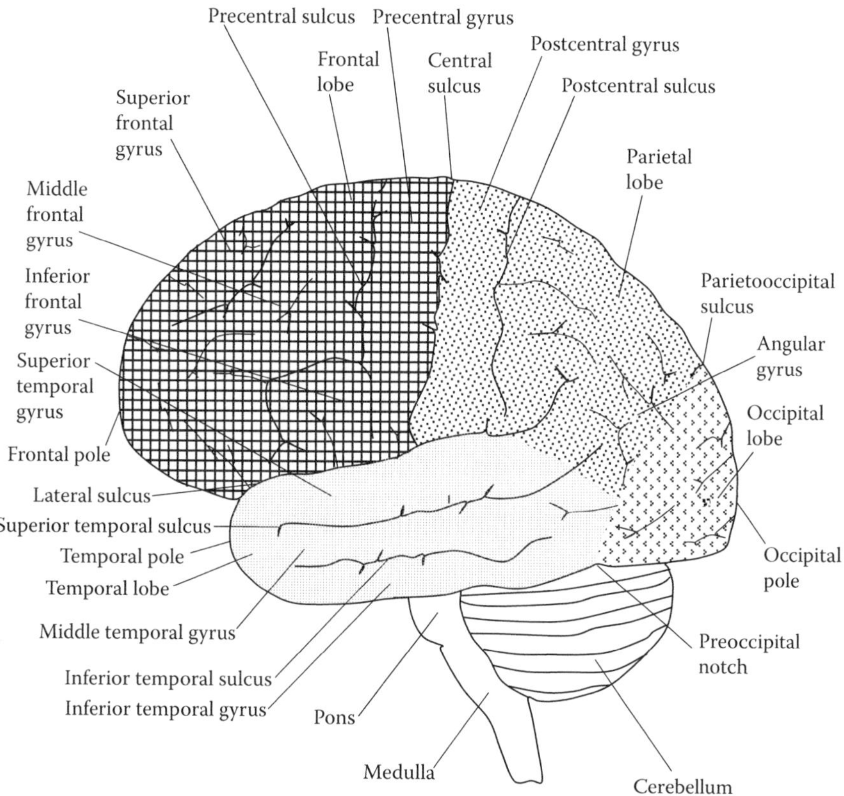

FIGURE 15.8
Lateral surface of the brain.

The lateral sulcus (Sylvian fissure) separates the temporal lobe from the frontal and parietal lobes, although part of the posterior border between the temporal and parietal lobe, as well as between the temporal and occipital lobe, is delineated by an arbitrary line. The occipital lobe is separated from the parietal lobe on the medial hemisphere surface by the parietooccipital sulcus. Imaginary lines connecting the parietooccipital sulcus with the preoccipital notch are the arbitrary boundaries on the lateral and inferior surfaces of the occipital lobe. The insular lobe (also known as insula or insular cortex) is the name given to a portion of the frontal, parietal, and temporal lobes buried deep within the lateral sulcus. Only if the folds of the lateral sulcus (known as the operculum), formed by overgrowth of the frontal, parietal, and temporal lobes, are drawn apart, will the insular lobe become visible. Sensory representations for items such as pain, taste, and balance are located there.

Olfactory nerves (I) mediate the sense of smell and directly enter the telencephalon when they synapse with neurons in the olfactory bulb. Axons of neurons in the olfactory bulb then project down the olfactory tract to several distinct regions, on the inferior surface of the hemispheres, collectively called the primary olfactory cortex. Primary cortices include visual, auditory, somatic sensory, motor, olfactory, and gustatory areas. The primary sensory regions of the somatic sensory, visual, auditory, and gustatory systems receive information directly, albeit via several relay nuclei (e.g., nuclei in thalamus), from peripheral receptors. The primary visual cortex, also known as striate cortex, is located on the banks of the calcarine fissure of the occipital lobe. The primary auditory cortex lies in the temporal lobe, specifically on Heschl's gyrus, while the primary gustatory cortex is located in the insular cortex and nearby operculum. The primary somatic sensory cortex lies on the postcentral gyrus and is involved in the initial processing of somatic sensations. Primary sensory areas are mainly devoted to the reception and initial processing of sensory information, and subsequently send this information to other areas for further processing. Names given to these other areas, such as higher-order, secondary, tertiary, association, and multimodal, can sometimes lead to confusion because they are not always used consistently across the literature. However, a system of dividing the cerebral cortex into over 50 divisions, known as the Brodmann's areas, allows cortical regions to be referred to unequivocally. The primary motor cortex, essential for the voluntary control of movement, lies on the precentral gyrus, and contains neurons that project directly to nuclei in the spinal cord and brainstem. Two well known language areas, Broca's area (important for speech) and Wernicke's area (important for understanding speech), occupy parts of the inferior frontal gyrus and posterior region of the superior temporal gyrus, respectively.

Having to carry out many functions, some type of intrinsic organization is necessary in the cerebral hemispheres. Somatotopy is the orderly mapping of the somatic sensory system and motor system on the cerebral cortex. The representation on the cortex is of the contralateral body parts. Using the mapping of the motor system as an example, in this type of organization the part of your brain that controls finger movements, located on the precentral gyrus, is right next to the part that controls your wrist movements, and the latter, in turn, is right next to the part that controls elbow movements. A similar correspondence exists for the mapping of the somatic sensory system, on the postcentral gyrus, between body surface and sensory perception. In determining the size of cortex dedicated to each body part, the size of the body part is less important than the density of receptors, for area of somatic sensory cortex, or innervation ratio (number of muscle fibers innervated by one motor neuron), for area of motor cortex, with a lower innervation ration generally allowing finer motor control of movements, and hence requiring more area of dedicated cortex.

At the medial aspect of the temporal lobe, and concealed by the parahippocampal gyrus, the cerebral cortex curls up to form the hippocampus. The hippocampus is a key component of the hippocampal formation (includes the hippocampus, dentate gyrus, and subiculum) and is important in learning and memory. The hippocampus is critically important in converting information from short-term memory into long-term memory storage. The subiculum, a continuation of the hippocampus, situated inferiorly and medially to the dentate gyrus, is a transitional zone between six-layered entorhinal cortex (anterior parahippocampal gyrus) and the three-layered hippocampus. The dentate gyrus receives input from the entorhinal cortex and contains neurons that project to the hippocampus. The fornix is a white matter tract that connects the hippocampus with the hypothalamus. The amygdala (or amygdaloid body or amygdaloid nuclear complex), a telencephalic structure that lies in the medial aspect of the temporal lobe, plays an important role in emotions and their behavioral expressions. It consists of three major subdivisions, the medial group, basal-lateral group, and central group. The medial group has extensive connections with olfactory bulb and olfactory cortex, whereas the basal-lateral group has major connections with the cerebral cortex. The central group is characterized by connections with the hypothalamus and brainstem.

Deep within the white matter of the cerebral hemispheres are a group of nuclei collectively known the basal ganglia. The basal ganglia are involved in regulation of stereotyped automatic muscle movements and muscle tone. Clinical findings from diseases involving the basal ganglia suggest that motor control is an important aspect of basal ganglia functions. Diseases involving the basal ganglia include Parkinson's disease (tremor and rigidity of movement) and Huntington's disease (erratic, involuntary movements). The nuclei in the basal ganglia can be divided into three categories: input nuclei, output nuclei, and intrinsic nuclei. The basal ganglia input nuclei receive connections from both basal ganglia and other (nonbasal ganglia) brain regions, and project to the intrinsic and output nuclei of the basal ganglia. Often the term striatum is used to refer to the input nuclei of the basal ganglia. There are three input nuclei in the basal ganglia, the caudate nucleus (participates in eye movement control and cognition), putamen (participates in limb and trunk movement control), and nucleus accumbens (participates in emotions). The caudate nucleus has a C-shape, consisting of a head, body, and tail. The output nuclei of the basal ganglia project to regions of the diencephalon (e.g., thalamus) and brainstem that are not part of the basal ganglia. There are three output nuclei, the substantia nigra pars reticulate (located in the midbrain), globus pallidus-internal segment, and ventral pallidum. In general, the intrinsic nuclei of the basal ganglia project within the basal ganglia. There are four intrinsic nuclei, the substantia nigra pars compacta (located in the midbrain, and contains dopamine projecting neurons that synapse widely at the striatum; a loss of these dopaminergic projection neurons is a feature of Parkinson's disease), ventral tegmental area, globus pallidus-external segment, and subthalamic nucleus (located in the diencephalon). Note that the globus pallidus (external and internal segments) and putamen are sometimes referred to as the lenticular nucleus (or lentiform nucleus) because their form is similar to a lens. The putamen is situated adjacent to the globus pallidus external segment, while the external and internal globus pallidus segments are adjacent to one another.

15.8 Supporting Structures of the Brain

There are three layers of meninges (connective tissue membranes) enclosing the brain. These are the dura mater, arachnoid mater, and pia mater. Together the arachnoid and pia mater are called the leptomeninges. The meninges surround both the brain and spinal cord, known as the cranial meninges and spinal meninges, respectively, and they are continuous with each other. These meninges surround the CNS, suspending it in cerebrospinal fluid (CSF), with the CSF cushioning the brain from contact with the skull during vigorous head movements. This protective space, filled with CSF, and known as the subarachnoid space, is located between the arachnoid and the pia mater (see Figure 15.9).

Dura mater is the thickest, toughest, and outermost of the three meninges and serves as a protective function. Concerning the cranial dura mater, it consists of an inner meningeal layer and outer periosteal layer, the latter attached to the inner surface of the skull, with the former reflected interiorly and partitioning brain tissue. The falx cerebri and the tentorium cerebelli are the two most significant partitions arising from the inner meningeal layer. The falx cerebri incompletely separates the two cerebral hemispheres, while the tentorium cerebelli separates the cerebellum from the cerebral hemispheres. The middle meningeal layer is called the arachnoid mater, and lies adjacent to the dura mater. These two meningeal layers are not tightly bound, allowing a potential space, known as the subdural space, to exist between them. Breakage of blood vessels in the dura mater can lead to subdural bleeding and the formation of a blood clot (subdural hematoma). The space between the arachnoid mater and pia mater is the subarachnoid space. Blood vessels are prevalent here, as is CSF. Arachnoid trabeculae cross the subarachnoid space, giving it a

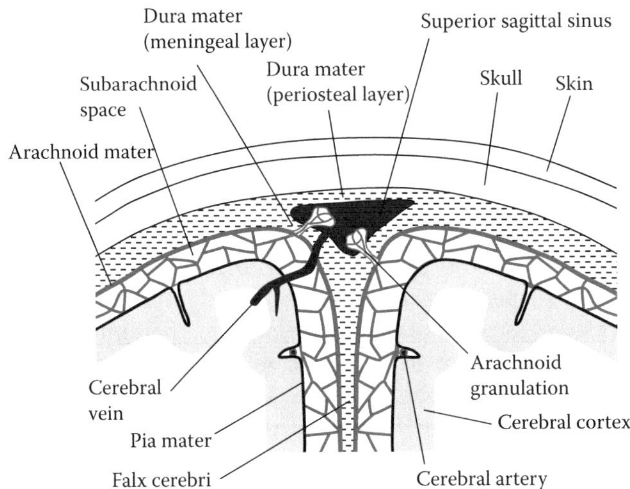

FIGURE 15.9
A schematic coronal section showing the superior sagittal sinus and associated structures, including the meninges (dura mater, arachnoid mater, and pia mater) and blood vessels.

web appearance. Arachnoid granulations are collections of arachnoid villi that protrude into the dural sinuses and also directly into certain veins. The arachnoid villi contain unidirectional valves through which CSF passes to the venous circulation. Dural sinuses are channels through which venous blood and CSF are returned to the systemic circulation. The very delicate pia mater is the innermost layer, and it adheres to the surface of the brain and spinal cord, following the brains contours closely and lining the various sulci.

Circulation and transportation of nutrients, chemicals, and waste products occurs in the CSF. The CSF circulates around the brain via the ventricles. CSF is produced by the choroid plexus, located in the ventricles, which includes specialized cells for the production of CSF. The choroid plexus consists of connective tissue membranes, small blood vessels, capillaries, and nerve fibers. A thin membrane, called the choroidal epithelium, coats the surface of the choroid plexus, and it is these specialized choroidal epithelial cells (also known as ependymal cells) that secrete CSF. It is also the site of the blood CSF barrier. The cells in this membrane are bonded by tight junctions, and contain enzymes specifically for the transport of ions and metabolites across the blood–brain barrier. CSF is also secreted by brain capillaries. This extrachoroidal source of CSF enters the ventricular system through ependymal cells that line the ventricles. The internal environment of most of the CNS is isolated from the blood and hence protected from potentially harmful substances in the blood by the blood–brain barrier. This barrier system, allowing only certain nutrients, hormones, and chemicals through, provides a stable and chemically optimal environment for neuronal function. There are two components to the blood–brain barrier: the blood–CSF barrier (discussed above) and the blood–ECF barrier. Extracellular fluid (ECF) is the fluid surrounding the brain cells (both neurons and neuroglia). The blood–ECF barrier resides in the CNS capillary bed, where the endothelial cells lining the capillaries are tightly joined, and prevent movement of compounds into the extracellular environment. To move compounds across the barrier these cells contain similar transport systems to those of the choroidal epithelium of the blood–CSF barrier. In addition to circulating through the ventricular system and subarachnoid spaces, CSF diffuses passively through the membrane lining the ventricles and enters the extracellular spaces. Hence, it adds to the ECF produced by the capillary bed and cell metabolism. In turn, excess ECF diffuses through the pia mater into the subarachnoid space, and then, as part of the CSF, follows the path through the arachnoid villi into the dural sinuses. This method of fluid drainage in the CNS compensates for the lack of lymphatics.

Cavities containing CSF within the CNS are called ventricles and are lined with ependyma (a thin membrane consisting of a singular layer of epithelial cells). There are two lateral ventricles, one within each cerebral hemisphere, consisting of a body (central part), anterior (frontal) horn, inferior (temporal) horn, and posterior (occipital) horn. The confluence of the three horns is known as the atrium. The lateral ventricles are the largest of the ventricles. A narrow channel, the interventricular foramen, connects each lateral ventricle with the third ventricle. The third ventricle is situated between the halves of the diencephalon. The third ventricle and fourth ventricle are connected via the cerebral aqueduct, with the fourth ventricle situated between the brainstem and the cerebellum. The ventricular system also extends into the spinal cord as the central canal. The CSF exits the ventricular system through apertures in the fourth ventricle, into the subarachnoid space, and from there bathes the surface of the CNS.

Blood supply for the brain is derived from two pairs of trunk arteries, the vertebral arteries (supplies posterior circulation) and the internal carotid arteries (supplies anterior circulation). The left and right vertebral arteries enter the cranium through the foramen magnum and then fuse together to form the single basilar artery. The basilar artery divides into two posterior cerebral arteries, which in turn branch off into the posterior communication arteries. The arteries emanating from the vertebral arteries supply blood to the brainstem, caudal portion of the diencephalon, cerebellum, occipital lobe, and aspects of the temporal lobes. The two internal carotid arteries enter via the base of the cranial cavity and supply blood to the rest of the cerebral hemispheres and diencephalon. The paired anterior cerebral arteries and middle cerebral arteries are the two main arterial branches of the internal carotid artery. The anterior cerebral arteries are linked by the anterior communicating artery. A ring of blood vessels is present on the base of the brain known as the Circle of Willis, and because the anterior and posterior circulations are connected through the Circle of Willis the circulation patterns can change, the brain being able to receive blood from either source. This is important in preventing serious interruptions to blood supply if either the anterior or posterior circulation becomes occluded. However, not all individuals have a fully functional Circle of Willis, due to them missing a component necessary to complete the circle. The spinal cord receives blood from the vertebral and radicular arteries.

Drainage of the blood from the CNS into the major vessels emptying into the heart can occur directly and indirectly. The spinal cord and the caudal medulla veins drain directly into the systemic circulation via a network of veins and plexuses. The rest of the CNS drains indirectly via veins that first empty into the dural sinuses before returning to the systemic circulation. The dural sinuses are valveless low pressure channels that return venous blood back to the systemic circulation, and they also act as a path for flow of CSF into the venous circulation. The dura mater surrounding the brain consists of an inner meningeal layer and outer periosteal layer, and the dural sinuses are located between these layers. The cerebral hemispheres are drained by either superficial or deep cerebral veins, with these veins in turn draining into dural sinuses. Prominent sinuses include the superior sagittal sinus, running along the midline at the superior margin of the falx cerebri, and the inferior sagittal sinus, running along the inferior margin of the falx cerebri. Eventually the venous blood drains into the two transverse sinuses. The transverse sinuses drain into the two sigmoid sinuses, which in turn return blood to the internal jugular veins.

Short Answer/Multiple Choice Questions

1. Which one of the below neural structures is considered part of the cerebrum?
   (a) Pons
   (b) Diencephalon
   (c) Cerebellum
   (d) Medulla oblongata
   (e) Midbrain
2. The cerebellum is considered very important in which function?
   (a) Pain control
   (b) Motor coordination
   (c) Hormone control
   (d) General arousal
3. Name the four surface lobes of the cerebral cortex.
4. What is the "fifth" lobe of the brain called?
5. Does the brain include the spinal cord?
6. The outer part of the cerebrum is what type of matter?
7. Which is the most correct definition of Rostral?
   (a) At or near the tail or hind parts
   (b) Relating to the front surface of the body
   (c) Situated toward the oral or nasal region
   (d) Toward or near the middle of the body
   (e) On or relating to the opposite side of the body
8. Running perpendicular to the longitudinal CNS axis is the:
   (a) Dorsoventral axis
   (b) Rostrocaudal axis
   (c) Superoinferior axis
   (d) Anteromedial axis
9. Name the three main anatomical planes in which brain sections are made.
10. A section through the spinal cord, at right angles to the longitudinal axis, is called a:
    (a) Midline section
    (b) Cross section
    (c) Horizontal section
    (d) Parasagittal section
    (e) Distal section
11. The dentatorubrothalamic tract terminates where?
    (a) Dentate nucleus
    (b) Red nucleus
    (c) Rubro nucleus

(d) Thalamus
(e) Rothalamo nucleus

12. Efferent tracts are axons carrying nerve impulses traveling where?
13. The telencephalon is the cerebrum minus the:
    (a) Diencephalon
    (b) Basal ganglia
    (c) Corpus callosum
    (d) Amygdala
    (e) Hippocampus
14. The hippocampus is well known to be involved in which function?
15. Which CNS division does the pons belong to?
16. What is the better known name for the mesencephalon?
17. What are the three main neural structures that comprise the brainstem?
18. The hypothalamus is part of the:
    (a) Metencephalon
    (b) Diencephalon
    (c) Telencephalon internals
    (d) Mesencephalon
    (e) Myelencephalon
19. The neural plate is derived from which of the three primary germ layers of an embryo?
20. What is the name of the process by which the neural plate region becomes committed to the formation of the CNS?
21. The process of formation of the neural tube from the neural plate is called what?
22. Failure of the caudal portion of the neural tube to close results in which crippling developmental abnormality?
    (a) Anencephaly
    (b) Spina bifida
    (c) Parkinson's disease
    (d) Hydrocephalus
    (e) Broca's aphasia
23. Ridges of gray matter in the cerebral cortex are known as what?
24. Fissures or slit-like incisures in the cerebral cortex are usually called what?
25. The posterior part of the superior temporal gyrus is part of
    (a) Broca's language area
    (b) Wernicke's language area
    (c) Striate cortex
    (d) Striatum
    (e) Limbic lobe

26. The central sulcus separates which two lobes?
27. The calcarine sulcus is located in what cortex?
28. The rhinal sulcus is an extension of which other sulcus:
    (a) Cingulate sulcus
    (b) Parietooccipital sulcus
    (c) Lateral sulcus
    (d) Collateral sulcus
    (e) Superior temporal sulcus
29. About how much of the cerebral cortex is exposed on the surface?
    (a) 1/3
    (b) 2/3
    (c) 1/4
    (d) 1/2
    (e) 3/4
30. Gyri and sulci have important functional significance. Is this statement true or false?
31. What brain section allows the limbic lobe to best be viewed?
32. The insular cortex is buried deep within the:
    (a) Longitudinal fissure
    (b) Superior temporal sulcus
    (c) Cingulate sulcus
    (d) Parietooccipital sulcus
    (e) Lateral sulcus
33. Collectively how many areas are termed the primary olfactory cortex?
    (a) Three
    (b) Four
    (c) Five
    (d) Two
    (e) Six
34. What is the major function of the occipital lobe?
35. What important functions is the limbic lobe associated with?
36. Incoming somatic sensory information is processed in which surface lobe of the cerebral cortex?
37. The best known divisions of the cerebral cortex are the:
    (a) Rexed's laminae
    (b) Brodmann's areas
    (c) Wernicke's areas
    (d) Broca's areas
    (e) Association areas

38. The orderly mapping of the somatic sensory system and motor system on the cerebral cortex is called what?
    (a) Somatotopy
    (b) Visuotopy
    (c) Tonotopy
    (d) Sensory homunculus
    (e) Motor homunculus
39. Multimodal convergent zones process:
    (a) Somatic sensory information only
    (b) Sensory information related to a single modality
    (c) Sensory information from more than one modality
    (d) Motor information only
40. Name the six primary cortices.
41. Which primary sensory area receives information from peripheral receptors that is NOT relayed via nuclei in the thalamus?
    (a) Gustatory cortex
    (b) Visual cortex
    (c) Auditory cortex
    (d) Olfactory cortex
    (e) Somatic sensory cortex
42. The cerebellum consists of how many lobes?
43. What is the name of the inner dura mater that separates the cerebellum from the overlying cerebral cortex?
    (a) Falx cerebri
    (b) Tentorium cerebelli
    (c) Vermis
    (d) Nodulus
    (e) Corona radiate
44. Folds of the cerebellar cortex are called
    (a) Folia
    (b) Lobules
    (c) Gyri
    (d) Flocculi
45. Which statement below best reflects the function of the flocculonodular lobe of the cerebellum?
    (a) Movement planning
    (b) Nonmotor functions
    (c) Eye movement control and maintaining balance
    (d) Control of limb and trunk movements

46. Which neuron projects the only axons to emerge from the cerebellar cortex?
    (a) Granule cells
    (b) Basket cells
    (c) Mossy fibers
    (d) Purkinje cells
    (e) Golgi cells
47. The actions of the axons of Purkinje cells are
    (a) Both excitatory and inhibitory
    (b) Excitatory only
    (c) Inhibitory only
48. The clusters of neurons deep inside the white matter of the cerebellum are known as what?
49. The cerebellum is connected to the brainstem by three pairs of what?
50. A classic sign or symptom of injury to the cerebellum is
    (a) Paralysis of limbs
    (b) The observation of a drunken gait
    (c) Experiencing constant headaches
    (d) Involuntary, erratic movements
    (e) Difficulty in hearing
51. The termination point of the spinal cord is known as the:
    (a) Conus medullaris
    (b) Filum terminale
    (c) Dorsal root ganglia
    (d) Medulla spinalis
52. How many pairs of spinal nerves are there?
53. The dorsal roots usually contain what type of information:
    (a) Outgoing information in the form of axons of motor neurons
    (b) Afferent sensory information
    (c) Both afferent and efferent information
54. The ventral horn consist of spinal cord:
    (a) White matter
    (b) Cerebrospinal fluid
    (c) Gray matter
    (d) Connective tissue
    (e) Epithelial tissue
55. Rexed's laminae I through VI forms the:
    (a) Dorsal horn of the spinal cord
    (b) Ventral horn of the spinal cord

(c) Posterior column of the spinal cord
(d) Anterior column of the spinal cord

56. The white matter on each side of the spinal cord is divided into which three regions?
57. The anterior spinothalamic tract carries:
    (a) Sensations of crude touch and pressure directly to the cortex
    (b) Sensations of crude touch and pressure to the thalamus
    (c) Motor information from the thalamus to peripheral effectors
    (d) Motor information from the cortex to ventral horn motor neurons
58. The medial lemniscus is a brainstem tract:
    (a) Containing primarily descending motor fibers for controlling axial and proximal limb structures
    (b) Seen as a pair of thick bands visible along the ventral surface of the medulla
    (c) Relaying sensory information from the thalamus to the primary sensory cortex
    (d) Containing axons traveling from dorsal column nuclei to the thalamus
59. Dermatomes:
    (a) Are the area of skin innervated by particular spinal nerves
    (b) Are two prominent bulges on the ventral surface of the medulla formed from corticospinal axons
    (c) Consists of a vast complex of loosely organized gray matter in the brainstem that contains embedded nuclei
    (d) Are mixed nerves, consisting of both afferent and efferent fibers, that are bound together to form a single spinal nerve
60. In what part of the brainstem are the pyramids located?
    (a) On the ventral side of the medulla after the descending tracts cross over
    (b) On the dorsal side of the medulla before the descending tracts cross over
    (c) On the ventral side of the medulla before the descending tracts cross over
    (d) On the ventral side of the pons after the descending tracts cross over
61. The inferior olivary nucleus is part of the:
    (a) Midbrain
    (b) Medulla oblongata
    (c) Pons
    (d) Thalamus
    (e) Spinal cord
62. The ascending medial lemniscus brainstem tract begins in the
    (a) Midbrain
    (b) Medulla oblongata
    (c) Pons
    (d) Cerebellum
    (e) Spinal cord

63. The locus ceruleus is "loosely" considered as part of the:
    (a) Reticular formation
    (b) Medial lemniscus
    (c) Cardiac reflex center
    (d) Pontine nuclei
    (e) Respiratory rhythmicity center
64. The pontine nuclei is located in the:
    (a) Midbrain
    (b) Cerebellum
    (c) Medulla oblongata
    (d) Hypothalamus
    (e) Pons
65. Neurons in the pontine nuclei
    (a) Participate in skilled movement control
    (b) Are involved in pain suppression
    (c) Keeps the cerebrum conscious, vigilant, and alert
    (d) Links the cerebellum with the medulla oblongata and spinal cord
66. Which is the shortest brainstem portion?
67. Neurons in the inferior colliculi are important:
    (a) In language processing
    (b) In controlling saccadic eye movement
    (c) In hearing
    (d) In the pain suppression circuit
    (e) In controlling breathing reflexes
68. Degeneration of the axon terminations from which nuclear mass is believed responsible for Parkinson's disease?
    (a) Periaqueductal gray
    (b) Substantia nigra
    (c) Red nucleus
    (d) Superior colliculi
    (e) The olives
69. Name the two largest structures of the diencephalon.
70. The pineal body produces which hormone:
    (a) Oxytocin
    (b) Vasopressin
    (c) Melanin
    (d) Melatonin
    (e) Thyroid hormones

71. Which hypothalamic nucleus participates in setting the normal sleep–wake cycle by controlling hormone production in the pineal gland?
    (a) Supraoptic nucleus
    (b) Suprachiasmatic nucleus
    (c) Lateral hypothalamic nucleus
    (d) Periventricular nucleus
    (e) Arcuate nucleus
72. Magnocellular neurosecretory neurons project their axons:
    (a) Directly into the anterior pituitary gland
    (b) Directly into the neurohypophysis
    (c) Close to blood vessels that take up chemicals released by the neurons, which then either promote or inhibit the release of hormones from secretory cells further down in the adenohypophysis
    (d) Close to blood vessels that take up chemicals released by the neurons, which then either promote or inhibit the release of hormones from secretory cells further down in the posterior portion of the pituitary gland
73. Which function is not generally true about the hypothalamus?
    (a) It is a control center for the autonomic system
    (b) It regulates reproductive development and function
    (c) The hypothalamus controls hormone production in the body
    (d) It helps regulate temperature, metabolic rate, water balance, appetite
    (e) It is involved in the processing of visual and auditory data
74. The thalamus is NOT a major relay center for sensory information going to the:
    (a) Olfactory cortex
    (b) Basal ganglia
    (c) Cingulate gyrus
    (d) Primary visual cortex
    (e) Primary auditory cortex
75. The medial geniculate nuclei of the thalamus:
    (a) Relays auditory information from the inferior colliculus to the primary auditory cortex.
    (b) Gets input from the retina and relays the visual information to the primary visual cortex
    (c) Projects visual information from superior colliculus to higher-order visual areas
    (d) Relays somatic sensory information to the primary somatic sensory cortex
76. Which thalamic reticular nuclei/nucleus only projects axons within the thalamus?
    (a) Lateral geniculate nuclei
    (b) Intralaminar nuclei
    (c) Reticular nucleus

(d) Pulvinar nucleus
(e) Lateral nuclei

77. How many pairs of cranial nerves are there?
78. Which of the below names does NOT belong to a cranial nerve?
    (a) Optic
    (b) Oculomotor
    (c) Vagus
    (d) Genu
    (e) Facial
79. The corona radiata
    (a) Is the portion of the subcortical white matter superior to the internal capsule
    (b) Is the portion of the subcortical white matter inferior to the internal capsule
    (c) Is the name of the input nuclei of the basal ganglia
    (d) Is a thalamic nuclei
80. Thalamic neurons whose axons project to the cerebral cortex send their axons primarily to which layer?
    (a) I
    (b) II
    (c) III
    (d) IV
    (e) V
81. The superior longitudinal fasciculus
    (a) Is a bundle of long association fibers
    (b) Is a minor commissure
    (c) Is a bundle of descending projection fibers
    (d) Is a nerve tract in the spinal column
82. The nuclei in the basal ganglia can be divided into three categories. Name the three categories.
83. Together the input nuclei of the basal ganglia are also known as the:
    (a) Extrastriate cortex
    (b) Lenticular nuclei
    (c) Striatum
    (d) Substantia nigra
    (e) Fornix
84. What are the two main types of nerve cells in the cortex?
    (a) Stellate cells and oligodendrocytes
    (b) Pyramidal cells and stellate cells
    (c) Pyramidal cells and oligodendrocytes

(d) Schwann cells and astrocytes
(e) Satellite cells and microglial cells

85. Giant pyramidal cells are known as
(a) Basket cells
(b) Stellate cells
(c) Chandelier cells
(d) Betz cells
(e) Oligodendrocytes
86. How many layers are there in the neocortex?
87. The cerebral cortex varies in thickness from about
(a) 1 mm to 2 mm
(b) 4 mm to 6 mm
(c) 2 mm to 4 mm
(d) 1 mm to 6 mm
(e) 1 mm to 10 mm
88. The neocortex makes up approximately how much of the cerebral cortex?
(a) 95%
(b) 100%
(c) 50%
(d) 75%
(e) 5%
89. The molecular layer:
(a) Is the innermost layer
(b) Is the outermost layer
(c) Is the middle layer
(d) Is the layer receiving input from thalamic neurons
90. The fusiform layer:
(a) Contains medium-sized pyramidal and stellate cells
(b) Contains spindle-shaped irregular pyramidal cells that project to the thalamus
(c) Contains few nerve cell bodies, but does contain glial cells and the tips of apical dendrites of pyramidal cells
(d) Is a dense layer composed of small pyramidal and stellate cells?
91. Name the three layers of connective tissue membranes completely enclosing the brain?
92. The subarachnoid space is filled with what?
93. Arachnoid granulations protrude into
(a) The ventricles
(b) The capillaries

(c) The dural sinuses
(d) The lumbar cistern

94. Cerebrospinal fluid (CSF) is
    (a) High in protein concentration
    (b) Is a clear colorless solution
    (c) Fills the subdural space
    (d) Does not surround the spinal cord
95. Name the ventricles in the brain.
96. How much cerebrospinal fluid is secreted by the choroids plexus every 24 hours?
    (a) 200 mL
    (b) 400 mL
    (c) 500 mL
    (d) 300 mL
    (e) 100 mL
97. Blockage of flow through the ventricular system or cranial subarachnoid space can result in what?
    (a) Parkinson's disease
    (b) Huntington's disease
    (c) Hydrocephalus
    (d) Spina bifida
    (e) Stroke
98. Name the two components of the blood–brain barrier?
99. The principal blood supply for the brain is derived from which two pairs of trunk arteries:
    (a) The vertebral arteries and the internal carotid arteries
    (b) The internal carotid arteries and the middle cerebral arteries
    (c) The middle cerebral arteries and the vertebral arteries
    (d) The posterior cerebral arteries and the anterior cerebral arteries
    (e) The basilar arteries and internal carotid arteries
100. Occlusion of which cerebral artery is the most common site of a stroke?

Answers to Questions

1. (b) Diencephalon
2. (b) Motor coordination
3. Frontal, parietal, occipital, and temporal lobes
4. The limbic lobe
5. No

6. Gray matter
7. (c) Situated toward the oral or nasal region
8. (a) Dorsoventral axis
9. The horizontal, coronal, and sagittal planes
10. (b) Cross section
11. (d) Thalamus
12. Traveling away from a particular structure—output
13. (a) Diencephalon
14. Memory
15. Metencephalon
16. Midbrain
17. Midbrain, pons, and medulla oblongata
18. Diencephalon
19. Ectoderm
20. Neural induction
21. Neurulation
22. (b) Spina bifida
23. Gyri
24. Sulci
25. (b) Wernicke's language area
26. Frontal and parietal lobes
27. Primary visual cortex
28. (d) Collateral sulcus
29. (a) 1/3
30. False
31. Median brain section
32. (e) Lateral sulcus
33. (c) Five
34. Primary visual processing
35. Emotions, learning, and memory
36. Parietal lobe
37. (b) Brodmann's areas
38. (a) Somatotopy
39. (c) Sensory information from more than one modality
40. Visual, auditory, somatic sensory, motor, olfactory, and gustatory cortical areas
41. (d) Olfactory cortex
42. Three
43. (b) Tentorium cerebelli
44. (a) Folia

45. (c) Eye movement control and maintaining balance
46. (d) Purkinje cells
47. (c) Inhibitory only
48. Deep cerebellar nuclei
49. Cerebellar peduncles
50. (b) The observation of a drunken gait
51. (a) conus medullaris
52. 31 pairs
53. (b) Afferent sensory information
54. (c) Gray matter
55. (a) Dorsal horn of the spinal cord
56. Posterior, anterior, and lateral white columns
57. (b) Sensations of crude touch and pressure to the thalamus
58. (d) Containing axons traveling from dorsal column nuclei to thalamus
59. (a) Are the area of skin innervated by particular spinal nerves
60. (c) On the ventral side of the medulla before the descending tracts cross over
61. (b) Medulla oblongata
62. (b) Medulla oblongata
63. (a) Reticular formation
64. (e) Pons
65. (a) Participate in skilled movement control
66. Midbrain
67. (c) In hearing
68. (b) Substantia nigra
69. Thalamus and hypothalamus
70. (d) Melatonin
71. (b) Suprachiasmatic nucleus
72. (b) Directly into the neurohypophysis
73. (e) It is involved in the processing of visual and auditory data
74. (a) Olfactory cortex
75. (a) Relays auditory information from the inferior colliculus to the primary auditory cortex
76. (c) Reticular nucleus
77. 12
78. (d) Genu
79. (a) Is the portion of the subcortical white matter superior to the internal capsule
80. (d) IV
81. (a) Is a bundle of long association fibers
82. Input nuclei, output nuclei, intrinsic nuclei

83. (c) Striatum
84. (b) Pyramidal cells and stellate cells
85. (d) Betz cells
86. Six
87. (c) 2 mm to 4 mm
88. (a) 95%
89. (b) Is the outermost layer
90. (b) Contains spindle-shaped irregular pyramidal cells that project to the thalamus
91. The dura mater, arachnoid mater, and pia mater
92. Cerebrospinal fluid (CSF)
93. (d) The dural sinuses
94. (b) Is a clear colorless solution
95. Two lateral ventricles, third ventricle, and fourth ventricle
96. (d) 300 mL
97. (c) Hydrocephalus
98. The blood–CSF (cerebrospinal fluid) barrier and the blood–ECF (extracellular fluid) barrier
99. (a) The vertebral arteries and the internal carotid arteries
100. Middle cerebral artery (MCA)

Bibliography

Fitzgerald MJT, Folan-Curran J. 2002. *Clinical Neuroanatomy and Related Neuro*science, 4th Edition. WB Saunders, Edinburgh.

Greenstein B, Greenstein A. 2000. *Color Atlas of Neuroscience: Neuroanatomy and Neurophysiology*. Thieme, Stuttgart.

Heimer L. 1995. *The Human Brain and Spinal Cord: Functional Neuroanatomy and Dissection Guide*, 2nd Edition. Springer-Verlag, New York.

Kahle W, Frotscher M. 2003. *Color Atlas of Human Anatomy, Vol. 3: Nervous System and Sensory Organs*, 5th Edition. Thieme, Stuttgart.

Kiernan JA. 2009. *Barr's: The Human Nervous System: An Anatomical Viewpoint*, 9th Edition. Lippincott Williams & Wilkins, Baltimore.

Martin JH. 2003. *Neuroanatomy: Text and Atlas*, Third Edition. McGraw-Hill, New York.

Martini FH, Nath JL. 2009. *Fundamentals of Anatomy and Physiology*, 8th Edition. Pearson Education, San Francisco.

Nieuwenhuys R, Voogd J, van Huijzen C. 2008. *The Human Central Nervous System*, 4th Edition, Springer-Verlag, Berlin.

Nolte J. 2009. *The Human Brain: An Introduction to its Functional Anatomy*, 6th Edition. Mosby, Philadelphia.

Purves D, Augustine GJ, Fitzpatrick D, Hall WC, LaMantia A-S, McNamara JO, White LE (Editors). 2008. *Neuroscience*, 4th Edition. Sinauer Associates, Sunderland, MA.

Siegel A, Sapru HN. 2006. *Essential Neuroscience*. Lippincott Williams & Wilkins, Baltimore.

Verkhratsky A. 2007. Introduction to the nervous system, Chapter 3. In: Petersen OH (Editor), *Lecture Notes: Human Physiology*, 5th Edition. Blackwell Publishing, Massachusetts, pp. 37–62.

16

The Biophysics of Sensation—General

Mark A. Schier and Andrew W. Wood

CONTENTS

16.1 General Principles

Sensation is an awareness or feeling of conditions on, surrounding or within the body. The traditional five senses of sight, hearing, taste, touch, and smell are augmented in some animals with sensing of magnetic and electric fields. In all of these classes of sensation there are specialized sensory receptors connected to sensory nerves involved. There are several principles that assist in setting the scene for understanding sensation which relate to the properties of the underlying structures and connections.

1. Sensory endings are the peripheral terminals of the afferent nerve fibers.
2. Sensory endings possess thresholds. In other words, some types of stimuli elicit afferent nerve impulses, whereas others will not at a given ending.
3. A sensory unit is a single primary afferent nerve fiber including all peripheral branches and central terminals, and its receptor cells.
4. A peripheral receptive field is the spatial area inside which a stimulus of the right strength and type will cause firing of the sensory unit. The threshold varies within the field, usually being the lowest near the center.
5. The peripheral branches of adjacent sensory fields overlap. The amount of overlapping depends on what area and to which modality the afferent fibers belong. The greater the degree of overlap, the greater the resolution within a given area. The number of sensory units having a receptive field within a given area is called the peripheral innervation density (or PID). This area has a representation on the cortex, equivalent to the number of sensory afferents (this is also known as the homunculus).
6. Different sets of nerve fibers, when active elicit different sensations by virtue of their central connections. The fibers elicit the same response whether stimulated naturally or artificially.
7. Sensory fibers differ from each other in the way they respond to a stimulus. For example a step function of sensory stimulus is applied to a receptive field. The activity in the primary afferent peaks shortly after the onset and then monotonically decreases. The decrease in output is called the adaptation of the fiber (Figure 16.1).

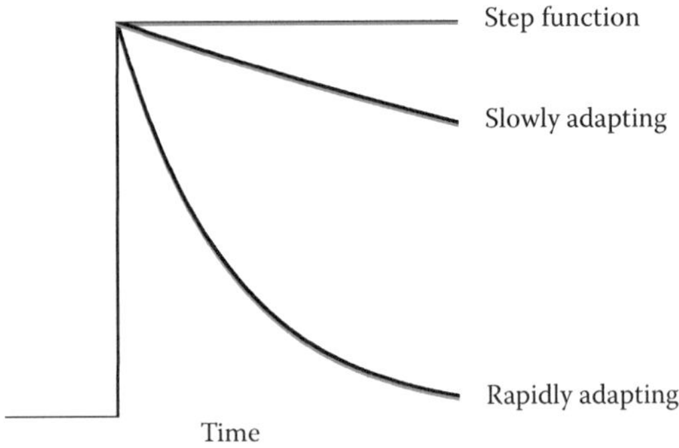

FIGURE 16.1
Adaptation: two levels of adaptation and a reference step function.

16.1.1 The Sensory Cell

We are aware of our environment through sensation. In addition to the traditional five senses of sight, hearing, smell, taste, and touch can be added the senses of heat and cold, movement and position, stretch, and pain. Many fish and monotremes such as the platypus are able to detect electric fields and birds and certain bacteria respond to the Earth's magnetic field. These environmental factors influence specific cells, or groups of cells. Although these cells have some features in common, ultimately there is a mechanism that confers specificity on each class of sensory cell, such that pain receptors do not respond to stimulation by light, for example. However, as we shall see, there is a certain amount of cross-sensitivity, such as the retina responding to pressure or electric current in the eyeball to give sensations of light ("phosphenes"). We will initially consider a generalized sensory cell shown in Figure 16.2. Internal current flows from top to bottom (apex to base) within the cell with the circuit made up by current in the external solution. A generalized stimulus (of a type depending on the specific cell) is applied to the apical regions, modulating the current to give rise to receptor (or generator) potentials in the membrane in this region (which attenuate the further they are from the point of stimulus). Further down the cell this depolarization may be sufficient to elicit action potentials, which are then propagated in an unattenuated manner down to the base, where neurotransmitter substances are released. In a sense, the apical part is similar to an axon hillock or a neuron soma, with temporal and spatial summation and the lower part is similar to an axon. Some receptors actually lack the "axon-like" portion.

There are three stages of the initial stimulus transduction.

(a) *Sensor mechanism*: Molecules in the membrane or the sensory region confer specific sensitivity, i.e., light of a particular wavelength, temperature or specific taste. These molecules convert the energy in the stimulus to the next stage. Some of these

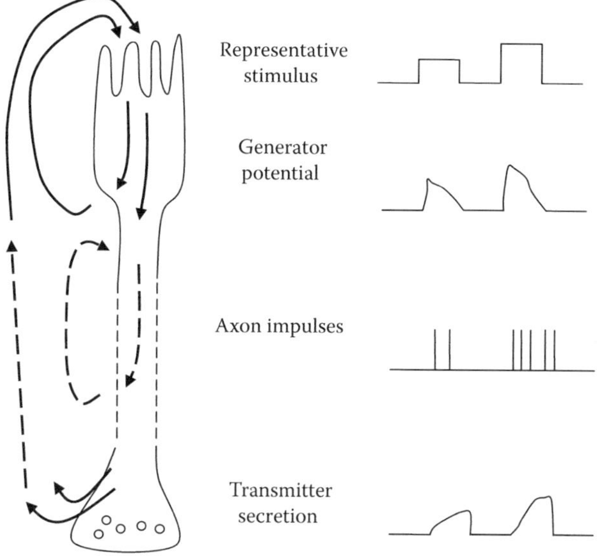

FIGURE 16.2
The signals at various stages of transduction in a generalized sensory cell. (With kind permission from Springer Science+Business Media: *Biophysics*, Fundamentals of transduction mechanisms in sensory cells, 1983, 657–666, Thurm, U. [Hoppe, W., Lohmann, W., Markl, H., Ziegler, H., eds.], Figure 15.23.)

molecules are extremely sensitive, such as responding to a single photon of light or a displacement of 10^{-10}m (less than the width of a "throat" of a K^+ channel). These small effects are amplified in some manner (ultimately relying on an external energy source).

(b) *Effector mechanism*: The amplified effects on the sensor molecules exert a control over membrane permeability via gated channels. Unlike axon Na^+ and K^+ channels, these are (relatively) insensitive to membrane potential. This modulation of membrane conductance gives rise to the change in receptor current (and hence generator potential) noted above. As in other membranes, the conductance changes show some degree of adaptation, as shown in Figure 16.1.

(c) *Environment mechanism*: In order for there to be a change in receptor current there needs to be a current there in the first place. This is linked to active ion transport and to cell metabolism.

16.1.1.1 Forms of Coupling

This can either be direct (in which there is only a 10-μsec or so delay between stimulus delivery and channel opening, or indirect (where the delay could be several orders of magnitude longer) involving a cascade of enzymes (amplification, see above). For example, the light sensing molecule is Rhodopsin, but the membrane regions where conductance is varied is several micrometers away.

16.1.1.2 Comparison with Axon Action Potentials

From the above description it may be hard to appreciate how sensory cells differ from more generalized neurons. Indeed, there are many similarities, but the following represent some degree of specialization. First, the sensory cells, in general, are part of an epithelial layer, held to adjoining cells by high-resistance tight junctions. The apical region also tends to protrude into the mucosal space above the brush border of the epithelium (see Figure 16.3), which tends to be Na^+-rich and K^+-poor. The receptor currents in the apical regions tend to be fixed in space, unlike the propagated changes in axons. Rather like a specialized soma, the apical region exhibits spatial and temporal summation giving rise to graded rather than all-or-none responses (which may occur in the more basal regions). Of course there are exceptions to the scheme in Figure 16.3, which will be dealt with below.

There is a wide diversity of receptor cells types, but many (such as retinal, ear hair-cells, touch receptors) have a specialized protuberance or "finger" extending into the external space, which shares some features of the hair-like cilia which are found on the apical surface of most epithelia. One particular feature is the ring of nine double tubules, which in cilia are "motile" or moveable, provide the molecular motor machinery to drive cilia to move, for example, mucus within bronchial tubes. In the hair cells in the cochlea and elsewhere this is used in reverse to generate current from mechanical motion.

16.1.1.3 Equivalent Electrical Circuit

The inward current (which is borne by both Na^+ and K^+ ions at the apical region arises from a K^+ leak, setting up a diffusion potential in the region beyond the tight junction (the Na/K pump can be electrogenic or electroneutral, it makes little difference to the argument). There is thus a potential gradient within the cell from the more positive apex to

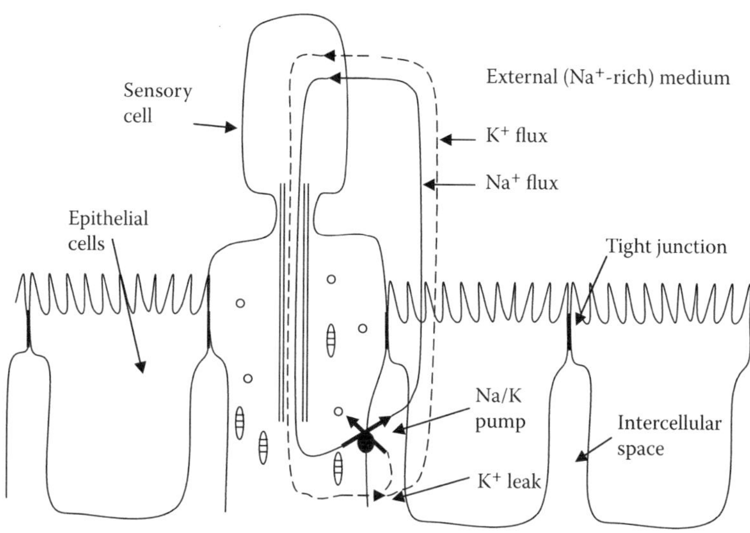

FIGURE 16.3
An illustration of the intracellular and extracellular currents of the sensory cell, specifically the potassium (K^+) and sodium (Na^+) fluxes.

the more negative base. The external regions above the epithelium tend to be negative, so current is drawn up through the adjacent cells to provide the circuit, but only if the apical membrane has sufficient permeability to allow it. The stimulus controlled permeability (or conductance) changes (which as we have seen may be somewhat remote from the sensor molecules) occur in this outer segment. Increased conductance will lead to an enhanced current and thus a depolarization of the membrane in these regions.

The stimulus is applied to a receptor site. The receptor site may be part of the nerve ending itself (e.g., bare nerve endings, olfactory tissue) or may be a nonneural cell that can activate a nerve cell (e.g., Merkel disk, taste bud). The energy of the stimulus is then transduced by the sensory receptor cell into either release transmitter across to the nerve ending or to cause a physiochemical change in the nerve terminal membrane. Following this stimulation of the nerve terminal membrane, there is a local permeability change in the surface membrane, which enables some charge transfer to occur, and a generator potential to be initiated. The generator potential is a nonpropagated, electrotonically spreading, graded potential, and will induce an action potential at some point in the nerve membrane.

The sequence of action is summarized: stimulus → local change in permeability → generator current (charge transfer) → local depolarization (generator potential) → conducted action potential (frequency is proportional to intensity).

The precise point at which the generator (or receptor) potential is transformed into an action potential varies between receptors, but two types can be distinguished. The *synaptic* type involves release of neurotransmitter substances in response to generator potentials (rather than action potentials) gives rise to postsynaptic potentials, which ultimately lead to action potentials in the postsynaptic cells. The other type is more similar to a neuron, in which action potentials develop within the axon of the sensory fiber itself. An example of this is the Pacinian corpuscle, which senses pressure in tissue, which still responds to mechanical stimuli if the "onion-like" layers are removed (see Figure 16.4). The response is a depolarization in the membrane in the first part of the fiber. Since this withdraws charge

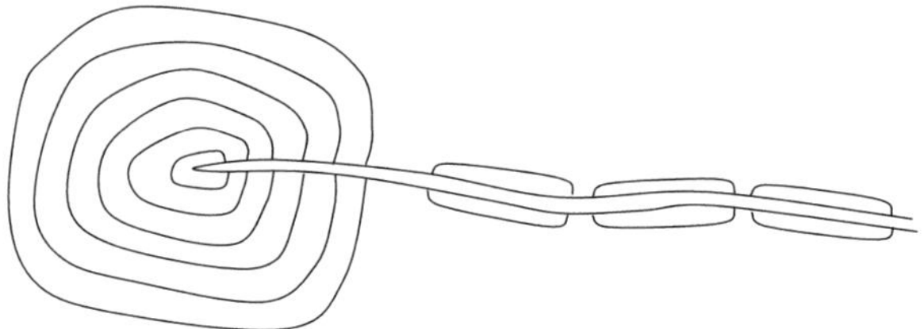

FIGURE 16.4
The structure of the Pacinian corpuscle showing the layered arrangement around the sensory ending.

from the first node of Ranvier, the action potential is elicited there (the impulse is then propagated by the usual saltatory conduction).

The responses to stimuli can also be differentiated into two types. The phasic (dynamic or differential) responses produce a generator potential proportional to the time rate of change of the stimulus (velocity in the case of mechanical receptors). Thus, an application and removal of a constant stimulus will give generator potentials at the start and finish of the application of the stimulus, but not during. Tonic (static or proportional) responses, on the other hand are proportional to the magnitude of the stimulus and will continue for all times that the stimulus is applied, although some diminution, or adaptation (see Figure 16.1) is usual.

Before the sensory information is conveyed to the CNS, there is some degree of "arborization," in which, like tree branches, the axons from several receptor cells will be joined together. There is this a "primary receptor field" over which two stimuli applied in this region simultaneously will be sensed as one. There can also be convergence in central neurons, where several afferents come together. In the cortex, there is further modification (including lateral inhibition) that contributes to the final subjective discrimination between two stimuli applied simultaneously. In mechanoreceptors, for example, the minimum discrimination distance is 2 mm–3 mm on the fingertips and tongue, but 65 mm on regions of the back.

16.1.2 The Coding of Stimulus Amplitude as Impulse Frequency

There can exist several types of response, and basically they all obey the same type of general law—that of a power function:

$$f = k(S - S_0)^n$$

where $n \geq 0$

f is the frequency of action potentials
S is stimulus intensity
S_0 is stimulus threshold
k is a constant

This function is logarithmic when $0 \leq n \leq 1$, linear when $n = 1$, and a power function when $n > 1$.

16.1.3 Divergence of Signals

When neurons entering the neuronal pool excite more neurons (one-to-many), this phenomenon is called divergence. There are two major types of divergence.

Same tract is where amplification occurs through successive stages. This is typical of motor neurons exciting many motor fibers (muscles) in parallel.

Multiple tracts, is where the signal splits, for example both into the cerebellum and the thalamus for somatosensory information (Figure 16.5).

16.1.4 Convergence

Convergence is essentially the opposite of the divergence processing and also appears in two types.

First, spatial summation of information at a summing terminal of an axon, so that thresholds are exceeded.

Second, convergence from mulitple sources, such as fibers from different receptor types yielding different signal types. Many examples of convergence in the cortex are easily found (Figure 16.6).

16.1.5 Mixed Messages

These occur as a result of combined excitatory and inhibitory outputs. The example shown below has an interneuron in one path that reverses the sense to make an inhibitory output to a separate neuron (Figure 16.7).

16.1.5.1 Further Processing in Sensory System

With all systems in the body, there are redundancies of information, so that we can better interpret the world around us. The redundancies become important when parts of the body are injured, or the brain is injured. They also allow us to "sacrifice" some information to improve the spatial localization of the stimuli. The particular aspect of redundancy of greatest interest is that of lateral inhibition.

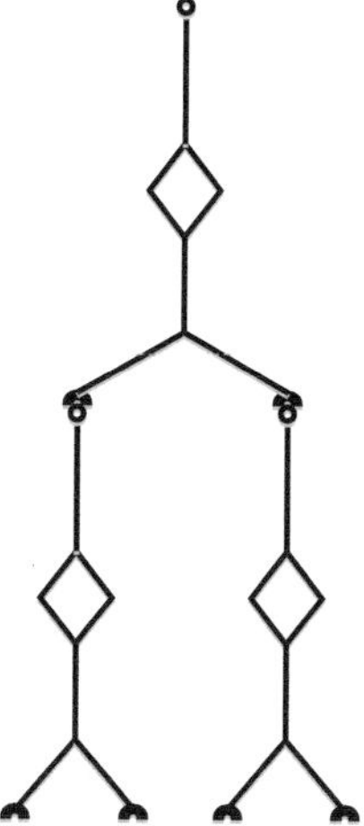

FIGURE 16.5
Example of divergence of neurons and pathways. Here, the neuron at the top of the figure branches with two axons and the second-order neurons branch similarly, giving a 1:4 divergence.

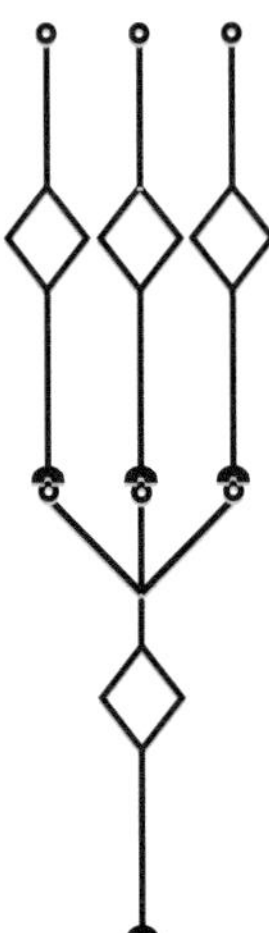

FIGURE 16.6
Example of convergence of neurons and pathways. Here, the three neurons at the top of the figure have input to the dendrites of a single axon, giving a 3:1 convergence.

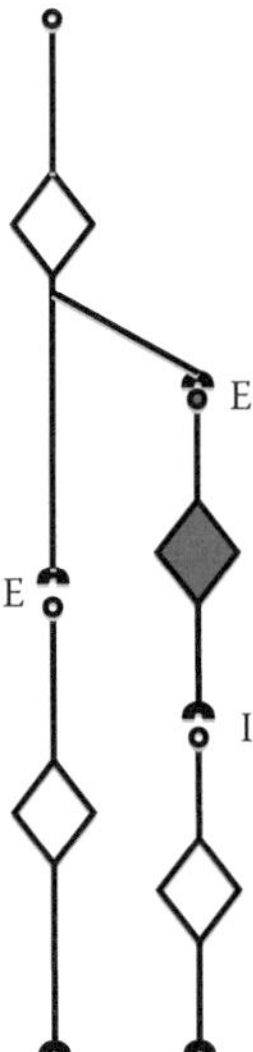

FIGURE 16.7
Example of a split neural pathway where one branch travels via excitatory neurons (labeled E), and the other via a small inhibitory interneuron (labeled I). This gives a mixed message and forms part of the system that is involved in lateral inhibition.

16.1.5.2 Lateral Inhibition

Lateral inhibition prevents lateral spread of information and increases the contrast occurring between adjacent neurons. Take the last example of the combined excitatory and inhibitory output of a neuron. The usual arrangement for lateral inhibition is that of the center of the receptive field depolarizes the next stage neuron, while the periphery hyperpolarizes the next stage neuron, making use of the mixed messages mechanism.

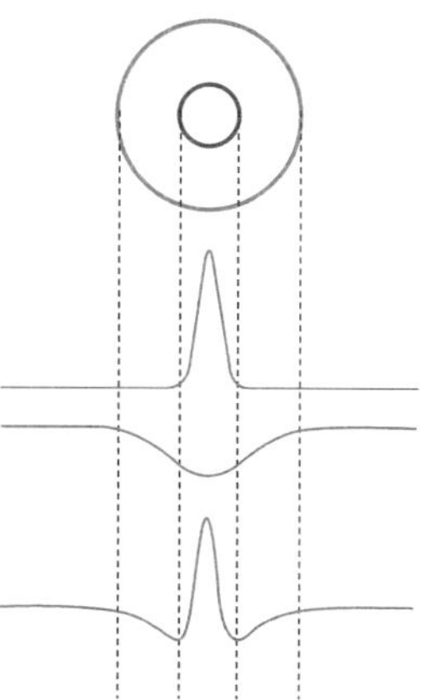

FIGURE 16.8
Excitatory and inhibitory receptive fields of a system connected to sensory receptors. The inner circle illustrates the excitatory, central field, while the outer circle illustrates the surrounding inhibitory field. The graphs below indicate respectively: the excitatory central response (Gaussian), the inhibitory peripheral response (inverted Gaussian), and the combined difference (difference of Gaussians).

The following diagram indicates what happens with weak and strong stimuli, and the next diagram indicates the extent of limiting the spatial information with lateral inhibition. The mathematical explanation relies upon differences between Gaussian curves (Figure 16.8).

16.1.6 Categories of Receptor

The receptors can be categorized as follows (with decreasing levels of consciousness).

Exteroreceptors. These respond to stimuli from surroundings including: touch, pressure, warmth, and cold.

Teloceptors. These are for special senses, and include rods and cones, olfactory mucous membranes of nose, taste buds of the tongue, and the cochlea of the ear. (Sometimes these are included with exteroreceptors).

Proprioceptors. These monitor the length of muscles, the tension of muscles, and the angle of joints and the overall position of the body (semicircular canals of the inner ear).

Interoceptors. Monitor internal organs, and include the baroreceptors (monitoring arterial pressure) and chemoreceptors (which essentially monitor blood pH).

Normally, the sensory information is combined and can yield very complex experiences, for example the sight, smell, taste, feel, color, temperature, and texture of a warm chocolate cake is readily drawn to the imagination.

How specific are receptors to the forms of energy they are expected to respond to? As we have seen, retinal cells will also respond to mechanical, electrical and magnetic stimuli. Cold receptors also respond to chemical stimulus (menthol, specifically) and hot receptors to calcium level. There is some degree of filtering to improve specificity, for example the vestibular system only responds to movements of less than 20-Hz characteristic frequency, whereas the cochlea responds to vibrations from 20 Hz to 20 kHz. The specific tuning characteristics of the cochlear membranes will be dealt with later.

16.2 Tactile Receptors

Fine touch and pressure receptors provide detailed information about a source of stimulation, including its exact location, shape, size, texture, and movement. These receptors are extremely sensitive and have relatively narrow receptive fields. The sensory information reaches our conscious awareness through the posterior column pathway.

Crude touch and pressure receptors provide poor localization and, because of relatively large receptive fields, give little additional information about the stimulus. These sensations ascend within the anterior spinothalamic tract, and the thalamus relays the information to appropriate areas of the primary sensory cortex. Tactile receptors range in complexity from free nerve endings to specialized sensory complexes with accessory cells and supporting structures (Figure 16.9).

1. Free nerve endings sensitive to touch and pressure are situated between epidermal cells. There are no apparent structural differences between these receptors and the free nerve endings that provide temperature or pain sensations. These are the only sensory receptors on the corneal surface of the eye, but in other portions of the body surface, more specialized tactile receptors are probably more important. Free nerve endings that provide touch sensations are tonic receptors with small receptive fields.
2. Pacinian corpuscles, or lamellated corpuscles (see also Figure 16.4), are sensitive to deep pressure. Because they are fast-adapting receptors, they are most sensitive to pulsing or high-frequency vibrating stimuli. A single dendritic process lies within a series of concentric layers of collagen fibers and supporting cells (specialized fibroblasts). The entire corpuscle may reach 4 mm in length and 1 mm in diameter. The concentric layers, separated by interstitial fluid, shield the dendrite from virtually every source of stimulation other than direct pressure. Pacinian corpuscles adapt quickly because distortion of the capsule soon relieves pressure

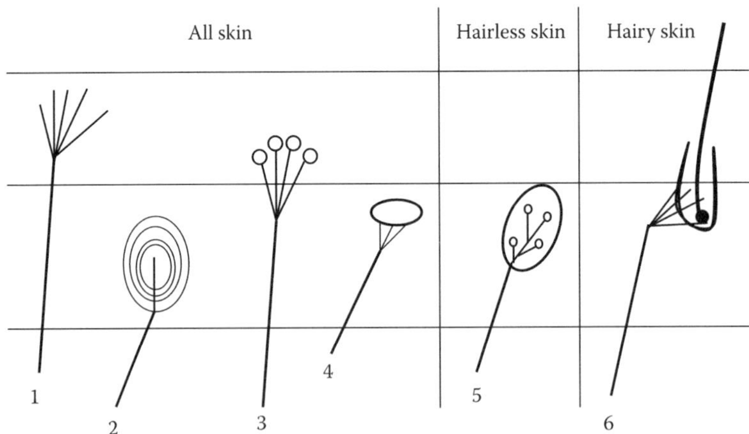

FIGURE 16.9
Receptors in skin: (1) free nerve ending, (2) Pacinian corpuscle, (3) Merkel disk, (4) Ruffini organ, (5) Meissner corpuscle, (6) hair follicle receptor.

on the sensory process. These receptors are scattered throughout the dermis, notably in the fingers, mammary glands, and external genitalia. They occur also in the superficial and deep fasciae, in joint capsules, in mesenteries, in the pancreas, and in the walls of the urethra and urinary bladder.

3. Merkel's disks are fine touch and pressure receptors. They are tonically active, extremely sensitive, and have very small receptive fields. The dendritic processes of a single myelinated afferent fiber make close contact with unusually large epithelial cells in the stratum germinativum of the skin.
4. Ruffini corpuscles are also sensitive to pressure and distortion of the skin, but they are located in the reticular (deep) dermis. These receptors are tonically active and show little if any adaptation. The capsule surrounds a core of collagen fibers that are continuous with those of the surrounding dermis. Inside the capsule, a network of dendrites is intertwined with the collagen fibers. Any tension or distortion of the dermis tugs or twists the capsular fibers, stretching or compressing the attached dendrites and altering the activity in the myelinated afferent fiber.
5. Meissner's corpuscles or tactile corpuscles, perceive sensations of fine touch and pressure and low-frequency vibration. They adapt to stimulation within a second after contact. Meissner's corpuscles are fairly large structures, measuring roughly 100 μm in length and 50 μm in width. These receptors are most abundant in the eyelids, lips, fingertips, nipples, and external genitalia. The dendrites are highly coiled and interwoven, and they are surrounded by modified Schwann cells. A fibrous capsule surrounds the entire complex and anchors it within the dermis.
6. Root hair plexus or hair follicle receptors. Wherever hairs are located, the nerve endings of the root hair plexus monitor distortions and movements across the body surface. When a hair is displaced, the movement of the follicle distorts the sensory dendrites and produces action potentials. These receptors adapt rapidly, so they are best at detecting initial contact and subsequent movements. For example, you generally feel your clothing only when you move or when you consciously focus on tactile sensations from the skin.

Summary Table of Properties of Mechanoreceptors

Free Nerve Endings	**Pacinian Corpuscle**	**Merkel's Discs**	**Ruffini Organs**	**Meissner's Corpuscles**	**Hair Follicle Receptors**
Hairy Glabrous	Hairy Glabrous	Hairy Glabrous	Hairy Glabrous	Glabrous	Hairy
$A\delta$ fibers Cool, sharp pain C fibers Warm Burning pain itch	$A\beta$ fibers Pressure Vibration (250 Hz–300 Hz)	$A\beta$ fibers Touch Pressure	$A\beta$ fibers Touch Pressure	$A\beta$ fibers Light touch Vibration (30 Hz–40 Hz)	$A\beta$ fibers Light touch Vibration (30 Hz–40 Hz)
Very slowly adapting	Very rapidly adapting	Slowly adapting	Slowly adapting	Rapidly adapting	Rapidly adapting
Pain Temperature Mechanoreception	Touch Vibration Acceleration	Touch Pressure Velocity	Touch Pressure Velocity	Touch flutter Velocity	Touch flutter Velocity

16.3 Thermal Receptors

In the skin exist some free nerve endings that provide information about temperature. These receptors have some special properties including: maintaining a discharge at a constant skin temperature (and the discharge rate is proportional to temperature), have a dynamic and phasic response, non responsive to nonthermal stimuli, very small receptive field (less than 1 mm^2), low conduction velocity (0.4 m/s–20 m/s). They are of two varieties: cold (13°C–36°C) and warm (41°C–47°C). Additionally, they respond to temperature changes within seconds.

They are distributed unevenly with a greater number of cold than warm receptors (cold 1–5/cm^2 compared with warm 0.4/cm^2), and the highest density is on the face (cold 9–16/cm^2). They are less numerous than touch receptors and tend to have large spatial thresholds. Cold receptors are also active at temperatures greater than 45°C and can signal falsely cold when hand placed in hot water (but only momentarily).

16.4 Psychophysics

Psychophysics is the study of the relationships between physical stimulus and the perceived subjective stimulus. It often focuses on intensity or magnitude of the physical stimulus and comparing this to the individual's perceived correlate of the intensity.

It consists of four separate but related aspects: detection, discrimination, recognition and scaling.

The quote below from Edgar Rice Burroughs *Jungle Tales of Tarzan* (1919) illustrates the four aspects reasonably well.

> Disturbed by the noise [detection] so close at hand, there arose from his sleep in a nearby thicket Numa, the lion. He looked through [discrimination] the tangled underbrush and saw the black woman and her young [recognition]. He licked his chops and measured the distance between them and himself [scaling]. A short charge and a long leap would carry him upon them.

In other words, yes, something exists; it exists against a background level (noise); it is (type of sensation); and is it strong/weak approaching or retreating?

To help us with the type of measurement, we need to understand basically how numbers can be used, or the scales of measurement.

16.4.1 Scales of Measurement

Nominal: The nominal scale exists as a numbering of cases (the numbers are convenient labels, for example, football numbers).

Ordinal: Determination of greater of less (the numbers provide information about big or small, for example hardness of minerals, pleasantness of odors).

Interval: The numbers provide information about a scale. The scale is special in that it is an absolute scale, for example temperature, energy, calendars.

Ratio: This is the most common type of measurement. The numbers mean something in relation to something else, for example length, loudness, pitch, weight, etc.

16.4.2 Categories of Measurements

Absolute thresholds: These are determined by way of when a person detects an event. The most usual method for determining absolute threshold is the method of limits.

Difference thresholds: This is to find the just noticeable difference, or "jnd." How much must be added or taken away before subject notices? Similar method of limits applies for absolute threshold.

Equality: Is this stimulus equivalent to that stimulus? For example subject is required to match the pressure on forehead with pressure on lips. This gives about a four-fold difference, with forehead requiring greater pressure.

Order: Subject ranks stimuli in order from largest to smallest. For some parameters, this will yield the same result for all subjects, but for others, "subjective" factors take hold, for example, most pleasing to least pleasing.

Equality of intervals: For example, making a gray halfway between white and black. Making gradations between marks halving, dividing by three etc.

Equality of ratios: Working out twice as loud.

Stimulus rating: Allowing for the system to distort the data. What is the velocity of that vehicle, weight of that pumpkin, length of that line in real units?

16.4.3 Method of Limits (Absolute and Difference Thresholds)

The method of calculating limits is described below:

A rough estimate of the threshold is established, and we want to make sure that our range of stimuli reach above and below this so that it is either unmistakably "on" or "off." Divide this range into several discrete levels and begin a procedure of increasing from the minimum of range until the stimulus is noted. Then take the stimulus to maximum range, and reduce until stimulus disappears. Repeat this procedure five times (i.e., five up and five down values of threshold). The next step is to find the median value of the threshold. This could be computed from SPSS, a graphics calculator, or graphically. The 50th percentile (median value is 5.8), and thus the absolute threshold is 5.8 units (Figure 16.10).

For difference thresholds, this is similar to absolute, but what we are interested in here is the just noticeable difference (jnd), and so we increase and decrease from the standard level. There are two median values: 3.6 and 12.3. To determine the threshold we take the average difference from the standard (8–3.6 and 12.3–8) and average = 4.35. This value is the difference threshold (Figure 16.11).

16.4.4 Method of Constant Stimuli

This method is similar to the method of limits but makes no assumptions about increasing or decreasing. The stimuli are randomly (or pseudorandomly) ordered and presented to the subject, who responds yes or no. The remainder of the analysis is identical to the method of limits.

16.4.5 Method of Adjustment

Here, the subject is in control of the stimulus. They increase and/or decrease the intensity until they can just notice it. It is in the style of fine-tuning the stimulus strength. It is, however, much more subjective and uncontrolled from the experimenter's point of view.

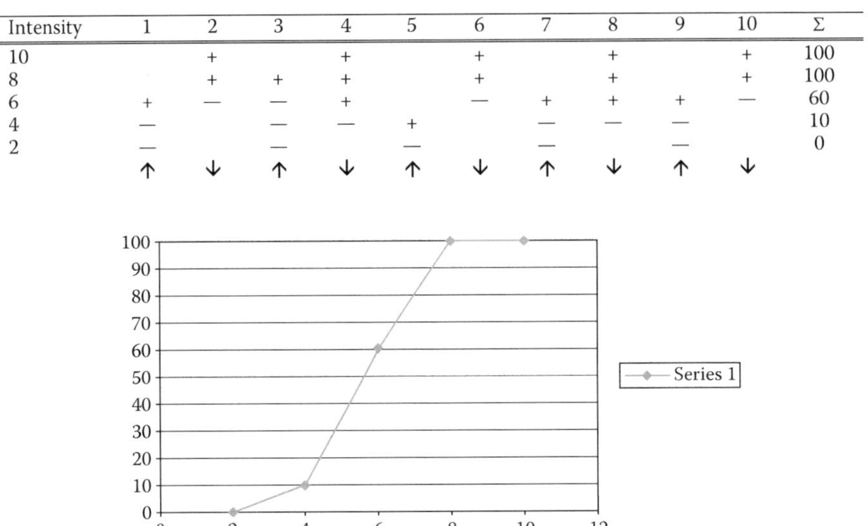

Intensity	1	2	3	4	5	6	7	8	9	10	Σ
10		+		+		+		+		+	100
8		+	+	+		+		+		+	100
6	+	—	—	+		—	+	+	+	—	60
4	—		—	—	+		—	—	—		10
2	—		—		—		—		—		0
	↑	↓	↑	↓	↑	↓	↑	↓	↑	↓	

FIGURE 16.10
Method limits for absolute threshold.

16.4.6 Automated Methods

Several methods exist for automatically adjusting the stimulus intensity, based upon the subject's response from the previous level. The most common of these is known as the ML-Pest method (maximum likelihood psychophysical estimation by sequential testing). The procedure follows a sequence that is based upon the shape of the psychophysical

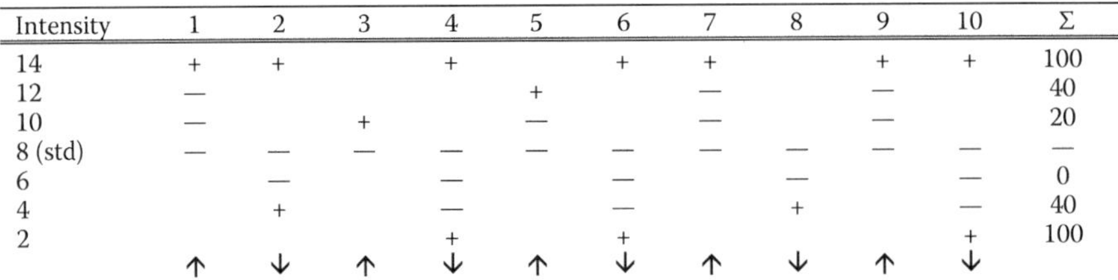

Intensity	1	2	3	4	5	6	7	8	9	10	Σ
14	+	+		+		+	+		+	+	100
12	—				+		—		—		40
10	—		+		—		—		—		20
8 (std)	—	—	—	—	—	—	—	—	—	—	—
6		—		—		—		—		—	0
4		+		—		—		+		—	40
2				+		+				+	100
	↑	↓	↑	↓	↑	↓	↑	↓	↑	↓	

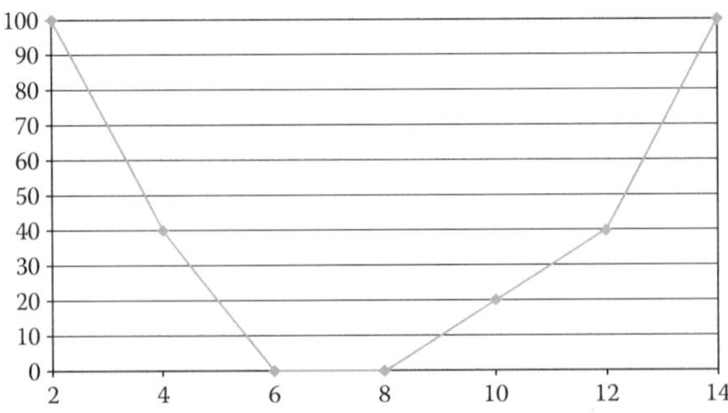

FIGURE 16.11
Method of limits for difference thresholds.

threshold function. Correct responses decrement the intensity by 0.3 log units, while incorrect result in an increment.

Our desire to understand how individuals (or groups) respond to stimuli dates back to Weber in the 1800s, when his observations of small changes in a static stimulus were only detectable when they exceeded a fraction of the static stimulus. This is usually known as the Weber fraction or Weber's law.

16.4.6.1 Weber's Law

This relationship is one of the ratio of the detectable difference is proportional to the absolute value of the stimulus intensity. If one plots ΔI vs I, we obtain a straight line with gradient k. This value of k is often termed the Weber fraction and is usually of the order of 5%–20%. The Weber law is expressed as

$$\Delta I = kI$$

or

$$k = \frac{\Delta I}{I}.$$

Unfortunately, things are never simple, and at very low levels of stimulus intensity, the relationship breaks down so that Weber's law is a poor description under some circumstances. An improved version is an expression related to the powers of the physical and behavioral quantities, known as Stevens' power law.

16.4.6.2 Stevens' Law

$$\psi = k\phi^{\beta}$$

where

ψ is the behavioral (subjective report) measure of stimulus intensity
ϕ is the physical value of stimulus intensity
β is the power (or index) of the function (if $0 < \beta < 1$ logartihmic; if $\beta = 1$ linear; if $\beta > 1$ power)
k is a constant

16.4.6.3 Distortion of Linearity

The nonlinearity of the sensory function gives rise to a nonlinear psychophysical interpretation. Line length estimation is reasonably linear. Some other modalities are nonlinear. For example, sound level estimation is logarithmic (or is compressed) with a β of 0.6, but electric shock has a power relationship of β = 3.5 (expansion).

Examples of Stevens' power law and diagrams come from the 1970 paper in *Science*. This is recommended reading and provides the required information about different powers or exponents and their interrelationship.

16.5 Signal Detection Theory

Apologies to all who have little interest in gambling, as most of the examples relate to probability, and hence gambling adds a good analogy for this section.

Briefly, the work will focus on the payoff matrix, observer bias and the discriminability function (the term also known as d′).

If we have a stimulus and wish to detect its threshold, then the subject may not behave truthfully (either deliberately, in an effort to please the experimenter, or to improve their "score" on some league table, imaginary or otherwise). To get around this, what we want is a method of determining when they are trying too hard or not reporting correctly.

If we put noise in the system, and have some trials that contain no stimuli, then we effectively engineer the experimental conditions to suit ourselves (as the experimenter). There are four possible pairs of stimulus and response: stimulus (present or absent) and response (detected or not detected). These form a "payoff" matrix, as illustrated in the following table:

Stimulus/Response	Yes	No
On (signal plus noise)	Hit	Miss
Off (noise only)	False alarm	Correct rejection

Alternative names exist for some of the cells of the payoff matrix.

- *False alarm* is also known as *error of commission* (analogous to a *Type I error*)
- *Miss* is also known as *error of omission* (analogous to a *Type II error*)

We can reward the subject if they perform correctly and punish them if they respond incorrectly.

We have two overlapping distributions of signal and signal plus noise separated by a distance d′ (Figure 16.12).

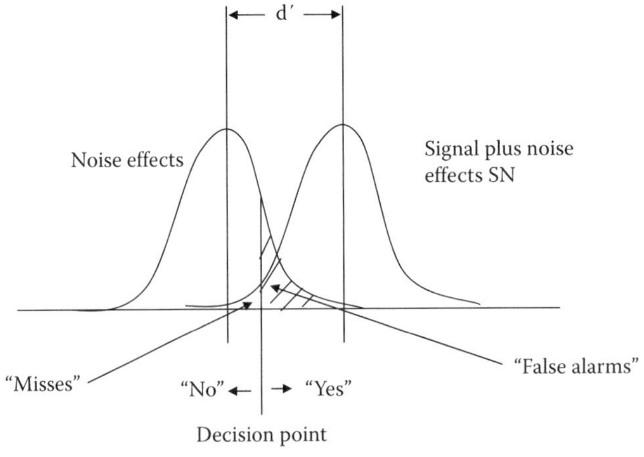

FIGURE 16.12
Distributions of the noise (left) and the signal plus noise (right). These distributions share the same variances but have different means separated by an amount d′. An overlap of misses and false alarms occurs in the gap.

The threshold can be moved according to the instructions given to the subject(s). For example an experiment where subjects have to respond to a tone presented at a low level, mixed with some background white noise.

	A	B	C
Hit	$100 win	$10 win	$10 win
Miss	$10 loss	$10 loss	$10 loss
False Alarm	$10 loss	$10 loss	$10 loss
C reject	$10 win	$100 win	$10 win

These three scenarios illustrate the manner in which experimental responses can be manipulated or biased by the experimenter. In all three cases, the subject will always respond yes if they are sure the stimulus is present, or no if they are sure the stimulus is absent. It affects the situation where they are unsure. In case A (so-called liberal bias), if they are unsure, then they are more likely to respond yes as the payout exceeds the penalty. In case B (so-called conservative bias), they are more likely to respond no as the opposite is true. In case C (so-called neutral bias), they will respond in the way somewhere between these two extremes.

Typical response values are

	A	B	C
Hit	98%	10%	75%
False alarm	90%	1%	10%

The steeper the curve, the more sensitive the subject is to the stimulus for a given stimulus level. The form of the curve depends upon the threshold and response bias (Figure 16.13).

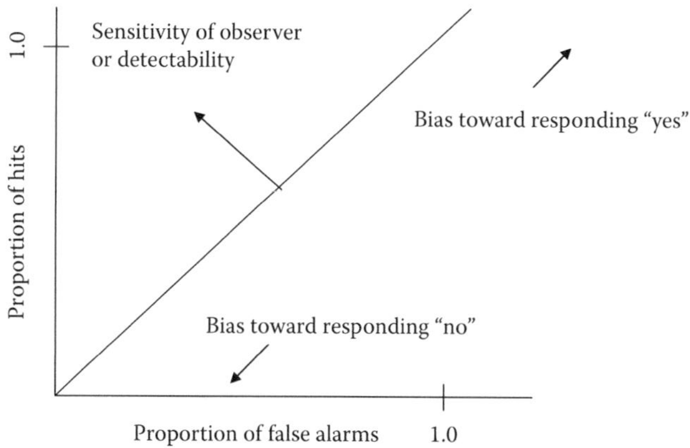

FIGURE 16.13
Plots of hits versus false alarms. The chance line is the 45-degree angled line. Point along this line will be determined by the bias that is made by the observer and are independent of observer sensitivity. Increased sensitivity will "stretch" the line into a curve indicated by the arrow pointing toward the ordinate axis.

16.5.1 Discriminability (d′)

To measure discriminability or the sensitivity of this curve, we take a 45-degree chance line and measure the distance to the curve. This distance is proportional to the discriminability and independent of the observer's response bias.

$$d' = \frac{\overline{X}_{SN} - \overline{X}_N}{\sigma_N}$$

where SN is the noise plus signal distribution and N is the noise only distribution. Essentially, it is the normalized difference between the means of the two distributions, as indicated in the previous diagram.

16.6 Information Theory

Let E be some event and P(E) the probability of the event occurring. After the event has occurred, we have received I(E) units of information. Where I(E) = log(1/P(E)) = –log(P(E)). The units used for information depend upon the base of the logarithm.

Log base	Information Unit Name	Conversion
$\log_2$	bit	1.00 bits
$\log_e$, *ln*	nat	1.44 bits
$\log_{10}$	Hartley	3.32 bits

If we have a source of information, such that successive symbols are statistically independent, and this is described by the source alphabet: $s_1, s_2, s_3, \ldots, s_r$

With each symbol having a probability associated with itself: $P(s_1), P(s_2), P(s_3), \ldots, P(s_r)$
The average information per symbol is given by the relation: $H(s) = \sum P(s_i)\, I(s_i)$
This quantity is known as the source entropy, and written as: $H(s) = -\sum P(s_i) \log_2(P(s_i))$

For example, with a standard six-sided die, the probability of any number appearing on the uppermost face is identical: P(S1) = P(S2) = P(S3) = P(S4) = P(S5) = P(S6) = 1/6 = 0.167

State	Pr	I (bits)
①	1/6	0.43
②	1/6	0.43
③	1/6	0.43
④	1/6	0.43
⑤	1/6	0.43
⑥	1/6	0.43
Σ	1	—
H(S)	—	2.58

And thus the entropy, H(S) = 2.58 bits.

As a second example, if we have a "loaded" die, where the number 6 is more likely to appear on the uppermost face (and the number 1 less likely as it is geometrically opposite). The probability of any face appearing is still 1, but the individual probabilities of each face appearing might now be: P(S6) = 0.32; P(S1) = 0.08; P(S2) = P(S3) = P(S4) = P(S5) = 0.15

State	Pr	I (bits)
①	0.08	0.29
②	0.15	0.41
③	0.15	0.41
④	0.15	0.41
⑤	0.15	0.41
⑥	0.32	0.53
Σ	1	—
H(S)	—	2.46

The entropy is now H(S) = 2.46 bits.

16.6.1 Information Channels

An information channel is described by giving an input alphabet and an output alphabet.

Input $A = \{a_i\}; i = 1, 2, 3, \ldots, r$

Output $B = \{b_j\}; j = 1, 2, 3, \ldots, s.$

These have a set of conditional probabilities: $P(b_j/a_i)$ for all i, j.

This is the probability that the output symbol b_j will be received, given the input a_i.

$$A \left\{ \begin{matrix} a_1 \\ a_2 \\ a_3 \\ a_r \end{matrix} \right. \Rightarrow P(b_j/a_i) \Rightarrow \left. \begin{matrix} b_1 \\ b_2 \\ b_3 \\ b_s \end{matrix} \right\} B.$$

We can describe the channel by a table. To do this, we would use some shorthand immediately:

$$P_{ij} = P(b_j/a_i).$$

Putting the input down the left side, and the output across the top, we end up with a matrix (called the channel matrix P).

$$\underline{P} = \begin{bmatrix} P_{11} & P_{12} & \cdots & P_{1s} \\ P_{21} & P_{22} & \cdots & P_{2s} \\ \cdots & \cdots & \cdots & \cdots \\ P_{r1} & P_{r2} & \cdots & P_{rs} \end{bmatrix}.$$

An information channel is described completely by its channel matrix

$$\sum P_{ij} = 1; i = 1, 2, 3, \ldots, r.$$

This means for any input we must have an output, and

$$\sum P(a_i) P_{ij} = P(b_j).$$

In other words, the probability of an event b_j occurring is the sum of all the products of $P(a_i)$ and P_{ij} for the appropriate j.

The operator $P(b_j/a_i)$ is known as the forward probability,
and hence, $P(a_i/b_j)$ is known as the reverse probability.

From Bayle's law,

$$P(a_i/b_j) = P(b_j/a_i) P(a_i)/P(b_j).$$

The numerator in this case is the probability of the joint event (a_i, b_j).

If we do not already know the output (a priori), then the entropy of A (the source alphabet) is:

$$H(A) = -\sum P(a_i) \log(P(a_i)).$$

If we do know the output (and it is b_j), then the a posteriori entropy of A is then:

$$H(A/b_j) = -\sum P(a_i/b_j) \log(P(b_j/a_i)).$$

The a priori entropy means the average number of bits needed to represent a symbol from the source alphabet.

The a posteriori entropy means the average number of bits needed to represent a symbol from the source alphabet given the arrival of b_j.

Taking the average of all the a posteriori entropies, i.e., H(A/B), is known as the equivocation of A with respect to B.

$$\begin{aligned} H(A/B) &= \Sigma P(b_j) H(A/b_j) \\ &= -\Sigma P(b_j) \Sigma P(a_i/b_j) \log(P(a_i/b_j)) \\ &= -\Sigma \Sigma P(a_i, b_j) \log(P(a_i/b_j)). \end{aligned}$$

16.6.2 Mutual Information

Originally, we needed an average of H(A) bits to specify an input symbol. If we know the conditional probabilities of the channel, we may calculate (via the intermediate steps—forward, backward, and joint probabilities), the equivocation. We need only an average of

H(A/B) bits of information. This means that there is information presented by the channel, and it is the difference of the a priori entropy and equivocation, or H(A) – H(A/B).

This quantity is known as the mutual information I(A/B) and given by the expression

$$I(A/B) = \sum_i \sum_j P(a_i, b_j) \log\left(\frac{P(a_i, b_j)}{P(a_i)P(b_j)}\right).$$

16.6.3 Channel Capacity

The channel capacity is the maximum amount of information we can have in the channel. It may be found by differentiating for maxima.

$$C = I(A/B)|_{max}\ P(a_i).$$

16.6.4 Information Coding in the CNS

Definition of terms:

- **x** the mean number of action potentials produced while stimulus is on,
- ν the frequency of action potentials,
- **y** the number of action potentials which pass through the channel,
- **t** the stimulus, (and observation) time.

$$x \propto \nu t.$$

Assuming the input range is

$$x_{min} \le x \le x_{max}.$$

then integrating yields

$$P(y_j) = \int P(y_j/x)P(x)dx.$$

and the uncertainty of the probability distribution:

The following equations are alternative representations of those in the previous section.

$$H(Y) = -\int P(y_j)\log(P(y_j))dy$$

$$H(Y/x) = -\int P(y_j/x_i)\log(P(y_j/x))dy$$

$$H(Y/X) = \int P(x)H(Y/x)dx$$

$$I(Y/X) = H(Y) - H(Y/X).$$

The work dealt with here has been using a frequency coding. This will not necessarily be true for every cell in an organism. For example, some neurons may use entirely frequency coding, others a mixture of interval, some unknown (Figure 16.14). These formulae are the maximum information content, and decoding may result in less information as higher order processing is integrative.

$$\text{Receptor} \rightarrow \text{nerve fiber} \rightarrow \text{CNS}.$$

Stimulus intensity is the information that is transmitted and subsequently received, decoded, and interpreted. Stimulus intensity is encoded into action potential frequency and the frequency of action potentials is interpreted as stimulus intensity in the CNS.

Information content is related to the number of states that can be distinguished following coding. If this is 0 or 1 impulses (two states), it can convey information about two stimulus intensities; with no impulses, the stimulus is below threshold (S0); with one, it is above. With more states, then this can allow more levels to be represented. If a neuron has a maximum of N impulses in response to stimulus, the receptor can signal N + 1 different intensity levels (or states, given by the numbers of impulses plus zero impulses) to an impulse counter in the CNS.

For an ideal receptor, N = f t.

The number of levels that can be represented is $N + 1 = (f_m\, t) + 1$, where f_m is the maximal firing rate of the receptor. Intensity increases with: the maximal firing frequency, and the observation time. The upper limit of the firing frequency is determined by the properties of the neuron (refractive period). Some receptors also have a resting discharge (f_0), and if this exists, then the frequency term f_m must be replaced by $(f_m - f_0)$.

The information content of a real system (Figure 16.5) can be computed:

$$
\begin{aligned}
I &= \log_2(n) \\
&= \log_2(N+1);\ \text{for the levels including zero} \\
&= \log_2(10) \\
&= \frac{\log_{10}(10)}{\log_{10}(2)} \\
&= \frac{1}{0.3010} \\
&\ 3.3\ \text{bits.}
\end{aligned}
$$

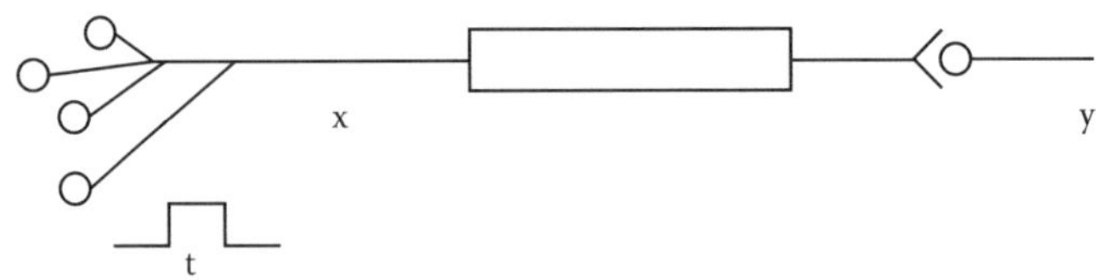

FIGURE 16.14
The diagram symbolizes a neuron with its receptive field, axon, and synaptic contacts. These form the input, the channel, and the output, respectively. The stimulus may vary in intensity and the frequency of the action potentials is related to this stimulus intensity.

- Information content for a receptor capable of responding to stimulation with a maximal frequency f_m

$$I = \log_2(f_m\, t + 1).$$

- I increases with both f_m and t. Hence, precision of the intensity of a prolonged constant stimulus increases with prolonged observation time. Even though it appears that increasing the observation time will yield greater I, in practice the CNS accumulates information about the afferent discharge over a limited time period and the real receptor does not achieve the theoretical number of stimulus-intensity levels due to the uncertainty and noisiness of the system (Figure 16.15). We can make an approximation based upon the (un)certainty of a staircase superimposed on the data (as illustrated in Figure 16.16).

For this example, we see that an eight-level staircase can be superimposed on the data. This contrasts with the ideal receptor that could discriminate 300 levels.

8 Levels	300 Levels
$\log_2 8$	$\log_2 300$
3 bits/stimulus	8.2 bits/stimulus

With this example, we have 8.2 – 3 = 5.2 bits of information are lost due to noise.

The largest value for a physiological example are found in the muscle spindles with a range of 4.8 – 6.3 bits per 1 sec stimulus.

In other systems there is inherent redundancy. For example, written English language contains more symbols than are required for the meaning to be understood. The excess of symbols equates to redundancy. The practical, measured information content of the 26-letter alphabet is measured to be 1.5 bits per letter. The theoretical value (based upon infinite combinations of letters) is 4.7 bits per letter ($\log_2 26$). The redundancy averages to 4.7 – 1.5 = 3.2 bits per letter. Although this seems like a waste its advantage is that is can it buffer disturbance or noise. In general, the greater the redundancy in the system, the more secure it is against disturbance.

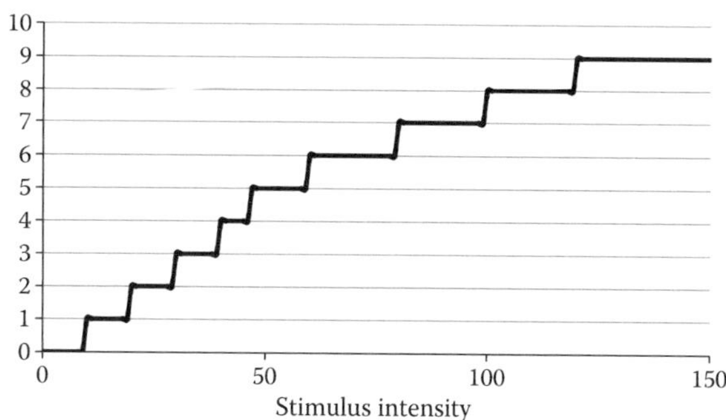

FIGURE 16.15
These data form a staircase function. There are 10 discrete levels (9 levels plus the zero value).

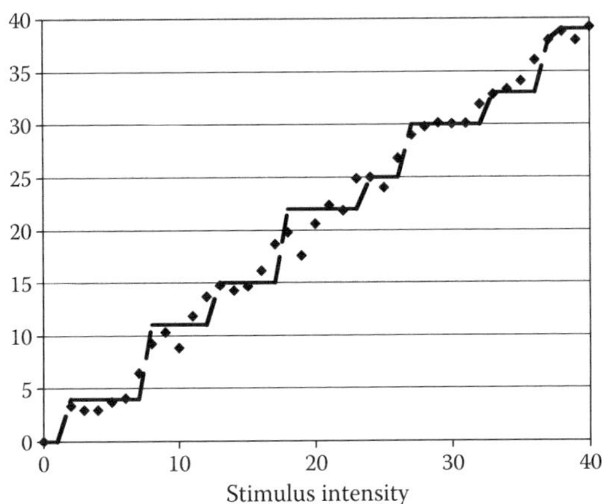

FIGURE 16.16
The plot of simulated firing frequency against stimulus intensity shows the raw data (diamonds) and the staircase fit indicating 8 levels plus zero or 9 intensity levels that can be adequately discriminated from these points.

16.6.5 Nervous System

If parallel systems or pathways exist, then there is protection against loss of information. In the body, parallel channels exist for transmitting information from receptors, and in the periphery, density of receptors is so high that even a point stimulus will excite several fibers. If we sum the information from each of the neurons, then we should have a better resolution based upon the number of impulses reaching the CNS.

PROBLEMS

1. A receptor with a linear relationship between firing frequency of action potentials and stimulus intensity is more likely to be found:
   (a) Muscle spindle (a proprioceptor)
   (b) Vibration receptor (e.g., Pacinian corpuscle)
   (c) Auditory receptor
   (d) Pressure receptor (e.g., Merkel disk)
2. A student experiment to find the threshold of touch by the method of limits yields the following results: Compute the threshold intensity level.

	Trial									
Intensity	1	2	3	4	5	6	7	8	9	10
70		+		+		+		+		+
60	+	+	+	+		+		+		+
50	—	—	—	+		—	+	+	+	—
40	—		—	—	+		—	—	—	
30	—		—		—		—		—	
Direction	↑	↓	↑	↓	↑	↓	↑	↓	↑	↓

3. A human participant can distinguish temperature with a resolution of 0.1°C, in the range of 21°C–40°C. How much information is within this range?

4. If the capacity of a nervous system is 5.3 bits, what temperature resolution could be obtained in range 0°C–20°C?

Bibliography

Ash RB. 1965. *Information Theory*. Interscience, New York.
Stein RB. 1967. The information capacity of nerve cells using a frequency code. *Biophys J* 7:797–826.
Stevens SS. 1970. Neural events and the psychophysical law. *Science* 170:1043–1050.
Zimmerman M. 1983. Cybernetic aspects of the nervous system and sense organs. In Schmidt RF and Thews G (eds.), *Human Physiology*. Springer-Verlag, Berlin.

17

Vision

David P. Crewther

CONTENTS

In this chapter, the anatomy, optics, physiology, and psychophysics of the human visual system will be summarized. The visual sense depends on the conversion of electromagnetic energy within a narrow range of wavelengths (from about 300 nm to 700 nm) into electrical energy within the retina. This is followed by a series of processing stages resulting in an output from the retina to several different parts of the midbrain and thalamus. Subsequently, signals are processed and may lead to conscious interpretation utilizing many regions of the cerebral hemispheres. Perhaps 50% of human visual cortex is visually sensitive.

17.1 What Does an Eye Need for Us to See?

The eye has developed a wonderful property—transparency. Thus, light can enter through several layers of epithelium, stroma, and endothelial membranes, comprising the cornea, traverse the anterior chamber, the crystalline lens, and the vitreous humor, pass through the retina and be absorbed at the level of the photoreceptor outer segments. This property of transparency is one of several seeming miracles of evolutionary development. How the collagen fibers of the corneal stroma and of the crystalline lens are maintained with spacing regularity that allows for visible light to pass through virtually undistorted is not a part of this chapter (for further reading, see Davson [1990]). A second, unrelated, miracle is also worth reporting. As well as passing undistorted through the eye, light is brought into focus, just in the same way as we might focus the rays from the distant sun onto a piece of paper with a magnifying glass. As can be seen from Figure 17.1, part of this refraction is brought about by the interface between air and cornea, which provides the largest part of refractive power. This power is added to by the crystalline lens, whose shape can be manipulated by fibers whose tension is controlled by the accommodation system through the ciliary body.

While the introduction pointed to the purpose of light absorption and visual sensation as a means to response to the environment, either to find food, to find a mate, to avoid danger, and so on, part of the machinery of the retina and surrounding membranes of the eye is also involved in adjusting the shapes of the eyeball in its growth so that focused images can be seen. Without such appropriate growth for optimal vision, the potential danger from a threat such as a leopard would not be available (see Figure 17.2).

Thus, across all vertebrate species studied to date, the shape of the eye and its refractive power are subject to change during early postnatal development. Normally, this results in a process called emmetropization—a progression toward zero refractive error, resulting in sharp images at the plane of the photoreceptors. However, if an animal such as a hatchling chicken is fitted with a defocusing lens, its eyeball ball can change by as much as 1 mm in axial length and quite rapidly refractively compensate to the applied to focus, whether defocus is provide by a convex (positive) lens, focusing light in front of the retina, or a concave (negative) lens, focusing light behind the retina. Further literature and theories of experimental refractive compensation can be found in reviews (Crewther 2000; Wallman and Winawer 2004).

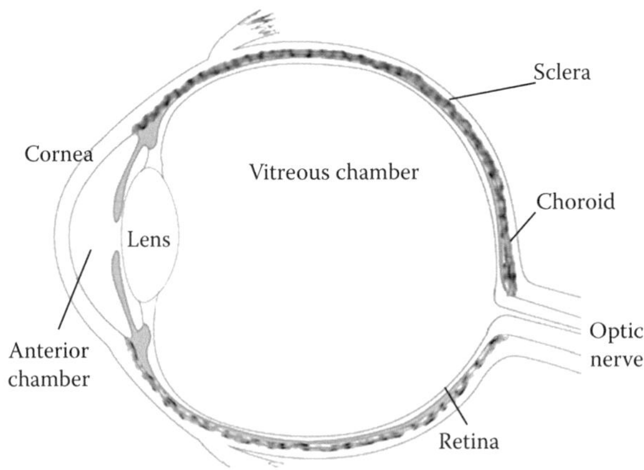

FIGURE 17.1
The eye and its components. Optically, the cornea and lens focus light onto the retina. The optic media are transparent to light. The amount of light transmitted to the retina is controlled by the size of the pupil.

FIGURE 17.2
A reclining leopard and the effects on recognition of applying Gaussian blur similar to that which occurs with optical defocus. Fine spatial vision is necessary for breaking the camouflage and seeing the spots.

17.2 Visual Sensation: The Photoreceptors

The retina is distinct from other peripheral sensory systems, in that the retina is part of the brain. That is, during development, the eye emerges as an outgrowth of the front part of the developing brain (the prosencephalon), invaginates to form the shape of the eye and the neural processes, and elements are thus strongly related to other neurons of the brain. The sensory elements are of two major forms—the rods and the cones (see Figure 17.3).

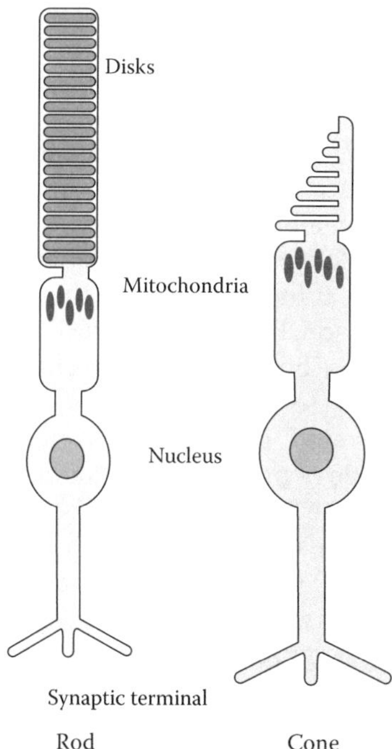

FIGURE 17.3
The two main photoreceptoral types—rods and cones. Cones come in three "flavors" (L, M, S) in humans differentiated by the peak absorption wavelength of their opsins (long, medium, and short wavelength). Rods come with only one pigment type.

Rods and cones differ in their distribution, their sensitivity, and their architecture. About 120 million rods are distributed uniformly across the primate retina, except for the fovea ("pit"), the visual center of the retina, which is rod-free. The cones, much fewer in number (about 6 million) are concentrated in the fovea and much less densely distributed in the retinal periphery. Rods are very sensitive in low light (scotopic conditions), such that single photon detection is possible in a fully dark-adapted state. Cones, on the other hand are effective under brighter (photopic) conditions. The rods and cones, together with various adaptive processes give visual sensitivity over an enormous range of light levels—approximately 10 log units.

17.3 Visual Transduction

Although it seems at first a backward process, the mammalian retina is actually working hard under no light conditions—i.e., in the dark. This so-called dark current depends on two processes. In the outer segment of the photoreceptor (the part containing the photosensitive membrane disks—see Figure 17.3), there is an inward current of sodium (Na^+) ions, flowing through cyclic GMP-gated channels. The second process is controlled by the electrogenic NaK_ATP-ase (a membrane bound sodium potassium pump that pumps out 3 Na^+ ions and pumps in 2 K^+ ions across the membrane). Working alone, this mechanism would result in a membrane potential of about –70 mV, the equilibrium potential for K^+. However, because the cGMP channels are open and Na^+ ions are flowing in, under dark conditions, the photoreceptor is depolarized with the membrane potential of around –40 mV.

When the photoreceptor absorbs light, there is a three stage process. We will use the rod as our example. The rod pigment called rhodopsin contains two parts—the *opsin* and the light absorbing part called *retinal*. When it is not activated, the 11-*cis* isomer of retinal is present and when retinal absorbs photons, it transforms into the all-*trans* configuration, and the retinal no longer fits into its binding site on the opsin molecule embedded in the disc membrane. The opsin transforms into a semistable molecular conformation called *metarhodopsin II*, which soon splits into the opsin and the all-*trans* retinal components.

In the second stage of transduction, the activation of the visual pigment by light absorption leads to a reduction in the concentration of the second messenger cGMP (cyclic guanosine 3′–5′ monophosphate), due to the activation of cGMP phosphodiesterase, which breaks down cGMP. When activated, the rhodopsin molecule diffuses within the disk membrane and activates hundreds of molecules of transducin. The light response cascade is terminated by two processes—the hydrolyzation of bound GTP by transducin, and rhodopsin is itself hydolyzed by a regulatory protein—arrestin.

When the concentration of cGMP is reduced by light absorption, there is a consequent closure of cGMP-gated Na^+ channels (which are also permeable to Ca^{2+} ions). Hence, less sodium flows across the membrane into the outer segment and the membrane becomes hyperpolarized. The process then transfers information about light absorption to the end-feet of the photoreceptors which synapse with bipolar (and horizontal) cells. Hyperpolarization of the membrane causes a reduction in the release of glutamate at the photoreceptor-bipolar interface (see Figure 17.3).

Phototransduction for the cones is very similar, with the acknowledgment that there are separate opsins for the L, M, and S cones and that the photoabsorptive membrane is continuous in the cone rather than being a series of floating disks.

17.4 Bipolar Cells and the Generation of Contrast Coding

As indicated above, in the dark, the photoreceptor cell is maximally depolarized and releases glutamate at the photoreceptor–bipolar synapse, and when light is absorbed in the outer segment, the photoreceptor hyperpolarizes and releases less neurotransmitter at the bipolar synapse. However, there are two major classes of bipolar cells that are distinguished by their response to glutamate. The so-called ON-bipolar cells are hyperpolarized by glutamate and the OFF-bipolar cells are depolarized by glutamate. In terms of the response to light the nomenclature makes sense: in response to light absorption by a photoreceptor and its hyperpolarization, glutamate release is reduced, and the ON-bipolar is depolarized via a sign-inverting synapse involving the mGluR6 receptor. The OFF bipolar interfaces with the photoreceptor through a sign-conserving synapse hyperpolarizing to the absorption of light by the photoreceptor (via the iGluRs receptor). Interestingly, each cone synapses with one ON-bipolar and one OFF bipolar cell, while the rod bipolars are only of the ON type (sign inverting).

The ON-bipolar cells connect, in most part, with ON-center retinal ganglion cells and the OFF-bipolar cells connect with OFF-center retinal ganglion cells.

Interestingly, the connections made by ON-bipolar cells are in a separate subdivision (sublamina b) of the inner plexiform layer (IPL) more proximal (toward the vitreous chamber) than those made by the OFF-bipolar cells (sublamina a). This raises the question, what is the purpose of such a division of processing of visual information? Early ideas about sensing onset or offset of light or perhaps the perception of brightness or darkness do not seem to hold up. More plausible, physiologically, is the idea that the separate ON and OFF

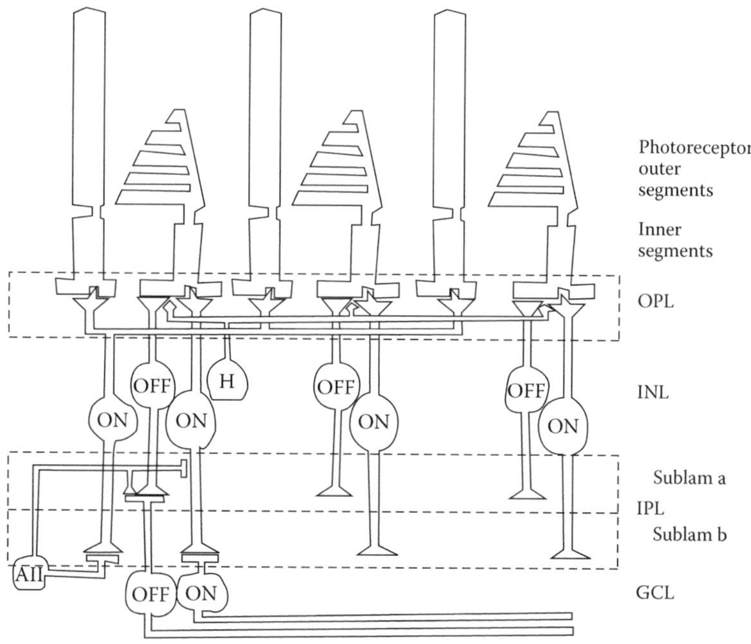

FIGURE 17.4
Retinal circuits for ON and OFF processing by rods and cones. For the L and M cones, contributions for the ON and OFF pathways separate at the photoreceptor bipolar synapse.

systems are there for the generation of antagonistic surround processes. However, this idea would require the convergence of ON and OFF bipolar processes—which in large has not been observed. The ON pathway and the OFF pathway maintain a separation from retina to cortex. The center-surround mechanism seems to be produced by lateral connections within the retina.

Another curious feature, observable in Figure 17.4, is the absence of rod OFF bipolar cells. While such an arrangement is rational, in terms of the role that rods play in very low light levels—things of interest are likely to brighter, and dark things will be hard to see, anyway, the retina has devised another way of feeding rod information into both the ON and OFF pathways of the cone bipolar system. Inspection of Figure 17.4 will show a synapse from the rod bipolar onto an amacrine cell—the AII amacrine cell that in turn synapses onto the OFF center retinal ganglion cell and onto the axon process of the ON cone bipolar that then synapses onto the ON center retinal ganglion cell.

17.5 Parallel Processing

The human retina possesses a dense array of photoreceptors—perhaps 120,000,000 rods and 6,400,000 cones. Why is there such an enormous number of detectors when the output stage of the eye—the retinal ganglion cells (RGC) number only 1.2 million? One explanation for economy in the degree of parallelism comes from answering the following question: If every one of these photoreceptors had an axon leading into the optic nerve, how big would the optic nerve be? The answer is quite staggering: if the average diameter of an axon is 1 μm, then 1 million would fit into 1 mm^2. Hence, the optic nerve would be roughly 13 mm in diameter—about half the diameter of the eye! Actually, for comparison, the human optic nerve possesses about 1.2 million axons and is about 2 mm in diameter. This is a very important fact when you consider how much you can move your eyes and the requirement of flexibility that this places on the optic nerve.

The solution taken by the mammalian retina appears to be one of redundant mapping by receptive fields of the different cell classes that have coarse filtering (few axons required) and fine filtering (excellent for acuity, but a heavy cost in terms of axon number). Thus, in the foveal region, there is a one-to-one mapping of L and M cones to midget ganglion cells, but in the periphery, an enormous number of rods plus cones will converge onto each peripheral RGC.

17.6 Retinal Specializations—The Fovea, Cell Density, and Visual Acuity

The compromise that evolved is a high foveal density of small receptive fields, with a lessening in density of ganglion cells and a corresponding increase in receptive field size toward the periphery. The visual system has optimized this patterning of the visual field in a neat fashion that supports perception without holes, but at the same time does not require too many ganglion cells. Thus, the overlap between neighboring ganglion cell receptive fields is approximately the same across the retina for each of the ganglion cell species. We are all familiar with the Snellen chart or Bailey–Lovie charts used to measure acuity clinically. Essentially, there is a progression of high contrast letters, which row by row become smaller and smaller. There is a point at which, standing, say 6 m from the chart, that you cannot recognize the individual letters. Across the normal populace this mean acuity is often represented as 6/6 acuity (or 20/20 in countries using feet and inches rather than metric systems). Someone with 6/6 vision can see (at 6 m distance from the chart) as clearly as the "normal" adult (at 6 m). The stroke of the letters used for these letters at threshold subtends about 1 min arc at the eye, or in terms of spatial frequency, 30 cycles per degree (cpd). Children, with clearer optic media, often score around 6/4.5 (their threshold line on the chart at 6 m is the same as a normal adult measured at 4.5 m).

With the advent of computers, single letter testing of acuity using better scientific testing basis (such as the Landolt C with 4 or 8 orientations of the gap in the ring—e.g., see the Freiberg acuity testing FrACT—http://www.michaelbach.de/fract/index.html). Such computer-based testing can rapidly find an unbiased threshold by forced choice testing, though they still must be used with appropriate care (e.g., testing within the spatial frequency range available).

While the circuitry of the primate retina appears to be enormously complex, in some ways, the outputs—that information flowing along the optic nerved from the eye toward the brain are quite simple, with luminance and/or color information enhanced by contrast information, with the central retinal area carrying much of the information, encoded at higher spatial frequencies, but with relatively simple receptive field structures. Eighty percent of the optic axis fibers are parvocellular (projecting from the midget ganglion cell type), 10% are magnocellular (projecting from the parasol type of ganglion cell), nearly 10% convey blue/yellow information into the koniocellular cells from the bistratified (with dendrites in both the ON and OFF sublaminae of the IPL) and large diffuse blue encoding cells. All of these cell types, (nearly 100%) have circular receptive fields, most of which are center/surround in nature.

It is worth taking a moment to convey information about the cells outside these classifications and their projection sites, particularly as most texts on vision play little attention to the retinal afferents beyond the lateral geniculate nucleus. Here we point to the major afferent midbrain and thalamic projections and point to their roles in overall vision and behavior (Table 17.1).

A brief inspection shows that in summation, these other roles of visual input—to help provide balance, ocular fixation, focus, and accommodation, to name a few, are essential to the optimal processing of visual information through the retinogeniculocortical system.

TABLE 17.1

Retinal Projections and Functions

Projection Site	Functions
Lateral geniculate nucleus (LGN)	Cortical visual processing; recognition; visual cognition; motion analysis and navigation
Superior collicullus (SC)	Oculomotor control; orienting response; multisensory integration
Pretectum	Input for pupillary control
Nucleus of the optic tract (NOT)	Stabilization of the eyes through optokinetic nystagmus (OKN)
Suprachiasmatic nucleus	Visual input to the diurnal rhythm
Accessory optic nuclei (AON—comprising medial terminal nucleus, lateral terminal nucleus)	Visual input into vestibular control; vestibulo-ocular reflex; visual input to cerebellum

To stress this point, let us ask a question: Consider what happens when you move your eyes steadily across a lecture theater. It is likely that you will recognize the lecturer and your classmates. That is fine, but it is also clear that the lecture theater does not move as you are moving your eyes. A quick consideration of the stimulation received by the retina indicates that this is strange, as the scene moving across the retina should stimulate motion detection systems. Another example is the commonly observed optokinetic nystagmus—a jerky back and forward movement of the eyes when, say, viewing a picket fence as you drive past it—yet as the observer of the picket fence you have no sensation of your eyes moving. Thus, we must conclude that stabilization of image is of utmost importance for maximizing perception and hence evolutionary success.

17.7 The Lateral Geniculate Nucleus—Parallel Inputs

The Lateral Geniculate Nucleus is thalamic nucleus found in the diencephalon of the brain and is one of the chief recipient zones of the optic nerve. In appearance it is immediately seen to be laminar in structure with two layers ventrally containing large cell bodies—the magnocellular layers, and four layers more dorsally—the parvocellular layers. The koniocellular cells were not initially observed, but appear in the intralaminar regions as well as throughout the magno and parvo laminae.

Why do these structures laminate? Developmentally, it is clear that the formation of laminae is rather late during development. A consideration of the differences in cell properties suggests certain organizational principles. There is a clear magno/parvo division. Thus, there is the possibility of cell recognition processes resulting in segregation. In addition, there are the inputs from the two eyes, causing a second dividing principle. Thus, from bottom to top, the laminae (in terms of M- or P- type neurons, and in terms of eye of origin—contralateral (C) or ipsilateral (I) are as follows: M(C); M(I); P(I); P(C); P(I); P(C). Why there are four parvo and two magno laminae is unknown. Some defining features of receptive fields, such as ON and OFF center polarity of ganglion cells does not result in sufficient plasticity for total segregation in primate, though it does in the ferret.

17.8 Function of Magno and Parvo Cells through Laminar Lesions

Two laboratories in the 1980s set out to systematically understand the contribution of the magno and parvocellular neurons of the LGN to visual processing. Using acrylamide lesions delivered to either the magnocellular layers (1 and 2) or to the parvocellular layers (3–6) of the primate LGN, Schiller et al. (2010) studied the responses to coarse and fine spatial information, color information and motion. Merigan and Maunsell (1993) selectively lesioned magnocellular layers with injections of ibotenic acid and found similar results. Clearly, the M cells show little color sensitivity, have higher contrast sensitivity at low contrasts, are not as responsive to high spatial frequencies at the P cells and show greater temporal frequency responses than do the P cells. P cells on the other hand provide all of our chromatic sensitivity, and our ability to see fine detail. Of course, P cells respond to achromatic as well as to colored stimuli.

17.9 Receptive Fields

It is clear that despite the simple nature of the primate retinal output—center-surround collections of receptive fields with ON-center and OFF-center persuasion, for the most part, the processing by the retina is far more sophisticated than that provided by the charge-coupled device that receives images in a video camera. However, even in the video camera, engineers have designed systems for automatic gain control (so that the screen adjusts to very bright or very dark conditions) and autofocus systems (generally based on maximizing local contrast across the pixel values).

A receptive field is a simple concept. For the somatosensory (touch) system, it is simply defined as the region of the skin for which touching/pressing/rubbing, causes a single receptor to fire. For the visual system, with the stimulus source generally external to the body, the principle generalizes quite easily. A solid angle of visual rays can be absorbed by a single photoreceptor. If we imagine playing a light source on a wall or screen, then there would a roughly circular region within which photo absorption occurs in the particular photoreceptor. As distinct from some other species (rabbit with direction selective receptive fields), most retinal ganglion cells (RGC) in primate have center-surround organization. By center-surround, we mean that there is a circular organization with the central part having one type of response (e.g., ON response) while there is an annular region—the surround that expresses an antagonistic response (in the current case, an OFF response).

Comparatively, the primate retinal output is also simple in the sense of receptive field properties. The humble amphibians, such as toads and frogs, have receptive fields that respond selectively to motion in a particular direction, particularly those with OFF center properties—possibly because of target food varieties such as darkly colored flies. The rabbit, a mammalian vertebrate has direction selective ganglion cell receptive fields that align with the directions of action of the extraocular muscles that move the eyes back and forth and up and down. It is when one progresses from retina to LGN and then to visual cortex that the complexity and variation in receptive filed (RF) structure suddenly blossoms in the primate.

As early as the 1960s, Hubel and Wiesel proposed that the orientation selectivity that they observed in simple cells of primary visual cortex might gain such selectivity via connections with a linear array of LGN neurons. Such a picture has remained remarkably intact, though the maps of orientation across cortical space certainly show plasticity and adaptation, possibly due to the feedback from striate cortex to the LGN—numerically about 70% of synapses on geniculate relay cells are of this nature.

While early models of cortical processing were based around segregation of M and P processing (Livingstone and Hubel 1988), this view has since been considerably modified (for a modern perspective, see Nassi and Callaway 2009). The diversity of receptive fields results partly from the convergence of information from the various neural types (M, P, K) largely segregated in retina and LGN, and clumping of RF properties occurs through the formation of columns, blobs, stripes and so on in primary visual cortex (V1) and extrastriate cortex (V2). The ON and OFF pathways begin to converge again, but other functional properties emerge including directional selectivity (necessary for motion processing), disparity selectivity (for stereopsis) and color and hue selectivity.

The descriptions simple, complex, and hypercomplex are still given to major receptive field classes, based on the work of Hubel and Wiesel and others:

Simple cells are characterized by having excitatory regions (ON and OFF) arranged side by side. Such an arrangement gives a clear orientation selectivity, if one imagines stimulating with a light bar on a darker background or vice versa. An antagonistically additive model of excitation would predict maximum response when the light bar is aligned over the elongated ON region, and lesser response would occur if stimulation occurred at different orientations.

Complex cells do not have spatially segregated ON and OFF regions. While tuning is not as sharp as with simple cells, complex cells still are orientation selective. In addition, most complex cells are direction selective, meaning that there is strong anisotropy in the firing rate for passing a stimulus in the preferred direction over the receptive field compared with the firing rate in the opposite direction.

Hypercomplex cells were originally described as a separate class of cells by Hubel and Wiesel (2009). They are characterized by a phenomenon known as end-stopping—namely, there appears to be a suppressive zone at one or both end of the receptive field such that if a stimulus bar extends into the zone, then firing rate is reduced. Detailed studies determined that the end-stopping property occurs in receptive fields that otherwise have simple or complex receptive field characteristics. Thus, hypercomplex cells give a means for the detection of angles or corners, perhaps the ends of objects.

This subject would be easy to understand if connections were all feed-forward in nature, but a remarkable feature of corticocortical connections is that they seem to possess a large number of reciprocal connections. Thus, Felleman and Van Essen in 1991 noted some 305 connections among the 32 visual areas that they analyzed. They surmised that 40% connectivity of all possible combinations of interarea communication was achieved.

However, even a detailed description of receptive fields cannot give a complete picture of how information is gathered for the purposes of perception. Differences in the nature and timing of neural spikes coming from retina to LGN to cortex must be taken into account. Here is it simply worth noting that a wave of early spikes communicates visual information rapidly across the cerebral hemispheres (see Section 17.14).

17.10 Ocular Dominance

One of the very interesting aspects of the visual system of the primate is understanding how the afferents from the two eyes—the axons, projecting from the retinal ganglion cells combine together to result in binocular vision, such as stereopsis. The process requires correct performance during several steps of development.

The first involves a process known as partial decussation that occurs where the optic nerves cross over. Rather than the almost total crossing that we are used to with the motor system—where left hemisphere drives the right side of the body (as exhibited so graphically with cerebral stroke), the visual system is broken into left and right visual hemifields, which project to the right and left cortical hemispheres, respectively. Simple optics indicates that with the two eyes pointing forward (say to a distant object), the parts of the object falling in left visual hemifield will optically project onto the nasal part of the left retina (the part near the nose) and the temporal part (toward the ear) of the right eye.

The fact that we see binocularly, without double vision, means that rather than being crossed, the optic chiasm where the two optic nerves join and separate again into the optic radiations mix the inputs from the two eyes. Thus, the right optic radiation will contain contralaterally projecting axons from the retinal ganglion cells in the nasal part of the left retina and ipsilaterally projecting axons from the retinal ganglion cells in the temporal part of the right retina.

In the first synaptic zone after the retina—the lateral geniculate nucleus, binocular connections do not occur. Rather, the half-maps from the hemiretinas sit one on top of the other, rather like a stack of pancakes. Another interesting feature becomes obvious—the neurons of these geniculate laminae segregate by cell type. Thus, the bottom two layers of cell nuclei appear to be rather larger (hence the term magnocellular layers) than those of the four layers above (termed the parvocellular layers). However, the third RGC type mentioned above—the konicocellular class also sends projections to the LGN. The cell bodies are relatively tiny and are thought to be distributed throughout the LGN, as well as in the intralaminar regions.

Why do the laminations form? Why is the order of the contralateral (C) and ipsilateral (I) as it is with a seeming order inversion between M and P layers? The answer must lie during development and have much to do with a process called Hebbian learning in terms of the conditions required for neural reinforcement.

Another very similar example occurs in the projection of LGN afferents to primary visual cortex, striate cortex, V1 (Brodmann area 17). Instead of arranging the input in terms of overlapping maps as in the LGN, the inputs from the left and right eyes arrives into layer IVc in a side-by-side fashion. Such an organizational feature was originally exhibited by the injection of radiolabeled, anterogradely transported tritiated amino acids into the vitreous chamber of one eye (in several various animal models), allowing sufficient time for transneuronal transport across the LGN synapse. This result has been confirmed in humans through the use of high-resolution functional magnetic resonance imaging. Here the stimulation of one eye (with the other eye covered) for 20 sec followed by the reverse procedures (stimulating the originally masked eye and putting a cover over the originally stimulated eye) gives a pattern of activity that is limited to visual area V1, in a series of side-by-side activations (note that this is close to the spatial resolution of the fMRI technique for whole body imaging).

17.11 Parallel Processing Streams of Visual Cortex

17.11.1 Old View

The simplest early ideas on how the different parallel pathways ramified into the more distant projections, soon proved to be inadequate. The simplest idea emerging around the start of the 1990s was that the magno projection synapses in layer IVCa and the parvo projection synapses in Layer IVCb followed by projections into layers II/II, which separate into the dorsal stream projection (toward parietal cortex) and ventral cortical stream (toward temporal cortex). The initial projection was thought to be conducted for parvo input to the newly discovered cytochrome oxidase rich blob regions of layers II/III and the interblob regions (of less color sensitive neurons) and then to Area V2, or of inputs from the magno afferent layer (4Ca to Layer IVB) and then directly to Area MT/V5. The segregation was thought to be maintained through extrastriate cortical area V2 with inputs from the blobs projecting to the V2 thin (cytochrome oxidase) stripes, interblobs projecting to V2 interstripes and layer 4B projecting to the thick stripes of Area V2 (as well as to MT/V5) (Figure 17.5).

17.11.2 Current View

With the advent of highly accurate neuronal tracing (including viral transfection techniques), the picture tended to become more complex. It was soon realized that Layer 4B

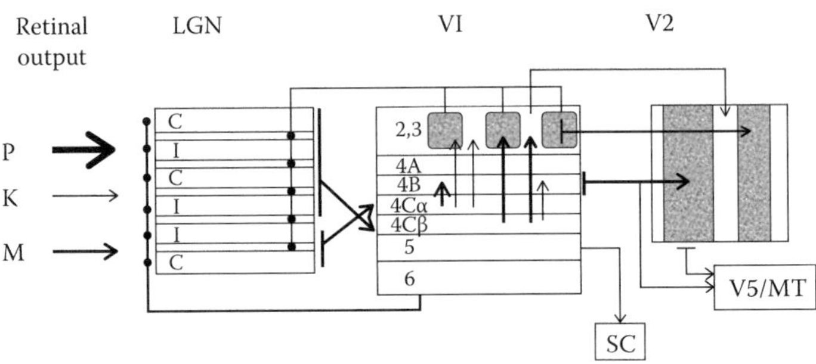

FIGURE 17.5
Postretinal processing of visual information by the geniculocortical system.

actually receives into from both 4Ca as well as 4Cb. Also, Layer 4C projects to blobs as well as interblob regions.

When these connections are considered together with those of Area V2 and MT/V5 it is clear that while the old view gives a reasonable sense of visual function—e.g., parietal neurons being relatively insensitive to chromatic stimuli fits with the notion of domination of inputs to parietal cortex by the M pathway, while inferotemporal cortical neurons are relatively insensitive to stimuli of coarse spatial features and not very motion sensitive, the additional connections referred to in the new model give access of the dorsal stream to color information in that MT/V5 neurons respond to moving edges that are separating isoluminant chromatic stimuli and similarly, our recognition of objects that are moving is presumably facilitated by the presence of some motion sensitivity in part of the lateral occipital complex called Area LO.

One example where may be tested is with the perception of moving objects that are either defined by luminance contrast or by texture contrast (see Figure 17.6). The figure shows two different sorts of gratings—one, a so-called first order (FO) contrast has bright and dark bars, while the other (second order—SO) (Figure 17.6b) produces form from the variation in the contrast of the random elements of texture. At low contrast, motor reaction

FIGURE 17.6
First- and second-order (FO and SO) grating stimuli created by luminance modulation and contrast modulation, respectively.

times to motion onset are 100 msec longer for the SO compared with the FO stimuli and visual evoked potential latencies are 20 msec longer.

However, at contrasts in the midrange (20% for VEP, 50% for MRT), the timing differences are no longer significant. Given that the M pathway responses saturate at around 10% to 20% contrast (from primate studies), but that P pathway responses do not saturate up to maximum contrast (black and white), these results suggest that the brain's motion system uses whatever information it can to respond, and is not solely wedded to purely M-pathway input.

17.12 Motion Processing—Area V5/MT

We process motion seemingly automatically and without thought, but when one compares the output stages of the cortical systems involved with orientation processing and edge detection, contour formation, closed contour recognition and hence object recognition, the end-point of which is a categorical name—something that taps into our semantic and linguistic systems as a noun, motion has a less strong grip on conscious classification—it if difficult to think of the processes of determining direction selectivity, speed, motion coherency, navigation, and so on, without starting to apply such measures to particular objects—and when one does so, they tend to be applied epithetically, as descriptors of objects—a car going fast, a police car heading north. Perhaps descriptions of invisible objects—for example, a strong wind, or a stiff current, give the strongest categorization of "motion objects." We depend on motion stimuli to a great extent—in evolutionary terms, motion breaks camouflage by causing coherent movement of one texture against another giving an advantage to the hunter. In modern times, automatic motion processing allows us to play ball games and drive cars. Motion processing seems also to be intimately related to brain circuitry underlying attention. Thus, MT/V5 is seen as the start of the dorsal stream, feeding regions such as intraparietal sulcus and superior parietal gyrus, known to be involved in the processing of complex geometrical transformations, navigation, and in attentional tracking of multiple objects.

17.13 Priority of Firing—The Magnocellular Advantage

From the electrical consideration of axon conduction velocity, different axons with considerably different axon thickness will result in different timing of impulses arriving at visual cortex (Area V1 = striate cortex). In primate, estimate of this time difference is 7 msec–20 msec, depending on the size of the monkey. In human, visual evoked potential measurements estimate the time to activation by the magnocellular and parvocellular projections to cortex as 25 msec–40 msec. Consideration of the cable properties of axons quickly indicates that large neurons with thick axons will conduct information more rapidly to cortex than will axons with small diameters.

Single cell neural recordings in monkeys show that M-pathway responses activate striate cortex earlier than P-pathway afferents by 7 msec–20 msec. Already by the LGN, the fastest M responses are 10 msec earlier than the fastest P responses and magnocellular lesion studies in primate confirm these timings. Confirmation of a magnocellular advantage in humans is important for the interpretation of visual psychophysics and for theories of recognition processes. Nonlinear visual evoked responses give an opportunity to separate the contributions of the M and P pathways to the cortical VEP by tapping into the different temporal coding properties of the M and P systems. Such techniques use pseudorandom stimulus sequences to measure neural recovery following stimulation, and under such conditions the nonlinearities generated by the M and P pathways have been shown to separate in the second order responses as a function of interaction time. In addition, the M-generated nonlinearity shows a contrast response function that has high contrast gain (growing rapidly at low contrast) and its amplitude saturates at around 35% to 40% contrast, whereas the main P-generated nonlinearity shows lower contrast gain and does not saturate.

These M and P nonlinearities show implicit time differences to the first major positivities of 25 msec–40 msec, depending on method. Such studies have direct application to studying the neural mechanisms of syndromes and behaviors that are defined only within human psychology/psychiatry—for example, autism or dyslexia.

Bullier proposed that there was sufficient time for the first wave of magnocellular afferents to project forward onto Area MT and into early parietal cortex and to retroinject information from these regions back onto area V1 and V2, which Bullier referred to as the active blackboard on the basis that these two cortical areas have the highest resolution of spatial information (i.e., smallest receptive fields) across all of the visual areas. This magnocellular temporal priority has been termed the "magnocellular advantage" (Laycock, Crewther, and Crewther 2007) and slowness of feedback from early dorsal cortical pathways may underlie several the abnormal visual perception seen in human conditions such as schizophrenia and autism (Dakin and Frith 2005).

17.14 The Feed Forward Sweep

Single unit recordings from primates have indicated a remarkable fact—information is carried for from the occipital cortex, where visual information first reaches the cerebral hemispheres to the front cortex in less then the blink of an eye. Meta-analysis of the timing of first action potential spikes evoked across cortical areas (Bullier 2001; Lamme and Roelfsema 2000) shows an extremely rapid input to frontal cortex. While such meta-analyses are characterized by different methods, stimuli, and animal species, they show a consistent trend. Table 17.2 gives a rough ordering of some 16 cortical regions of interest ranked by earliest spike firing.

In humans, event related potentials show a similar and early forward sweep of activation (Foxe and Simpson 2002), with earliest occipital activation at 56 msec with dorsolateral frontal cortical activation by 80 msec, in line with the increased brain size of human compared with the primate species sampled. This feed-forward sweep has been used as the basis of a model for recognition (Roelfsema 2006) based around the process of grouping. Roelfsema uses the term "base-grouping" to connote the feed-forward process whereby neurons that are tuned to particular features can respond very rapidly conveying the information of the scene through these selective filters. His second component is called "incremental grouping" and uses processes of horizontal connections and feedback to reinforce the elements that emerge as the received percept.

TABLE 17.2
Earliest Spike Firing in Primate Visual Cortical Regions

Area	Earliest Spikes (m)
V1	36
MT	41
MST/FST	43
FEF	44
7a	46
SMA	48
V3	50
PreFr	51
V2	55
PreM	57
TPO	61
V4	61
Tem	65
Tea	65
OrbFr	80
EntRh	100

Note: Sixteen regions chosen from meta-analyses (Bullier 2001; Lamme and Roelfsema 2000) shown in order of firing of the first spikes to visual stimulation. It is clear that frontal regions such as frontal eye fields (FEF), supplementary motor area (SMA), and prefrontal cortex (PreFr) are activated very rapidly—even before response is generated in the extrastriate cortical area V2 and V4. Dorsal cortical regions tend to be activated considerably earlier than ventral cortical regions such as orbitofrantal cortex (OrbFr) and entorhinal cortex (EntRh).

17.15 Ventral Processing and Visual Recognition

The brain performs an amazing transformation on the enormous amount of parallel information that is imposed on it. We have seen how this information is initially processed in parallel, and also is processed on a number of scales, in terms of receptive field size, and in terms of various modalities, such as color, orientation, motion, and so on. From this seeming chaotic information, a perception arises that is smooth and uncluttered. It consists of relatively few objects that may have associated motion or intention, and that may draw our attention closer. The extraction of object features and associations do not appear to depend on either perceptual load nor on intention, indicating that such processing has occurred prior to the time in which conscious selection can be made, but also allowing for break-in of stimuli that must be attended to—e.g., threatening stimuli. Imaging studies in monkeys have shown an ordered array of columns of neurons within monkey inferotemporal cortex that best code for particular kinds of complicated stimuli. When responding to a particular stimulus, certain of these columns are activated and others are not. Interestingly, when a stimulus can be decomposed into various parts—e.g., silhouette, internal features (eyes, mouth), body, and so on, there is typically common columnar activation for stimuli containing common parts.

17.16 Faces and Places, Chairs, and Motor Cars

While the organization of neural structures in early visual cortex is set up before or around birth, the existence of regions that respond to particular features such as faces (fusiform cortex, sometimes called the fusiform face area, FFA) and places (parahippocampal cortex, sometimes called the parahippocampal place area, PPA) must rely on the process of visual experience and learning to create the specialization required for recognition (say, of a loved one). Are such areas preordained, or are they built up through a process of establishing expertise. A vigorous argument in the literature has shone much light on this question, without a final answer.

Activation in such regions seems to correlate with conscious percept, as demonstrated by binocular rivalry experiments. When one views a face through one eye and a scene through the other, the resultant percept is an alternation between face and place with fMRI event-related activation showing corresponding fluctuations in FFA and PPA.

It should be clear, however, that much more is to be learned before conscious vision is fully understood.

References

Bullier J. 2001. Integrated model of visual processing. *Brain Res Brain Res Rev* 36:96–107.

Crewther DP. 2000. The role of photoreceptors in the control of refractive state. *Prog Retin Eye Res* 19: 421–457.

Dakin S, Frith U. 2005. Vagaries of visual perception in autism. *Neuron* 48:497–507.

Davson H. 1990. *Physiology of the Eye*, 5th Edition. Pergamon.

Felleman DJ, and Van Essen DC. 1991. Distributed hierarchical processing in the primate cerebral cortex. *Cereb Cortex* 1:1–47.

Foxe JJ, Simpson GV. 2002. Flow of activation from V1 to frontal cortex in humans. A framework for defining "early" visual processing. *Exp Brain Res* 142:139–150.

Hubel DH, and Wiesel TN. 2009. Republication of The Journal of Physiology (1959) 148:574–591: Receptive fields of single neurones in the cat's striate cortex. 1959. *J Physiol*, 587:2721–2732.

Lamme VA, Roelfsema PR. 2000. The distinct modes of vision offered by feedforward and recurrent processing. *Trends Neurosci* 23:571–579.

Laycock R, Crewther SG, Crewther DP. 2007. A role for the "magnocellular advantage" in visual impairments in neurodevelopmental and psychiatric disorders. *Neurosci Biobehav Rev* 31:363–376.

Livingstone M, Hubel D. 1988. Segregation of form, color, movement, and depth: Anatomy, physiology, and perception. *Science* 240:740–749.

Merigan WH, and Maunsell JH. 1993. How parallel are the primate visual pathways? *Annu Rev Neurosci* 16:369–402.

Nassi JJ, Callaway EM. 2009. Parallel processing strategies of the primate visual system. *Nat Rev Neurosci* 10:360–372.

Roelfsema PR. 2006. Cortical algorithms for perceptual grouping. *Annu Rev Neurosci* 29:203–227.

Schiller PH. 2010. Parallel information processing channels created in the retina. *Proc Natl Acad Sci USA* 107:17087–17094.

Wallman J, Winawer J. 2004. Homeostasis of eye growth and the question of myopia. *Neuron* 43:447–468.

18

Audition and Vestibular Sense

Joseph Ciorciari

CONTENTS

In this chapter, the fundamentals of the anatomy and physiology of the human auditory and vestibular systems will be covered briefly. The focus of this chapter will be to outline the wide range of measurements that can be made and the appropriate tests that can be performed to measure the integrity of the neurophysiology associated with audition and vestibular function. The disorders of these systems will be characterized in these terms.

Auditory anatomy, physiology, impedance matching qualities, and information theories will be covered in Sections 18.1–18.7 while the vestibular system (bony labyrinth) will be covered in Section 18.8. The neuroanatomical pathways for both systems are reviewed, focusing on techniques for measuring the various reflexes, the cortical and brainstem responses, and typical clinical tests.

18.1 Introduction

The human auditory system is sensitive to frequencies from approximately 20 Hz to 20 kHz. Unfortunately, our hearing capacity decreases as we age. Within this range, the ear is most sensitive between 1 kHz and 5 kHz. This is due to the resonant qualities of the ear canal and the ossicles of the middle ear and is also the range associated with speech. According to Australian statistics, however, hearing decreases rapidly from the age of 51 with males displaying more loss. Also one in five of the Australian population suffers from some hearing impairment* and it is estimated that this will be one in four by 2050.† In 2009, 15% of the USA had hearing deficits (National Health Interview Survey) similar to the UK.‡ According to the World Health Organization, it estimates that approximately 278 million people worldwide have hearing deficits (WHO 2010).

The human hearing system consists of a conduction system, impedance matching system and a transduction system; collectively relaying sound information to the auditory cortex of the brain. Simply put, the sound waves enter the *ear canal* and cause the ear drum (*tympanic membrane*) to vibrate in synchrony with the sound waves, with energy transferred through the *ossicles* to cause the same vibrations to enter and travel through the *perilymph* of the *cochlea* (see Figures 18.1a,b and 18.2). The *basilar membrane* is displaced in synchrony with the sound waves and causes hair cells to react to the movement because of the tension changes on the *kinocilium* and release neurotransmitter which then elicits action potentials in the auditory pathway through to the temporal cortex. This process of converting sound wave energy into electrical impulses is referred to as *transduction* and arrives at the sound processing areas of the cortex in approximately 10 ms.§

The major anatomical features associated with the hearing system are the organ of Corti, round and oval windows, vestibular canal (scala vestibule), cochlear duct (scala media), tympanic canal (scala tympani), basiliar membrane, and the receptor hair cells, all of which are housed or part of the cochlea, a coiled snail shell-like structure. As illustrated by Figure 18.2, the cochlea receives sound information via the *ossicles* (malleus, incus, and stapes), three very small bones that act as an mechanical interface between air and the fluid filled chambers of the cochlea.

* http://www.vicdeaf.com.au/statistics-on-deafness-amp-hearing-loss.
† http://www.hearing.com.au.
‡ http://www.earhelp.co.uk/facts-figures-about-hearing-problems.html.
§ As measured with a Auditory Brainstem Evoked Response (ABER).

18.2 Overview of Ear Physiology

Sound waves enter the auditory canal and hit the eardrum (*Tympanic membrane*). The membrane vibrates in synchrony with waves. The waves travel to *malleus, incus* then the *stapes* (*ossicles*). The stapes vibrates the oval window. The *Stapes* transmits energy to the perilymph of the *scala vestibuli* (vestibular duct). Stapes oscillates backward and forward driving the

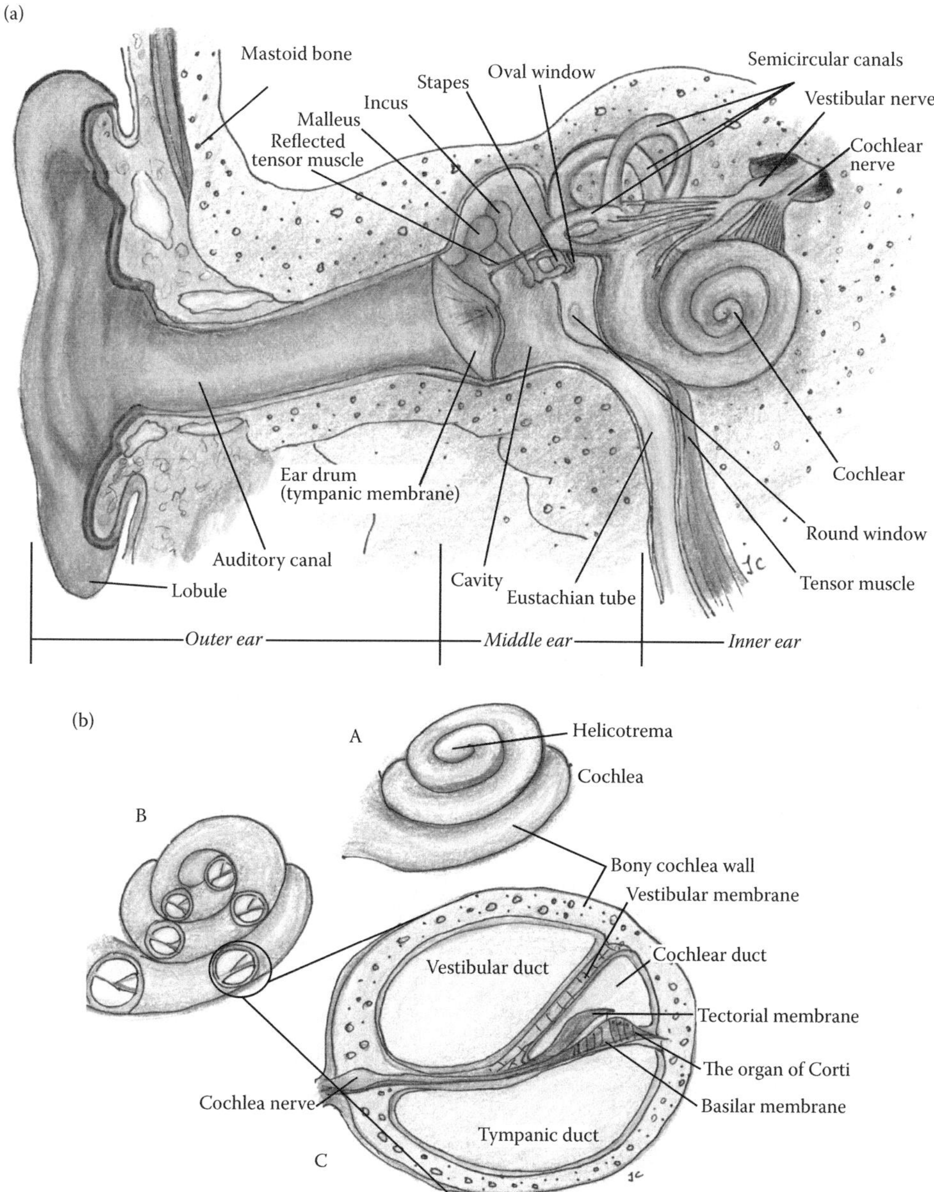

FIGURE 18.1
(a) Anatomy of the hearing and balance system. (b) Anatomy of the cochlea and the hair cells: A, cochlea; B, cross sections through cochlea; C, cross section of cochlea illustrating the ducts and organ of Corti.

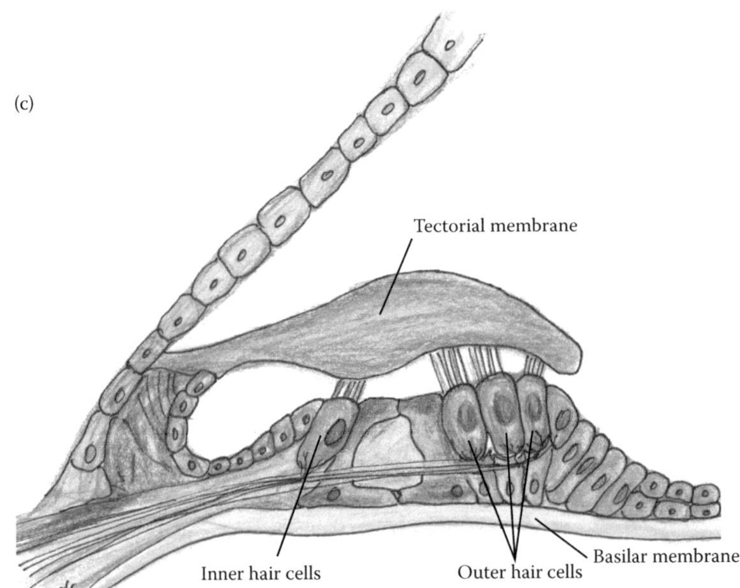

FIGURE 18.1 (Continued)
(c) Organ of Corti. This illustration identifies the location of the inner and outer hair cell rows.

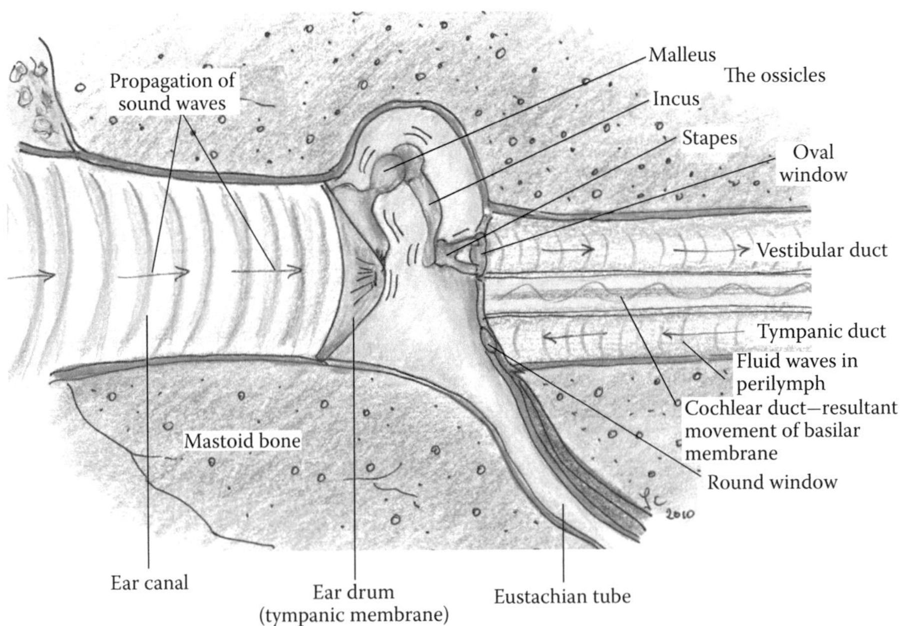

FIGURE 18.2
Middle ear sound wave conduction. Sound waves travel down canal and vibrate the ear drum. The ossicles transmit the vibrations to the oval window thereby causing traveling waves through perilymph fluid of cochlea and subsequently the basilar membrane and hair cells.

incompressible fluid (corresponding movement of the *round window*). *Receptors* (hair cells) are activated by vibrations set up by *traveling waves* of pressure changes in fluid. The point of maximum amplitude of the excursions of the *basilar membrane* moves progressively toward the apex (*helicotrema*) as the frequency of the stimulus decreases. Even with a pure tone a wide stretch of the basilar membrane will oscillate and a number of *hair cells* will be excited. This is also the basis for the Place Theory; points of maximum displacement maps each frequency. Shearing the *cilia* (of the *hair cell*) constitutes the adequate stimulus of the hair cells. Note that there is a three-dimensional movement of the basilar membrane, which may code for other auditory sensitivities. This will be discussed later in Section 18.4.

There are inner and outer rows of hair cells enclosed in the organ of Corti. The inner row consists of a single row of approximately 3500 hair cells with 90% having nerve endings. The outer rows consist of three rows, of approximately 12,000 hair cells with only 10% having nerve endings. All hair cells found in the organ of Corti are called secondary sense cells because neuron endings connect to them from elsewhere in the neuroanatomical pathways. Approximately 30–40,000 afferent fibers run in the auditory nerve. Efferents are also found in this nerve. These may have a modulatory role.

18.2.1 Neuroanatomical Pathways

This *coded* information is transferred through a number of sites in the *central nervous system* via the VIIIth cranial nerve; the *auditory vestibular nerve*. This pathway is illustrated in Figure 18.3. From the organ of Corti, the auditory information travels to the cochlear nuclei (dorsal and ventral nuclei). After some processing, this information is projected bilaterally to both superior olives. The left and right *superior olives* will receive information from ipsilateral and contralateral sides (i.e., from both ears). At this level, some "calculation" of the direction of the sound is achieved. This processed information is then projected to the *inferior colliculi* via the *lateral lemniscus*, to be projected to the *medial geniculate nuclei* of the *thalamus*.

From the *lateral lemniscus* sound related information is also passed on to the *cerebellum* and the *reticular formation* (medullary nuclei). It is thought that this is to elicit appropriate postural reflexes in the event of a startling sound. A heightened sense of vigilance is also elicited by the *reticular formation*. By 10 ms the information has arrived at the auditory cortex. The integrity of these pathways can be tested using a *Brainstem Auditory Evoked Response* (BAER or ABR) test. This is illustrated in Chapter 21, Figure 21.8. Note that the pathways consist of both inhibitory and excitatory pathways.

18.3 The Electrical Activity of the Ear

In this section, the process of *transduction* will be examined. The process of transduction is the process associated with converting sound wave information into electrical information or a neuronal signal. We will also further examine the electrophysiology associated with the higher order areas of the auditory neuroanatomical pathways.

If a microelectrode is used to measure the micropotentials in the inner ear (with the Vestibular Duct as reference point), the cochlear duct (scale media) exhibits a positive potential (+80 mV) (see Figure 18.1b). The organ of Corti is negatively charged (–ve mV). During the absence of sound, these potentials are referred to as *Standing Potentials*. With sound however (see Figure 18.4), other potentials can be recorded from the cochlea and round window. These include the

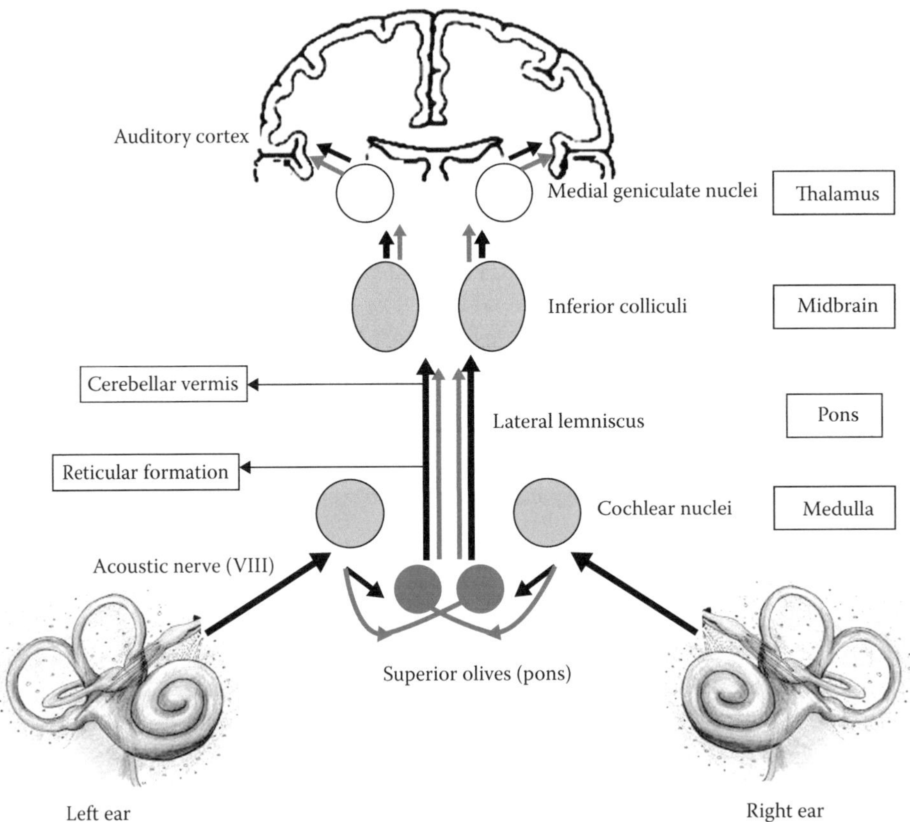

FIGURE 18.3
Auditory neuroanatomical pathways. It takes approximately 10 ms for the sound information to arrive at the auditory cortex.

cochlear microphonic, hair cell potentials, endocochlear potentials and action potentials of the auditory nerve. These potentials will now be discussed separately.

18.3.1 Cochlear Microphonic

This potential can be recorded from the *round window. It reproduces closely* the fluctuations in sound pressure. Unlike an action potential, this potential essentially has no latency, no refractory

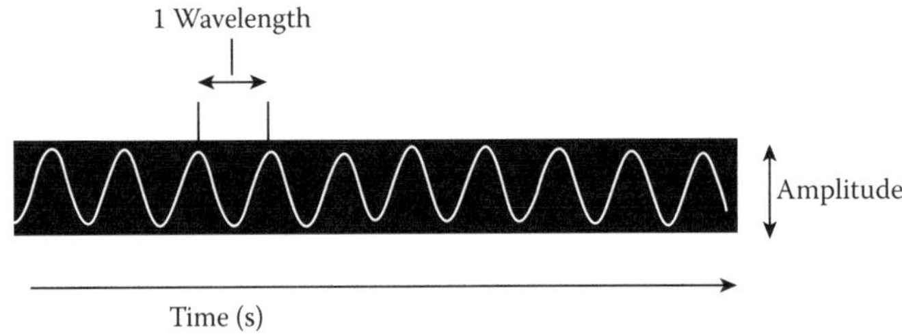

FIGURE 18.4
Sound waves. A sinusoidal waveform (tone), with amplitude and wavelength illustrated.

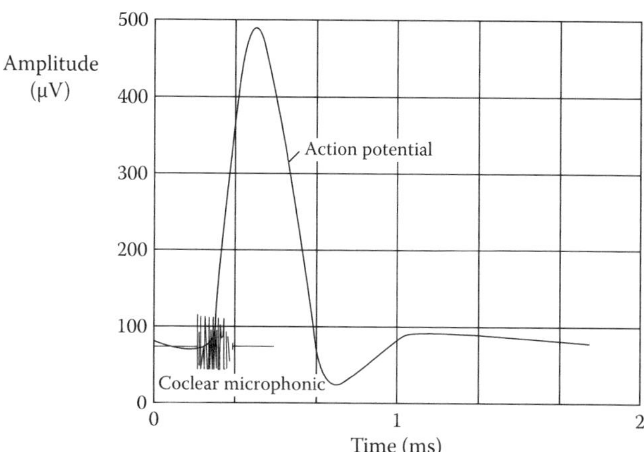

FIGURE 18.5
Action potential (individual hair cell) and cochlear microphonic (hair cell and duct ion activity).

period, no measurable threshold, and is not fatigueable. It is associated with the summation of all hair cell activity. Its amplitude is normally 100 μV with duration of less than 1 ms.

18.3.2 Endocochlear Potential

With the movement of the basilar membrane, a shear force is created and a small potential is generated due to difference in potential between the endolymph (scale media +80mV) and the perilymph (vestibular and tympanic ducts –70 mV). Acoustic stimulus produces a simultaneous change in conductance at the membrane of the receptor cell. Because there is a steep potential gradient (150 mV), changes in membrane conductance are accompanied by rapid influx and efflux of ions which in turn produce the receptor potential. This has previously been referred to as the *battery hypothesis*. The receptor potential for each hair cell causes a release of neurotransmitter at is basal pole, which elicits excitation of the afferent nerve fibers.

18.3.3 Action Potentials

These potentials signal excitation of the hair cells to the central nervous system and their responses are associated with the temporal and loudness qualities of the sound. The duration of sound is coded by duration of neuronal activity. The frequency of sound is coded by the frequency of neuronal activity. These potential scan be recorded at round window using microelectrodes. Whereas the amplitude of cochlear microphonic is under 100 μV in amplitude and less than 1 ms in duration, the amplitude of an action potential is 500 μV with duration of 1 ms (see Figure 18.5).

18.4 Information Processing

18.4.1 The Place Theory

With sound the *stapes* transmits energy to the perilymph of the *scala vestibuli* (vestibular duct). It does this by oscillating backward and forward driving the incompressible fluid

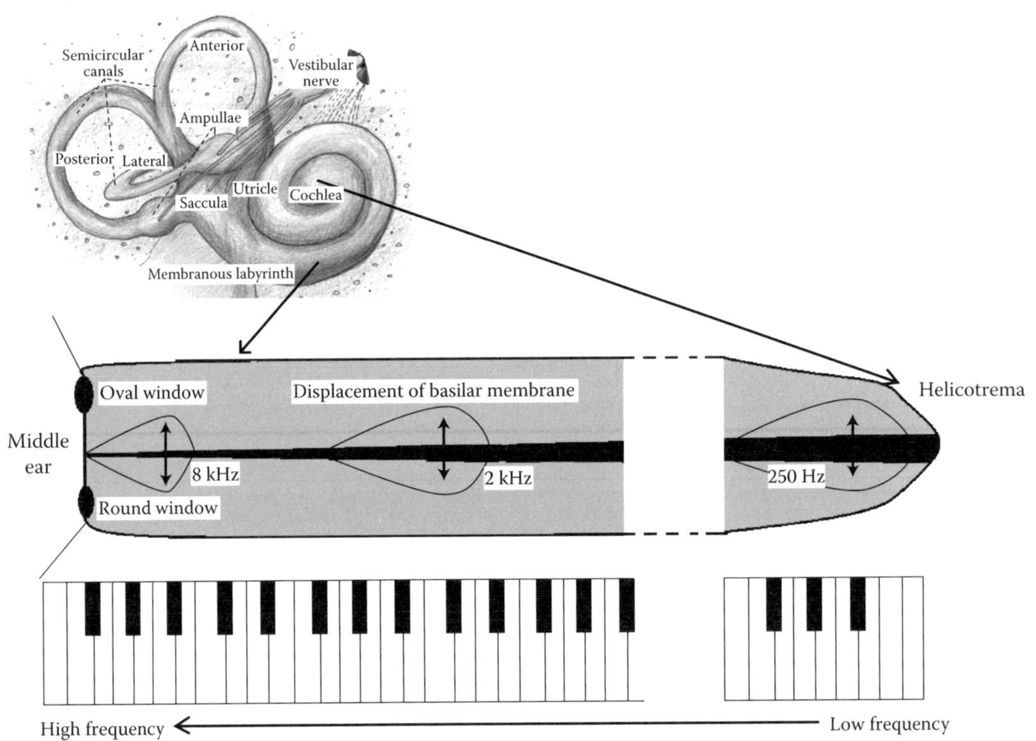

FIGURE 18.6
Frequency mapping the basilar membrane. Displacement of the membrane excites hair cell, however, the shape and size of the membrane has areas where maximum displacement which are frequency dependent. In this illustration, the higher tones cause maximum displacement of the membrane closest to the oval and round windows; analogous to a reversed keyboard.

(perilymph) and is noted with the corresponding movement of the *round window*. Figure 18.6 illustrates this process. This movement causes displacement of the *basilar membrane and s*tarts a wave of movement moving toward the *helicotrema*. Each frequency will cause vibrations of the particular regions of the basilar membrane. Shearing the *cilia* (of the *hair cell*) constitutes the adequate stimulus of the hair cells. Therefore, each frequency will excite different sensory cells located at the points of maximum displacement. This is also referred to as the tonotopic organization of the primary auditory neurons. This process is explained as the Place theory.

18.4.2 Coding of Frequency Information

The relationship between stimulus frequency and loudness and each subsequent nerve fiber as we move through the neuroanatomical pathway can be quantified or profiled by producing *response threshold curves* for single nerves also known as *tuning curves*. These are associated with the APs of groups of hair cells (see Figure 18.7). Each nerve has a characteristic *threshold frequency* and a responsivity range associated with loudness of the stimulus and frequency range. Outside these ranges, it does not respond to the sound stimulus at all. Basically, each neuron has its own individual response. For example, consider a neuron with a critical frequency of 500 Hz. It may be able to fire in synchrony at 500 Hz (impulses/s), but if it cannot fire this quickly it may fire at divisions of 500 Hz, say 250, 125, or 100 Hz. (fundamental plus its harmonics). Therefore, intervals between

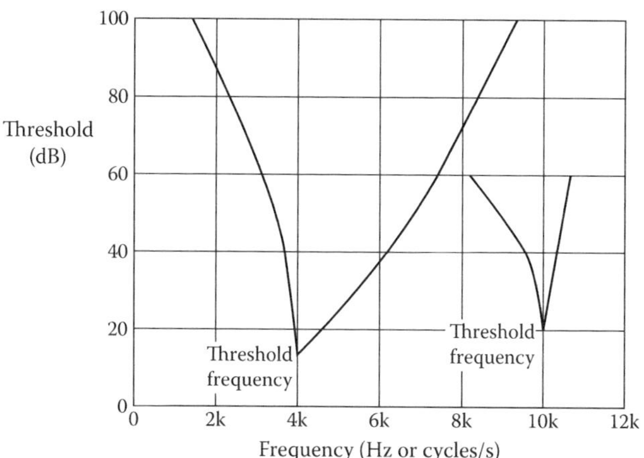

FIGURE 18.7
Response curves for two auditory nerve fibers. One has a threshold frequency of 4 kHz while the other has a threshold frequency of 10 kHz.

firing may also be a form of information about the original sound frequency. If a hair cell fatigues, a neighboring cell will fire in its place. Auditory nerves can follow signals as high as 3 kHz–4 kHz, but an individual cannot.

18.5 Coding of Intensity Information

The neural responses to changing intensities of sound are less complex than those associated with frequency related responses. If stimulus frequency of sound is constant and loudness (intensity) increases, then response rate of each neuron increases. For example, an individual neuron may fire at 200 Hz and increase to 300 Hz with an intensity increase of 10 dB, but then shows no change as the loudness increases. This characteristic is also illustrated by the threshold response curve.

18.5.1 The Volley Principle

The volley principle (previously *volley theory*) is an information scheme used to describe a possible neurological encoding process for sound information in human hearing. The action potentials generated by excited auditory neurons are able to transmit information about the sound by the way they respond to the sound stimuli. For example, it is possible for a neuron to have an excitation rate consistent with the frequency of the stimulus. However, to compensate for higher rates of stimulus frequency which a single neuron cannot respond to, neighboring neurons may become excited or they may take turns. Figure 18.8 illustrates this behavior. The neuron responses are phase locked to the sound stimulus frequency characteristics. A number of neurons phase locked to the stimulus frequency can encode this information.

18.5.2 Auditory Cortex

Information about intensity and frequency arrive at auditory cortex together with information about the direction and location of the sounds. Information about pitch and timbre

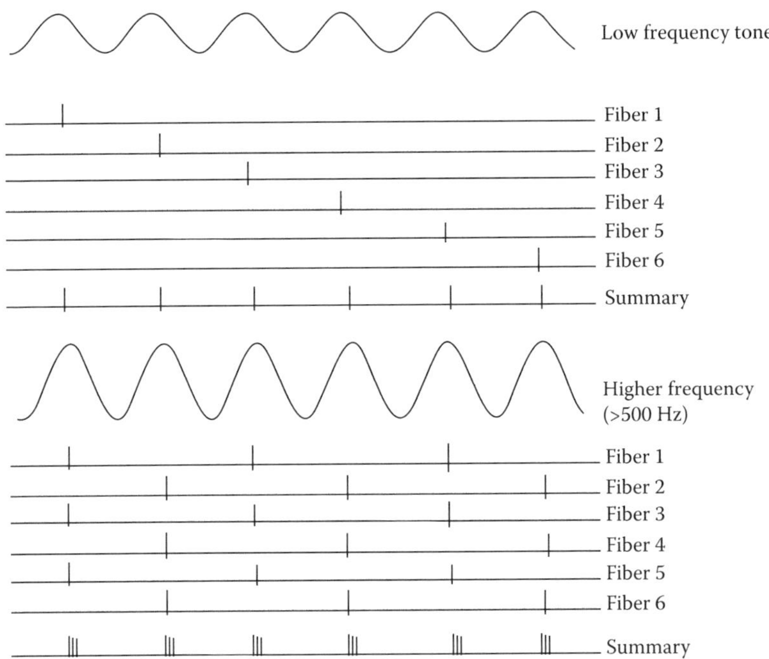

FIGURE 18.8
The volley principle. This illustration demonstrates how fibers respond in synchrony with the stimulus frequency. When the intensity of the stimulus increases, fibers may discharge at a faster rate although still in synchrony with the original stimulus frequency (harmonic).

are also encoded and are further processed in the primary auditory cortex (Brodmann's areas 41, 42). In the cortex, there are cells which response to particular changes in sound. There are those particularly suited to processing human speech. In fact 40% of the cortical cells will respond to pure tones while 60% will only respond to particular changes in sound. For example, certain cells will increase their rate of firing only when a sound first occurs (excitatory response), and others respond when sound is diminished (inhibitory response). This response may be associated with a decrease in firing rate. Some cells function in an interactive mode, inhibiting or exciting other cells, at specific frequencies. Some of these responses are illustrated in Figure 18.9.

18.6 Impedance Properties of the Ear

The major role of the middle ear is to allow the efficient transfer of sound wave energy from one medium to another; sound waves traveling through air to sound waves traveling through perilymph fluid within the cochlea. The middle ear is the interface between air and perilymph (both having fluid characteristics; fluid mechanics). The pinna, the auditory canal, the eardrum and round window acoustic properties also allow the efficient transfer of sound information, without loss or reflection. All components have resonance or acoustic qualities. We can calculate the resistance to sound waves by examining the impedance qualities of the middle ear.

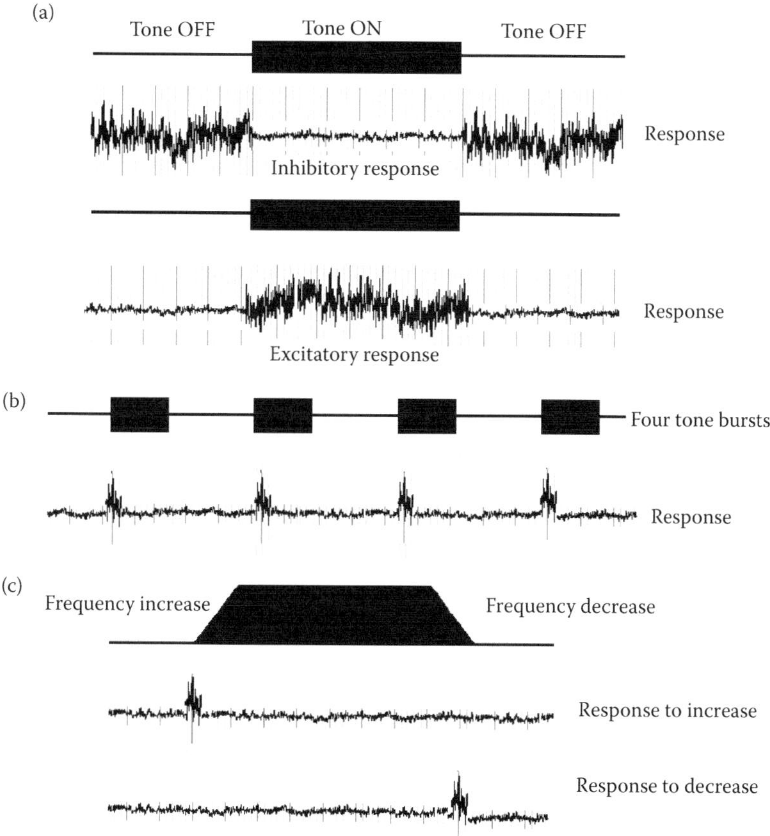

FIGURE 18.9
Neural responses with sound and frequency changes. (a) An illustration of ON and OFF response. (b) Bursts of activity associated with the presentation of the tone. (c) Responses associated with an increase and decrease in stimulus frequency.

18.6.1 The Gain of the Middle Ear System

A number of physical characteristics associated with the middle ear can be used to calculate the gain of the system, utilizing basic physics principals. For example there is an *area gain* when you compare the area of the eardrum and the stapes foot plate (oval window). Also, the ossicles act like a lever (see Figure 18.10) and produce a *pressure gain*. Other factors that help with the transmission of the sound waves are the mass and elasticity in the transmission chain, the curvature and oscillatory properties of the eardrum as well as the shape of pinna, helix, and external meatus and ear canal.

18.6.2 Calculating the Gain in Pressure

If we use the principal, that pressure is equal to force per unit area (Press = Force/Area), we can estimate the potential pressure gain just by the differences in areas; area of eardrum (A_d) = 0.55 cm^2, and the oval window (A_o) = 0.032 cm^2, then A_d/A_o gives a pressure gain of 17.

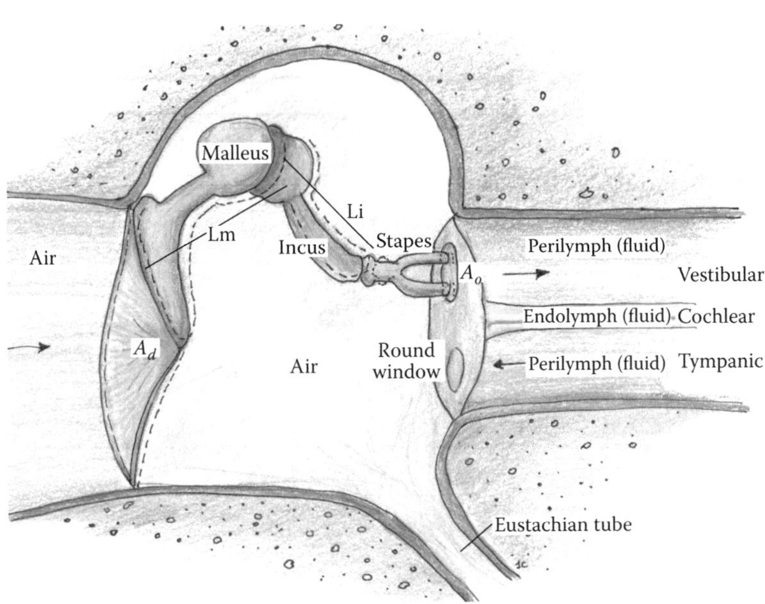

FIGURE 18.10
Middle ear–impedance matching. (See text for description.)

Therefore, in summary,

$$\text{Pressure} = \text{Force/Area}$$

$$\text{Assuming force is constant then pressure gain} = A_d/A_o = 17 \text{ times.}$$

If we also include the lever properties of the ossicles, length of the incus (l_i) and the length of the malleus (l_m), then the overall pressure gain is multiplied by 1.3 because the malleus is longer than the incus by 1.3.

Therefore, in summary, overall pressure gain is

$$17 \times 1.3 = 22 \text{ times.}$$

We can convert this to sound pressure level (SPL) which is measured in decibels (dB) by using the following relationship;

$$\begin{aligned} \text{SPL} &= 20\log_{10}(\text{Pressure change}) \\ &= 20\log_{10}(22) \\ &= 27 \text{ dB.} \end{aligned}$$

Therefore, we have a resultant improvement in sound hearing of 27 dB. Of course, if the ossicles or eardrum movement is impeded (say by infection or wax build-up), then a dramatic loss up to 27 dB in hearing can manifest.

18.6.3 Impedance Matching

The middle ear acts like an impedance matching device, that is, it allows the faithful transmission for information (sound waves) from one fluid medium (air) to another (perilymph) in the cochlea. By using the following relationship, we can calculate the impedance properties of the middle ear,

$$\text{Impedance (Z)} = \text{Applied Pressure (P)/Resultant Velocity(V)}.$$

If we assume that

$$P_{drum} = A_o \times l_i$$
$$P_{oval} = A_d \times l_m.$$

And that the resultant pressure is given by

$$P_{drum}/P_{oval} = (A_o \times l_i)/(A_d \times l_m)$$

and that the ratio of velocity is approximately equal to ratio of lengths

$$V_{drum}/V_{oval} = l_m/l_i.$$

Then the impedance ratio is

$$Z_{drum}/Z_{oval} = \left(P_{drum}/P_{oval}\right)\times\left(V_{drum}/V_{oval}\right)$$
$$= A_o/A_d(l_i/l_m)^2$$
$$= 0.035\ (30\ \text{dB}).$$

Note the input impedance of the cochlea is 5.6×10^4 rayls, while the input impedance for the drum is 1960 rayls. This equates to a further 30 dB (approx) increase in hearing sensitivity. Together with the gain calculated for pressure, the middle ear improves hearing by approximately 60 dB (27 + 30).

18.6.4 Sound Localization

The brain is able to use sound information regarding sound intensity, timing (phase) and sound spectrum (frequency) to create a perception of a three-dimensional image of the acoustic landscape. It is able to do this by comparing the differences in sound arrival with the use of the ears. The positions of the ears are not for convenience but are crucial to detection of the source of sound in a 3D landscape. It seems there are at least two mechanisms are involved. The human ear is sensitive to intensity and frequency differences at both ears. These are referred to as the *interaural level difference* (ILD) for intensity of sound, while frequency or phase difference is referred to as *interaural time difference* (ITD). So differences in arrival time of same sound wave can be interpreted and the direction of the sound source sensed. This mechanism is particularly sensitive for human speech.

For low tones (2 kHz or less), error for the azimuth is approximately 5–10 degrees, whereas this is higher for 2 kHz–4 kHz tones (20–25 degrees). One of the mechanisms can

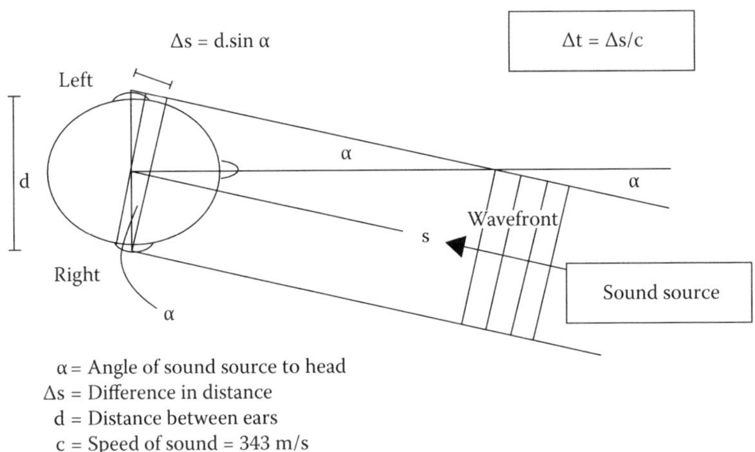

FIGURE 18.11
Sound localization for low frequencies. Calculation of conduction time differences in directional hearing. Using the above formula, delays as little as 3×10^{-5} s can be reliably detected.

detect sound intensity differences (high-frequency system) while the other is sensitive for phase or delay between sound reaching both ears (low-frequency system).

For example (see Figure 18.11), sound arrives earlier in right ear than the left. The brainstem auditory nuclei compare sounds from both ears and are able to "calculate" differences in the sound coming from the same source. Using the example illustrated in Figure 18.11, the time delays as little as 3×10^{-5} s can be detected. This mechanism is associated with the superior olives (first site for binaural interaction) and the interaction between lateral and medial nuclei.

At higher frequencies (above 3 kHz), cells in the lateral nucleus of the superior olive are excited by the ipsilateral input. These inputs are finely balanced and have a higher sensitivity for detecting differences in intensity of sound arriving at both ears. The medial superior olive is also associated with mapping sound in 3D space. To do this it must also have accurate information about head position and head movement to construct map. Therefore, the vestibular system (discussed in Section 18.8) provides this information to the superior olives.

In an engineering perspective, understanding these phases and intensity detection mechanisms in humans have helped to design cinema sound systems to give viewers the perception of realistic sound.

18.7 Pathology and Clinical Tests

Pathologies of hearing can be grouped into three categories. These include the disturbances of sound conduction, disturbances of sound sensation, and retro cochlear damage. Some changes can be improved whereas some are permanent. Each category will be examined separately with discussion about symptoms, methods of detection, and potential actions.

18.7.1 Disturbances of Sound Conduction

Generally, disturbances of sound conduction involve some change to the mechanisms of the middle ear. This could be due to infection and the subsequent build up of fluid within the middle ear chamber. With infection, the Eustachian tube may block, creating a sealed chamber which develops a negative pressure with respect to the external atmospheric pressure. Unable to equalize pressure, negative pressure in chamber causes fluid build-up. The normally pearly white eardrum is sucked inward and then bulges out as fluid pressure builds. On observation, the eardrum may appear reddish in color. This condition is reported as *otitis media*. The fluid may also impede the movement of the ossicles. This condition may be chronic and may require surgical intervention.

The ossicles can be damaged by physical trauma or by changes to bone health in conditions such as *otosclerosis* (bone growth on stapes) or calcium depletion as seen in *osteoporosis*. Some individuals also have a problem with a build up of ear canal wax which can impede the movement of the drum. In some case a loss of 20 dB–30 dB is possible.

By simple observation of the eardrum, one can see whether there is evidence of infection by looking for discoloration and condition of the eardrum. However, if you wanted to quantify the changes, you would use audiometry to measure sound loss (see Figure 18.12) or measure the stiffness (compliance) of the eardrum using a tympanogram (see Figure 18.13). Bacterial based infections could be improved with the use of antibiotics, however chronic conditions where the Eustachian tube is permanently blocked, then a biocompatible grommet needs to be fitted into the eardrum to equalize pressure within the middle ear and allow for discharge. In some cases a change in diet may be usual in reducing allergies which may lead to an overproduction of mucus in the sinuses which may lead to Eustachian dysfunction.

18.7.2 Disturbances of Sound Sensation

This condition is usually associated with changes to hair cell function. In some cases, transduction or neurotransmitter production and release may be altered. High sustained noise trauma, especially high-intensity, high-frequency sounds will damage hair cells. Table 18.1 lists the sound pressure levels associated with everyday items and conditions, illustrating those which can cause potential damage.

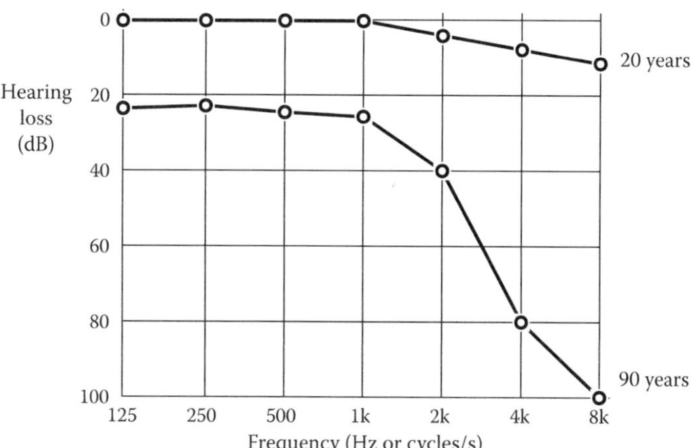

FIGURE 18.12
An audiogram. This graph illustrates the typical hearing range of a healthy 20-year-old compared to that of a 90-year-old. Note the loss of the higher frequency ranges.

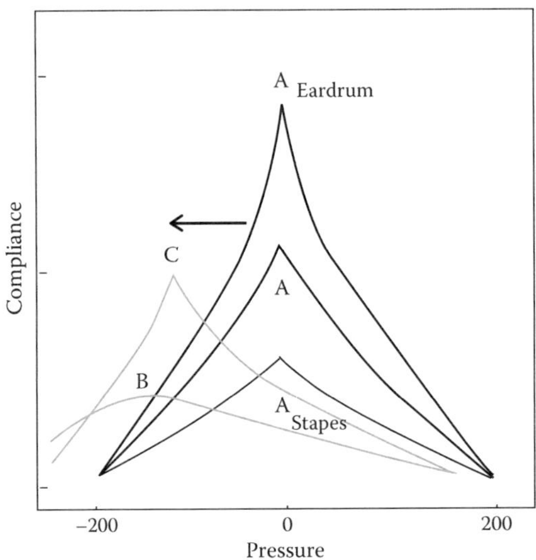

FIGURE 18.13
Ear drum compliance. This is an illustration of a tympanogram. (A) Normal compliance curve with two of lesser compliance associated with progressive stiffing of eardrum. (B and C) There is negative pressure in the middle ear. The tympanogram is shifted to the left with these conditions.

Other factors which lead to hair cell damage or dysfunction are the use of ototoxic drugs—certain antibiotics/diuretics. Hair cell loss with may occur with age and changes to cardiovascular-perfusion of this area. Some may occur because of neurodevelopmental disruptions affecting the growth and development of these systems. Meningitis can cause these disruptions. Generally air and bone conduction audiometry techniques are used to establish the degree of hearing perception loss. In some cases if the auditory nerve is

TABLE 18.1
Sound Level Comparisons

Level (dB)	Example	Dangerous Time Exposure
0	Lowest audible sound	
30	Quiet library; soft whisper	
40	Quiet office; living room; bedroom away from traffic	
50	Light traffic at a distance; refrigerator; gentle breeze	
60	Air conditioner at 20 feet; conversation; sewing machine in operation	
70	Busy traffic; noisy restaurant	Some damage if continuous
80	Subway; heavy city traffic; alarm clock at 2 feet; factory noise	More than 8 hours
90	Truck traffic; noisy home appliances; shop tools; gas lawnmower	Less than 8 hours
100	Chain saw; boiler shop; pneumatic drill	2 hours
120	"Heavy metal" rock concert; sandblasting; thunderclap nearby	Immediate
140	Gunshot; jet plane	Immediate danger
160	Rocket launching pad	Hearing loss inevitable danger

intact, a *bionic ear* may be surgically implanted to electrically stimulate the nerves and return hearing perception.

18.7.3 Retrocochlear Damage

With this condition, the middle and inner ear (cochlea) are intact, but there are neurological changes to the auditory neural pathways to the auditory cortex. There may be problems associated with the central/primary afferent nerve fibers. Brain tumors or acoustic neuroma can lead to partial or total deafness, sensation of vertigo or tinnitus. The swelling of the endolymphs such as occurs in *Labyrinthitis* can lead to all of these symptoms as well. The condition can be quantified and diagnosed with a Brainstem Auditory Evoked Response (see Chapter 27) or with a structural imaging technology such as Magnetic Resonance Imaging (MRI). Standard audiometry can be done to measure changes over time. The BAER may be of value clinically and is useful in demonstrating structural brain stem changes that may be associated with coma, multiple sclerosis, posterior fossa masses, and related disorders.

18.7.4 Overcoming Hearing Loss

Apart from the procedures mentioned in previous text, a number of aides are available to improve hearing in relatively intact yet damaged systems. Hearing aides have a number of characteristics designed to improve perception of sound. Some of these include, specific frequency band amplification, selective for human speech, they can be powered, and individually fitted to overcome specific hearing range loss. In the 1970s, the bionic ear was invented by Professor Graeme Clark and a team of scientists in Melbourne, Australia. This device primarily consists of a receiver and stimulus unit that is surgically placed in the mastoid bone behind the ear with a very fine electrode array wound through the cochlea so that the auditory nerves can be stimulated electrically by passing the faulty hair cells. A transmitter and sound decoder are worn externally by the recipient. More information about this great bioengineering feat can be found at the following website: http://www.bionicear.org/bei.

18.8 The Vestibular System: The Mechanical Sense of Equilibrium

The vestibular sense refers to the system that detects the position and the movement of the head. It also directs compensatory movements of the eye and helps to maintain balance. The vestibular organ is adjacent to the cochlea. In maintain balance, three systems are integrated. These include the visual system, the proprioceptive system and the vestibular system.

18.8.1 Anatomy of the Vestibular System

The major sensory components of the vestibular organ or the *membranous labyrinth* are illustrated in Figure 18.14. The vestibular organ consists of two *otolith* organs (the *saccule* and *utricle*) and three semicircular canals. The otolith organs have calcium carbonate particles (*otoliths*) that activate hair cells when the head tilts by shearing. The three semicircular canals are oriented in three different planes and filled with a jellylike substance that activates hair cells when the head moves. The semicircular canals consist of the superior, lateral and posterior canals. Within each canal is a structure called the *ampullae*. Within

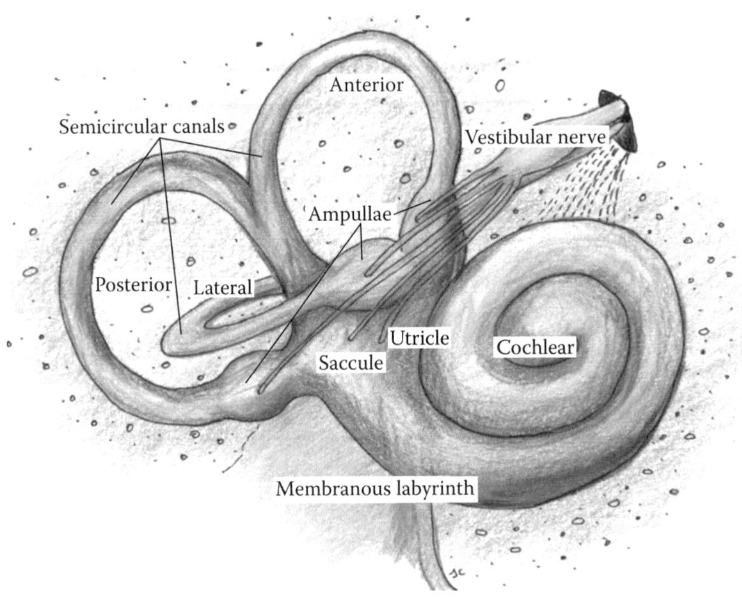

FIGURE 18.14
The membranous labyrinth. This may also be called the bony labyrinth. For illustrative purposes, this diagram shows the major components of the vestibular system; the macula organs (saccule and utricle) and the three semicircular canals, the posterior, anterior, and lateral (or horizontal).

each ampullae, is the *cupola* which forms the mechanical detection system associated with rotation, while the macula utriculi (utricle) and macula sacculi (saccule) are the tilt detectors. Even though hair cells are the main sensory receptor they are housed in different ways. The shearing of the kinocilium initiates a change in electrical discharge.

The different sensory mechanisms are illustrated in Figure 18.15. Two morphologically different hair cells types (Types 1 and 2) are found in the vestibular system (Figure 18.15a). Both, however, have cilia and a kinocilium sensitive to movement. These are secondary sense cells and have no neural process of their own. They are innervated by afferent from nerve cells in the vestibular ganglion. In the saccule and utricle, the hair cells are imbedded in a gelatinous mass which has otoliths (composed of calcite crystals) on the upper side, making the structure top heavy and creating shearing forces when the head is tilted (see Figure 18.9b and c). Another structure houses the hair cells in the semicircular canals. Within the ampullae, a bell-like structure called the cupola has hair cells embedded in it. With rotational movement, the endolymph fluid displaces the cupola, moving the kinocilium of the hair cells indicating rotational acceleration (see Figure 18.9d). For all systems, the direction of the shear will determine the output discharge of the hair cell.

18.8.2 Detecting of Movement and Acceleration

Hair cells in saccule/utricle macula organs respond to linear acceleration, or tilt of head in a vertical and horizontal direction. The saccule and utricle are oriented at right angles to each other to detect vertical and horizontal movement. They are also referred to as gravity detectors. However, hair cells in the cupula respond to angular acceleration (rotational). Both produce a high neuronal discharge at rest and increase or decrease discharge rates depending on the direction of displacement or movement. From the

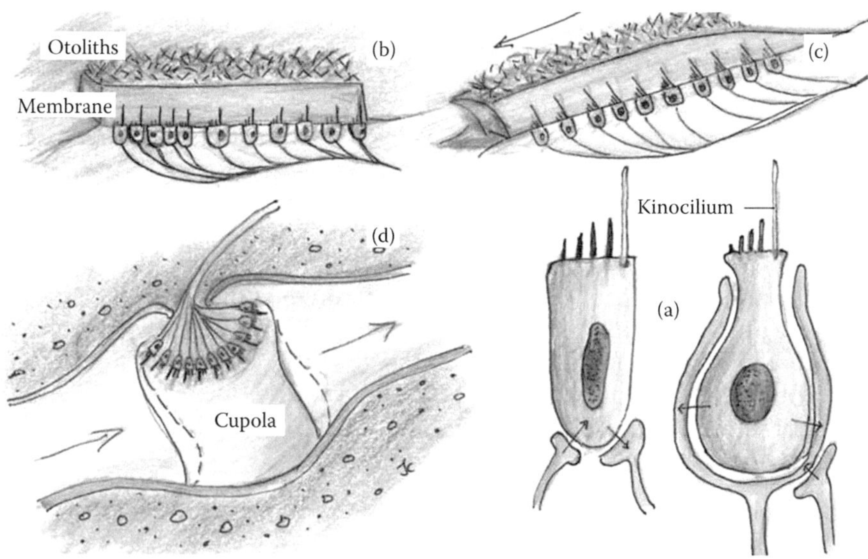

FIGURE 18.15
Cells of the membranous labyrinth. (a) Two morphologically different cells. (b) Calcite crystals (otoliths) and the gelatinous mass. (c) The shearing of the hair cells with tilt of the gelatinous mass. (d) Semicircular canals: movement of the cupola, shearing the kinocilium of the hair cells.

three discharge patterns from the semicircular canals, the brain is able to determine the direction of angular acceleration, together with pitch and roll. When turning the head, there is a slight movement of the cupola detecting angular acceleration, but returns to rest position. However, with prolonged rotation the cupola is now deflected by the movement of the endolymph fluid. This may also occur with a vestibulo-ocular reflex (VOR) called *nystagmus* (see Figure 18.16).

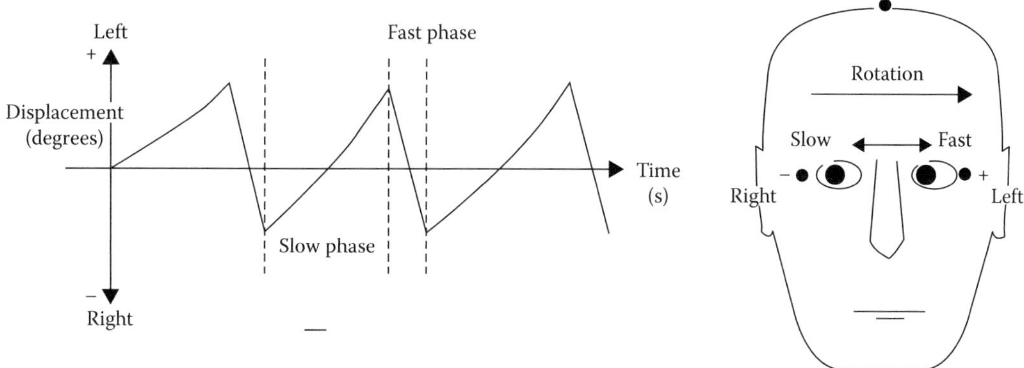

FIGURE 18.16
Nystagmogram. Movement of eyes compensate for rotational movement of a chair. An EOG trace illustrating fast and slow eye movements during rotation. This reflex is evidenced as nystagmus.

18.8.3 Vestibular Pathways—Reflex Arcs

There are two main reflexes associated with the vestibular system. Both are associated with maintaining gaze and posture regardless of whether the individual is moving or the environment is moving around the individual. These include the *static reflex*, rolling of eyes to compensate for movements, and the statokinetic reflex, to compensate for body movements in order to maintain direction of gaze. These reflexes work in conjunction of the proprioreceptor and visual systems. By observing eye movements during rotation, one can generally check the health of the reflexes. Figure 18.16 illustrates the compensatory eye movements associated with rotation. In order to maintain gaze, the reflex compensates by moving the eyes quickly in the direction of rotation followed be a slow phase. These involuntary eye movements are referred to as *nystagmus*.

There is some degree of filtering to improve specificity, for example the vestibular system only responds to movements of less than 20 Hz characteristic frequency, whereas the cochlea responds to vibrations from 20 Hz to 20 kHz.

18.8.4 Vestibular Pathways and Networks

The vestibular information about movement of head and rotational acceleration are crucial pieces of information for maintaining posture. It is also an essential component in the production of motor responses critical for daily function and survival. This importance is illustrated by the number of connections from the vestibular nuclei (see Figure 18.17).

From the vestibular sensory organs, primary afferent nerve fibers arrive at the vestibular nuclei (in the medulla). These are neurons in the brainstem responsible for receiving, integrating, and distributing information that controls motor activities (eye movements, head movements, postural reflexes). From these nuclei, projections go to the spinal regions, the

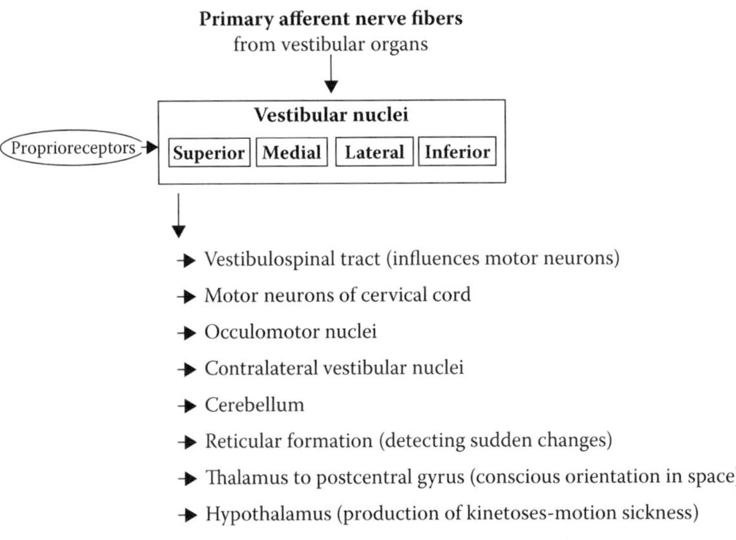

FIGURE 18.17
Vestibular pathways.

cervical motor neurons, the oculomotor nuclei which are responsible for mediating eye movements. There are also projections to the cerebellum, the reticular formation (vigilance) and the thalamus (conscious orientation) and the hypothalamus (associated with motion sickness).

These can be summarized as five networks.

- *Peripheral receptor apparatus* (head motion and position)
- *Central vestibular nuclei* (eye movements, head movements, postural reflexes)
- *Vestibulo-ocular network* (controlling eye movement)
- *Vestibulospinal network* (controls postural reflexes)
- *Vestibulo-thalamo-cortico network* (conscious perception of movement and spatial orientation)

An example of such a network is the vestibulo-ocular reflex (VOR). This network for compensating eye movements during body movement so as to maintain gaze is illustrated in Figure 18.18. There are three forms of VOR; rotational VOR, linear VOR, and nystagmus. Each is elicited by a specific direction of rotation or movement. With rotation, the horizontal semicircular canals and utricle are responsible for controlling horizontal eye movements. The vertical semicircular canals and saccule are responsible for controlling vertical eye movements, while torsional rotation is controlled by the vertical semicircular canals and the utricle.

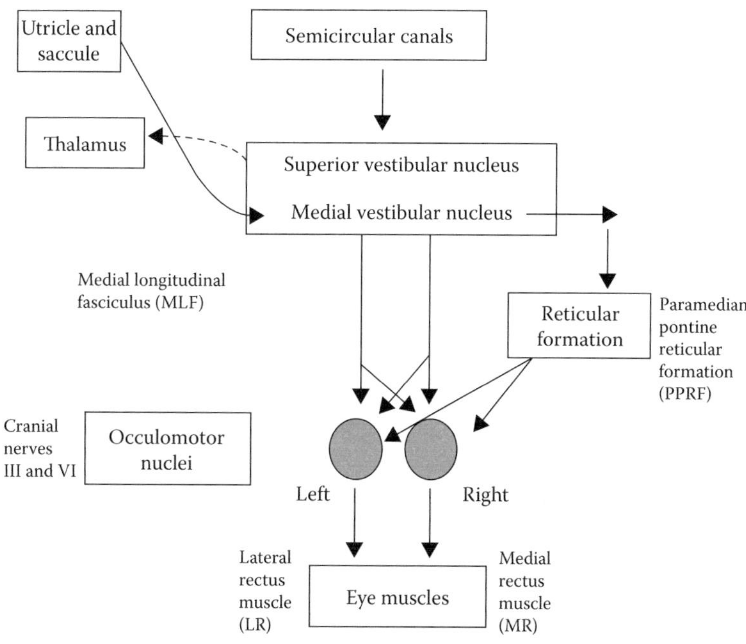

FIGURE 18.18
The vestibulo-occular reflex (VOR).

18.8.5 Testing the Vestibular System

A number of techniques can be used in clinical conditions to test the integrity and health of the vestibular systems and the related postural and eye movement control networks. The technique for measuring nystagmus is called *nystagmography*. This usually involves a special rotating chair where eye movements can be monitored. With large head movements (e.g., 360° body rotation), a number of compensatory eye movements occur.

(a) *Nystagmus:* combination of slow and fast phases of involuntary eye movements.
(b) *Slow phase:* vestibulo-ocular reflex (VOR) directs the eyes slowly in the direction opposite to the head motion.
(c) *Fast phase:* when the eye reaches the limit of how far it can turn in the orbit, it springs rapidly back to the central position, moving in the same direction as the head.

Electronystagmography (ENG) can be used as diagnostic indicator of vestibular integrity. The main purpose of the ENG clinical test is to determine whether or not dizziness may be due to inner ear disease. By monitoring eye movement compensation during tracing, moving the head to different positions and implementing the *caloric test*, the integrity of the system can be quantified.

The caloric test measures responses to warm and cold water (7°C above or below assumed body temperature) circulated through a small, soft tube in the ear canal. This effectively sets up a convection current in the endolymph fluid of the horizontal semicircular canal, and the participant feels they are rotating. Involuntarily, compensatory eye movements (nystagmus) occurs which can be measured with the EOG (electrooculogram). As soon as the participant opens their eyes, this reflex is suppressed. The amplitude of the nystagmogram is used to calculate the suppression index. However, if the nystagmus cannot be effectively suppressed then this indicates an abnormal vestibular system. This may then explain why dizziness and vertigo symptoms are occurring.

Normal stimulation of the vestibular system will have equal responses on each side. It does not assess vertical canal function or otolith function.

18.8.6 Vestibular Disorders

The main symptoms associated with disorders of the vestibular system are dizziness and vertigo. The dizziness may include spatial disorientation which may or may not involve feelings of movement. These symptoms may also be accompanied by nausea and or postural instability. However, other disorders may contribute to these symptoms and not be necessarily vestibular in origin. Examination of the eyes during these symptoms may also reveal nystagmus.

Clinically, by administering a *rotation test* or *caloric test*, the sensitivity of the system can be quantified. Vertigo can also be elicited optokinetically if the visual surroundings are rotated while the body remains stationary.

One of the most common vestibular disorders is *benign position vertigo*. It is characterized by brief episodes of vertigo coinciding with changes in body position. It can be triggered by rolling over in bed, getting up in the morning, bending over, or rising from a bent position. Posterior canal abnormalities in the ampullae are implicated. Other possible explanations include the otoconial crystals from utricle separating from the otolith membrane, or an increased density of cupula where abnormal cupula deflections occur when the head changes position relative to gravity.

There are a large number of possible causes of vertigo. Generally, any disease or injury that damages the acoustic nerve can cause vertigo. This may include congenital disorders (present at birth), physical trauma (car accident), rubella, blood vessel disorders with hemorrhage (bleeding), clots, atherosclerosis of the blood supply of the ear, cholesteatoma and other ear tumors, some poisons, ototoxic (toxic to the ear nerves) medications, including aminoglycoside antibiotics, some antimalarial drugs, loop diuretics, and salicylates.

Questions on Auditory Function

Auditory Multiple Choice

Identify the choice that best completes the statement or answers the question.

1. An average healthy adult can hear pitches in what frequency range?
   (a) 1 Hz–5,000 Hz
   (b) 15 Hz–20,000 Hz
   (c) 100 Hz–100,000 Hz
   (d) 1000 Hz–20,000 Hz
2. The eardrum vibrates at
   (a) A much higher frequency than the sound waves that hit it
   (b) Half the frequency of the sound waves that hit it
   (c) The same frequency as the sound waves that hit it
   (d) A constant frequency regardless of the frequency of the sound
3. Three small bones connect the tympanic membrane to the oval window. What is the function of these bones?
   (a) They hold the tympanic membrane in place
   (b) They convert airwaves into waves of greater pressure
   (c) They spread out the air waves over an area of larger diameter
   (d) They change the frequency of air waves into lower frequencies that can be heard
4. The tympanic membrane is to the ________ as the oval window is to the ________.
   (a) Anvil; hammer
   (b) Stirrup; anvil
   (c) Inner ear; middle ear
   (d) Middle ear; inner ear
5. Why is it important for sound vibrations to be amplified as they pass through the ear?
   (a) The inner membrane gets less sensitive with age
   (b) More force is needed to create waves in fluid
   (c) Much of the vibration is lost in the eardrum
   (d) Too much is lost through friction
6. To what does the stapes (stirrup) connect?
   (a) Tympanic membrane
   (b) Scala media

   (c) Oval window
   (d) Basilar membrane
7. Which of the following are presented in the correct order when describing some of the structures that sound waves travel through as they pass from the outer ear to the inner ear?
   (a) Pinna, tympanic membrane, oval window, cochlea
   (b) Tympanic membrane, pinna, cochlea
   (c) Pinna, stapes, eardrum
   (d) Malleus, tympanic membrane, oval window, pinna
8. The scala tympani makes up part of the:
   (a) Tympanic membrane
   (b) Middle ear
   (c) Cochlea
   (d) Ossicles
9. A cross-section of the cochlea shows:
   (a) Three long fluid-filled tunnels
   (b) One long fluid-filled tunnel
   (c) Three small bones collectively known as ossicles
   (d) Calcium particles called otoliths
10. How do sound waves ultimately result in the production of receptor potentials?
   (a) The tectorial membrane squeezes the auditory nerve
   (b) The basilar membrane releases neurotransmitters
   (c) Hair cells in the cochlea vibrate, causing ion channels to open in their membrane
   (d) The scala vestibuli has receptors that create action potentials

Vestibular System

1. Describe how the vestibular apparatus functions. Include in your answer reference to the structure and function of the tests used to assess damage to the vestibular apparatus.
2. Describe the vestibulo-ocular system. In your answer, describe the major pathways and functions.
3. Describe the types and orientations of semicircular canals found in other animals. Are there differences? Why?

Internet

Australian Hearing. http://www.hearing.com.au.

Burke J. 2010. Hearing Loss- Demographics-Deafness Statistics http://deafness.about.com/cs/earbasics/a/demographics.htm.

EARHELP (UK). http://www.earhelp.co.uk/facts-figures-about-hearing-problems.html.

The Bionic Ear Institute. http://www.bionicear.org/bei.
The Victorian Deaf Society (Vicdeaf) http://www.vicdeaf.com.au/statistics-on-deafness-amp-hearing-loss.
World Health Organization (WHO) 2010. http://www.who.int/whosis/en/.

Bibliography

Summary Health Statistics for U.S. Adults: National Health Interview Survey, 2009. Series 10, No. 249.

19

Chemical Senses

John Patterson

CONTENTS

19.1 Introduction

From an evolutionary perspective, chemical sensing is perhaps the most ancient of biological senses not dependent on mechanical touch. In the original aquatic environment, it was a form of remote sensing as even single-celled organisms could/can "taste" compounds in water. Such a sense was/is essential for the detection of food and the avoidance of chemical hazards. In this way even simple animals can remove themselves from an environment too low in salt, or too high in salt. It is important to distinguish between the sensing of something chemical for the benefit of the whole organism and the use of chemical receptors for internal processes. With the evolution of multicellular organisms (which allowed for the partitioning of the internal and external environment on a larger scale, and for tissue/organ specialization), the basic concept of chemical sensing was internalized, as well as remaining in its external locations. The creation of tissues meant the specialization of dedicated specific receptor systems; more correctly sensors. Once the aquatic environment was left, external sensors on the surface of the entire organism became two important variants of chemical detection: *olfaction* (smell) and *gustation* (taste). Over the same evolutionary time period, the processor for handling the information needed to develop as well. The simple organisms reacted to the sensed molecules: they either moved towards the source and engulfed it (food), or they moved away from it (not food, possible threat). As the organisms became more complex, they had more complex behaviors and systems to process, requiring a "brain." Once the brain and the chemical sensors were inextricably linked the rest of the evolutionary story is one of continual development of the processing and use of chemical sense information. The chemical senses become exquisite in their detection and provide some remarkable adaptations for certain animals.

There was a parallel chemical sense which developed and made use of certain chemicals for the purposes of controlling certain animal behaviors. This is the "pheromonal" sense which relies on a particular sense organ in the nose. It is similar to olfaction in that it depends on volatile compounds, but it differs from olfaction in that it can only produce programmed behavioral changes and is not, as far as is known, used for identifying harmful or harmless molecules. It will not be mentioned further as its relevance in people is still being debated.

Some chemical sensing relies on *stereochemical* structure. All molecules have a three-dimensional shape based on their volume occupancy of space, as well as the arrangement of their electrical charge on the surfaces of the molecules. Some chemical sensing is more direct, with molecules triggering depolarization of target cells directly. In other circumstances the receptor and the molecule which interact must be matched in some way. Although several molecules may interact with a single receptor to some extent, for optimal functioning they need to interact completely. In doing so they become a new molecule, temporarily. Importantly, the persistence of this molecule is determined by the affinity of each for each other. An interaction which exhibits a high level of affinity will characteristically have the molecule stay on the receptor for longer periods than for a low affinity interaction. The affinity is determined by the physicochemical nature of the two components. There has to be, however, a way—if the interaction is to act as a biological signal system—for the interaction to be time-limited. This may be as a result of the affinity, or, a change in the effective concentration of the molecule. The latter can be modulated by removing the molecule from the region of the receptor, either by breaking it down to inactive components, or removing it by dilution, migration or displacement. Every natural receptor has characteristic ways of limiting the interaction time. Such dynamics are termed receptor kinetics.

We term a molecule which interacts with a receptor, as the *agonist*, or *ligand*, for that receptor. There may be many such potential agonists for a receptor or there may be only one. There are molecules which can block an agonist from attaching to the receptor and they are termed antagonists. Such interactions are the purview of pharmacology, and not chemical senses, although much of the dynamic process is the same in both, and the study of receptor kinetics is shared. The characteristics of gustation and olfaction are determined by the different circumstances in which they operate. Olfactory agonists are called olfactants, or odors, or, commonly, smells; gustatory agonists are technically referred to as tastants, sapids, or, commonly, tastes.

Olfaction is confined to molecules which are naturally volatile, while gustation requires the molecules to be to some extent water miscible. By nature, olfaction is more remote—in terms of the necessary proximity of the source to the person—as a sense, than gustation which requires some ingestion or direct exposure of tissues to the source. Functionally, the olfactory system is intended to provide advanced information of the nature of the material before it is ingested to the extent that it cannot be eliminated. Gustation requires that the molecule dissolves in the saliva. It can be removed from the body but not as readily as for olfaction. If the molecule is harmful to epithelial tissues, then harm may have already occurred. The purpose of the chemical sensing is also different for the two processes in the range of different sensations which can be obtained.

The effect of odors and tastants are linked to our preferences* for food, and to our protection from unpleasant experiences. We can be both aware or unaware of these molecules. We need to be aware of them for us to express a preference. We do not need conscious detection to occur for the chemical to have an influence on our behavior. Olfactory signals are detectable at lower levels (these may be measured as a series of dilutions, or dilution

* Preference is how we express our enthusiasm, or not, for a smell or taste.

steps, or levels, and quoted as such, rather than the more convenient concentration measure) generally than the detection levels for gustatory compounds. Olfactory agonists are much more diverse than gustatory ones. Gustatory agonists are limited in the range of like and dislike they cause. Our enjoyment of food and drink, as well as of the environment, is derived from both our olfactory and gustatory senses. There are, however, deeper purposes reliant on our chemical sense as it can affect behavior, biological rhythms and physiology. Structurally, our two external chemical senses have similarities, but they are also quite different. The location of the olfactory system is in the airflow when we breathe and is found in the nasal cavity; the gustatory system is restricted to the mouth (oral cavity).

19.2 Olfaction

Located under the base of the skull (Figure 19.1), the olfactory epithelium sits as a specialized region in the general nasal epithelium from which it is different in structure and function. Recent work has suggested that the historical location may be incorrect, with a more anterior situation than previously thought. Such a difference could be very important in

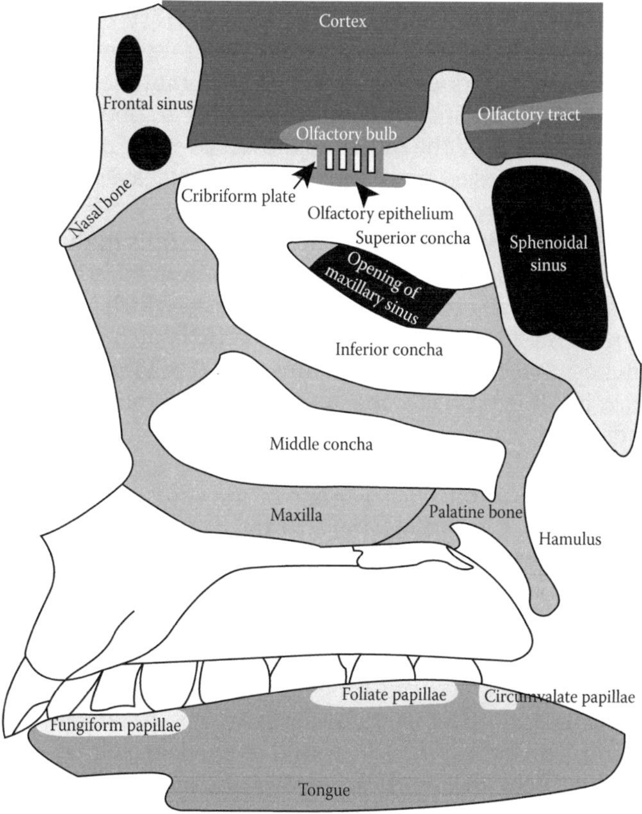

FIGURE 19.1
A diagrammatic representation of the face showing the approximate locations of the olfactory tract, the olfactory bulb, the olfactory epithelium, and the approximate sites of the tongue papillae.

clinical circumstances if the epithelium needs to be examined in full. More recent studies have used expression of DNA markers and stains to reveal its location. It is a small area in man, but much larger, both actually and relatively, in other animals. Its size determines olfactory sensitivity. It is located such as to ensure that air which we breath in (orthonasal) flows over the epithelium; however, it is also positioned such that aromatic components which have already been consumed may be detected a second time by a retronasal process with perceptual differences. If we wish to increase the amount of the odor we expose to our olfactory epithelium, we sniff. This accelerates the air and when forced by the shape of the roof of the nasal cavity to turn, the odors continue in a straight line as a result of their momentum and impact on the olfactory epithelium. At lower velocities, they turn with the air and only those odorants on the periphery of the airflow come in contact with the epithelium. The effects of this odor delivery system and the mucus has been modeled with good agreement with empirical findings.

The olfactory epithelium is a single layer of epithelial cells. The thin layer of connective tissue under the olfactory epithelium lies on the bone and attaches the epithelium. Nutrients and waste materials are transported via the blood vessels. The olfactory epithelium is regenerative and readily repaired. It has serous and mucous glands that produce a protective covering over the fragile epithelial cells. The main protection is to prevent the epithelium from drying out. The mucus is needed as it is less likely to just "drip" away and will stick to the surface more reliably than a more watery serous secretion. Quoted as being from 5 μm–20 μm to 30 μm thick over the olfactory epithelium, it is produced continuously. As it moves along the olfactory epithelium it eventually is drained away in the nasal cavity, removing as it does any entrapped odorant molecules and any particles which were brought in with the inspired air. It is replaced about every 10 min in normal circumstances. It is not clear how the mucus is moved over the olfactory epithelial region: it has been suggested that this section of the whole mucous layer is simply transported as a consequence of the movement of the contiguous material on the nasal epithelium; it has also been suggested that it may be moved by non-sensing cells in the olfactory epithelium.

The olfactory epithelium (Figure 19.2) in humans is estimated to have an area of between 5 cm^2 and 10 cm^2. A typical dog may have an epithelium which is approximately 170 cm^2 by comparison. The number of olfactory receptor cells is also dramatically different: in humans it is estimated that there are approximately 10×10^6 receptor cells; in dogs, the figure is up to 200×10^6. It is probably the latter difference that endows dogs with such a spectacular sense of smell.

The structure of the olfactory epithelium consists of the *lamina propria** in intimate contact with the *cribriform plate* of the *ethmoid* bone of the skull and the olfactory epithelium sitting on the *lamina propria*. It is in the latter that the vascular bed is found along with the base of the glands most commonly thought to be responsible for producing the protective mucous secretion: *Bowman's glands* (or olfactory glands). The duct from these glands extends through the olfactory epithelium to the surface. The epithelium is also termed the neuroepithelial layer as it consists of brain (neural) cells and the supporting epithelial cells. There are three well studied cell types in this mixed neuroepithelial layer: basal cells; sustentacular cells and olfactory receptor cells. The basal cells are found near the margin with the lamina propria and so are found at the deepest part of the layer. They are known to be a type of "stem cell" and can replace the damaged/aged olfactory receptor

* *Lamina propria* means "own, or special layer," referring to the fact that it is special for that overlying epithelium; which if a mucosa, it would be *lamina propria mucosae*.

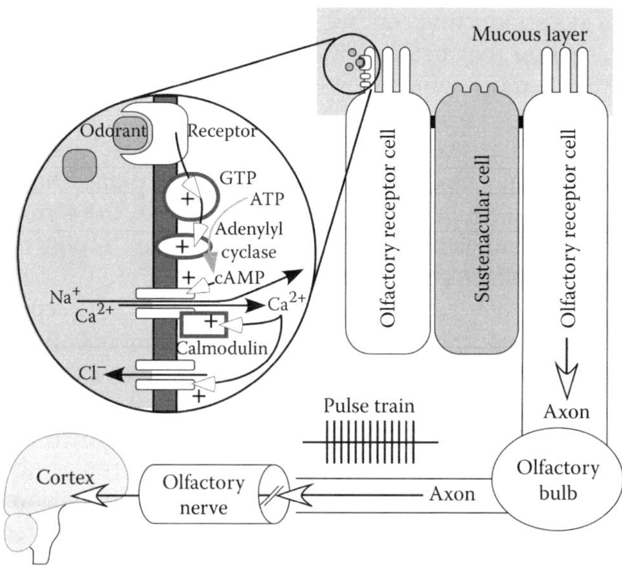

FIGURE 19.2
Relationship between the olfactory epithelium, nerve, and bulb. The signaling within specific cell types in the epithelium is described in the text.

cells. This is a remarkable process which will be treated more later. The sustentacular* (also sustenacular) cells are columnar in type (narrow base relative to height), they secrete a mucopolysaccharide which contributes to the mucous layer, and they are the main cell type in the epithelial layer except where there are olfactory receptor cells and the duct of Bowman's glands. The ducts are surrounded along their length by sustentacular cells on their way to the surface. In a similar fashion, the olfactory receptor cells pass from where they traverse the lamina propria through the cribriform plate, to the brain. There are other cell types (perhaps as many as five) in the epithelium, but their functions are not yet well understood.

The olfactory receptor cells have a cell body—located roughly in the middle of the neuroepithelial layer—a dendrite, and communicate via an axon which travels to the olfactory bulb of the brain. The dendrites of the olfactory receptor cells extend from the cell body to the surface of the neuroepithelial layer. In fact they extend a little way beyond the surface of the sustentacular cells and terminate in a knob. Radiating from these knobs are between eight and 20 olfactory cilia which tend to lie parallel to the surface of the epithelium. The cilia are not motile (that is they do not move and are not thought to move the mucous layer) and these intertwine in a loose meshwork completely covered by the mucous layer. Another feature of the knobs is the numerous mitochondria which are found in them, strongly suggestive of a high level of metabolic energy required in this region of the olfactory receptor cells. The cilia are where the detection of the odorant molecules occurs and where the signal transduction process takes place. The sustentacular cells probably have a mechanical supporting function, but may also phagocytose the olfactory receptor cells once they are to be replaced; additionally, they may act like neuroglial cells (a neural-tissue

* *Sustentacular* is from the old Latin *sustentaculm* meaning "support" and refers to cells found in a number of epithelia.

support cell) and form a supporting conduit for the replacement cells. They must act as an epithelium as well, sealing the surface with tight junctions around their rim. They also have microvilli on their surface which may be one of the ways in which the mucous layer is moved.

The olfactory nerve cells are surrounded in groups, or fascicles, by a series of olfactory ensheathing cells as they pass through the cribriform plate. These are thought to be another form of glial cell. While not in any way involved in the signal processing of the olfactory epithelium, these olfactory ensheathing cells seem to play a critical role in the next topic: replacement of olfactory receptor cells.

It has long been known that the olfactory receptor cells can be replaced. Every four to eight weeks, they are replaced by maturation of basal cells, but not all at once. Replacement is done in regions, or zones; this ensures that we are not anosmic (no sense of smell) while the replacement cells develop. Not only do the replacement cells develop the necessary cilia and the receptor molecules, but they produce an axon which grows back into the olfactory bulb of the brain where it makes new connections replacing the old cell's synapses. The olfactory ensheathing cells maintain the pathway for the new neurones to grow in precisely the way they were previously connected, and to the correct destination. For many years this was the sole known example of natural repair mechanisms involving brain tissue. Research has been carried out with these basal cells extracted from the neuroepithelium, and they have been shown to be able to develop into nerve cells in other locations, such as the spinal cord. Interestingly, the olfactory ensheathing cells are needed to make this process very effective and might be able to allow repair in the spinal cord without the need for olfactory basal cells.

Situated on the cilia are the receptor molecules which must also be replaced. The genetic basis for this has been revealed with the olfactory receptor genes the largest superfamily of genes so far encountered. This was first discovered in the rat (Buck and Axel, the 1991 Nobel Prize for Medicine recipients), then humans, and more recently in the mouse. The genes are estimated to code for perhaps 1000 receptors in the mouse, but only around 350 olfactory receptors in humans; this represents nearly 1% of the total genes in vertebrates. While there is an expectation that humans have some homology with mice (shared genes), this discrepancy in receptor numbers is thought to represent our loss of olfactory ability. Such genes encode for proteins which are part of the G protein-coupled receptor family. These chemical-detecting receptor molecules are the first step in a complex process which permits us to be aware of a smell. The name comes from a shortening of *guanine nucleotide-binding protein*. This family of proteins are part of the *second messenger* system of internal communication in many cells which are activated by an external ligand.

Olfactory receptor cells have proved to be highly variable across species in terms of their resting membrane potentials. The range is from as much as –90 mV to as little as –30 mV; the frog olfactory epithelial cells have been determined to have a –75 mV potential. When the accumulation of the various graded membrane potentials exceeds the threshold at the axon hillock, or critical segment, of the olfactory receptor cell, an action potential is generated and these continue to be produced as long as the cell potential exceeds the threshold. In the newt, olfactory epithelial cells, the threshold value is approximately –30 mV and the resting membrane potential is –70 mV. Other work has demonstrated that the dendrite itself could be the site for generation of action potentials, at least in the salamander, measured using isolated dendrites obtained by solubilization with alkaline Ca^{2+} solution.

Transduction is the conversion of one signal (energy type) to another signal (energy type) and in the case of the olfactory and gustatory systems, this means converting a chemical stimulus to a biological one. For the olfactory system it provides a way of detecting odor molecules and generating a specific neural (electrical) signal. This detection, as stated previously, occurs on the olfactory receptor cell cilia. Because the number of odorant molecules may not be very large, there needs to be a mechanism which can "amplify" the effect of each interaction. The end result of the process is the depolarization of the membrane of the ofactory receptor cell. There are two processes which achieve this outcome: second messenger activation, and the creation of an action potential in the axon of the receptor cell. Second messengers are internal cellular signaling systems. They start through the activation of the membrane receptor by the agonist. There are a large number of second messenger systems involved in a large number of control systems in the body. In olfactory signaling, the main second messenger system involves G protein. This very large membrane-bound system switches from guanosine diphosphate (GDP) to guanosine triphosphate (GTP) when activated by the presence of the odorant (ligand) on the odorant binding protein embedded in the cell membrane of the cilia. The GTP in turn switches on the cytoplasmic adenylate cyclase enzyme (sometimes referred to as adenylyl cyclase, or adenyl cyclase) associated with this complex. The effect is to cause the conversion adenosine triphosphate (ATP) to cyclic-adenosine monophosphate (cAMP). The increase in cAMP in the cytoplasm opens specific cation gates located in the cell membrane: these are sodium (Na^+) and calcium (Ca^{2+}) specific and the influx of these ions raises the internal positive charge in the cell, initially creating a local proportional membrane voltage change, but, eventually—should there be enough agonist on enough receptors—leading to a depolarization of the axon.

Observations on patch-clamped olfactory receptor cells have revealed a very large potential amplification in the cAMP-mediated process. There is another second messenger system which is involved is that of the phopholipase-C linked mechanism. This is also activated by the G-protein and is the first step in a series which eventually can open Ca^{2+} channels in the endoplasmic reticulum increasing the intracellular levels of this cation. The enzyme acts on a phospholipid—phosphatidylinositol 4,5-bisphosphate (PIP2)—which is broken into diacyl glycerol (DAG) and inositol 1,4,5-triphosphate (IP3). DAG is not free in the cytoplasm but remains attached to the cell membrane, but IP3 is released into the cytoplasm where it can diffuse freely. It will bind to any IP3 receptors including specific calcium channels. The last of the second messengers involves cyclic guanosine monophosphate (cGMP). This is thought to be involved in adaptation of the olfactory receptor cells; it is also known to be a more potent form of second messenger. In mice it has been shown that a protein in the neurones can alter the response to odors in the cilia *via* a change in cAMP kinetics. This suggests that there may be even more *in situ* processing of the signal. The resultant action potential from this process travels the length of the olfactory receptor cell axon until the neurone synapses in the olfactory bulb.

Not all of the receptor genes are necessarily expressed and functional. From the human genome project, we have about 400 functional genes for olfactory receptors with the remaining 600 considered to be pseudogenes.* This explains, in part, the diversity of the odors we can detect which are variously estimated to be around 4000; but this is far more than the number of identifiable receptors, so there is more to this story. There is even evidence of detection of natriuretic peptide hormones by a subset of receptors. The approximately 400 genes are thought to control production of the G protein-coupled receptor molecules of which around seven have been identified which consist of between 300 and 350 amino acids.

* Pseudogenes are inactive variants of known genes with no protein-coding capacity, or, are no longer expressed.

The discrepancy between the receptor numbers, the gene numbers, and the number of odors we can detect is explained as follows. It is not generally accepted that each receptor codes for, that is detects, only one odor molecule, but rather it encodes for a "class" of molecule. Additionally it is believed that any odor molecule may actually bind to more than one receptor. So, olfactory receptors can bind to a number of subtly different molecules, and different molecules can activate more than one receptor. The combinations and permutations available are very large, and more than enough to cover the known number of detectable odorant molecules. This system also allows for classes of molecules to be detected which means that there are many—even synthetic—molecules which may be detected that are not necessarily common, or even previously experienced by the individual. A combinatorial coding is how this works. As long as the combinations of activated olfactory receptor molecules is "unique" for a given compound and this combination of action potentials occurring on the nerves leaving the epithelium is transmitted to the olfactory bulb, then we will distinguish that molecule from any other molecule which does not activate that combination. A slightly different molecule, or mixture, may activate a slightly different combination of receptors and so present to the olfactory bulb another unique signature. This infers that molecules activating an identical combination of receptors will not be separately distinguishable. Any of the group will produce the same awareness. There is one very important fact which needs to be considered at this point, no mention has been made of concentrations, or amounts, of odorant molecules. In this we come to the simple physical constraints of the olfactory system. First, the odor has to be volatile, but not all are equally volatile at the same temperature; so ambient temperature, or the temperature of the source is important. The more easily the material is vaporized, if not in a gas state to begin with, the more readily it will get to the olfactory epithelium at a high concentration. Second, not all odorant molecules "dissolve" or diffuse equally in the mucous layer. This will alter the rate of access of different molecules to the receptors. Those which rapidly penetrate to the receptors will be detected more rapidly, and may require lower concentrations (gradients) for an effective stimulus than less "soluble" molecules. They will be detected before any less soluble components. Consequently we have a time-based, or "temporal" component to olfactory detection. While the physical chemistry of this will be complex, large molecules may be less mobile than smaller ones in the mucus even with similar solubilities. Third, the rate of flow of the inspired air over the olfactory epithelium will determine the "contact" time for these molecules to the mucous layer. Gentle breathing may not sweep odorant molecules to the olfactory epithelium particularly well, but if we sniff, it may raise the effective concentration.

Once we breathe a mixture of odors, or a single odor, the output of the olfactory epithelium is a number of activated axons sending action potentials onto the next stage: the olfactory bulb. At this point we have no evidence that any awareness occurs, only the possibility of detection. From this point on, i.e., the olfactory bulb, the processing is much more complex. Before explaining what is thought to happen we need to understand the structure of the olfactory bulb.

The olfactory bulb, which connects to the brain by the olfactory nerve (Cranial Nerve I), is where the first of the significant neural processing of the olfactory signal occurs and it essentially has five layers or regions in which particular cell types are found. It has been described as a form of olfactory cortex, or olfactory thalamus such is its processing power. These are: the mitral (or tufted) cell layer; an external plexiform layer, the glomerular layer; the internal plexiform layers, and the granular layer (these are listed from the edge of the bulb nearest the cribriform plate towards the inner part of the bulb connected to the olfactory nerve). If we examine the progress of a signal arising in an olfactory receptor cell, it will arrive at the olfactory bulb and will enter the glomerular layer and synapse on a dendrite. This is the dendrite of a mitral cell, tufted cell, granule cell or periglomerular cell. With only around 2000 glomeruli receiving as many as 10,000 olfactory receptor cell outputs, this is an example of a form of neural signal processing called convergence. The extent of this is even more dramatic when it is realized that it is believed that there are only around 25 mitral cells in total to receive all these converged signals. In effect there is around a 700:1 ratio of sensory cells to glomeruli. This is a remarkable level of convergence, which at first might suggest a loss of discrimination or resolution. In fact the pattern of combinations is still very large and the number of levels of signal detection is very large: these lead to identification and discrimination of odor molecules and to a capacity to handle different intensities of signals giving a measure of the concentration of the odorant. For example, furaneol (described as having a caramel, sweet odor) can be detected by people at concentrations as low as 5 parts in 1 billion (5 ng mL^{-1}); or, methyl mercaptan which is reported to be as low as 2 parts per billion (2 ng mL^{-1}) which may be added to odorless natural gas to allow detection of leaks. The output of the olfactory bulb is contained in the axons of the mitral and tufted cells which form the lateral olfactory tract. This projects (a neurophysiological expression for a functional connection) onto nuclei in the primary olfactory cortex and the temporal lobe of the brain (Figure 19.3). Consequently, the glomerulus is considered to be the basic odor map which is capable of further modification. Cells other than the mitral cells and tufted cells are important to this process and these are collectively termed juxtaglomerular cells. Included in this label are short-axon cells and periglomerular cells. These are thought to be involved in feedback from the cortex or by another neural process termed lateral inhibition which is able to adjust the extent of the signal being processed in very simple terms. Granule cells, in the granular layer, are typical of a type of cell found in the central nervous system, particularly the brain. Characteristically, these cells are very small and in the olfactory bulb have no axon (not true of all granular cells). Functionally, the periglomerular and granule cells are inhibitory. The former synapses with the mitral and tufted cells as well as the olfactory sensory neurones (by dendrodentritic connections); the latter make dendrodentritic synapses only on the mitral and tufted cells. The short axons cells are excitatory to these cells. Last, the organization of the olfactory epithelium and the olfactory bulb is by way of zones, or regions. In the olfactory epithelium, there are four zones and any olfactory receptor cell is found only in one zone. Each of the nerve cells in these zones connects to one or two glomeruli in one of four zones in the bulb. Optical imaging using reflection of specific wavelengths which are sensitive to oxyhemoglobin levels shows that similar odors activate different clusters of glomeruli. This is so specific that even isomers* of some, but not all, odorants can activate different patterns of areas.

While the olfactory bulb is primarily a sensory device with the majority of the neurones afferent in nature, the last major cell type is efferent and descends into the bulb from higher cortical centers. Termed centrifugal efferents, these project onto mitral, tufted and granule cells. They function to "tune" the olfactory bulb in terms of output intensity and may be associated with aspects of odor memory. They release neurotransmitters such as noradrenaline, serotonin, or acetylcholine. A second group of these centrifugal efferents arise from the olfactory cortex, and because that region in turn receives inputs from non-olfactory processing centers of the brain, the centrifugal efferents arising from there can bring information to the olfactory bulb concerning cues which might be related

* Isomers, from the Greek *isos* meaning "equal" and, *méros* meaning "part," are compounds with the same atomic composition but with a different physical arrangement of these atoms.

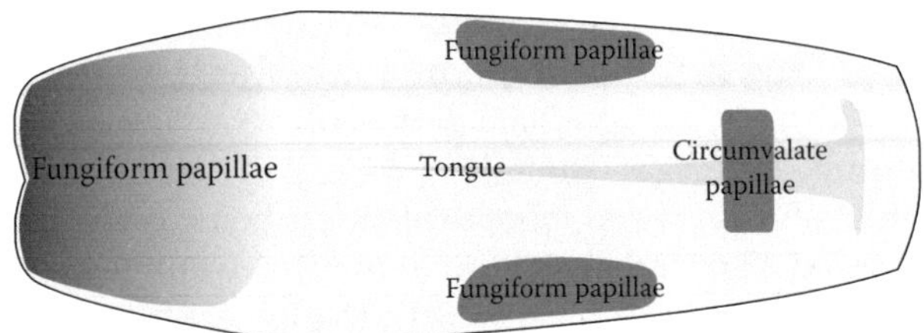

FIGURE 19.3
Dorsal view of the tongue showing the likely locations of papillae involved in tongue gustation. The area where fungiform papillae are found is filled with a gradient indicating the progressive reduction in density of these papillae.

to associations between smell and other senses (such as vision, hearing) and behavioral states (feelings, emotion).

The lateral olfactory tracts carry the axons of the mitral and tufted cells to the ipsilateral (same side) cortex, but olfactory information can cross over to the contralateral hemisphere *via* the anterior commissure. There are two cortical olfactory areas on each side of the brain: the primary olfactory cortex and the secondary olfactory cortex. Anatomically, the primary olfactory cortex is found in the piriform lobe (also piriform cortex, and sometimes spelt "pyriform"); the secondary olfactory cortex is in the orbitofrontal cortex. The piriform lobe, or piriform cortex, is a remnant of the paleopallium—or "ancient" layer—which confirms the ancient nature of smell as a function. It lies ventrally in each hemisphere in humans towards the midline, physically quite close to the brain stem. The orbitofrontal cortex (OFC) derives its label from being above the eye sockets and in the front of the hemisphere. It is part of the "association" cortex of the brain concerned with cognitive processes such as decision-making. It receives projections from a number of significant cortical centers, and functions in emotional states involving some form of reward. The "satisfaction" and "pleasure" which may derive from an odor arises from activation of these areas. Brain pathways for olfactory signals project onto a number of important areas in terms of their functionality, all of which align with our known experiences of a smell. We have areas which provide conscious perception (awareness and recognition). We have areas which engage mood and behavioral changes. These regions are all interconnected to a greater or lesser extent so the cortical tissues involved in a full range of responses to lateral olfactory tract signals are much greater than just the primary and secondary olfactory cortices. A very brief summary of the major regions and their roles in olfactory processing are

- Orbitofrontal cortex—sensory integration and discrimination;
- Thalamus/hypothalamus (limbic system)—discrimination and emotional responses;
- Hippocampus—conscious odor perception;
- Amygdala—emotional responses;
- Entorhinal cortex—integration of all sensory modalities.

The processing of individual olfactory signals—that is from an odor—is complex with some of the above areas having many cells processing a single odor, or, having few cells processing several odors. This processing is not at all fully understood. For example, it has been observed that two odors will activate several olfactory receptor cells each. Each odor has neurones which project onto a single glomerulus. In turn the output from these glomeruli project onto the piriform cortex: very straightforward and in keeping with the general description, presented previously. But it has been noted that the projection on the piriform cortex for these two odors is not to a single cell. Instead, each odor has a discrete area, or areas, where the piriform cortical cells are activated by the odors. There may be several of these and importantly they do not overlap. In another part of the piriform cortex various target cells are activated but the two odors cause a pattern more reminiscent of scattering. Here there is overlap and it is thought that this piriform cortex activation—unique to a mixture of two odors—assists us in discriminating mixtures of smells, or, more complex odors.

Because the neural processing of odors is so complex and abstract, there is a problem of communicating what we perceive to others who may perceive the same stimulus in a totally different way. Our individual likes and dislikes are complex and affected by experience, gender, age, ethnicity, and culture. For many years attempts have been made to try to develop an objective description of odor quality. Ideally if one person describes the odor has having a pineapple quality, this would be everyone's description. The problem is caused by the subtlety of our sense of smell and the ability to respond to very small differences in odor quality and our varying sensitivity to different odors. Add to this the problems of our language and the issue of experience and the issue of subjective senses, and then the complexity may become more obvious. One early theory (although given that really early attempts go back several hundred years at least, this is only relative early, posed as it was in 1962), was that odor perception was only based on stereochemistry. This theory fits for many situations, but falls down when very small variations in structure give different odors, or when very different structures are perceived to have the same "odor." The total number of different odors are estimated, from anecdotal reports and many specific empirical trials, to be 40,000 with only around 20% of these considered to be pleasant. Stereochemistry, by itself, is not the whole answer, but is certainly an important part of the system and the full explanation of how odors and receptors interact may not fit other "receptor" models.

While our sense of smell is fundamentally a physiological phenomenon, the involvement of language and emotion in its effects, introduces behavioral issues. Opinions and subjective effects are then of great interest. This is the arena of psychophysics where relationships between stimulus strength and perception are examined. These do not necessarily have a linear relationship as most sensory systems in the body have a logarithmic response. There are several aspects to our sense of smell which need to be considered: adaptation, habituation, masking, discrimination, and threshold. All these require a verbal, or non-verbal (marking an intensity scale, choosing some presented descriptor) response from the person and so fall under psychophysics. Both *adaptation* and *habituation* are used to describe a phenomenon in which a previously recognized odor (any sense can be affected) temporarily reduces in intensity over time, or with repeated exposures. They are distinguished from each other by the effect of a second stimulus presented during the period of the reduced perception. If the original odor is still perceived at a low intensity, then the process is one of adaptation. If, however, it be can be detected again, an habituation has occurred. Changes in intensity by these processes can be as much as a 50% reduction in perceived strength in a single second. In some experiments looking at long-term exposure the reductions in sensitivity have persisted for up to two weeks postexposure. Part of this can be central (the centrifugal inhibitory influences mentioned previously), part of it can be peripheral at the receptors. There is certainly some cognitive input, as a perception of the risk associated with an odor may alter the adaptation. There are times when these two different labels are used interchangeably, this should be avoided; most likely "adaptation" is the effect being discussed in most of our personal circumstances. The issues involving both are similar so we will concentrate on adaptation.

Adaptation in which one odor alters the sensitivity to another odor also can occur (*cross-adaptation*); this is particularly the case if they are closely related in structure and function. Odor fatigue is a version of adaptation termed self-adaptation, in which the strength of the odor apparently declines with successive stimuli. We are very familiar with adaptation as we become unaware of a smell if we are exposed to it for some time; we will very likely become aware of it again if the concentration changes suddenly. There are several factors affecting adaptation: intensity of the odor and length of exposure to the odor. The magnitude of the adaptation is commonly directly proportional to the odor concentration. The longer the exposure time the greater the adaptation. These are not absolutes as a very unpleasant odor may cause the person to remove themselves well before any habituation can occur. This behavior is protective. *Self-adaptation* can also occur in which continued or repeated exposure to an odor can result in a loss of sensitivity to the stimulus. While the common technique used to determine adaptation is a psychophysical one (the participant is asked to rate the odor intensity on a scale of 1 to 10), it can be also determined in an objective way by recording the electrical changes in the receptor cells (termed an electro-olfactogram), or in the various brain centers (termed the electroencephalogram) caused by the odor.

Masking occurs when one odor predominates over other odors. Generally the weaker odors are masked. This occurs through a combination of central and peripheral effects. The extreme case is when one odor is not detected in the presence of the other.

Discrimination is a measure of how well we can distinguish one odor from another. We do not need to be able to identify the odors, just to be able to compare the two odors when they are presented, and say they are different. If the process is extended to three odors we can have a forced-choice triangle test, which is a very common psychophysical tool, and also one used for smell determination in clinical circumstances and in food technology studies.

The process of *identification* is one which is potentially complex. Very many factors alter identification, not the least of which is experience. It is not likely that an unknown odor could be identified unless there were many non-olfactory cues provided. For example, if someone, who has not ever experienced the smell of flower before, but can visually identify the flower, is asked to identify the smell and the only smell they are exposed to is of that flower, they will get it right. The flower is the clue, not the smell. You can change the perception and response of someone to an odor by showing them a contrasting image at the same time. This is a process of presenting the person with a so-called *incongruent* condition. In odor identification, there will need to be some prior linkage between an odor and a label. This will involve memory, and may be one of the reasons why some elderly people struggle with odor identification tests. Not only is the identification of odors potentially hard, if there are several components in a mixture it gets very hard, very quickly to distinguish the separate components. The limit may be as few as three odors, but such experiments are greatly affected by the odors used, their interactions (cross-adaptation) and the various thresholds. The effects of age reveal one of the largest changes when analyzed in relation to identification. There are smaller differences between males and females—gender differences—in identification.

One of the more common psychophysical and physiological tests is that of *threshold* testing. There is a high degree of variability among people in terms of threshold levels for odors. We define threshold very carefully as it is not a straightforward parameter. This is because there are three thresholds: detected and identified; detected but not identified; and subconsciously detected. These follow a descending concentration as listed, with a stronger concentration usually needed for the first and the last occurring at the weakest level. Threshold levels vary with odor, they vary with age, they vary with gender, and they are affected by many environmental, cultural, experiential, and medical circumstances. So, testing for thresholds for odors has resulted in the creation of a series of procedures involving bottles, odor impregnated strips, odor impregnated pens, and olfactometers. There is a strong interest in threshold testing in otorhinolaryngology and geriatric clinical areas and there are many references to variations on the simple step-test, or forced-choice triangle testing. Commercial kits exist for testing a variety of olfactory abilities (for example, the "Sniffin' Sticks" test from Burghart Medizintechnik, Wedel, Germany). There are also commercial olfactometers—devices designed to deliver a series of known concentrations (automated dilutions) of odors (for example, the "Dynascent" from Environodor Australia Pty Ltd, Rosebery, New South Wales, Australia; the OM2-OM8b from Burghart Medizintechnik, Wedel, Germany). Many laboratories have developed their own systems for research purposes.

In general, the determination of conscious threshold requires a participant to indicate when they notice the odor. This is usually done by presenting them with increasing concentrations of the odor, starting with the strongest solution which is expected to be easily detected by all who are not *anosmic** in the form of a serial dilution (. . . 0.1, 0.2, 0.4, etc.) until they detect its presence. The normal procedure is then to reverse the direction and go down in the series until they indicate they no longer can detect the odor. This reversal process is undertaken a variable number of times (depending on the author), and the average value is taken from these staircase iterations. There are other approaches used: extrapolation or interpolation is used on the intensity response (category scale) provided to different odor (suprathreshold) concentrations. These values are then fitted to an equation which can then allow the "zero" value to be inferred. The technique, based on the methods of limits, has a number of alternative approaches: ascending concentrations from a subthreshold level; descending concentrations from some suprathreshold value. Another alternative is to provide the participant with two samples they compare to determine if one is stronger, or detected/not detected. For olfaction, the ascending methods—single ramp or staircase—are preferred as the impact, such as carry-over, or inhibition, by a strong concentration on a weaker one is a confounding issue.

Obviously, being able to determine subconscious detection of an odor cannot be done with a system which depends on a conscious response. This brings us to the electrophysiology of olfaction. There are two ways of determining the responses to an odor: recording directly from the epithelium, or from the central brain area. The former, known as an electro-olfactogram, has been recorded while participants were—by their lack of response—not aware of the odor, as a negative signal after the (subthreshold) odor has been delivered. This has now been undertaken using a noninvasive approach with electrodes on the outside of the nose. The use of the abbreviation—EOG—is unfortunate for physiology as it is already used for eye movements in the electro-oculogram. As this chapter is only about chemical senses it will be used here for the less common measure of olfactory epithelial activity. EOG work in animals has emphasized the synchronization of olfactory function with respiratory cycles. This is expected as there is little point in attending to the information from the olfactory epithelium when there is no airflow. It must be noted that there is

* *Anosmia* is the absence of a sense of smell; it can be restricted to a single chemical, or a total absence of any olfactory chemosense.

also a well-recognized retronasal detection of smell. This form of smelling may only apply to food-chemicals and not to non-food chemicals. It is also quite feasible to obtain chemosensory event-related potentials (ERPs: Rombaux et al. 2006a; also CSERPs or CERPs) and continuous electroencephalograms (EEGs) in response to odor delivery. The use of ERPs requires very precise delivery of the odor as the averaging of the EEG traces is dependent on a fixed reference point. Any "jitter" in the reference time in relation to the cortical events and the signals will be decreased rather than enhanced. To overcome the flow problems associated with normal respiration, that is variable flow rates affecting the time at which the odor arrives at the epithelium from its release at the external nares, artificial flows have been used which require the participant to breath abnormally. The impact on such an artificial system in light of the comments previously has not yet been addressed. The consistency of CSERPs recorded with different techniques suggests that they really exist and can be used to identify clinical olfactory dysfunction. They have been compared to auditory and visual ERPs and the odor induced ones are different from those from the other two modalities. Although the expected N1, P2, and P3 can be discerned in CSERPs, it is interesting that the correlation between the different modalities and their EP amplitudes only showed a relationship ($p < 0.001$) between the CSERP P2 and the P2 of the other modalities using an average of all electrode sites; for the P3 there were correlations between the visual P3 and the CSERP P3 at electrode location Pz, and the auditory P3 at the same location. For the Cz electrode site, only the visual P3 correlated with the CSERP P3 signal. The potential delays in odor diffusion and transduction of odors was not mentioned in the analysis, but the data are reasonably typical of midline CSERPs. For more details of the nature and the recording of EEG potentials, see Chapter 27.

CSERPs have been shown to decline with age in a linear fashion in a study using age groupings of 18–35, 36–55 and 56–80 in which the first age step had a "better" CSERP than the second age grouping; gender related differences were also observed. The duration of the odor stimulus has been shown to affect the CSERP but only for strong stimuli (in comparison to a "weak stimulus") when a longer duration stimulus elicited a larger CSERP amplitude.

As indicated by the CSERPs, there are changes to olfactory capacity with age with steady deterioration throughout middle age with over 62% of people aged between 80 and 97 years of age showing impairment, even though the self-reports (awareness of a loss) is low (between 12% for women and 18% for men). Men had a greater level of impairment. Although the impact of life-long environment and work-place issues has not been clarified, in the short-term a low-hazard working environment may not have a deleterious effect on olfactory abilities.

The higher central connections of the olfactory system have been investigated for some time with the clear revelation that many emotion-associated centers of the brain receive olfactory signals. Using rCBF to monitor activity changes in PET studies, aversive stimuli changed blood flow in the amygdala and left orbitofrontal cortex. Less aversive stimuli still caused rCBF changes in the orbitofrontal cortex, but not in the amygdala. Subjective ratings of aversion correlated well with rCBF changes in the amygdala. The impact of circumstances—other attentional tasks—when the odor is being perceived have been shown in rats to affect the firing pattern of the olfactory bulb. This is empirical confirmation of our own experience that we may not notice an odor if we are distracted.

While our sense of smell is not the most sensitive or the most discriminating, it does provide a great deal of information, much of which we are unaware of, but all which must add to our enjoyment of a lot of things, and, may, still play a role in defending us against undesirable things. It is certainly a complex and still being unravelled process. The critical and complex nature of the brain connections used by olfactory signals is revealed in the loss of olfaction in normal aging and those with Alzheimer's disease.

19.3 Gustation

While the olfactory system is a long-range detection system, the purpose of gustation is to "check" the nature of ingested materials before swallowing. It goes further and may be a process by which we are able to determine the content of the food to satisfy biochemical needs. So, unlike olfaction which is primarily telling us about the overall, or very specific, chemical nature of the food or drink—but not much about its hydrophilic components—gustation is linked to our "needs" at a lower, or dietary, level. We cannot separate the two processes from the perspective of homeostasis, they both do their parts, but gustation has capacities not found in olfaction. Although far simpler in structure and nature of receptors, gustation can tell us about the quantity of some essential non-organic materials in our diet. From this we can determine how much to ingest to satisfy needs. These essential non-organic materials are needed for all electrophysiological, renal, and liver function: they are our salts. Of equal value is the need for water. Satisfying thirst by drinking until the blood pressure rises due to an increase in blood volume would be a poor way to regulate this critical aspect. Or drinking until the osmotic pressure of our blood falls through dilution would be equally bad. These adjustment processes would occur too slowly to stop intake before the delays would be detrimental. A process of "integrating" the intake while it happens is the most effective way of stopping intake when a desirable level is reached. This would require water and salt detectors in the mouth, so do these exist?

To answer that question it is necessary to look at the anatomy of the mouth and the gustatory receptors found there. First, gustatory receptors are not only found on the tongue, as most believe. There are "taste" receptors in the cells of the brush border of the rat small intestine for sweet substances, for example; including, in the rat, for sweeteners which are thought to be nutritionally neutral or inactive.

The concept that gustation has three forms of signal outcome processing has been put forward: an identification step; an *ingestive motivation* stage and a *digestive preparation* process. This brings into perspective the delayed consequences of eating as well as the immediate aspects, and links these to the maintenance of an adequate diet. It has been proposed that we can actually specifically taste free fatty acids, something which was disputed for a long time. If this can be extensively verified and not just be a result of some other sensation (mouth-feel, olfaction, direct action on membranes), it will increase the number of basic tastes which are recognized. Fatty acids are an important part of our diet and it was always curious that it was believed we could not detect them. Traditionally, it has been known that the taste buds or receptors are broadly distributed in the mouth but are more concentrated on the tongue. Further, the way the highest concentration of the receptors is distributed on the tongue may have inadvertently produced a concept that these receptors are only found in certain tongue regions. The concept of a "tongue map" received serious damage in 1974 when a study showed that the spread of receptors is fairly even over the tongue. Collings studied the psychophysics of the spread of receptors and she was able to confirm earlier observations in general and showed that the soft palate was more sensitive to quinine and urea than the tongue, but not for other tastants. There have been efforts to propose a different approach to the issues of tastes and perhaps to remove the constraints imposed by older ideas and look to a form of collective coding to improve our understanding of taste. Gustation may prove in the future to be as complex as olfaction, but at present, it is accepted that we have five taste modalities: salt, sweet, acid, bitter and, the most recent addition (not yet including fatty acids), umami (or savouriness). Water receptors are also recognized, but a traditional taste has not been associated with these.

The location of the structures on the tongue and palate associated with tasting are the taste buds, the anatomy and location of which have long been described (Figure 19.3). More modern mapping of the functionality of the taste buds suggests a broadly tuned set of tastes in terms of tongue anatomy (Figure 19.4).

The receptors for taste are not like those for smell. They are discrete (modified epithelial) cells which are attached to, and synapse with, the nerve terminal of cranial nerves (the Facial, VII; Glossopharyngeal, IX; Vagus, X, nerves). Although they are epithelial in basic form and origin, these taste receptor cells (TRC) have some properties associated with neurones in that they can depolarize and have similar cell markers to neurones. These TRCs are collected in groups called taste buds found in the tongue, soft palate, uvula, epiglottis, pharynx, larynx, and oesophagus. The innervation is such that the chorda tympani branch of the facial nerve collects information from the anterior of the tongue; the glossopharyngeal nerve from the posterior of the tongue; and the larynx *via* the vagus nerve. The taste buds have different structural characters in different locations, although all are discrete egg-shaped volumes or papillae. On the tongue, the buds are of three kinds: fungiform, foliate, and circumvallate (Figure 19.5).

The process of transduction of each of the five recognized tastes has yet to be fully clarified, but for some, the picture is clear. First, though, it has been shown that the receptor interaction process which depolarizes the TRC may involve complex G-protein second messenger systems, or, may be a direct depolarization caused by the tastant acting on ion-channels (Roper 2007). So, taste transduction may be a simple direct action or a process similar to that for olfaction: second messenger cascades.

Once the TRC is depolarized it releases a messenger the nature of which is not at all clear. Confusing the issue is the lack of certainty as to the anatomy of taste cells. Up to four types of cells have been described from electron microscopy, but the variability of these is high and their individual function unclear (Roper 2007). At present there is some agreement to the concept that there are two distinct cells involved in taste reception: a taste sensory cell and a presynaptic cell which can synapse. The evidence for this is that the former have phospholipase C, a critical sensory cell component, while the latter have

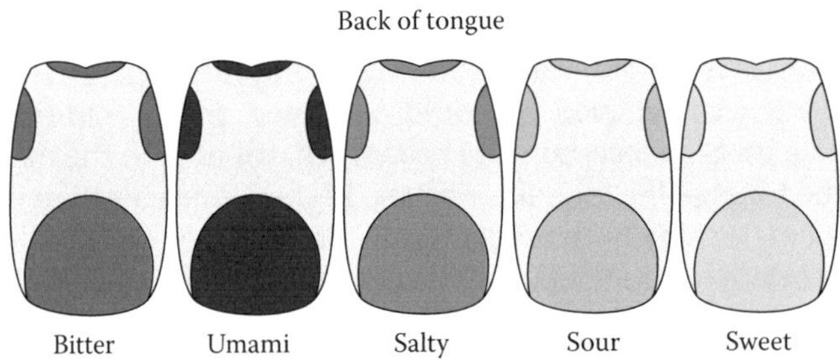

FIGURE 19.4
In contrast to the view from the traditional "tongue map," all taste modalities are now thought to be detected by any of the papillae.

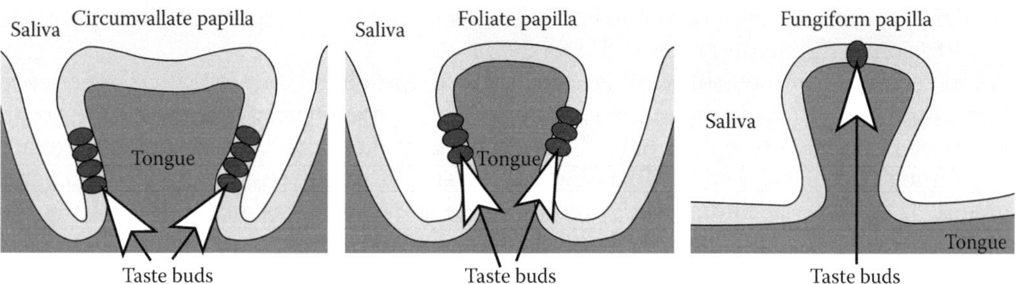

FIGURE 19.5
The structure of the three taste papillae of the tongue shown in cross section.

synaptosomal-associated protein 25 (SNAP25) which is believed to be required to have synapses. This may explain the difficulty that electron microscopists had in clearly identifying cells which were structurally able to detect and to synapse. As for the neurotransmitter released, adenosine triphosphate has been shown to be released by receptor cells in mice response to tastants and that 5-hydroxytryptamine 5-HT, serotonin) has been shown to be released by the presynaptic cells. Some of these cells may co-release noradrenaline. This has led to a model of transduction of tastants in which the sensory cell releases ATP in response to the presence of the tastant on its receptors; this ATP can stimulate the afferent (sensory) nerve directly, or, it can stimulate the adjacent presynaptic cell to release 5-HT which is presumed to activate the sensory nerve. The role of co-released noradrenaline is yet to be revealed, but the presence of it might contribute to the complex coding which is believed, but not proven, to be associated with taste. Several models of this have been proposed based on either the idea that one receptor cell processes one single tastant, the labeled line model; or, that they have multiple tastant reception, or multiple innervations (both collapsing the coding of the neurone across tastants), the across-fiber concept. How the coding is arranged and how it is linked to the two neurotransmitters is not yet clear. There is support for a simple single relationship between the five tastants and cell depolarizations in studies on "knockout" mice.

Of the five agreed taste modalities, transduction processes for sour and salty are yet to be fully clarified. For sweet, bitter, and umami, the situation concerning how they are transduced is much clearer.

Sour tastes vary according to the sour tastant with differences occurring according to concentration of the chemical, but is becoming accepted that the main moiety involved in sourness is the proton (H^+), or protonated material. The effects of this may be directed to the outside of the receptor cell or may, alternatively, occur inside the receptor cell, as pH changes have been observed in both locations when acids are administered. Mineral acids and organic acids may work through different mechanisms. There may even be involvement of epithelial sodium channels along with around eight other channels implicated in sour-taste transduction (Roper 2007).

For salt, the situation should be simpler with sodium channels ubiquitous in cells, but, in fact, the way we detect saltiness is not clear. Early work suggested that the epithelial sodium channels (ENaC) were located on the tongue, but while these work for NaCl, they may also play a part in sensing KCl. Not only do we have to explain the sensing of the cation, but also the dissociated anion (Cl^-, for example) as different sodium salts differ in "saltiness"; an issue with such important cations is that they permeate epithelia readily and the effects cannot be assumed to be surface-mediated only. Larger anions may impede the diffusion of the smaller cations (Roper 2007).

A wide variety of molecules, other than sucrose, can elicit a sweet-tasting experience from dilute NaCl to proteins, so the range of sweet tastants is very extensive. One group of G-protein coupled receptors (GPCR) had been identified: the T1Rs of which there are now three (T1R1, T1R2, and T1R3), two of which are the putative receptors for sweet and umami. There may be undiscovered receptors for some sweet tastants, but certainly the combination of T1R2 and T1R3 working as a receptor is sweet sensing (Roper 2007).

Receptors for bitter tastes have been surprisingly wide-ranging with up to 30 suggested. In 2000, two independent groups published the discovery of bitter taste receptors. The currently accepted receptor is yet another of GPCR: T2R (Roper 2007; originally named by Adler et al. 2000) which is coupled to gustducin, a G-protein subunit. How this family transduces bitter tastes is not known as the extracellular N-terminal part of the molecule is quite short and may restrict access to agonists (Roper 2007).

The most recently accepted taste modality is umami, an experience traditionally linked to the ingestion of monosodium glutamate. There are many other compounds which elicit this experience from various foods such as tomatoes, cheeses, and meats. The receptors for these have been found to be GPCRs yet again with a combination of T1R1 and T1R3 as one, and a version of mGluR1, the other (Roper 2007).

One aspect for which there seems no doubt in the transduction of taste is the involvement of second messengers, but, yet again, the precise details for various modalities are yet to be understood. Both cAMP and Ca^{2+} mediated processes have been indentified.

Of the three types of cells in taste buds which are thought to be involved in gustation, the receptor cells and the presynaptic cells are electrically excitable. The cation, mainly sodium, channels have been identified as being of tetrodotoxin-sensitive types: SCN2A, SCN3A, and SCN9A. These have been implicated in the initiation of the taste–cell action potentials. For the presynaptic cells the depolarization process is thought to involve Na^+ and Ca^{2+} currents *via* voltage-gated channels, while repolarization is K^+ based. For the taste receptor cells there appears to be no voltage-gated Ca^{2+} channels. The role of Cl^- in repolarization is not definite. The firing of individual taste buds from isolated fungiform papillae of the mouse show patterns which differ for the different tastants tested; recordings from the Chorda Tympani nerve (from the facial nerve VII) to the same tastants had similar, but not identical, firing patterns. Further, the array of taste receptor cells in the taste bud were variably responsive to the tastants, with 67% responding to one tastant only, 30% to two tastants and 3% to three or more tastants. This further suggests that the encoding is not necessarily 1:1 between tastants and receptor/presynaptic cells, but that it can be.

The lifespan of taste receptors cells is not clearly defined, although for many years their replacement has been well documented. It has been suggested that there are two sources of replacement taste cells, one each for the taste receptor cell and the presynaptic cell and that they may be replaced every 10 days, and, in the rat, both types die by apoptosis.

The density of taste buds vary in different people as do their distribution. The buds are found in association with lingual papillae of which there are four recognized types: filiform, fungiform, foliate, and circumvallate. Only the filiform papillae are not known to contain taste buds. Of the other three the circumvallate have the largest number of taste buds at an average of 240, but there are usually only eight or nine such papillae on the whole tongue: along the caudal border of the tongue. The taste buds are arrayed on the sides of the central column of the papilla, facing the circular groove from which the papillae get their name. The foliate papillae have fewer buds (around 100) but are more numerous than the circumvallate. They are found on the lateral aspects of the tongue. While not all fungiform papillae contain taste buds, and those which do only have fewer than five, they are dispersed over the anterior portion of the upper tongue surface. All the taste buds are serviced by extensive salivary secretion from the tongue and mouth. This rinses the mouth but ensures solubilization of the tastants.

19.4 Central Connections for Gustation

The traditional locations for central connections of taste are the nucleus of the solitary tract in the medulla, the thalamus and the cortex (See Chapter 18: CNS). This region is the destination of the afferents traveling in the facial, glossopharyngeal, and vagus nerves. The geniculate nuclei taste target cells which have two broad functions and subsequent projections to higher centers: the higher order circuit for discrimination and homeostasis; and, a local medullary control circuit for immediate responses to tastants. From the medulla, projections occur to the ventroposterior nucleus of the thalamus, and from there to the primary and secondary gustatory cortices. Not unexpectedly, aversive—unpleasant—tastants have been shown to activate many of the higher centers associated with emotion: the amygdala and orbitofrontal cortex. Other regions were involved which are on the pathway to these higher centers: medial thalamus, pregenual, dorsal anterior cingulate, and hippocampus. Another not-unexpected concept is that of the importance of the temporal and spatial processing of tastants. Unraveling the way tastants are coded would seem to require subtle processing from receptors to the higher cortical centers, all of which are not equally distributed, nor work at identical times. The gustatory circuitry has been proposed as the site of sensory integration of intraoral stimuli. The impact of prestimulus activity centered in the gustatory cortices on the interpretation of tastants is suggestive that a simple coding between the receptor and the outcome does not exist; or, is not explicable with present data.

Measuring the activity of the gustatory system is easier than for the olfactory system as most of the components are more accessible: the tongue, the cranial nerves, and the medulla are all more readily accessed with less damage than accessing the olfactory bulb and deep brain centers. For this reason more detailed work has been achieved with isolated individual receptor cells, ganglion cells, axons, and the geniculate nuclei of the medulla. Further it is more appropriate to deliver taste stimuli artificially as, unlike with smell when breathing may be an important aspect. Stimulating the tongue electrically is also a simpler process than for olfaction.

Electrogustometry is a process in which electrical currents are imposed on the tongue to determine thresholds. It has clinical uses in determining empirical loss of, or reductions in taste, but requires careful application. While current intensity—around 50 μA—is readily determined by using the correct stimulator, it has been argued that it is the current density which is important. There are non-clinical outcomes from the application of currents to the tongue to reveal the nature of taste reception, such as the impact of simultaneous electrical stimulation and tastant application. When applied together, cathodal currents inhibit, specifically, salty taste, with no affect on sweet receptors. Both anodal and cathodal polarizations on the tongue surface induce taste experiences. The issue with application of currents is that electrolysis can occur and normal bathing anions and cations in the saliva are likely to move, thus altering the local milieu of the receptors. A further confound is that, particularly anodal stimulation may stimulate trigeminal nerve axons directly. A common experience caused by electrical simulation of the tongue is that of a "metallic" taste. Stimulation across different regions of the tongue have shown that the thresholds for electrical stimulation by electrogustometry inversely follow the distribution of the fungiform papillae; low thresholds occur where fungiform papillae are numerous. Tastes induced by electrical stimulation are not described in a directly comparable way to those created by actual chemicals chosen to be similar to sour, bitter, or metallic experiences.

Combining electrical stimulation and EEG recording has produced clear gustatory event-related potentials, once the stimulus artefact was removed by processing. This confirmed the projection of signals to the primary taste cortex. Magnetic field recording from the human cortex (magnetoencephalography: MEG) during taste stimulation has confirmed the location of the primary gustatory cortex (between the parietal operculum and the insular cortex) with subsequent activation of other regions, including the insula, superior temporal sulcus and the hippocampus. Similar regions have been observed to be altered in fMRI studies. Combined EEG for gustatory evoked potentials and MEG has shown that the first EEG deflection (some 130 ms after the stimulation, in this case with NaCl solution) is synchronous with activation of the primary gustatory cortex.

While gustatory ERPs have been recorded, the variation in tongues, delivery issues, and removal of tastants have all introduced variation, making replication of experiments difficult. GERPs have been recorded to different taste modalities.

19.5 Gustatory Psychophysics

While the same psychophysical concepts which apply to olfaction also apply to gustation—adaptation, thresholds, discrimination, identification, and masking—the details are quite different. Taste shows very strong adaptation which appears to occur centrally. Cross-adaptation can occur depending on the taste receptors involved. Coincidental processing of two tastants by the same receptor system may not result in both tastants being detected. Thresholds, as have been indicated, are very tastant dependent, but also may be affected by circumstances surrounding the experience. As with olfaction, any maladies which affect the environment of the mouth will impact on thresholds. One striking difference between smell and taste thresholds is the genetically-based differences in ability to detect certain tastants. Two compounds are known to be tasted at markedly different concentrations by different people, so much so they can be genetic markers. Phenylthiocarbamide (PTC) and 6-*n*-propylthiouracil (PROP) are chemicals to which some experience nothing, others a slight taste, and yet to others, an extreme experience; these have proved to be invaluable tools for the psychophysics of taste. People who have a disproportionate number of fungiform papillae and taste pores on their tongue, and who react very strongly to many bitter tastes, including to PROP, have been described as "supertasters." Women are more likely to be supertasters. It is also known that supertasters have different liking/disliking of other tastes such as sweet and high-fat foods. Because the studies on this all involved psychophysics there have been discrepant results. Supertasters are now, however, a well recognized group in the community. Cross-cultural differences are known to occur, but most cultures have around 30% of the population who are nontasters for PTC and PROP. As an addendum to this, most supertasters have heightened pain and trigeminal responses making them more sensitive than non-tasters or partial tasters.

19.6 Chemosensory Disorders

The complexity of the olfactory system and the gustatory systems means that they are prone to functional failures. Each sense may be diminished, distorted, or absent. A loss of the olfactory sense is anosmia; that of taste is ageusia: these may be partial, specific, or complete. Hyposmia and hypogeusia are conditions where the senses are reduced in the ability to detect the chemical. Dysosmia and dysgeusia are conditions where the response to the chemical is abnormal: such as aversion to something that others find appealing. Hyperosmia and hypergeusia are when the person has an increased ability to smell or taste a chemical. If there is a perception of a chemical, but no actual chemical present, then it may be a case of phantosmia, or phantogeusia. These are all compared to "normal" responses, termed normosmia, or normogeusia. That such a condition may not actually exist in an individual means that such "normal" values need to be population averages. There are also agnosias, or the inability to classify, contrast or verbally identify smells and tastes in which the process is totally central, unlike the other conditions which can be central or peripheral in origin.

Trauma to either system is an example of peripheral influences. Both the olfactory and gustatory systems are on the edge of the protective mechanisms of the body and damage to the olfactory nerves during frontal head impacts can debilitate our flavor sense. There are many common, and many uncommon comorbid conditions, which somehow result in altered smell and taste ability. Pregnancy, influenza, hay fever, treatment with some medications, some psychiatric disorders, endocrine disorders, and some genetic disorders can all be linked, in some circumstances, with changes in chemosense. Self-reporting is not a reliable measure of determining dysfunction, so the reliability of psychophysical and physiological methods are essential.

19.7 Flavor and Its Sources

When commonly used, "flavor" has little direct relationship to our senses, unlike "tones" or "colors." It is an expression which defies simple definition. It varies according to the nature of the source of the flavor: whether food, drink, warm, cold, bitter, sweet, odorous, and its texture. All these aspects can influence the flavor which is an experience dependent on all of our senses, but not always all at once. It will be very strongly influenced by our own preferences, experiences, and culture. What represents a preferred flavor will be mostly individually determined by threshold levels, sensor status, and cognition. Along with mechanical senses, even pain and temperature—other parts of our somatosenses—can be involved in its derivation. It is not only humans who ingest bitter and possibly painful materials as part of their diet (Roper 2007). The multi-modal nature of flavor ensures that it is a highly variable and hedonic experience for which, if all the chemosenses play their part, safety and satisfaction are ensured when we consume food and drink. Studies using umami have further confirmed the integration of smell and taste, with single neurones combining signals about the taste of glutamate with the olfactory, oral texture, and temperature as well as visual stimuli. As a monitor of the presence of valuable protein, glutamate taste stimulates a reward response via the orbitofrontal cortex. Any pleasantness of umami is derived from areas removed from just taste and odor detection, or the pregenual cingulate or orbitofrontal cortices (Rolls 2009). This confirms an assumed purpose for the olfactory and gustatory systems, that of ensuring proper food intake.

For such an essential aspect to our lives, the lack of information about the working of our chemosensory systems is curious. While a considerable amount of research continues there are still some fundamental questions to be answered.

Tutorial Problems

1. The surface area of the human olfactory epithleium is estimated to be ____?

   Answer: Between 5 cm^2 and 10 cm^2.

2. The olfactory epithelium is attached to ____ bone?

   Answer: Cribriform plate of the ethmoid bone.

3. The olfactory epithelium contains ____ cells which have a known function?

   Answer: Three.

4. The olfactory epithelium may contain as many as ____ cell types?

   Answer: At least five.

5. The olfactory bulb has ____ layers?

   Answer: five layers.

6. There are two communication directions involving the olfactory bulb; what are they and why are they important?

   Answer: Afferent and efferent (to and from the brain). The sensory flow takes the primary olfactory signals to the brain from the bulb and the efferent (reciprocal) signals are thought to provide feedback control of the bulb by the rest of the brain.

7. How often are olfactory receptor cells replaced?

   Answer: Around every eight weeks.

8. What is the meaning of convergence and how does it impact on the signal processing of olfactory signals in the bulb?

   Answer: When many signals synapse onto a smaller number of target neurones, the output is effectively the sum of all inputs in the temporal sense. Providing the information does not collide in time on the target neurone or neurones, the target cell still can process all the input information. This can then be sent to areas of the brain where there are an excess of cells available to process all the information.

9. Why are all-or-none action potentials different from proportional signals in excitable cells?

   Answer: Action potentials cannot be summed in amplitude, they are all of the same size in any given neurone. If it is necessary to summate signals to determine if they exceed a threshold, then a mechanism for handling proportional signals is required (such as on the cell body before the axon hillock where the influence of very many dendrites can be summed).

10. What is the maximum resting membrane potential which has been recorded for olfactory receptor cells?

    Answer: Currently, –70 mV.

11. How many cells are thought to be involved in the signal processing path for an odor signal to be converted into a train of action potentials?

    Answer: Three, not counting the olfactory nerve cells.

12. How many mitral cells are estimated to be present in the human olfactory bulb?

    Answer: 25.

13. Which of the following regions of the brain is NOT considered to be immediately involved in processing of olfactory signals:

    - Orbitofrontal cortex;
    - Limbic system;
    - Hippocampus;
    - Amygdala;
    - Entorhinal cortex;
    - Cerebellum.

    Answer: Cerebellum.

14. What percentage of the known odors is thought to be unpleasant?

    Answer: 80%.

15. What is a CSERP?

    Answer: Chemosensory event-related potential.

16. The recognized taste modalities are ____?

    Answer: Sweet, sour, acid, bitter, salt, and umami.

17. What is a membrane-bound molecule commonly associated with taste transduction?

    Answer: Gustducin, or a G protein–coupled receptor.

18. The umami detection may involve a receptor for a more common tastant which is ____?

    Answer: The glutamate receptor.

19. Supertasters can detect at low concentrations which chemical?

    Answer: 6-*n*-Propylthiouracil, PROP.

20. What is an anatomical or structural explanation for the extra sensitivity of supertasters?

    Answer: Unusually large numbers of fungiform papillae.

Bibliography

Adler E, Hoon MA, Mueller KL, Chandrashekar J, Ryba NJP, Zuker CS. 2000. A novel family of mammalian taste receptors. *Cell* 100:693–702.

Chandrashekar J, Hoon MA, Ryba NJP, Zuker CS. 2006. The receptors and cells of mammalian taste. *Nature* 444:288–294.

Collings VB. 1974. Human taste response as a function of locus of stimulation on the tongue and soft palate. *Percept Psychophysics* 16: 169–174.

Finger TE, Danilova V, Barrows J, Bartel DL, Vigers AJ, Stone L, Hellekant G, Kinnamon SC. 2005. ATP signaling is crucial for the communication from taste bud to gustatory nerves. *Science* 310:1495–1499.

Haase L, Cerf-Ducastel B, Buracas G, Murphy C. 2007. On-line psychophysical data acquisition and event-related fMRI protocol optimized for the investigation of brain activation in response to gustatory stimuli. *J Neurosci Methods* 159(1):98–107.

Mace OJ, Affleck J, Patel N, Kellett GL. 2007. Sweet taste receptors in rat small intestine stimulate glucose absorption through apical GLUT2. *J Physiol* 582(1):379–392.

Matsunami H, Montmayeur JP, Buck LB. 2000. A family of candidate taste receptors in human and mouse. *Nature* 404:601–604.

Owen CM, Patterson J, Simpson DG. 2002. Development of a continuous respiration olfactometer for odorant delivery synchronous with natural respiration during recordings of brain electrical activity. *IEEE Trans Biomed Eng* 49(8):852–858.

Pun RYK, Kleene SJ. 2004. An estimate of the resting membrane resistance of frog olfactory receptor neurones. *J Physiol* 559(2):535–542.

Rolls ET. 2009. Functional neuroimaging of umami taste: What makes umami pleasant? *Am J Clin Nutr* 90:804S–813S.

Rombaux P, Weitz H, Mouraux A, Nicolas G, Bertrand B, Duprez T, Hummel T. 2006. Olfactory function assessed with orthonasal and retronasal testing, olfactory bulb volume, and chemosensory event-related potentials. *Arch Otolaryngol Head Neck Surg* 132(12):1346–1351.

Roper SD. 2007. Signal transduction and information processing in mammalian taste buds. *Pflugers Arch—Eur J Physiol* 454:759–776.

Zhang X, Firestein S. 2002. The olfactory receptor gene superfamily of the mouse. *Nature Neurosci* 5(2):124–133.

Part VI

Systems and Signals

20

Physiological Signal Processing

Peter J. Cadusch

CONTENTS

The field of signal processing is vast and generally well served by specialist textbooks, so rather than attempting a broad coverage, this chapter will concentrate on basic ideas and techniques.

20.1 Signals

A signal is something that conveys information or from which information can be extracted. In practice, the physical signal usually takes the form of a quantity, such as voltage or current, which varies in time. In *analog signal processing* the continuously varying signal is directly manipulated by physical devices; in *digital signal processing*, the continuously varying signal is first sampled, usually at regular intervals, to produce a sequence of numbers which are then processed by computer or special purpose digital processors (Figure 20.1).

A useful example to keep in mind is the electrocardiogram (ECG). This signal has its origin in the electrical activity of the heart, which produces small, characteristically varying currents throughout the body that in turn produce time-varying potential differences between sites on the surface of the body. In a typical modern processing chain, these potential differences are amplified and filtered by special purpose analog devices to produce electrical potentials that an analog-to-digital converter (ADC) converts to a sequence of numbers.

20.1.1 Signals in the Time Domain

The number of different forms the signal took in the example above is by no means unusual, and a description of signals which tried to include all of the features at each stage would be cumbersome. For processing purposes what is physically varying is not particularly important and a mathematical abstraction of the signal is usually sufficient. The ADC output in the ECG example provides a useful starting point; the result of the recording process is just a list of numbers, $\{x_1,x_2,x_3,\ldots,x_N\}$ giving the voltage at successive sampling times. The times $\{t_n;\ n = 1,2,3,\cdots,N\}$, at which each value was recorded are also known. Thus, a possible representation of a (discrete) signal is as a set of ordered pairs of numbers $\{(t_n,x_n);\ n = 1,2,3,\cdots,N\}$. In mathematical terms this set of ordered pairs is a function. A signal defined at a discrete set of times is often called a *discrete time signal* or just a *discrete signal*.

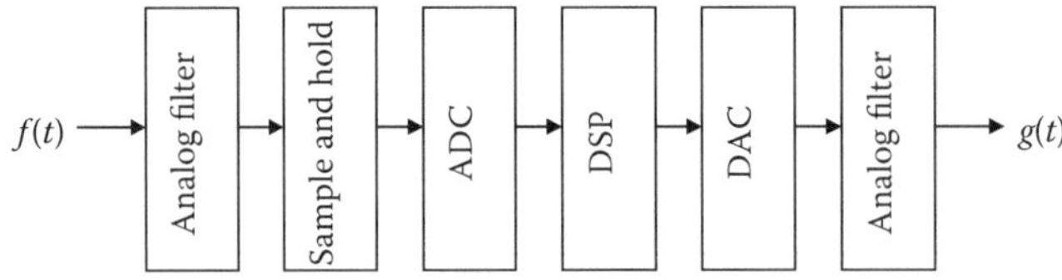

FIGURE 20.1
Typical digital signal processing chain. The analog filter at the front end limits the range of frequency components entering the chain, and the filter at the end of the chain smoothes the output of the digital to analog converter to remove the effects of sampling.

The idea generalizes to *continuous time signals* for which the set of times at which the signal is defined forms a continuum. In practice discrete-time signals usually arise as sampled continuous time signals. For idealized signals it may also be possible to describe the relation between t and $x(t)$ by a simple formula, for example, $x(t) = A\sin(\omega t + \phi)$, but this is not usually the case.

To summarize: the abstract mathematical model of a signal is a real valued function $x(t)$, of a single continuous or discrete real variable (usually called time, t). A signal represented directly in this way is often referred to as the signal in the time-domain.

20.1.2 Signals in the Frequency Domain

In the early part of the 19th century, Fourier showed that continuous-time periodic signals can be written as sums of *harmonic functions* such as

$$f(t) = \sum_{n=-\infty}^{\infty} F_n e^{jn\frac{2\pi}{T}t}$$

where the *Fourier coefficients* are given (where T is the length of the fundamental period in time) by

$$F_n = \frac{1}{T} \int_{-\frac{1}{2}T}^{\frac{1}{2}T} f(t) e^{-jn\frac{2\pi}{T}t}\, dt$$

(In signal processing, it is conventional to use the symbol $j = \sqrt{-1}$.)

For nonperiodic signals that go to zero sufficiently rapidly as $t \to \pm\infty$, the idea can be extended to represent the signal as an integral over a continuous frequency variable (ω):

$$f(t) = \frac{1}{2\pi} \int_{-\infty}^{\infty} F(\omega) e^{j\omega t}\, d\omega$$

where the *Fourier transform* $F(\omega)$ of the function $f(t)$ is given by

$$F(\omega) = \int_{-\infty}^{\infty} f(t) e^{-j\omega t}\, dt.$$

Note that negative frequencies appear in these expressions due to the choice of complex exponentials as the functions in which to expand the general function $f(t)$. For real valued functions the value of the Fourier transform at $-\omega$ is related to its value at ω by

$F(-\omega) = (F(\omega))^*$ where the star denotes complex conjugation.

Subsequently Helmholtz and others realized that representing the signal by its Fourier transform or Fourier series coefficients greatly simplified the description of how signals are transformed by physical systems, and also provided a way of extracting information about the processes that gave rise to the signal.

The signal represented by its Fourier transform (or Fourier coefficients) is referred to as the signal in the frequency domain. It is convenient to use the short hand notation $f(t) \leftrightarrow F(\omega)$ to represent the relation between time-domain and frequency domain representations of the same signal.

Example 1.1:

For the rectangular pulse signal $P_\tau(t) = \begin{cases} 1; & |t| \le \tau \\ 0; & |t| > \tau \end{cases}$ the expression for the Fourier transform is

$$\int_{-\infty}^{\infty} P_\tau(t) e^{-j\omega t}\, dt = \int_{-\tau}^{\tau} e^{-j\omega t}\, dt = \frac{1}{j\omega}\left[e^{j\omega\tau} - e^{-j\omega\tau t}\right].$$

Using Euler's formula $e^{j\theta} = \cos\theta + j\sin\theta$ this is easily recast as $2/\omega \sin(\omega\tau) = 2\tau\,\mathrm{sinc}(\omega\tau)$, where we have introduced the common shorthand $\mathrm{sinc}\, x \equiv \sin x / x$. (Note that sinc $0 \equiv 1$.) Thus, $P_\tau(t) \leftrightarrow 2\tau\ \mathrm{sinc}(\omega\tau)$.

20.1.3 Sampled Signals

In modern signal processing a continuous time signal is usually converted to a discrete time signal by sampling an analog to digital conversion (ADC), processed by a general purpose computer or specialized digital signal processor (DSP) and then, if necessary, reconstructed as a continuous time signal by a digital to analog converter (DAC) (Figure 20.1).

An important question is whether any information is lost in this process. A fundamental mathematical result, *Shannon's theorem*, ensures that a continuous-time signal can be sampled without loss of information provided it does not vary too rapidly between sampling instants. The problem is to be precise about what is meant by "vary too rapidly."

A *band limited* signal is one that requires only a finite range of frequencies to represent. For a signal $f(t)$ with a Fourier transform $F(\omega)$ (or Fourier series F_n) this implies that

$$F(\omega) = 0;\ |\omega| > \omega_c \quad \text{or} \quad F_n = 0;\ \left|n\frac{2\pi}{T}\right| > \omega_c.$$

The upper limit of frequencies required to represent such a signal, ω_c is called its *cutoff frequency*. Shannon's theorem guarantees that any band limited signal may be sampled and reconstructed exactly, provided it is sampled at a rate greater than twice its cutoff frequency. (Twice the cutoff frequency of a signal is called its *Nyquist rate*.) Another way of stating this is to insist that the cutoff frequency is less than half the sampling frequency. (Somewhat confusingly, half the sampling frequency is called the *Nyquist frequency*.)

Example 1.2:

The duality theorem of Fourier analysis can be used to show that $\text{sinc}(\Omega t) \leftrightarrow \pi/\Omega P_{\Omega}(\omega)$. Thus, a signal of the form $f(t) = A\text{sinc}(2\pi t)$ is band-limited, with a cut-off frequency $\omega_c = 2\pi s^{-1}$, or equivalently 1 Hz. If this signal is sampled at faster than 2 Hz, it can in principle be exactly reconstructed from its samples. If, however, the signal is sampled at 1.8 Hz, then the Nyquist frequency is 0.9 Hz, which is less than the cutoff, so some information would be irretrievably lost in the sampling process.

In practice most physiological signals are at least approximately band-limited, but it can happen that the signal has frequency components higher than the Nyquist frequency of the sampling system. The error introduced by such *undersampling* is as if the components with frequencies above the Nyquist frequency "wrap around" and appear as additional components at frequencies lower than the Nyquist frequency. Alternatively, the sampled signal can be associated with an "aliased spectrum" made up of the Fourier transform ($F(\omega)$) of the original signal plus frequency shifted images ($F(\omega - n\omega_s)$) of the transform:

$$\tilde{F}(\omega) = \sum_{n=-\infty}^{\infty} F(\omega - n\omega_s).$$

The information available from the sampled signal is just that available from the aliased spectrum. If the images do not overlap then the original signal can be recovered from the sampled signal by an appropriate filter. Any overlap of the images results in a difference between the aliased spectrum and the original spectrum on the frequency interval $-\omega_c \le \omega \le \omega_c$ which results in a distortion of the reconstructed signal. To reduce aliasing error most digital signal processing systems include an analog *antialiasing filter* before the analog to digital converter to limit the range of frequencies to less than the Nyquist frequency.

Example 1.3:

The signal $f(t) = \text{sinc}^2(\Omega t/2)$ has as its Fourier transform the triangular pulse function

$$F(\omega) = \frac{2\pi}{\Omega} Q_{\Omega}(\omega) = \frac{2\pi}{\Omega} \begin{cases} \left(1 - \frac{|\omega|}{\Omega}\right); & |\omega| \le \Omega \\ 0 \quad ; & |\omega| > \Omega \end{cases}.$$

Clearly, $f(t)$ is band-limited with cutoff frequency $\omega_c = \Omega$. For simplicity take the cutoff to be at $f_c = 1$ Hz (so the angular frequency of the cutoff is $\omega_c = 2\pi f_c = 2\pi$). If this signal were to be sampled at $f_s = 1.5$ Hz for example, the components above $f_{Nq} = 0.5 f_s = 0.75$ Hz would wrap around and add to the components below 0.75 Hz as shown in Figure 20.2c. To extend the frequency range over which the signal is undistorted an anti-aliasing filter with cutoff at the Nyquist frequency (0.75 Hz) could be used (Figure 20.2d–f). In practice the difficulty in making filters with very sharp cutoffs means that the anti-aliasing cutoff would need to be somewhat below the Nyquist frequency.

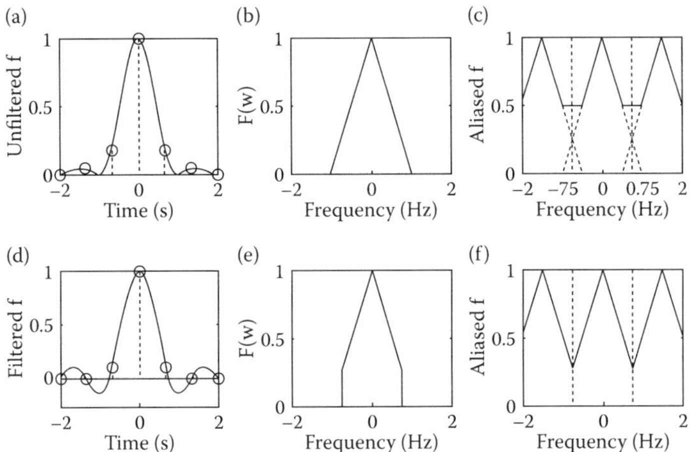

FIGURE 20.2
(a) Unfiltered continuous time signal and samples at 1/1.5 or 0.67 sec intervals. (b) Fourier transform. (c) Spectrum showing original transform and two images (dotted) as well as the aliased spectrum (solid). (d) Antialias filtered signal and samples. (e) Fourier transform and (f) aliased spectrum of filtered signal. The dotted vertical lines mark the position of the Nyquist frequency.

20.1.4 Digitized Signals

Another complication arises in DSP due to the fact that numbers generated by ADCs have finite precision, i.e., have a finite number of bits (B) per sample (typically in the range 8 bits to 16 bits). A rough guide to the error associated with this quantization of the sampled values is given by the root-mean-square (rms) error due to quantization when sampling a full scale sine wave. The resulting signal-to-quantization noise ratio (SQNR) can be shown to be approximately

$$SQNR = 20\log_{10}\left(\frac{\text{Signal rms amplitude}}{\text{Quantization rms error}}\right) \approx 1.76 + 6.02B \text{ dB}.$$

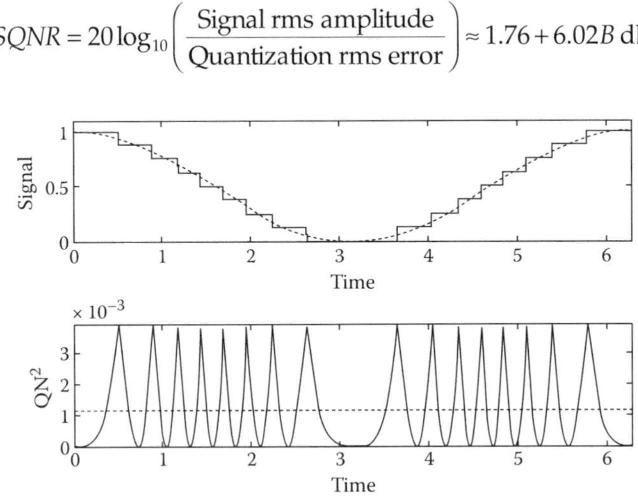

FIGURE 20.3
A 3-bit quantization of full scale cosine (upper) and the resulting squared quantization error (lower). The horizontal dotted line in the lower panel indicates the mean squared quantization error. The measured SQNR is close to the value of 20 dB predicted by the equation above.

Each extra bit improves the SQNR by around 6 dB, the downside is that more bits generally require more time to quantize and/or more expensive ADCs.

Example 1.4:

If a full scale signal is quantized at 3 bits the $SQNR \approx 1.76 + 6.02 \times 3 \approx 20$ dB. Figure 20.3 shows the actual quantization error for a raised cosine signal $f(t) = 0.5(1 + \cos(t))$ with rms amplitude $f_{rms} = 0.5/\sqrt{2} = 0.354$; the measured rms amplitude of the quantization error is $QN_{rms} = 1.1 \times 10^{-3}$ so the SQNR is given by $SQNR = 20\log_{10}(0.354/1.1 \times 10^{-3}) = 20.6$ dB. At 8 bits the SQNR is around 50 dB and at 16 bits it is closer to 98 dB.

20.2 Processes and Signals

Signals are usefully categorized by the nature of their sources as well as by the general nature of their temporal variation. Two very two basic types of source are introduced in this section, deterministic processes and stochastic (or random) processes, together with an overview of the ways in which stochastic sources and signals may be characterized.

20.2.1 Deterministic Processes and Signals

If the process that generates a signal is predictable, in the sense that repeated experiments in the same conditions would always produce the same signal, then the process is said to be *deterministic*. More strictly, a deterministic process has a time course that is determined by its initial conditions; if the initial conditions are repeated exactly, the same signal will result each time. Not many biological signals are strictly deterministic, but features of many signals can be explained by deterministic processes. In practice, deterministic signals are generated by relatively simple processes and, to a degree, the future behavior of the signal can be predicted from known values up to the present time.

Example 2.1:

The synthetic ECG signal shown in Figure 20.4a was generated using the model of McSharry et al (IEEE Trans. Biomed. Eng., 50(3), 289–294, 2003). The model consists of three coupled ordinary differential equations of the general form:

$$\frac{dx}{dt} = \alpha(x,y)x - \omega y$$

$$\frac{dy}{dt} = \alpha(x,y)y + \omega x$$

$$\frac{dz}{dt} = -z - \sum_{i=1}^{5} f_i(x,y) + A\sin(\omega_2 t)$$

(For details, see McSharry et al. A MATLAB® function to solve this system of equations, ecgsyn.m, written by Patrick McSharry and Gari Clifford is freely availble from Physionet—http://www.physionet.org/.)

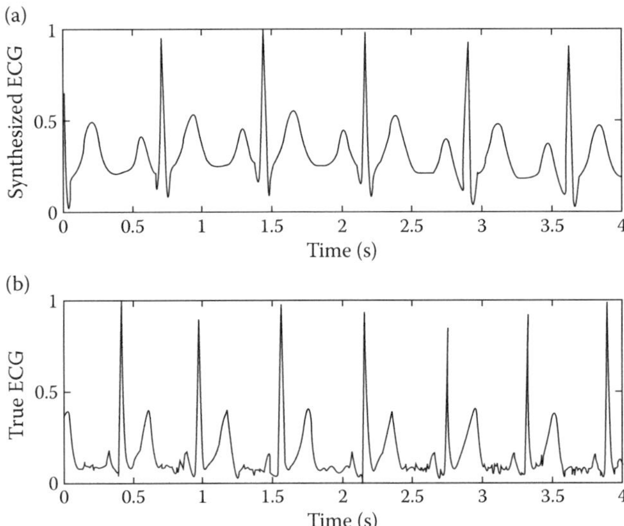

FIGURE 20.4
Comparison of idealized deterministic ECG signal (a) synthesized by a set of three coupled differential equations, with recorded ECG (b).

20.2.2 Stochastic Processes and Signals

If a process that produces a signal is unpredictable, in the sense that repeated experiments in the same conditions would produce different signals at random, then the process is said to be *stochastic*. The signals produced tend to have a random temporal character as well so that it is harder to predict future values from past values than for deterministic signals. The most common source of stochastic signals is thermal noise, but many large-scale physiological signals, such as the EMG and ongoing EEG, have an apparently stochastic character, at least on some time scales (Figure 20.5). The description of stochastic signals is largely statistical.

For stochastic signals, averages are more useful than instantaneous values since repeated experiments will yield different exact values at corresponding times after the start of the each of the experiments, but the statistical properties of the values at each instant may be quite stable. There are two types of average to be distinguished, *ensemble averages* and *time averages*.

20.2.2.1 Ensemble Averages

An ensemble average is an idealized average over the signals, $\{x_k(t); k = 1,2,3,\cdots\}$, generated in a theoretically infinite number of repeated experiments. Three important ensemble average quantities are the *mean* value of a stochastic signal at a given time t (measured from the start of each experiment) given by

$$\bar{x}(t) = E\left[x(t)\right] \equiv \lim_{K\to\infty} \frac{1}{K} \sum_{k=1}^{K} x_k(t) \equiv \int_{-\infty}^{\infty} xp(x,t)\,dx,$$

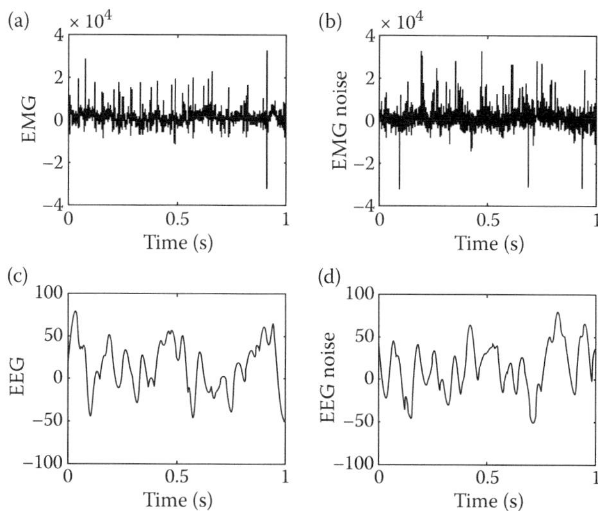

FIGURE 20.5
Examples of stochastic signals. (a) Raw EMG data. (b) Filtered noise with similar statistical properties to the EMG. (c) Raw EEG data. (d) Filtered noise with similar statistical properties to the EEG.

where K is the number of experiments, the *variance* of the signal at t given by

$$\sigma^2(t) = E\left[\left(x(t) - \bar{x}(t)\right)^2\right] \equiv \lim_{K\to\infty} \frac{1}{K} \sum_{n=1}^{K} \left(x_n(t) - \bar{x}(t)\right)^2 \equiv \int_{-\infty}^{\infty} p(x,t)\left(x - \bar{x}(t)\right)^2 dt,$$

and the autocovariance of the signal at two times t and $t + \tau$, given by

$$\begin{aligned} C(t, t+\tau) &= E\left[\left(x(t) - \bar{x}(t)\right)\left(x(t+\tau) - \bar{x}(t+\tau)\right)\right] \\ &\equiv \lim_{k\to\infty} \frac{1}{K} \sum_{k=1}^{K} \left(x_k(t) - \bar{x}(t)\right)\left(x_k(t+\tau) - \bar{x}(t+\tau)\right) \\ &\equiv \int_{-\infty}^{\infty}\int_{-\infty}^{\infty} \left(x - \bar{x}(t)\right)\left(x' - \bar{x}(t+\tau)\right) p_2(x, x', t, t+\tau)\, dx\, dx'. \end{aligned}$$

In each of these definitions, the expression involving the sum gives the formal ensemble average over the infinite set of signals obtained in repeated experiments. In practice, of course, such an average can only be taken over a finite number of repetitions (see Figure 20.6).

The second form of each definition involves the probability density function (pdf), $p(x,t)$, for the signal amplitude at a given time or the joint probability density function, $p_2(x,x',t,t')$, for the signal values measured at two different times. The quantity τ is usually called the *lag time*. If the pdfs of the process producing the signal are independent of the time t (although the joint distributions still depend on τ) the process is said to be *stationary*. To

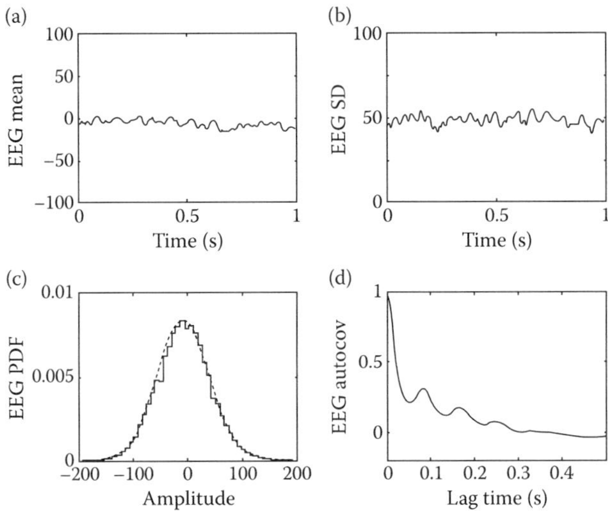

FIGURE 20.6
(a) Mean and (b) standard deviation of the EEG signals of Figure 20.5, based on an ensemble of 100 trials, each of 1 s duration. In this case the ensemble averaged quantities are approximately constant in time suggesting that the processes are stationary, at least on the time scale displayed. (c) Pdf of EEG signal estimated directly from data (solid) compared with a Gaussian pdf with the same mean and standard deviation (dotted). (d) Normalized autocovariance functions the EEG signals based on 100 trials. The periodic character of the EEG shows up in the peaks at multiples of the period of the alpha rhythm (frequency 8 Hz–12 Hz).

calculate these averages in practice one needs to assume a form for the distributions, for example, for a stationary Gaussian process

$$p(n) = \frac{1}{2\pi\sigma^2}\exp\left(-\frac{(n-\bar{n})^2}{2\sigma^2}\right)$$

and

$$p_2(n_1, n_2, \tau) = \frac{1}{2\pi\sigma^2\sqrt{1-\rho^2}}\exp\left(-\frac{1}{2(1-\rho^2)\sigma^2}\left((n_1-\bar{n})^2+(n_2-\bar{n})^2-2\rho(n_1-\bar{n})(n_2-\bar{n})\right)\right)$$

where $\bar{n}$ and σ are constants and the correlation coefficient $\rho(\tau) = C(\tau)/\sigma^2$, which is proportional to the covariance, depends on the lag time. The usual way of estimating the parameters that appear in these distribution functions is to compute them from measured signals.

20.2.2.2 Time Averages

A time average is simply an average of a given signal over time T and if the process is stationary (and *ergodic*—which means that a single long recording is statistically representative of the process as a whole) then time averages can be used to estimate ensemble averages, and vice-versa. The three main averages, estimated as time averages (for continuous and discrete processes, respectively), are then given by

$$\bar{x} = \lim_{T\to\infty} \frac{1}{T} \int_{-\frac{1}{2}T}^{\frac{1}{2}T} x(t)\,dt = \lim_{N\to\infty} \frac{1}{2N+1} \sum_{n=-N}^{N} x(n)$$

$$\sigma^2 = \lim_{T\to\infty} \frac{1}{T} \int_{-\frac{1}{2}T}^{\frac{1}{2}T} (x(t)-\bar{x})^2\,dt = \lim_{N\to\infty} \frac{1}{2N+1} \sum_{n=-N}^{N} \left(x(n)-\bar{x}\right)^2$$

$$C(\tau) = \lim_{T\to\infty} \frac{1}{T} \int_{-\frac{1}{2}T}^{\frac{1}{2}T} (x(t)-\bar{x})(x(t+\tau)-\bar{x})\,dt$$

$$C(m) = \lim_{N\to\infty} \frac{1}{2N+1} \sum_{n=-N}^{N} \left(x(n)-\bar{x}\right)\left(x(n+m)-\bar{x}\right).$$

In practice, the integrals or sums can only be carried out over a finite time interval, so the limit is not strictly possible. T or N needs to be reasonably large for the time averages to be accurate estimates of the ensemble averages; how large depends on the signal.

The pdf introduced above that depends on the signal at a single time is actually a *marginal pdf*, i.e., a probability density for the signal amplitude at a given time independent of the values the signal takes at other times. For stationary, ergodic processes the marginal pdf $p(f)$ can also be estimated from the fraction of time the signal is found in each amplitude bin (i.e., amplitude between x and $x + dx$):

$$p(x)dx \approx \frac{\text{Total duration for which } x \le X < x+dx}{\text{Duration of signal}}.$$

Example 2.2:

The pdf of a signal can sometimes be used to distinguish between a stochastic signal and the same signal corrupted by interference. For example, the pdf of the EEG signal is significantly different from that of the same signal corrupted by 50 Hz interference as shown in Figure 20.7.

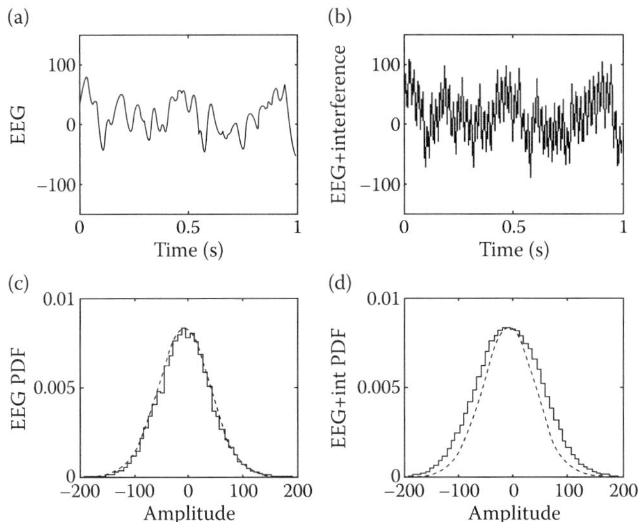

FIGURE 20.7
Comparison of pdfs of EEG signals and EEG corrupted by 50 Hz interference. The dotted curves show a Gaussian pdf with the same mean and standard deviation as the uncorrupted signal.

20.3 Fourier Spectra

A frequency spectrum (or more correctly an energy spectral density [ESD] or power spectral density [PSD]) measures the distribution of power in a signal over frequency. ESDs or PSDs can be defined for both deterministic and stochastic signals. Deterministic signals tend to have characteristic spectral features, often with most of the signal energy at lower frequencies, occasionally with narrow spectral peaks representing the contribution of periodic processes to the signal. For example, the PSD of a typical ECG would show a series of peaks at harmonics of the pulse frequency with the relative amplitudes of the peaks determined by the shape of a single cycle of the ECG signal. The spectral peaks would also be broadened due to slight irregularities of the pulse rate over time as well as peak-to-peak variation in the pulse shape (see Figure 20.8a and b).

Stochastic signals often show relatively broad-band spectra, reflecting their noise-like character, or a spectrum with broad peaks at unrelated frequencies. For example, the eyes-closed resting EEG signal for humans usually shows a prominent spectral peak (the α rhythm peak) in the vicinity of 10 Hz, with a width of a few hertz.

It should be noted that, in general, the interpretation of Fourier spectra is not a trivial matter. A direct interpretation of spectral features in terms of underlying system properties is only possible for linear time-invariant systems, and then only when the input signal is known, at least in general form. There are also problems in calculating a reliable spectrum from recorded data which will be discussed in a later section.

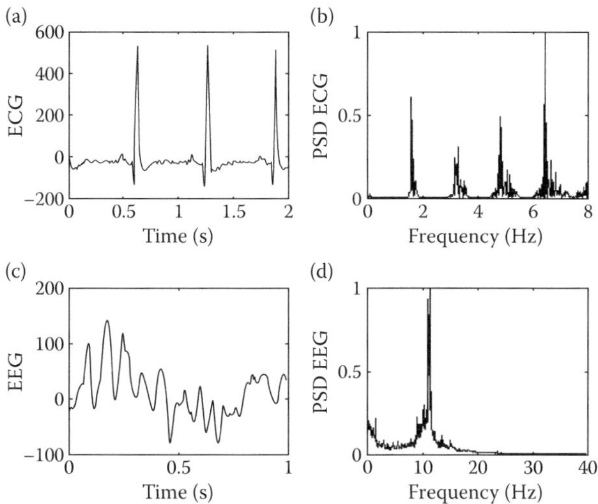

FIGURE 20.8
(a) Waveform and (b) normalized power spectral density (PSD) for ECG. (c) Waveform and (d) PSD for resting, eyes closed EEG.

20.3.1 Spectra of Finite Energy Signals

In signal processing, the instantaneous power of a signal $x(t)$ is defined as the square of the signal, $p(t) = x^2(t)$ (since the real power of a signal is often proportional to the square of the physical signal, e.g., voltage). The energy in the signal is then given by

$$E = \int_{-\infty}^{\infty} |x(t)|^2 \, dt \quad \text{and} \quad E = \sum_{n=-\infty}^{\infty} |x(n)|^2$$

for continuous time signals and discrete time signals, respectively. If the energy is finite then the signal is said to be a *finite energy signal*. Signals of finite duration or signals with a region outside of which the signal decays rapidly to zero are typical finite energy signals (but strictly any truly physical signal must be finite energy).

Example 3.1:

The Gaussian pulse $x(t) = Ae^{-\frac{1}{2}\left(\frac{t-t_0}{\sigma}\right)^2}$ is finite energy since

$$\int_{-\infty}^{\infty} |x(t)|^2 \, dt = |A|^2 \int_{-\infty}^{\infty} e^{-\left(\frac{t-t_0}{\sigma}\right)^2} dt = |A|^2 \frac{\sigma}{\sqrt{2}} \int_{-\infty}^{\infty} e^{-\frac{1}{2}s^2} ds = |A|^2 \frac{\sigma}{\sqrt{2}} \sqrt{2\pi} < \infty \text{ (for real values of } \sigma).$$

If a continuous time signal is finite energy then it has a *Fourier transform*:

$$X(\omega) = \int_{-\infty}^{\infty} x(t) e^{-j\omega t} \, dt$$

where ω is the angular frequency related to frequency in hertz by $\omega = 2\pi f$.

For a discrete time signal the closest equivalent to the Fourier transform is the *discrete time Fourier transform* (DTFT), and inverse, defined respectively by

$$X(\omega)=\sum_{n=-\infty}^{\infty}x(n)e^{-j\omega n};\quad 0\le\omega<2\pi\quad x(n)=\frac{1}{2\pi}\int_{0}^{2\pi}X(\omega)e^{j\omega n}\,d\omega$$

where ω is now the *digital frequency* and is a continuous variable related to the frequency in hertz by $\omega=2\pi\frac{f}{f_s}=2\pi fT$, where $f_s=\frac{1}{T}$ is the sampling frequency. (The notational convention used in relating the sampled values to the original signal is $x(nT)\rightarrow x(n)$.)

The DTFT is periodic ($X(\omega+2\pi)=X(\omega)$), so digital frequency needs to range only over the values 0π to 2π. The sampling frequency corresponds to a digital frequency of 2π and the Nyquist frequency corresponds to π. Frequencies between π and 2π correspond to negative frequencies since $X(2\pi-\omega)=X(-\omega)$. For real signals the information at negative frequencies is redundant ($X(-\omega)=X^*(\omega)$) so discrete spectra are usually only plotted for digital frequencies in the range 0 to π. If the sequence $x(n)$ results from sampling a band-limited function at greater than the Nyquist rate, then the DTFT at the digital frequency ωT is proportional to the Fourier transform of the original signal at ω: $X_{FT}(\omega)=TX_{DTFT}(\omega T)$;

Example 3.2:

A decaying one-sided harmonic signal

$$x(t)=\begin{cases}0 & ;\quad t<0\\ e^{-t}\cos(30\pi t); & t\ge 0\end{cases}$$

is sampled (without antialias filtering) at 100 Hz (sampling interval $T=1/100$ s). The digital frequency corresponding to the frequency of the oscillatory component of the signal is $\omega_p=30\pi/100=0.3\pi$. The sampled sequence is finite energy and its DTFT is well defined.

$$X_{DTFT}(\omega T)=\sum_{n=0}^{\infty}e^{-0.01n}\cos(0.30\pi n)e^{-j\omega Tn}=\frac{1}{2}\sum_{n=0}^{\infty}\left[e^{-(0.01+j(\omega T-0.3\pi))n}+e^{-(0.01+j(\omega T+0.3\pi))n}\right]$$

$$TX_{DTFT}(\omega)=\frac{1}{200}\left[\frac{1}{1-e^{-\left(0.01+j\left(\frac{\omega}{100}-0.3\pi\right)\right)}}+\frac{1}{1-e^{-\left(0.01+j\left(\frac{\omega}{100}+0.3\pi\right)\right)}}\right].$$

Since the original signal is not band-limited the DTFT is at best only an approximation to the Fourier transform of the original signal, which is given by

$$x(t)\leftrightarrow\tfrac{1}{2}\left[\frac{1}{1+j(\omega-30\pi)}+\frac{1}{1+j(\omega+30\pi)}\right].$$

The approximation is fairly accurate in the vicinity of the frequencies of the oscillatory part of the signal (Figure 20.9) as demonstrated by the following MATLAB code:

```
freq = linspace(0.1,50,1001);
w = 2*pi*freq;
Fa = (1/200)*(1./(1 - exp(-(0.01 + 1j*(w/100 - 0.3*pi)))) ...
            + 1./(1 - exp(-(0.01 + 1j*(w/100 + 0.3*pi)))));
Fx = 0.5*(1./(1 + 1j*(w - 30*pi)) ...
        + 1./(1 + 1j*(w + 30*pi)));
loglog(freq,abs(Fa),'k',freq,abs(Fx),':k');
xlabel('Frequency (Hz)');
ylabel('|F(w)|');
legend('DTFT','FT','Location','NorthWest')
```

The basic result on which spectroscopy relies is Parseval's relation:

$$\int_{-\infty}^{\infty} |x(t)|^2\, dt = \frac{1}{2\pi} \int_{-\infty}^{\infty} |X(\omega)|^2\, d\omega.$$

Physically, this means that the total energy in a signal is equal to the sum of the energies in the separate frequency components, without any cross terms. This leads to the definition of the energy spectral density as the square of the magnitude of the corresponding Fourier transform

$$S(\omega) = \frac{1}{2\pi} |X(\omega)|^2.$$

Note that, the spectrum here is really a density, being proportional to the energy per unit frequency contained in components with frequencies close to the given frequency, rather than the energy at a given frequency. So, for infinitesimal $d\omega$, $S(\omega)d\omega$ is proportional to the energy in components with frequencies in the interval $[\omega, \omega + d\omega]$. This signal is referred to as a two-sided spectrum since the energies in +ve and −ve frequencies are represented separately. In fact, for real signals it can be shown that the ESD is an even function of frequency, $S(-\omega) = S(\omega)$, so no extra information is given by the −ve frequencies. It is also common to define a one-sided spectrum which lumps the energies at +ve and −ve frequencies together:

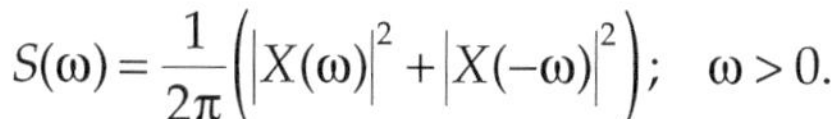

$$S(\omega) = \frac{1}{2\pi}\left(|X(\omega)|^2 + |X(-\omega)|^2\right); \quad \omega > 0.$$

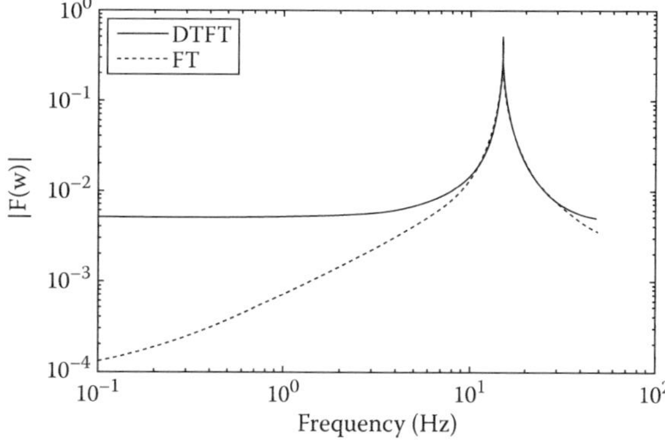

FIGURE 20.9
Comparison of the magnitude of the exact Fourier transform of the decaying cosine (dotted line) with the magnitude of the DTFT of the sampled sequence.

Also, since spectra are often normalized in practice, it is common to drop the factor of $1/2\pi$ in the definition. (The factor $1/2\pi$ is due to the use of angular frequency $\omega = 2\pi f$ as the frequency variable, it is absent if one uses frequency f [in hertz] as the transform variable.)

Example 3.3:

The one-sided (or "causal") exponential: $x(t) = \begin{cases} 0 & ; t < 0 \\ e^{-t/\tau}; & t \geq 0 \end{cases}$ is finite energy and has a Fourier transform $X(\omega) = \dfrac{\tau}{1 + j\omega\tau}$.

The 2-sided ESD is thus $S(\omega) = \dfrac{1}{2\pi}\left|\dfrac{\tau}{1 + j\omega\tau}\right|^2 = \dfrac{1}{2\pi}\dfrac{\tau^2}{\left(1 + (\omega\tau)^2\right)}$.

Since $S(-\omega) = S(\omega)$, and $\int \dfrac{dx}{1 + x^2} = \operatorname{atan} x$, the fraction of the total energy in the pulse with frequencies below the 3 dB point $\omega_{3dB} = \dfrac{1}{\tau}$ is given by

$$\int_0^{\omega_{3dB}} S(\omega)\,d\omega \bigg/ \int_0^{\infty} S(\omega)\,d\omega = \int_0^{\omega_{3dB}} \frac{1}{1 + (\omega\tau)^2}\,d\omega \bigg/ \int_0^{\infty} \frac{1}{1 + (\omega\tau)^2}\,d\omega = \frac{\operatorname{atan}(1)}{\operatorname{atan}(\infty)} = \frac{\frac{1}{4}\pi}{\frac{1}{2}\pi} = 0.5.$$

20.3.2 Spectra of Finite Power Signals

While all physical signals are ultimately finite energy it is convenient to idealize some signals, for example, periodic signals and noise signals, as finite power signals. The expressions for average power in a signal follow from those for the energy in a signal. For continuous and discrete time signals, respectively,

$$P = \lim_{T\to\infty} \frac{1}{2T} \int_{-T}^{T} |x(t)|^2\,dt \qquad P = \lim_{N\to\infty} \frac{1}{2N + 1} \sum_{n=-N}^{N} |x(n)|^2.$$

If these quantities are finite, the signals are called *finite power signals*. The expressions for the two-sided spectra follow in a similar fashion:

$$P = \frac{1}{2\pi} \lim_{T\to\infty} \frac{1}{2T} \left| \int_{-T}^{T} x(t) e^{-j\omega t}\,dt \right|^2 \qquad P = \frac{1}{2\pi} \lim_{N\to\infty} \frac{1}{2N + 1} \left| \sum_{n=-N}^{N} x(n) e^{-j\omega n} \right|^2.$$

Example 3.4:

A cosine of frequency Ω has a two-sided power spectrum given by

$$P(\omega) = \frac{1}{2\pi} \lim_{T\to\infty} \frac{1}{2T} \left| \int_{-T}^{T} \cos(\Omega t) e^{-j\omega t}\,dt \right|^2 = \frac{1}{2\pi} \lim_{T\to\infty} \frac{1}{2T} \left| \int_0^T \left[\cos((\omega - \Omega)t) + \cos((\omega + \Omega)t)\right] dt \right|^2$$

$$P(\omega) = \frac{1}{2\pi}\lim_{T\to\infty}\frac{T}{2}\left|\frac{\sin((\omega-\Omega)T)}{(\omega-\Omega)T} + \frac{\sin((\omega+\Omega)T)}{(\omega+\Omega)T}\right|^2$$

$$= \frac{1}{4\pi}\lim_{T\to\infty} T\left|\operatorname{sinc}\big((\omega-\Omega)T\big) + \operatorname{sinc}\big((\omega+\Omega)T\big)\right|^2$$

All the power in the signal is concentrated in the vicinity of the frequencies of the cosine since $P(\omega) = 0$; $\omega \neq \pm\Omega$ (at least in the limit as $T \to \infty$). The explicit expression in the vicinity of $\omega \approx \pm\Omega$ can be written $P(\omega) = \frac{1}{4\pi}\lim_{T\to\infty} T\operatorname{sinc}^2\big((\omega \mp \Omega)T\big)$, which behaves like a delta function since $\int_{-\infty}^{\infty} T\operatorname{sinc}^2(\omega T)\,d\omega = \pi$ and $\lim_{T\to\infty} T\operatorname{sinc}^2(\omega T) = 0;\ \omega \neq 0$. Thus,

$P(\omega) = \frac{1}{4}\big[\delta(\omega-\Omega) + \delta(\omega+\Omega)\big]$.

The total power, which is the integral of $P(\omega)$ over all ω, yields 1/2, as expected for a harmonic signal.

20.3.3 Spectra of Stochastic Processes

If the signals $x(t)$ are stochastic the expressions given above result in stochastic spectra; repeated experiments will produce different spectra. To define a stable estimate of the power distribution for these signals one needs to average the results of the individual experiments over a large number of experiments. The average is usually written symbolically as the expected value, E[*], of the appropriate spectrum:

$$P(\omega) = E\left[\frac{1}{2\pi}\lim_{T\to\infty}\frac{1}{2T}\left|\int_{-T}^{T} x(t)e^{-j\omega t}\,dt\right|^2\right] \quad \text{or} \quad P(\omega) = E\left[\frac{1}{2\pi}\lim_{N\to\infty}\frac{1}{2N+1}\left|\sum_{n=-N}^{N} x(n)e^{-j\omega n}\right|^2\right].$$

The spectrum without the average is a surprisingly poor estimate of the "true" spectrum, so some form of averaging is essential (Figure 20.10).

20.3.3.1 Weiner–Kinchine Theorem

The autocorrelation function of a process (which is equal to the autocovariance function defined in 2.2.1 for signals with zero mean, i.e., if $E[x(t)] = 0$), is defined as

$$R(t, t+\tau) = E[x(t)x(t+\tau)].$$

For stationary processes the autocorrelation function depends only on the lag time, τ, and the Weiner–Kinchine theorem states that the PSD of a stationary process is the Fourier transform of the autocorrelation function of the process.

$$R(\tau) \leftrightarrow P(\omega).$$

Thus, the spectrum of a signal is an alternative way of displaying the statistical connection between successive values of the signal. A broad spectrum implies a narrow autocorrelation function, i.e., little relation between signal values at different times. Conversely a narrow

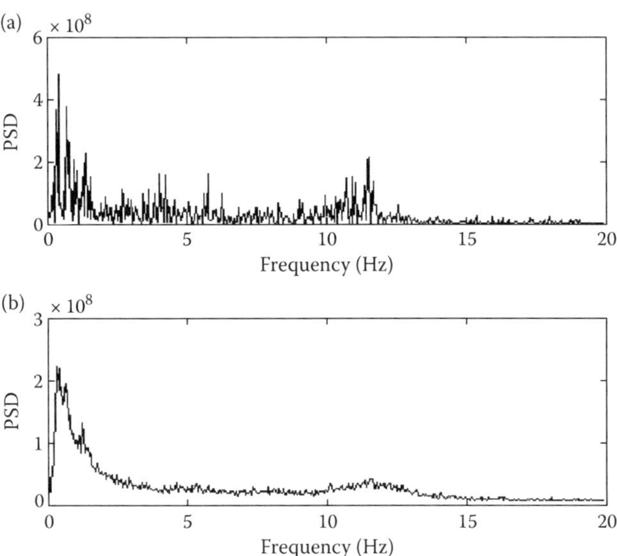

FIGURE 20.10
Comparison of PSD of human EEG: (a) based on a single 25-sec recording sampled at 173.61 Hz, and (b) based on the average of 100 such recordings.

spectrum implies a broad autocorrelation function, so signal values separated by considerable intervals in time can be strongly correlated.

Example:

If white noise (a zero-mean stochastic signal with a flat spectrum) is passed through a single-pole low-pass filter with a 3-dB point at frequency Ω, the resulting signal has a PSD of the form: $P(\omega)=\dfrac{A}{\left(1+(\omega/\Omega)^2\right)}$, where A is a constant. The inverse Fourier transform of this is $R(\tau)=\dfrac{A\Omega}{2}\exp(-|\Omega\tau|)$. Since $R(0)=E[x^2(t)]=x_{rms}^2$ the constant A can be related to the rms amplitude and bandwidth of the filtered noise thus: $A=2x_{rms}^2/\Omega$.

For deterministic signals a similar relation holds between the PSD of the signal and its autocorrelation function defined by

$$R(\tau)=\int_{-\infty}^{\infty} x(t)x(t+\tau)\,dt \leftrightarrow S(\omega)$$

for finite energy functions, and by

$$R(\tau)=\lim_{T\to\infty}\frac{1}{T}\int_{-\frac{1}{2}T}^{\frac{1}{2}T} x(t)x(t+\tau)\,dt \leftrightarrow P(\omega)$$

for finite power signals.

20.3.4 Calculating Spectra

The equations defining the Fourier transform are formal in the sense that, except for fairly simple signals, one cannot use the expressions to actually calculate the corresponding spectra. At some stage approximations must be introduced and it is here that many of the problems associated with spectral analysis arise. In particular what is the "best" numerical estimate of a spectral density?

In earlier times power spectra were estimated by passing the signal through a filter which responded strongly to a narrow band of frequencies. The full spectrum could be obtained by using a variable filter and repeating the measurement at a range of frequencies or by using a bank of filters. In modern practice, especially for physiological signals, spectra are nearly always calculated numerically based on sampled data.

20.3.4.1 The Discrete Fourier Transform

The most common algorithm is based on the Discrete Fourier Transform (DFT), mainly because a very efficient implementation (the Fast Fourier Transform: FFT) exists. The DFT differs from the DTFT introduced in Section 3.1 in that it is defined only for finite length sequences and only at a finite set of frequencies. The DFT needs to be used carefully because, although it is often a good approximation to the sampled Fourier transform, there are situations in which it may yield misleading spectra.

The DFT is only defined for sequences of finite length and by convention both the time index (n) and the frequency index (m) run from 0 to $N - 1$. With this convention the DFT is given by:

$$X(m) = \sum_{n=0}^{N-1} x(n) e^{-j\frac{2\pi}{N}nm}; \quad m = 0, 1, 2, \cdots, N-1.$$

This is clearly equivalent to the DTFT of a finite sequence evaluated at the digital frequencies $\omega = m\frac{2\pi}{N}$. The corresponding "true" frequencies are given by $\omega_m = m\frac{2\pi f_s}{N}$, where f_s is the sampling frequency in hertz. The DFT invertible and it can be shown that the inverse DFT is given by

$$x(n) = \frac{1}{N}\sum_{n=0}^{N-1} X(m) e^{j\frac{2\pi}{N}mn}; \quad n = 0, 1, 2, \cdots, N-1.$$

Efficient implementations of these formulae are available for most programming languages.

20.3.4.2 Aliasing Errors

If the restrictions on the domains of n and m are relaxed, it can be shown that both $x(n)$ and $X(m)$ are periodic sequences, $x(n + N) = x(n)$ and $X(m + N) = X(m)$. One consequence of this is that the negative frequencies are represented in the DFT by the second half of the sequence since $X(-m) = X(N - m)$. It is evident then that the DFT cannot represent frequency components greater than half the sampling frequency $f_N = \frac{1}{2}f_s$ (the *Nyquist*

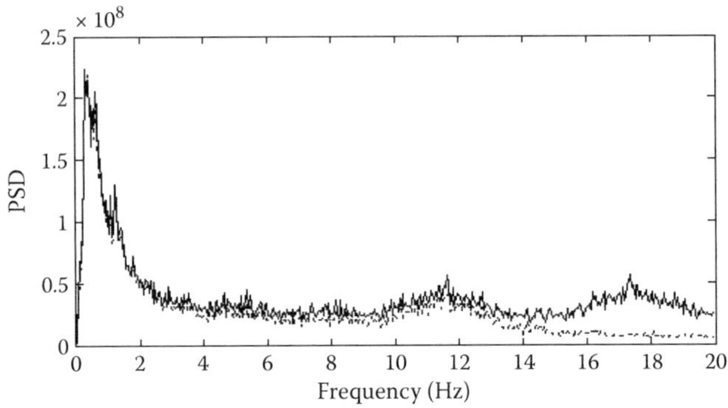

FIGURE 20.11
Comparison of EEG spectra. The dotted curve is the true spectrum, the solid curve is the spectrum based on a signal sampled at 30 Hz without an antialiasing filter. The apparent peak at around 18 Hz in the under-sampled spectrum is, in fact, due entirely to aliasing error.

frequency). This leads to the first restriction on the use of the DFT: the signal must contain no components above the Nyquist frequency, if it does these should be removed **before** the signal is sampled. In practice a margin of error needs to be applied and the sampling rate needs to be somewhat greater than twice the highest frequency in the signal. The errors that arise when a signal has frequency components above the Nyquist frequency are the *aliasing errors* discussed earlier in the section on the sampling theorem (Figure 20.11).

20.3.4.3 Windowing

The fact that the DFT is defined only for finite length sequences means that ongoing signals need to be truncated in order to use the DFT in calculating the spectrum. In effect the spectrum calculated is that of the signal

$$x_T(t) = w(t)x(t)$$

where $w(t)$, referred to as a *window function*, is a function which is nonzero only over a finite interval (we say that $w(t)$ has *compact support*). For example, we might choose a *rectangular window*

$$w(t) = \begin{cases} 1; & |t| \leq \frac{1}{2}T \\ 0; & |t| > \frac{1}{2}T \end{cases}.$$

For a finite energy signal, the Fourier transform of a product becomes a *convolution* of the Fourier transforms of the two functions.

$$w(t)x(t) \leftrightarrow \frac{1}{2\pi} W * X(\omega) = \frac{1}{2\pi} \int_{-\infty}^{\infty} W(\omega - \omega')X(\omega')d\omega'.$$

A convolution can be thought of as a type of moving average in which the value of $X(\omega)$ at ω is replaced by the weighted average of its values in the vicinity of ω with $W(\omega)$ serving as the weighting function.

This leads to a pair of related problems. First, since the main lobe of the weighting function has a width of the order of $1/T$, the frequency resolution (how close in frequency two features of the spectrum can be and still be individually seen) is limited to roughly $1/T$, where T is the duration of the segment used in calculating the spectrum. Second, since the weighting function decays to zero relatively slowly (for example, as $1/\omega$ for the rectangular window) a prominent feature in the spectrum could alter the power observed at frequencies distant from the feature, a problem called *spectral leakage*. A common way of reducing the problem of leakage is to use a window function $w(t)$ which does not go so quickly to zero, this reduces the side lobes of the weighting function $W(\omega)$, at the expense of producing a wider main lobe. Thus, the problems are linked, a window which produces good resolution (such as the rectangular window) also produces severe leakage, and a window that reduces the leakage problem (such as the Hamming window—a common choice), does so at the expense of poorer resolution (Figure 20.12).

20.3.5 Spectra of Stochastic Processes

Spectral analysis is particularly useful for stochastic processes since their temporal behavior is often largely unpredictable. The main problem in estimating spectra for these signals is the need to average the spectrum over many experiments.

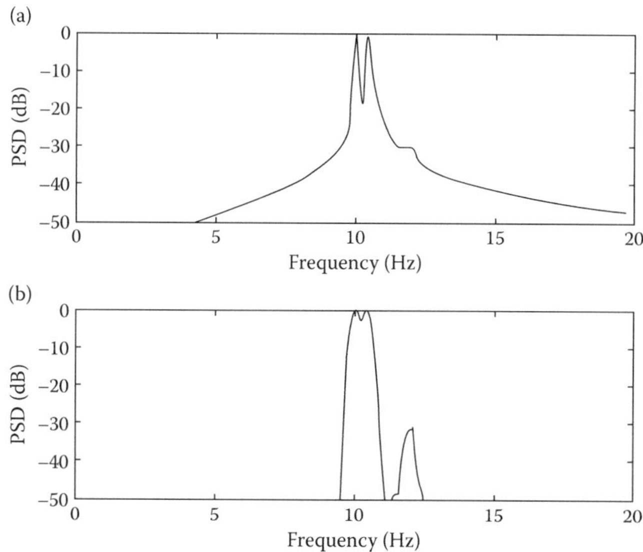

FIGURE 20.12
Comparison of spectra of a signal with two strong peaks (0 dB at 10 Hz and 10.5 Hz) and one weak peak (−30 dB at 12 Hz) calculated using the DFT of approximately 4 sec of data sampled at roughly 4 kHz. (a) Rectangular window and (b) Hamming window. The rectangular window allows the two close lines to be resolved but leakage obscures the weak line, whereas the Hamming window coalesces the two close lines but allows the weak line to be easily distinguished.

If repeated measurements, $\{x_k(t);k = 1,2,3,\cdots,K\}$ are available then a simple average is usually a good option: calculate the PSDs for each experiment, $\{S_k(\omega) = |X_k(\omega)|^2; k = 1,2,3,\cdots,K\}$, in the usual fashion (using an appropriate window) and then average the resulting spectra:

$$S(\omega) = \frac{1}{K}\sum_{k=1}^{K} S_k(\omega).$$

In many situations repeated experiments are not done, but one may have a long continuous recorded signal. If the process is *stationary* and *ergodic*, it may be possible to calculate a reliable estimate of the spectrum of the process by splitting the long signal into a number of shorter signals (thus sacrificing resolution) and averaging the PSDs of these segments. The segments may be nonoverlapping, but there is some advantage in using overlapping segments (typically 50% overlap). The resulting spectrum is called the *Welsh periodogram* and is the method implemented, for example, in the MATLAB *pwelch* command.

20.4 Signal Estimation

20.4.1 Artifacts, Interference, and Noise

A measured signal, $x(t)$, is usually a corrupted version of an information containing signal, $f(t)$. As shown in Figure 20.13, the corruption may occur in more than one way, The following is a partial list:

- *Artifacts* are signals originating from the source but due to phenomena which are extraneous to the process of interest. For example, in recording the ECG, signals due to muscle activity, or to the effects of, say, respiration, may be added to the real ECG signal. Another common example occurs in the electroencephalogram (EEG) where the effects of eye movements and blinks on the recorded signal can be large, especially at frontal sites.

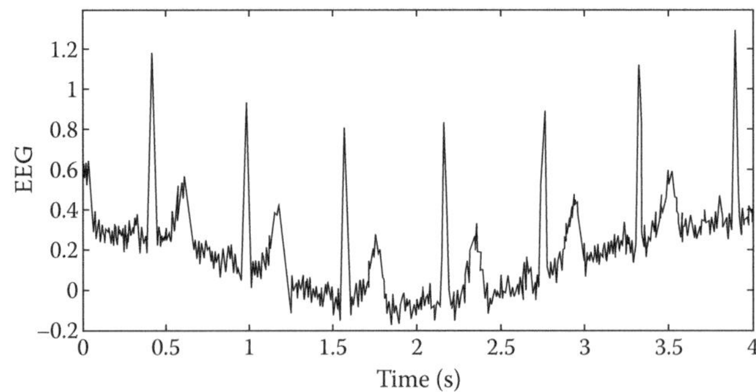

FIGURE 20.13
A 4-sec ECG record showing baseline drift, artifact, mains frequency interference, and noise.

- *Instrumental effects* are due to the limitations of transducers and recording devices. As a result the recorded signal is at best an approximation to the signal produced by the source.
- *Drift* is a general term for instrumental effects which produce a slow drift in the baseline of a signal.
- *Interference* is due to extraneous signals that have been added to the true signal. The most common type of interference is mains interference, a narrow band signal at the power line frequency and usually also at integer multiples of this frequency.
- *Noise* is a "random" signal which accompanies all signals. It may, for example, be due to thermal effects which generate voltages in the components of an amplifier or transducer.

The simplest model of the corrupted signal, ignoring instrumental effects, and the one most used in practice, is

$$x(t) = f(t) + g_I(t) + g_T(t) + g_A(t) + n(t)$$

where $f(t)$ is the "true" signal, $g_I(t)$ is due to interference, $g_T(t)$ is the trend due to drift, $g_A(t)$ is the artifact and $n(t)$ is the noise.

Often the $g(t)$ terms are omitted in formal treatments since their character is closer to that of the true signal than to noise and they are thus lumped with the "signal" part of $x(t)$, to give the simpler additive noise model

$$x(t) = f(t) + n(t).$$

Signal estimation is the problem of recovering the best estimate of the true signal from a recording that usually includes noise and interference.

20.4.2 Noise Reduction by Averaging

At the base of most signal estimation and detection techniques is the fact that the signal component, $f(t)$, is not correlated with the other components of the measured signal. For the case of a signal corrupted with additive zero-mean noise, estimation usually relies on the fact that the power in the averaged noise signal decreases to zero as the number of terms in the average (N) increases. To see this start with the definition of the variance of the averaged noise, i.e.,

$$\mathrm{var}(\overline{n}) = E\left[\left(\frac{1}{N}\sum_{k=1}^{N} n_k(t)\right)^2\right].$$

Write the square as a product of two summations

$$\left(\frac{1}{N}\sum_{k=1}^{N} n_k(t)\right)^2 = \left(\frac{1}{N}\sum_{k=1}^{N} n_k(t)\right)\left(\frac{1}{N}\sum_{k'=1}^{N} n_{k'}(t)\right) = \frac{1}{N^2}\sum_{k=1}^{N}\sum_{k'=1}^{N} n_k(t)n_{k'}(t)$$

and use the linearity of the expectation value

$$E\left[\frac{1}{N^2}\sum_{k=1}^{N}\sum_{k'=1}^{N} n_k(t)n_{k'}(t)\right] = \frac{1}{N^2}\sum_{k=1}^{N}\sum_{k'=1}^{N} E\left[n_k(t)n_{k'}(t)\right].$$

The noise in different sweeps is uncorrelated so that $E[x_k(t)x_{k'}(t)] = 0$ if $k \neq k'$, and the only terms that contribute to the double sum are those with $k = k'$. Thus,

$$\frac{1}{N^2}\sum_{k=1}^{N}\sum_{k'=1}^{N} E\left[n_k(t)n_{k'}(t)\right] = \frac{1}{N^2}\sum_{k=1}^{N} E\left[n_k(t)n_k(t)\right] = \frac{1}{N^2}\sum_{k=1}^{N} E\left[n_k^2(t)\right].$$

Since the noise is assumed to have the same properties in each trial $E[n_k^2(t)]$ is independent of k, and is equal to the mean square amplitude of the noise and is the instantaneous power of the un-averaged noise signal at time t, written as n_{rms}^2 for convenience. Thus,

$$\mathrm{var}(\bar{n}) = \frac{1}{N^2}\sum_{k=1}^{N} n_{rms}^2(t) = \frac{Nn_{rms}^2(t)}{N^2} = \frac{n_{rms}^2(t)}{N} \to 0 \text{ as } N \to \infty.$$

Thus, the root-mean-squared amplitude of the noise component of the averaged signal is smaller than that in the original signal by a factor $1/\sqrt{N}$. The signal component is the same in each experiment so that the average of the signal component is just equal to the signal component. The net effect is to increase the amplitude signal-to-noise ratio by a factor $\sqrt{N}$. (Note: the signal-to-noise ratio is usually based on some measure of the signal power or energy, for example

$$\left(\frac{S}{N}\right)_{power} = \frac{f_{rms}^2}{n_{rms}^2},$$

and the amplitude signal-to-noise ratio is the square root of this. Conventionally the ratio is stated in dB thus: $\left(\frac{S}{N}\right) dB = 10\log_{10}\frac{f_{rms}^2}{n_{rms}^2} = 20\log_{10}\frac{f_{rms}}{n_{rms}}$.

A similar noise reduction occurs if the noisy signal is averaged over a short time interval, T, around each time, t

$$\bar{x}(t) = \frac{1}{T}\int_{t-\frac{1}{2}T}^{t+\frac{1}{2}T} x(t')dt' = \bar{f}(t) + \bar{n}(t) \quad \text{(Continuous time signals)}$$

$$\bar{x}(n) = \frac{1}{2M+1}\sum_{n'=n-M}^{n+M} x(n') \quad \text{(Discrete time signals).}$$

The power in the noise component, $\bar{n}(t)$, of the time averaged signal, tends to reduce by a factor $\frac{\tau_c}{T}$ where τ_c is a measure of the lag time beyond which the autocorrelation function of the noise is negligible. The signal component, $\bar{f}(t)$, of the time average about time t is approximately equal to the value of the signal, $f(t)$, at that time, provided T is not too large, so that the amplitude signal-to-noise ratio increases by a factor $\sqrt{\frac{T}{\tau_c}}$. If the window duration, T, is too large, the error (called *bias*) introduced into the signal component by the average increases and eventually dominates, so there is a limit to the length, T, of the average beyond which the total error (noise plus bias) in the processed signal begins to rise (Figure 20.14).

Example:

Assuming a signal corrupted by additive noise such that the signal-to-noise ratio (SNR) is 20 dB, the number of trials required to increase the SNR to 40 dB can be found from:

$$\left(\frac{S}{N}\right)(dB) = 20\log_{10}\left(\frac{S}{N}\right)_{amp} = 20\log_{10}\left(\sqrt{N}\left(\frac{S}{N}\right)_0\right) = 10\log_{10}(N) + 20\log_{10}\left(\frac{S}{N}\right)_0$$

$$= 10\log_{10} N + \left(\frac{S}{N}\right)_0 (dB).$$

Thus, $40 = 10\log_{10} N + 20 \Rightarrow N = 100$.

If the noise had a correlation time of 1 ms, the length of a moving window average required to achieve the same improvement in SNR, ignoring the effect of bias, would be given by $T/\tau_c = N \Rightarrow T = N\tau_c = 100 \times 0.001 = 0.1$ s. If the signal component changed significantly over times of the order of 0.1 s then the bias term would dominate and

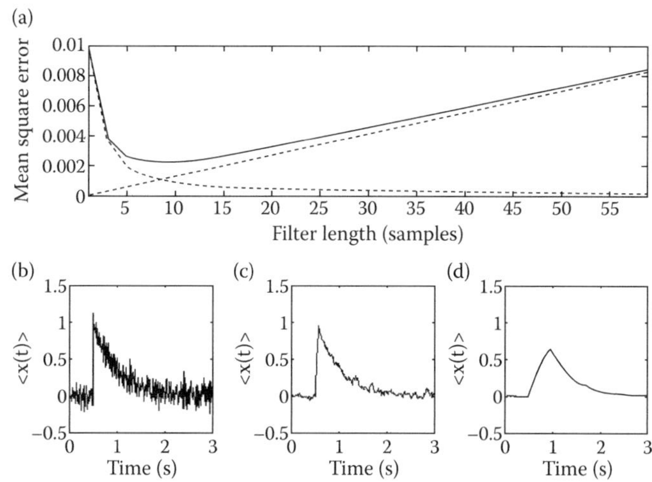

FIGURE 20.14
(a) Total mean square error (solid) in a moving window averaged signal as a function of window length. The dotted lines represent the separate contributions of the noise and bias. (b) Unsmoothed signal. (c) Smoothed signal with minimum mse. (d) Smoothed signal with bias error approximately equal to the noise error in the unsmoothed signal.

a moving window would probably not be able to achieve the required improvement without unacceptable signal distortion.

20.4.3 Moving Window Averages

The time averages introduced above are members of a wider class of processing schemes, called moving window averages, which introduce a weighting function, $w(t)$, into the average so that not all points in the window are treated equally:

$$\bar{x}(t) = \int_{-\infty}^{\infty} w(t'-t)x(t')dt' \quad \text{or} \quad \bar{x}(n) = \sum_{n'=-\infty}^{\infty} w(n'-n)x(n').$$

The area under the weighting function is usually 1, i.e. $\int_{-\infty}^{\infty} w(t)dt = 1$, and the window generally has *compact support*, so that, for example, $w(t) = 0$ for $|t| > \frac{1}{2}T$.

For example, the window functions corresponding to the simple averages of the last section are of the form

$$w(t) = P_{\frac{1}{2}T}(t) = \begin{cases} 1: & |t| \leq \frac{1}{2}T \\ 0: & |t| > \frac{1}{2}T \end{cases}.$$

The main reason for preferring a moving window with a shape other than the rectangular pulse above is that with an appropriate choice of window shape, for a given noise reduction the bias error can be smaller. A near optimal window in this respect is the parabolic (or Welch) window:

$$w(t) = \begin{cases} \frac{3}{2T}\left(1-\left(\frac{2t}{T}\right)^2\right); & |t| \leq \frac{1}{2}T \\ 0 \quad ; & |t| > \frac{1}{2}T \end{cases}.$$

A commonly used quick-and-dirty discrete time moving window average is the 1/4 – 1/2 – 1/4 filter for which

$$\bar{x}(n) = \frac{1}{4}x(n-1) + \frac{1}{2}x(n) + \frac{1}{4}x(n+1).$$

If more noise reduction is required the 1/4 – 1/2 – 1/4 filter can be applied repeatedly to the data.

Example:

To estimate the potential improvement in SNR from a single pass of a 1/4 – 1/2 – 1/4 filter, assume that the noise values at successive sample times are independent and identically distributed (iid) random variables (the discrete time equivalent of white noise). The noise variance in the filtered signal follows directly:

$$\mathrm{var}(\bar{\nu}) = \left[\left(\frac{1}{4}\right)^2 + \left(\frac{1}{2}\right)^2 + \left(\frac{1}{4}\right)^2\right]\mathrm{var}(\nu) = \frac{3}{8}\mathrm{var}(\nu).$$

Thus, the SNR of the filtered signal (ignoring bias) in dB is

$$\left(\frac{S}{N}\right)(dB) = 10\log_{10}\left(\frac{S}{\mathrm{var}(\bar{\nu})}\right) = 10\log_{10}\left(\frac{8}{3}\right) + 10\log_{10}\left(\frac{S}{\mathrm{var}(\nu)}\right)$$

$$= 4.26 + \left(\frac{S}{N}\right)_0 (dB).$$

Thus, the best possible improvement in SNR is around 4.26 dB.

In practice the noise may not be white and the effect of bias will not be entirely negligible so the figure just calculated is an optimistic upper limit on the improvement. Further passes produce smaller improvements, in effect because the earlier passes introduced correlations between successive samples which were not present in the original signal.

20.4.4 Filters

The moving window average is an example of a filter, i.e., a linear device or process that selectively alters the spectrum (or more correctly the Fourier transform) of a signal. To see this first define the time reversed window function $h(t) = w(-t)$, and then write the moving window in terms of $h(t)$ rather than $w(t)$

$$\bar{x}(t) = \int_{-\infty}^{\infty} h(t-t')x(t')\,dt' \quad \text{or} \quad \bar{x}(n) = \sum_{n=-\infty}^{\infty} h(n-n')x(n').$$

In this form, the expression can be recognized as a *convolution* and a theorem of Fourier analysis is that the Fourier transform of the convolution of two functions is the product of the Fourier transforms of the two functions:

$$\bar{X}(\omega) = H(\omega)X(\omega).$$

Thus, the Fourier transform of the averaged signal is just the transform of the original signal multiplied by the transform of the time reversed window function. The quantity $h(t)$ is the *impulse response function* of the filter (the response of the filter to a very short unit-area pulse), and the quantity $H(\omega)$ is called the *system function* of the filter. A complication is that $H(\omega)$ is a complex valued function of frequency, and it is usual to think of its operation in terms of the effects it has on the amplitude and phase of the transform of the input signal separately. Thus, write $H(\omega) = A(\omega)e^{j\phi(\omega)}$ where $A(\omega)$ is the *amplitude response* and $\phi(\omega)$ is the *phase response* of the filter. For example, the output of a filter in response to a sinusoidal input $f(t) = \sin(\omega t)$ of frequency ω is $A(\omega)\sin(\omega t + \phi(\omega))$. In other words, each Fourier component of a signal has its amplitude changed by a factor $A(\omega)$ and its phase shifted by an amount $\phi(\omega)$.

A filter which selectively removes the higher frequency components of a signal is called a *low-pass filter* and moving window averages are usually low-pass filters (Figure 20.15). This provides a different perspective on why moving window averages improve the signal-to-noise ratio of a signal, since the true signal component usually has a restricted range of frequencies on which the spectrum is concentrated, whereas noise signals typically have a broad spectral range. Removing the high frequencies selectively removes noise components without much effect on the signal components, thus the ratio of signal power to noise power is greater in the averaged signal.

Example:

A single-pole analog filter has a system function proportional to

$$H(\omega) = \frac{T}{1 + j\omega T} = \frac{T}{\sqrt{1+(\omega T)^2}}\exp\left(-j\,\mathrm{atan}(\omega T)\right).$$

The amplitude response is $A(\omega) = \dfrac{T}{\sqrt{1+(\omega T)^2}}$ and the phase response is $\phi(\omega) = -\mathrm{atan}(\omega T)$.

Clearly, the high frequencies are attenuated more strongly than the low frequencies hence the system is a low pass filter.

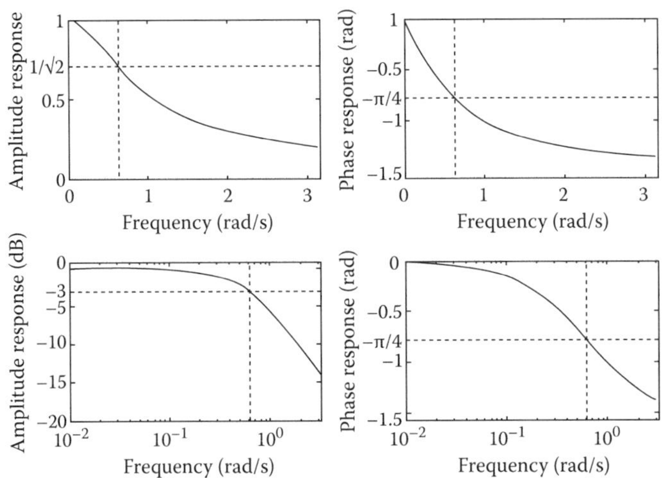

FIGURE 20.15
Amplitude and phase responses for a single-pole low-pass filter. The pass-band edge frequency (where the attenuation falls is 3 dB) is set at 0.1 Hz. The upper panels show the responses on linear axes, the lower panels show Bode-plots in which the amplitude response in dB is plotted with a log frequency axis. The latter is the more common display for filter responses.

20.4.5 Finite Impulse Response Filters

One of the advantages of discrete signal processing is that it is much easier to design and implement filters that approximate ideal behavior than in analog processing. It can be shown that the system function of a discrete filter is essentially the DTFT of the impulse response sequence, so the design problem is to choose the correct values for $h(n)$ to produce the desired system function.

The easiest discrete filters to understand are those that have only a finite number of nonzero terms in their impulse response—called *finite impulse response* or *FIR* filters. The time-domain response of such a filter to an input sequence $x(n)$ is most conveniently written in the form

$$\bar{x}(n) = \sum_{n'=N_1}^{N_2} h(n')x(n-n')$$

where $N_2 - N_1 < \infty$. Provided $N = N_2 - N_1 + 1$ is not too large, the filter can also be realized in this form, i.e., the program that performs the discrete filter simply calculates this expression repeatedly, once for each output value required.

For a filter to operate in real time the output at the nth sample time cannot depend on values of the input at future times, so $h(n') = 0$ for $n' < 0$. Filters which satisfy this condition are said to be *causal*. In practice ideal filters (such as an ideal low-pass filter) can only be realized as noncausal systems, but they can be approximated by causal filters with an appropriate time delay between input and output. A time delay, τ, between input and output transforms to a change in the phase of the signal by an amount $\Delta\phi = -\omega\tau$. A filter whose entire phase response is like this is called a *linear phase filter* and is about a close to the ideal filters as one can get. It can be shown that it is possible to approximate most amplitude responses with linear phase FIR filters (Figure 20.16). This accounts to a large extent for the popularity of FIR filters in practice. The downside is that the length of the impulse response may need to be quite long to realize the desired amplitude response.

Example:

An ideal (zero-phase) low-pass filter has the system function:

$$H(\omega) = P_{\Omega}(\omega).$$

The inverse DTFT of this yields the appropriate impulse response sequence

$$h(n) = \frac{1}{2\pi}\int_{-\pi}^{\pi} P_{\Omega}(\omega)e^{j\omega n}\,d\omega = \frac{1}{2\pi}\int_{-\Omega}^{\Omega} e^{j\omega n}\,d\omega = \frac{1}{2\pi jn}\left[e^{j\Omega n} - e^{-j\Omega n}\right] = \frac{\Omega}{\pi}\operatorname{sinc}(\Omega n).$$

This is noncausal and clearly not FIR. One way to generate a causal linear-phase FIR filter with approximately the same amplitude response is to truncate the sequence symmetrically and add a delay to ensure $h(n) = 0$ for $n < 0$:

$$h_{FIR}(n) = P_N(n-N)h(n-N) = \begin{cases} \dfrac{\Omega}{\pi}\operatorname{sinc}\big(\Omega(n-N)\big); & n = 0,1,2,\cdots,2N. \\ 0 & ; \text{ otherwise.} \end{cases}$$

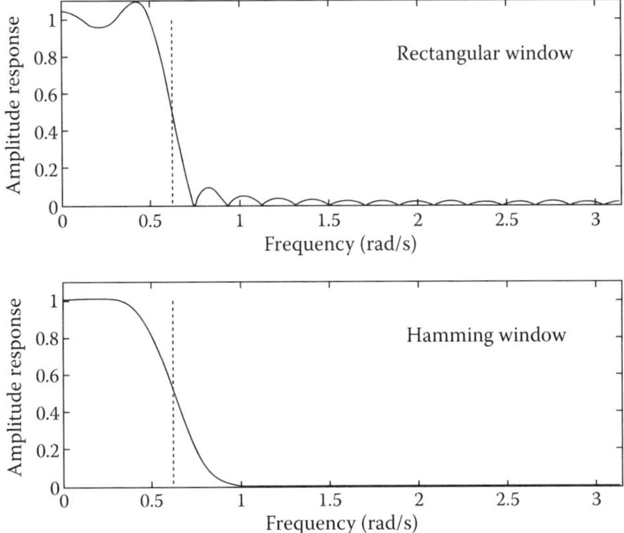

FIGURE 20.16
Amplitude response for FIR Type I linear phase low-pass filters of length 33 samples. The upper panel shows the amplitude response when the ideal impulse response sequence is simply truncated (or equivalently windowed with a rectangular window). The lower panel shows the amplitude response when the sequence is truncated using a Hamming window. The superior pass-band and stop-band performance has come at the expense of a broader transition band.

This is an example of a Type I linear-phase filter. Unfortunately, the amplitude response corresponding to the truncated filter can show relatively large oscillations in the vicinity of the pass-band edge, i.e., near Ω, (this is called Gibb's phenomenon—a common problem when a signal is truncated rapidly). Also, the stop band attenuation is adversely affected by the rapid truncation. To overcome this, at the expense of a broader transition band, a window with a more gradual transition to zero may be used, for example, $h_{FIR}(n) = w(n - N)h(n - N)$ where $w(n)$ is a Hamming window

$$w(n) = P_N(n)[0.54 + 0.46\cos(\pi n/N)].$$

20.4.6 Infinite Impulse Response Filters

If the number of terms in the impulse response function is infinite the filter is referred to as an *infinite impulse response* filter. It might be thought that it is impossible to implement such a filter in practice, but it is possible if one allows feedback from the filter output. Such designs are said to be *recursive*.

The main advantage of IIR filters is that they usually require far fewer operations (additions and multiplications) for each output than an equivalent FIR filter. A disadvantage is that it is much harder to control the phase response of the filter. Also, FIR filters are stable by design (do not produce unbounded outputs for bounded inputs) and relatively insensitive to small errors in the coefficients and calculations whereas IIR filters may be unstable and their simplest implementations are very sensitive to coefficient and calculation errors.

Example:

Suppose at time n that the input to a filter is $x(n)$ and the output is $y(n)$.

Consider the filter generated by adding a fraction, α, of the output at time $n - 1$ to a fraction $(1 - \alpha)$ of the input at time n to form the output at time n: Thus,

$$y(n) = \alpha y(n - 1) + (1 - \alpha)x(n).$$

Repeated substitution for $y(n - k)$ on the RHS yields:

$$\begin{aligned} y(n) &= \alpha\left[\alpha y(n-2)+(1-\alpha)x(n-1)\right]+(1-\alpha)x(n) \\ &= \alpha^2 y(n-2)+(1-\alpha)\left[x(n)+\alpha x(n-1)\right] \\ &= \alpha^3 y(n-3)+(1-\alpha)\left[x(n)+\alpha x(n-1)+\alpha^2 x(n-2)\right] \\ &= \alpha^{k+1} y(n-k-1)+(1-\alpha)\sum_{r=0}^{k}\alpha^r x(n-r). \end{aligned}$$

Continuing to $k = \infty$ and assuming that $|\alpha| < 1$ so that $\lim_{k\to\infty}\alpha^k = 0$, yields $y(n) = (1-\alpha)\sum_{r=0}^{\infty}\alpha^r x(n-r)$.

Thus, the impulse response sequence for this system is

$$h(n) = \begin{cases} 0 & ; \quad n < 0 \\ (1-\alpha)\alpha^n ; & n \geq 0 \end{cases}$$

which is clearly infinite in length. Thus, due to the recursion, the finite system realizes an IIR filter.

The system function for this filter can be found by taking the DTFT of the impulse response:

$$H(\omega) = (1-\alpha)\sum_{n=0}^{\infty}\alpha^n e^{-j\omega n} = (1-\alpha)\sum_{n=0}^{\infty}\left(\alpha e^{-j\omega}\right)^n = \frac{1-\alpha}{1-\alpha e^{-j\omega}}.$$

This acts as a low-pass filter for band-limited signals as can bee seen from Figure 20.17.

20.4.7 Uses of Filters

Noise reduction using low-pass filters is only one use of filters, in a similar fashion notch filters with a sharp rejection band (or a set of such bands) can also be used to reduce mains-frequency interference. As seen earlier, analog filters are also essential in the sampling of continuous time signals and in the reconstruction or synthesis of continuous time signals from sampled values. A common use of digital (and analog) filters is to separate different components of a signal; for example, in communications several signals which occupy nonoverlapping parts of the frequency spectrum can be simultaneously sent over the same transmission line; filters play an important role in the separation of these signals at the receiver. Similarly, the EEG is often studied by splitting it into a set of nonoverlapping

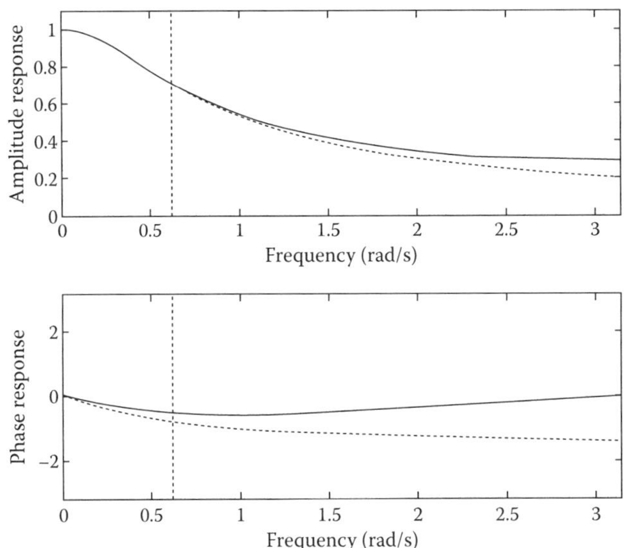

FIGURE 20.17
Amplitude and phase responses of a single pole IIR filter (solid lines) compared with the responses of an analog single pole filter with the same 3-dB point (dotted lines).

frequency bands, either directly using a digitally calculated spectrum or, for real time monitoring, by a bank of (typically digital) filters. As described in the next section, filters are also used in signal detection.

20.5 Signal Detection

Signal detection is the problem of detecting the presence and location in time of a known (or possibly unknown), usually deterministic waveform in a complicated combination of signals and noise. Measurements involving time delays in reflection (RADAR, SONAR, Ultrasound, etc.) are common examples in which detection of a signal of known form is important.

20.5.1 Matched Filters

The conceptually simplest approach to the problem of signal detection is to design a filter which responds strongly to the target signal but only weakly to other signals. A threshold detector on the output of such a filter can provide a reliable signal detector, and the time at which the output of the filter is largest can accurately locate the detected signal in time. The expression which produces the maximum signal-to-noise ratio for a signal of known form, $y(t)$, in a signal, $x(t)$, consisting of a delayed version of y corrupted with white noise, is the cross-correlation between the known signal form and the input signal:

$$R_{yx}(\tau) = \int_{-\infty}^{\infty} y(t)x(t+\tau)\,dt.$$

This can be shown to be equivalent to a filter with impulse response equal to the time reversed known signal:

$$R_{yx}(\tau) = \int_{-\infty}^{\infty} y(t)x(t+\tau)\,dt = \int_{-\infty}^{\infty} y(t'-\tau)x(t')\,dt' = \int_{-\infty}^{\infty} y(-(\tau-t'))x(t')\,dt' = h * x(\tau)$$

where $h(t) = y(-t)$.

This is an example of a *matched filter*. The expression is a little more complicated if the spectrum of the noise is not flat, but the basic principle is similar. Similar expressions hold for the case of discrete signals.

If $x(t) = y(t - T) + n(t)$ then, because the signal and noise are uncorrelated,

$$E[R_{xy}(\tau)] = \int_{-\infty}^{\infty} y(t'-\tau)y(t'-T)\,dt'$$

and it is not difficult to see that the maximum expected value occurs when $\tau = T$, i.e., when the lag in the autocorrelation matches the delay in the signal.

20.5.2 Phase-Sensitive Detector

A special case of a matched filter of great practical utility occurs when the expected signal is a harmonic signal of known frequency, as expected when a driving signal is harmonic in form. Adapting the expression for a matched filter to the case when the expected signal is finite power rather than finite energy, gives

$$R_{xy}(\tau) = \lim_{T\to\infty}\frac{1}{2T}\int_{-T}^{T} y(t-\tau)x(t)\,dt = \lim_{T\to\infty}\frac{1}{2T}\int_{-T}^{T} \cos\big(\Omega(t-\tau)\big)x(t)\,dt$$

$$= \cos(\Omega\tau)\lim_{T\to\infty}\frac{1}{2T}\int_{-T}^{T}\cos(\Omega t)x(t)\,dt + \sin(\Omega\tau)\lim_{T\to\infty}\frac{1}{2T}\int_{-T}^{T}\sin(\Omega t)x(t)\,dt.$$

The advantage of this form is that the full cross correlation for any lag can be synthesized from the two readily measured quantities

$$X = \lim_{T\to\infty}\frac{1}{2T}\int_{-T}^{T}\cos(\Omega t)x(t)\,dt \quad \text{and} \quad Y = \lim_{T\to\infty}\frac{1}{2T}\int_{-T}^{T}\sin(\Omega t)x(t)\,dt$$

$$R_{xy}(\tau) = X\cos(\Omega\tau) + Y\sin(\Omega\tau).$$

From this expression it can be seen that $R_{xy}(\tau)$ is the projection of the vector $\mathbf{R} = \hat{\mathbf{x}}X + \hat{\mathbf{y}}Y$ in the direction of the unit vector $\mathbf{n} = \hat{\mathbf{x}}\cos\theta + \hat{\mathbf{y}}\sin\theta$ where $\theta = \Omega\tau$. The maximum of $R_{xy}(\tau)$

will occur when the two vectors are parallel. Thus, the maximum occurs when $\Omega\tau$ = atan2(Y, X) and the maximum value is $R_{xy}(\tau_{max}) = \sqrt{X^2 + Y^2}$, so the delay (modulo the signal period) and the detected signal strength can be calculated from X and Y directly. This forms the basis of a *phase-sensitive detector.*

Problems

1. A periodic sequence has the property that there is an integer p (the period of the sequence) such that $x(n + p) = x(n)$ for all integers n. Show that the sequence obtained by sampling the signal $f(t) = \cos(2\pi t)$ sampled at intervals $T = 0.1$ s is a periodic sequence. What is the period of the sequence? If the sampling interval were $T = 1.1$ s would the sequence be periodic? If so what would be the period?
2. Generalize the results of question 1 to show that the sequence would be periodic if the sampling interval were any rational number (i.e., could be written as a ratio of two finite integers).
3. Find the Fourier series coefficients for the rectangular pulse train, $f(t) = \sum_{n=-\infty}^{\infty} P_\tau(t - nT)$, of pulse width 2τ and period T (assume $T > 2\tau$). Show that, for the case of a square wave ($T = 4\tau$), there are no even harmonics (i.e., show that $F_{2n} = 0$ for all nonzero integers n).
4. The Fourier transform has a number of general properties which can be derived from the defining equation. For example, the *shift property* states that, if $f(t)$ has a Fourier transform $F(\omega)$, then the shifted signal $f(t - T)$, where T is a constant, has a Fourier transform $e^{-j\omega T} F(\omega)$ (so a shift affects only the phase of the signal). In shorthand we would write $f(t) \leftrightarrow F(\omega) \Rightarrow f(t - T) \leftrightarrow e^{-j\omega T} F(\omega)$. This can be demonstrated by a simple change of variable in the definition as follows:

$$f(t-T) \leftrightarrow \int_{-\infty}^{\infty} f(t-T)e^{-j\omega t}\,dt = \int_{-\infty}^{\infty} f(t')e^{-j\omega(t'+T)}\,dt' = e^{-j\omega T}\int_{-\infty}^{\infty} f(t')e^{-j\omega t'}\,dt' = e^{-j\omega T}F(\omega)$$

In a similar fashion, demonstrate that the following important properties follow from the definitions: (assume all functions involved go to zero as $t \to \pm\infty$)

(i) *Scaling*: $f(t) \leftrightarrow F(\omega) \Rightarrow f(at) \leftrightarrow \frac{1}{|a|}F\left(\frac{\omega}{a}\right)$ for $a \neq 0$

(ii) *Duality*: $f(t) \leftrightarrow F(\omega) \Rightarrow F(t) \leftrightarrow 2\pi f(-\omega)$ (Hint: use the inverse transform.)

(iii) *Derivative*: $f(t) \leftrightarrow F(\omega) \Rightarrow \frac{df(t)}{dt} \leftrightarrow j\omega F(\omega)$ (Hint: integrate by parts.)

(iv) *Hermitian*: $f(t) \leftrightarrow F(\omega)$ and $f(t)$ real $\Rightarrow F(-\omega) = (F(\omega))^*$.

(v) *Convolution*: $f(t) \leftrightarrow F(\omega)$ and $g(t) \leftrightarrow G(\omega) \Rightarrow \int_{-\infty}^{\infty} f(t')g(t-t')\,dt' \leftrightarrow F(\omega)G(\omega)$

(Hint: Swap the order of integration and use the shift property.)

5. Use the duality property to show that the Fourier transform of the *sinc* function is given by: $\text{sinc}(\Omega t) \leftrightarrow \frac{\pi}{\Omega} P_\Omega(\omega)$.
6. A band-limited signal has a cutoff frequency of 200 Hz and is sampled at 380 Hz. What is the Nyquist rate for the signal? What is the Nyquist frequency for the sampling system? Explain why information is lost in sampling the signal. What cutoff would you recommend for an antialiasing filter in order to minimize the effect of aliasing on the spectrum of the sampled signal? If no antialiasing filter were used what would be the highest frequency at which the sampled signal could be used to estimate the spectrum of the signal?
7. The convolution of two signals $f(t)$ and $g(t)$ is a function of t defined by the expression $f * g(t) = \int_{-\infty}^{\infty} f(t')g(t-t')dt'$. Use the properties of the Fourier transform to show that, if $f(t)$ has a Fourier transform, then the convolution of $f(t)$ with a function $g(t) = \text{sinc}(\Omega t)$ is band-limited.
8. Reconstructing a signal from its samples is equivalent to removing all frequency components above the Nyquist rate from the aliased spectrum of the signal. A signal that has the Fourier transform $F(\omega) = P_{200\pi}(\omega)$, is sampled at 180 Hz and then reconstructed. Sketch the form of the aliased spectrum and of the form of the spectrum of the reconstructed signal.
9. The reconstruction step in Shannon's theorem results in the following expression for the reconstructed signal in terms of the sampled values:

$$f_R(t) = \sum_{n=-\infty}^{\infty} f(nT)\,\text{sinc}\left(\frac{\pi}{T}(t-nT)\right).$$

Verify that this signal is band-limited (even if the original signal sampled is not band-limited) and that it exactly reproduces the values of the original signal at the sampling times ($t = mT$; $m = \cdots -1,0,1,2,3,\cdots$).
10. An idealized stochastic signal can be constructed from a set of independent identically distributed random variables $\{x_n;\ n = \cdots,-1,0,1,2,3\cdots\}$ by treating them as samples of a random signal thus: $x(t) = \sum_{n=-\infty}^{\infty} x_n\,\text{sinc}\left(\frac{\pi}{T}(t-nT)\right)$. Assume that the samples have zero mean and that they are uncorrelated, so that $E[x_n] = 0$ and

$$E[x_n x_m] = \begin{cases} \sigma^2; & n = m \\ 0; & n \neq m \end{cases}.$$

(i) Show that the signal is band-limited. What is the cutoff frequency?
(ii) Show that $x(t)$ has zero mean for all time t. (Hint: $E\left[\sum a_n x_n\right] = \sum a_n E\left[x_n\right]$ for any set of random variables $\{x_n\}$ and constants $\{a_n\}$.)
(iii) Calculate the autocovariance function for the process and show that it depends only on the lag time $\tau = t_2 - t_1$.

(Hint: the identity, $\text{sinc}\left(\frac{\pi}{T}(t_2 - t_1)\right) = \sum_{n=-\infty}^{\infty} \text{sinc}\left(\frac{\pi}{T}(t_2 - nT)\right)\text{sinc}\left(\frac{\pi}{T}(t_1 - nT)\right)$, which follows from Shannon's theorem, will help.)

(iv) What is the rms amplitude of the signal?

11. A zero-mean, stationary, ergodic Gaussian random process has an rms amplitude of 10mV. Estimate the fraction of time for which the magnitude of the signal is greater than 50mV. (A table of the cumulative normal distribution function, or a program to calculate this function will be needed.)
12. The pdf of deterministic signals can be defined in the same way as the pdf of ergodic processes, i.e., p(x)dx = fraction of time the signal spends in the interval [x,x+dx). Use this definition to find the pdf for a triangular wave of unit amplitude:

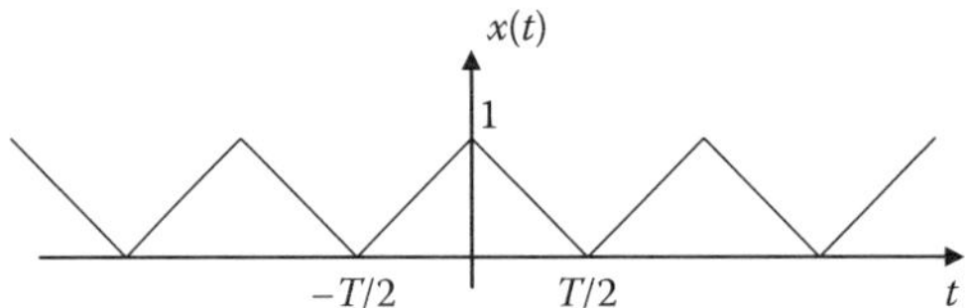

13. For a differentiable deterministic signal $x = f(t)$ show that the pdf as defined in problem 12 is given by $p(x) = \lim_{T\to\infty} \frac{1}{T} \sum_n \frac{1}{|f'(t_n)|}$ where $f'(t)$ is the derivative of f with respect to its argument and t_n are the solutions of $f(t_n) = x$ on the interval $-\frac{T}{2} \le t < \frac{T}{2}$.
14. Which of the following signals are finite energy? which are finite power? which are neither?

(i) $x(t) = P_1(t)$ (ii) $x(t) = \sin(t)$ (iii) $x(t) = \begin{cases} 0 \ ; & t < 0 \\ e^{-t}; & t \ge 0 \end{cases}$

(iv) $x(t) = |t|$ (v) $x(t) = \frac{1}{1+t^2}$ (vi) $x(t) = e^{-t^2}\cos(t)$

15. Find the DTFT of the two sided sequence $x(n) = a^{|n|}$ where a is a constant of magnitude less than 1.
16. If a band-limited signal is sampled faster than the Nyquist rate it can be written in terms its sampled values as given in problem 9. Use this expression to prove that the DTFT of a sampled band-limited signal is proportional to the Fourier transform of the continuous time signal. (Hint: Use the expression given together with the shift and duality properties of the Fourier transform.)
17. Use Parseval's relation to prove that $\int_{-\infty}^{\infty} \text{sinc}^2(t)\,dt = \pi$.
18. Show that the energy spectral density of a real finite energy signal is an even function of frequency, i.e., that $f(t) \leftrightarrow F(\omega)$ and $S(\omega) = \frac{1}{2\pi}|F(\omega)|^2 \Rightarrow S(-\omega) = S(\omega)$ (Hint: $|F(\omega)|^2 = F(\omega)(F(\omega))^*$, where * is the complex conjugate.)

19. Show that the energy spectral density (ESD) of a finite energy signal is the same as the ESD of the signal shifted in time by an arbitrary amount, i.e., that $ESD\left[f(t)\right]=S(\omega)$ and $ESD\left[f(t-T)\right]=\tilde{S}(\omega)\Rightarrow S(\omega)=\tilde{S}(\omega)$ (Hint: $|e^{j\theta}|=1$ for real θ.)
20. Show that the power spectral density of a sequence of zero-mean independent identically-distributed (iid) random variables is flat, i.e., is independent of frequency. (Hint: for zero mean iid variables $E\left[x_n\right]=0$, $E\left[x_n x_m\right]=0$ if $n\neq m$, and $E[x_n^2]=\sigma^2$ for all n.)
21. Band limited white noise, which has a flat spectrum up to a cutoff and zero spectral power beyond the cutoff, is often used as an approximation to nonwhite noise. The equivalent bandwidth of a noise signal is defined as the cutoff required for band-limited white noise to have the same spectral power density at 0 Hz and the same total power as the original noise. Show that the equivalent bandwidth for a noise with PSD $P(\omega)$ is given by $\Omega_{eq}=\int_0^\infty \frac{P(\omega)}{P(0)}d\omega$.
22. White noise passed through a first order low-pass filter with time constant τ, has the PSD $P(\omega)=\frac{P(0)}{1+(\omega\tau)^2}$. Find the Equivalent bandwidth for such a noise signal. (Hint: $\int\frac{1}{1+x^2}dx=\text{atan}(x)$.)
23. Find the DFT of the sequence [0, 1, 0, 0, 0, 0, 0, 1].
24. An EEG signal is sampled at 80 Hz and 12 bit quantization without prefiltering, if the signal were corrupted by mains frequency interference at 50 Hz, at what frequency would this appear in the spectrum calculated from the sampled data? Assume that the rms amplitude of the interfering signal is roughly equal to the rms amplitude of the EEG signal. What attenuation (in dB) would be required for an antialiasing filter to reduce the error due to the interference to levels comparable with that due to quantization?
25. A continuous Hamming window of length $2T$ has the form $w(t)=P_T(t)(0.54+0.46\cos(\pi t/T))$

    where $P_T(t)=\begin{cases}1; & |t|\leq T\\ 0; & |t|>T\end{cases}$ is the rectangular window of the same length.

    Find the Fourier transform of the Hamming window and for the case T = 0.5, plot its magnitude in dB as a function of frequency in hertz. Compare this plot with a similar plot for the rectangular window. (If you have access to MATLAB, the *wintool* command can be used to systematically study the properties of a range of windows.)
26. The signal-to-noise ratio for a single event related potential may be 0 dB. How many trials would have to be averaged to increase the ratio to 30 dB? Assuming that the noise in a single trial has a correlation time of 1 ms, estimate the length of a rectangular moving window needed to increase the SNR of a single trial by 30 dB? Given that the typical length of features in an event related potential are of the order of (say) 30 ms, is a moving window average practical in this situation? Explain.
27. Show that two passes of a $\left(\frac{1}{4}\ \frac{1}{2}\ \frac{1}{4}\right)$ moving window average is equivalent to a single pass with the window $\left(\frac{1}{16}\ \frac{1}{4}\ \frac{3}{8}\ \frac{1}{4}\ \frac{1}{16}\right)$. Hence, estimate

the improvement in SNR for two passes of the $\left(\frac{1}{4}\ \frac{1}{2}\ \frac{1}{4}\right)$ window. (Hint: write $y(n)=\frac{1}{4}x(n-1)+\frac{1}{2}x(n)+\frac{1}{4}x(n+1)$ for the output of the first pass and $z(n)=\frac{1}{4}y(n-1)+\frac{1}{2}y(n)+\frac{1}{4}y(n+1)$ for the output of the second; then substitute for y in terms of x and collect terms.)

28. A simple low-pass filter can be obtained by using the RC circuit shown:

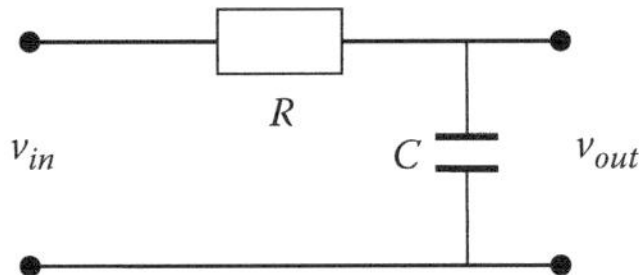

The relation between the input and output is $RC\frac{dv_{out}}{dt}+v_{out}=v_{in}$.
Find the system function and impulse response for this system. (Hint: the system function is the ratio of the Fourier transform of the output to that of the input: $H(\omega)=\frac{V_{out}(\omega)}{V_{in}(\omega)}$, and the impulse response is the inverse Fourier transform of the system function.)

29. Find the amplitude response of the $\left(\frac{1}{4}\ \frac{1}{2}\ \frac{1}{4}\right)$ filter as a function of digital frequency.

30. Show that the recursive implementation defined by $y(n)=y(n-1)+\frac{1}{N}(x(n)-x(n-N))$ is equivalent to the FIR filter $y(n)=\frac{1}{N}\sum_{r=0}^{N-1}x(n-r)$. (Hint: start with the FIR filter and calculate $y(n)-y(n-1)$.) What is the main advantage of the recursive implementation over the nonrecursive one?

21

Bioelectrical Signals: The Electroencephalogram

Joseph Ciorciari

CONTENTS

The field of electroencephalography (EEG) is a well researched and published field. Rather than discuss all aspects of the field, this chapter will endeavour to cover the main methodological issues associated with recording the EEG.

21.1 Introduction

Electroencephalography (EEG) is the study of the electrical activity associated with the brain. In humans, EEG has been a useful measure of cognition, attention, vigilance, and a way of quantifying states of consciousness such as sleep (Pfurtscheller and Lopes da Silva 1989, Regan 1989, Niedermeyer and Lopes da Silva 2004, Andreassi 2007). It has also been applied to the study of neurological disorders such as epilepsy (Hughes 2008) and psychiatric disorders such as the *psychoses*.* The author has also published EEG studies examining the effects of trauma (Cook et al. 2009), consciousness (White et al. 2009), dissociation and brain laterality (Ashworth et al. 2008) and personality (Stough et al. 2001).

It has been over 100 years since an English physiologist known as Richard Caton was the first to measure voltages on the surfaces of animal brains. In 1924, using a double-coil Siemens recording galvanometer, Dr. Hans Berger (1873–1941) made the first EEG recording in humans—the "electroencephalogram" (Berger 1924). He was first to describe the oscillatory activities; rhythms or waves, in both normal and abnormal brains. Berger's wave, also known as the alpha rhythm, was identified as a waveform which was suppressed when a human participant opened their eyes. This activity was also replaced by faster smaller waveforms known as beta waves (refer to Figure 21.1). Berger also expanded his work by also examining disease, and was possibly the first to record epileptic activity. For these reasons, Berger has been bestowed with the honour of being the "Father of Electroencephalography" (Regan 1989). The recording of EEG is considered to be a noninvasive process; being relatively expeditious to set-up; unlike the more invasive imaging techniques which also carry a significant *ionizing*† radiation burden. The EEG also has a superior temporal resolution but many argue that there is too much information and until the advent of the Fast Fourier Transform there was an inadequacy in techniques for analyzing all these data. EEGs are routinely done in patients simply to identify whether the disorder is due to an organic component; epilepsy, multiple sclerosis (MS), and other demyelinating diseases.

21.2 Origins of the EEG

This electrical activity is associated with the discharge of nerve cells in the cortex (synaptic excitation), principally the pyramidal cells and their associated dendrites (white matter).

* Abnormal mental state associated with brain dysfunction.
† Ultraviolet, X-rays, and gamma rays.

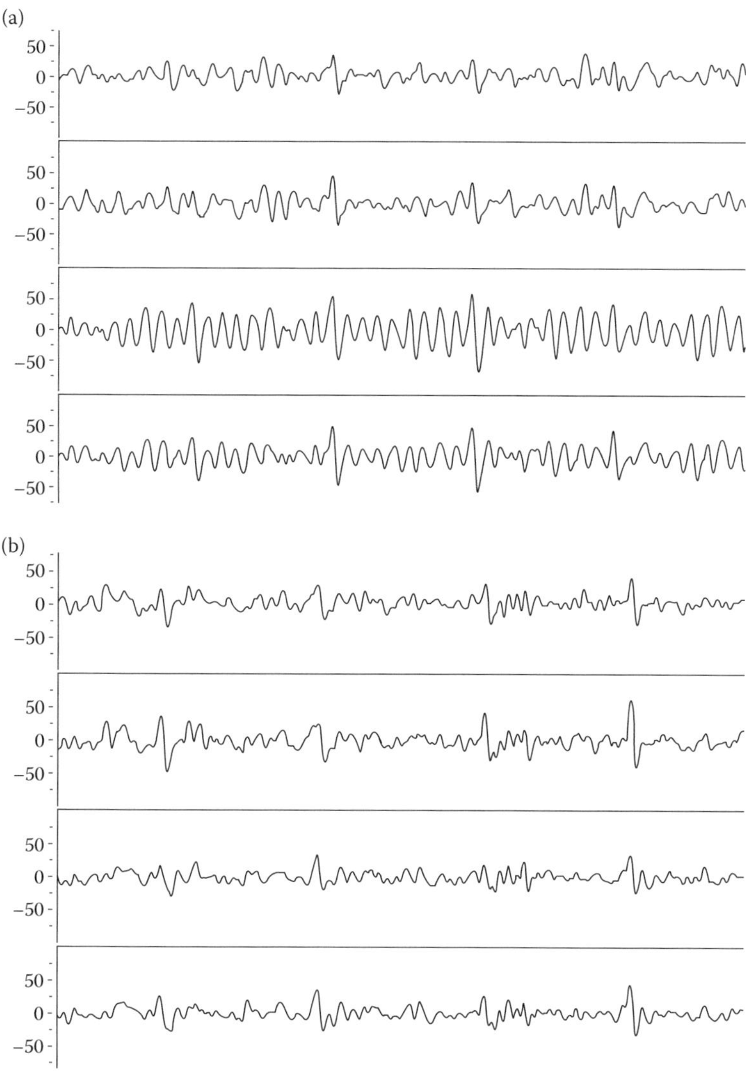

FIGURE 21.1
(a) Ongoing EEG (time varying voltage changes) recorded from four scalp electrodes at International 10/20 sites C3, C4, O1, O2 with participant with *eyes closed*. Note the dominant alpha wave in the last two traces (regular or synchronized activity). Units: Amplitude (μV) versus time (4 s). (b) Ongoing EEG recorded from four scalp electrodes at International 10/20 sites C3, C4, O1, O2 with participant with *eyes open*. Note the absence of synchronized activity with more irregular activity present. Units: Amplitude (μV) versus time (4 s).

These structures make up approximately 80% of the brain's mass. Ion current flows produced by thousands of these cells (see Figure 21.2a) are arranged perpendicular to the surface, allows the generated potential to migrate via volume conduction (current source smearing) to the scalp surface to be recorded as electrical activity (Nunez 1981). It follows the path of least resistance through the supporting tissue of the brain, skull, and scalp, where it can be recorded noninvasively from the scalp and is recorded as an electroencephalogram (Volume Conduction theory). On the scalp, the EEG has a voltage ranging

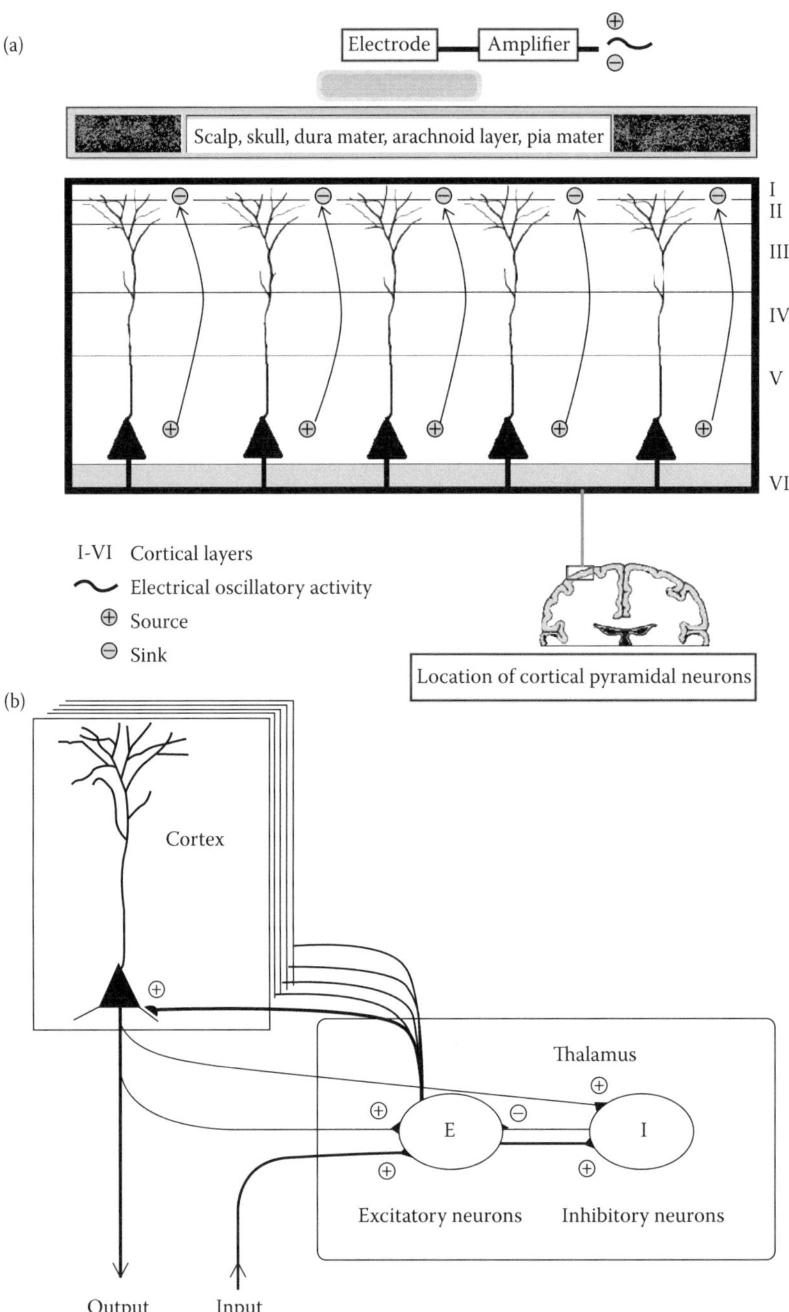

FIGURE 21.2
(a) Origin of scalp recorded electrical activity. Thousands of pyramidal cells contribute to the EEG signal recorded at the scalp electrode. This diagram simplistically illustrates how a very small section of the neocortex pyramidal cell population contributes to the EEG recorded from the scalp. The cortical layer V acts as a source while layer I acts as a sink. (b) Brain electrical activity is also driven by thalamic activity of the excitatory cells; an example of the pacemaker interpretation of rhythmic activity changes. E = excitatory, I = Inhibitory. The input (afferent) pathways come from other parts of the neocortex, and subcortical regions and the output (efferent pathways) project back to these areas.

from 1 μV to 200 μV whereas recorded from the *neocortex** using indwelling electrodes, the voltage range is 0.1 mV–1.5 mV (Gevins and Remond 1987).

Ongoing spontaneous brain electrical activity can be classified into two forms; irregular and synchronized. When pyramidal neurons fire de-synchronously this is represented as irregular electrical activity. But while the neurons fire synchronously this activity is represented as synchronized activity, characterized by large amplitudes and regular cycles. An example of this is waveform characteristic is when alpha appears as the dominant waveform while the participant's eyes are closed. Figure 21.1a clearly demonstrates the dominance of the alpha wave during a recording of a participant's EEG when their eyes are closed when compared to the "eyes open" EEG (Figure 21.1b). The alpha wave is more prominent when recorded from occipital sites. This synchronized activity can also be due to the thalamus excitatory cells driving the cortical pyramidal cells (see Figure 21.2b).

21.3 EEG Developments

Historically, a major development in electroencephalography was the announcement by Jasper to the World Health Organization the idea of a standard electrode locating and placement system. All EEG recording laboratories could now follow a standard system; the International 10/20 system (Jasper 1957). These electrode sites were selected on the basis of anatomical positions and brain lobes. These sites were labeled as O-occipital, F-frontal, P-parietal, T-Temporal and nonlobe determinations; C-Central. The 10/20 system utilizes measures (10% and 20%) from the *nasion*† to *inion*‡ and right and left *preauricular*§ sites to determine electrode sites, and interelectrode distances (see Figure 21.3).

However, the most important development in electroencephalography was the advent of the digital and computing age. Analyses of large quantities of data are now possible, extracting qualities of the EEG in both time and frequency domains and displaying them in near to real time. As an example, the Cooley and Tukey's Fourier transform algorithm could be utilized to calculate the energy or power associated with each of the various frequency bands or brain waves (See Table 21.1) with comparative ease, allowing better quantification of the energy or power associated with the rhythmic EEG activity. Other techniques can now be readily downloaded and implemented from internet based EEG resource sites (Delorme and Makeig 2004).

21.4 Oscillatory Activity

Brain oscillatory or spontaneous electrical activity has been categorized into several frequency bands which now include delta, theta, alpha, beta, and gamma (See Table 21.1 for detailed description). Activity is not limited to one or more bands at any singular time

* The outer layer of the cerebral hemispheres, and made up of six layers, labeled I to VI (refer to Figure 21.2a).

† Area directly between the eyes, just above the nose.

‡ Process of the occipital bone at the lower rear of the skull.

§ Location where top of the ear joins to the scalp.

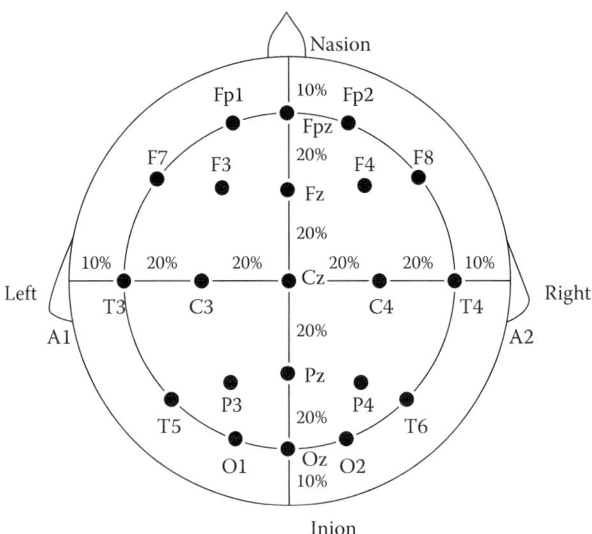

FIGURE 21.3
The 10–20 International Electrode System for determining positioning of electrodes on scalp (A1–2 *mastoid** reference sites). Note division of the left–right ear distance and nasion–inion distance into 10% and 20% segments. Circumferential spacing uses a combination of both distances for example P3 is 30% of nasion–inion distance from inion and 20% of left–right ear distance.* Bony prominence of the temporal bone (behind the ear).

period. In fact, the ongoing EEG may consist of proportions of all frequency bands and that the proportions may vary with changes in cognitive and sensory processes.

A few seconds of recorded ongoing EEG can be analyzed with a Fourier transform algorithm to reveal the energy or power associated with each bandwidth. Figure 21.4 illustrates the spectral EEG distribution and the relative amounts of power for each frequency band; total area in each bandwidth. Two measures can be made for each band, *absolute power*: which is the actual area under the curve within the band limits or *relative power* which is the normalized area under the curve (total power) divided by the total area (Total power density $\mu V^2/Hz$). EEG rhythms can be associated with cortico–cortico neuron interactions (emergent) or could also be driven by thalamic excitatory cells (pacemaker). This process is illustrated in Figure 21.2b.

TABLE 21.1
Normal Frequency Range of Ongoing EEG Subdivided into Five Bands (in Hertz, Cycles per Second)

Band	Symbol	Lower (Hz)	Upper (Hz)	Amplitude Range (μV)
Delta	(δ)	0.5	4	20–200
Theta	(θ)	5	7	20–100
Alpha	(α)	8	14	20–60
Beta	(β)	15	30	2–20
Gamma	(γ)	30	50[a]	5–10

Source: Hughes JR, *Epilepsy Behav* 13, 1, 25–31, 2008.
[a] 40 Hz is typically used, although some have reported that gamma can range up to 100 Hz.

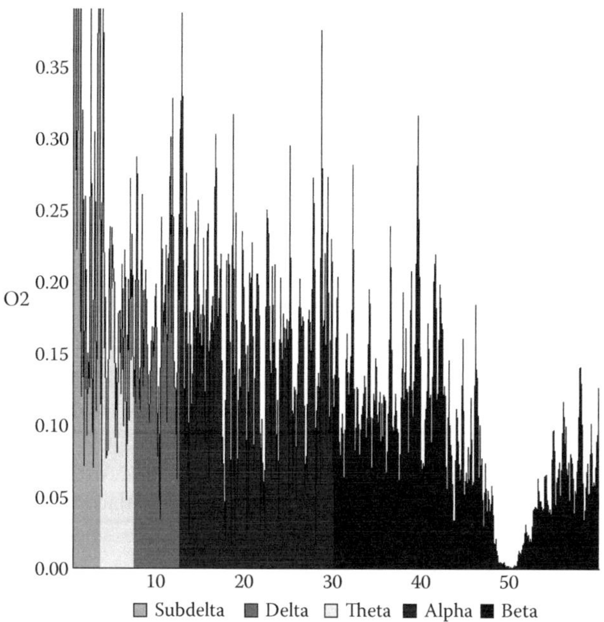

FIGURE 21.4
EEG power spectral data (of a 1-s epoch) with a 50-Hz notch filter. The unit are spectral power density (μV² per Hz) versus frequency range. Unlike a previous spectral graph in Chapter 26, this graph illustrates an unwindowed (No Hamming or Hanning filter applied) spectra with a more "spiky" distribution of frequency with a resolution of 0.004 Hz.

21.4.1 Spectral EEG

The features associated with each band vary according to location of recording electrode site, mental activity, age, gender, and level of consciousness. Five of the main bandwidths are examined here briefly. Quantitative characteristics of each of the five bands are summarized in Table 21.1.

21.4.1.1 Delta

This frequency demonstrates higher amplitude (up to 200 μV) and is the slowest of the waveforms (lowest frequency). It is generally readily seen during stages of sleep (in particular Stages 3 and 4) and drowsiness. Its presence may also indicate a subcortical or diffuse lesion if present as the dominant waveform during wakefulness in adults. However, it is the dominant waveform in infants under one year of age. Its distribution across the scalp will also vary depending on age; it is generally distributed frontally in adults and posteriorly in children. In many recordings, delta may actually be associated with eye movement artifact (blink, horizontal, and vertical movement). A diagram illustrating the contamination of the EEG with eye movement is seen in Figure 21.6b. In this example, rhythmic movements of eyes moving from left to right resembles delta waveform.

21.4.1.2 Theta

This rhythmic activity is also described as "slow" activity and is present in children and during states of drowsiness in younger adults. In adults it is less frequent but has been reported to occur during meditative states and memory consolidation. The amplitude for this waveform is generally reported to be around 100 μV. In brain pathologies where it is dominant, it is associated with generalized subcortical brain damage and cases of epilepsy. It is generated by the hippocampus or cortically. It is also associated with *sensorimotor** processes, learning and memory, and present during deep stages of sleep.

21.4.1.3 Alpha

The alpha wave was noted by Berger to be one of the dominant waveforms in adults, especially if they are relaxed. Subsequently, it has been shown to be associated with cognitive processes, consciousness and degree of vigilance. With eyes closed this dominant waveform is distributed bilaterally in posterior regions of the head. It decreases in amplitude and power when eyes are opened. It has also been reported to decrease in amplitude and power during tasks which require higher cognitive demand. This phenomenon is known as alpha desynchronization or *alpha blocking*. During this process, beta becomes the more dominant frequency. With motor or intended motor activity, blocking can also occur. This motor sensitive waveform is known as the *Mu rhythm* (8 Hz–13 Hz) and is distributed centrally and anteriorly.

21.4.1.4 Beta

This irregular smaller amplitude waveform is generally associated with higher cognitive processes and sensorimotor activity. In vigilant adults, it is distributed frontally and bilaterally. It is the dominant waveform when an individual is alert or anxious. It is also present while recording the EEG during a Rapid Eye Movement (REM) sleep cycle.

21.4.1.5 Gamma

Researchers have suggested that this high frequency waveform may be associated with synchronizing different parts of the brain: a network or *binding* frequency. This binding connects regions associated with processing information about a task at hand. Although, some studies have indicated that it may be predominantly electromyographic (EMG) contamination or muscle artifact.

21.5 Methodological Issues: Sources of Artifact in Studies of Cognition

According to the electrophysiological literature, it has been suggested that various methodological issues associated with recording the EEG (Gevins and Cutillo 1986, Maurer and Dierks 1991) may lead to misleading information or interpretations. Some are endogenous (artifact) and some exogenous (due to recording environment).

The EEG represents predominantly cortical activity with few possible subcortical events allowing recording. Unfortunately, it is also highly vulnerable to artifact. Artifact such as eye movement, head and neck muscle activity, speech, tongue movements, and other upper body muscle activity can contaminate the EEG, making interpretation of the record difficult. Figure 21.6b illustrates how artifact in the form of muscle activity (recorded as EMG)

* Neurological systems associated with coordinated motor and sensory function.

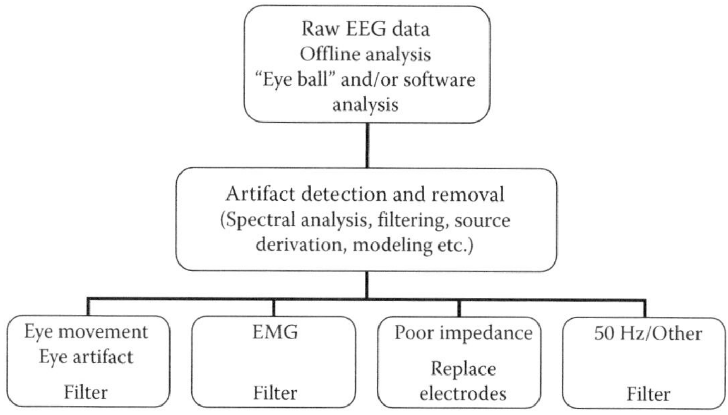

FIGURE 21.5
EEG can be analyzed by "eye balling" or visually inspecting the EEG trace to remove artifact contaminated epochs (time periods) of data. Preferably one can use a variety of signal analysis techniques so that artifact can be detected and removed. This flowchart illustrates what needs to be done to the EEG before it can be used to quantify cognitive and alertness correlates.

and eye movement (recorded as the electro-occulogram or EOG) can contaminate the EEG. When analyzing the EEG, the contribution of any potential artifact must be quantified. A number of commercial analyses systems will allow the user to identify and remove artifact contamination from the EEG. A number of signal analysis techniques (mathematical algorithms) can be employed to do so (See previous chapter). A suggested flowchart for an analysis method is illustrated in Figure 21.5.

Some of these principal artifact problems and contributions while recording the EEG are summarized in Table 21.2. The table also summarizes the remedies to improve the signal

TABLE 21.2
Endogenous and Exogenous Common Artifacts and Suggestions to Minimize or Remove

Sources of Artifact	EEG Bandwidth Affected	Location of Artifact	Possible Solutions
Endogenous			
Eye blink	Delta (δ)	Frontal	Rest participant
Eye movement—horizontal	Delta (δ)	Frontal	Provide a fixation point
Eye movement—vertical	Delta (δ)	Frontal	Provide a fixation point
Eyelid activity	Alpha (α)	Frontal	Rest subject, lower fixation point
Forehead muscle contraction due to stress or anxiety	Beta (β)	Frontal	Rest subject, lower fixation point
Jaw muscle movement or verbalization (or subverbalization during cognitive task)	Beta(β)	Temporal	Have subject open mouth slightly/stop talking
Neck strap muscle contraction (Depends on electrocap type)	Beta (β)	Occipital	Rest/reposition subject
Exogenous			
Poor electrode contact or impedance, worse if reference electrodes affected	Any	Any electrode	Reapply gel or change electrode/lead
50 Hz interference–power sources, lack of shielding	Any/all	Any/all electrodes	Shielding/grounding check

to noise ratio. Some of the other issues also include the use of a standard electrophysiological stable reference, variability in the 10/20 international electrode placement due to technician error if not using a set electrode cap, and possible instrumentation faults. Variations in electrophysiological recordings can also be attributed to by changes in emotional states (anxiety) performance, motivation and attention, underlying neurological conditions, and mental illness. If the EEG is to be used for the study of cognitive processes in normal individuals and those with mental illness, then these issues as well as the "state" of the subject requires addressing. The use of controls; their age, and education may also be confounding variables.

21.5.1 Controlling for the Endogenous Electrical Artifact

The degree of artifact that may be recorded may vary considerably if you are dealing with special patient types. In psychoses (particularly in schizophrenia), it is quite well documented, that the degree of muscle and eye movement artifact is considerably greater than in nonpsychotics and therefore does pose a substantial problem for recording EEGs. Instrumentation designed to filter the artifact, in particular eye movement and eye blink, cannot remove all contamination without removing wanted signals/data as well. Movement artifact may also be a contributing factor as some neurological patients can also exhibit uncontrollable body movement during the recording session. Perspiration could also be a problem especially if the test room is inadequately ventilated. This has the possibility of contributing to recording slow wave activity or delta. Therefore is it a possibility that if a patient feels anxious, this may contribute to an increase in delta because of artifact associated with perspiration?

To properly control for some of these artifacts, then it is imperative that they be recorded directly. Figure 21.6a, illustrates the electrode sites required for recording EOG, and EMG during an EEG recording. Electrodes placed around the neck, eyes, and ears are necessary for this purpose. Yet with the anxiety reported by researchers in mental health, the added burden of applying more electrodes would be undesirable.

21.5.2 Use of a Standard Electrophysiological Stable Reference

It is quite possible that not all published EEG studies recording from patient and healthy *cohorts** have used the same electrode arrangement or montage. More importantly, not all studies use the same reference system. Some have used linked ears, the tip of the nose, the forehead or the average reference (see Figure 21.6a). This makes comparisons between and across studies problematic. Each reference system has its own level of stability as well. The issue of using a stable reference is possibly more crucial to *topographic*† mapping because of the possibility of varying spatial distribution of brain electrical activity. It is also a very important consideration if comparisons across studies and similar testing paradigms are to be made.

For example, a number of researchers have examined EEG coherence in schizophrenia. Contradictory findings, however, have been reported across these studies, and it has been suggested that these inconsistencies were probably due to the use of different reference systems rather than the differences associated with the disorder itself. This reference

* Groups of participants with similar demographics.
† Surface distribution of features.

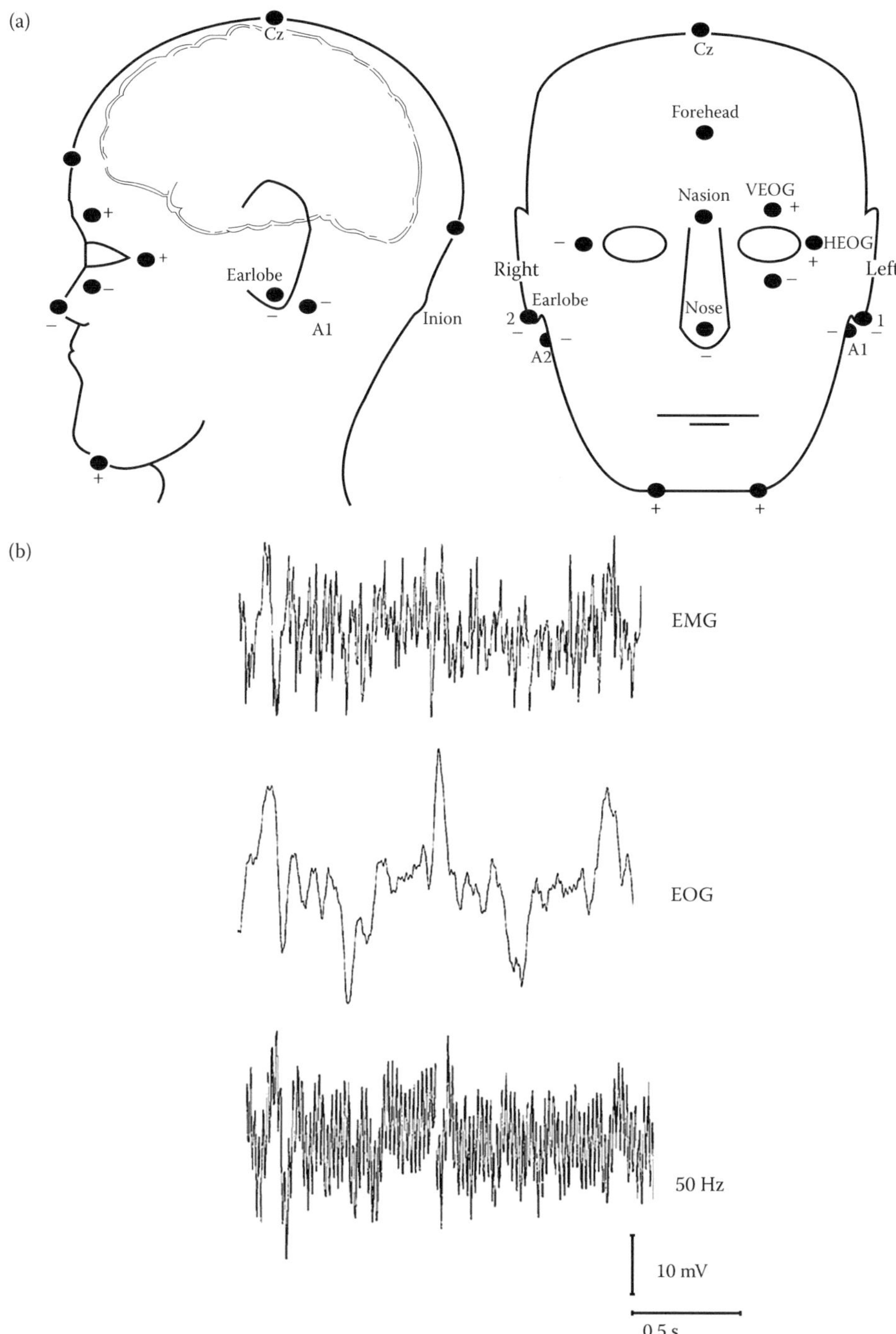

FIGURE 21.6
(a) Reference and ground electrodes for recording artifact; EMG and EOG during an EEG recording. Possible grounds include Cz, forehead, nasion, and nose. Common reference (inverting) sites include A1, A2, and earlobes 1 and 2 (linked). Electrode sites used to record neck/jaw EMG are shown. These can be reference to A1 and A2. The electrode sites for measuring are situated above and to the sides of the eye (canthi) and labeled VEOG (vertical) and HEOG (horizontal). (b) Examples of artifacts embedded in the electroencephalogram.

problem was also examined in a simulation study and subsequently demonstrated that a stable recorded reference system is the best technique to use.

It has been pointed out that not only the EEG coherence values or topographies can be altered by the reference, but the very wave shape, and amplitude of the recorded EEG signal can be distorted as well. In 1951, Stephenson and Gibbs published laboratory notes about a reference electrode system, which would not distort wave shapes or amplitudes, especially eliminating temporal sites distortions; the balanced noncephalic reference electrode. The problem with this reference system, though, is the increased cardiac artifact which must be monitored and balanced "out." In practice, it may be difficult to completely eliminate all electrocardiogram activity. Stephenson and Gibbs (1951) also demonstrated how a slow wave temporal focus can be distorted if an ear electrode is used as reference; especially the ear closest to the focus. By using a balanced noncephalic reference the focus is more clearly differentiated from other sites. Also, Nunez (1981) indicated that linked earlobes may make an asymmetric EEG feature appear symmetric.

EEG asymmetries which are generally related to psychopathologies may even be affected by cranial and brain parenchyma asymmetries. Electrode placement over inconspicuous cranial deformities (plagiocephaly) and *parenchymal** brain asymmetries can distort electrophysiological signals. Some studies have demonstrated that changes in alpha distribution in the parieto-occipital regions were due to slant cranial deformity, and fluid shunting due to fluid in the subgleal space or at dural levels, rather than cognitive related asymmetries.

21.5.3 Variability in the 10/20 International Electrode Placement due to Technician Error

Figure 21.3 is a representation of the standard electrode sites used by EEG technicians, as indicated by Jasper (1957). This system is referred to as the International 10/20 system because 10% and 20% of inion to nasion measures are used to site each electrode on the scalp. According to Maurer et al. (1991), it would take an experienced technician about 20 minutes to affix all these electrodes to a scalp. In some of the studies reviewed, as few as one active electrode to over twenty has been used to record the EEG from patients. Therefore one of the variations across studies would be the preparatory time each of the various groups has had to "wait" before actual recording. It has been shown in number of studies, that time could increase the anxiety felt by the patients and thereby affecting the recorded EEG.

With an increase in the number of electrodes used for recording, in the past few years, predominantly because of instrumentation and computer power available for topographic mapping, it has become more crucial that electrodes be placed correctly (equally low values of impedance) and sited accurately. Poor contacts (high impedances) can also affect the wave shape, amplitude, and artifact content. Good contact (low impedance) can be made by preparing the electrode site with abrading gel followed by wiping the surface with an alcohol wipe. Once dry, the electrode with conductive gel or paste can be placed on the prepared skin or scalp site.

* Tissue structure.

21.5.4 Possible Instrumentation Faults

Since each electrode is connected to a dedicated preamplifier, there is still the possibility of faults with the instrumentation, differences in gain, filtering; differing high cuts, low cuts, and dB roll-off, sampling rate variability, aliasing and so on, which can all affect the electrical signals to be analyzed (Maurer et al. 1991). We assume that all technicians and researchers show a degree of competency in locating faults, and exhibiting high degree of quality control for their equipment. Still, technical faults due to human error may still occur; say recording and measuring the power of beta activity with a sampling rate far below the Nyquist rate.

Commercial systems differ greatly in the number of electrodes, amplifier characteristics, hardware for detection and rejection of artifact and display techniques. Variability in the types of analysis of the ongoing data may also differ; some account for reference distortion, volume conduction while others do not. This difference though is more applicable to the physics of topographical mapping and will therefore be covered later. For these reasons, making comparisons across some studies may prove difficult but not impossible.

Researchers or technicians must discriminate between what is real biological signal and what is artifact and eliminate or minimize the artifact from the records. Next they may perform an analysis to determine the spectral band characteristics from the background EEG, the fast Fourier transform for converting data to spectral domain. The variability of the spectral characteristics are then assessed by visually examining real time spectral data to determine whether the parameters are intermittent or not (if 50 Hz is present or not). If no artifact is present in the spectral data, then these data can then be used to construct the spectral or wave shape distribution over the scalp (with topographic mapping). A statistical procedure can be utilized to then determine whether the effects are significant or not. Many commercially available analysis software packages have many of these analysis features. Comparing different EEG studies, at each one of these stages, may be problematic as there may be some variation in procedure making it difficult to identify common findings. Another source of confounding variables, are the participants themselves.

21.5.5 Variations in Electrophysiological Recordings Attributed to by Changes in Emotional States

The EEG is also subject to levels of alertness, consciousness, and cognitive activity. Therefore, this would be even more of an important issue if one was recording from patients with mental illness. The majority of schizophrenia patients have difficulty in maintaining attention; "incapable of holding the train of thought in the proper channel" (Bleuler 1951), and would therefore affect any state; or attempting to maintain some consistency across experiments and patients. The state of anxiety will also greatly increase the amount of artifact generated by the psychotic subject; a common fear is that of electrocution. In the paranoid patient this apprehension can be extremely overwhelming.

Drowsiness can also be a problem and may be difficult to exclude, especially if the recording period is long and perhaps a task being performed by the subject may be too simple, repetitive, and consequently boring. Most of the EEG studies reviewed require participants to relax with eyes closed for long periods of recording time. This may relax normal controls to the point of drowsiness but may make paranoid patients quite anxious. Some medications taken by patients will also affect their mood and thereby directing or indirectly their EEG characteristics. Alcohol can also have the same effect.

21.5.6 Possible Confounding Gender Influences

According to the neuropsychological literature, males and females perform differently on neuropsychological tasks; language and visuo-spatial skills, and more complex attentional tasks. This is an important finding; especially if we consider that the majority of EEG studies have participants performing some task while EEGs are recorded. This is usually done to investigate cognition and identify the associated electrophysiological correlates. There is evidence that visuo-spatial performance variability as well as other factors, are affected by the menstrual cycle effects, and cerebral asymmetry. Gender is certainly an issue in cognitive neuroscience because of the possible influence of sex endocrines on the development of the brain and the subsequent hemispheric specialization.

The electrophysiological evidence demonstrates definite differences between males and females while performing various neuropsychological-type tasks (Petsche, Rappelsberger, Pockberger 1989). Two features were noted by Petsche et al., a higher power in the beta range as well lower beta coherence. Interestingly, this finding has also been replicated in another study by Jonkman, Veldhuiszen and Poortvliet (1992) where sex differences in *Quantitative Electroencephalography* (QEEG) parameters of normal adults were noted; high beta power in females, parieto-occipital alpha power higher in men, temporal theta power higher in females. The authors concluded by stating, "only subjects of the same sex should be included in the group of interest and the reference group." Therefore it seems that gender differences in EEG may involve differing absolute power and topography.

In the light of this evidence it seems methodologically inappropriate to mix gender in an EEG and cognition study (Allison, Wood and Goff 1983). This issue is not only a concern for EEG studies but a wide range of studies involving other neuroimaging techniques such as regional cerebral blood flow, magnetoencephalography and magnetic resonance imaging.

21.5.7 Other Factors Associated with Neurological and Mental Illness

Another factor which may influential in confounding electrophysiological characteristics directly may involve the patient history itself. It may be difficult to group together patients with identical psychotic characteristics and background history for group averaging. Most studies will control for age, but may not account for the number of years on medication or the years since onset of the illness, or the severity or variability of the illness. Therefore, in a patient group, of a badly constructed study, we may have patients with duration of illness varying from say two months to ten years or more. The course of treatment may also not have been considered.

21.5.8 Use of Controls, Matching for Age and Education

Just as it is important to have a consistent baseline to compare EEG characteristics to, the type of control group for comparison is just as vital. Matching for age, level of education, gender, are all very important factors. It would not be methodologically sound to compare the EEG of a group of patients with an average age of 20 years with a control group of say 60 years of age. Age related changes in EEG characteristics are quite well documented. There is also evidence that education may also need to be matched between patient and control group. DeMyer and colleagues illustrated how structural changes in the brain using MRI were also influenced by education levels in patients with schizophrenia (DeMyer, Gilmor, Hendrie, DeMeyer, Augustyn and Jackson 1988). It was noted that the size and neuronal density of the frontal regions of schizophrenia brains were related to years of education. Could this poor performance be related to their disorder or perhaps to a lack of education and training due to the early disruption to normal education years by the onset of the illness? Studies of healthy cohorts have certainly demonstrated that the years of education have direct bearing on a range of cognitive problem solving skills.

21.5.9 The Resting State

The focus of most of the electrophysiological research has been to identify distinctive electrophysiological features which can aid in the identification of possible cognitive processes in participants with and without illness. Any condition which can increase the chance of a successful identification is of great value. The state in which the participant presents during testing is very important and should be controlled; at least to maintain uniformity across a control cohort. The EEG characteristics of a fully awake and vigilant participant, is quite different to that of a drowsy subject where increased slow wave activity may be recorded. In terms of identifying differences associated with cognitive processes, an active state is more reliable than the unreliable "rest" state or condition where an individual is asked to simply rest with eyes open or closed. There is evidence that a participant can be in any state when told by researchers to "think of nothing." Some may actually become more anxious, especially with their eyes closed. This "resting state" is regarded as the baseline state of which comparisons are made between cognitive states and noncognitive states. However, a well constructed paradigm can have a low demand active state to be the baseline task. Studies have demonstrated that as the task difficulty increases; in this case a visual task, the amount of parietal occipital alpha decreased and the power of left frontal theta increased in association with the amount of mental processing.

The use of a simple task "state" for a baseline state for comparison is regarded as a better controlled experimental procedure, minimizing for intersubject variability, and age related variability. In healthy or nonhealthy cohorts, the use of a low demand task as baseline, is a far better method to maintain uniformity (Gevins and Bressler 1989, Gruzelier et al. 1989).

A stimulus driven EEG for examining cortical function has been identified with the use of externally driven Evoked Potential (EP) techniques. Any modality (vision, auditory, somatosensory, chemosensory-taste, and motor can be stimulated to elicit a subcortical and cortical evoked response which can be recorded with EEG techniques. Some of those techniques and their associated methodological constraints will be briefly covered next.

21.6 Evoked Potentials

Evoked potentials are the responses recorded to some form of stimulation, i.e., evoked by a physical stimulus. They are typically recorded from the scalp, although they may be recorded from within the brain, or in the cortex. The response shows a time-locked relationship to the stimulus. There are different types of evoked potentials that can be recorded for different sensory systems (modalities), and different scalp locations. The optimal sites will depend on the different modalities. For example, the auditory evoked potential gives the largest results when recorded over the contralateral temporal region (see Figures 21.7 and 21.8). Somatosensory EPs are best recorded over the contralateral central region while Visual EPs are best recorded over the occipital region, closest to the visual cortex.

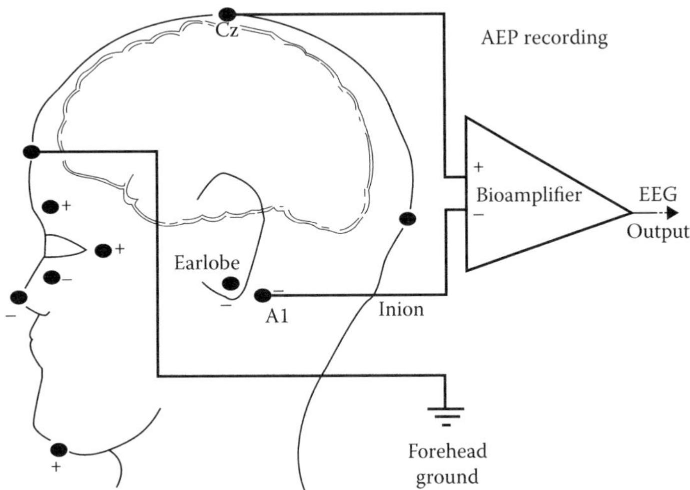

FIGURE 21.7
A typical recording setup to record the Auditory Brainstem Response (ABR) and Auditory Evoked Potentials (AEPs) or cortical responses. If the headphone presenting the "click"stimulus is one the same side as the recording electrodes then it is referred to as an ipsilateral recording. With the headphone on the opposite side then the cortical potentials are referred to as contralateral recordings. Note the electrode sites for monitoring eye movement, jaw, EMG, and alternate references are also shown for separately recording potential artifact.

21.6.1 Labeling Schemes

There are a number of conventions for labeling evoked potentials. *For example; the p*olarity: (positive/negative), roman numerals: (brainstem), peak labeling (time after stimulus) or peak order (P1-first peak, P2-second peak, P3-third peak). Figure 21.8, illustrates the differing types associated with the Auditory Evoked Potentials which are cortical in origin and early potentials which are brainstem in origin. Roman numerals have been used to

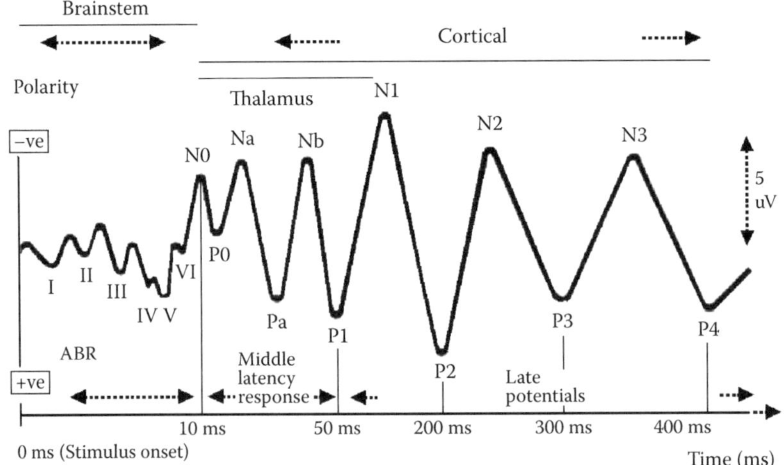

FIGURE 21.8
ABRs and AEPs on the same trace to demonstrate the timing and source of the potential. (Timeline not to scale so as to illustrate all potential waveforms in one diagram.)

label peaks and/or troughs of early brainstem evoked potentials, and two systems are shown in Figure 21.7. For the Auditory brainstem response (ABR) waveform, the positive components are labeled with the roman numerals I to VII. Their latencies are fixed and can therefore be used in clinical scenarios (such as multiple sclerosis) to determine delays in processing associated with the auditory components of brainstem sensory processing. For the peak and latency system, both the polarity of the peak (positive or negative), and the approximate latency in milliseconds denote a peak or trough, thus for example P2 becomes P200 in this system. Historically, the nomenclature used for evoked potential studies has generally used negative is up; the graph polarity is reversed (negative up and positive done). However, the student should be aware some laboratories have adopted the intuitively correct positive is up.

21.6.2 Recording Techniques

Single trials of evoked responses are very noisy; contaminated with other EEG and artefact interference. One method of reducing this interference or noise is to average together a number of individual responses or trial epochs to form an averaged evoked response. In the case of an ABR, at least one thousand trials (or much more-depending on participant condition) may be necessary. For cortical EPs, at least 200 trials are sufficient. Most averaging analysis is centered on the fact that the noise reduces with the square root of the number of averages (n). Therefore one would expect at least twice the improvement in signal to noise with 128 averages compared to 32 averages. It is assumed that the signal to noise will improve with the number of trials (epochs) associated with each stimulus event (n). This can be summarized by the following relation;

Evoked Potential Amplitude (n)/EEG amplitude (Square root n) = Gain

In other cases for example, EPs are extracted from single trials using other algorithms based on the *Hilbert's transform.** This improves the speed in acquiring the potential and removes the need to repeat trials so many times inconveniencing the participant or patient. Another technique useful for extracting the EP from noise is the use of the artificial neural networks method (a pattern recognition method). A number of paradigms are also available to elicit cognitive potentials, such as the oddball paradigm. These are discussed later in Section 6.5.

21.6.3 Early Potentials

The early potentials are those which occur in the period prior to about 100 ms or so. They may be very early (short latency) such as the auditory potentials prior to 10 ms which are known as the ABR; somatosensory prior to 20 ms (depending upon the distance from the brain the stimulus takes place); and may be prior to about 60 ms in the visual system, when the information reaches the brain.

(i) *Auditory Brainstem Response (ABR):* The integrity of the auditory pathways can be assessed and quantified with the ABR testing paradigm. By placing recording electrodes on sites illustrated by Figure 21.7, and getting the participant to listen to a thousand or so "clicks" or tones of set duration, the EEG is recorded for a period

* Hilbert transform is a linear operator which produces a function with the same domain.

time locked to the sound stimulus. Each epoch is then averaged to produce the ABR. The presence of any form of blockade (tumor) can be identified within either the mechanical (middle ear) or neural systems associated with hearing (inner ear and auditory neuroanatomical pathways to the brain). If all the potentials are delayed, then the problem is likely to be in the transduction, before the sound is converted to neural impulses. The peaks are generally associated with the processing at various structures along the auditory neural pathway to the cortex.

(ii) *Somatosensory:* There are a number of associated structures from which the early latency SSEP responses are believed to emanate before finally reaching the sensory cortex (20 ms). The individual peaks are due to synapses occurring in the various regions of the brainstem, and spinal cord. Median nerve response when stimulated at the wrist. This can be recorded at various points along the nerve path until it enters the spinal cord (at level C6), and from there propagates to the cortex.

(iii) *Visual:* The first visual signals which can be recorded are at about 35 ms when the VEP can be recorded in the visual cortex. The time taken for the VEP to reach the cortex is longer for the visual system as it passes through many more structures (it is more highly developed than the other senses), as well as traversing a number of parallel pathways from the retina to the cortex. Changes in the amplitudes in the VEP can be due to the physical properties of the actual stimulus; whether they are transient, sustained, or have differing spatial frequencies (such as a *checkerboard** stimulus). Intensity, duration, complexity, and position of stimulus can also affect the amplitude and latency of these early potentials.

21.6.4 Mid-Latency Potentials

The mid-latency potentials are generally considered 100 ms–200 ms and are mostly directly related to sensory processing. There are studies which show that changes in the amplitude and latency of these potentials may be affected by attention; larger amplitudes in some peaks. In a classic study, Hansen and Hillyard (1980) recorded event related potentials to an attentional task which involved selectively attending to one of three locations while fixating on a central location. Attention was maintained at a location by requiring the subject to find randomly occurring target stimuli (at that location) which differed from the ordinary stimuli. When examining the response in the occipital scalp (at recording sites 01 and 02), there was increased amplitude of the evoked potential components P135 and N185. This paradigm is very useful in studying attention in patients who have attention deficits.

21.6.5 Late Potentials

These potentials are in the 200 ms–300 ms (and later) latency, and are very susceptible to the effects of attention, novelty of the stimulus, and other cognitive aspects of the stimulus. Exogenous potentials are those elicited by external (or exogenous) stimuli. All the examples which we have seen up until now are exogenous evoked potentials. Endogenous potentials are those elicited by internal stimuli. A distinction between the terms, Event-related Potential (ERP) and Evoked Potential (EP) can be made. Event-related potentials are small phasic potentials elicited in conjunction with internal sensory, cognitive, and

* Alternating black and white checks as seen on a checkered flag.

motor decisions and manifest about 300 ms after stimulus presentation. They are generally known as cognitive potentials. Whereas, EPs are elicited by the physical characteristics of the visual, auditory or tactile stimuli are not cognitively related.

21.6.5.1 Cognitive Potentials (P300, N400)

The P300 is a positive peak of latency between 250 ms and 400 ms (see Figure 21.8), which appears if a change in an expected event occurs. The auditory oddball paradigm is a well documented protocol for eliciting this cognition related (event) waveform. The oddball can consist of two tones which are presented every second. One tone (low frequency tone or nontarget) may occur 85% of the time while the other (high frequency tone) appears less often (15%). The participant is requested to attend and count the high tone stimuli (target tone). The elicited responses to the high and low tones are averaged separately. The total number of tones presented should be around 200–300 to produce a satisfactory response for the "target" average and "nontarget" averages. The "target" response will produce a peak at about 300 ms (positive), and the nontargets show a very small peak (or none) at the same latency. This P300 is referred to as a cognitive potential. It is also affected by the state of the participant. The P300 can also be recorded in other modalities with a similar oddball design.

Another cognitive potential is the negativity N400. This has been referred to as a language potential because it can be elicited by listening to sentences with semantic incongruities. It is not elicited when the meaning of a sentence is congruous (Hink, Hillyard and Benson 1983).

21.7 Brain Electrical Activity Mapping: Methodological Considerations

> The brain produces an ever-changing, electric-magnetic field—not packs of line tracings. Therefore, the adequate way to deal with brain field data is to view them as series of maps (Lehmann 1990).

The use of computer processing power, the internet and open source codes has made it possible to process brain signals in an infinite number of ways, extracting useful diagnostic information. Mapping and analysis algorithms programs are available to the community of researchers from a variety of sites; one notably the EEG Lab website (http://sccn.ucsd.edu/eeglab/.) Mapping can be used to summarize vast amounts of data, displaying them as efficiently as possible so as to allow better opportunities for detecting distribution of activity, event-related changes (see Figure 21.9a and b) to abnormalities associated with clinical groups. Even though this field is extremely attractive to both researchers and clinicians because of the advantages over previous quantification of clinical EEGs there are some unique problems associated with topographical mapping of the EEG and EP. For example how many data sources or electrodes are necessary for constructing topographies of electrical activity? Are there differences in mapping EPs and EEG? What mapping routines are the best to use? Can varying reference electrode sites cause distortion of the topography? Does volume conduction alter topographies? Is the 10/20 System an adequate electrode placement system for mapping? These methodological issues will be discussed here briefly.

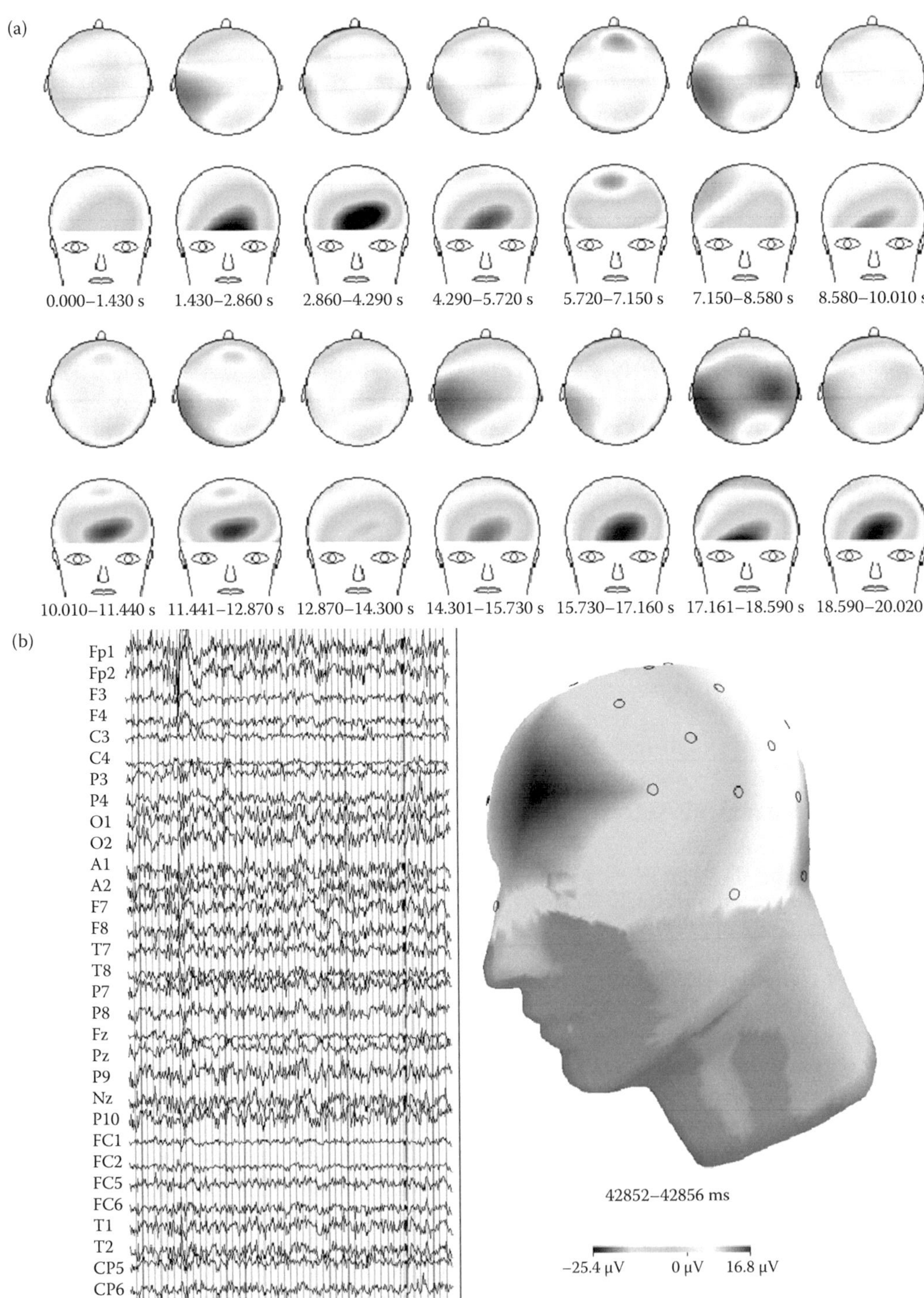

FIGURE 21.9
(a) Topographical mapping of 20 s ongoing EEG using spherical spline interpolation. Plotting second for second maps of activity can help with visualization of topographical interactions. (b) Topographical distribution of activity at one moment in the EEG. This can help visualize the distribution of activity relative to a key moment.

21.7.1 Number and Distribution of Electrodes

The number of electrodes needed for topographical mapping of brain electrical activity mainly related to the issue of spatial aliasing (related to the Nyquist sampling theorem); the relationship between the characteristics of the original signal to be recorded, and the position; or more importantly the interelectrode distance (Gevins and Bressler 1989). For maps of the topographic distribution of the Visually Evoked Response (VER), 2.5 centimeters separation of electrodes is required. For the somatosensory evoked potential (SEP) electrode distance should be kept to under 3 centimeters. The interelectrode distance is 7 centimeters with the international 10/20 system therefore well above the required amount for EEG topography. For most clinical recording of widespread signal fields or for clinical EP recording, the 10/20 is adequate. At least 19 electrodes are recommended for pharmacological-EEG correlation mapping (Herrmann 1989). However, this number would not detect transients such as spikes (Maurer et al. 1991). As early as the 1990s, some laboratories began using 124 electrodes, to achieve good signal to noise and have enough data points for interpolation of data for a map reconstruction of the distribution of the EEG. The expense of this technology was a reason why most research centres had not adopted the large multichannel or high density systems. Still based on the 10/20 system, new inter-electrodes were introduced to improve sampling sources for topography mapping. Figure 21.10b illustrates a typical augmented 10/20 electrode arrays used for mapping.

Localization of EEG sources can be hindered by a lack of the appropriate electrode numbers. It can also be affected by volume conduction effects of the scalp, skull, and

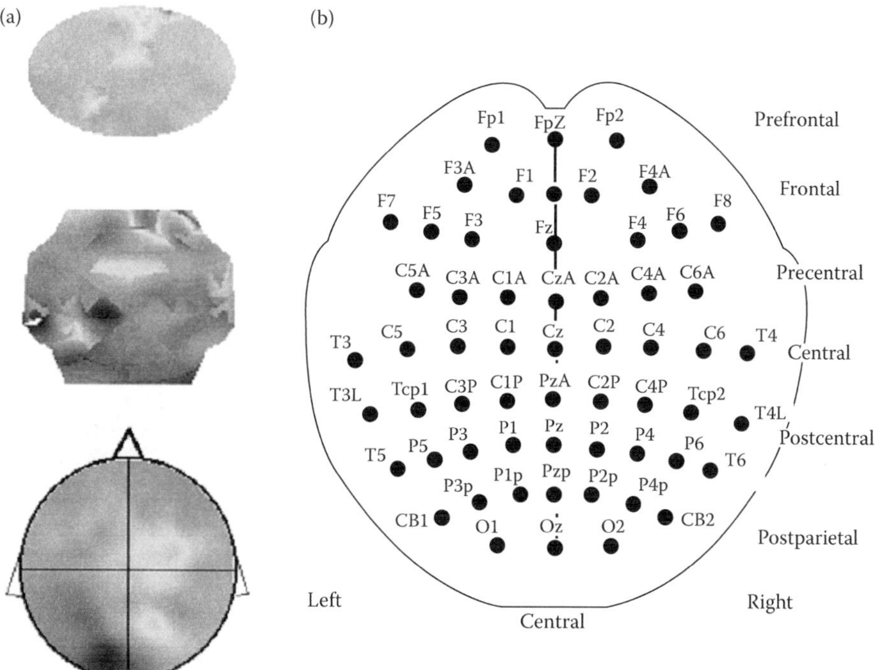

FIGURE 21.10
(a) Interpolation techniques: linear (top), quadratic spline (middle), laplacian spherical spline (bottom). (b) Additional electrodes based on 10/20 System.

cerebrospinal fluid surrounding the brain (Nunez 1981). It was suggested that localising any electrical event is quite a task perhaps erroneous with just a few electrodes and with simple mapping techniques (Fender 1987). However, combining *multidensity** electrode monitoring with various mathematical models for mapping, it has been possible to compensate for some of the volume conduction and localization problems. In particular, mathematical techniques which use the Laplacian spline for fitting topography, dipole estimation, deconvolution techniques and the estimation of spatial properties for finding sources have been applied to EEG topographical mapping (Lehmann and Michel 1990, Maurer and Dierks 1991).

The *Laplacian spline techniques*† have been applied to giving the topography a smoother appearance in the distribution of brain activity on the scalp; improve the *spatial resolution*.‡ Figure 21.10 illustrates the smoothing superiority of spherical spline Laplacian method in mapping spectral data. This technique has often been reported to be a spatial "deblurring" technique.

Source estimation techniques are based on knowledge of the volume conduction characteristics of the scalp, skull, cerebral spinal fluid, the vasculature (blood vessels), and the meninges. *Spatial deconvolution* algorithms offer to overcome the distortion of the EEG from volume conduction effects. These mathematical algorithms or models attempt to recreate and map the distribution of the sources as if no volume conduction effects have occurred. Another area of mathematics attempts to represent the generators of the EEG as fixed or moving *dipole*§ sources. Information pertaining to the various models and algorithms can be found at the EEGLAB website (http://sccn.ucsd.edu/eeglab/). Another functional imaging technique which involves a linear inverse solution which can model sources of EEG in three dimensions is known as *LORETA*¶ (low resolution electromagnetic tomography). Currently, one of the most utilized strategies for exploring the neural substrates associated with human cognitive processes combines LORETA with a spatial "deblurring" technique such as Laplacian spline to produce images comparable to fMRI images, but representative of electrical sources. *High density electrode arrays*** are required for this form of high resolution EEG.

21.7.2 Application of Various Topographical Mapping Algorithms and Techniques

According to various early reviews (Maurer et al. 1991), a variety of mapping algorithms were in use and so in an attempt to standardize the use of topographical mapping of the electrophysiology, the International Pharmaco-EEG Group made recommendations. These recommendations included guidelines associated with map formation; projection techniques, electrode numbers, pixel representations, equal *dimensionality*†† between maps and original scalp, and interpolation techniques incorporating at least three nearest electrodes (Herrmann 1989).

There are three features involved in selecting an appropriate projection geometry: (a) simplicity and ease in interpretation, (b) accuracy in visualization of EEG feature relative to scalp topography, and (c) the electrode positions as part of the geometrical model. One of the earliest projection models which incorporated these features was reported by Estrin and Uzgalis (1969). Electrodes were located by a grid system mapped onto a sphere

* High-resolution EEG.
† Spherical, ellipsoidal, and realistic geometry spline Laplacian algorithms have been used.
‡ Directly related to the number of electrodes available.
§ A closed circulation of electric current.
¶ http://www.uzh.ch/keyinst/loreta.htm.
** High number of equally spaced electrodes with a separation of under 2 cm.
†† Spatial 3D characteristics.

by secant cone projection. A number of projection systems have since been used; mercator projections, equal area projections, and ellipsoid model projections. Each has had some inherent problem or series of assumptions (Gevins and Bressler 1989). It is not the intention of this chapter to cover this in depth but only to make the reader aware of possible methodological concerns.

Interpolation schemes have also differed; ranging from simple linear three nearest neighbor electrodes inverse interpolation schemes, linear 4 nearest neighbors interpolation), to quadratic interpolation schemes; Chebyschev and cubic interpolation to those requiring ten nearest neighbor electrodes and smoothing of data using sophisticated spline interpolation techniques. Some examples are illustrated in Figure 21.10a, where an example of simple linear, quadratic (Ciorciari, Silberstein and Schier 1988), and Laplacian spline data (Cadusch, Breckon and Silberstein 1992) are presented. This illustration demonstrates, some "real" features of the EEG can be distorted or totally excluded by some of the simpler linear techniques. Another problem associated with interpolation schemes and choice of nearest neighbors for interpolation is edge distortion; a prominent distortion of the boundaries of maps, where a lack of suitable electrodes are most noticeable. Some researchers have used simple linear extrapolation techniques to "fill in the gaps"; inverse square law, which inherently accentuates high or low amplitude activity. Laplacian spline techniques have been used by some to overcome this problem (Cadusch et al. 1992, Nunez, Silberstein, Cadusch and Wijesinghe 1993).

21.7.3 Variation in Skull Shape and Thickness across Individuals and Volume Conduction

It cannot be assumed that the topographies of the EEG or EPs represent solely neuronal activity; instead it represents some volume conduction effect as well due to the skull, cerebrospinal fluid, and dura. Successive waves of depolarization and hyperpolarization over the whole scalp may simply be due to a focus of neuronal activity at some point in the cortex. Of course variations in skull and brain tissue all contribute to variations to volume conduction effects within individuals and groups. Even though this was mentioned in Section 5.3, the effects of skull thickness and tissue variations can especially affect the accuracy of mapping the EEG.

Cranial and brain parenchyma asymmetries may contribute to the inaccurate representation of the distribution of EEG and Evoked Potentials with mapping. This may lead to misinterpretations of the EEG. As early as 1981, solutions to this problem were sought by Ary and colleagues (Ary, Klein and Fender 1981). The authors proposed a model which took into account the variations in skull and scalp thicknesses. More recently, combining both neuroimaging (MRI) with EEG/EP mapping can control for this problem most satisfactorily. Another solution would be to use indwelling electrodes spaced evenly across the cortex. This solution, is not necessarily viable or available as an option for studies of normal participants. Techniques such as LORETA are able to improve on these issues.

21.7.4 Use of a Standard Electrophysiological Stable Reference

The issue of reference is one of the other important issues which can significantly affect the accuracy of mapping EEG data. As discussed previously, there are many types of reference systems, each with their own methodological issue which affects mapping. Some common reference systems are listed in Table 21.3 together with the problem they may introduce. Some of these problems involve the introduction of ghost field potentials, attenuation, artifact and interference. Figure 21.11 illustrates the effects of two referencing systems on the amplitude of a potential. The isopotential lines are the same but the zero lines shift. These isopotential lines may vary with linked earlobe reference system. The preferred reference will depend on a number of factors; the type of study, number of electrodes, and mapping techniques used (Rosenfeld 2002).

Questions

Multiple choice: select one of the best answers for each question.

1. Theta activity
   (i) Contains frequencies greater than 20 Hz
   (ii) Is best seen over the frontal region with eyes open
   (iii) Is associated with meditation
   (iv) Is observed during a REM sleep state
2. Relative EEG power spectra
   (i) Allows comparison between subjects
   (ii) Allows accurate measurements of activity
   (iii) Is the absolute total EEG
   (iv) Allows comparison between subjects in the range 0.1 Hz–45 Hz
3. Which is not a methodological issue associated with EEG mapping?
   (i) Number and distribution of electrodes
   (ii) Application of various topographical mapping algorithms and techniques
   (iii) Use of tin electrodes
   (iv) Variation in skull shape and thickness across individuals

TABLE 21.3

Reference Types and Artifact Characteristics Specifically Affecting the Mapping of the Distribution of Scalp Activity

Reference System	Problems in Mapping
1. Ear or mastoid (A1/A2)	• Electrically active
2. Linked ears or mastoids	• Electrically active • Attenuates amplitude • Potential difference effects which minimize asymmetries
3. Nasion, chin, forehead	• Electrically active
4. Common average reference	• Ghost potential fields • Distortion due to focal pathology
5. Local average reference (source derivation/Laplacian)	• Ghost potential fields • Equal distance electrodes needed • Extra interference
6. Balanced noncephalic	• ECG artifact and vertical dipole effects

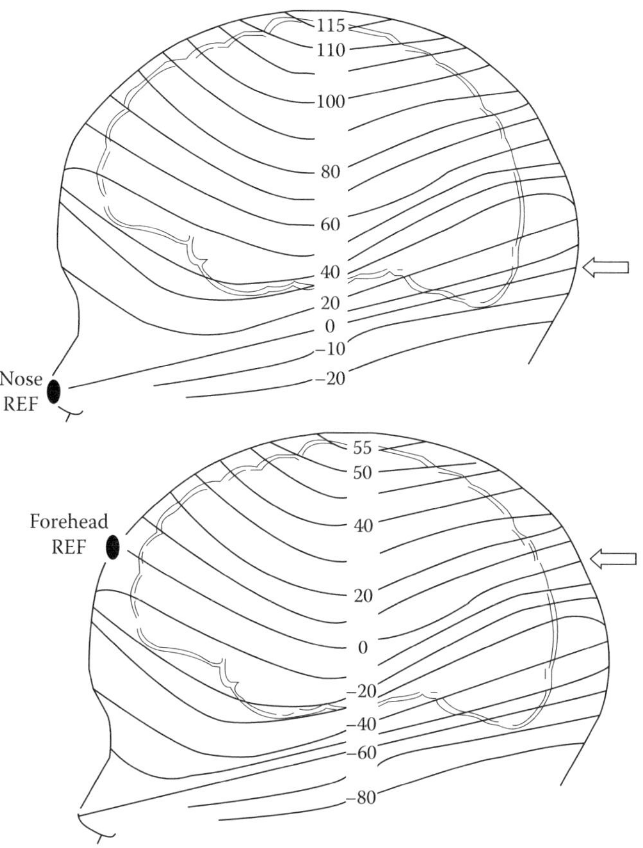

FIGURE 21.11
The topography amplitude variations when using different referencing; varying the reference point. (Top) Isopotential map of a nose reference. (Bottom) Using a forehead reference.

4. Which source of artifact in posterior EEG signals is a problem?
   (i) Neck muscle activity
   (ii) Heart activity
   (iii) Ocular activity
   (iv) Facial muscle activity
5. Alpha activity
   (i) Contains frequencies from 10 Hz to 20 Hz inclusive
   (ii) Is best seen over the occipital region with eyes open
   (iii) Cancels all delta activity
   (iv) Is best seen over the occipital region with eyes closed
6. Which source of artifact in EEG signals is rarely seen?
   (i) Muscle activity
   (ii) Heart activity
   (iii) Ocular activity
   (iv) Myographic activity

7. Electrodes located over the somatosensory cortex would be
   (i) C3, C4, T5, T6
   (ii) C3, P3, T5, O2
   (iii) O1, O2, Oz, Pz
   (iv) F3, F8, F4, F7
8. An example of a "monopolar" recording is
   (i) F3-A1A2, F4-A1A2
   (ii) F3-Fz, Fz-F4
   (iii) O1-P3, O2-P4
   (iv) Cz-Fz, Fz-Fpz
9. Alpha activity
   (i) Contains frequencies from 10 Hz to 20 Hz inclusive
   (ii) Is best seen over the occipital region with eyes open
   (iii) Cancels all delta activity
   (iv) Is best seen over the occipital region with eyes closed

Questions to Research

1. Briefly discuss the common forms of endogenous and exogenous artifacts associated with recording the EEG. What can be done to minimize their effects?
2. What are the benefits of using EEG for studying human brain function?
3. Discuss the advantages of computer based data acquisition systems as against amplifier/chart recording systems.
4. Discuss how electrophysiological techniques may be useful in the diagnosis of neurodegenerative disorders. When are they not useful? Your discussion should also make reference to some clinical examples.
5. Explain the main differences between emergent and pacemaker accounts of the genesis of rhythmic activity in the electroencephalogram (EEG).

References

Allison T, Wood CC and Goff WR. 1983. Brainstem auditory, visual, and short-latency somatosensory evoked potentials: Latencies in relation to age, sex, and brain and body size. *EEG and Clinical Neurophysiology* 55:619–636.

Andreassi JL. 2007. *Psychophysiology: Human Behaviour and Physiological Response*, 5th Edition. Lawrence Erlbaum, Hillsdale, NJ.

Ary JP, Klein SA and Fender DH. 1981. Location of sources of scalp evoked potentials: Corrections for skull and scalp thicknesses. *IEEE Trans Biomed Eng* 28:447–452.

Ashworth J, Ciorciari J and Stough C. 2008. Psychophysiological correlates of dissociation, handedness and hemispheric lateralization. *J Nervous Mental Dis* 196(5): 411–416.

Berger H. Über das elektroenkephalogramm des menschen. 1924. *Archiv Psychiatrie Nervenkrankheiten* 87: 521–570.

Bleuler E. 1951. *Textbook of Psychiatry*. (Translation J. Zinkin), Dover, New York.

Cadusch PJ, Breckon W and Silberstein RB. 1992. Spherical splines and the interpolation, deblurring and transformation of topographic EEG data. Pan Pacific Workshop on Brain Electric and Magnetic Topography, Melbourne, Australia.

Ciorciari J, Silberstein RB and Schier MA. 1988. A high resolution mapping technique for brain electrical activity. *Neurosci Lett Suppl* 30:S146.

Cook F, Ciorciari J, Varker T and Devilly G. 2009. Changes in long term neural connectivity following psychological trauma. *Clin Neurophysiol* 120:390–314.

Delorme A and Makeig S. 2004. EEGLAB: an open source toolbox for analysis of single-trial EEG dynamics. *J Neurosci Methods* 134:9–21.

DeMyer MK, Gilmor RL, Hendrie HC, DeMeyer WE, Augustyn GT and Jackson RK. 1988. Magnetic resonance brain images in schizophrenic and normal subjects: Influence of diagnosis and education. *Schizophr Res* 14(1):21–32.

Estrin T and Uzgalis R. 1969. Computerized display of spatio-temoral EEG patterns. *IEEE Trans Biomed Eng* BME-16(3):192–196.

Fender DH. 1987. Source localization of brain electrical activity. In: *Methods of Analysis of Brain Electrical and Magnetic Signals*. Gevins AS and Remond A (eds). Elsevier, 355–403.

Gevins AS and Cutillo BA. 1986. Signals of cognition. In *Clinical Applications of Computer Analysis of EEG and other Neurophysiological Signals. Handbook of Electroencephalography and Clinical Neurophysiology: New*. FH Lopes da Silva, W Storm van Leeuwen and A Remond (eds). Series V2, Elsevier, Amsterdam, 335–384.

Gevins AS and Bressler SL. 1989. Functional topography of the human brain. In: *Functional Brain Imaging*. G Pfurtscheller and FH Lopes da Silva (eds). Hans Huber Publishers, Toronto, 149–160.

Gevins AS and Remond A. 1987. Methods of analysis of brain electrical and magnetic signals. In: *EEG Handbook* (revised series Volume 1.). AS Gevins and A Remond (eds). Elsevier.

Gruzelier J, Seymour K and Wilson L. 1989. Topographical mapping of electrocortical activity in schizophrenia during nonfocussed attention recognition memory, and motor programming. In: *Functional Brain Imaging*. G Pfurtscheller and FH Lopes da Silva (eds). Huns Huber Publishers, Toronto.

Hansen JC and Hillyard S. 1980. Endogenous brain potentials associated with selective auditory attention. *EEG Clin Neurophatol* 49:217–290.

Herrmann WM (Chairman) International Pharmaco-EEG Group (IPEG). 1989. Recommendations for EEG and evoked potential mapping April 5 1990. *Neuropyschobiology* 22:170–176.

Hink RF, Hillyard SA and Benson PJ. 1978. *Biological Psychology* 6:1–16.

Hughes JR. 2008. Gamma, fast, and ultrafast waves of the brain: their relationships with epilepsy and behavior. *Epilepsy Behav* 13(1):25–31.

Jasper H. 1957. Report of committee on methods of clinical exam in EEG. The Ten Twenty Electrode System of the International Federation. Proceedings of the General Assembly held on the occasion of the IVth International EEG Congress. July 24, 1957, 371–375.

Jonkman EJ, Veldhuiszen RJ and Poortvliet DCJ. 1992. Sex differences in QEEG parameters of normal adults. X.5 p.65 ISBET Conference Proceedings PS6. Amsterdam, 31.

Lehmann D. 1990. Past, present and future of topographic mapping. *Brain Topogr* 3(1):191–202.

Lehmann D and Michel CM. 1990. Intracerebral dipole source localization for FFT power maps. *EEG Clin Neurophysi* 76:271–276.

Maurer K and Dierks T. 1991. *Atlas on Brain Mapping: Topographic Mapping of EEG and Evoked Potentials*. Springer-Verlag, Berlin.

Niedermeyer E and Lopes da Silva F. 2004. *Electroencephalography: Basic Principles, Clinical Applications, and Related Fields*. Lippincott, Williams and Wilkins.

Nunez PL (ed). 1981. *Electric Fields of the Brain*. Oxford University Press, Oxford.

Nunez PL, Silberstein RB, Cadusch PJ and Wijesinghe R. 1993. Comparison of high resolution EEG methods having different theoretical bases. *Brain Topogr* 5:361–364.

Petsche H, Rappelsberger P and Pockberger H. 1989. Sex differences in the ongoing EEG: probability mapping at rest and during cognitive tasks. In: *Functional Brain Imaging*. G Pfurtscheller and FH Lopes da Silva (eds). Hans Huber Publishers, Toronto.

Pfurtscheller G and Lopes da Silva FH. 1989. *Functional Brain Imaging*. Hans Huber Publishers, Toronto.

Regan D. 1989. *Human Brain Electrophysiology: Evoked Potentials and Magnetic Fields in Science and Medicine*. Elsevier, New York.

Rosenfeld JP. 2002. Theoretical implications of EEG reference choice and related methodology issues. *J Neurother* 4(2):77–87.

Stephenson WA and Gibbs FA. 1951. A balanced non-cephalic reference electrode. *EEG Clin Neurophysiol* 3:237–240.

Stough C, Donaldson C, Scarlata B and Ciorciari J. 2001. Psychophysiological correlates of the NEO PI-R openness, agreeableness and conscientiousness: preliminary results. *Int J Psychophysiol* 41:87–91.

White D, Ciorciari J, Carbis C and Liley D. 2009. EEG correlates of virtual reality hypnosis. *Int J Clin Exp Hypnosis* 57(1):1–23.

22

Magnetic Stimulation and Biomagnetic Signals

Andrew W. Wood

CONTENTS

This chapter deals with two aspects of the interaction of magnetic fields to biological tissue: (i) the use of rapidly changing magnetic fields (caused by rapidly changing electrical currents in coils) to stimulate excitable tissue, and (ii) the measurement of the tiny magnetic fields associated with naturally occurring current loops within the body. There is a reciprocal relationship between the two phenomena: the same result is obtained if we treat a particular small current loop within the body as a *source* and we *measure* a magnetic field (or induced current in a loop of wire on the surface, which is called magnetometry) as when we have a loop of current on the surface of the body as a *source* (magnetic stimulation) and we *measure* the induced current at the same location in the body where the source loop was previously. Of course, in the first situation, the magnetic fields concerned are around 10^{-12} those associated with the second. In both cases, the currents concerned are those associated with neural activity (spontaneous or perhaps evoked in the first case and sufficiently strong to evoke responses in the second case).

22.1 Magnetic Stimulation

When current in a wire (particularly a coil of wire) changes, there is a current induced in adjacent media. This is used in a transformer, shown in Figure 22.1, where the meter will indicate a brief current when the switch is opened or closed, but not at other times. The size of the induced current in the secondary depends on how rapidly the magnetic field in the primary changes. This can be applied to living systems, where, if the magnetic field in

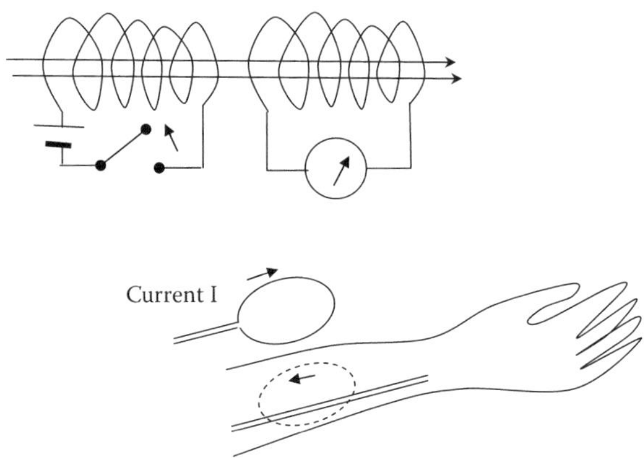

FIGURE 22.1
Top: a transformer. Current flows in the secondary (the coil on the right) when the switch is opened or closed, but not at other times. Bottom: a circular coil placed above the forearm will induce a current in the opposite direction to the current in the coil, when this current is time-varying.

the coil is made to change fast enough, there will be enough current induced in tissue (the forearm, in the case of Figure 22.1) that nerve stimulation will ensue. As we will see, this offers several advantages over the more conventional means of stimulating nerves, using surface electrodes.

22.1.1 Varieties of Magnetic Stimulation

Transcranial magnetic stimulation (TMS) involves placing coils close to the scalp to stimulate specific regions within the cortex. Typically, regions of the motor cortex can be stimulated to elicit specific responses (such as finger movement) for diagnosis. TMS can also be used to interfere with cognitive processing tasks to elucidate the pathways the brain uses to perform these tasks. It can also be used as an alternative to electroconvulsive therapy (ECT) to treat major depression and other illness affecting the brain. Another variety, shown in Figure 22.1 lower, is for peripheral nerve magnetic stimulation. This is used mainly in experimental studies on animals rather than in routine clinical practice.

22.1.2 Principles of Magnetic Stimulation

As shown in Figure 22.1, the induced current follows the path of the magnet current, but in the opposite direction. What happens when we put a current through a coil of wire? A familiar situation is a cylindrical coil wrapped around a soft iron core (a solenoid, as shown in Figure 22.2). The ability of the soft iron to attract small metallic items, such as nails, depends on (1) the current in the coil, (2) the number of turns in the coil, (3) the coil geometry, and (4) the nature of what is inside the coil (iron, air, etc.). These factors determine the magnetic flux density, more commonly referred to as the magnetic field. The formula for estimating the magnetic field inside a solenoid is given in Figure 22.2, where μ is the magnetic permeability of the material in the core. This is often expressed as $\mu_r\mu_0$, where μ_r is the permeability *relative* to vacuum (which can have a value of several thousand for soft iron) and μ_0 is the permeability of vacuum, which, by definition, is $4\pi \times 10^{-7}$ in SI units. As already mentioned, if we have a second coil (as in Figure 22.1), we can generate a current in the second coil (secondary), only if the current in the first, primary, coil is

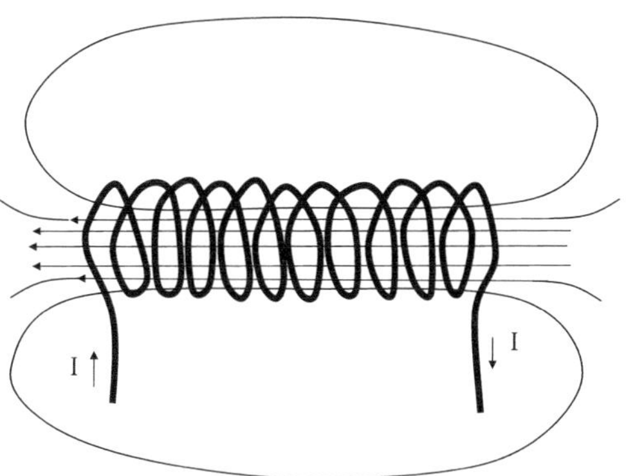

FIGURE 22.2
A solenoid, showing the magnetic field lines. The number of lines per unit area is an indication of the magnetic flux density B. The value of B is given by μnI, where n is the number of turns per meter, I the current, and μ the magnetic permeability, which for air is $4\pi \times 10^7$. Note the magnetic flux lines are at a nearly uniform density inside the solenoid and elsewhere they are highly divergent.

changing. However, a force exists on susceptible items all the time a current is flowing in the primary.

The unit of magnetic flux density is tesla: the magnitude of the Earth's magnetic field, for example, is approximately 50 μT and for the main magnet of a magnetic resonance imaging (MRI) system commonly 3 T. The concept of flux density is similar to current density: inside the core (Figure 22.2) the imaginary lines of magnetic flux are crowded together, but outside of the coil the lines are much further apart. In terms of SI base units, tesla is in kg s^{-2} A^{-1}. The symbol is usually B. The older unit for magnetic field is the Gauss (1 G = 10^{-4} T). For a flat coil:

$$B\ (\text{tesla}) = \mu\ N\ I/(2R) \tag{22.1}$$

where μ is the permeability ($4\pi \times 10^{-7}$), for air in this case, N is the number of turns, I is the current (Amps), and R is the coil radius (m). So, for N = 10, I = 10 A; R = 0.1 m, we get 630 μT.

In a two-coil system, the changing magnetic field causes a current to flow in the other coil (or more precisely a *voltage* across the terminals). This voltage depends on (1) how rapidly the magnetic field changes, (2) the area enclosed by the other coil (πR_2^2), and (3) how many turns in the other coil (N_2). For a flat coil:

$$V = (dB/dt) N_2\ \pi R_2^2. \tag{22.2}$$

So, if the 630 μT changes to zero in 10^{-6} s, and if $N_2 = 1$ and $R_2 = 0.1$ m, then V = 20 volts! The length of the other coil is $2\pi R_2 = 0.6$ m, so the *electric field* induced along the coil is 20/0.6 = 33 V/m.

By convention, the direction of magnetic field lines follow the right-hand thumb rule, that is, if the right hand were to grasp a long current-carrying straight wire, with the thumb pointing in the direction of the current, then the fingers would point in the direction of the lines of magnetic field (or flux) circling around the wire (Figure 22.3). For a loop of wire, the field lines are as shown in Figure 22.4. If we have two loops side by side, with currents flowing in opposite directions, then the field lines are as shown in Figure 22.5 upper. When the coils are brought into contact, forming a "figure-of-eight" arrangement,

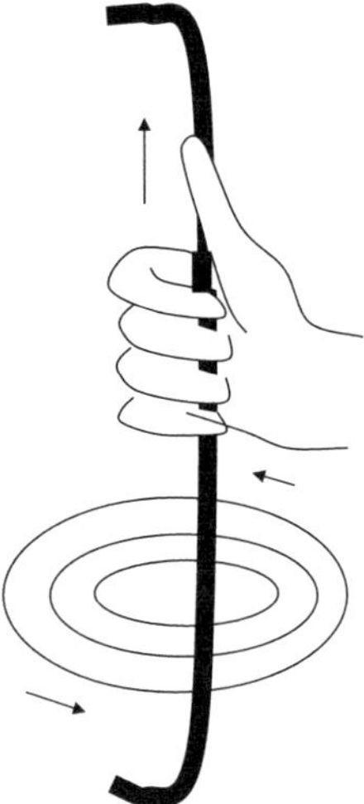

FIGURE 22.3
The right-hand thumb rule. If the thumb indicates the direction of current, the fingers indicate the direction of the magnetic field lines.

there is summation of field lines in the individual coils from the contributions of the individual coils (Figure 22.5 lower). Viewed from the side (with a slightly altered initial condition) when the initial coaxial arrangement of coils is moved as shown by the light colored arrows to the final figure-eight position (Figure 22.6). The field lines are particularly dense below the point at which the two coils come into contact (in fact, the windings are often interleaved, so the two coils intersect). The induced current path is as shown by the dotted circle and again is maximal below the point if intersection. From electrophysiological studies, the current density (J) is the important parameter to determine whether a nerve will fire or not. Experiments show that a tissue averaged value of J of about 2 A/m^2 (2 μA/mm^2) is required to stimulate a nerve. This is related to the induced electric field by the following:

$$J = \sigma E \tag{22.3}$$

where σ is tissue conductivity (0.1 S/m) and E is the induced field in V/m.

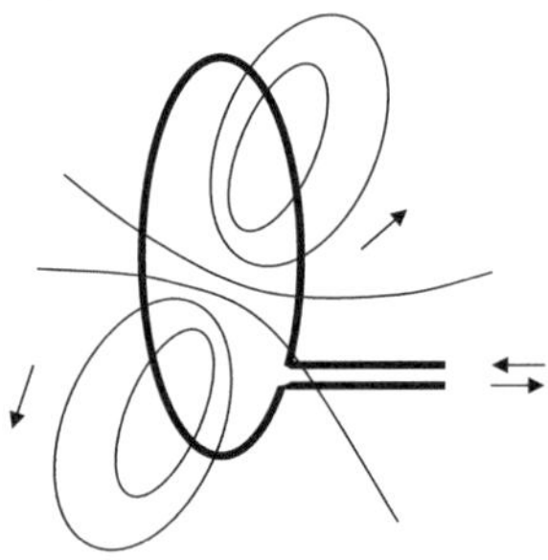

FIGURE 22.4
Magnetic field lines associated with a loop of wire carrying a current.

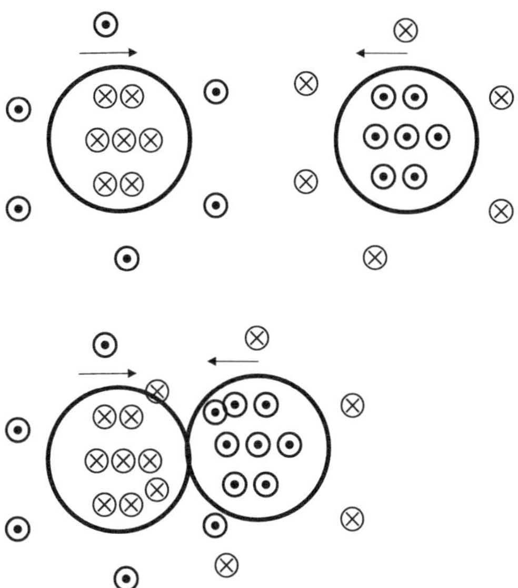

FIGURE 22.5
The magnetic field lines associated with two loops brought together in a "figure-of-eight" configuration. The arrows denote the direction of current and by convention, the symbols ⊗ and ⊙ denote field direction into and out of the page, respectively.

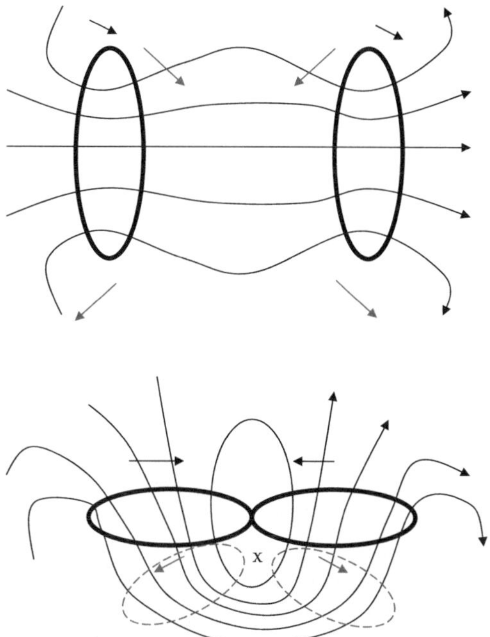

FIGURE 22.6
Field lines in a "figure-of-eight" coil. The dotted circles indicate the direction of the current induced in tissue. Note that at point x the induced current density is maximal and represents a withdrawal of positive charge from the surface of excitable tissue cells. Light colored arrows in upper diagram denote the way the coils would be moved to achieve the configuration in the lower diagram.

In the earlier example we get J = 0.1 × 33 = 3.3 A/m^2, above the stimulation threshold. By combining the two equations we get

$$J = \frac{1}{2}\sigma R_2(dB/dt). \tag{22.4}$$

In our previous example of dB/dt = 630 T/s, R_2 = 0.1 m and σ = 0.1 S/m, we can easily see that a similar result is obtained from this equation.

How is a nerve stimulated? We need withdraw a specified amount of positive electric charge from a small region of the nerve surface: this is equivalent to depolarizing the nerve membrane at a particular location (see Chapter 6). The amount is actually 0.1 millicoulomb per m^2 (1 coulomb is the charge accumulated when 1 A flows for 1 s). When stimulating a nerve using electrodes, this charge will be withdrawn under the negative electrode (cathode) for a positive-going pulse (positive charge flows toward the current sink). With magnetic stimulation the arrangement is somewhat different, but by analogy this is referred to as a "virtual cathode."

In a "figure-of-eight" coil the virtual cathode is at a point where the induced currents are beginning to diverge (Figure 22.7). This represents regions where the surface charge on nerve membranes can be taken away from the neurons rather than just being transferred from one region to another. By withdrawing this charge from bends or at the ends of the nerve, depolarization can be achieved (see Figure 22.8). In other words, the current flowing parallel to the membrane surface will not remove charge until a discontinuity is encountered.

This can be demonstrated in a relatively simple experiment by setting up a toad or frog sciatic nerve in a large Petri dish as shown in Figure 22.9. Here, the nerve is set up with a right-angled bend by the use of a piece of cotton fixed to the bottom of the Petri dish (the nerve comes up toward the reader in the figure). A "figure-of-eight" magnetic stimulating coil is placed below the Petri dish. To measure the response, a pair of Ag/AgCl electrodes is placed under the nerve as shown, with the measured signal amplified and displayed on an oscilloscope. The lower diagram indicates the averaged response to 32 magnetic stimuli (a stimulus artifact is visible just after the origin). The action potential of <1 mV can be seen between 1.5 ms and 3 ms after the stimulus starts. If there is no bend in the nerve, the response is not seen.

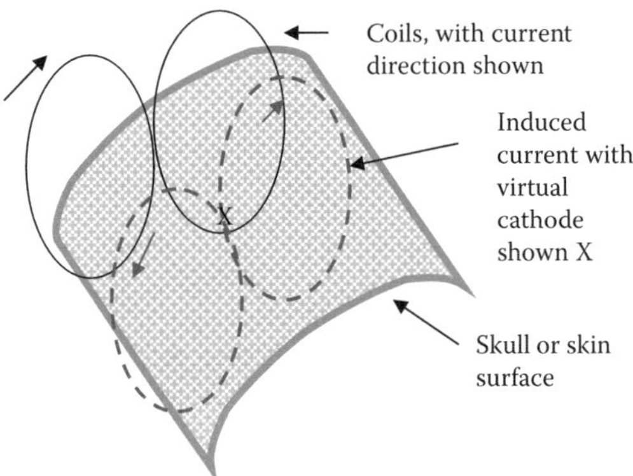

FIGURE 22.7
Illustrates how the "figure-of-eight" coil placed above the scalp will induce currents within the cerebral cortex below the skull. The current density falls away with distance from the skull–cortex interface, but is maximal near point X in the diagram. The volume of tissue for which the current density is sufficient to cause nerve stimulation may be quite small (a few mm^3).

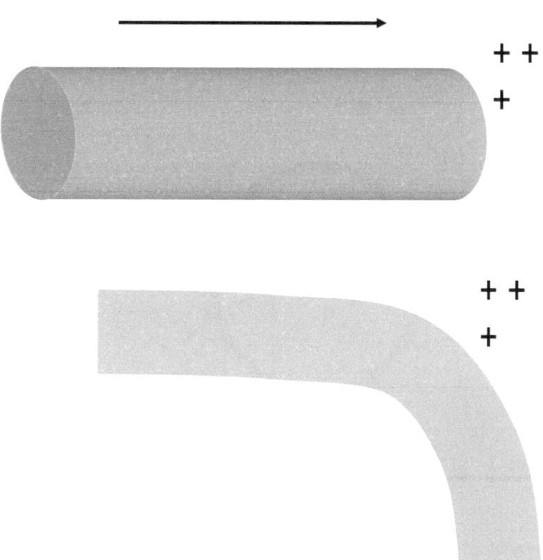

FIGURE 22.8
The cylinder represents a nerve axon and the arrow the direction of an electric field (or voltage gradient) induced in the bulk tissue. Note that the positive charge will be removed from the end of the axon, or if the axon bends around, at the bend.

22.1.3 Magnetic Stimulation in Clinical Studies

The main advantage in magnetic stimulation is that there is no pain or discomfort from high-current density below stimulating electrodes (because there are no electrodes). There is no contact with skin (which is more hygienic). When applying magnetic stimulation to the head the high resistivity of skull is not a problem, since the magnetic fields are not attenuated by bone or other tissues. Since the induced currents can be concentrated by using a figure-of-eight stimulator, it is possible to arrange for currents to be above threshold in very small regions of cortex, leading to stimulation of specific groups of fibers (such as those controlling muscles in the little finger, for example). The induced current density is maximal in the cortical regions immediately below the "virtual cathode" (see above) then it declines the further we go away from the coil. The magnetic field distribution can be quite easily computed for ideal conditions, that is, for a loop consisting of N turns, including off-axis positions. Resources are available via the Internet (http://www.netdenizen.com/emagnet/offaxis/iloopcalculator.htm) to do these calculations. Fields due to a figure-of-eight coil can be computed by vector superposition of two loops of N turns, centers separated by a diameter. Within the brain, the magnetic permeability μ has the same value in all tissues, so the flux density (B) value at a particular location will not be influenced by the specific tissue type. Using this approach, the magnetic field lines within the head can be plotted. The induced electric fields (or current density) will always be at right angles to these field lines and can be calculated using Maxwell's equations. A rule of thumb is that the induced currents in the tissue tend to follow the same path as the current in the coil, but in the opposite direction. This is illustrated in the upper part of Figure 22.10. Inside the head, electric fields are also induced because of build up of charge at interfaces between tissues (such as between the skull and underlying CSF), so there is an extra component.

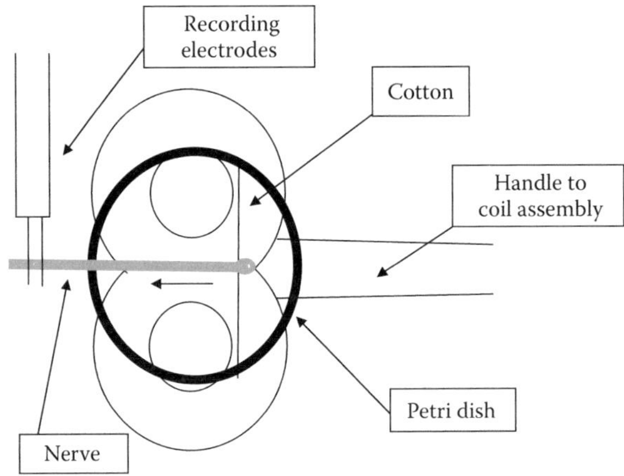

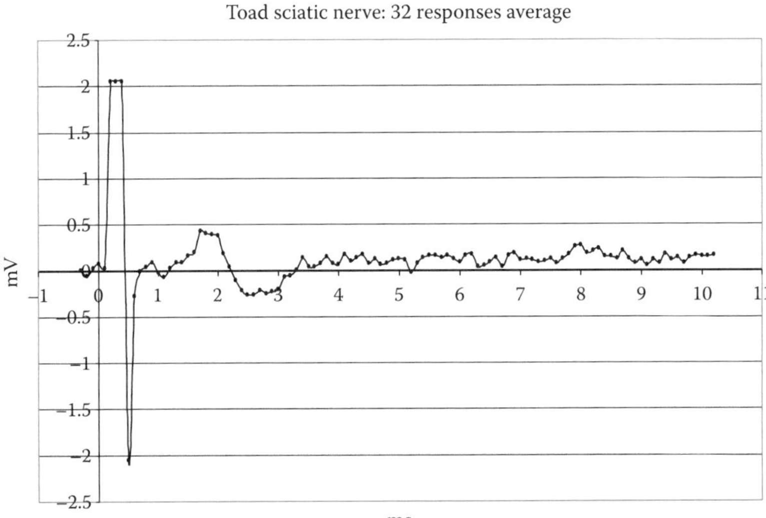

FIGURE 22.9
The use of a "figure-of-eight" stimulator to elicit an action potential (lower diagram) from a toad sciatic nerve. The experimental setup is shown in the upper diagram with the sciatic nerve bent 90° around a strand of cotton. The arrow shows the direction of induced current in the nerve.

This is illustrated in Figure 22.10 (lower), which shows that for an off-axis coil, the field due to interfacial charge build up is quite significant.

Induced fields can be measured experimentally in head-like phantoms, for example a cylinder of saline with approximate dimensions of the head. Figure 22.11 shows results from a head phantom made from a hollowed watermelon filled with saline. The induced electric field (and hence current density) is measured using a calibrated dipole probe placed at various locations and at different orthogonal directions within the saline. The plot in the lower part of Figure 22.11 is beneath the "virtual cathode" on the stimulator coils. The maximum value corresponds to 210 V/m. Experiments of this type reveal the strong directionality of the induced current. Since, as was pointed out above, individual axons are stimulated at bends or terminations, the direction of the current in relation to regional microscopic anatomy will determine which axons are stimulated. Figure 22.12

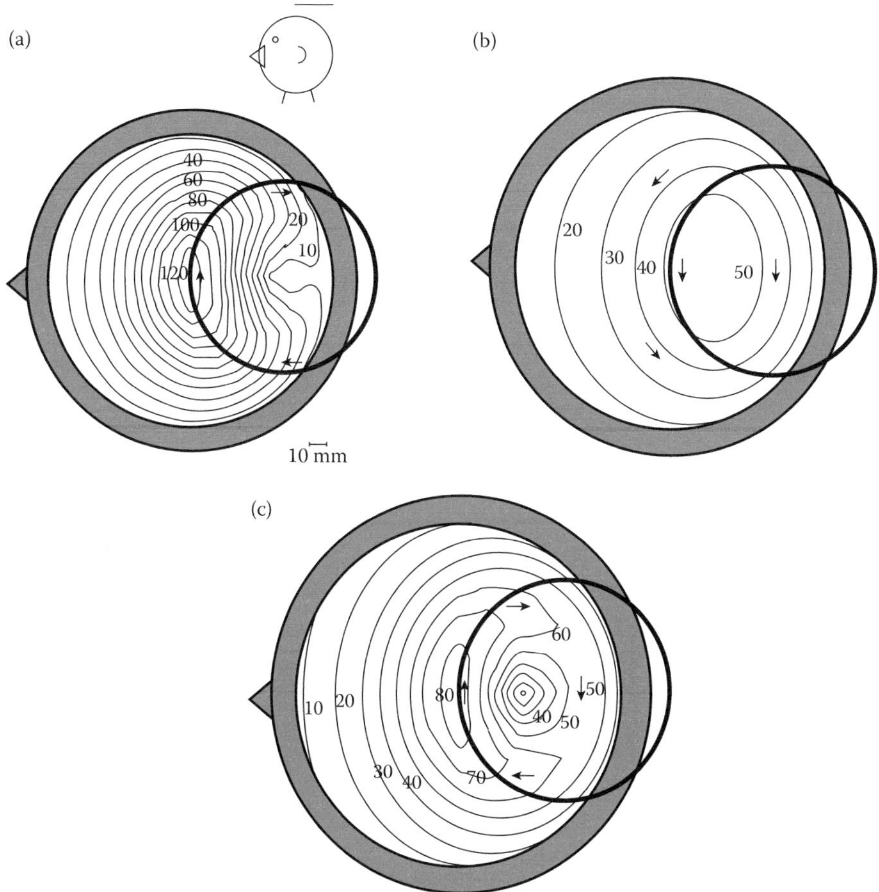

FIGURE 22.10
Induced electric fields inside a spherical model of the head. A circular magnetic stimulating coil (excitation 10 kT/s) is placed toward the back of the head as shown. The electric field due to induction E_A is shown at top left (maximum value of around 130 V/m under the coil edge). The diagram at top right shows values of electric field E_ϕ due to charge buildup at bone/tissue interfaces (maximum value just over −50 V/m). High values are associated with regions where the contours are closest together in the top left diagram. Bottom (c): Total electric field, by adding together the contributions from the top two diagrams. For a coil placed symmetrically over the head, the value of E_ϕ is zero and the E-field contours are concentric to the coil. (Diagram from Roth et al. *Electroencephalogr Clin Neurol* 81:47 (1991), used with permission.)

illustrates how a particular physiological response (in this case contraction of the thenar muscle, which controls the thumb) only takes place effectively if the handle of the figure-of-eight coils is within a small range of angles.

Commercial magnetic stimulators offer a variety of coils for stimulating peripheral nerves or the cortex (TMS—see above). Both applications can consist of single or of multiple stimulating pulses. In the case of the cortex, multiple stimulations are referred to as repetitive TMS, or rTMS. The coils can be single or figure-of-eight, and the latter can be planar or angled. A typical single pulse waveform is shown in Figure 22.13. The current is a single sinusoid leading to a magnetic field with a 1.2-T peak-to-peak amplitude (see Equation 22.1). The period of 0.36 msec represents an underlying frequency (f) of 2.78 kHz. Since the maximal rate of change of field (dB/dt) for a sinusoid is given by $2\pi f B_0 = 2\pi \times$

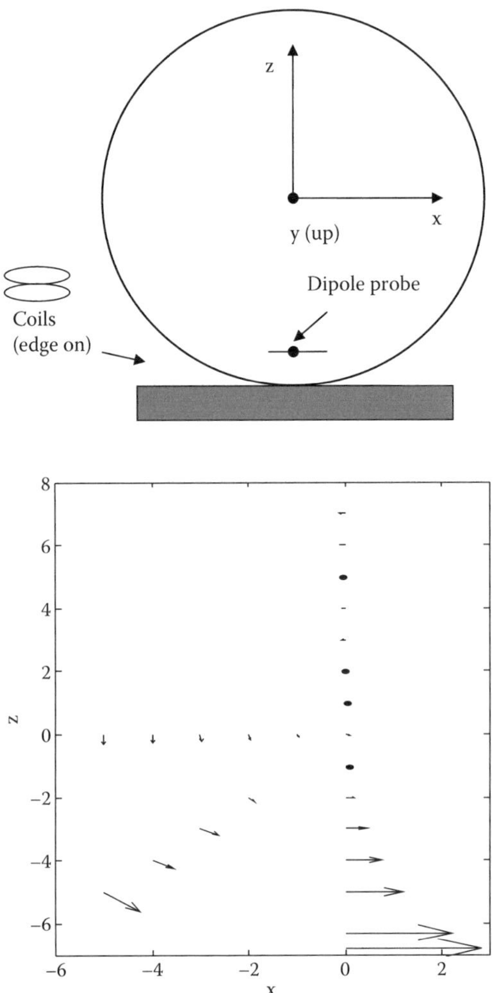

FIGURE 22.11
Simulation of the human head using a cylinder filled with saline. The induced electric field is measured using a dipole probe (in this case measuring E_x). The lower diagram shows the resultant field $\sqrt{(E_x^2 + E_z^2)}$ in the plane at the intersection of the two coils in the "figure-of-eight" pattern. In the x-y plane (not shown) the resultant fields follow the "figure-of-eight" pattern, as expected.

2,780 × 0.6 = 10.5 kT/s, this will be the effective stimulating waveform, shown at the bottom of Figure 22.13. There are thus two reversals in direction in the induced current. However, the first, positive-going, portion of the waveform is designed to initiate the action potentials constituting the response. The induced current density, which is given by Equation 22.4, is expected to be approximately 40 A m^{-2} (assuming a tissue conductivity of 0.1 S/m and a radius of 80 mm) and the induced tissue electric field up to 400 V/m (Equation 22.3).

There are two main applications of magnetic stimulation in clinical practice: first, as a method for investigating cognitive processes in experimental neuroscience, and second, as a therapeutic modality. In the first application a cognitive stimulus is presented, such as an image on a computer monitor where the participant has to make a judgment and then make a response, such as pressing a certain key on the computer. The TMS is given

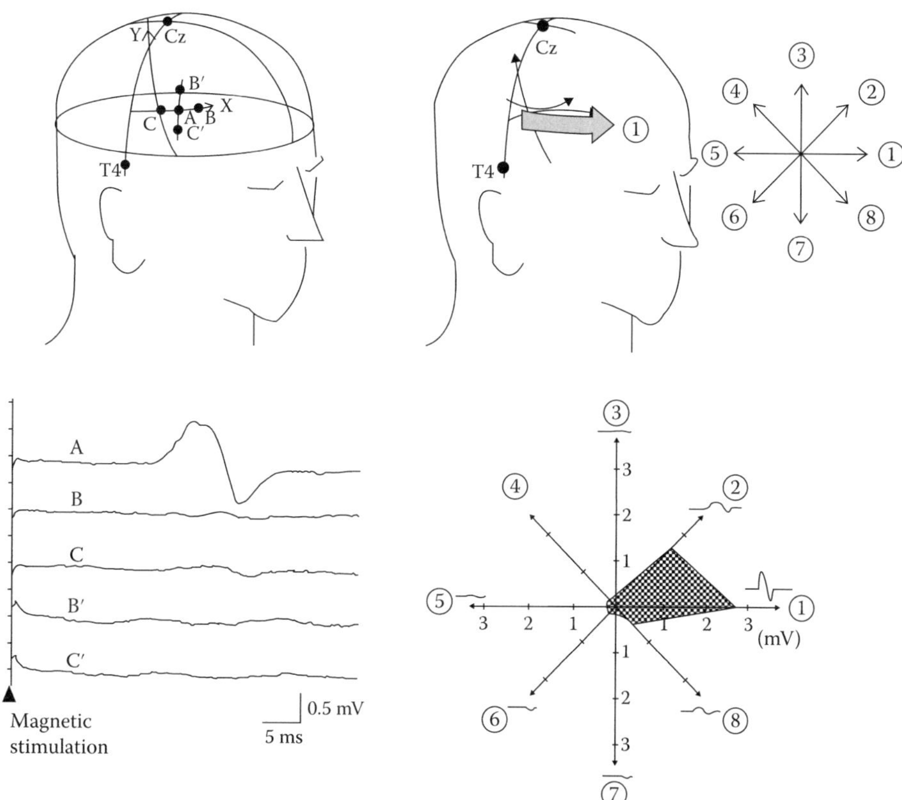

FIGURE 22.12
The importance of position (at left) and direction (at right) of "figure-of-eight" stimulating coils in TMS. The letters A–C′ refer to positions 5 mm from each other and the numbers 1–8 refer to the direction the handle of the figure-of-eight coils are pointing in. The shaded area represents response amplitudes for the various directions. (From Ueno et al. *IEEE Trans Magn* 26:1539 (1990).) The handle is similar to that shown in Figure 22.9 upper.

at certain times between visual stimulus presentation and response in such a way that specific parts of the cognitive pathways (those involved in making the judgment) and the ensuing response are interrupted or otherwise interfered with. The TMS may alter speed or accuracy of responding. This then forms the basis of inferring the regions of the brain involved in the information processing.

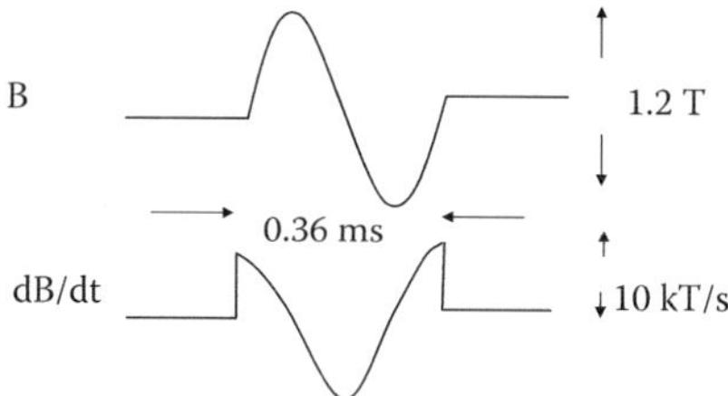

FIGURE 22.13
Top: the single sinusoidal waveform used in a commercial Transcranial Magnetic Stimulation (TMS) machine, with an amplitude of 1.2 T and a cycle length of 0.36 ms. The lower diagram is the first differential of the upper diagram and is proportional to the induced electric field within the tissue.

The use of rTMS in treatment of (i) schizophrenia, (ii) major depression, and (iii) bipolar disorder is an emerging application. For many years, the efficacy of electroconvulsive therapy or ECT has been well documented for at least the second and third of these illnesses. ECT involves heavy sedation and care has to be taken to avoid injury arising from the convulsion. TMS does not have these disadvantages, but although there is good evidence for efficacy, it still has to be accepted as a mainstream therapy. Some recent reviews are included in the list of references at the end of this chapter.

22.2 Biomagnetometry

22.2.1 Varieties of Biomagnetometry

Just in the same way that there are many types of electrical signal that can be recorded using skin surface and other electrodes, there are analogous measurements involving magnetic field monitoring. These include magnetoencephalography (MEG) and magnetocardiography (MCG), which are the magnetic equivalents of EEG and ECG, respectively. In addition, magnetic susceptance measurement in lungs and liver can be carried out, also axonal currents in-vitro can be monitored by threading the nerve fibers through a ferrite pick-up coil. Another area of investigation is the measurement of embryo fields from avian eggs during development.

Table 22.1 compares the magnitudes of the biomagnetic signals from various sources.

22.2.2 Measuring Femtotesla in a Microtesla Environment

Given that most of these signals are around 8 orders of magnitude less than the Earth's field, there is a great challenge to detect them, particularly since there are both temporal and spatial variations in the Earth's field. There are a number of sensitive magnetic field detectors, but by far the most practical are detectors based on Superconducting QUantum Interference Devices, or SQUIDs, for short. These effectively count quanta of magnetic flux, where a single quantum is given by the formula

$$\Phi_0 = h/(2e) = 2\ \mathrm{fTm}^2.$$

Thus, for a single turn of wire, area 2 cm^2 (2×10^{-4} m^2), this quantum will correspond to 10 pT of flux density, since $B = \Phi_0/A$. The sensitivity is much less than this, because of the

TABLE 22.1

Range of Magnetic Flux Densities (in fT)

MEG	10^3–10^4 (i.e., in the pT range)
MCG	10^5
Lung contaminants	10^6
Retinogram	10^2
Evoked cortical activity	10^2

Notes: The relevant frequency ranges are similar to corresponding electrical signals, except for lung contaminants, which are a DC signal. 1 fT = 10^{-15} T: for comparison, the Earth's field is 50×10^{-6} T or 5×10^{10} fT.

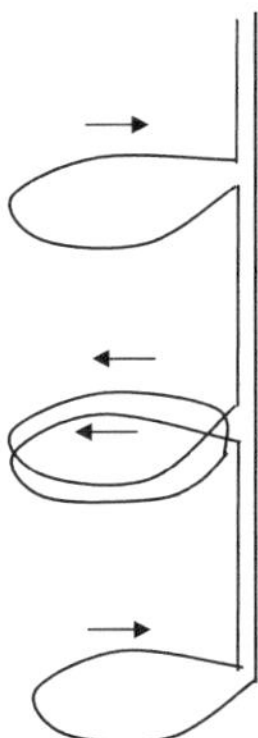

FIGURE 22.14
The second gradient pickup coils inside the "tail" of the SQUID biomagnetometer. The voltages induced in the coils are such that these will cancel for steady magnetic fields and for uniform field gradients. However, fields that die away rapidly with distance (as those produced by cortical currents do) will be detected.

quantum interference effects exploited in the measurement. To attenuate the effects of the Earth's field, extra loops of wire in the sensor can be used to measure the second gradient (d^2B/dz^2), that is, the sensor will only sense fields that have a nonlinear spatial gradient. The way this is accomplished is shown in Figure 22.14, where two sets of windings are opposed as shown. This way steady fields (such as the Earth's) or constant gradients in field are not sensed. More recent designs with multi-channel sensors rely more on efficient magnetic shielding of the measurement laboratory. This will often involve 'active shielding', which is the use of field sensors to control the currents through coils designed to nullify the external fields and field gradients.

22.2.2.1 Measuring Brain Currents

A magnetic field is generated whenever a current flows (see Figure 22.1). When nerve axons "fire" this represents a current in the extracellular fluid toward the depolarized region. If many axons are firing simultaneously, there is a local region of high current density, with the return current loops are more dispersed (and thus the density is far less). In net terms, the region can be represented by a short current density vector J, surrounded by concentric magnetic field lines. This is called an equivalent current dipole. Figure 22.15 shows that for current dipoles tangential to the surface of the cortex (due to axonal currents mainly in the crevices or sulci) give rise to magnetic field lines leaving the scalp at one point and entering at an adjacent point. The deeper the current dipole, the further apart will be those two points. On the other hand, radial current dipoles give rise to field lines that would not be detected by the sensor coils. Thus, the MEG only records the tangential component of current dipoles. Note these radially directed dipoles are detected by EEG, so there is value in combining both measurements. However, spatial resolution of EEG is much poorer than MEG: the skull acts as a low-pass spatial filter for EEG because of its low conductivity. Combined measurements are achieved by requiring patients or study participants to wear a "bathing cap" containing an array of small EEG electrodes, while simultaneously measuring from multiple sites immediately above the scalp using a multichannel SQUID.

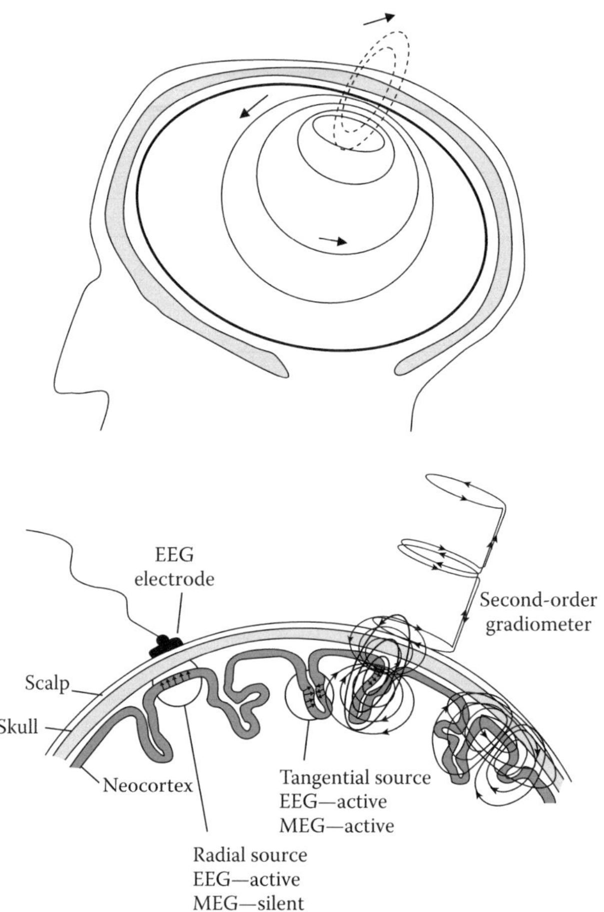

FIGURE 22.15
Upper: lines of current due to a region of high current density within the cortex and directed tangentially to the cortical surface. Note directions of magnetic field lines, out of the scalp at one point and in at an adjacent point. Bottom: currents associated with tangential and radial current dipoles. In the case of the latter, the magnetic field lines are contained within the cortex and the gradiometer does not "see" these sources.

22.2.2.2 Detecting Magnetic Quanta

Since the SQUID, by definition is a superconducting device, the whole detector system has to be kept at liquid helium temperatures (<4 K). The second gradient of the magnetic flux from the body ($d^2\Phi_{body}/dz$) produces a voltage in the sensor coils that are then coupled to the actual SQUID sensor at a point further away from the body. The flux induced in this secondary coil, Φ_{ext}, influences the voltage across the Josephson Junctions within the SQUID. A Josephson junction is effectively a small section of insulator between two superconductors (as illustrated in Figure 22.16a). The current-voltage (I-V) characteristic is modulated by magnetic flux Φ with a period of one flux quantum Φo (2 fT m^2). The detection actually consists of a feedback circuit (Figure 22.16b): a bias current, is initially adjusted to keep a zero voltage across the two superconducting Josephson junctions. An

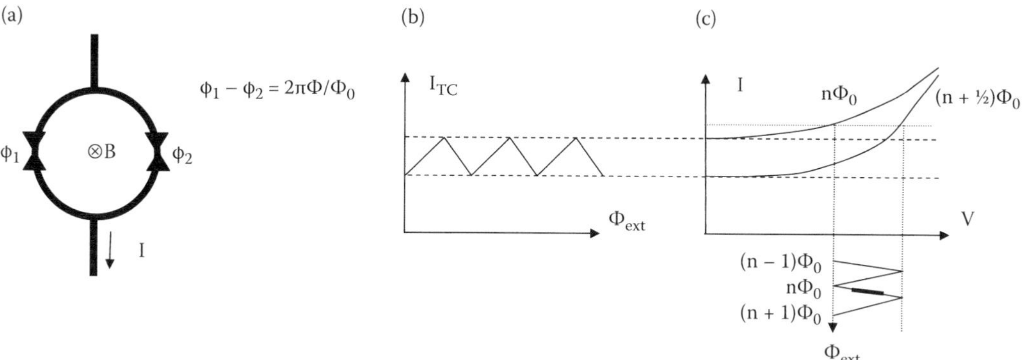

FIGURE 22.16
(a) The SQUID sensor: two Josephson junctions with phases ϕ_1, ϕ_2 of the wave function describing the junction. (b) The addition of a magnetic field B (from the gradient coils) introduces a difference in the two phases dependent on the ratio of the added flux Φ (= B × area) to the flux quantum $\Phi_0 = h/2e$. As the flux Φ_{ext} (due to B) varies, so does the current I_{TC} at zero bias voltage. (c) The current–voltage relationship varies between two limiting characteristics, corresponding to $n\Phi_0$ and $(n + ½)\Phi_0$, as shown. In practice, a lock-in amplifier feeds current to a feedback coil to keep the measured voltage within the region shown thickened in the lower part of the diagram. It also feeds in a (typically) 100 kHz signal to facilitate the lock-in measurement, because if the net value of Φ_{ext} moves into the "V" of the characteristic, the 100 kHz signal will become half-wave rectified and thus the actual 100 kHz component will disappear. The output from the lock-in amplifier is essentially the measure of external magnetic field.

added flux Φ_{ext} causes a phase difference (interference) between these junctions (hence the name "quantum interference device").

A 100-kHz modulating current drives the "phase-locked loop" to achieve maximum sensitivity (Figure 22.16c). A specialized Dewar, with a thin "tail" to allow close position of measuring coil is filled with liquid helium to maintain the temperature of the sensing system below 4 K. An early single-channel SQUID magnetometer is shown in Figure 22.17 (upper). A map of brain-induced magnetic field can be obtained by giving repeated identical stimuli and recording responses from a number of sites in sequence. It is assumed that the brain does not adapt during this quite lengthy acquisition time. More recent magnetometer systems use multichannel SQUID magnetometer (or MEG) systems, typically with 64 or more sensors arranged in a honeycomb pattern to fit closely to the head. Figure 22.17 (bottom) shows an array of around 100 square pick-up coils, each feeding to an individual SQUID sensor. The Dewar system is shaped to accommodate the head and the whole system can be moved to allow comfortable data acquisition with the patient (or volunteer participant) in a seated position, perhaps viewing a visual stimulus.

22.2.2.3 Source Localization

For single current dipole sources, it is possible to characterize them using the MEG data. Using electromagnetic field theory it is possible to derive the following:

$$d = \Delta/\sqrt{2}$$

$$Q_t = (32.6)d^2B_{max}/\mu_0.$$

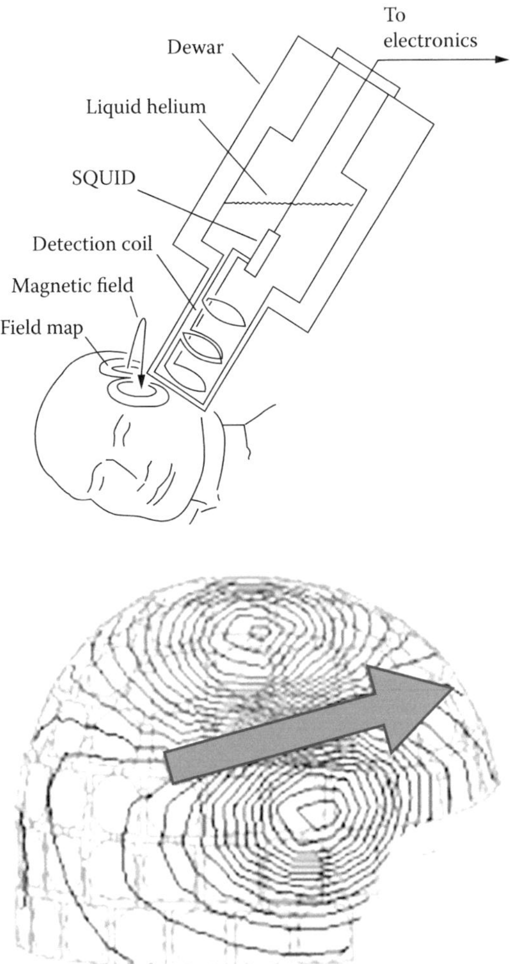

FIGURE 22.17
Upper: A single channel SQUID gradiometer within a Dewar tail. Bottom: Array of magnetometers and gradiometers with the dipole current source indicated by the large arrow. (Taulu and Hari: *Human Brain Mapping*. p. 1254. 2009. Copyright Wiley-VCH Verlag GmbH & Co. KGaA. Reproduced with permission.)

Here, d represents the distance below the skull surface that the current dipole is located and Q_t its dipole moment. Figure 22.18 shows some data obtained from the author's laboratory several years ago. The stimulus was a semicircular checkerboard image, where the black and white checks alternated in color in time temporally (13 Hz) as well as spatially. The top diagram shows the isofield contours (in fT) and the middle diagram the phase (relative to the stimulus) measured at various locations on the scalp above the visual cortex (point 0, 0 represents the inion, the small bony projection at the lower rear of the skull).

Note that the contours do not discriminate between positive and negative values of magnetic field (here the field refers to the tiny additions to or subtractions from the Earth's field). The arrows representing phase angle suggest that 1.5, 1.5, and 1.5, 6.0 are different

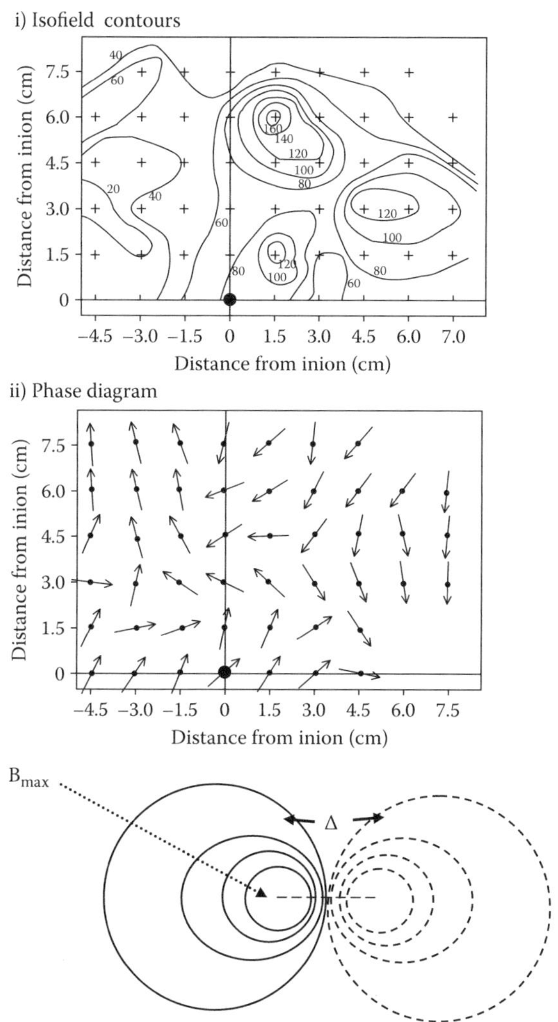

FIGURE 22.18
Top: Magnetic field intensity over the visual cortex due to a steady-state stimulus. Middle phase between sinusoidal stimulus and response. Bottom: Separation of maxima D, with maximum value B_{max} can be used to estimate magnitude, and position of current dipole.

by 180°, thus $\Delta = 6.0–1.5 = 4.5$ cm. B_{max} is around 140 fT, thus the dipole is located at $4.5/\sqrt{2}$ or 3.2 cm below the scalp surface and Q_t in this case is around 4 nA m.

Bibliography

Couturier JL. 2005. Efficacy of rapid-rate repetitive transcranial magnetic stimulation in the treatment of depression: a systematic review and meta-analysis. *J Psych Neurosci* 30:83–90.

Fitzgerald P. 2008. A Randomized-Controlled Trial of Bilateral rTMS for Treatment-Resistant Depression. *Progress in Neurotherapeutics and Neuropsychopharmacology* 3: 211–226.

Papanicolaou AC. 2009. *Clinical Magnetoencephalography and Magnetic Source Imaging*. Cambridge University Press, Cambridge.

Reilly PJ. 1998. *Applied Bioelectricity: From Electrical Stimulations to Electropathology*. Springer-Verlag, New York.

Reilly PJ, Diamant AM. 2011. *Electrostimulation: Theory, Applications, and Computational Models*. Artech House, Norwood, MA, USA.

23

Medical Imaging

Andrew W. Wood

CONTENTS

23.1 Introduction

Imaging consists of spatial mapping of a certain quality of an object at a given point. A normal photographic image taken by a digital camera records the amount of reflected light from a scene within a particular range of wavelengths. Objects which look blue in the image are those which strongly absorb yellow and red light from a white light source and reflect blue. A blue object illuminated by a red light looks black.

There is a range of imaging systems used in biomedicine. These can be used both to delineate structure and to infer function. Some have associated health concerns if over-used, whereas others are safer. This chapter will serve as a brief introduction to medical

TABLE 23.1

Common Imaging Modalities

Type of Imaging System	Field or Radiation Involved	Detector	Quality of Object Recorded
Photography, thermography, video	Visible light, UV, IR	Film, photoconductors, charge-coupled devices (CCDs)	Reflectivity or emissivity as a function of wavelength
X-rays, computed tomography	Electromagnetic (E/M) waves of wavelength 10–100 pm	Photographic film, ionization chambers, thermoluminescent devices, solid-state detectors	Absorption (attenuation) coefficient (atomic number), thickness
Nuclear imaging and single-photon emission tomography (SPECT)	Gamma and occasionally beta rays	NaI crystal + photomultiplier, Li-drifted Ge detectors	Affinities of particular tissues to "tagged" substances: compartments available to diffusion
Positron emission tomography	Gamma rays associated with positron annihilation	As above	As above
Nuclear magnetic resonance imaging (MRI)	Strong static magnetic field + radiofrequency (RF) E/M radiation	RF coils	Hydrogen and other elemental concentration. Relaxation times in response to pulsed stimuli
Ultrasound	Sound pressure waves (wavelength around 1 mm)	Piezoelectric crystals	Reflection of waves from interfaces (which depends on compressibility and density)

imaging systems. Table 23.1 is a summary of some of the more common imaging modalities. There are also some less common modalities, including electrical impedance imaging and terahertz imaging, the latter using radiation bordering on the IR region and finding applications in security screening, the location of tumors and dental imaging.

23.2 Ultrasound

23.2.1 Basic Physics of Ultrasound

Ultrasound is sound whose frequency is above the audible range for humans, i.e., above 20 kHz. The range used in clinical medicine is mostly 1 MHz–10 MHz but with some applications up to 100 MHz. Unlike electromagnetic radiation, the wave is longitudinal, and involves the displacement of molecules from their equilibrium positions (displacement in the direction of wave propagation: hence longitudinal). For the one dimensional case we can write a wave equation:

$$c^2\partial^2\xi/\partial z^2 = \partial^2\xi/\partial t^2;$$

where ξ represents the displacement (in meters) from the equilibrium position (Figure 23.1). The cumulative distance is given by z and t is time. The wavespeed c is given by $c^2 = \kappa/\rho$, where κ is the bulk modulus of the medium (that is, the change in pressure caused by a fractional change in volume) and ρ is the (average) density of the medium. At 5 MHz, the wavespeed is approximately 1530 m/s in distilled water, a little greater than this in most tissues except fat (1450 m/s) and in bone is higher still (2500–4700 m/s). In air, the wavespeed is 331 m/s.

The wavelength, given by $\lambda = c/f$ (i.e., $1500/(5 \times 10^6) = 0.3$ mm), is smaller than most structures we may be interested in and ensures that adequate resolution can be obtained when obtaining images. At these frequencies the sound beams obey simple optical principles, remaining essentially nondispersed (that is, retaining a "pencil" waveshape) over relatively long distances. The intensity or power density of the beam (in W/m^2) can be estimated from the following formulae.

$$I = \frac{1}{2} p_0^2/(\rho c) = \frac{1}{2} \rho c \omega \xi_0^2 = \frac{1}{2} \rho c v_0^2$$

where ρ_0, ξ_0, and v_0 refer to pressure, displacement, and velocity amplitudes, respectively. Note that velocity here refers to particle displacement velocity rather than wave velocity (or wavespeed). In fact the use of these equations is usually the other way round, to estimate the pressure, displacement and velocity amplitudes for a given beam intensity, as shown in Table 23.2.

Figure 23.1 is a diagram of layers of particles in a material in their equilibrium positions (full line) and when displaced due to an ultrasonic wave passing through the medium (dotted line) showing areas of compression and rarefaction.

Ultrasonic imaging is based on detecting echoes from interfaces within the tissue. The size of echo depends on the amount of acoustic mismatch between the media on either side of the interface. This in turn depends on the difference in acoustic impedance ($Z = \rho c$) for the two media in question. The ratio of echo (or reflected) intensity to the incident intensity for normal incidence (I_r/I_i) is given by the following formula:

$$(I_r/I_i) = ([Z_2 - Z_1]/[Z_2 + Z_1])^2 = (A_r/A_i)^2;$$

where A_r, A_i refer to reflected and incident amplitudes (pressure, displacement, or signal voltage), respectively, and 1 and 2 refer to the layer from which and to which the beam is passing. The remainder of the power will of course be transmitted into the second layer. This transmitted intensity I_t is given by:

$$(I_t/I_i) = 4Z_1Z_2/(Z_1 + Z_2)^2.$$

You can check that $I_i = I_r + I_t$. Typical values for muscle are: $\rho = 1070$ kg/m^3, $c = 1590$ m/s; and for blood $\rho = 1060$ kg/m^3, $c = 1570$ m/s. This gives I_r/I_i of 1.2×10^{-4}, i.e., −40 dB. In other

TABLE 23.2

Typical Values for I = 10 mW/cm^2 (or 100 W/m^2), c′= 1540 m/s and ρ = 1000 kg/m^3 (Using the Formulae Given in the Text)

f (MHz)	Particle Displacement (m)	Particle Velocity (m/s)	Pressure Amplitude (kPa)
2.5	7.3×10^{-10}	0.011	17.6
10	1.8×10^{-10}	0.011	17.6

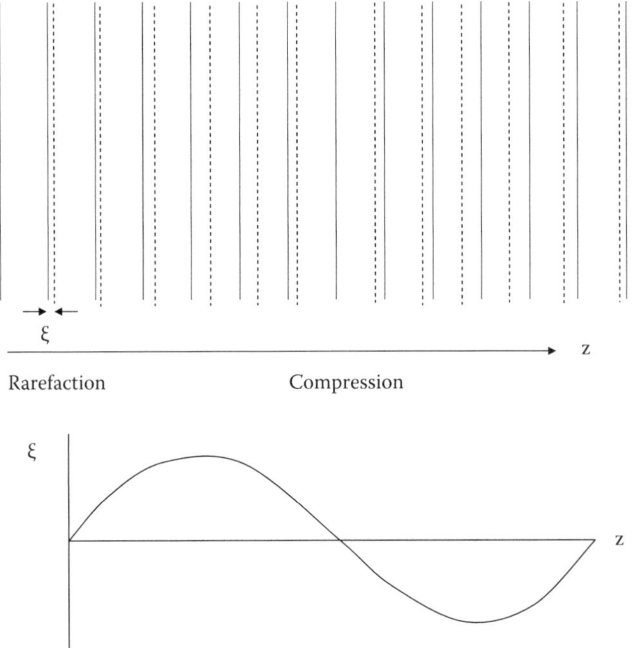

FIGURE 23.1
Features of longitudinal pressure waves (such as ultrasound). Upper diagram shows the normal positions of particles (full lines) compared to the displaced positions (dotted lines). The displacement of each layer from its undisturbed position is given by the distance ξ.

words, ultrasonic echoes from the internal chambers of the heart are tiny in comparison to the incident intensity, so high-gain amplifiers need to be used to detect them. Any sudden changes in density within a particular organ with give rise to echoes, hence the speckled appearance of ultrasound images.

For nonnormal incidence, after each Z in the above formulae add a $\cos\theta$ term, where θ refers to the angle the beam makes with the normal to the interface in that medium. Note that $\sin\theta_1/\sin\theta_2 = c_1/c_2$ (analogous to Snell's law) and the angles of incidence and reflection are equal.

Finally, the radiation force of the beam is given by $F = IA/c$, where A is the cross-sectional area of the beam. A beam of 100 W/m^2 intensity would exert a force of 6 μN on a 1 cm^2 fully-absorbing target. Even though this force is small (equivalent to the weight of a 0.6 mg mass), it can be measured using sensitive balances and can be used to calibrate the beam intensity.

There are a number of interesting features of these values: (a) the displacement is about the same as the distance between Na^+ and Cl^- ions in a salt crystal and that this displacement decreases with increasing frequency, (b) that particle velocity is around 1 cm/s, and (c) that the pressure variations (2 × 17.6 kPa) are rather large and are in fact around 1/3 of an atmosphere in this case.

23.2.2 Safety of Ultrasound

Although the ultrasound radiation is nonionizing and therefore unlikely to cause chromosomal or genetic damage at low levels, the pressure variations just referred to exert mechanical stress on biological material. This manifests itself in three ways: cavitation, due to the appearance and collapse of small bubbles in the biological fluid; fluid streaming, due to the movement of organelles within a cell and third heating, due to the friction of particles in motion. Although the average intensities used in diagnostic ultrasound are low (around 10 mW/cm^2) the beams are often pulsed, with maximal (spatial peak temporal averaged) intensities above 1 W/cm^2. Indeed, the use of ultrasound therapeutically relies on the warming effect within tissues. High power ultrasound is used to de-scale teeth and dentures and to clean jewelry. The World Federation for Ultrasound in Medicine and Biology (WUFMB) evaluate the safety of ultrasound for use in diagnosis and issue recommendations from time to time (http://www.wfumb.org/44-reports.htm).

23.2.3 Generation of Ultrasound

In biomedical applications ultrasound is generated by exploiting the piezoelectric effect in small ceramic crystals, usually a solid solution of lead zirconate and lead titanate (abbreviated PZT). The crystal thickness (z') is determined by the desired frequency of operation f (such that $z' = c/2f$, where c is the wavespeed in PZT). For example, a 2.5 MHz transducer would need to be 0.75 mm thick, if the wavespeed in PZT is 1875 m/s. The faces of the crystal have a thin metallic layer, across which a voltage pulse can be applied. The pulse is somewhat similar to striking a bell, in that the crystal will resonate at its natural frequency and emit sound (displacement of particles from their equilibrium position) from its surfaces as shown in Figure 23.2.

A 100-V pulse will produce a change in thickness in the order of 10s of nm. The crystal, or ultrasonic transducer, is normally mounted in an acoustic insulator or backing material, such that the acoustic pulse emanating from the left hand surface in the diagram is absorbed. The transducer behaves much like a small loudspeaker in a mobile phone handset. There is also a thin layer on the right hand surface which performs the dual role of impedance matching and protection. The thickness of this layer is a quarter of a

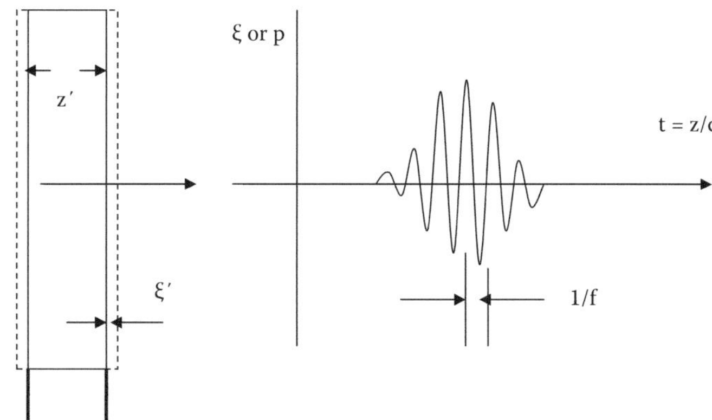

FIGURE 23.2
Left: ultrasonic generator, consisting of a piezoelectric crystal z' thick. The faces have conducting layers deposited on them, so the application of a voltage pulse will give rise to a displacement of the crystal lattice to give an additional thickness of $2\xi'$. Right: displacement ξ or pressure p variations within the medium into which the ultrasound in directed (as a function of time).

wavelength in that particular layer, and ensures that returning echoes are transmitted into the crystal, rather than being reflected off the surface.

When a returning echo impinges on the crystal surface, it behaves rather like a microphone, in that the sonic displacements generate a small voltage across the crystal faces (the inverse piezo-electric effect). Pressure amplitudes of 17.6 kPa (due to an intensity of 10 mW/cm^2) will produce around 300 mV. As we saw above, the typical echo will only be 10^{-4} of the intensity (10^{-2} of the amplitude) associated with the incident beam, thus 3 mV would be typical.

23.2.4 Attenuation of Ultrasound Beams

As the ultrasound beam passes through biological or other media it becomes attenuated. This can be due to absorption of energy because of frictional forces in the media, but it could also be due to scattering off point objects or by being reflected from interfaces between different tissues. The absorption process can be modeled as a simple exponential fall-off as follows:

$$A_z = A_0 \exp(-\mu z)$$

where A_z and A_0 are the pressure (or particle displacement) amplitudes on either side of a slab of material, thickness z. The dimensions of μ are the reciprocal of the units of z (cm^{-1}, for example). A similar equation could be used to describe the fall-off of beam intensity

$$I_z = I_0 \exp(-\mu' z)$$

where $\mu' = \mu/2$, and the factor of 2 recognizes that $I \propto A^2$. However, because the voltage signal from the transducer is proportional to A, the first form is preferred. Two other measures of absorption are used: α (in dB/cm) and $l_{1/2}$ or "half-value layer thickness." The conversions are as follows: $\alpha = 8.69\ \mu$ (the factor 8.69 is in fact $20(\log_{10}(e))$) and $l_{1/2} = 0.693/\mu = 6.02/\alpha$.

In general, $l_{1/2}$ decreases with increasing frequency: in fact the relationship is $l_{1/2} \propto 1/f^h$, where h is slightly greater than unity (Table 23.3). Although higher frequencies are associated with better spatial resolution, the depth of penetration is poorer, so cannot be used for imaging deep structures. Note that because the amount of absorption in air is high ($l_{1/2}$ low) forming images of lung tissue is difficult: this also implies that the transducer needs

TABLE 23.3

$l_{1/2}$ Values (in cm) for Various Media

	1 MHz	5 MHz	20 MHz
Air	0.25	0.01	-
Bone	0.2	0.04	-
Perspex	7	1.4	0.35
Muscle	1.5	0.3	-
Olive oil	116	6.3	0.5
Fat	5	1	0.25
Blood	17	3	1
Water	1400–3000	54	3.4

some coupling gel between it and the skin to ensure that there is minimal loss. The high absorption of bone means that visualizing organs in the chest must be done between the ribs. Because of low absorption by water, it is standard practice to perform uterine examinations when the patient has a full bladder.

A typical value for absorption is soft tissues is 1 dB/cm ($l_{1/2}$ of 6 cm) at 2 MHz. The ultrasound pulse is attenuated on both the outward path and on its return as an echo. Thus, an echo which is received from a blood/muscle interface 10 cm from the transducer will be −60 dB (10 dB attenuation for each 10 cm traversed and a further 40 dB attenuation from the interface as was determined above). The typical echo would be further attenuated from the 3 mV value given above, to 0.3 μV.

23.2.5 Shapes of Ultrasound Beams

The transducer can be though of as a piston, projecting alternating compressions and rarefactions into the media it is in contact with. The width of the beam is determined by the diameter of the transducer, with the beam remaining parallel for a certain distance (near field) before beginning to diverge (far field) (Figure 23.3). The near-field distance and the divergence angle in the far-field is shown in the diagram below. There is a trade-off in trying to make the beam too narrow, in that if D is too small the near-field distance will be small and the divergence angle θ will be large.

Typical values for D are 20 mm at 1 MHz and 5 mm at 10 MHz and for θ: 2°–5°. Note that these values correspond to near-field distances of 70 mm at 1 MHz and 40 mm at 10 MHz.

The intensity of the beam varies in a complicated way with distance in the near field, but in the far field falls off approximately as $1/z^2$.

One way to make the beam narrower is to use a curved crystal, producing a beam shape as shown in Figure 23.3 (lower).

23.2.6 Ultrasound Modes

Historically, medical ultrasound has been divided into a number of modes, each one exploiting a different aspect of the echo signal. These modes were briefly discussed in Chapter 11 and are: A, B, Real-time, M (TM), and Doppler.

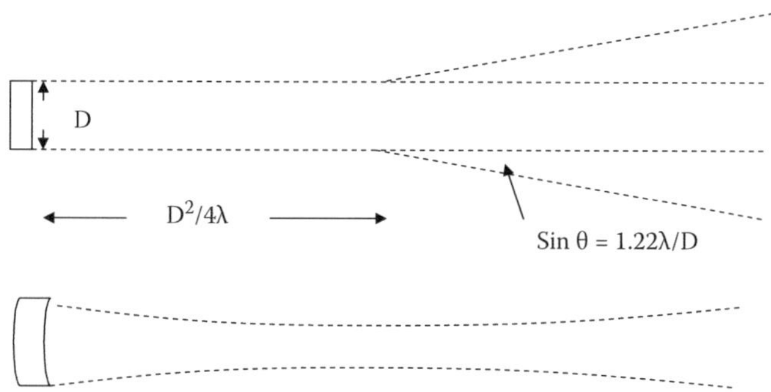

FIGURE 23.3
Top: divergence of an ultrasound beam from a flat transducer in the far field, with the distance to the far field given by the formula shown. Bottom: concave-faced transducer leading to beam focusing as shown. Maximal resolution will be obtained at this focal point.

23.2.6.1 A (or Amplitude) Mode

This is a simple display of the ultrasound echo signal on the y axis of an oscilloscope, with the timebase set on 0.01 ms per division. Since sound travels at approximately 1500 ms^{-1} in tissue, the time taken for the ultrasound pulse to reach the muscle/blood interface (assuming this to be 5 cm from the transducer) will be 0.067 ms. Assuming the next interface is a further 5 cm away, then the second echo will be another 0.067 ms later. Note that the second echo is attenuated in relation to the first.

In Figure 23.4a, the regions M and B indicate the muscle and blood layers, respectively. Note that the transducer acts as both as a transmitter (loudspeaker) and receiver (microphone). The input stage of the amplifier needs to be protected somehow from the pulse used to excite the transducer crystal (which could be 30 V in amplitude and a few μs in duration.

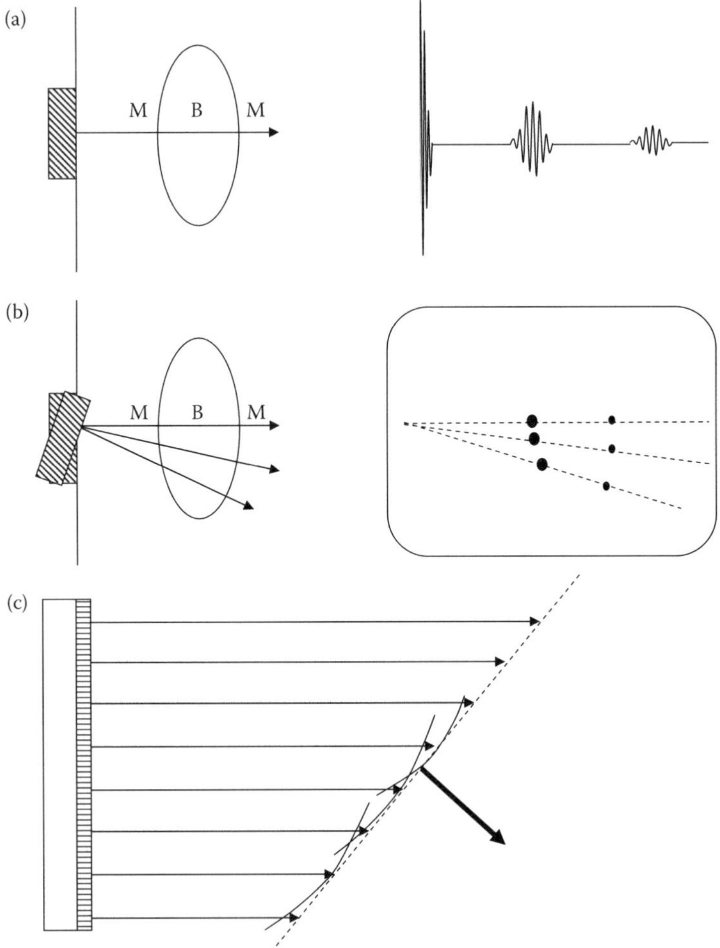

FIGURE 23.4
(a) Simple A-mode display, showing echoes from the muscle/blood interfaces: the two echoes are 0.067 s apart. (b) Bright spots corresponding to the position of the interfaces. (c) Multitransducer array with constant delay between beams. Some of the wavefronts are shown. The dotted line represents the advancing resultant wavefront moving in the direction indicated.

23.2.6.2 B (Brightness) Mode

In this mode the intensity of the "spot" on the display screen modulated by intensity of echo: the stronger the echo, the brighter the spot. The line joining the spots is made to correspond to the direction of the beam within the patient's body. If the direction of the beam alters, then the line joining the spots will alter, too, as shown in Figure 23.4b.

Each time the transducer sends out a pulse, a new line of spots will appear on the screen. A typical pulse repetition rate is 3 kHz, so pulses are sent out at 0.33 ms intervals (remember that the echoes are taking around 0.15 ms to return to the transducer). To collect a "frame" of 256 lines (in a fan formation) a time of 84 ms is required. Remember that this represents a 2D "slice" into the chest in a plane determined by the direction the transducer is moved. Scanning can be done manually, mechanically or (as is more usual in modern instruments), electronically. Electronic scanning uses a phased array of tiny transducers to artificially construct a wave-front. Figure 23.4c shows 8 such beams, each beam progressively more delayed from bottom to top. The arrowheads represent where each pulse has got to: from the Huygens principle, the individual pulses can be thought of as a plane wave advancing in the direction shown. By altering the relative delays between pulses, the direction of the beam can be altered, to produce a range of angles.

By changing the beam direction really quickly, it is possible to do a fan-shaped scan over a particular area in around 50 ms. Images can be updated faster than the "flicker fusion frequency," so appear as a movie in real-time. By having a 2D array of transducers a 3D image can be obtained from "slices."

WORKED EXAMPLE

A multielement ultrasonic linear array is placed (with coupling gel) on one surface of a test block as shown in Figure 23.5 (A is air, B is Perspex, and C is water)

	Density (kg/m)	Wavespeed (m/s)	Absorption dB/cm
Air	1.0	330	14.4
Perspex	1200	2700	4.95
Water	1000	1500	0.0174

(i) If the transducer is producing a plane wave, what would the (B mode) image look like?

(ii) Calculate the relative heights of the two main echoes produced by beam X (ignore multiple reflections and losses at the transducer interface). (You may express your answer in dB if you wish).

Answer

(i) The B-mode image would look similar to the lower part of Figure 23.5, with echoes from the upper and lower sides of block B and the bottom of C, but with only part of A being imaged, because of the curving surface. Part of the lower interface may also be in C's shadow.

(ii) The first echo will come from the front surface of block B. Since the question asks the relative height of the two echoes we need to consider the attenuation of each. For the first echo, the attenuation produced by a two-way journey through material C will be 6 × 0.0174 = 0.10 dB. The attenuation produced by the reflection from the interface will be $(\rho_C c_C - \rho_B c_B)/(\rho_C c_C + \rho_B c_B)$ = (1000 × 1500 – 1200 × 2700)/(1000 × 1500 + 1200 × 2700) = (–) 0.367 or –8.7 dB, or –8.8 dB in toto. The second echo results from the following processes: two trips through C, two transmissions through the C/B interface, two trips through B and then a reflection from the B/A interface. The attenuation through 3 cm of C and 4 cm of B is 6 × 0.0174 + 4 × 4.95 = 19.9 dB. The attenuation from reflection from the bottom surface is (1200 × 2700 – 330)/(1200 × 2700 + 330) = 0.9998 or 0 dB and the amount transmitted across the C/B interface is found by considering the power flux. The fraction of power reflected is 0.367^2 (see above), so the power transmitted is 1 – 0.367^2 = 0.865. The amplitude ratio is found by taking the square root, but since the amplitude is reduced by the same amount on the return journey, the amplitude attenuation is $(\sqrt{0.865})^2$ = 0.865 or 1.25 dB attenuation. This the total attenuation is 19.9 + 1.3 + 0 = 20.5. Thus, the difference in height is 20.5 – 8.8 = 11.4 dB.

Chapter 11 describes the use of ultrasound Doppler to measure blood flow. This information can also be obtained, either "continuous" (giving information about flows at all distances from the transducer) or "pulsed," where the returning echoes can be "gated" to select only those returning within a small time interval (and therefore from a particular depth in the tissue). The Doppler information is often displayed as a color-coded overlay to the 2D or 3D image.

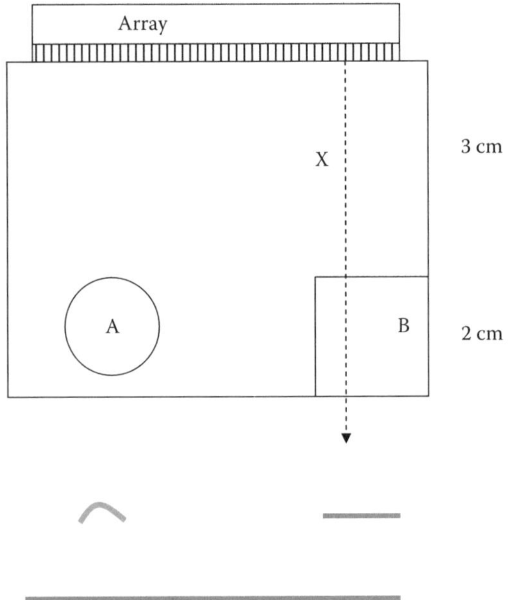

FIGURE 23.5
Top: test object described in worked example; Bottom: answer to part (i).

23.2.6.3 M (or TM: Time-Motion) Mode

This is mainly used to follow heart valve motion. For this, the beam direction is fixed, but the display scrolls to give a "history" of where the structure is. Suppose the MB interface oscillates to-and-fro (as it does when the heart is beating).

Displays are usually rotated through 90°, so that downwards represents deeper into the tissue and L-to-R indicates the passage of time. Heart valves have a characteristic "M" (or "W") shaped signature, and the maximum slope gives the closing or opening velocity of the valve leaflet (which will be less in disease).

23.2.7 Resolution and Artifacts

There are two aspects to resolution (that is, the smallest distance apart two features will appear as separate echoes). Axial resolution is related to the wavelength and also the shape of the echo signal, with better resolution associated with higher frequencies. For example, at 1 MHz and 10 MHz the axial resolutions are 4.5 mm and 0.5 mm, respectively. On the other hand, lateral resolution is related to beam shape and the amount of focussing (Figure 23.3 [lower]). This also relates to the crystal diameter. Lateral resolution of 2 mm at 10 MHz would be typical.

Artifacts are mainly to do with multiple reflections from interfaces with large acoustic impedance mismatch, tissue/bone or tissue/air, for example. This gives rise to "ghost" structures in the image. It is also possible for echoes to reach the transducer having been reflected from two nonnormal interfaces. In fact, echoes reaching the transducer due to refraction from a nonnormal interface can give the impression of being directly below the transducer, but being in fact some distance away from the initial beam direction.

23.3 X-Ray Techniques—Plain X-Rays and Computed Tomography (CT)

23.3.1 Basic Principles of the Generation and Detection of X-Rays

X-ray generation can be thought of as being the inverse of the photoelectric (PE) effect—that is, electrons are accelerated against a metal target (as shown in Figure 23.7 below) to give up some of their energy as electromagnetic radiation (the PE effect involves the release of electrons from a metal surface due to the absorption of electromagnetic radiation, typically light). There are two forms of electromagnetic radiation emitted in the former circumstances: bremsstrahlung, or "braking radiation" due to some of the kinetic energy being converted to electromagnetic radiation (the remainder to heat) and characteristic radiation, due to the ejection of one of the metal atom's inner electrons and subsequent electromagnetic radiation of a specific wavelength being emitted as another electron falls into the vacated position from a higher orbit (Figure 23.6). The wavelength λ is given by

$$hc/\lambda_{ij} = E_i - E_j$$

where h is Planck's constant, c the velocity of light and E_i and E_j the energy levels of the two electron orbital involved. Bremsstrahlung wavelengths are mainly in the range

10 pm–100 pm, with a minimum value determined by the voltage between anode and cathode. The minimum wavelength is in fact given by the formula:

$$\lambda_{min} = 1.24 \times 10^{-9}/V,$$

where V is the accelerating voltage, in kV. In the case of a typical medical X-ray operating voltage of 80 kV, the value of λ_{min} is thus 15.5 pm. This wavelength corresponds to all of the kinetic energy of the electron being absorbed to form bremsstrahlung and no heating takes place. The longer wavelengths correspond to progressively more of the KE being absorbed as heat (Figures 23.6 and 23.7).

The characteristic radiation is determined by the values of the energy levels of the metal target, and for tungsten the main emissions are at 58 keV and 68 keV, where eV refers to the "electron volt" a measure of energy (in fact, the maximum energy the electron attains when accelerated by 80 kV is 80 keV: note that 1 eV is equal to 1.6×10^{-19}J). Since it is important to have monochromatic X-rays (that is, at a single wavelength, to avoid dispersion effects) it is usual to filter out much of the low-energy or "softer" X-rays. This is known as "beam hardening" and is accomplished by the use of thin aluminum disks placed in the exit window, as shown in Figure 23.7.

The grid between the patient and the recording device consists of thin strips of lead interspersed with less absorbent material, such as perspex (Lucite). The purpose of this is to prevent any rays produced by secondary emissions (scattering) from being detected. In order to prevent a "Venetian blind" artifact in the image, the grid is oscillated to-and-fro, thus smearing out any such artifact. There will be some loss in recorded intensity (due to absorption by the lead strips) but this is outweighed by the decreased "fogging" due to secondary emissions (backscattered radiation).

Figure 23.7 shows as a recording system a cassette containing a photographic film. Although these analog systems are still in use in many locations, they are progressively being replaced by digital recording devices.

The main determinants of the ability of material to absorb X-rays are the density and atomic number of the material and the principal energy of the X-rays. Absorption

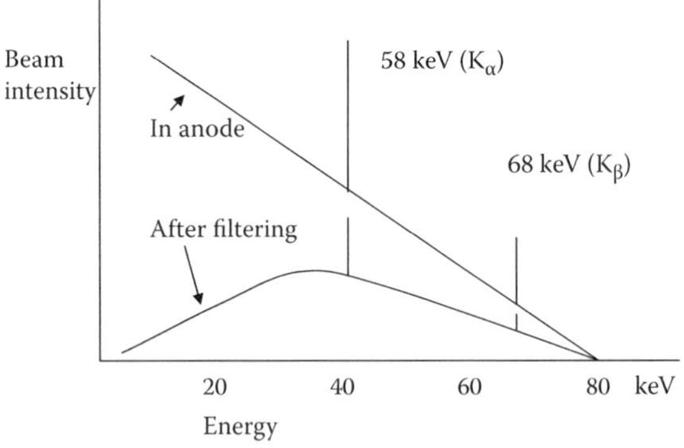

FIGURE 23.6
Spectrum of intensity of beam from a Cu target (top) and after filtration through a thin Al plate (bottom). The lines represent characteristic radiation and the rest of the spectrum is due to Bremsstrahlung. The filtration removes much of the "soft" radiation and ensures that the beam is concentrated into a smaller range of energies (or wavelengths).

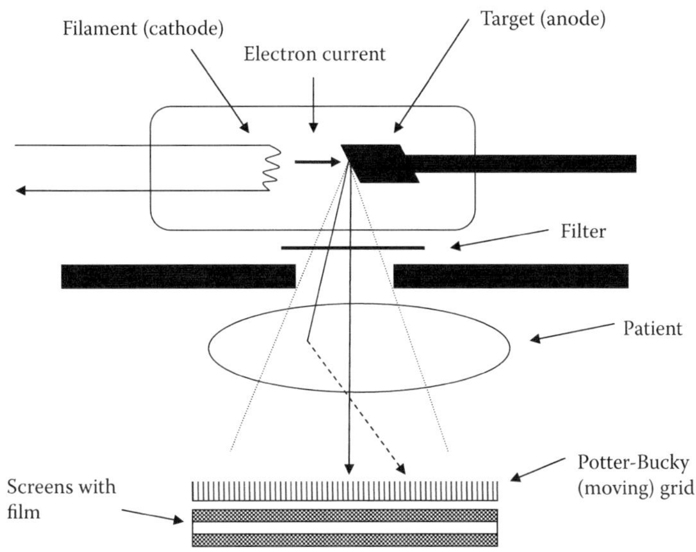

FIGURE 23.7
The production of X-rays from accelerated electrons colliding with the target anode. The film records direct X-ray beams and the grid is designed to absorb scattered X-ray beams (shown with a dotted arrow). The screens in film cassettes are designed to fluoresce when the X-rays are absorbed in them. The fluorescence (light) is then recorded by the film, which is far more sensitive to light than to X-rays.

coefficients are defined in terms of the decrement in intensity, rather than amplitude as in the case of ultrasound. Thus:

$$I_z = I_0 \exp(-\mu z).$$

Because of the expected variation with density (ρ), a mass absorption coefficient (in m²/kg) is defined as

$$\mu_m = \mu/\rho$$

and the variation with atomic number Z is $\mu_m \propto Z^3$, approximately. This explains why (at 50 keV) the value of μ_m for bone, which is composed mainly of Ca and P (Z = 20 and 15, respectively) is approximately 4 times more than muscle, which is similar to water (for O, Z = 8). Lead has a value around 24 times that of water. At 100 keV, the higher photon energy of the X-rays means that the radiation will penetrate further (the values of μ_m are less) but there is less of a difference between bone and muscle (hence the contrast in the image is less).

The dark areas in a conventional film thus represent regions where little absorption has taken place before the X-ray intensity I_z has been recorded (by the film or by a discrete photon detector) and conversely, the regions of the patient where there is a large amount of bone will greatly attenuate the intensity I_z, thus the image will appear pale below those regions. The precise relationship between the darkness of the image and the value of I_z depends on the characteristics of the image detector, but in general, there is a relationship between the optical density (OD) of the image and the X-ray exposure (X-ray intensity [I_z] × time of exposure) of the form shown below in Figure 23.8. The OD is the $\log_{10}$ of the

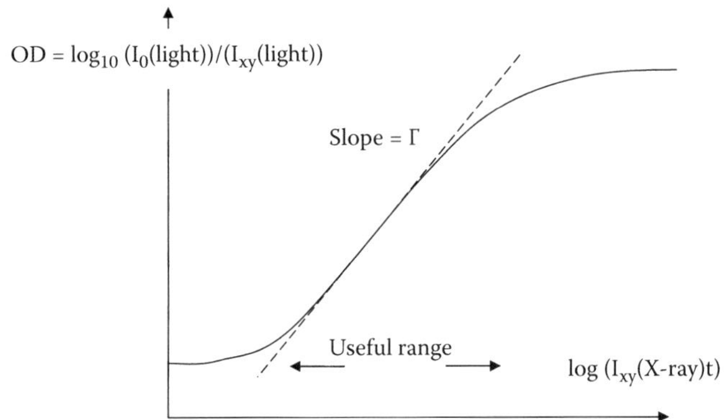

FIGURE 23.8
The characteristics of X-ray images, recorded on film or on a computer screen. The y-axis represents the darkness of the film (the darker the film, the higher the OD). The x-axis represents the X-ray "dose" which produced that darkening. This is a product of the intensity of X-rays impinging on the detector (film, for example) and the time over which this happens (exposure time).

ratio of the intensity of the "white" portion of the image to the intensity at a given location in the image (if X-ray images are being viewed on a light-box, this is the ratio of the light intensity from the light-box before the film is put into place to the intensity from given regions of the surface of the film after it has been put into place. The slope of the central region is known as the "gamma" of the imaging system, and optimally, all regions of the film (or image generally) should be within the range indicated (known as "latitude") (Figure 23.8).

WORKED EXAMPLE

The following data were obtained by measuring the amount of light transmitted through an X-ray film taken of a copper step wedge with each step 0.1 mm thicker than the last. The incident illumination can be assumed to be spatially uniform. The X-ray absorption coefficient at 60 keV is 1500 m^{-1}. Estimate gamma for the film.

Thickness (mm)	Light Intensity (mW/m^2)
0.1	2.7
0.2	4.5
0.3	7.4
0.4	11.6
0.5	20.1
Background	0.2

Remember, the more X-rays absorbed the lighter the film and the greater the transmitted intensity. First, the background is subtracted from each value, then the value of $\log_{10}$ of *light* intensity found. High values of OD = $\log_{10}(I_0/I)$ are associated with blacker regions of the film, but higher values of I (lighter film) correspond to a greater X-ray absorption and hence a lower exposure (lower X-ray intensity). From the formula $I'_z = I'_0 \exp(-\mu z)$ where I′ here refers to *X-ray* intensity, we note that there will be a relationship between μz and the log of light intensity. However, since the optical density of film (OD) is defined in terms of $\log_{10}$ the slope of the graph of OD versus μz will need to be multiplied by ln(10) to adjust for this difference.

μz	$\log_{10}$ (Light Intensity, Background Corrected)
0.15	0.4
0.3	0.63
0.45	0.86
0.6	1.06
0.75	1.3

These figures can be plotted via a spreadsheet and the slope estimated. The value is 1.49. Thus, gamma is 1.49 × 2.303 = 3.43.

23.3.2 Computed Tomography

A tomographic image is one that isolates a single plane or slice within the object, in this case the human body. Before the era of electronic computing, tomographic images were obtained by simultaneously moving the X-ray tube and the image detector around a pivot point within the body. The movement smeared out all objects above and below the pivot plane in the image, so the image was clear for one plane only.

The earliest computed tomography systems consisted of a X-ray source, producing a thin or "pencil" beam whose intensity was registered by a detector, as shown in Figure 23.9. The source and detector were moved together in the x direction in the diagram. The registered intensity at a given point x (N(x) is given by the formula

$$N(x) = N_0 \exp(-\int \mu(x,y)dy)$$

where N_0 is the intensity (or number of photons/s) being emitted from the source (which is assumed to be constant) and μ(x,y) refers to the X-ray absorption coefficient at a specific location in the x,y plane. For a uniform structure $N = N_0\exp(-\mu y)$. The integral is along the line from the source to the detector. If the body has a high quantity of bone along line x_1 the value of N(x) will be low and conversely will be higher where only soft tissue is encountered (x_2).

Having determined these "line integrals" for locations along the x-axis, the x and y directions are then rotated through a few degrees and the process repeated (this involves mounting the source and detector on a gantry which can be rotated around the patient) Figure 23.10. After 90 such rotations, each 2° more than the last, the body will have been "viewed" from sufficient directions to remove any ambiguities that might otherwise exist regarding underlying structures. In general, for an angle θ′ made by n rotations (each θ_0 from the initial position), it is possible to define a "transmission" $T_{\theta'}(x')$ as follows:

$$T_{\theta'}(x') = -\ln(N(x')/N_0) = \int \mu(x',y')dy'.$$

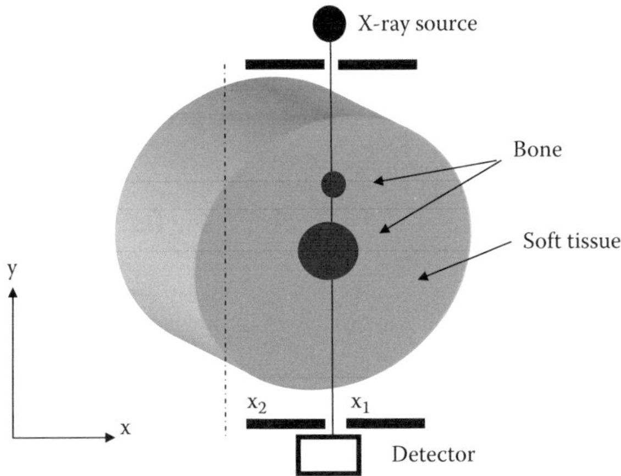

FIGURE 23.9
The principle of X-ray computed tomography (CT). The source and detector are moved together in the x-direction to take a series of line attenuation measurements over the body (a "projection").

In other words, this represents the weighted sum of μ values along a particular line through the body at a particular location on the x′ axis. N.B. $T_{\theta'}(x')$ is dimensionless.

If, instead of X-rays, a strong light source was used and the object was semitransparent, then the value of N(x′) as a function of x′ is the value of intensity along a line on a projection screen placed on the far side of the object. The functions $T_{\theta'}(x')$ are often referred to as "projections," although of course the regions of greatest absorption are those with the highest value of $T_{\theta'}(x')$. For example, the $T_0(x)$ for $\theta = 0$ (the top diagram) would look something like Figure 23.11a and the corresponding "projections" for $\theta = 45°$ and 90° something like Figure 23.11b and c.

Instinctively, it should be possible to work back from the $T_{\theta'}(x')$ values for each value of θ to reconstruct a map of the actual values of absorption coefficient in each location in the x,y

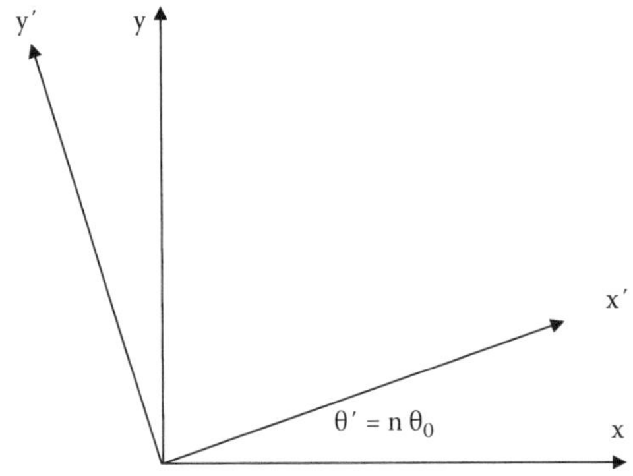

FIGURE 23.10
CT imaging. The source and detector are then rotated by a small amount ($\theta_0 = 2°$, say) and then the process repeated (new axes are represented by x′ and y′ after n rotation steps).

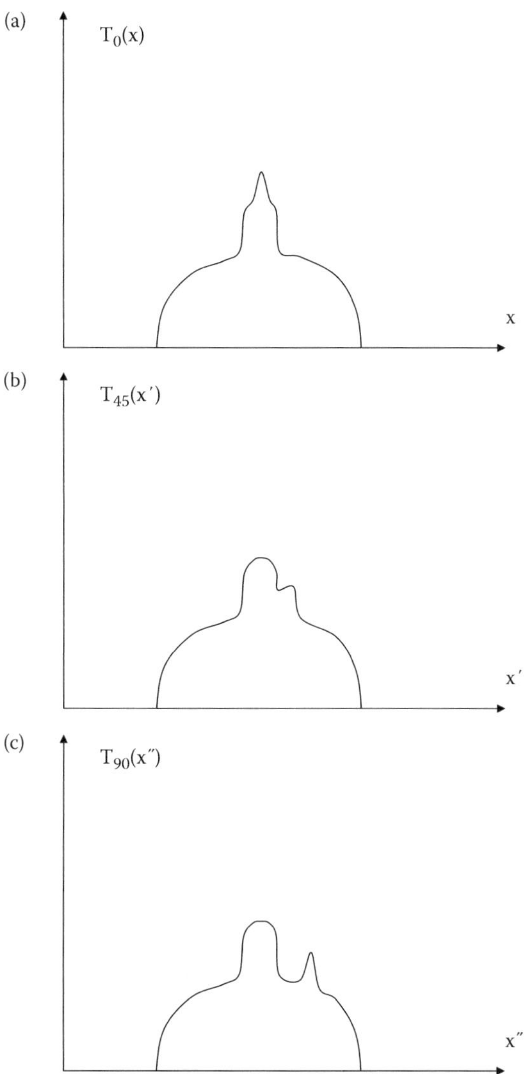

FIGURE 23.11
Three CT projections $T_\theta(x)$, for $\theta = 0$, 45, and 90°, for the bony features shown in a simplified torso in Figure 23.9.

plane ($\mu(x,y)$). For illustrative purposes only, we will consider first the simplest reconstruction method, successive iteration. Suppose there are only two projections, T_0 and T_{90}, both with only two values 1 cm apart as follows:

$$T_0\text{: } 0.4, 0.6$$
$$T_{90}\text{: } 0.7, 0.3.$$

We make an initial guess based on the average values: 2.5 cm^{-1} in each cell. This can be analyzed as follows (Figure 23.12a):

(a) First guess at the 4 μ values: comparing against T_{90} values:

			Measured	**Calculated**	**Error**	**Error/cell**
2.5	2.5	→	0.3	2.5 + 2.5 = 5	4.7	2.35
2.5	2.5	→	0.7	5	4.3	2.15

Corrected guess: subtract error/cell from each 1st guess, then compare to T_0 values:

2.5 – 2.35 = 0.15	2.5 – 2.35 = 0.15
2.5 – 2.15 = 0.35	2.5 – 2.15 = 0.35
↓	↓

Measured	0.4	0.6
Calculated	0.15 + 0.35 = 0.5	0.15 + 0.35 = 0.5
Error	0.1	–0.1
Error/cell	0.05	–0.05

Now: The corrected second guess gives the observed $T_\theta(x')$ values from the μ values:

0.15 – 0.05 = 0.1	0.15 – (–0.05) = 0.2
0.35 – 0.05 = 0.3	0.35 – (–0.05) = 0.4

(b)

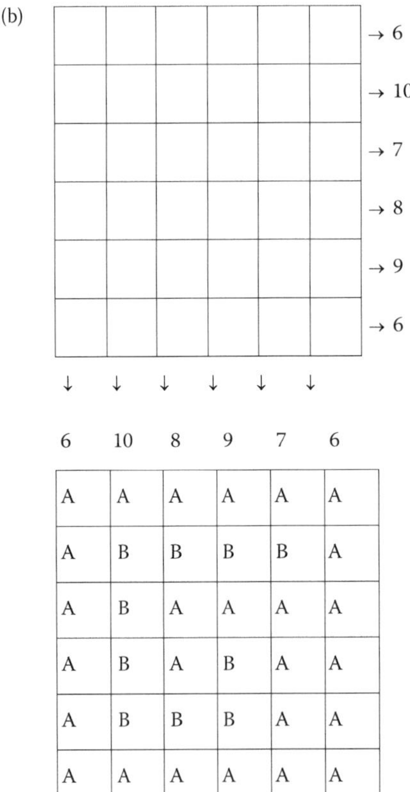

FIGURE 23.12
(a) A simple iterative procedure for "discovering" the μ values in four contingent cubes of differing materials from data two projections T_0 and T_{90}. (b) An even simpler procedure where the 5 × 5 array consists of just two materials of known m values. See worked example.

This method can of course be extended to 90 projections, each with, say 128 measurements along each x′ axis. The computing time could prove to be a limiting factor, and certainly was in early days of CT imaging.

WORKED EXAMPLE

A test object is made from 1 cm cubes of two substances, A and B, where the absorption coefficients are 1 cm^{-1} and 2 cm^{-1}, respectively. The object is then X-rayed from the top and the side (see Figure 23.12b) and the values of "transmission function" $-\ln(I/I_0)$ are measured (these values are shown to the right and below the diagram). Which of the blocks are A and which B?

Answer

Since $-\ln(I/I_0) = \int \mu(x', y')dy' = \Sigma \mu_i \Delta y_i$ and $\Delta y_i = 1$ cm, the maximum and minimum values must be 12 and 6, respectively. Thus, the top and bottom rows and first and last columns must be all "A." The remaining blocks in row 2 and column 2 must therefore be "B." Following similar argument, it is possible to identify the other blocks as shown in the diagram in the lower part of Figure 23.12b.

23.3.2.1 Filtered Back-Projection

This procedure effectively replaces each element by the initial projection value, then the values from each subsequent projection are added together. To the example above an additional T_{45} is included, with three values, 0.3, 0.1 + 0.4 = 0.5 and 0.2 (this ignores, for the moment, the factor of $\sqrt{2}$, for the extra distance along the diagonal) (Figure 23.13).

If we now subtract 1 from each cell then divide by 3 we get the original values. However, there is a problem with this approach—which is illustrated by the need to subtract an amount (in this case 1) from each of the computed values. This arises out of the unwanted contributions to neighboring cells, best illustrated by the following example of a 5 × 5 grid with a single object (absorption value 1) at the center. Again, ignoring the factor $\sqrt{2}$ for the diagonals, the sums of the distributed projection values would be as shown in (Figure 23.14).

As more angles of projection are added (like spokes of a wheel), adjacent projections will tend to overlap, particularly near the central cell creating a volcano-like effect. Thinking in terms of the analogy mentioned earlier, if the projections were to be set up in the directions they were initially created in and "back-projected" using a lamp behind each projection (represented as a film transparency), then the tracks of light on a table-top viewed from above would give a rather blurred representation of the original 2D object. Alternatively, if each projection were represented by an array of LEDs, whose brightness represented $T_\theta(x')$, placed so that the pencils of light grazed the surface of a table, then the intersecting modulated beams would give a similar representation of the object. This is illustrated in Figure 23.15 (left).

The problem then is to modify the projections $T_\theta(x')$ in such a way that when they are recombined, the "spokes" disappear and a clear image emerges. The method relies on the fact that if a large number of projections are superimposed the spokes merge into a "volcano-like" artifact surrounding each point pixel in the image and that this has the form $I = k/r$, where r is the radial distance from the center of the pixel. The values of $T_\theta(x')$ will be modified to give a new set of projections $G_\theta(x')$, derived from a mathematical transformation which will be described shortly. Before doing this however, an approximate method will

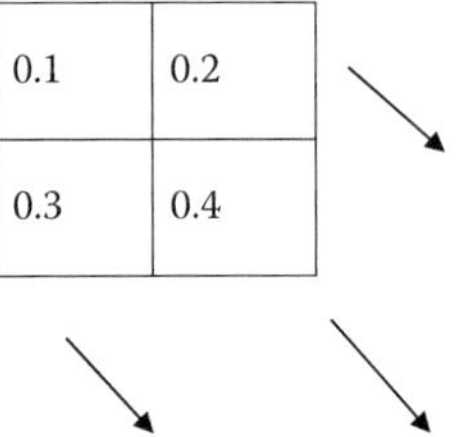

Adding T_{90} and T_{45} gives

0.3 + 0.5 = 0.8	0.3 + 0.2 = 0.5
0.7 + 0.3 = 1.0	0.7 + 0.5 = 1.2

Then adding in T_0

0.8 + 0.4 = 1.2	0.5 + 0.6 = 1.1
1.0 + 0.4 = 1.4	1.2 + 0.6 = 1.8

Last, add in T_{-45}, whose values are: 0.1, 0.2 + 0.3 = 0.5, 0.4

1.2 + 0.1 = 1.3	1.1 + 0.5 = 1.6
1.4 + 0.5 = 1.9	1.8 + 0.4 = 2.2

FIGURE 23.13
This procedure "discovers" the m values by superimposing (back-projecting) the projection values T_θ for 0, 45, 90, and –45. This procedure is closer to the "filtered back-projection" methods used in practical reconstructive imaging.

1	0	1	0	1
0	1	1	1	0
1	1	4	1	1
0	1	1	1	0
1	0	1	0	1

FIGURE 23.14
See worked example.

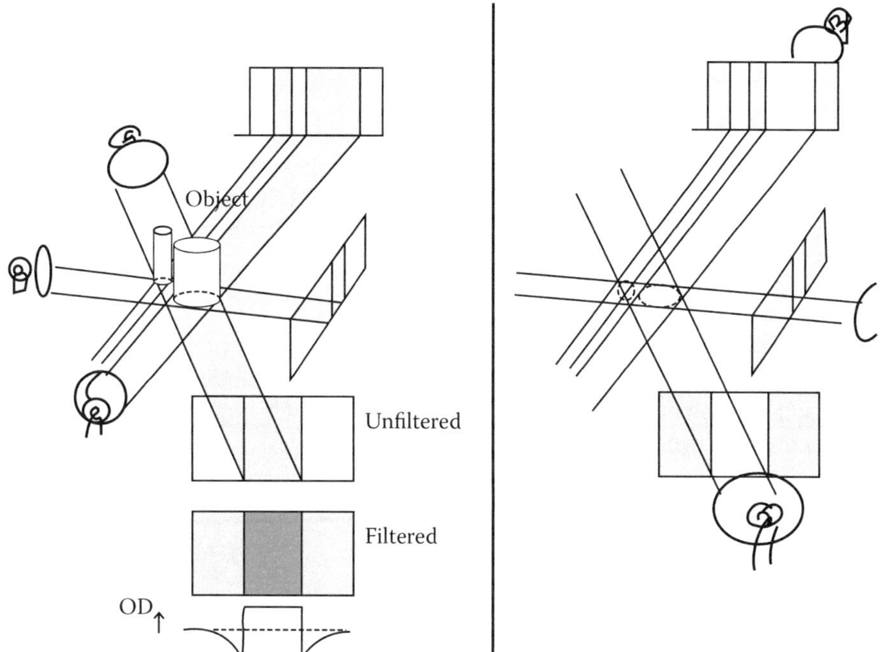

FIGURE 23.15
A cartoon representation of filtered back-projection. On the left the objects produce characteristic shadows when light is projected in the directions shown. These could be captured on film as a transparency. If the objects are removed, then the circular patches of light on the flat surface (shown at right) are representations of the objects when the transparencies are illuminated from behind (back-projection). Unfortunately, there are also star-shaped streaks of light surrounding the circular representation of each object. These can be much attenuated by "doctoring" the transparencies (this involves darkening the edges, as shown at the very bottom of the diagram on the left). This is equivalent to filtering the back-projections.

be described, to provide some insight into how the procedure operates. If, instead of the transparencies being black and white they are grey, black, and white (as shown) the black regions in the tracks of light on the table will tend to block the addition of the grey regions surrounding the white regions.

The precise form of the "touching up" of the projections can be derived by considering the spatial frequency content of the projections and of the 1/r artifact.

23.3.2.2 Fourier Methods

The Fourier Transform (FT) of the projection can be found as follows:

$$F[T_\theta(x')] = \int T_\theta(x')\exp(-ju'x')dx = \mathscr{T}_\theta(u')$$

where u′ is the spatial frequency in m^{-1} parallel to the x′ axis. From the definition of $T_\theta(x')$ given above we can write:

$$\mathscr{T}_\theta(u') = \iint \mu(x',y')\exp(-ju'x')dx'dy'$$

This can be compared to the 2D FT of the absorption coefficients $\mu(x',y')$

$$F[\mu(x',y')] = \iint \mu(x',y')\exp(-j(u'x' + v'y')dx'dy'$$

where v' is the spatial frequency parallel to the y axis.

This equation when compared to the equation immediately prior, shows that the FT of a projection is the central section (i.e., with $v' = 0$) of the 2D FT of the absorption coefficients:

$$F[T_\theta(x')] = F[\mu(x',y')]_{v'=0}.$$

As we saw above, the values of $\mu(x',y')$ obtained on reconstruction ($\mu_R(x',y')$) are the true values ($\mu_T(x',y')$) convolved with a blurring function, which has the form $1/r$. Using the symbol * for convolution:

$$\mu_R(x',y') = \mu_T(x',y') * (1/r).$$

Now, given that the FT of the function $1/r$ is $1/w$, where w is a distance in frequency space ($w^2 = u^2 + v^2$), and also remembering that the FT of a convolution of two functions is equal to the product of the FTs of the individual transforms:

$$F[\mu_R(x',y')] = F[\mu_T(x',y')](1/w) = F[T_\theta(x')_{v'=0}](1/w).$$

What we can now do is estimate values for modified projections ($G_\theta(x')$) such that when these are back-projected, instead of giving blurred estimates of absorption coefficient, give the true values $\mu_T(x',y')$. That is, when the modified projections $G_\theta(x')$ are convolved with the blurring function $1/r$ they give the true rather than the observed (blurred) values for μ. Analogous to the equation above, the values for $G_\theta(x')$ must satisfy:

$$F[\mu_T(x',y')] = F[G_\theta(x')]1/w = F[T_\theta(x')].$$

Or, taking inverse FTs

$$G_\theta(x') = T_\theta(x') * F^{-1}(w).$$

The inverse FT of w is of the form $(\sin(x))/x$ and the maximum frequency w_{max} (corresponding to the spatial resolution) needs to be defined.

$$G_\theta(x') = w_{max}T_\theta(x') - \Sigma\, T_\theta(x'')\sin^2(\pi w_{max}(x' - x'')/(\pi(x' - x'')^2.$$

Or, since $T_\theta(x')$ will consist of discrete readings $T_\theta(n\Delta x)$ this can be approximated to

$$G_\theta(i) = (w_{max}/2)T_\theta(i) - (2/\pi^2)\Sigma\, T_\theta(j)/(i - j)^2.$$

For j odd, where $s = 1/2w_{max}$ is the smallest spacing observable in the image.

WORKED EXAMPLE USING MATLAB TO PRODUCE A RECONSTRUCTION VIA BACK-PROJECTION

Suppose a tophat shaped solid object were to be X-rayed from the side (see Figure 23.16). Since $T(x) = \ln(I_0/I(x)) = \Sigma\mu_i y_i$ we can estimate the path lengths at various points along the x axis (the center is at x = 33 and the diameter of the hat is 32 units. If we assume that $\mu = 1$ for the hat and 0 for air, then the path length will be $2\sqrt{(16^2 - (x-33)^2)}$. The following m-file (tophat.m) estimates T(x) (and hence how the inverse density of X-ray film would look—white in the center and black at the edges). Since the object is symmetrical about the center, the T(x) would look the same, whichever angle in the x,y plane the object were X-rayed from.

```
tophat.m

% generates a tophat function
for x = 1:65
z = (16^2 - (x-33)^2);
if z<= 0
```

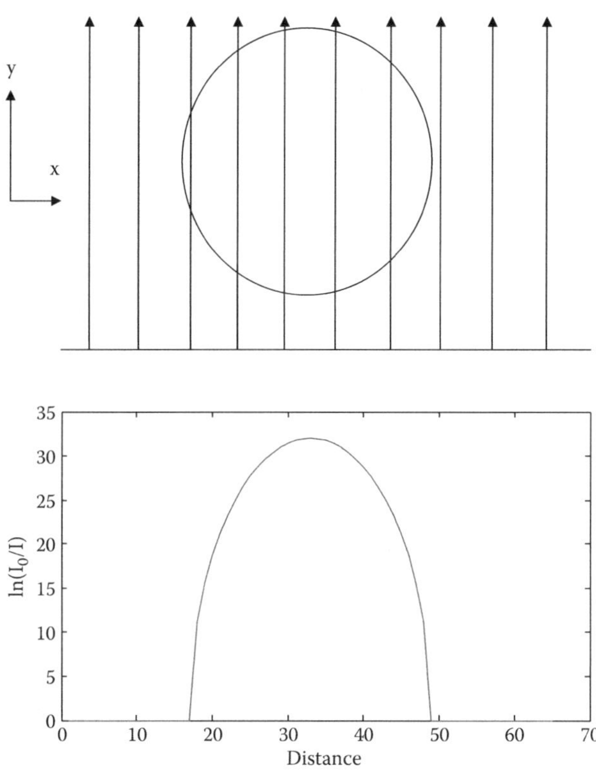

FIGURE 23.16
Top: a cylindrical object X-rayed from the side. Notice that since $I = I_0\exp(-\int\mu dx)$, we can estimate these values from the path length (obtained from geometry) through the object at each position (bottom).

```
    tophat(x) = 0;
else
    tophat(x) = 2*sqrt(z);
end
end
```

The next stage is to convolve this with a filter, in this case the Ramachandran Lakshminarayanan filter (see Figure 23.17), which is generated by the following code:

```
ramacfil.m

function Pstar=ramacfil(P)
% Ramachandran- Lakshminarayanan filter for back projection
for I = 1:65
Q=P(I)*2.467401;
JC= 1 + rem(I,2);
    for J = JC:2:65
    Q=Q - P(J)/(I - J)^2;
    end
Pstar(I) = Q/(pi*.3333*50);
end
```

This has the same number of x-axis positions, but the value P* is such that when the back-projection is performed, the "star-shaped" artifacts are greatly attenuated. The next m-file (backprl.m) performs this back-projection.

```
backpr1.m

function F=backpr1(y,phi)
% does the back projection on the output from Ramacfil
```

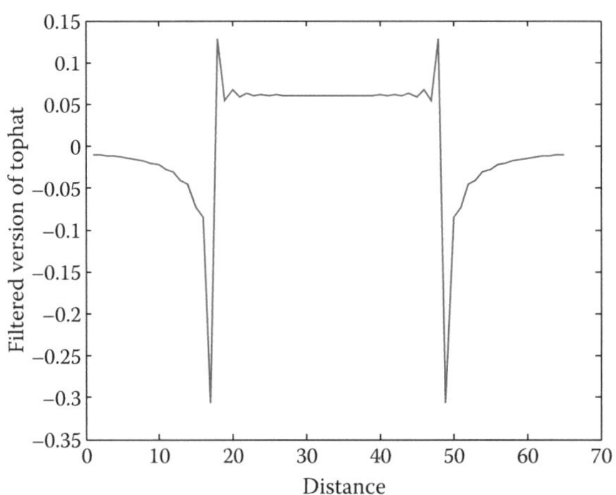

FIGURE 23.17
A digital spatial filter for modifying the projections in such a way that the "star-shaped" artifacts will be greatly reduced.

```
Image=zeros(1,4225);
    for J = 1:65
    imin = J*65 - 32 - floor(sqrt(1024-(33-J)^2));
    imax = (2*J - 1)*65 - imin + 1;
    X = 33 + (33 - J)*sin(phi) + (imin - J*65 + 31)*cos(phi);
    for I = imin:imax
        X = X + cos(phi);
        IX = floor(X);
        Image(I) = Image(I) + y(IX) + (X - IX)*(y(IX + 1) - y(IX));
        end
    end
for i = 1:65
    for j = 1:65
    F(i,j) = Image(65*(i-1) + j);
    end
end
```

Finally, the whole sequence is performed by the following m-file (reconst.m), which first of all reads the path integrals $\Sigma\mu_i y_i$ from the m-file "tophat," then generates a filtered version y by calling the filtering algorithm "ramacfil." These values are then back-projected for 30 angles from approximately 0° to 360°, generating reconstructed values μ_R (note how the various contributions are added at I = I + z) in the code below. Finally, the final I values are plotted as a mesh plot (Figure 23.18). Note a certain amount of adjustment needs to be done to make the values 0 in the surrounding area and 1 where the cylindrical top has is. Arbitrary shapes could be used instead of the tophat.

```
reconst.m

tophat;
% displays a reconstructed tophat by 1) generating a tophat
projection
% 2) doing a Ramachandran Lakshminarayanan filter on this & 3) doing
a
```

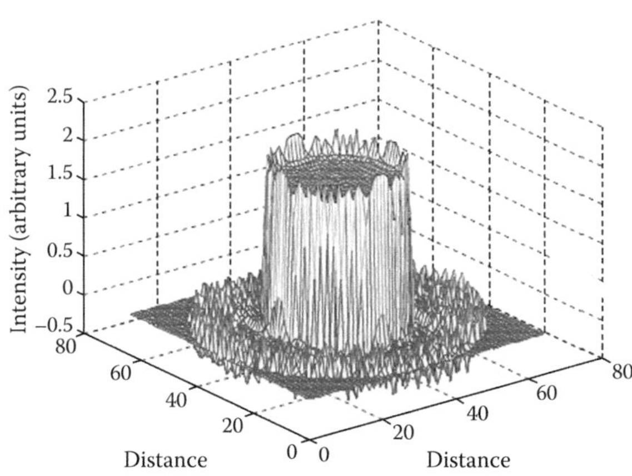

FIGURE 23.18
The reconstructed "tophat" (the object shown in Figure 23.16 [top]) using the m-files described in the text.

```
% back projection from 30 angles between 0.1 and 3.1 radians
y=ramacfil(tophat);
I=zeros(65,65);
for phi=0.1:0.1:3.1
    z=backpr1(y,phi);
    I=I + z;
end
mesh(I)
```

As an exercise, modify the m-file reconst.m so that the reconstruction is of the raw, unfiltered, back projections.

23.3.3 Modern CT Systems

23.3.3.1 Magnetic Resonance Imaging

Magnetic Resonance Imaging or MRI was developed in the 1970s and 1980s as a convenient method for obtaining 3D tomographic images without the use of ionizing radiation. MRI is based on a phenomenon known since the 1940s called nuclear magnetic resonance (NMR), which was originally exploited as a technique for analyzing chemical structure, via spectrometry. Basically, a nucleus has a *magnetic moment* due to the equivalent of a current flowing in the nucleus, due to nuclear spin. This is referred to as a *magnetic dipole*, by analogy with an electric dipole (see Chapter 8.3), even though in the magnetic case there is no equivalent to point charges separated by a certain distance. The magnetic moment μ is the current flowing multiplied by the loop area, thus $\mu = I.A = I\pi r^2$. Actually, μ is a vector quantity (**μ**), with the direction determined by the direction of the current and the right-hand rule, as shown below.

Figure 23.19a shows the nuclear magnetic moment and energy in a magnetic field B. The magnetic field exerts a turning force or torque on the magnetic dipole which is maximum when μ is parallel to the magnetic field **B** and zero when perpendicular. A subscript N will be added to the symbol μ to denote nuclear (as opposed to electron) magnetic moment. The corresponding potential energy U is given by

$$U = -\boldsymbol{\mu}_N \mathbf{B} = -\mu_N B \cos\phi.$$

The energy required to "tip" a nucleus from being aligned in the same direction as B (parallel) ($\cos\phi = 1$) to opposite (antiparallel) ($\cos\phi = -1$) is thus $2\,\mu_N B$. Actually, the values of $\cos\phi$ allowed by quantum restrictions are ±1/2. In addition, a so-called "g-factor" has to be included, which for a hydrogen (H) nucleus (which is a single proton) is 5.586, so the energy difference is $g\mu_N B$, where μ_N is given by the formula

$$\mu_N = eh/4\pi M = 5.05 \times 10^{-27}\ \mathrm{J/T}$$

where e is the electronic charge, h is Planck's constant and M the mass of a proton. This energy can be supplied by an electromagnetic source, whose quantum energy is $h\nu$, where ν is the frequency of the radiation. Thus, in this case, to supply this energy, the frequency (the Larmor frequency) needs to be $2 \times 5.586 \times 5.05 \times 10^{-27}/6.63 \times 10^{-34}$ = 42.58 MHz per T. This is within the nonionizing portion of the electromagnetic spectrum, within the radiofrequency (RF) range, in fact. The reason that NMR became such a useful tool in chemical analysis is that the protons experience a local magnetic field B′ =

$B_0 - \delta B$, where δB represents the "screening effect" due to the immediate environment of particular protons, which varies according to which chemical group the protons find themselves in. NMR chemical analyzers thus give a spectrum, where the strength of the absorption at a particular frequency is a measure of the preponderance of particular chemical end-groups (such as $-CH_3$ and CH_2OH groups).

However, in order to produce an image of an object, the effect that gives rise to the spectrum just referred to is not used. Instead, the magnetic field B is varied across the object: in this way the resonant absorption frequency varies from position to position. In the diagram below the static magnetic field varies as $B = B_0 + G.y$, where B_0 is the magnetic field at point $y = 0$ and G is the gradient in T/m. The strength of absorption is thus an indication of the proton concentration at different positions (Figure 23.19b).

Since the most likely place to find protons (H^+ nuclei) is in water (H_2O), the absorption pattern maps water content. The very first image to be obtained using this technique was that of an orange, in 1975. Obtaining a 3D image is, of course, more complicated, but the steps are as follows:

1. *Dynamics of RF absorption*

   When material is placed in a static field (in the z direction, say), it will acquire a net magnetization (M_z) as the spinning ^1H nuclei (and others that we are less interested in) align with the field. Most align parallel, but some align antiparallel, as we have seen. The application of RF as a pulse at the resonant frequency will cause some to change alignment ('flip') from antiparallel to parallel. The moment each chooses to flip varies throughout an ensemble of H nuclei. As the RF is applied, the net magnetization of the ensemble of nuclei precesses, like a top, progressively further and further away from the direction of the applied field (z). In fact, if the RF pulse is applied for $\pi/(2\gamma B_1)$ seconds (where B_1 is the strength of the RF field) the magnetization will be 90° from the z direction (in the xy plane). Detector coils in the z direction will pick up no signal at that instant, but coils at right angles (in the xy plane) will pick up a sinusoidal signal at the Larmor frequency. At the cessation of the RF pulse, this signal will decay away with a time constant determined by the speed at which the individual nuclei become out of

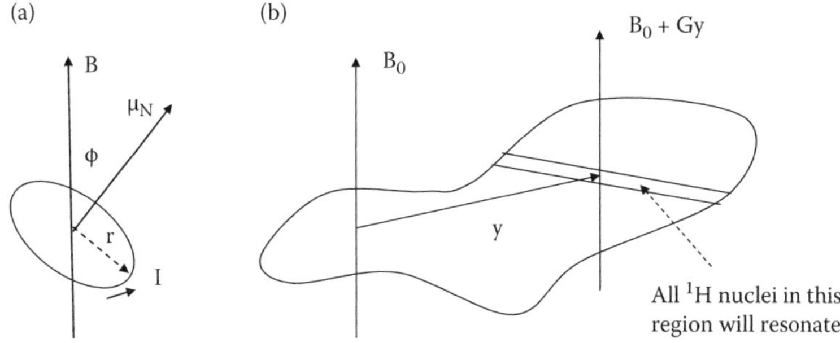

FIGURE 23.19
(a) Nuclear magnetic moment μ_N and its interaction with an applied static magnetic field B_z. (b) Magnetic resonance imaging (MRI). There is a gradient in static field in the y direction. At a fixed radiofrequency (RF) excitation frequency, only those ^1H nuclei at a particular position in the y direction will resonate. Altering the frequency of the RF will alter this position, so the whole object can be scanned.

phase with each other, due to interactions with their surroundings. This signal is known as the Free Induction Decay (or FID) and the rate of decay is known as T_2 (or actually T_2^*, which is slightly different, but can be converted to the intrinsic T_2 by a standard procedure). This is called the transverse or "spin-spin" time constant. At the same time, the signal in the z coils grows exponentially back to the initial value (M_z). The time constant for this process is T_1, which is called the longitudinal or "spin-lattice" decay constant. MRI images can display T_1, T_2 or "proton concentration" (M_z) values. Each of these emphasizes different features in the tissue and gives complementary information.

2. *Field gradients*

   We saw earlier that when a gradient in static field (G) is applied, the resonant or Larmor frequency depends on position, so spatial resolution can be produced. This is employed both in the excitation and the readout phases of the FID. For example, in the diagram above, if a RF 90° pulse is applied of frequency f′, a slice perpendicular to the y direction within the body will resonate and emit a FID. If now a gradient is applied in the x direction, a range of frequencies (close to f′) will appear in the FID, due to the concentration of protons in the different regions at right angles to the x direction, as shown. These lines are similar to the projections referred to in CT imaging. The Fourier Transform of the FID, since it has frequency on the abscissa (which is an indication of positions in the x direction) will be equivalent to a "projection." By applying gradients in different directions in the xz plane, a series of projections will be obtained at different angles. These can be "back-projected" using software algorithms to give point by point values of the quality (T_1, T_2, or M_z) being imaged, in the same way that point-by-point values of absorption coefficient μ can be obtained in CT.

3. *Spin sequences*

   In practice, T_1 and T_2 cannot be obtained from a single FID. Instead both 90° and 180° pulses are applied with varying times between them to obtain estimates of these parameters. The details of these pulse sequences are outside the scope of this book and the interested reader is referred to a detailed text (Woodward 2001).

4. *Functional MRI (fMRI)*

   Since MRI uses nonionizing radiation (RF and static fields) to produce an image, its use is not limited to diagnosing and monitoring disease, where the use of ionizing radiation has to be justified on the basis of a net benefit over the risk of harm from the radiation. Safety issues for MRI are limited to the possibility of tissue heating and of injury due to the movement of ferromagnetic objects into the strong field. Death has resulted from the rupture of major vessels due to the movement of artery clips, but these mishaps are extremely rare and patients or volunteer participants are routinely screened for implants. Studies are increasingly using volunteer participants because of the safety of this imaging modality. In particular, the discovery by three independent groups in 1992 that the T_2 values in particular are affected by the degree of oxygenation of the blood (the so-called BOLD or Blood-Oxygen Level Dependent signal) has given rise to many neurophysiological studies, aiming to localize which area of the brain is devoted to which cognitive or other brain tasks. The underlying assumption is that the metabolism associated with neural activity is indicated by a local drop in blood oxygen in venous blood, giving rise to a greater contrast of these regions compared to other regions or of that region prior to the onset of neural activity. The BOLD signal thus provides the basis for three-dimensional mapping of brain *function* at a time resolution of a few seconds, hence the acronym fMRI.

5. *Other MRI techniques*

   There are several other physiologically relevant signals which can be derived from MRI signals. For example, Diffusion-weighted Imaging (DWI) is a measure of the diffusion coefficient of water in particular locations within the body. A pulsed gradient is applied, which causes several Larmor frequencies to get excited, then an equal but opposite gradient is applied which causes imperfect refocusing of the spins due to diffusion of protons. The reduction in signal is related mathematically to the diffusion coefficient. Because this is nonisotropic, pulsed gradients are applied in three orthogonal directions to produce a Diffusion Tensor Image (DTI).

23.3.4 Emission Techniques: PET and SPECT

In emission techniques, a radioactive substance (radionuclide) is injected into the body and its emitted radiation is recorded using a detector. The radionuclide is often concentrated in particular organs due to it being conjugated with a biological agent which for a particular organ has affinity.

23.3.4.1 Positron Emission Tomography (PET)

Positrons (positive electrons) are emitted from certain radionuclides as they decay. An example is an isotope of Fluorine, ^{18}F, which has a half-life of 1.87 hours and decays to ^{18}O, emitting positrons as it decays. These positrons soon encounter normal electrons (within a few mm of their creation) and as they do so they become annihilated. The energy of the annihilation event is carried away from the point of annihilation by two γ-rays at 180° to each other (Figure 23.20). These are then detected by a pair of detectors from a ring-shaped array of detectors surrounding the patient. The pair of detectors which detect the two annihilation γ-rays within a very short time of each other (so-called coincidence events) serve to identify a line along which the initial annihilation event must have occurred (see Figure 23.20). By a filtered back-projection method or other image reconstruction algorithm similar to those just discussed, the location of the ^{18}F within a narrow slice can be determined to an accuracy of a few mm. Using multiple rings can give a stack of 2D images to produce a 3D construction.

The ^{18}F is usually joined (conjugated) with another molecule which is used to identify certain functional characteristics. For example, to study glucose metabolism in the brain, ^{18}F-labeled 2-fluorodeoxy-D-glucose, or 2FDG, is used. The brain regions where this substance accumulates correspond to those regions where there is a high demand for glucose, because of the high amount of cognitive activity there. Other positron emitters, such as ^{15}O and ^{11}C are also used in these types of study, to measure blood flow and the expression of certain neurotransmitter substances.

Because these radionuclides have such short half-lives, they need to be generated on site. This is usually done using a cyclotron, which is an electromagnetic device for accelerating charged particles (such as protons) to high energy, where they can interact with a target to produce the desired radioactive product. PET is often combined with other

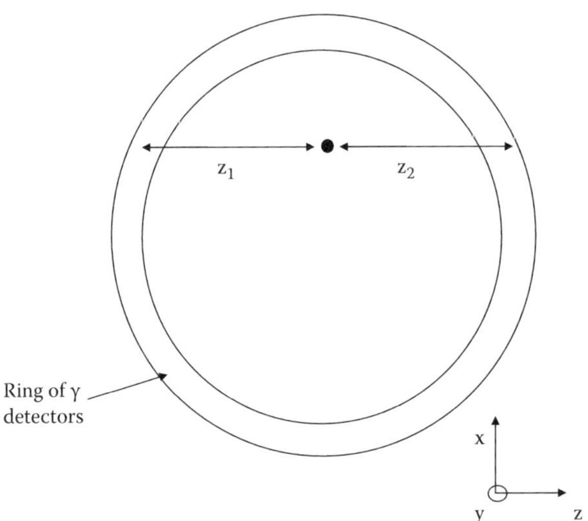

FIGURE 23.20
Positron emission tomography, showing an event in which a 18-F nucleus decays by emitting a positron (positive electron) which annihilates with an electron within a few mm, emitting two gamma-ray particles, each carrying around 0.511 MeV, at 180° to each other. The γ rays are captured by two detectors almost in coincidence and this can be used to locate the line within which the original β^+ emission occurred. If we represent the number of disintegrations per second per mL by g(x,y), then, because of the self-absorption by the tissue, the coincidence count rate is given by $_0\int^L g(x,y)\exp(-\mu z_1)\exp(-\mu z_2)dz(\Delta y\Delta x) = \exp(-\mu L)_0\int^L g(x,y)dz(\Delta y\Delta x)$, where $L = z_1 + z_2$. An external radioactive source with energy close to 0.51 MeV is used to correct for the $\exp(-\mu L)$ term.

imaging modalities such as CT and MRI, to provide additional structural or functional information. The former is particularly useful, because the CT measurements give estimates of attenuation coefficients, which are then used in the PET image reconstruction to compensate for differential absorption in the two paths between the annihilation event and detection.

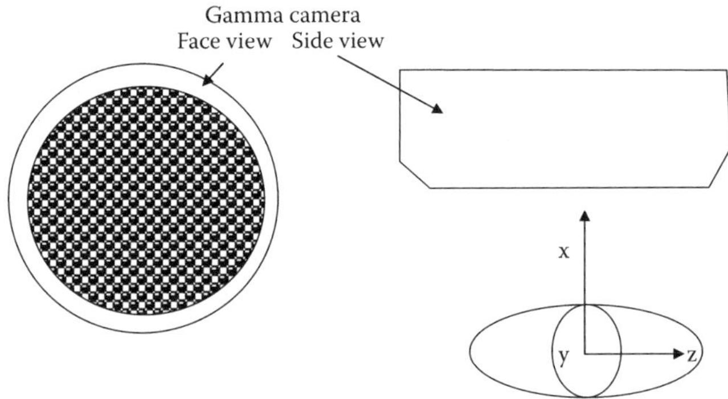

FIGURE 23.21
Single photon emission computed tomography (SPECT). The gamma camera is rotated around the center line of the body (y axis) at typically 64 positions. These "views" are then back-projected to "discover" the distribution of γ-emitter concentration g(x,z). As with PET, a correction factor is used to compensate for tissue self-absorption.

23.3.4.2 Single Photon Emission Tomography (SPECT)

This is a method for producing 3D images from 2D images of injected radionuclide in the body. The 2D image is obtained by using a device such as a "Gamma camera" (or "Anger" camera after its inventor). This consists of a large (0.5 diameter) crystal of NaI, which emits flashes of light when γ-rays impinge on it. A metal plate with parallel holes drilled through it select only those γ-rays traveling in lines normal to the surface of the crystal, ensuring a 2D representation in the crystal of where the radionuclide is located. These flashes of light are then detected by an array of photomultiplier tubes, which give location and photon energy information. By rotating the Gamma camera around the patient, a series of projections are obtained, which can then be combined into a 3D image in a similar manner to CT and MRI. Double and triple camera systems are often used to speed up data acquisition. SPECT is used frequently in cardiology, to gauge the severity of myocardial disease. Image acquisition is aided by the use of ECG gating (called MUGA, or MUltiple Gated Acquisition), which entails data from corresponding epochs from several cardiac cycles being aggregated to improve image quality.

23.3.5 Analysis of Image Quality

The detection of an abnormality in a medical image requires a trained observer. It is possible to make quantitative comparisons between the ability of individuals to detect abnormalities by studying receiver operating characteristics (ROC). This entails using a set of test images in which the presence or absence of abnormality has been determined by subsequent surgery or other method of verification. A curve can be constructed (a ROC curve) which shows the characteristics for a particular individual. It plots the rate of true positives against false positives (that is, the rate at which a verified abnormality was correctly identified versus the rate that which images were classified as abnormal, but were in fact normal). The plot consists of a number of points, each one with a different criterion for accepting images with borderline abnormalities as abnormal. Criteria could range from strict (we will only accept for surgery patients with images showing definite abnormality) to cautious (we will send patients for further tests if there is the slightest hint of abnormality). Another way of doing this is to ask the individual to rate, on a scale of 0–100, the likelihood, in their opinion, of an abnormality being present. The cumulative probability of true and false positive identification to a particular percentile is equivalent to adopting a particular criterion. Figure 23.22 (upper) shows a graph of the responses of a particular individual in which they had to rate over 1000 images on the following scale: 1 = 80% – 100% certain, 2 = 60% – 79% certain and so on to 6 = 0% – 5% certain. It was known that say, 50% of images actually contained an abnormality. The graph is constructed from the cumulative probabilities of the true and false positive rates for each category 1-6.

Differing ROCs need not imply that some individuals are better than others at detecting abnormalities. On the other hand it does reflect the differing mental criteria that individuals have in making decisions on normal/abnormal. This can be factored in to overall quality assurance. It can also be used to study the differences between individual imaging devices or systems. Devices with poorer imaging capacity (in which the signal-to-noise ratio is low) will produce a ROC with a greater false positive rate for a given true positive rate, as shown by the dotted line in Figure 23.22 (upper).

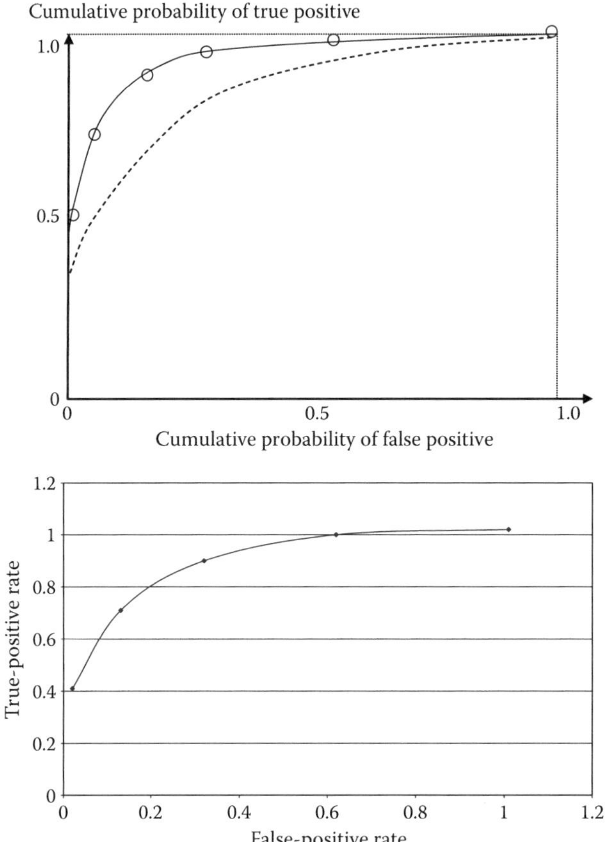

FIGURE 23.22
ROC curves. Top: high (full) versus low (dashed) signal-to-noise ratio. Bottom: plot of data from worked example.

WORKED EXAMPLE

An imaging specialist was asked to rate a set of 400 ultrasonic images of cases according to a scale of certainty that an abnormality was present. Following surgery, it was established that 200 of these cases actually had an abnormality present. The following table gives the number of times in each rating category that were actually abnormal or normal. Construct a ROC for this specialist.

Rating Category	No. Actually Abnormal	No. Actually Normal
Very definite	81	4
Definite	59	21
Likely	38	37
Unlikely	19	60
Very Unlikely	3	78

Answer

First, we estimate the true positive rate by dividing the number of abnormal in each category by 200, and similarly the false positive rate by doing the same with the normals. The cumulative probability is obtained by adding the individual probabilities.

True Positive Rate	Cumulative Probability	False Positive Rate	Cumulative Probability
0.41	0.41	0.02	0.02
0.30	0.71	0.11	0.13
0.19	0.90	0.19	0.32
0.10	1.00	0.30	0.62
0.02	1.02	0.39	1.01

Note that due to rounding errors the cumulative probability is greater than 1 (which is should not be). However, plotting the second and last columns against one another gives the required ROC (see Figure 23.22 [bottom]).

Tutorial Questions

1. (a) A 4 × 4-cm phantom for use with a CAT scanner is shown below (Figure 23.23):

   X-ray absorption coefficients: A 3 m^{-1}; B: 1 m^{-1}; C: 0.5 m^{-1}

   Sketch the projections T_θ for $\theta = 0°$, 90°, and 45°, calculating actual values for $\theta = 0°$ and 90°.

   Describe why, in reconstructing an image from projections, the $T\theta$ have to be modified if a back-projection reconstruction method is to be used.

   (b) Briefly describe how the principle of nuclear magnetic resonance can be used to obtain an image of the brain. What qualities of the tissue can be imaged by this method?

2. (a) Describe how an image of the liver can be obtained using ultrasound. Discuss factors affecting the resolution in the image so obtained.

   (b) Transducers 1 and 2 are placed either end of the perspex (Lucite) block shown. Transducer B is connected to a CRO, which is triggered by the transmitted

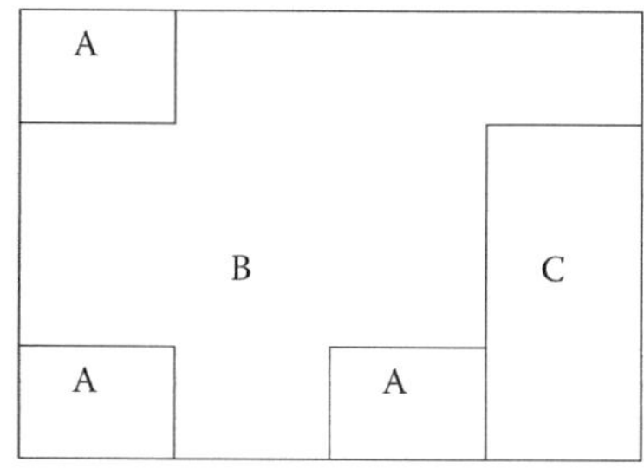

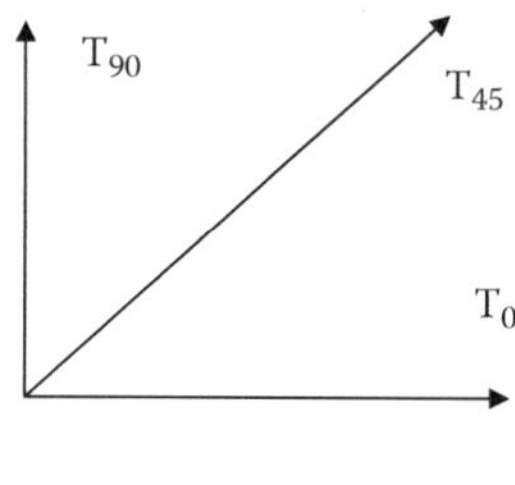

FIGURE 23.23
See Tutorial Question 1.

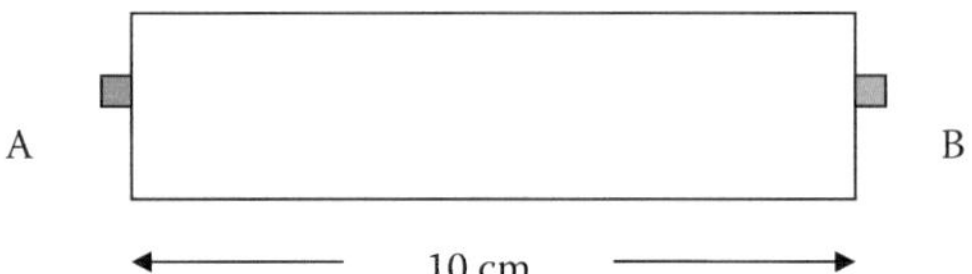

FIGURE 23.24
See Tutorial Question 2.

pulse from A. The received pulse is detected 37 μs from the start, with an amplitude of 1.7 mV. Transducer 2 is now removed and transducer 1 is used to both transmit and receive (Figure 23.24).

Calculate (i) the speed of sound in perspex; (ii) the amplitude of the echo received by transducer 1 (assuming it to be identical to transducer 2).

Data: Perspex: ρ = 1180 kg/m^3; amplitude attenuation coefficient = 0.57 cm^{-1}

Air: ρ = 1.2 kg/m^3; speed of sound = 330 m/s

3. The diagram below represents a phantom test object of 1 cm cubes of 3 substances of known absorption coefficient (values in cm^{-1}). X-rays are beamed in directions 1, 2, and 3 shown, and scanned across in a direction perpendicular to these directions (Figure 23.25).
   (a) What are the half-value layer thicknesses of the 3 substances?
   (b) Draw graphs of (i) the X-ray intensity and (ii) the transmission $T_\theta(x)$ along the lines A, B, and C indicated (give actual values, with units in ii only).
   (c) Explain how these projections (with others) can be combined to reconstruct an image of the original test object, explaining especially why the projections have to be modified in the filtered back-projection method?

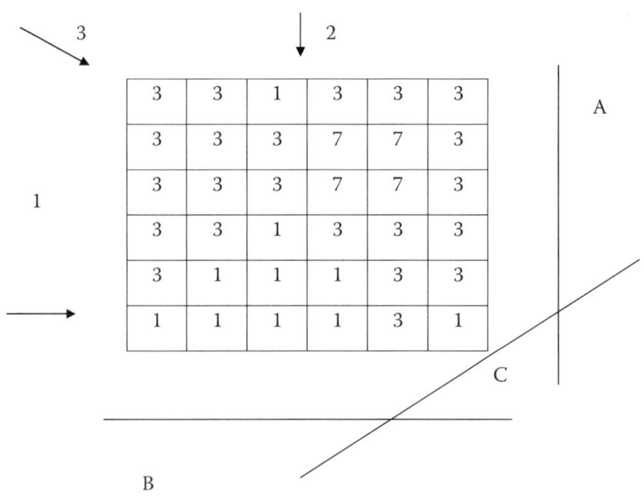

FIGURE 23.25
See Tutorial Question 3.

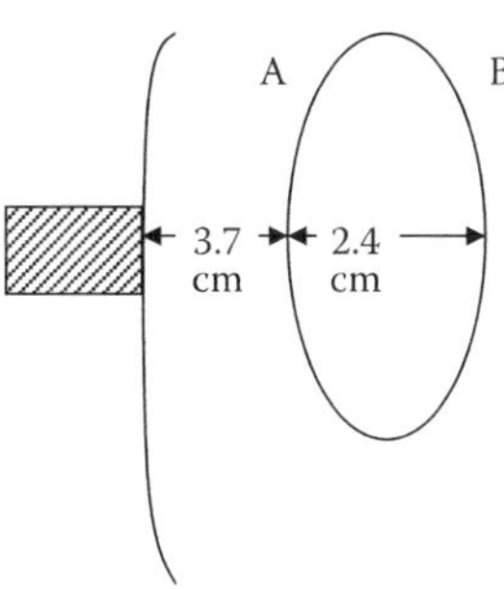

FIGURE 23.26
See Tutorial Question 6.

4. Explain how the phenomenon of nuclear magnetic resonance is used to produce medical images. What exactly do the images portray?
5. How is resolution and image quality checked in the case of (a) X-rays and (b) ultrasound?
6. (a) What properties of tissue are exploited in ultrasound imaging? Further, what are the essential design features for a transducer and associated beam steering apparatus for ultrasonic investigations of the liver?
   (b) An ultrasound transducer is directed through the left ventricle as shown (Figure 23.26).

   The received echo from interface A is 1 mV in amplitude:

   (i) How long after the initial pulse is transmitted would the echo from interface B be received?
   (ii) What will the amplitude of the received echo from this interface B be?
   (iii) What amount of depth compensation (in dB/cm) would be necessary to ensure that both of these echoes would be the same amplitude?

   Data: Blood density 1060 kg/m^3; wavespeed 1570 m/s; attenuation 0.04 cm^{-1}

   Muscle density 1070 kg/m^3; wavespeed 1590 m/s; attenuation 0.46 cm^{-1}
   (c) A duplex ultrasound scanner can give real-time images and blood flow information simultaneously. Explain how this can be accomplished.

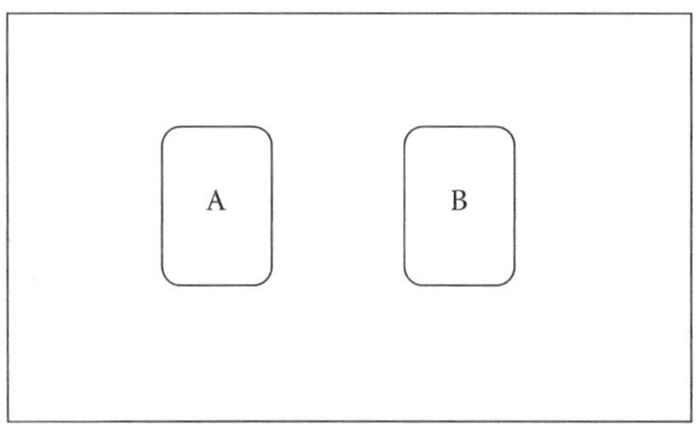

FIGURE 23.27
See Tutorial Question 8.

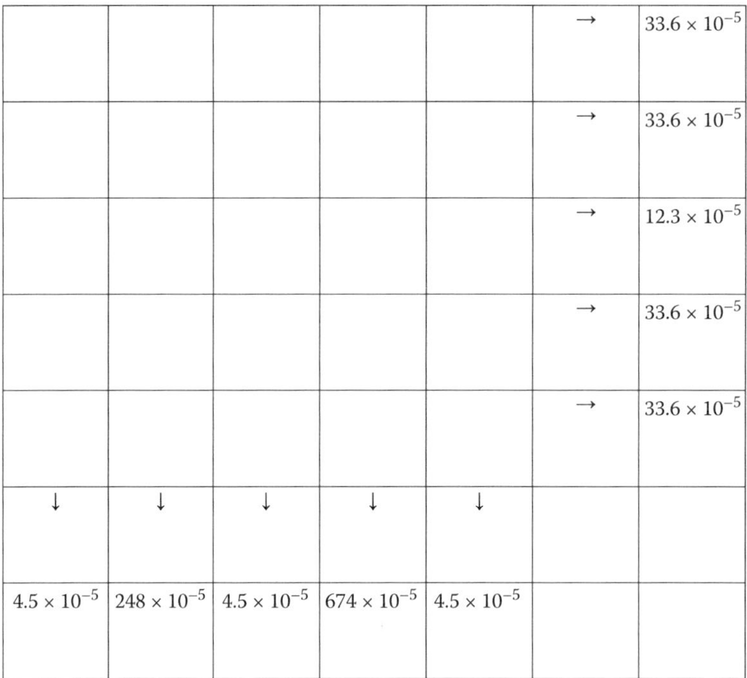

FIGURE 23.28
See Tutorial Question 10.

(d) Outline some of the ways that the arterial image and the arterial flow waveform can be analyzed to give diagnostic information on the state of the artery walls (particularly in the aorta).

7. (a) In an X-ray image the smallest change in Optical Density distinguishable by the eye is 0.07. What implication does this have in terms of X-ray dose in diagnostic imaging?

   (b) Explain how a Receiver Operating Characteristic curve could be constructed for a particular clinician. How would signal-to-noise ratio affect this?

8. (a) Explain how, in X-ray imaging, (i) a monochromatic beam of X-rays can be produced and (ii) a clear image of blood vessels within the brain can be obtained.

   (b) The diagram represents an X-ray image of two objects, both 20 mm thick. A is pure copper (μ = 1200 m^{-1}) and B is aluminum (μ = 70 m^{-1}) completely coated with platinum (μ = 9000 m^{-1}). Calculate the thickness of platinum, given that on the X-ray film both areas (A and B) appear equally dark (Figure 23.27).

   (c) A is replaced by another block of copper, this one 21 mm thick. The light intensity transmitted through region A on the illuminated screen is now 50% of the previous value. Calculate "gamma" for the film.

9. (a) Explain how, in a gamma camera, a final image with a resolution of a few mm can be produced, despite the size of individual photomultiplier tubes.

   (b) Compare SPECT and PET in terms of physiological functions that can be imaged, and in terms of image construction techniques.

10. A 5-cm^2 test block made up of lead and copper cubes is X-rayed from the sides. If the absorption coefficients of lead and copper are 2 cm^{-1} and 1 cm^{-1} respectively, and the emerging X-ray intensity (relative to the incident intensity) is as given below, identify which of the cubes are copper, and which are lead (Figure 23.28).

Bibliography

Brown BH, Smallwood RH, Barber DC, Lawford PV, Hose DR. 1999. *Medical Physics and Biomedical Engineering*. IOP Press, Bristol, UK.

Hobbie RK, Roth B. 2007. *Intermediate Physics for Medicine and Biology*, 4th Edition. Springer, New York.

Webb S. 1988. *The Physics of Medical Imaging*. Adam Hilger, Bristol, UK.

Woodward P. 2001. *MRI for Technologists*. 2nd Edition. McGraw Hill, New York.

24

Microscopy and Biophotonics

Andrew W. Wood

CONTENTS

24.1 General Aspects of Image Display

The first microscope consisted of a fused glass sphere attached to a brass plate by a screw and was constructed by Anthonie van Leeuwenhoek in about 1670. Until the mid-19th century, all microscope image recording was by careful drawing of the image observed by eye. The invention of photography allowed more accurate image recording (photomicrography). In the 1980s, the development of computer vision devices such as frame grabbers, together with image sensors, such as charge-coupled devices (CCDs) have marked a shift from the use of visual inspection and photographs to the use of digital software to display, analyze and record images. At the same time, some of the software used for the 3-D reconstruction and rendering of medical images such as ultrasound, CT, and MRI has been adapted to microscopic images. A particularly useful software package, produced under the auspices of the National Institutes of Health in the USA, *ImageJ*, is in the public domain and can be downloaded, with full supporting documentation from http://rsbweb.nih.gov/ij/.

One of the most exciting developments in recent years is confocal microscopy, so called because the image is formed from light emanating from a single focal point in the specimen, any other light being removed by optical means.

24.2 Confocal Microscopy

24.2.1 Introduction: The Concept

In conventional microscopy, the specimen is typically mounted on a glass slide and then a coverslip placed to sandwich the specimen between it and the slide. For biological samples, the tissue is often stained to visualize particular cell components and then embedded in paraffin wax. This destroys cell function, although there are techniques, for example, using light polarization properties for visualizing fresh tissues or cell suspensions without the use of stains or embedding, which allow live-cell imaging. Since, in conventional microscopy, the specimen is illuminated by focusing a light via a condenser lens, there is no way to avoid shadowy images of structures above and below the focal plane from appearing in the viewed image. This is illustrated in Figure 24.1a. The light path from the condenser to the objective lens is shown in Figure 24.2a, which emphasized that the regions above and below the focal plane are illuminated and will contribute to the final image.

The essence of the confocal microsopy technique, on the other hand, is that light from above or below the focal plane does not play a part in image formation. This means that in forming an image of an array of cells, only those structures in the focal plane will be imaged. The rest will not give rise to an out-of-focus blur seen in conventional microscopy. This is accomplished by using an optical device for preventing the rays of light emanating from the cones A and B from forming the final image. The most convenient means of doing this is to use pinholes for both the illuminating beam and the reflected beam (which carries the image information), so that only information from the focal plane (or as shown in Figure 24.2b, the focal spot) will reach the detector which forms the image. In this example, the image is formed by scanning the focal spot in a raster pattern (like a TV image) over the focal plane. The detector is typically a photomultiplier tube (PMT) or a charge-coupled device (CCD) similar to that found in digital cameras. The image is thus a digital image of light intensity in 512 × 512 or more pixels.

24.2.2 Confocal Laser Scanning Microscopy (CLSM)

Sometimes this is referred to as Laser Scanning Confocal Microscopy (LSCM). As the name suggests, a CLSM uses a laser beam as a light source, because the monochromatic lines in a laser source (such as an argon-ion laser, which has lines at 351, 454.6, 457.9, 465.8, 476.5, 488.0, 496.5, 501.7, 514.5, and 528.7 nm, but with strongest emissions at 488 nm and 514.5 nm)

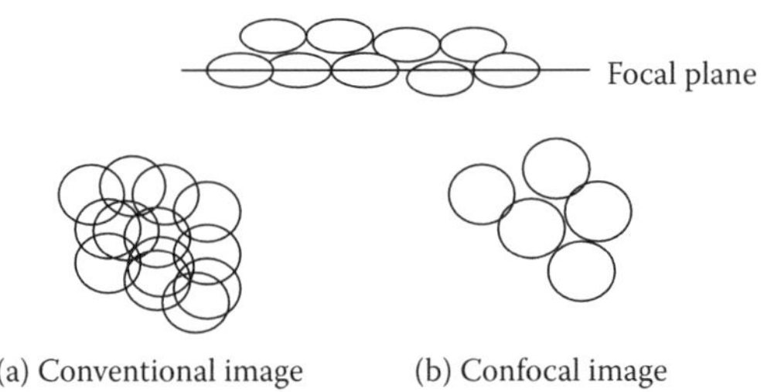

FIGURE 24.1
The appearance of a stack of cells imaged (a) by a conventional microscope and (b) a confocal microscope. Because of the removal of out-of-plane information by the confocal technique, the other layers of cells do not appear in the confocal image.

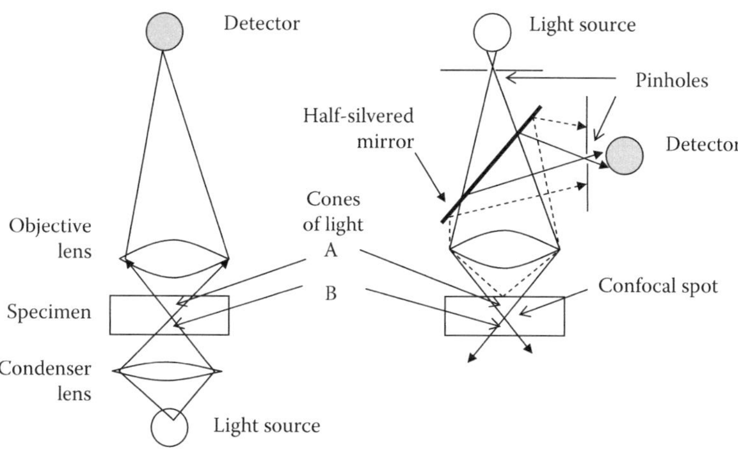

FIGURE 24.2
Ray paths for conventional (left) and confocal microscopes. Conventional microscopes illuminate the specimen via a light source, usually below the specimen stage. This illumination is of two conical regions (A, B) above and below the focus. In confocal imaging, the light source and detector are on the same side of the specimen. Rays toward and away from the focal spot follow the same path (hence the name "confocal"), but rays emanating elsewhere from the cones of light are not confocal and are unable to focus within the exit pinhole. As a consequence, very little of the light from these other regions reaches the detector.

can be used to excite specific tissue dyes. The information recorded by the detector is thus usually fluorescent amplitude rather than reflectance, although the latter has been used in studies of silver particles in nerves.

The focal plane can in fact be less than a micron thick. If the focal plane is gradually moved through the cell (by moving the stage on which the specimen is placed toward the objective), a stack of two-dimensional image "slices" can be taken. Showing these in rapid succession gives an impression of moving down through the cell. Software can be used to "render" these 2D images as a 3D shape, which can be rotated to give a very clear impression of cell morphology (see "Volume Viewer" plug-in in ImageJ) (Figure 24.3).

Because the focal spot is scanned across each focal plane, the acquisition time for one frame has to be taken into account. In modern CLSM systems each image can be acquired in 0.1 s, so a stack of 10 images would take 1 s. However, this depends on the degree of fluorescent intensity. Often, several frames have to be taken to provide an acceptable image. Thus, cell processes which take place on a millisecond timescale normally cannot be imaged, although there are special techniques to allow this, which will be explained in a later section.

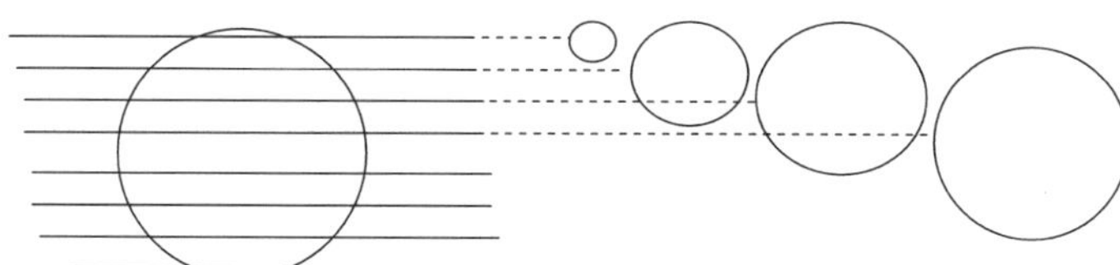

FIGURE 24.3
A series of eight confocal planes through a spherical cell giving an image "stack," which can be reconstructed into a three-dimensional representation via image rendering software.

24.2.3 A Brief History of Confocal Microscopy

The invention of confocal microscopy is of interest. It was invented in 1955 and patented two years later by Marvin Minsky, who at that time had finished doctoral studies on nervous connections in the brain. In his own words (http://web.media.mit.edu/~minsky/papers/ConfocalMemoir.html). "The serious problem is scattering. Unless you can confine each view to a thin enough plane, nothing comes out but a meaningless blur. Too little signal compared to the noise: the problem kept frustrating me." His invention is shown in the diagram from the patent in Figure 24.4. In this version the detector measures transmitted light and rather than scanning the focal spot, the specimen is moved around in fixed position. Nevertheless, the use of pinholes to eliminate the nonconfocal rays is clearly used. Minsky went on to make significant contributions to artificial intelligence, developing, with Seymour Papert, the first LOGO "Turtle," an early precursor to computer graphics software.

24.2.4 Performance

The resolution of a confocal system is determined by the numerical aperture (NA) of the objective lens: the strict definition of NA is the product of the refractive index n of the medium in which the lens is working (often oil with $n \sim 1.5$) and the sine of the angle θ of the maximum cone of light the lens can accept. Very expensive lenses can have a NA of 1.4. In the xy plane (see below) the resolution is typically 0.3 microns; in the z-plane it is 1 micron. The depth resolution is thus not quite as good as in the focal plane. The full equations for these resolutions are given in Figure 24.5.

Movement in the z-direction can be accomplished by a stepper motor (to produce micron steps) or by a piezoelectric device. The single pinhole in the original Minsky design has been replaced by a series of pinholes on a spinning disk (Nipkow disk), with the holes set to produce the scanning motion across the sample, or more recently, the delivery of light to and from a mechanical scanning head is via an optical fiber—the small acceptance angle at the end of the fiber is equivalent to a pinhole and is far less susceptible to vibration.

24.2.5 Use of Fluorescent Dyes in Confocal (and Conventional) Microscopy

In confocal imaging, the tissue is fully functional and cellular processes (such as the development of axonal connections) can be viewed in real time. The simplest method of

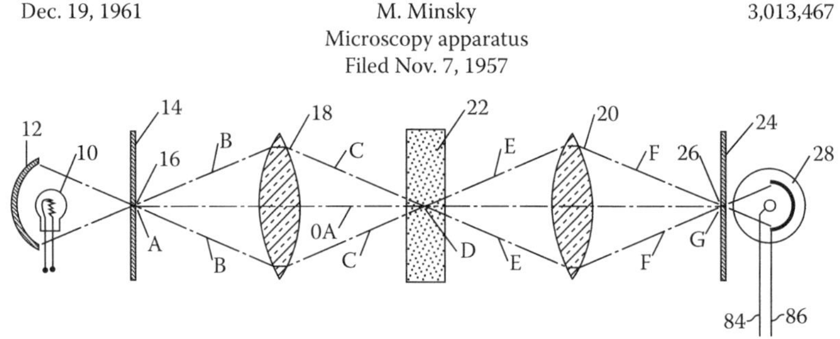

FIGURE 24.4
Diagram from Marvin Minsky's patent, showing the confocal principle used to image single layers within tissue (labeled 22), which is moved in relation to the focal spot (D). The pinholes (14, 24) are shown.

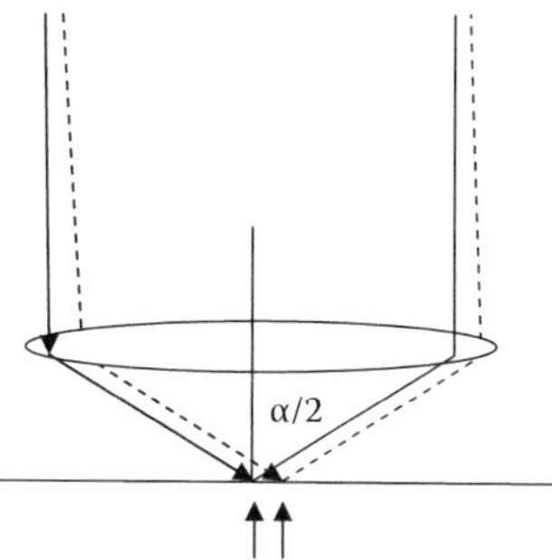

$I_L = [2J_1(V)/V]^4$, where $V = 2\pi r.\sin(\alpha/2)/\lambda = 2\pi rNA/(\lambda n)$
$NA = n.\sin(\alpha/2)$, where n is the refractive index of the object and
α is the angle subtended by the cone of light entering the objective
J_1 is a Bessel function of the 1st kind
r is radial distance from focal point in lateral direction
r for $(I_L/I_0) = ½$ is of the order of 0.3 μm, for NA = 0.95

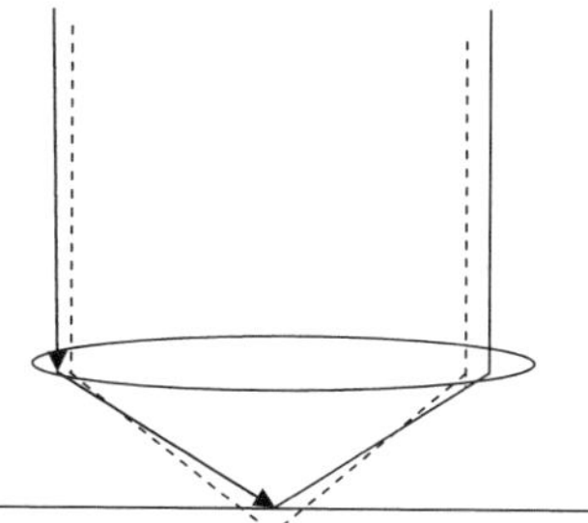

$I_A = [\sin(u/2)/(u/2)]^2$
Where $u = 8\pi z.n.\sin^2(\alpha/2)/\lambda$, where z is vertical distance.
z for $I_A/I_0 = ½$ is of the order of 1 μm, for NA = 0.95

FIGURE 24.5
The lateral (L) resolution and axial (A) resolution for confocal systems are determined by the Numerical Aperture (NA) of the lens and wavelength λ of the light. I_L and I_A represent the light intensity as a function of radial and axial distance respectively.

obtaining an image is to use a reflectant, such as colloidal silver or gold (40 nm particles), but the commonest technique is to use fluorescence imaging, using nontoxic dyes to respond to particular cell components. The majority of these dyes are based on fluorescein (which is excited optimally at around 490 nm, or the 488 nm blue line of an argon-ion laser) or rhodamine (which is excited at around 510 nm, or the green argon-ion laser line). These dyes are conjugated (i.e., joined on to) chemical species with affinity for the quantity to be studied (calcium, DNA, pH, membrane potential, and so on). There are several suppliers of these fluorescent probes, for example, the US-based company Invitrogen (http://probes.invitrogen.com/handbook/) has an extensive catalogue of such dyes (and other products) with full descriptions and application notes for each.

It is necessary at this point to give a brief background to the process of fluorescent emission. A fluorescent dye (or fluorophone) will absorb light (excitation) and then will emit light at a longer wavelength (emission). Between these two processes, the dye molecule undergoes an internal conversion, in which some of the quantum energy is absorbed without emission of radiation. Since the emitted radiation is of lower quantum energy (shorter arrow) than that

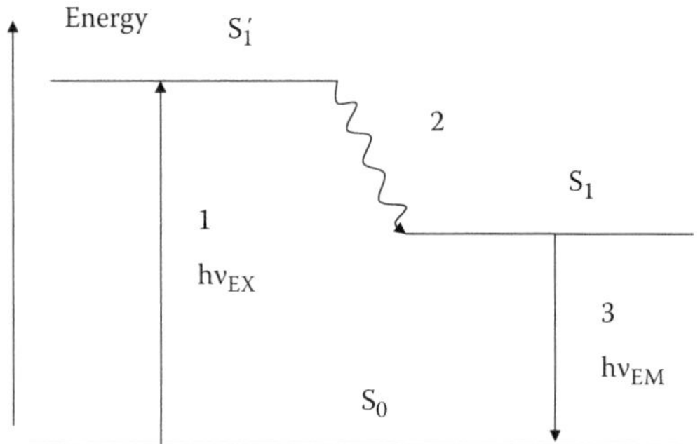

FIGURE 24.6
Jablonski diagram: S_1' and S_1 represent excited energy levels for the fluorophore. Process 1 indicates the absorption of a quantum of light (excitation), energy $h\nu_{EX}$, or hc/λ, where h is Planck's constant, ν is the frequency, λ the wavelength, and c the velocity of light; 2 represents a radiationless internal conversion to a lower energy state and 3 the subsequent emission (EM) of fluorescent light, at a longer wavelength than the absorbed light.

absorbed in excitation, the former has a longer wavelength. If the wavelength of light being absorbed does not exactly match the excitation wavelength (which is characteristic of the species of dye), there will still be some absorption, but not optimally.

The fluorescent intensity depends on a number of factors, but in particular, the amount of the target (calcium, say) which has associated with the dye. An example of this is the dye fluo-3, which "chelates" calcium. "Chelation" refers to a claw-like region of the molecule which has a specific affinity to a metal ion, in this case calcium. The substance ethylene diamine tetraacetic acid (EDTA) is used in a variety of industrial and medical application for lowering calcium concentration in solutions. The dye fluo-3 has two parts to it: the dye part related to fluorescein and the chelator part related to EDTA. The amount of fluorescence depends on the amount of calcium present in solution, which determines the percentage of dye molecules with chelated calcium ("% bound"). The variation of % bound with calcium concentration in solution is shown in Figure 24.7. The concentration of fluo-3 is such that the chelation of calcium has minimal effect on the overall calcium

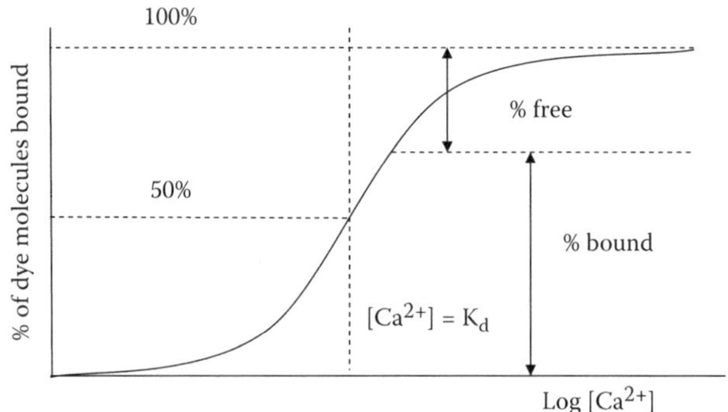

FIGURE 24.7
Percentage of calcium chelator dye molecules bound to calcium varies with ambient $[Ca^{2+}]$. For $[Ca^{2+}] = K_d$ for the dye (390 nM for fluo-3), half of the molecules are bound. The intensity of fluorescence is proportional to % bound. The range of concentration between no fluorescence and saturation of fluorescence is quite small, typically 2 orders of magnitude.

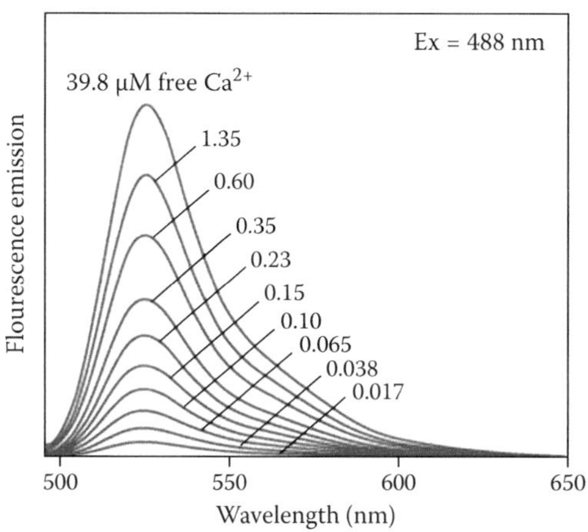

FIGURE 24.8
Fluo-3 emission spectrum for excitation at 488 nm with values of $[Ca^{2+}]$ shown in μM. For $[Ca^{2+}]$ at the K_d value for fluo-3, 0.39 μM, the emission peak is at approximately 50% of maximum, when Ca^{2+} is in excess. (Illustration from Invitrogen on-line catalogue http://www.invitrogen.com/site/us/en/home/Products-and-Services/Applications/Drug-Discovery/Target-and-Lead-Identification-and-Validation/g-protein_coupled_html/cell-based-second-messenger-assays/fluo-3-calcium-indicator.html.)

concentration in solution (in a cell any drop in cytoplasmic calcium will be quickly restored to equilibrium values from internal stores or from elsewhere).

Fluo-3 can be excited by argon-ion laser 488 nm line (and the fluorescence measured at >515 nm), but suffers the disadvantage that as imaging of a particular region progresses, the dye becomes photobleached (that is, the amount of fluorescent intensity decreases with time) (Figure 24.8). However, if a low-intensity laser beam is used for excitation, photobleaching is often minimal. $[Ca^{2+}]$ at particular locations in the tissue can be estimated from the equation (which is derived and explained in the next section):

$$[Ca] = K_d(F - F_{min})/(F_{max} - F)$$

where F, F_{min}, F_{max} refer to fluorescent intensity; respectively: at the moment of interest, when dye fluorescence is quenched, when $[Ca^{2+}]$ is allowed to saturate the dye.

WORKED EXAMPLE

In an experiment on lymphocytes, fluo-3 (K_d = 390 nM) is taken up by the cells where it becomes florescent. Image analysis software is used to measure the fluorescence in a particular cell, where the average intensity value (the grayscale value 0–255 in 25 pixels representing the cell) taken over a particular 5-min period is 2387. At the end of the experiment, an ionophore is added, which floods the cell with Ca^{2+}. The intensity value now becomes 4,978. Mn^{2+} is then added to quench the fluorescence, but there is still a residual intensity of 89. Estimate $[Ca^{2+}]$ during the 5 min measurement period.

Answer

We can take F, F_{max} and F_{min} as 2387, 4978, and 89, respectively. Using the above equation, we can calculate $[Ca^{2+}]$ = 390 (2387–89)/(4978–2387) = 345 nM.

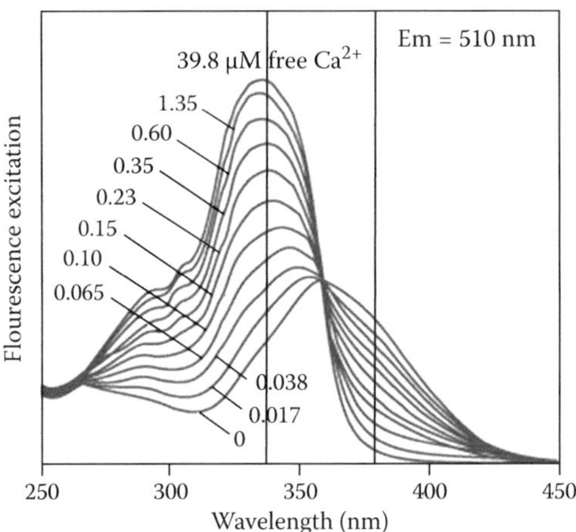

FIGURE 24.9
Fura-2 excitation spectra for emission at 510 nm. When excited at wavelengths of 340 and 380 nm (see vertical lines shown), it can be seen that the ratio of emitted light (in response to these excitation wavelengths) varies enormously as $[Ca^{2+}]$ rises from 0 μM to 39.8 μM. (Diagram from Invitrogen catalogue, http://www.invitrogen.com/site/us/en/home/References/Molecular-Probes-The-Handbook/Indicators-for-Ca2-Mg2-Zn2-and-Other-Metal-Ions/Fluorescent-Ca2-Indicators-Excited-with-UV-Light.html.)

However, because of the uncertainties in many of these quantities, variations in $[Ca^{2+}]$ (rather than absolute values) are often estimated from $\Delta F/F_0$, where F_0 refers to the initial fluorescence and ΔF the change due to change in $[Ca^{2+}]$.

Some of the most useful dyes are the so-called ratiometric dyes, in which the amount of the quantity to be measured is proportional to the ratio of fluorescent intensities at two different wavelengths. Occasionally, this ratio is of a single fluorescent wavelength but two different excitation wavelengths (such as Fura-2 for calcium imaging). Fura-2 has the disadvantage of needing two UV wavelengths to activate it, which until recently were not available from laser sources. However, the overwhelming advantage is that photobleaching affects both wavelengths similarly, so that the ratio remains the same, despite the amount of bleaching (Figure 24.9).

24.2.6 Theory of Measurement

First, for a dye such as fluo-3 which acts as a Ca^{2+} chelator. The equilibrium reaction can be represented as

$$\text{Fluo-3} + Ca^{2+} \underset{k_d \text{ dissociation}}{\overset{k_a \text{ association}}{\leftrightarrow}} \text{Fluo-3}Ca^{2+}.$$

Let c_f be the concentration of free fluo-3, c_b, the concentration of bound fluo-3 (fluo-$3Ca^{2+}$). At equilibrium:

$$k_a\, c_f\, [Ca] = k_d c_b.$$

Or, $c_f[Ca]/c_b$, $= K_d$, putting $k_d/k_a = K_d$, the dissociation constant.

Given that fluorescent intensity F is proportional to % of dye in the bound form ($\%c_b$), we can put

$$F = S\ \%c_b + F_{min}$$

where the second term reflects the amount of nonspecific fluorescence which may still be present when $[Ca^{2+}]$ is zero. We can substitute from the equation above, given that $\%c_b + \%c_f = 100$:

$$F = S[Ca]\%c_f/K_d + F_{min}.$$

By examining Figure 24.7, we can see that the % free ($\%c_f$) is proportional to $F_{max} - F$.

$$\%c_f = (F_{max} - F)/S.$$

Thus,

$$F = [Ca]\ (F_{max} - F)/K_d + F_{min}.$$

Rearranging we get:

$$[Ca] = K_d(F - F_{min})/(F_{max} - F)$$

which allows us to estimate calcium concentration, given fluorescent intensity F at a certain position in the image. F_{max} is estimated at the end of the imaging period by flooding the cell with calcium. This is accomplished by adding an ionophore, which effectively punches holes in the cell membrane. After this, the ion Mn^{2+} is used to displace Ca^{2+} from the chelation sites. This renders the dye nonfluorescent, leaving the non-specific fluorescence, F_{min}.

The derivation of the equation for ratiometric determinations (that is by examination of the intensity at either two different excitation, or two different emission wavelengths) is a little more involved.

Let fluorescent intensity be F_1, F_2, at 2 wavelengths 1 and 2.

$$F_1 = S_{f1}c_f + S_{b1}c_b$$

$$F_2 = S_{f2}c_f + S_{b2}c_b$$

where $S = I\varepsilon Q\Delta x e_c$, I is the incident intensity, ε is the extinction coefficient ($cm^{-1}M^{-1}$), Q is quantum efficiency, Δx path length, and e_c is the collection efficiency.

Fluorescence ratio $R = F_1/F_2 = \{S_{f1} + S_{b1}[Ca]/K_d\}/\{S_{f2} + S_{b2}[Ca]/K_d\}$;

When $[Ca] = 0$, $R = S_{f1}/S_{f2} = R_{min}$

When [Ca] is in excess, $R \approx S_{b1}/S_{b2} = R_{max}$

Thus, $[Ca] = \beta K_d\{R - R_{min}\}/\{R_{max} - R\}$, where $\beta = S_{f2}/S_{b2}$.

K_d for Fura-2 is 200 nM approximately. Cellular [Ca] ranges from 0.1 μM to 10 μM. Fura-2 is usually added to give a final concentration of around 10 μM.

WORKED EXAMPLE

(a) From the data given in the table below estimate the cytoplasmic calcium concentration at instants T1 and T2 (which is 30 min after T1). The data are of fura-2 fluorescence when excited by UV light at 330 nm and at 380 nm.

	T1	T2	Ionophore Added	Mn^{2+} Added
F_1(330 nm)	6200	3100	4100	900
F_2(380 nm)	1800	900	600	2100

Further data $K_d = 200$ nM; $\beta = S_{f2}/S_{b2} = 2100/600$

Answer

Let us deal with T1 first. Since we define $R = F_1/F_2$, the values for R, R_{max} and R_{min} are 6200/1800, 4100/600, and 900/2100, respectively. We are also given values for the other two parameters in the equation, so we have:

$$[Ca^{2+}] = 200.(2100/600)\{6200/1800 - 900/2100)/(4100/600 - 6200/1800)\} = 200 \times 3.5(3.01/3.39) = 621 \text{ nM}.$$

For T2, the ratio F_1/F_2 is the same, so the value of $[Ca^{2+}]$ will be unchanged, despite a fall in the overall fluorescence. This illustrates the way a ratiometric dye is far less sensitive to photobleaching effects.

24.2.7 Loading Techniques

In order to get fura-2, fluo-3, or other fluorophores into the cytoplasm of the cells under study, three main methods are used: (i) microinjection using a micropipette, (ii) using the acetomethoxy (AM) form of the dye (this form is lipid soluble, but is nonfluorescent), or (iii) by using fluorescent molecules which are manufactured in situ, in a way similar to naturally occurring bioluminescent molecules such as luciferase. Microinjection involves a controlled bleed off of pressure produced by a microsyringe, so that the liquid displaced by the micropipette is of the order of 10^{-10} mL, in other words, far less than the volume of a cell. The AM-dye method involves conjugating the dye with the AM moiety, in which form the complex diffuses into the cell. Next, esterases in the cytoplasm are able to cleave off the AM part of the dye. The reaction can be represented as shown in Figure 24.10. Once inside the cell, and having cleaved off the AM, the dye is trapped and can no longer pass through the membrane. It is now in its fluorescent form.

The third method involves transfecting the tissue with an engineered baculovirus (a specialized virus) containing DNA which codes for a fluorescent protein. Typical proteins include Cameleon (a calcium indicator) and green fluorescent protein (GFP).

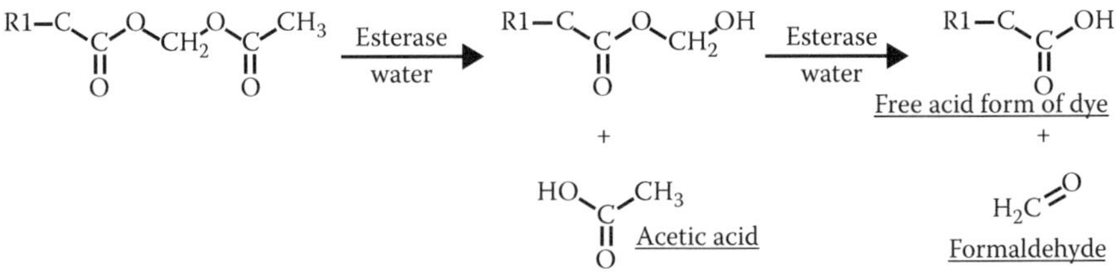

FIGURE 24.10
The dye is represented by R1.C.CO.O, conjugated to the AM moiety CH_2.O.CO.CH_3. This permeates the cell membrane and once in the cytoplasm, native esterases remove, in stages, the acetic acid and then the formaldehyde parts of the AM moiety. The free acid form of the dye is fluorescent and trapped inside the cell.

We have concentrated mainly on dyes which measure cell [Ca^{2+}]. There are a vast range of other ion concentrations which can be measured using specific dyes, including [H^+], which of course is related to pH. Dyes can also estimate changes in membrane potential: some are Nernstian (i.e., the log of the fluorescence ratio inside and outside the cell gives an estimate of membrane potential, e.g., TMRE) others sit in the membrane and respond directly to changes in potential via altering the dipole moment of the molecule (di-8-ANEPPS is a typical example). The latter responds in milliseconds to changes whereas the former takes several seconds. In addition to these applications, specific dyes are attracted to different components of the cell (nucleus, microfibrils and tubules, cell membranes). By using multiple dyes, each with nonoverlapping excitation and emission spectra, it is possible to build up a composite image of different cell components, each component represented by a different color (which may not be the actual emission color—images are often enhanced by the use of computer-generated pseudocolor).

24.2.8 Multiphoton Imaging

In Figure 24.6, we saw that a photon of energy $h\nu = hc/\lambda$ can lift the dye molecule into an excited state. Very occasionally, two photons of half this energy (or, with wavelength of twice that of the original) can accomplish the same transition. More rarely still, three photons of three times the original wavelength will also suffice. One way of increasing the chances of these multiphoton events is to use very high intensity beams. If, instead of continuous illumination by a laser beam, a modulated beam with high pulse intensity is used, this phenomenon will occur. The photon source is typically a femtosecond laser, that is, a laser with a pulse of around 80 femtoseconds (80×10^{-15} s) in duration and with a 80-MHz repetition rate. The "duty cycle" is thus approximately $(1/80 \times 10^6)/80 \times 10^{-15} = 1.6 \times 10^6$. Thus, a 1-mW/m^2 average intensity is concentrated into 1.5 kW/m^2 during the brief pulses. This is sufficient to cause significant multiphoton events. If infrared pulses are used at 700 nm, two-photon absorption is equivalent to single photon absorption at 350 nm (which is in the UV). Because of nonlinear absorption properties, two-photon absorption occurs as a function of the square of intensity, so near the focal spot (where the intensity is greatest) the chances of two-photon absorption increases dramatically, but elsewhere the chances are far less. In fact, the fluorescent intensity (which is of light of wavelength somewhat greater than 350 nm) falls off as $1/r^4$ and is thus very localized. Because of this localization (and the absence of fluorescence from the cones of light referred to in Figure 24.1), there is no need for pinholes. In fact, the detector can be a pair of light-detecting devices either side of the sample (designed to collect as much of the fluorescence from the focal spot as possible). Remember, if the excitation is IR, it will be invisible—the only visible light will be from the fluorescence in the spot. The benefits of two-photon imaging include imaging of UV (ratiometric) fluorophores using conventional visible light optics (two red photons being equivalent to one UV photon); the use of red light excitation is less damaging to cells than blue light excitation; red light excitation can image deeper into tissues than blue light; photobleaching is reduced in the sample since excitation is confined to the focal point.

24.2.9 Developments in Confocal Imaging

A particularly useful technique is the so-called line scan mode. Rather than the light spot scanning in a x,y plane to produce a 2D image, the line is scanned in the x direction along a particular point in the y direction. Subsequent scans along this line are displaced downward on the display, but now the lines represent subsequent moments in time, around 1 ms apart. If there is a variation in dye fluorescence at a point along the line with time, then this will be seen as an alteration in line intensity at this particular point on the line from one line to the next. In particular, if there is a sudden release of, say, calcium from a localized region of the cells, then these events (known as "puffs" or "spikes") will show up in some of the lines but not others. This is also very useful when using voltage-sensitive dyes, to follow changes in membrane potential in, for example, cardiac myocytes (isolated muscle cells). The variation with time where the scanning line crosses the membrane can be seen from the fluorescent intensity variations.

A second useful technique (or group of techniques) consists of overcoming the diffraction limit. Conventional microscopy has a resolution determined by diffraction of the object, which is estimated from the Rayleigh Criterion ($\Delta x = 0.61\lambda/NA$), where Δx is the smallest separation which can be distinguished in the image, λ is the wavelength (400 nm, say) and NA is the numerical aperture, usually around 1.4 for a good objective lens. Thus, it is unlikely to be able to image structures less than 500 nm apart. The diffraction or "Rayleigh limit" can be overcome the following way (shown in simplified form in Figure 24.11). This uses an overlapping doughnut-shaped secondary beam to de-excite all but the central region of the exciting beam. Thus, the subsequently emitted fluorescence will only come from that central region, which can be as small as 50 nm. Using this technique, it has been possible to image individual vesicles containing fluorescent neurotransmitters. The technique is known as STED (stimulated emission depletion). There are several other acronyms applied to new confocal imaging techniques. These include FRAP (fluorescence recovery after photobleach), which measures the dye diffusion rate from adjoining areas after the dye in the target region has been bleached (rendered nonfluorescent) by an intense light beam, FRET (fluorescence resonance energy transfer), which measures molecular proximity on a nanometer scale, because resonant transfer is highly distance dependant, OCT (optical coherence tomography), which is like ultrasound imaging, but uses IR light.

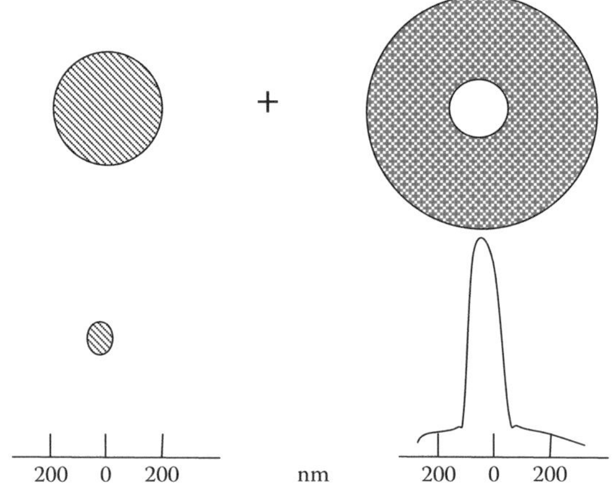

FIGURE 24.11
Stimulation emission depletion (STED) microscopy. Top left shows the normal diffraction-limited excitation spot of around 300 nm in diameter (typically blue light, producing orange fluorescence). Superimposed on this spot is the doughnut-shaped orange beam shown at top right. This suppresses the fluorescence in the periphery by a process of stimulated emission, effectively reversing the fluorescence process (see Figure 24.6). This leaves just the region in the "hole" of the doughnut, which may be less than 100 nm or 50 nm resolution. (From Willig et al., *Nature* 440:935, 2006.)

This last technique is particularly good for retinal imaging and like ultrasound, gives an image of the plane the beam directions lie in. There is also a related technique, OPT (optical projection tomography), which is like CT but uses visible light and is good for developmental biology of translucent organisms.

An important specialized technique is TIRFM (total internal reflectance fluorescence microscopy) which has been called "patch clamping with light" (see Chapter 6). This utilizes a beam of light coming from below the sample at a critical angle which is totally reflected into the glass slide. A small amount of light (the evanescent wave) penetrates a few tens of nanometers into the sample and can produce fluorescence in this narrow region. The patch clamping arises because it is possible to quantify calcium movements using highly regional fluorescence measurements from large cells (such as frog eggs, or oocytes) in response to voltage pulses (Demuro and Parker 2005).

24.3 Laser Techniques in Medicine

24.3.1 Photodynamic Therapy

This amounts to treating tumors with laser light, leaving normal tissue unaffected. It arises from two properties of certain substances: first, some photosensitive pharmaceuticals are *selectively* taken up by tumor tissue, and second, these pharmaceuticals produce *toxic free radicals* when activated by laser light, which then destroy the tumor tissue. A free radical is a molecule with an incomplete shell of valency electrons, but is not a free ion. These are very reactive and examples include $OH^{\bullet}$, superoxide $O_2^{\bullet -}$ and lipid radicals (the dot $^{\bullet}$ indicates that there is an incomplete shell). The procedure is as follows: the photosensitive substance is injected (not necessarily into the tumor) and then three days later, the tumor is illuminated with intense laser light (typically 630 nm, >100 J/cm^2), perhaps via optical fibers. The tumor will by this time have accumulated the photosensitive substance and the subsequent release of free radicals leads to localized necrosis. Typical photosensitive pharmaceuticals include: hematoporphyrin derivative, or HpD, which is related to hemoglobin; rhodamine 123 (a common dye); phthalocyanines; 5-aminolaevulinic acid (which is naturally occurring); hypericin (from the plant St John's wort) and other metal complex porphyrins and bacteriochlorin. Photodynamic diagnosis (PDD) is similar, but laser intensity is less strong: precancerous (and cancerous) cells give a fluorescence and can be located prior to PDT. The technique is particularly useful for cancers of internal surfaces, such as the bladder. However, the technique is still considered experimental (Figure 24.12).

24.3.2 Endoscopy and Endomicroscopy

Essentially, modern systems consist of a *coherent bundle* of optical fibers (that is, the ordering of the fibers is identical at both ends), individually clad. Most are in a flexible sheath, with steerable sections. The overall diameter is around 1 cm or less, but over 1 m in length. There are various types suited to: gastroscopy, colonoscopy, bronchoscopy, cystoscopy (bladder), laparoscopy (peritoneum), arthroscopy (knee and other joints), and angioscopy (arteries and veins). The individual optical fiber diameters are 10 μm –20 μm (typical), with 1.5 μm–2.5 μm cladding to minimize cross-talk, giving a resolution of around 0.02 mm. Lenses control working distance and depth of focus and there is provision both for direct

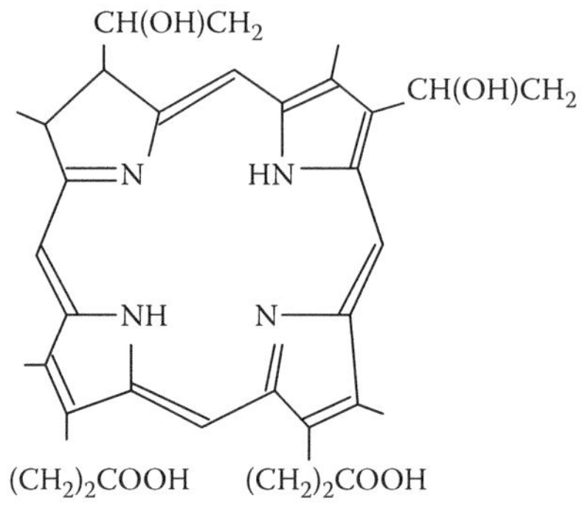

FIGURE 24.12
Structure of hematoporphyrin (Sigma Aldrich).

viewing, monitor viewing and recording via VCR/DVD. Contained within the endoscope tube are additional optical fibers to provide illumination, a mechanical device for taking biopsies and a tube to allow the introduction of gas to clear an area for viewing. Some models have a confocal microscope incorporated, with the scanning mechanism at the tip. This takes the resolution down two orders of magnitude and with the help of suitable dyes can provide in situ histological diagnosis (Figure 24.13).

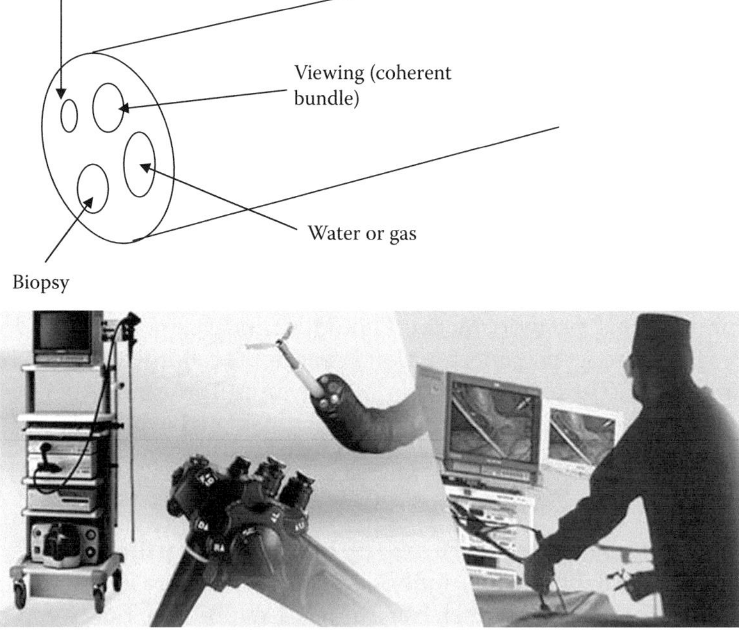

FIGURE 24.13
Diagram of a flexible endoscope showing the components within the sheath. The light is to illuminate the area, the coherent bundle to capture the image, the water or gas conduit to give a clear area for viewing and a mechanical device to take biopsies. The various segments of the sheath are steerable as shown in the lower diagram. (http://www.olympus.co.jp/en/mesg/endoscope/.)

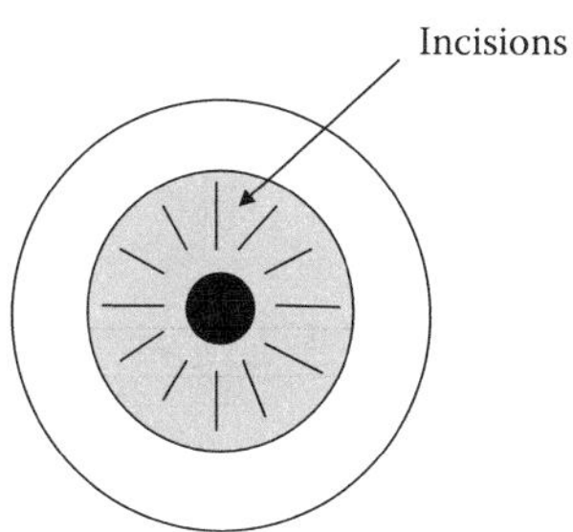

FIGURE 24.14
Radial keratotomy using an excimer laser. The incisions shown change the focal length of the cornea.

24.3.3 Laser Surgery

Lasers have their place in treating regions where small volumes of tissue need to be removed. The mechanisms whereby this is accomplished are thermal or boiling of tissue water (vaporization), denaturation of proteins (leading to coagulation of tissue) then ablation (physical removal of tissue). If the laser is pulsed, this provides a means of tissue ablation, carbonization, coagulation, and desiccation. High intensity short pulses produce photomechanical disruptions of tissue whereas longer pulses produce photochemical reactions. Corneal sculpting (radial keratotomy or keratectomy) is used to correct defects of vision using excimer (or excited dimer) lasers, typically with a wavelength of 193 nm with an intensity of 350 kJ/m^2 or thereabouts (see Figure 24.14). Water absorption increases dramatically above 2 μm wavelength, so CO_2 lasers (wavelength around 10 μm) are useful for rapid tissue heating (absorption dominated). Ablation, on the other hand occurs more at lower wavelengths, for example, the neodymium–yttrium–aluminium–garnet or Nd-YAG laser at 1 μm. Argon ion 0.5 μm wavelength is absorbed by red blood cells so is used for angioplasty (or removal of atheromatous plaque)—however, other methods are considered superior for this. The removal of port-wine stains on the skin and the treatment of Barratt's esophagus (where gastric epithelial cells migrate abnormally into the oesophagus) are also treated by laser.

Tutorial Questions

1. (a) You have been requested to measure calcium levels in single muscle fibers during electrical stimulation of the fiber. Explain:
   - (i) The essential requirements for selecting a fluorescent dye to achieve this purpose
   - (ii) How you would obtain a three-dimensional representation of calcium distribution at a set time (e.g., 10 ms) after stimulation
   - (iii) How you would follow the progress of calcium waves along a particular muscle fiber

   (b) Using image analysis software, the intensity of fluo-3 fluorescence at a particular time and in a particular region of interest in the muscle is 1425 counts. At the end of the experiment an ionophore is used to flood the cell with [Ca^{2+}],

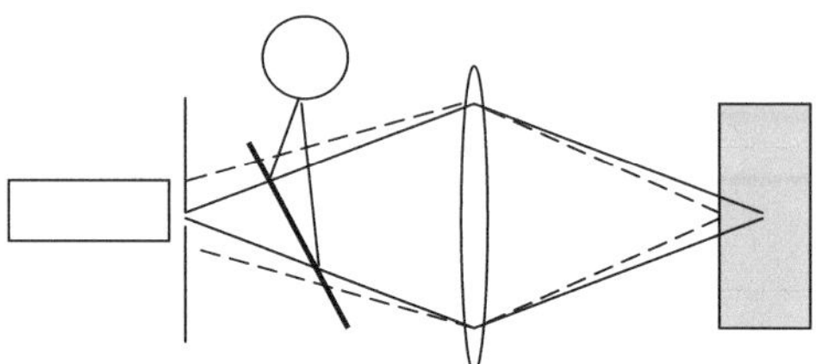

FIGURE 24.15
See Tutorial Question 2.

and the count in this region rise to 5322. The fluorescence is then quenched with Mn^{2+}, and the residual fluorescence is 57 counts. Determine the calcium concentration, given that K_d for fluo-3 is 390 nM. Why would a dye like fura-2 give a more reliable estimate?

2. (a) By referring to the ray paths in Figure 24.15, briefly explain the principles of confocal microscopy.

   (b) Further, explain how it is possible to produce (i) two-dimensional and (ii) three-dimensional views of cells.

3. (a) Explain how, using a laser scanning confocal microscope (LSCM), it is possible to image a 1-μm slice within a sample of biological tissue several tens of microns thick.

   (b) Further, explain how it is possible to produce a three-dimensional view of cells, for example, a group of interconnected neurons using LSCM.

   (c) Fluo-3 AM is taken up by tumor cells. In one particular cell, an average fluorescence value of 150 units is recorded. At the end of the experiment, the cell is treated in such a way that calcium enters the cell in such quantities that fluo-3 is saturated. The fluorescence value is then 830 units. Manganese is then added, which displaces the calcium from the fluo-3. The fluorescence is then 80 units. Given that K_d for fluo-3 is 390 nM, estimate cell $[Ca^{2+}]$.

   (d) What is multiphoton microscopy, how does it work, and what are its advantages?

Bibliography

Demuro A, Parker I. 2005. "Optical patch-clamping" single-channel recording by imaging Ca^{2+} flux through individual muscle acetylcholine receptor channels. *J Gen Physiol* 126:179–192.

Hibbs A. 2004. *Confocal Microscopy for Biologists*. Springer, New York.

Pawley JB. 2006. *Handbook of Biological Confocal Microscopy*. Springer, New York.

Sheppard CJR, Shotton DM. 1997. *Confocal Laser Scanning Microscopy*. BIOS Scientific, Oxford.

Part VII

Systems Integration

25

Physiological Modeling

Andrew W. Wood

CONTENTS

Mathematical modeling of physiological and general biological systems has proved to be a powerful tool for understanding mechanisms in health and disease, also in predicting behaviors in possibly dangerous environments, such as deep-sea diving or space-walking. Broadly, models can be divided into two types: mechanistic and heuristic. The first (*mechanistic*) takes a "bottom-up" approach, integrating basic mechanisms at the cellular, subcellular or even molecular level, and then integrating these to describe a more complex system. The Hodgkin-Huxley model of the nerve, which was introduced in Chapter 6 is an example of this. The second type, *heuristic*, is a "top-down" approach in which no specific mechanism is assumed for the process or subprocess, which is described in terms of how the output varies for a given set of inputs. This is a "black box" approach, in which the mechanisms are inferred or discovered (hence "heuristic") by judicious probing from the outside. The various black boxes representing the subprocesses or subsystems are then integrated to form a more complex model. In practice, most mechanistic models contain black boxes (consisting of relationships determined empirically rather than theoretically) and the integrating of black box subsystems represents an understanding of mechanistic relationships.

There is interplay between model and experiment. The parameters in a model need to be determined by experimental data, or if none exist, values are guessed, then systematically increased or decreased until the model gives a realistic behavior. The model will identify anomalies and curiosities which will form the basis for further specific experiments, which will then inform the modelers of needed improvements.

So, in summary, if we have a model that works, it tells us about how the system operates (we can then begin to understand the mechanism better). If the model is a good one, we can predict what will happen for example in disease or in extreme environments; it can help in designing medical treatment for disease (for example radiotherapy planning) and it is part of a cycle where model is compared with experimental data and further refined to fit data better. The model will suggest further experiments to produce more refinement. We can also study such properties as equilibrium, dynamic responses (that is, how thing will change with time), and stability.

What follows is a brief history of the application of modeling to physiological systems. In the 1930s, WB Cannon, who was professor of Physiology at Harvard University, studied homeostasis and control mechanisms in living systems, publishing "Wisdom of the Body" in 1932. The prevailing paradigm of the time was that certain physiological systems had some affinity with mechanical control system, in particular servo systems, using negative feedback to correct errors in functioning. The next two decades moved on to consider flow of materials between discreet compartments and also to the analogies which can be drawn between electrical circuits and physiological systems. More recent developments, made possible by superfast computing, have included the flow of information within individual cells (the so-called "omics"—genomics, proteomics, metabolomics, and cellomics).

The area of physiological modeling is vast and this chapter can only skim the surface. The various sections should be regarded as preliminary only and reference should be made to the recommended reading at the end of the chapter for further information. The topics covered in this chapter are biological reactions (enzyme kinetics), systems analysis, compartmental systems, glucose and iodine system models, thermoregulation, biological oscillations, electrical analogs of arteries, integrated models of biological systems, electrical properties of tissue, and frequency analysis of biological systems.

25.1 Law of Mass Action and Enzyme Kinetics

An appropriate starting point for this chapter is to consider the "law of mass action." This refers to chemical reactions, but by analogy, the concept can be applied to many physiological systems. The book by Keener and Sneyd (1998), is an excellent resource for additional information on this topic (see Bibliography).

Let chemicals A and B react to produce C

$$A + B \xrightarrow{k} C.$$

The *reaction rate* is the rate of production of C (i.e., d[C]/dt), the square brackets denoting concentration. This reaction rate is proportional to both $[A]$ and $[B]$, so

$$d[C]/dt = k\,[A]\,[B] \quad \text{(Law of mass action)}$$

k is the rate constant for the reaction and A *or* B could be constant also.

Actually, most reactions can go in both directions, so

$$A + B \underset{k_-}{\overset{k_+}{\rightleftarrows}} C.$$

Here k_+ and k_- are rate constants for forward and reverse reaction, respectively (N.B. they have different units). Considering now the rate of consumption of A:

$$d[A]/dt = k_-\,[C] - k_+\,[A]\,[B] \quad \text{(mol/L/s)}.$$

Often $k_+ >> k_-$ (or vice versa) so one can be ignored.

At equilibrium d$[A]$/dt = 0, so $[C^{eq}] = (k_+/k_-)\,[A^{eq}]\,[B^{eq}]$. As A decreases, C increases. In fact $[A] + [C] = A_0$ (if A and C are not involved in other reactions). Eliminating $[A]$ gives

$$[C] = A_0[B]/(K_{eq} + [B])$$

where $K_{eq} = k_-/k_+$ is the equilibrium constant of the reaction (whose units are concentration). Note that if $K_{eq} = [B]$, then $[A] = [C] = A_0/2$.

Now let us specifically consider enzymes, which catalyze the conversion of a specific substrate S into a product P. More specifically S combines with a free enzyme E to form a complex ES which later decomposes to form product P together with E again, which is unaltered (there may be other small molecules such as CO_2 or OH^- formed, but these do not affect the analysis). We can represent this two step process as follows:

$$E + S \underset{k_{-1}}{\overset{k_{+1}}{\rightleftarrows}} ES \underset{k_{-2}}{\overset{k_{+2}}{\rightleftarrows}} E + P.$$

We can usually ignore the back reaction k_{-2}. We can write the rate of change of substrate concentration as

$$d[S]/dt = -k_{+1}[E][S] + k_{-1}[ES]$$

$$d[ES]/dt = k_{+1}[E][S] - (k_{-1} + k_{+2})[ES]$$

$$d[P]/dt = -d[S]/dt = k_{+2}[ES].$$

Now $[E] + [ES] = \text{constant} = [E_T]$.

Thus, we can rewrite the second equation as

$$d[ES]/dt = k_{+1}[E_T][S] - (k_{-1} + k_{+2} + k_{+1}[S])[ES].$$

If $[S] >> [E_T]$ then $d[ES]/dt \to 0$; giving

$$[ES] = k_{+1}[E][S]/(k_{-1} + k_{+2} + k_{+1}[S]).$$

The velocity of reaction (V) is the same as the rate at which both the product increases and the substrate decreases, i.e.,

$$V = d[P]/dt = -d[S]/dt = k_{+2}[ES] = k_{+2}k_{+1}[E][S]/(k_{-1} + k_{+2} + k_{+1}[S]).$$

And if we write $K_m \equiv (k_{-1} + k_{+2})/k_{+1}$, we get

$$V = k_2[E_T][S]/(K_m + [S]).$$

Note that as $[S] \to \infty$, $V \to k_{+2}[E_T] \equiv V_{Max}$, thus:

$$V = V_{Max}[S]/(K_m + [S]) = V_{Max}/\{1 + K_m/[S]\}.$$

This is called the *Michaelis-Menten* equation, and K_m is known as the Michaelis constant. Note that K_m is actually the same as the substrate concentration at which $V = V_{Max}/2$. We assume the reaction is in steady state, that is the reaction rate is constant with time.*

Experimentally K_m and V_{Max} can be deduced from the slope and intercept of a Lineweaver-Burke plot in which 1/V is plotted against 1/[S]. Rearranging the above equation:

$$1/V = K_m/(V_{Max}[S]) - 1/V_{Max}.$$

Many enzymes consist of subunits which need to cooperate to allow the reaction to proceed. It can be shown that in this case the velocity is given by

$$V = V_{Max}[S]^n/(K_m^n + [S]^n).$$

* Note that the equation was derived by Michaelis and Menten for equilibrium, with K_m defined as k_{-1}/k_{+1}. The extension to steady-state was derived by Biggs and Haldane 12 years later.

This is called the *Hill Equation* and n is the degree of cooperativity, which can be as high as 5, but is often not an integer. n can be found experimentally by plotting $\ln(V/(V_{max}-V))$ versus ln([S]) (Hill plot). n is the slope.

WORKED EXAMPLE

The spreadsheet in Figure 25.1 (lower) tabulates partial oxygen pressures (pO_2) (column A) and % O_2 saturation in the blood (column C). If column is plotted against A a sigmoidal curve is obtained (this was first shown by AV Hill, who gave his name to the equation in question). Here, the velocity of reaction is interpreted as the % of hemoglobin molecules carrying oxygen molecules (up to a V_{max} of 100%). Columns B and E compute ln[S] and $\ln(V/(V_{max} - V))$, respectively. Plotting these against each other will give a straight line, slope 2.3, which is the value for n in this case.

Notice that if $[S] \ll K_m$ then $V \to V_{Max}[S]^n/K_m^n$; i.e., $V \propto [S]^n$, which is the same as the "law of mass action" if n = 1. In this case,

$$V = d[P]/dt = -d[S]/dt = k[S].$$

Note, too, that as the degree of cooperativity increases, the graph of V versus [S] becomes increasingly sigmoidal in shape (Figure 25.1 (upper)).

Finally, if we consider the situation where n = 1 and $[S] \ll K_m$. Here we can use linear approximation:

$$V = d[P]/dt = V_{max}[S]/K_m.$$

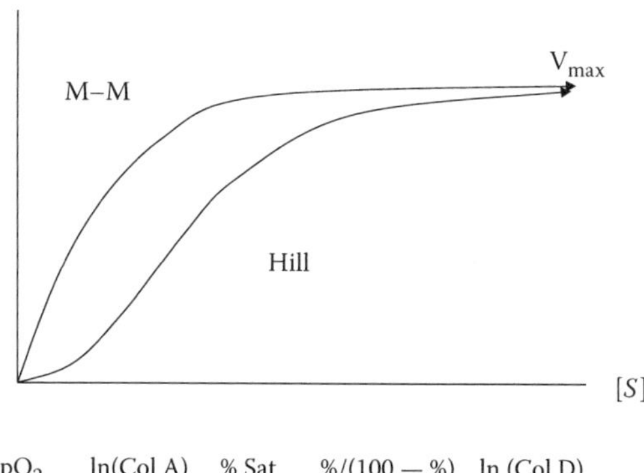

pO_2	ln(Col A)	% Sat	%/(100 — %)	ln (Col D)
20	1.30103	36	0.5625	−0.24988
40	1.60206	74	2.846153846	0.454258
60	1.778151	88	7.333333333	0.865301
80	1.90309	93	13.28571429	1.123385

FIGURE 25.1

Enzyme kinetics. Upper diagram: plot of reaction rate V versus substrate concentration [S] comparing Michaelis-Menten kinetics (top curve) with Hill kinetics (bottom curve). The curve becomes more sigmoidal the greater the degree of cooperativity between groups. Lower diagram: human data of oxygen saturation (%Sat: 3rd column) and blood partial O_2 pressure (pO_2: 1st column). The final column is the quantity ln (%Sat/(100 – %Sat)). Entering these data into a spreadsheet and plotting the second and last columns gives an approximate straight line whose slope is the Hill exponent *n*.

We now assume that the product P is being broken down with rate constant k_+ and assume the law of mass action gives the rate of breakdown (in mol/L/s) as $-k_+[P]$.

We can thus write:

$$d[P]/dt = V_{max}[S]/K_m - k_+[P].$$

In other words, rate of change of concentration of product equals rate of production minus rate of breakdown. In general, for a fixed volume, the masses (*p, s*) rather than concentrations ([P], [S]) are conserved, but the importance of this will only become apparent when we consider compartmental systems, in Section 25.4.

In general, we can write the previous equation in terms of masses as

$$dp/dt = a_1 s - a_2 p,$$

where a_1 and a_2 are constants. More generally still, if we have n components, whose masses are $x_1, x_2, \ldots x_i, \ldots x_n$, we can write a series of simultaneous first-order ordinary differential equations (ODEs) as follows:

$$dx_1/dt = a_{11}x_1 + a_{12}x_2 \ldots + a_{1i}x_i \ldots + a_{1n}x_n$$

$$dx_2/dt = a_{21}x_1 + a_{22}x_2 \ldots + a_{2i}x_i \ldots + a_{2n}x_n$$

$$\ldots$$

$$dx_n/dt = a_{n1}x_1 + a_{n2}x_2 \ldots + a_{ni}x_i \ldots + a_{nn}x_n.$$

This is a very important way of representing creation-breakdown reactions. As we will see this form of equations can be applied to a wide range of physiological systems and they are easily solved (for simple systems at least) using some special functions within MATLAB®.

25.2 Systems Analysis: Application of Formal Control Theory Concepts to Physiological Systems

Our focus now shifts to feedback control systems, such as the way we sweat more on a hot day in order to lose heat by evaporation and thus keep core temperature constant. Feedback control involves the *modification* of a process or system by its *results* or *effects*. It may be helpful to think of physiological control mechanisms as consisting of two relationships: one in which the output is the *controlled* variable (e.g., core temperature, plasma osmotic pressure, blood pressure, blood glucose, and so on—that is, the *results* or *effects*) and the other has an output (which depends on the controlled variable) which exerts control or *modification* (e.g., hypothalamic neural activity controlling sweating, ADH secretion, baroreceptor activity, insulin secretion rate). The first relationship is known as the forward, or "feed-forward" relationship, the second, the feedback relationship. Most physiological systems involve negative feedback, which is the basis for *homeostasis*, or the maintenance of constancy as maintained by physiological processes. Some systems, such as the generation of the action potential, the clotting of blood and the immune reaction involve positive feedback, at least in early phases. In negative feedback systems the feed-forward and the feedback relationship can be plotted on the same set of axes (as a first example ADH secretion rate versus plasma osmotic pressure [OP]). In this example the controlled variable is OP and ADH is the factor that exerts control. In the feed-forward relationship increased ADH secretion causes *decreased* OP through water retention. In the feedback relationship raised OP is sensed by osmoreceptors to cause *raised* ADH secretion. This is shown in Figure 25.2. Denoting a relationship in which *increase* of the independent variable gives rise to a *decrease* of the dependent variable by a dotted arrow and a relationship in which a *rise* in one gives a *rise* in the other by a full arrow, we can write for this particular system:

$$\text{ADH} \dashrightarrow \text{OP (forward reaction)}$$

$$\text{OP} \rightarrow \text{ADH (feedback reaction)}.$$

We can think of these as two curves which cross at some point, the equilibrium point (or operating point, if we draw an analogy with operational amplifiers, see Chapter 2).

In reality, there are many intermediate steps, and a circle of relationships A → B → C → D → E → F → A is closer to actuality. Of course, once we can represent these steps by functional relationships (i.e., theoretical or empirical formulae) we can reduce the six relationships down to just two, and then just one relationship. It is often better to leave it at two relationships, so that the distinction between controlled and controlling system can be drawn.

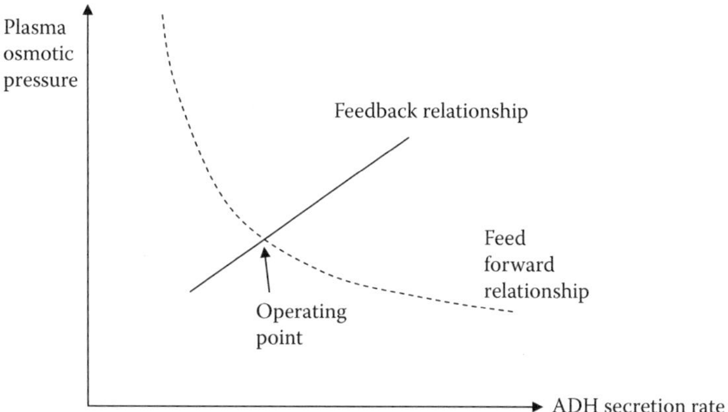

FIGURE 25.2
Illustrates a general negative feedback control system. As ADH secretion increases, this leads to lowered plasma osmotic pressure (OP), because of water retention by the kidney ("feed forward" relationship shown dotted). The osmoreceptors in the brain, however, provide feedback by stimulating the neurohypophyseal regions of the brain to release ADH, if the OP becomes elevated. For simplicity, we can assume this relationship to be linear (full line). The point where the two relationships cross is the equilibrium or "operating point" of the system.

WORKED EXAMPLE

Suppose an analysis of experimental data following trends similar to those shown in Figure 25.2 give the following relationships (in appropriate units, which need not concern us at the moment)

$$\text{Feed-forward: OP} = 20.9/\text{ADH}$$

$$\text{Feedback: ADH (secretion rate)} = 2 \times 10^{-4} \times \text{OP}$$

Substituting the second equation into the first gives OP = 20.9/(2 × 10^{-4} × OP), or OP = $\sqrt{20.9/(2 \times 10^{-4})}$ = 323 OP units. Using the second equation gives 0.065 for the rate of ADH secretion. This represents the operating point for this system.

In a complex biological system, were it is difficult to decide at the outset what factors need to be considered, it is useful to draw a "symbol and arrow" diagram based on what is known about the physiological system in question. The stages of this exercise, which help in the initial organization of information, are as follows: (i) identify what the variables are in a control loop; (ii) identify what depends on what (i.e., direction of *causality*); (iii) draw an arrow from the *independent* variable(s) towards the *dependent* variable(s) for each step (in the direction of causality); (iv) if the dependent variable increases as the independent variable decreases, draw the arrow *dotted* (otherwise a full arrow); and (v) identify loops of variables, and for each loop ignore arrows *pointing away* from variables in that loop. For a negative feedback system there has to be an odd number of dotted arrows in this representation of the loop (otherwise changes would accumulate, as in positive feedback).

An example of a loop of relationships is the pupillary reflex, which was described in Chapter 1. Here there are four relationships in the overall control pathway: The variables are p, the flux of light incident on the retina, f_p the rate of nervous impulses to the CNS from the retina associated with this reflex, f_m the rate of nervous impulses from the CNS to the ciliary muscles and A the pupil area. L is light intensity. As f_m increases A decreases, so this is the dotted arrow step. The "symbol and arrow" diagram is shown in Figure 25.3 (top). The single dotted arrow in the loop ensures negative feedback.

An alternative way of representing this system is a block diagram (Figure 25.3, bottom). This is the next stage on from the initial "symbol and arrow" diagram. Here the blocks represent *processes* or often specific organ systems. Each box or block contains a representation of the actual relationship that exists between the independent variable (x axis) and the dependent variable (y axis). The dependent variable of the first block serves as the independent variable in the next block, and so on. The precise relationships are a mixture of experimental data and informed guesswork. In the pupillary example, an initial attempt at specifying relationships could yield the following:

$$p = A.L \quad \text{(from definitions of flux and intensity)}$$
$$f_p \approx k_1\sqrt{(p - p')} \quad \text{(by looking at experimental data)}$$
$$f_m = k_2 f_p + k_3 \quad \text{(guesswork)}$$
$$A = k_4/f_m \quad \text{(guesswork).}$$

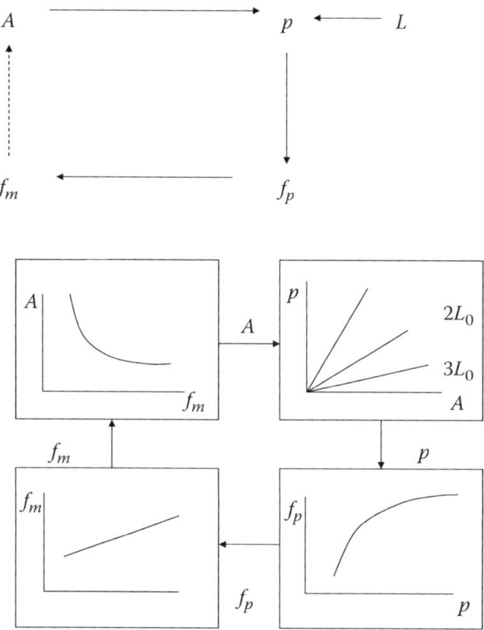

FIGURE 25.3
Top: symbol and arrow diagram and bottom: block diagram for the pupillary control system. See text for explanation.

These can be represented by a composite graph shown in Figure 25.4.

We can now choose an arbitrary starting point and note that with a very few iterations the equilibrium point is reached. The final values are shown as circles and the same point is reached whether arbitrary high or low starting values of the parameters (A in this example) are chosen (Figure 25.5).

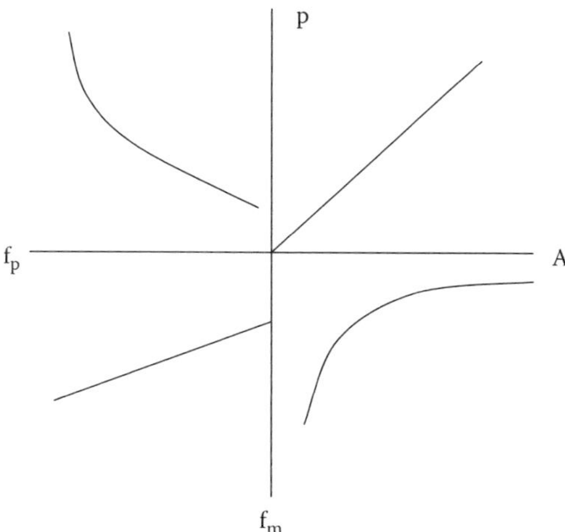

FIGURE 25.4
Relationships illustrated in the block diagram of Figure 25.3 displayed on four sets of axes "back-to-back."

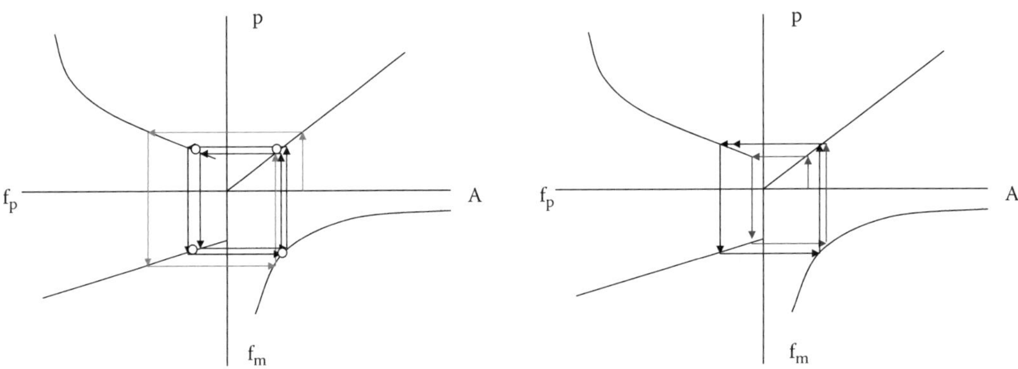

FIGURE 25.5
Equilibrium reached from high and low initial values of A (left and right, respectively).

We can, of course, reduce the four equations to just two (feed-forward and feedback) by substitution. For example, the first three could be combined to yield:

$$f_m = k_2 k_1(\sqrt{[AL - p']} + k_3$$

and the last equation can then be use to eliminate f_m and to discover the "operating point" shown graphically in Figure 25.5.

25.3 Control Theory and Open Loop Gain

It is now time to make reference to the more formal representations of feedback control systems as presented in engineering texts. The top part of Figure 25.6 shows this, with a reference level signal (ideal blood pressure, for example) denoted by "r" and a controlled output by "c." The value of "c" is determined by some process which compares "r" against "b" (which is directly related to "c") to produce an error signal "e." If "e" is zero, then the actual blood pressure is the ideal pressure and there is no need to exert any further control via "c." G is the gain in the relationship between "e" and "c." It may just be a number, like amplifier gain, or it could be a more complicated relationship. H represents the way the controlled output is monitored to produce a feedback signal "b." Suppose H is a simple fraction (such as 0.3).

The system equations are as follows: $c = Ge$; $b = Hc$; $e = r - b$.

Thus,

$$c = (r - b)G = (r - Hc)G,$$

or

$$c(1 + GH) = rG.$$

Thus:

$$c/r = G/(1 + GH) = \text{output change/input change}.$$

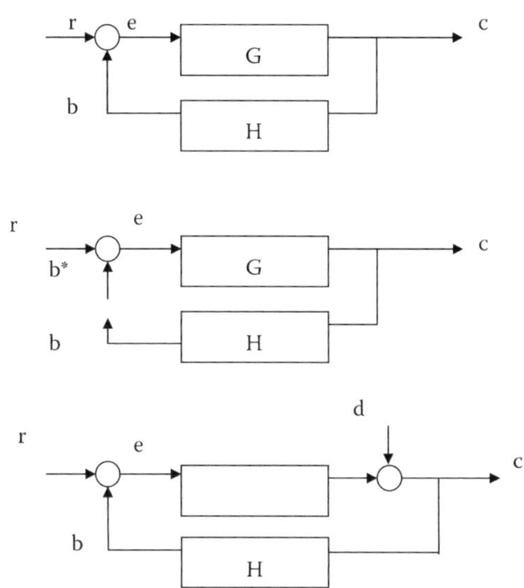

FIGURE 25.6
Formal representation of feedback systems: G is forward gain and H is feedback fraction (or feedback characteristic). Middle diagram: opened loop, with a gain of G × Hr: reference level; e: error signal; c: controlled output; d: disturbance.

The middle part of Figure 25.6 represents an artificially opened controller: we treat b* as input with b as output. Now:

$$\text{output change/input change} = GH = \textit{open loop gain.}$$

The significance of open loop gain (OLG) can be appreciated by considering the effects of a disturbance d (such that $c' + d = c$), and the change in output Δc due to a change in reference Δr.

Consider input/output of box G: output is actually $\Delta c' = \Delta c - d$, so

$$\Delta c - d = eG = (\Delta r - b)G = (\Delta r - H\Delta c)G.$$

Rearranging gives

$$\Delta c = G\Delta r/(1 + GH) + d/(1 + GH).$$

The first term is the change in output due to the change in reference and the second is the change due to the disturbance. If $\Delta r = 0$ then $\Delta c = d/(1 + GH)$. In the absence of feedback the change Δc would be equal to the disturbance d, so

$$\frac{\text{change in output in closed loop situation}}{\text{change in output in open loop situation}} = \frac{1}{(1+GH)}.$$

The reduction is considerable for large G. Now, returning to the top diagram of Figure 25.6, if we replace r by x and c by y, we get for the feed forward and the feedback relationships the following:

$$y = Gx$$

$$x = r - Hy.$$

In terms of a symbol and arrow diagram we get x → y for the first relationship and y ⇢ x for the second, thus fulfilling the requirement for negative feedback. Note that for the first relationship dy/dx = G and for the second dx/dy = –H. Thus, the open loop gain (OLG) G × H is the modulus of the *product of the slopes*. In general, if we represent the two relationships as follows:

$$y = f(x)$$

$$x = g(y)$$

then $|(f'(x)g'(y))|$ is a measure OLG, and for the reasons outlined above measures the effectiveness of control of the system, since it measures how well the system minimizes the effects of disturbance. In the case of the plasma osmolarity example above, $f'(x)$ is $-20.6/ADH^2$ and $g'(y)$ is just 2×10^{-4}. The OLG at the "operating point" (323, 0.065) is thus $-20.6 \times 2 \times 10^{-4}/(0.065)^2 = 0.98$. Of course this is a highly artificial example. Typical values of OLG are 5.5 for the respiratory controller, 5 for the glucose controller (see Hobbie and Roth, 2007). The OLG for the pupillary control system in Section 25.2 can be found algebraically by multiplying the differentials of the equations reduced to two $|(df_m/dA)(dA/df_m)|$. Actually the same results would be obtained by multiplying the four differentials $|(dA/dp)(dp/df_p)(df_p/df_m)(df_m/dA)|$. However, it is important to go around the loop in the direction of causality: going around the other way gives the inverse result.

25.3.1 Problems of Applying Formal Control Theory to Biological Systems

It is important to realize the limitations of applying these rather crude techniques, which are really for the purpose of preliminary rather than in-depth analysis. These include

- Inherent nonlinearities imply that because superposition may not hold, Fourier analysis (which is important for in-depth analysis) may not be appropriate.
- The comparison of reference level and feedback signal (so-called comparator) may not be a separate entity. In fact, usually nothing corresponds easily to a "reference input" (there is no part of the body that provides a "standard 37°C," for example).
- Usually there are several inputs for one output. This is important if the influences controlling these other inputs are not included or not understood.
- Systems often cannot be considered in isolation—for example, the phase of breathing affects the heart rate.
- Difficulty of opening loops to study gain. There are some specialized ways of doing this which will be mentioned later.

25.3.2 Example of OLG for a Simple Model of the Complete Circulation

This example uses the model in Khoo and assumes that the circulation can the split into two components: the heart and the circulation (Figure 25.7). The main loop variable are cardiac output (= venous return) Q and right atrial pressure P_{RA}. It is based on the Guyton analysis described in Chapter 10. First, the "feed-forward" relation is the Starling law of the heart. For the "linear" region of the Starling curve:

$$Q = fC_D(P_{RA} - C_S P_A/C_D - P_{PL}).$$

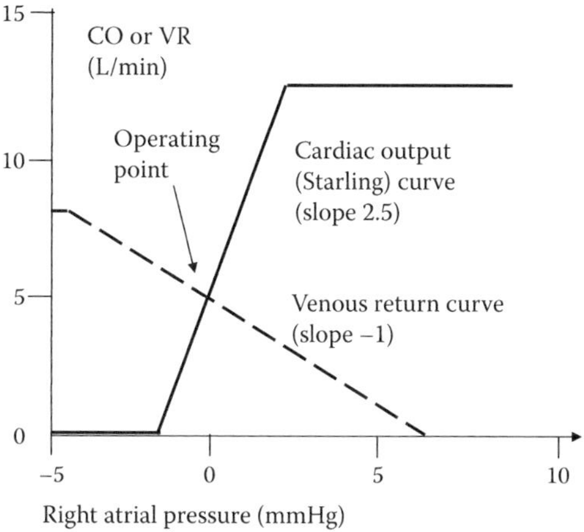

FIGURE 25.7
The heart and circulation as a "feed-forward" and "feedback" loop. The product of the slopes is 2.5 (ignoring sign) which is the Open-Loop Gain or OLG. (Adapted from Khoo: *Physiological Control Systems*. 2000, Copyright Wiley-VCH Verlag GmbH & Co. KGaA. Reproduced with permission.)

Essentially this is Q = heart rate (f) times the stroke volume, C_S, C_D are systolic and diastolic capacitances and P_{PL} is pleural pressure. P_A is arterial pressure. The slope dQ/dP_{RA} is simply fC_D.

The feedback relationship is the determination of P_{RA} by the total flow Q through the circulation. This follows the Guyton concept of P_{ms}, the mean systemic pressure (see Chapter 10). The following relationship for venous return Q_R is obtained (remember $Q_R = Q$):

$$Q_R = (P_{ms} - P_{RA})/(R_V + R_A/19)$$

where R_V and R_A are the arterial and venous resistances. The 19 appears because the venous capacity is 18 times bigger than the arterial capacity. This can be rearranged to make P_{RA} the dependant variable (and dropping the R subscript in Q_R):

$$P_{RA} = P_{ms} - Q(R_v + R_A/19).$$

Thus, $dP_{RA}/dQ = -(R_V + R_A/19)$ and hence, $OLG = fC_D(R_V + R_A/19)$.

No values are quoted, but examining Figure 3.11 in Khoo gives a value of around 2.5.* Note that as heart rate, cardiac compliance and vascular resistance increase, these all contribute to better control.

The Operating Point as a function of P_{RA} is given by

$$P_{RA} = \{P_{ms} + A(C_S P_A/C_D + P_{PL})\}/(1 + A)$$

where $A = fC_D(R_{v+} + R_A/19)$.

* Khoo reserves the term "Open Loop Gain" for the gain of the feed-forward relationship (fC_D). What is referred to as "OLG" above, he terms "Loop Gain" (LG).

25.4 Compartmental Systems: General Multicompartment Systems

Compartmental analysis is particularly useful in studying the effects of drugs, anesthetics, hormones, toxins, and other mobile substances in the body. Ion transport and the movement of respiratory gases are also amenable to this approach, as indeed is the study of body temperature. Compartments are assumed to be well-stirred and have a defined volume V_i. The volume may vary with time, but everywhere within that volume (at a given time) the quantity under consideration (amount of substance, temperature, pH etc.) is assumed not to alter. In simple compartmental systems with fixed volumes the variation with time can be found by obtaining an exact (or analytical) solution. For many compartment systems, numerical methods, such as a provided by MATLAB functions, are particularly useful, but because these are approximate solutions, accuracy has to be assessed. Using Laplace transforms and matrix methods some more complicated systems can be solved exactly, but a decision has to be made on whether this increased precision is in line with natural experimental variation.

Figure 25.8 shows a single compartment, which is able to exchange material (or heat etc.) with other compartments. We can label it as the *ith* compartment (i.e., compartment *i*). In addition to the n − 1 other compartments that can exchange with this compartment, there is material coming in from any *source* of the material in question (such as an enzyme producing it or a system infusing it) and there may be loss of the material to a *sink* (considered to be infinite in size and producing no back exchange (this could by enzymic breakdown or excretion).

Let the *amount* in compartment i be x_i, the amount in compartment j be x_j and so on. Only positive or zero amounts are allowed. Each compartment has a defined volume V_i, V_j, etc. We define the flow from compartment i to j as f_{ij}, and that from j to i as f_{ji}. Typical units of f are mass per unit time, such as kg/s or mole/min. If we now consider the rate of change of mass in i we get:

$$dx_i/dt = \sum f_{ji} - \sum f_{ij} + f_{oi}(t) - f_{io}(t).$$

Here the summations are taken from j = 1 up to n, but excluding j = i. f_{oi} is the flow from the source and f_{io} is the flow from the *ith* compartment to the sink. The source can often be represented by a step function $u_i(t)$ or an impulse function $\delta(t)$. The sink can be modeled (using the law of mass action) as: $f_{io}(t) = k_{io}x_i(t)$. Here, k_{io} is a rate constant whose units are time^{-1}.

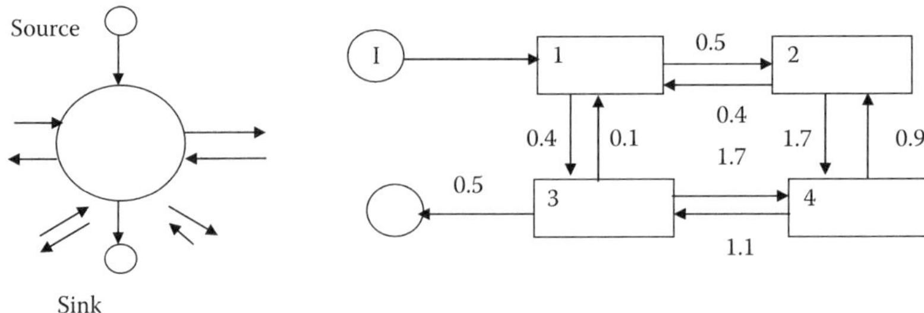

FIGURE 25.8
Left: a single compartment showing exchanges with other compartments. Right: a four compartment system (with source and sink compartments) with numerical values for the exchange between compartments (in min^{-1}).

Other flows represent *transits* and can be similarly expressed as

$$f_{ji} = k_{ji}x_j(t) \quad \text{and} \quad f_{ij} = k_{ij}x_i(t).$$

Thus we can write:

$$dx_i/dt = \sum k_{ji}x_j(t) - x_i(t) \sum k_{ij} + u_i.$$

Here the second summation includes j = 0; i.e., the sink, but excludes j = i. In words:

Rate of change of x in compartment i equals rate at which material arrives in i minus rate at which material leaves compartment i. We can also write this as

$$dx_i/dt = k_{1i}x_i(t) + k_{2i}x_2(t) + k_{3i}x_3(t) + \ldots + k_{ii}x_i(t) + k_{ni}x_n(t) + u_i$$

which is of the form that was given at the end of Section 25.1 and as we will see can be solved using MATLAB and other programs which solve simultaneous first order ("ordinary") differential equations (ODEs).

We can also express these equations in terms of concentrations

$$c_i(t)V_i dc_i/dt = \sum k_{ji}V_jc_j(t) - V_ic_i(t) \sum k_{ij} + u_i.$$

The k_{ij} values have units of reciprocal time (i.e., s^{-1}). They can also be expressed as "clearance" q in liters per second (or similar) where $q_{ji} = k_{ji}V_j$. Many of the k_{ij} will be zero and the nonzero k_{ij} can be estimated from controlled experimental data.

WORKED EXAMPLE

The lower part of Figure 25.8 represents a compartmental system, with the numbers representing the transfer coefficients k_{ij} (in min^{-1}) of a substance between compartments. Write down a set of equations to describe the rate of change of mass (m_i) in compartments 1–4. I = 25 g/min.

Answer

Following the form of the equation shown above we can write down the following:

$$dm_1/dt = 25 - 0.9m_1 + 0.4m_2 + 0.1m_3$$

$$dm_2/dt = 0.5m_1 - 2.1m_2 + 0.9m_4$$

$$dm_3/dt = 0.4m_1 - 2.3m_3 + 1.1m_4$$

$$dm_4/dt = 1.7m_2 + 1.7m_3 - 2m_4$$

Suppose the volume of compartment 1 is 10 mL and the remaining 3 compartments 5 mL. Modify the equation for compartment 1 in terms of concentrations c_i (in g/mL) in it and the adjacent compartments.

Answer

Remember that m = Vc and substitute for the appropriate values in the first equation.

$$10dc_1/dt = 25 - 10 \times 0.9c_1 + 5 \times 0.4c_2 + 5 \times 0.1c_3$$

$$\text{so } dc_1/dt = 2.5 - 0.9c_1 + 0.2c_2 + 0.05c_3$$

Estimate the rate of change in concentration (stating units) in compartment 1 at the moment in time when concentrations in compartments 1, 2, and 3 are the same, 1 g/mL.

Answer

$$dc_1/dt = 2.5 - 0.9 + 0.2 + 0.05 = 1.85 \text{ g/mL/min}$$

25.4.1 Special Case: State Equations for a Closed 2-Compartment System

The total amount of material $A_o = A_1(t) + A_2(t)$ remains fixed for all times.

$$dA_1/dt = k_{21}A_2 - k_{12}A_1 = -dA_2/dt$$

$$dA_1/dt = k_{21}A_o - (k_{12} + k_{21})A_1(t).$$

This can be integrated to give:

$$[A_1(t) - k_{21}A_o/(k_{12} + k_{21})] = [(A_1(0) - k_{21}A_o/(k_{12} + k_{21})]\exp[-(k_{12} + k_{21})t]$$

and since $C_1(t) = A_1(t)/V_1$, and as $t \to \infty$ $dA_1/dt \to 0$, so

$$k_{21}A_2(\infty) = k_{12}A_1(\infty); \text{ and since } C_1(\infty) = C_2(\infty)$$

$$A_1(\infty)/V_1 = A_2(\infty)/V_2; \text{ so } k_{12}/k_{21} = V_2/V_1$$

$$\text{and of course } C_1(\infty) = A_o/(V_1 + V_2) = C_2(\infty)$$

now putting $q = k_{12}V_1 = k_{21}V_2$ (units: L/s)

$$\begin{aligned} C_1(t) &= A_1(t)/V_1 \\ &= qA_o/(V_1V_2(q/V_1+q/V_2)) + [A_1(0)/V_1 - qA_o/(V_1V_2(q/V_1+q/V_2))]\exp[-(q/V_1+q/V_2)t] \\ &= A_o/(V_1 + V_2) + [A_1(0)/V_1 - A_o(V_1 + V_2)]\exp[-q(1/V_1 + 1/V_2)t] \end{aligned}$$

$$C_1(t) = C_1(\infty) + [C_1(0) - C_1(\infty)]\exp[-q(1/V_1 + 1/V_2)t] \text{ and}$$

$$\begin{aligned} C_2(t) &= C_2(\infty) + (C_2(0) - C_2(\infty)]\exp[-q(1/V_1 + 1/V_2)t] \\ &= (A_o - V_1C_1(t))/V_2 \end{aligned}$$

WORKED EXAMPLE

The data below represents the fluorescent intensity I(t) of a dye in blood samples (each the same volume) taken at 100 second intervals. Initially, 10 μg of this dye was added to the blood, which was assumed to rapidly mix within the circulatory blood volume, 6 L. It is also assumed that the dye exchanges with a single second compartment. Note that the ratio of initial and equilibrium intensities of blood fluorescence $(I(0)/I(\infty))$ is 4.0.

Time	100	200	300	400	500
Intensity (counts)	989	721	672	663	661

(a) Show that the volume of the other compartment is 18 L. **Answer:** The ratio of fluorescent intensities is the same as the ratio of initial and final concentrations in the blood. The mass of dye has been diluted to ¼ of its original value so the total volume $(V_1 + V_2)$ must be four times the original volume the dye was in, V_1V_2 must be $3 \times V_1$, hence 18 L.

(b) What are the initial and final concentrations (in μg/L) of dye in the two compartments? Initially the concentration of dye in the blood is 10/6 = 1.67 μg/L, then finally it is 1.67 × ¼ = 0.42 μg/L, the same as the concentration in compartment 2 **Answer:** (because it is by then in equilibrium). We can assume that initially there was no dye in compartment 2.

(c) What is the value of the clearance q between compartments (and in what units)? **Answer:** To get the clearance, we need the slope of the graph of $\ln[I(t) - I(\infty)]$ versus time (this can be seen by remembering that I is proportional to C and by taking natural logs of both sides of the equation above).

I(t) – I(inf)	328	60	11	2	0
ln[I(t) – I(inf)]	5.793014	4.094345	2.397895	0.693147	

The estimate of the slope is –0.017 and this is equal to $-q(1/V_1 + 1/V_2)$ from the equations above. This gives 0.077 L/s for q.

(d) Given that the permeability of capillary walls to the dye is 3.7×10^{-5} m/s estimate the total surface area capillary walls in the circulation (assume all of exchange occurs across capillary walls). **Answer:** The relationship between permeability and clearance is simply P = q/A. We first need to change q to m^3/s: remembering that $1\ L = 10^{-3}\ m^3$ gives $A = q/P = 0.077 \times 10^{-3}/3.7 \times 10^{-5} = 2\ m^2$.

25.4.2 Open 2-Compartment System

This is similar to the example above, except that material can exit from compartment 1 to the environment (which in this case will be labeled "3"). We assume there is no flow from compartment 3 back into 1 (the environment is a "sink"). This corresponds to a situation where a drug is injected into the blood (1) which then exchanges with another compartment labeled "2" (ECF, say) but is simultaneously metabolized or broken down in the blood (sink) (Figure 25.9).

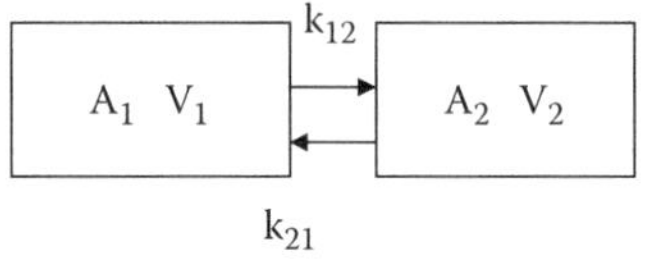

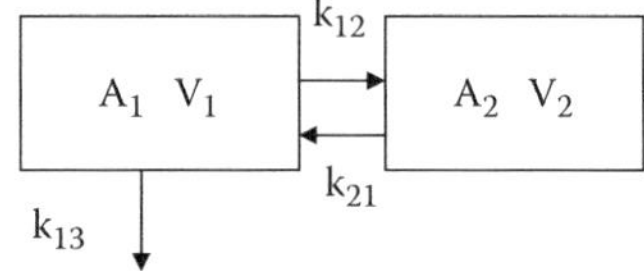

FIGURE 25.9
Two compartment systems. Top: Closed and bottom: open. Symbols as follows: A_i is amount; V_i volume; k_{ij} rate constants for transfer from i to j.

The two equations now become:

$$dA_1/dt = -(k_{12} + k_{13})A_1 + k_{21}A_2$$

$$dA_2/dt = k_{21}A_1 - k_{21}A_2$$

and in terms of concentrations (note the addition of volume ratios):

$$dc_1/dt = -(k_{12} + k_{13})c_1 + (V_2/V_1)k_{21}c_2$$

$$dc_2/dt = (V_1/V_2)k_{12}c_1 - k_{21}c_2.$$

25.5 Glucose and Iodine Compartmental Models

As we saw in the previous section, many physiological systems are best analyzed in terms of *compartments*. The first example is the Bolie model for plasma glucose. This does not fit readily into the usual compartmental systems analysis, because it is really two single compartments, considering two different quantities, glucose and insulin, which we could represent by y_1 and y_2, but for the sake of clarity we represent these by G and I. These symbols represent the *amounts* of these quantities in the total volume of plasma. The compartments are shown in Figure 25.10. Since both represent total body plasma they both have the same volume V.

Usually B(t) = 0, unless insulin is introduced as an injection or infusion. A(t) could be in the form of a meal or an infusion. For the Glucose Tolerance Test, 100 g of glucose is given as a drink after a period of fasting. The form of A(t) in this case depends on how quickly this is taken up from the gut. For the sake of simplicity, we will assume it to be a step function, but it could also be represented as an impulse.

The two equations are

$$dG/dt - k_4G + k_5 - k_6I + A(t)$$

$$dI/dt = k_1I + k_2 + k_3G + B(t).$$

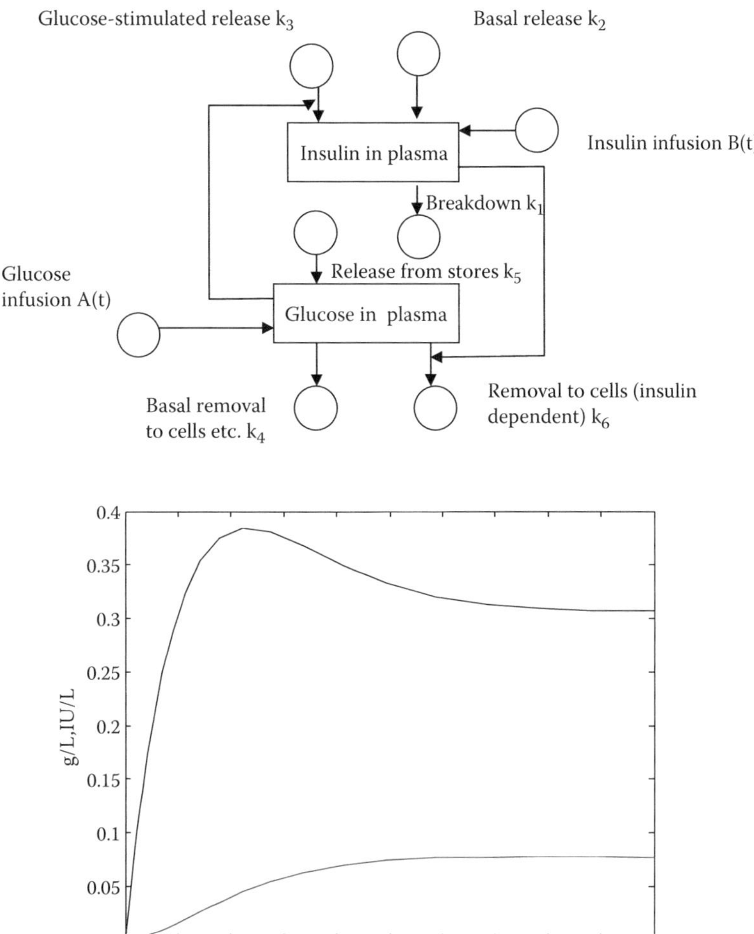

FIGURE 25.10
Top: a simple model for insulin control of plasma glucose. Bottom: output from a Simulink™ implementation of the model shown at top. Response to a step input of glucose. Upper curve is [glucose] change from resting values and lower is [insulin].

Please note that k_2 and k_5 will have different units from the others, even if both G and I are in grams (actually Insulin is usually measured in "international units" or IU).

Considering an equilibrium state.

$$dG/dt = 0 = k_4G_o + k_5 - k_6I_o + 0$$

$$dI/dt = 0 = k_1I_o + k_2 + k_3G_o + 0.$$

Thus, we can put $k_5 = k_4G_o + k_6I_o$ and $k_2 = k_1I_o - k_3G_o$.

Thus, $dG/dt = -k_4G + k_4G_0 + k_6I_0 - k_6I + A(t)$; $dI/dt = -k_1I + k_1I_0 - k_3G_0 + k_3G + B(t)$

If we now put $i = I - I_0$; and $g = G - G_0$, the equations now become:

$$dg/dt = -k_4g - k_6i + A(t)$$

$$di/dt = + k_3g - k_1i + B(t).$$

And we can set A(t) = au(t) (u(t) is a step input in glucose, which = 1 for t => 0 and 0 for t < 0)

$$B(t) = 0.$$

If we assume the effective volumes of the glucose and insulin compartments to be the same, we can interpret the symbols g and i as concentrations rather than amounts (we need to do this with caution).

G is in g/L and G_0 is specifically 0.8 g/L
I is in International Units/L, and I_0 is 0.005 IU/L
Typical values: k_1 is 0.8 h^{-1}; k_3 is 0.2 IU/h/g; k_4 is 2 h^{-1}; k_6 is 5 g/h/IU; a is 1 g/L/hr.

25.5.1 Analytic Solution

Actually, in this case we can get an exact solution. If we put x = i and y = g we get

$$dy/dt = -k_4y - k_6x + au(t)$$

$$d^2y/dt^2 = -k_4(dy/dt) - k_6(dx/dt) + adu(t)/dt$$

and since $dx/dt = -k_1x + k_3y$, we can substitute in the above equation, giving (after some algebra):

$$d^2y/dt^2 + (k_1 + k_4)(dy/dt) + (k_1k_4 + k_3k_6)y = k_1a + adu/dt.$$

This has a standard solution, which can found in a standard textbook. Remember that y (= g) is the difference from the equilibrium value for glucose concentration. Both methods of solution give similar curves, which have a resemblance to Glucose Tolerance Test (GTT) curves (shown in Figure 25.10: lower).

25.5.2 Numerical Solution

We have 2 simultaneous Ordinary Differential Equations and the method of solving these in MATLAB will now be described. An appropriate function is called ODE23. ODE stands for Ordinary Differential Equations and the 23 refers to the precise form of the algorithm (called a Runge-Kutta routine) used. There is also ODE45, but the difference need not concern us just now. ODE23 is called from the "command prompt" in the following way:

```
>>[t,y] = ODE23(myfile,[0 5], [0 0]);
```

Here, t represents time, y a matrix containing the values of the variables (in this case 2) at each time point and "myfile" is the so-called m-file (see below) which specifies the particular set of equations. The numbers in the first set of square brackets represent the start and stop time for the time points and those in the second set of brackets the initial values of the variables (in this case 2). The equations themselves are entered into the Editor of MATLAB in the following way:

```
function dydt = gluc(t,y)
dydt(1) = -2*y(1) - 5*y(2) + 1
dydt(2) = +0.2*y(1) - 0.8*y(2)
```

Here y(1) is *g* and y(2) is *i*. These three lines are saved as "gluc.m". Actually they can be entered even more concisely as follows:

```
function dydt = gluc(t,y)
dydt = [-2 -5; 0.2 -0.8]*y + [1 0]';
```

then type the following at the MATLAB prompt:

```
>> [t,y] = ode23(gluc,[0 5],[0 0]);
>> plot(t,y)
```

And a plot similar to the one at the lower part of Figure 25.10 should appear.

An alternative approach is to use the MATLAB toolbox "Simulink," where the equations can be represented by circuit elements. The following is a representation of our equations, but in integral form: the glucose equation (on the right of the diagram) is

$$g = \int [-2^*g - 5^*i]\mathrm{d}t + \int \delta(t)\,\mathrm{d}t = \int [-2^*\mathrm{g} - 5^*\mathrm{i}]\mathrm{d}t + 1.$$

It can easily be seen that the "*g*" signal is multiplied by –2, the "*i*" signal by –5 then added to unity from the step function (for t > 0) before integrating to get g as an output, which is then fed back (as above) and also fed into the "i" equation. The g(t) and i(t) values are displayed on autoscale graphs. Simulink has a useful facility of being able to group similar circuits as single blocks, which can be joined as appropriate, then individually specified (Figure 25.11).

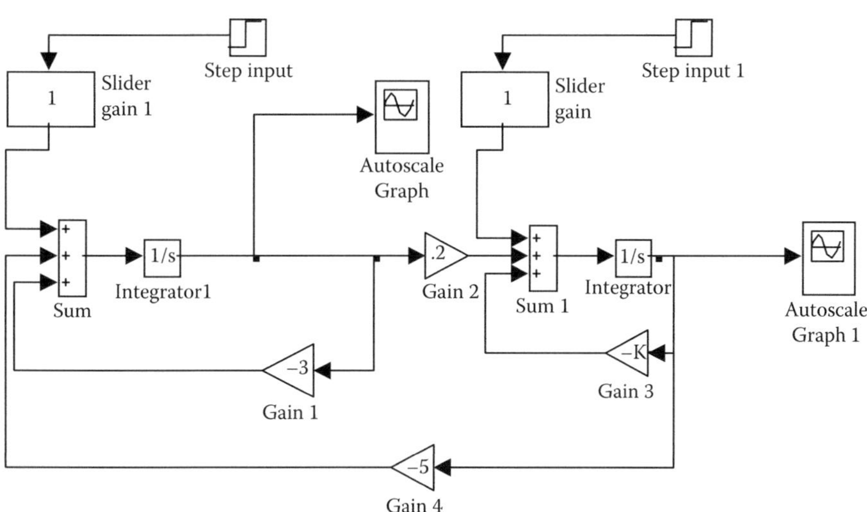

FIGURE 25.11
Simulink™ solution of the glucose-insulin model described in the text. The glucose equation is on the left and insulin on the right. The two autoscale graphs plot the concentrations of *g* and *i* (see Figure 25.10: lower). Ingestion of glucose is simulated by adjusting the slider gain on the left (the one on the right is set to zero, but could be used to simulate insulin infusion). Gain 3 is –0.8.

25.5.3 The Riggs Iodine Metabolism Model

In his book *Control Theory and Physiological Feedback Mechanisms*, Riggs (1970) presents the model shown in Figure 25.12.

The 3 state equations are:

$$\frac{dI}{dt} = -(k_1 + k_4)I + k_3H + B_1(t)$$

$$\frac{dG}{dt} = k_1I - k_2G$$

$$\frac{dH}{dt} = k_2G - (k_3 + k_5)H + B_3(t).$$

Now put $k_I = k_1 + k_4$ and $k_H = k_3 + k_5$ and label I, G, and H as x_1, x_2, and x_3. Substituting numerical values:

$$dx_1/dt = -2.52x_1 + 0.08x_3 + 150$$

$$dx_2/dt = 0.84x_1 - 0.01x_2$$

$$dx_3/dt = 0.01x_2 + 0.1x_3 + 0$$

If the iodine input (B(1)) remains at 150 μg/d the amounts in the three compartments stay the same over time, 81.2, 6821, and 682 μg, respectively. If the intake were to suddenly drop to 10% of the normal value (15 μg/d) the amounts would fall and since there are three compartments, three exponentials would be expected.

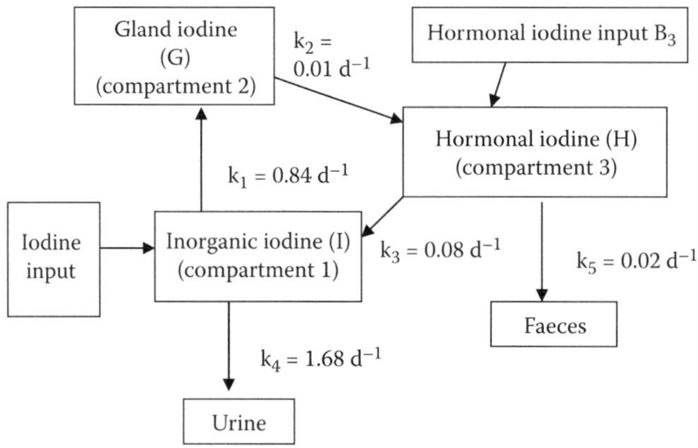

FIGURE 25.12
The Riggs model for iodine metabolism. Iodine equilibrates between the three compartments shown. The model has the provision of administration of thyroxine (hormonal iodine) which would be necessary to treat certain thyroid conditions (hypothyroidism). Note that the rate constants are net values (i.e., $k_1 = k_{12} - k_{21}$).

Using exact analysis of this equations (via Laplace transforms, which will not be dealt with here), the time course of the amounts in the three compartments is given by the following equations:

$$x_1 = 8.1 + 53.6\exp(-2.52t) - 1.58\exp(-0.103t) + 21\exp(-0.00712t)$$

$$x_2 = 683 - 17.9\exp(-2.52t) + 14.3\exp(-0.103t) + 6141\exp(-0.00712t)$$

$$x_3 = 68.1 + 0.074\exp(-2.52t) - 47.5\exp(-0.103t) + 661.2\exp(-0.00712t).$$

Note that at t = 0

$$x_1 = 8.1 + 53.6 - 1.58 + 21 = 81.1$$

$$x_2 = 683 - 17.9 + 14.3 + 6141 = 6820$$

$$x_3 = 68.1 + 0.074 - 47.5 + 661.2 = 681.9$$

which is close enough to the initial conditions given; and at $t = \infty$

$$x_1 = 8.1$$

$$x_2 = 683$$

$$x_3 = 68.1$$

(all values in µg): in other words about 10% of the initial values. Note the same exponentials apply to all three compartments. With the aid of MATLAB this can computed fairly easily using the following m-file and command line.
m-file "riggs.m"

```
function yp = riggs(t,y);
yp = [-2.52 0 .08;.84 -.01 0;0 .01 -.1]*y + [15 0 0]';
```

command lines

```
>> [t,y] = ode23("riggs",[0 100],[81.2 6821 682]);
>> plot(t,y)
```

The plots are more easily understood if the values (y) are normalized to the initial values (this is an exercise for the student).

25.6 Thermoregulation: Control of Human Body Temperature

Heat loss and heat maintenance processes are "turned on" by (respectively) high or low blood temperature as sensed by the hypothalamus in the brain. There are also temperature sensors in the skin. Control is exerted via: Evaporation (for heat loss); Endocrine control of basal metabolism; Neural control of shivering activity (for heat maintenance) and Vasomotor control of blood distribution (for both heat loss and maintenance). There are heat diffusion processes through the various layers of the body. The diffusive processes give rise to first order ODEs and are thus solved easily in MATLAB.

25.6.1 Modeling of the Controlled Process (Temperature in Various Layers of the Body)

Basic equation: if we have several layers Δx thick, then the difference in rate of heat flow (q) in adjacent layers (n, n + 1) must be equal to the rate of storage of heat in a given layer:

$$q_n - q_{n+1} = \rho c A \Delta x (d\theta_n/dt);$$

where ρ is density, c is specific heat, A is cross-sectional area and θ_n is the temperature in layer n. We can write this as

$$q_n - q_{n+1} = m_n c_n (d\theta_n/dt); \text{ where } m_n = \rho_n A_n \Delta x, \text{ the mass of layer n.}$$

The rate of heat flow q is given by the thermal diffusion equation: $q_{n,n-1} = kA_n(\theta_n - \theta_{n-1})/\Delta x$; where k is the thermal diffusion coefficient 0.03 J/(m °C s) for tissue. A simple model of the human is a 1.7-m-high set of three concentric cylinders, as shown in Figure 25.13 (this follows the model of Stolwijk and Hardy, as described in Milsum, 1966).

Other quantities are

M_b: Basal metabolism; F_r: Respiratory heat loss; M_m: Muscle metabolism; F_c: Convective heat loss; F_e: Evaporative heat loss; F_{rad}: Radiative heat loss

First order ODEs are as follows:

Core:

$$m_c c_c d\theta_c/dt = M_b - F_f - q_{cm} - q_{cs} \text{ where } q_{cm} = kA_c(\theta_c - \theta_m)/(r_c) \text{ etc.}$$

Muscle:

$$m_m c_m d\theta_m/dt = M_m + M_x + q_{cm} - q_{ms}.$$

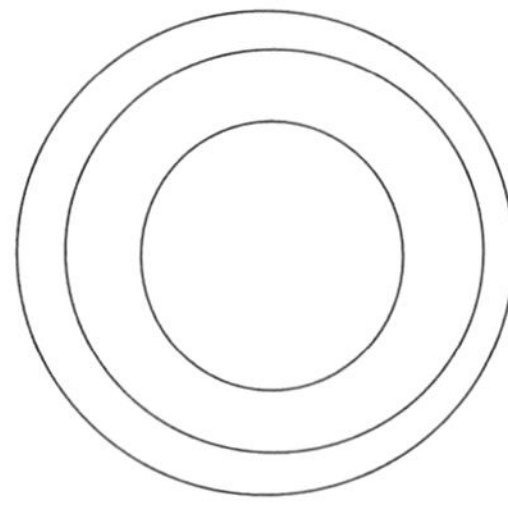

FIGURE 25.13
The Stolwijk–Hardy cylindrical model of the human body. The three layers represent core; muscle, and skin. The radii are 114; 154, and 160 mm, respectively. The temperature is the same throughout each layer (i.e., 3 compartments). The temperatures of these shells are represented by θ_c, θ_m, θ_s, their masses as m_c, m_m, and m_s, and c_c, c_m, and c_s their specific heats.

Skin:

$$m_s c_s d\theta_s/dt = -(F_c + F_e + F_{rad}) + q_{cs} + q_{ms}.$$

The next task is to specify the values of the Fs and Ms in terms of core and skin temperature.

25.6.2 Heat Flux Control Equations

Evaporation control: $F_e = k_1(\theta_c - \theta_c^*)$ for $\theta_c > \theta_c^*$ and $F_e = F_{eo}$ for $\theta_c^* < \theta_c$; where θ_c^* is a reference level.

$$\text{Muscle Shivering: } M_m = 250(\theta_c^* - \theta_c) + 18(\theta_{SO} - \theta_c);$$

where θ_{SO} is a skin reference level.

Vasomotor action: Equivalent thermal conduction constant

$$k_v = k_0[1 + k_6(\theta_b - \theta_b^*) + k_7 d\theta_b/dt] \text{ for } \theta_b > \theta_b^*,$$

where θ_b^* is a body temperature reference level

$$\text{i.e., } q_{cs} = k_v A_{cs}\, c(\theta_c - \theta_s)/(r_s - r_c).$$

Other equations (skin surface)

$$\text{Convection: } F_c = hA(\theta_s - \theta_a);$$

where h is heat transfer coefficient; A is area; θ_s is skin temperature, and θ_a is ambient temperature

$$\text{Radiation: } F_{rad} = \sigma A'(\theta_s^4 - \theta_e^4) \approx 4\sigma A'(\theta_s - \theta_e)\theta_e^4$$

where σ is the Sefan–Boltzmann radiation constant, A′ is the effective black body area and θ_e is the effective black body temperature of the environment.

All parameters in the model are now specified and it is possible to solve for skin and core temperature θ_s and θ_c, respectively, and also to compare these to experimental measurements. This has been done for a situation in which an unclothed person moves from a hot to a cold room and begins to shiver. Modeling revealed a paradoxical phenomenon, that of the core temperature rising initially before falling, which was subsequently found in careful measurements.

25.7 Biological Oscillations and Population Models

Biological oscillations cover a wide range of cycle periods, ranging from a few tens of milliseconds in the CNS, 1 s for the heart, 4 s–5 s for breathing, one day for various parameters such as temperature and melatonin, one month for the menstrual cycle and one year for the breeding cycle.

The nature of each of these oscillators is somewhat different, but most have been modeled and are able to exhibit self-sustained oscillations. This section covers two examples of these: calcium oscillations in single cells and the circadian (24 h) rhythm. These will only be described briefly, but m-files are provided to allow the student to interact with them. Fully explanations can be found in associated research papers. First, though, two simple oscillating systems will be described, along with the use of MATLAB to study these.

25.7.1 The van der Pol Equation

This is the equation for simple harmonic oscillation with an extra term in the middle:

$$d^2y/dt^2 + \varepsilon(y^2 - 1)dy/dt + y = 0.$$

Note that if $\varepsilon = 0$, then a sinusoidal solution results. If we want to use the MATLAB ODE command, we need to eliminate the second order term. This is done quite easily by splitting the equation into two, with a second variable the differential of the first. We thus get:

$$dy_1/dt = -\varepsilon(y_2^2 - 1)y_1 - y_2$$

$$dy_2/dt = y_1$$

where $y_1 = dy/dt$ and $y_2 = y$ (hence the second equation). In the form of an m-file this can be written very compactly as

```
function yp=vanderp(t,y)
% solves the van der Pol equation
u=3;
yp=[(1-y(2)*y(2))*u -1; 1 0]*y;
```

which is then called from a MATLAB command line

```
>> [t,y]=ode23("vanderp",[0 50],[0 1]);
>> plot(t,y)
```

The plot shown in the upper part of Figure 25.12 then results. There is a slight similarity between the waveform and the membrane potential from the pacemaker region of the heart. Coupled van der Pol oscillators have also been used to simulate peristaltic movements in the colon and electrical activity of the brain.

25.7.2 The Lotka–Volterra Equations

This is an early example of an ecological model in which two species inhabit the same region, but there is interspecies competition. The lynxes depend on hares for their survival and hares eat grass, which is abundant. If the lynxes eat too many hares, then ultimately lynx numbers will begin to fall. Again the change in numbers is the difference between birth and death rate.

Absolute birth rate of *hares* depends on number of hares, but death rate depends both on number of hares and number of lynxes.

$$dH(t)/dt = k_1H(t) - kH(t)L(t).$$

On the other hand, the birthrate of *lynxes* depends both on the number of parents (lynxes) and the availability of food (hares). The death rate depends only on the number of lynxes.

$$dL(t)/dt = k'H(t)L(t) - k_2L(t).$$

If we represent H(t) by y_1 and L(t) by y_2 we can write the equations (the Lotka–Volterra equations) as follows:

$$y_1' = k_1(+1 - \alpha y_2)y_1$$

$$y_2' = k_2(-1 + \beta y_1)y_2$$

where $\alpha = k/k_1$ and $\beta = k'/k_2$.

These are a pair of Ordinary Differential Equations (ODEs) as before. There is a built-in function "lotkademo" in MATLAB, which illustrates these equations, which are further explained in MATLAB "help" under "Numerical Integration of Differential Equations." For output, see Figure 25.14 (bottom).

```
function yp = lotka(t,y)
yp = diag([1 - .01*y(2), -1 + .02*y(1)])*y;
%   To simulate the differential equation defined in LOTKA over the
%   interval 0 < t < 15, we invoke ODE23:
t0 = 0;
tfinal = 15;
y0 = [20 20]';    % Define initial conditions.
% [t,y] = ode23('lotka',t0,tfinal,y0);
```

Figure 25.15 illustrates the Cuthbertson–Chay model for calcium dynamics in a single cell (see Cuthbert and Chay [1991]). The calcium concentration in the cytoplasm interacts with a receptor in the membrane surrounding the endoplasmic reticulum (the IP3 receptor) in such a way that as calcium rises, the probability of the IP3 channel opening (and releasing Ca^+) also increases. The three main variables are: cytoplasmic calcium concentration (c), Ga-GTP (g) and DAG/IP3 (d). These can be modeled using first order differential equations as in the worked example in Section 25.4. The rates of synthesis are given by Hill-type equations as outlined in Section 25.1. The equations are coded in a MATLAB m-file "cuthbert.m" as shown below:

```
function yp=cuthbert(t,y)
% Model of cell calcium dynamics from Cuthbertson & Chay, Cell Calcium 12:97
% Parameter values
rg=5; hg=10; Kp=40; Kc=100; kd1=500; hd=4; Id=0.3; Kd=5; Kg=50;
hc=0.4; Ks=5; Ic=80; kc=770;
% g is y(1); d is y(2) and c is y(3): use these initial values
% g=15; d=1; % c=200
```

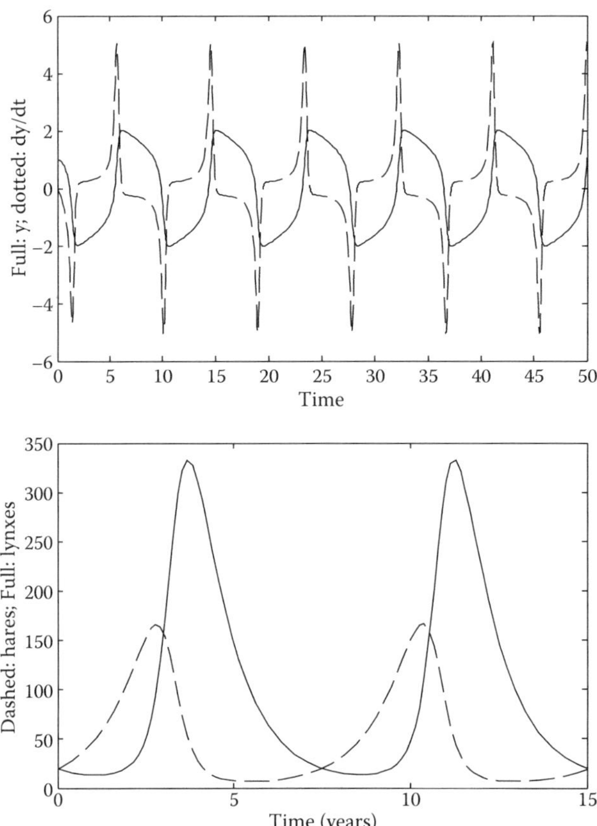

FIGURE 25.14
Top: the van der Pol oscillator, with the parameter u set at 3. Bottom: solution of the Lotka–Volterra equations. Hare and lynx population over a 15-year period.

```
% The following two code lines represent the "Hill" terms; and the last line
% represents the three differential equations dg/dt, dd/dt and dc/dt
r1=1/((1+Kp/y(2))*(1+Kc/y(3)));
r2=1/((1+(Kd/y(2))*(Kd/y(2)))*(1+Kg^4/(y(1)^4)));
yp = [rg-hg*r1*y(1); kd1*r2-hd*y(2)+Id; kc/(1+(Ks/y(2))*(Ks/y(2))*(Ks/
y(2)))-hc*y(3)+Ic];
```

The oscillatory behavior of cell calcium (c) over a period of 100 seconds can be studied using the following script file:

```
%script file for cuthbertson model
[t,y]=ode23("cuthbert",[0 100],[15 1 200]');
plot(t,y)
```

The final model in this section is for studying the variation of the hormone melatonin over a 24-hour period, this hormone being linked to the "biological clock." The rhythm can either be free-running (where the period is actually a bit longer than 24 h) or entrained, by a light–dark cycle.

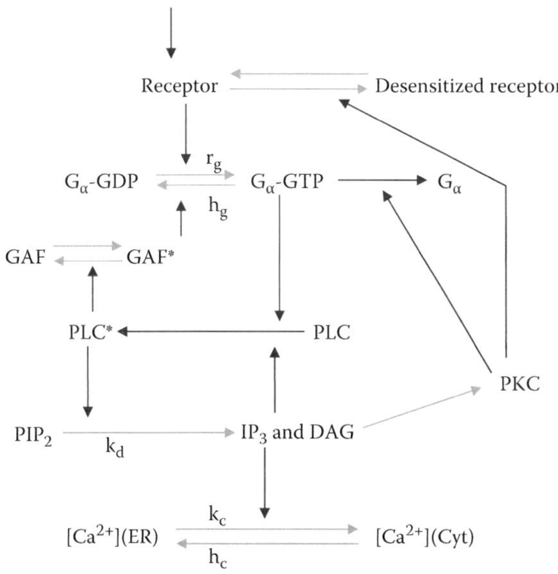

FIGURE 25.15
The Cuthbertson and Chay (1991) model of Calcium dynamics within a singe cell.

The following minimal model of Thompson et al. (2001) demonstrates both conditions and has seven differential equations, as briefly described in the comments lines in the m-file "melat.m" below.

```
function yp = melat(t,y)
% This m-file solves the 7 equations of Thompson et al. Physica A
% The variables are concentrations as follows: y1 is Product ; y2 is
% Melatonin; y3 is Serotonin; y4 is specific RNA messenger; y5 is free
% enzyme; y6 is second messenger and y7 is enzyme-substrate complex
% Fixed parameter values
a=.02; k3=3.65; d=3.65; u=78; w=.117; q=26; g=.26; KA=2;
n=6; h=26; KR=1; N=3; k=.52; km=1; kp=3; k2=260; kp1=3.38;
km1=13; r=.52; m=3; v=26;

y1 = y(1); y2 = y(2); y3 = y(3); y4 = y(4);
y5 = y(5); y6 = y(6); y7 = y(7);

% System equations
dy1dt=k2*y7 - k3*y1;
dy2dt=k3*y1 - d*y2;
dy3dt=u - w*y3 + km1*y7 - kp1*y5*y3;
dy4dt=q*y6^n/(1 + KA*y6^n) - g*y4;
dy5dt=h*y4^N/(1 + KR*y4^N) - k*y5 + (km1 + k2)*y7 - kp1*y5*y3;
dy6dt=v*km/(km + kp*y(2)^m) - r*y(6);
```

```
dy7dt=kp1*y5*y3 - (km1 + k2)*y7;

yp=[dy1dt
    dy2dt
    dy3dt
    dy4dt
    dy5dt
    dy6dt
    dy7dt];
```

Again, the behavior of these parameters over 100 hours can be studied by using the following script file:

```
% Script file for Thompson melatonin model
y]=ode23("melat",[0 100],[11.6552  12.1186  175.9831   0.0247   0.0723
0.0085   0.1575]);
 plot(t,y)
```

25.8 Electrical Analogs of Arteries: Transmission Line Theory Applied to Arterial Segments

This treats the arteries as a series of (arbitrarily) short segments, each characterized by a resistance (R) related to blood viscosity, capacitance (C) related to wall compliance, inductance, (L) related to blood mass and conductance, and (G) related to vessel wall permeability. This last factor is zero for arteries and arterioles but is significant for capillaries. The "transmission line" is shown in Figure 25.16 and is similar to diagrams in standard engineering texts (see, for example, Ramo, Whinnery and van Duzer 1994).

The resistance per unit length is a measure of blood viscosity; specifically,

$$R/\Delta x = 8\mu/(\pi r^4) \text{ (Poiseuille's law, with } \mu \text{ as blood viscosity).}$$

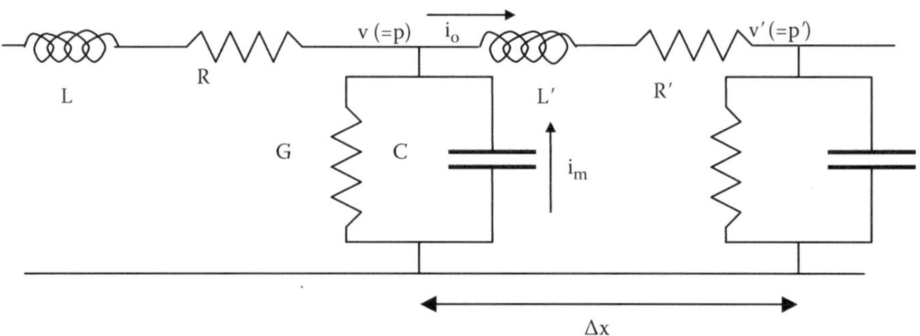

FIGURE 25.16
The transmission line model for an artery. The artery is divided into arbitrary segments each Δx long, and each segment has its particular values of inertia L, viscous resistance R, wall compliance C, and wall permeability G. In normal arteries G will be zero.

The inductance (or "reluctance") is a measure of the mass or blood in the vessel. The electrical equivalent for inductance is voltage/rate of change of current. In fluid mechanical terms (see Chapters 9 and 11) this is

$$L/\Delta x = (dp/dx)/(dQ/dt) = (dp/dx)/(Adu/dt),$$

where p is pressure, Q is volume flow rate, u is velocity and A is cross-sectional area (πr^2). Now since the equation of motion for fluid systems can be written as $dp/dx = \rho du/dt$ (see Chapter 9), this equation becomes:

$$L/\Delta x = \rho/A.$$

Similarly,

$$C/\Delta x = dA/dp,$$

which is analogous to $C = dq/dv$ in the electrical case. Since the distensibility of artery walls is defined as $D = (dA/A)/dp$, we can put

$$C/\Delta x = DA.$$

Also $G/\Delta x$ = mass flow across wall/net pressure = $J/(\Delta p - \sigma\Delta\pi)$, where π refers to osmotic pressure difference across the wall and Δp the pressure gradient across the wall. As mentioned above, this is zero for arteries.

The circuit shown in Figure 25.16 is the standard representation of an electrical transmission line such as a coaxial cable. Consulting a standard book on electromagnetics, such as

WORKED EXAMPLE 1

Show that typical values for L and R in a 1 cm length of the aorta (r = 8 mm) are 5.3×10^4 and 3.1×10^4 SI units, respectively. Assume μ is 5×10^{-3} Pa s and ρ is 1070 kg/m^3.

Answer

$R/\Delta x = 8\mu/(\pi r^4) = 8 \times 5 \times 10^{-3}/[3.14 \times (8 \times 10^{-3})^4] = 3.1 \times 10^6$. However, this would be for 1m, so dividing by 100 gives the required result. $L/\Delta x = \rho/A = \rho/(\pi r^2) = 1070/[3.14 \times (8 \times 10^{-3})^2] = 5.3 \times 10^6$, which has to be divided by 100, as before. Note that the units of $L/\Delta x$ are kg/m^5 and $R/\Delta x$ are kgm^{-5}s^{-1}.

WORKED EXAMPLE 2

In a capillary, the flow rate of plasma across the capillary wall is 0.08 μm/s. Calculate $G/\Delta x$ given that the net filtration pressure ($\Delta p - \sigma\Delta\pi$) is around 5 mmHg.

In SI units J is 8×10^{-8} m/s (that is 8×10^{-8} m^3/s per m^2 of wall area) and the denominator 665 Pa (which is $5 \times 13.3/100$ kPa). Again, a factor of 100 has to be introduced to estimate the value for 1 cm, so G/Dx is 1.2×10^{-12} m/(sPa). In terms of mass flux, if we assume a density of 1000 kg/m^3, then we get 1.2×10^{-9} s for a 1 cm length.

Ramo, Whinnery and van Duzer will reveal that the voltage v (or in this case the pressure p) will obey the following:

$$d^2p/dt^2 = \gamma^2 p,$$

where

$$\gamma^2 = (R + j\omega L)(G + j\omega C) \equiv (\alpha + j\beta)^2.$$

Comparing real parts we get:

$$\alpha^2 - \beta^2 = RG - \omega^2 LC,$$

which if we substitute from the above equations gives:

$$= 0 - \omega^2(\rho/A)(dA/dp) = -\omega^2\rho D.$$

The wavespeed c in an elastic tube such as an artery is given by the formula

$$c^2 = 1/(\rho D),$$

so substituting this in the above gives us

$$\alpha^2 - \beta^2 = -\omega^2/c^2 = -k^2,$$

where k is the familiar wavenumber ($2\pi/\lambda$).

Equating imaginary parts gives:

$$2\,\alpha\beta = 8\omega\mu D/r^2 = (8\mu\,/r^2)(\omega/(\rho c^2)).$$

Thus, $\beta^2 = \frac{1}{2}k^2(1 \pm \sqrt{(1 + \varepsilon)})$, where $\varepsilon = 64\mu^2/(\rho^2\omega^2 r^4)$

And when $\mu \to 0$, $\beta \to k$ and $\alpha \to 0$

We can represent the sinusoidal Fourier components of the arterial pressure waveform as

$$p = p_0 \exp(-\alpha t)\exp(-j(\omega t - \beta x))$$

the first exponential being a representation of attenuation, the second, oscillation. Thus, each Fourier component will have its own values of p_0, α, β, and ω. Since $\alpha \to 0$ in arteries, there is little attenuation there.

WORKED EXAMPLE

This gives us a means of estimating the arterial capacitance $C/\Delta x$ in the model in Figure 25.16, given a typical wavespeed of 6 m/s, since $c^2 = 1/(\rho D)$ and $C/\Delta x = DA = A/(\rho c^2) = 3.14 \times (8 \times 10^{-3})^2/[1070 \times 6^2] = 5.2 \times 10^{-9}$. Thus, for a 1 cm length this would be 5.2×10^{-11} SI units.

25.8.1 Characteristic Impedance

For an infinite line in electromagnetics:

$$v = A\exp(-\gamma x) + B\exp(+\gamma x)$$

$$i = (A/Z_0)\exp(-\gamma x) + (B/Z_0)\exp(+\gamma x)$$

where $Z_0 = \sqrt{[(R + j\omega L)/(G + j\omega C)]} \rightarrow \sqrt{(L/C)} = \sqrt{(\rho C)}/A = \sqrt{(\rho/(DA^2))}$ using the substitutions from above. Now since the wavespeed in the transmission line (and thus the artery wall) $c^2 = 1/(\rho D)$, we can write $Z_0 = \rho c/A$.

Now for branches in arteries, like branches in transmission lines we get Figure 25.17.

It is more convenient to deal with admittances $Y = 1/Z$ when analyzing this situation, so $Y = A/(\rho c)$. These equations can be derived by considering the boundary conditions across the junction. These are: the pressure cannot change abruptly across the junction and that the sum of the flows in the branches must equal the flow in artery 0. For the moment though, we will take a simple analogy with the electromagnetic wave case with branching transmission lines. The following equations can be found in a standard text.

The ratio of reflected (R) to incident (I) pressure amplitude is given by

$$p_R/p_I = [(Y_0 - (Y_1 + Y_2))/(Y_0 + (Y_1 + Y_2))]$$

and for the transmitted amplitudes in branches 1 and 2:

$$p_{T1}/p_I = p_{T2}/p_I = 2Y_0/(Y_0 + (Y_1 + Y_2)).$$

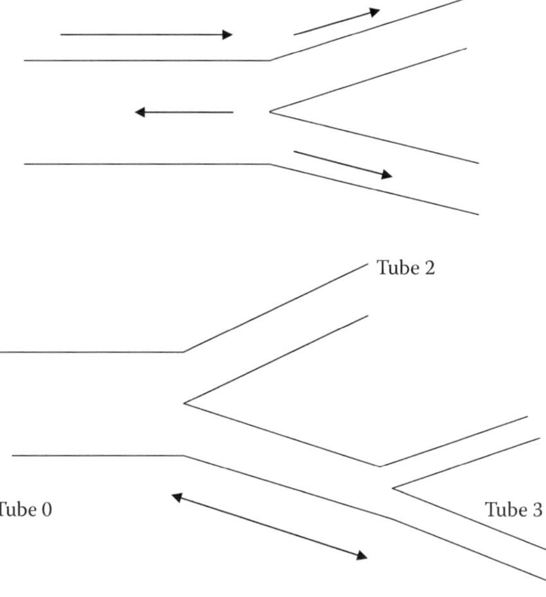

FIGURE 25.17
Branching arteries. Top: each arterial segment is assumed to be infinitely long. Bottom: a finite length l between two bifurcations, otherwise the segment is infinitely long.

Since power is proportional to amplitude squared:

$$\% \text{ power reflected} = 100 \times [(Y_0 - (Y_1 + Y_2))/(Y_0 + (Y_1 + Y_2))]^2 = 100 \times R^2$$

and thus power transmitted = $1 - R^2$, which divides into branches 1 and 2 in the ratio

$$Y_1/(Y_1 + Y_2) \text{ and } Y_2/(Y_1 + Y_2), \text{ respectively.}$$

The pressure in tube 0 is made up of the sum of the instantaneous pressures of the incident and the reflected wave. These could, of course, cancel in certain locations giving a standing wave pattern. In fact, it can be shown that if tube 1 divides further after a distance l the effective admittance of this tube becomes frequency dependent. This will be explored further below.

25.8.2 Arterial Branches: Derivation of Equations Starting from Fluid Dynamic Considerations (Lighthill 1986)

25.8.2.1 Infinitely Long Tubes

Consider the following branch in which the parent tube 0 splits into two daughters 1 and 2, all three tubes being considered long. Suppose a pulse wave approaches the junction from the left: some of this will be reflected; the remainder will enter the tubes 1 and 2. In each tube the characteristic impedance can be calculated from the expression $\rho c/A$, with ρ the fluid density, c the wavespeed, and A the cross-sectional area, πr^2.

Pressures must be same on each side of branch: N.B. p refers to instantaneous amplitudes of sinusoidal Fourier components

$$p_I + p_R = p_{T1} = p_{T2} \quad \text{(analogous to voltages on branching electrical line).}$$

Similarly, flow rates through junction should balance

$$Q_I - Q_R = Q_{T1} + Q_{T2} \quad \text{(analogous to current).}$$

The characteristic impedance of each branch is given by $\rho c/A$. In fact, the admittance $Y = A/(\rho c)$ is more useful. We can write $Q = Yp$ for each wave.

From the above equations:

$$Y_0(p_I - p_R) = Y_1 p_{T1} + Y_2 p_{T2} = (Y_1 + Y_2)(p_I + p_R)$$

and after some rearrangement we get the same equations as above:

$$p_R/p_I = \{Y_0 - (Y_1 + Y_2)\}/\{Y_0 + (Y_1 + Y_2)\} = R \quad \text{(as previous)}$$

and

$$p_{T1}/p_I = p_{T2}/p_I = 2Y_0/\{Y_0 + (Y_1 + Y_2)\} = T.$$

Power ratios: % power reflected given by $100 \times R^2$;

$$\% \text{ power transmitted} = 100 \times (Y_1 + Y_2)T^2/Y_0.$$

Note that the % power in T1 is $100 \times Y_1 T^2/Y_0$, and the % power into T2 $100 \times Y_2 T^2/Y_0$ which is equivalent to the previous expressions.

Note too, that if $Y_1 + Y_2 = Y_0$, there is no reflected wave. This would be the case if the sum of the cross-sectional areas in the daughter tubes is the same as the parent tube (wave-speed and density of fluid being the same). In Chapter 10 we saw that wavespeed is given by the Moens Korteweg equation $c^2 = E(h/d)/\rho$, where h/d is the ratio of artery wall thickness to diameter and E is the Young's modulus of the artery wall.

WORKED EXAMPLE

The aorta (0) branches into two smaller arteries (1) and (2):

The ratio of wall thickness to diameter in each segment is the same, 0.05, but the Young's (elastic) modulus for arteries 1, 2 is 20% higher than in 0, where it is 0.9 MPa. The radii are as follows: 0: 9.0 mm; 1: 2.0 mm; 2: 3.0 mm. $\rho = 1070\ \text{kg/m}^3$ and $c^2 = (Eh/(\rho d))$.

Consider the tubes to be infinitely long. Calculate

(i) Percentage of incident energy reflected from junction.
(ii)Percentage transmitted into artery 2.

Answer

Note that $Y = A/(\rho c) = \pi r^2/[\rho\sqrt{(E(h/d)/\rho}]$. Since $R = [Y_0 - (Y_1 + Y_2)]/[\,Y_0 + (Y_1 + Y_2)]$ many factors cancel. In fact

$$R = [r_0^2/\sqrt{E_0} - [r_1^2/\sqrt{E_1} + r_1^2/\sqrt{E_1}]]/[r_0^2/\sqrt{E_0} + [r_1^2/\sqrt{E_1} + r_1^2/\sqrt{E_1}]]$$

It is also not necessary to convert from mm to m, since the factor of 10^3 cancels.

$$R = [9^2/\sqrt{0.9} - [2^2/\sqrt{(0.9 \times 1.2)} + 3^2/\sqrt{(0.9 \times 1.2)}]/[\,9^2/\sqrt{0.9} - [2^2/\sqrt{(0.9 \times 1.2)} + 3^2/\sqrt{(0.9 \times 1.2)}]$$

Note that the factor 0.9 also cancels and $\sqrt{1.2} = 1.1$, so

$R = [81 - (4 + 9)/1.1]/[81 + (4 + 9)/1.1] = 69/101 = 0.68$. $R^2 = 0.47$, so 47% reflected.

Of the 53% transmitted, this divides according to the fractional area, i.e., $3^2/(2^2 + 3^2) = 9/13$, so 37% goes into artery 2.

25.8.2.2 Effect of Finite Lengths

Figure 25.17 (lower) shows a finite length arterial segment between two bifurcations. Suppose ℓ is the length of tube 1 and k is the wave number $= 2\pi/\lambda$. Here λ refers to the wavelength of the Fourier components of the incident wave, and $f\lambda = c_1$, where f is the relevant frequency. The effective admittance of tube 1 is given by

$$Y_{1(\text{eff})} = Y_1\{(Y_{3(\text{eff})} + Y_{4(\text{eff})}) + jY_1 \tan(k\ell)\}/\{Y_1 + j(Y_{3(\text{eff})} + Y_{4(\text{eff})})\tan(k\ell)\}.$$

Where $Y_1 = A_1/(\rho c_1)$, as before. $Y_{3(\text{eff})}$ and $Y_{4(\text{eff})}$ are determined in a similar fashion, and so on to the smallest vessels, where admittance can be estimated from the Poiseuille formula, $1/R = \pi r^4/(8\mu\ell)$.

Consider some special cases: if $k\ell = n\,\lambda$; $\tan(kl) = 0$, thus $Y_{1(\text{eff})} = Y_{3(\text{eff})} + Y_{4(\text{eff})}$. This is equivalent to $\ell = n\,\lambda/2$. The system behaves as if tube 1 were not there.

If, on the other hand, $k\ell = n\,\lambda/2$, with n odd, $\tan(k\ell) = \infty$, so $\ell = n\,\lambda/4$, and

$$Y_{1(eff)} = Y_1^2/(Y_{3(eff)} + Y_{4(eff)});$$

which can be large or small depending on the relative magnitudes of Y_1, $Y_{3(eff)}$, and $Y_{4(eff)}$.

Modeling of the entire arterial system is possible by assuming that the terminal arterioles have impedance values determined by the Poisieulle formula, then taking each generation of bifurcations back to the aorta, where the entrance impedance can be estimated. A number of groups have made experimental determinations of the entrance impedance (or admittance) as a function of sinusoidal frequency (in the range 1 Hz–25 Hz) for both systemic and pulmonary arterial trees and have been able to correlate the shape of this characteristic (with particular resonant frequencies) with lengths between branches.

25.9 Integrated Models of Biological Systems

The segmented model for an artery can equally be applied to longer segments: for example the whole of the aorta can be represented by characteristic values of R, L, and C (since

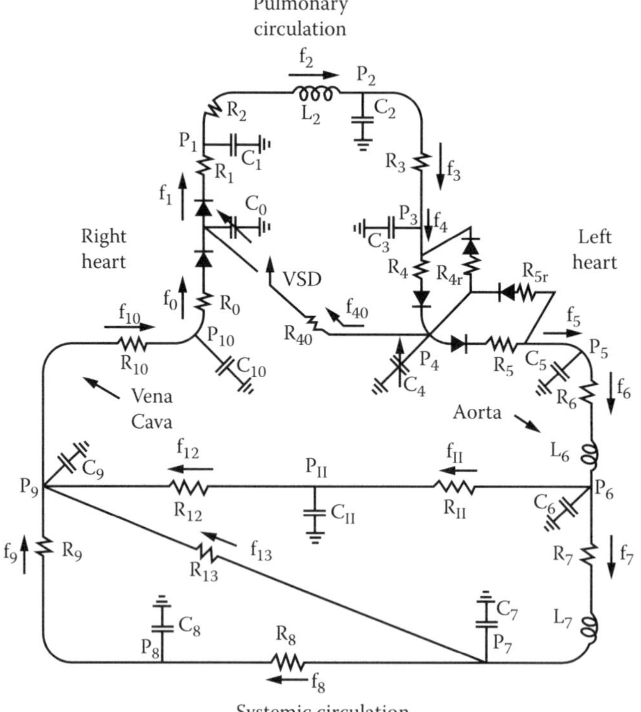

FIGURE 25.18
The Rideout model of the complete circulation based on circuit analogs. Note segments corresponding to regions as follows: 0: R Atrium; 1: R Ventricle; 2 Pulmonary arteries; 3: Pulmonary capillaries; 4: L Atrium; 5: L ventricle; 6: Aorta; 7: Major arteries; 8: Capillaries; 9: Major veins; 10: Vena Cava; 11: Renal arteries; 12: Renal capillaries; 13 Portal (liver) capillaries. (Adapted from VC Rideout, *IEEE Trans Biomed Eng* 19:101, 1972.)

these were initially defined for unit length, the total length of each class of blood vessel needs to be included). The model of Rideout shown in Figure 25.18 is an example of this. Each segment is represented by two simple differential equations, one for flow and the other for pressure, in terms of the electrical parameters. The system is driven by varying the compliance C in the ventricles and atria. The form of this variation is a half sinusoid, lasting 0.3 s then a pause of 0.5 s before the next one (corresponding to systole and diastole). The behavior of the complete model is remarkably close to reality, with aortic and major arterial pressure waveforms displaying the characteristic dicrotic notch. The effects of various illnesses, such as atrial septal defect, can be simulated. The model was originally implemented as an undergraduate teaching aid. Students are referred to the full paper (see caption to Figure 25.18) for further details.

25.10 Electrical Properties of Tissue

25.10.1 Conductivity of Electrolyte Solutions

When studying the effects of endogenous or exogenous electrical currents within or upon the body, an accurate model of the electrical properties of tissue is required. This section will give an introduction on how this can be done. There are now some highly sophisticated electrical models of the human body available for research purposes (see, for example, http://www1.itis.ethz.ch/index/index_humanmodels.html).

The conductivity σ of a simple electrolyte (like NaCl) is given by the following:

$$\sigma = C[\lambda_+ \nu_+ + |\lambda_-| \nu_-]$$

where λ_+, λ_- are valencies and $\nu_+ \nu_-$ are molar conductivities (see typical values below).

If C (concentration in mol/m^3) is small; if C is large ν_+ varies as $(1 - constant \sqrt{C})$

Molar conductivities (at 25°C) in Sm^2 mol^{-1} (S is siemens):

K^+: 7.35×10^{-3}

Na^+: 5.011×10^{-3}

Cl^-: 7.634×10^{-3}

NO_3^-: 7.14×10^{-3}

Thus, a 0.1 M solution of NaCl (100 mol m^{-3}) would have a σ value of 1.26 S/m, approximately. Actually, σ varies significantly with temperature (≈ 2%/°C).

25.10.2 Homogeneous Model of Tissue

For a simple cylindrical model of the chest the resistance R, and conductance G are given by

$$R = L/(\sigma \pi r^2)\ (\Omega); \quad \text{alternatively } G = (\sigma \pi r^2)/L\ (S).$$

A saline phantom of the torso would have a resistance of a few ohms ($0.4/(1.3\pi 0.15^2) = 4.4\ \Omega$): the chest is actually about 30 Ω at 5 kHz: chest resistance is determined by tissue cellular structure rather than constituents. In the GHz region chest impedance falls to a few ohms.

25.10.3 Lumped Model

For a cylindrical model with uniform cross-sectional structure along the length, conductance is given by

$$G = A_0 \sum f_i \sigma_i / L;$$

where f_i are the volume fraction of each tissue type, σ_i is the conductivity of that tissue and A_0 is the total cross-sectional area. L is the length, perpendicular to the slice shown in Figure 25.19.

It is important to identify the frequency at which σ_i has been estimated, and also the physiological state of tissue (that is, whether it is dehydrated or wet).

25.10.4 Cellular Models

Cells are often ordered into layers with close adhesion between cells (skin, blood vessels, alveoli, cardiac muscles, etc.). At high frequency a greater proportion of current goes through the cells rather than between them. The diagram shows an epithelial layer (skin or stomach lining) with a pathway between cells and another through cells (Figure 25.20).

Cell membranes consist of lipid bilayers with variable conductance represented by ion-specific channels (pores) embedded in the layer (Figure 25.21). The conductance of normal cells is 0.1 – 10 S/m², rising above 100 S/m² in excited cells. The capacitance is ≈1 μF/cm² (0.01 F/m²).

25.10.5 Dispersion Curves

The term dispersion relates to variation of σ (conductivity) and ε (permittivity) with frequency: note that in general, conductivity increases with frequency and permittivity tends to fall. Conductivity is related to the imaginary part of the complex dielectric constant (see below) which is in turn related to wavespeed and refractive index: the processes are similar to optical dispersion. Conductivity is also related to energy dissipation via $\sigma \boldsymbol{E}^2$ (W/m³), where $\boldsymbol{E}$ is the local electric field.

From Maxwell's equations:

$$(\boldsymbol{n}^*)^2 = \mu_r \mathbf{K}' - j\, \mu_r \sigma/(\omega \varepsilon_0)$$

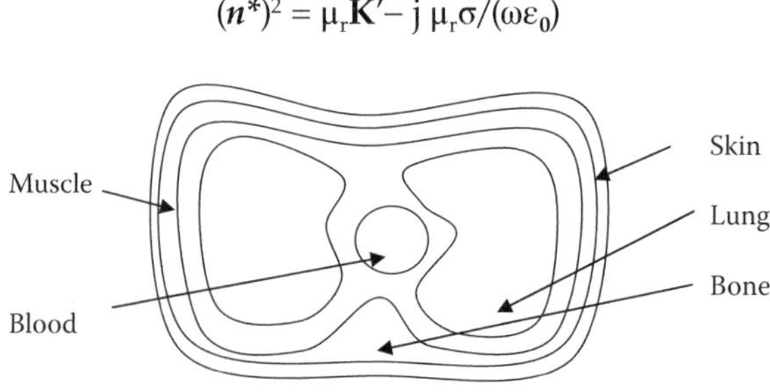

FIGURE 25.19
Cross-section of chest showing approximate anatomy and representation by five tissue types.

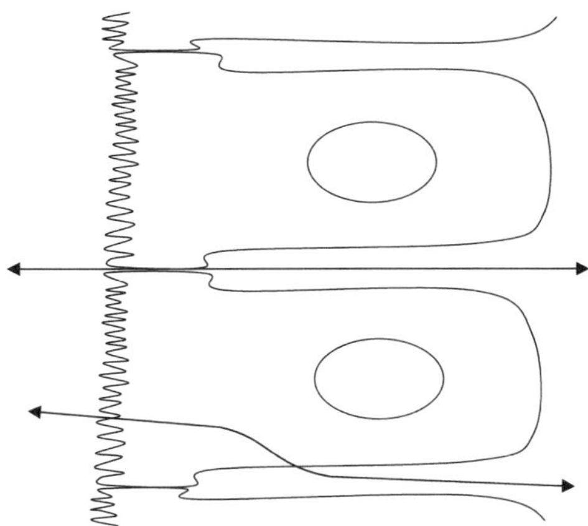

FIGURE 25.20
Epithelial tissue layer showing intercellular (top arrow) and intracellular (bottom arrow) current pathways.

where
n^* = complex refractive index; μ_r = relative magnetic permeability; K′ = dielectric constant; ε_0 = free-space permittivity and $\omega = 2\pi f$. For biological tissues $\mu_r = 1$, and

$$(n^*)^2 = K^* = K' - j\sigma/(\omega\varepsilon_0) = K' - jK''.$$

25.10.5.1 Cole–Cole Plots

This is a standard way of investigating biological tissue; and is essentially plotting the real versus the imaginary part of the tissue impedance as a function of frequency.

As an example, if an epithelial or other cell layer membrane is placed between two chambers containing physiological solutions (e.g., Ringer's), the voltage across the membrane in response to a sinusoidal current can be monitored (Figure 25.22) to measure the (complex) impedance of the membrane (in practice the impedance of the electrodes has to be taken into consideration to ensure that the measurement is just of the membrane).

$$\text{Let } |Z| = V_m/I_m = \sqrt{(R^2 + X^2)}$$

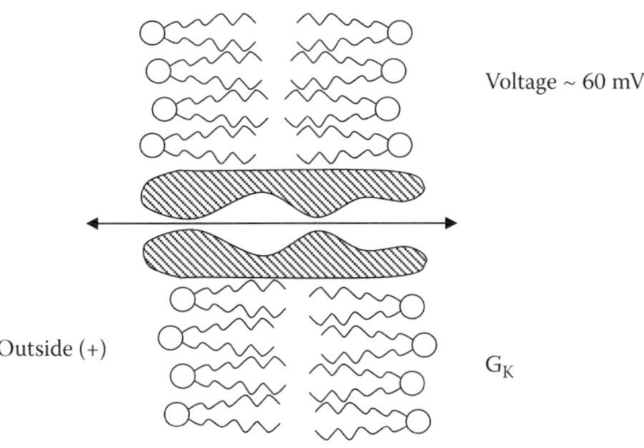

FIGURE 25.21
A representation of a water-filled ion (K^+) channel embedded in a lipid bilayer.

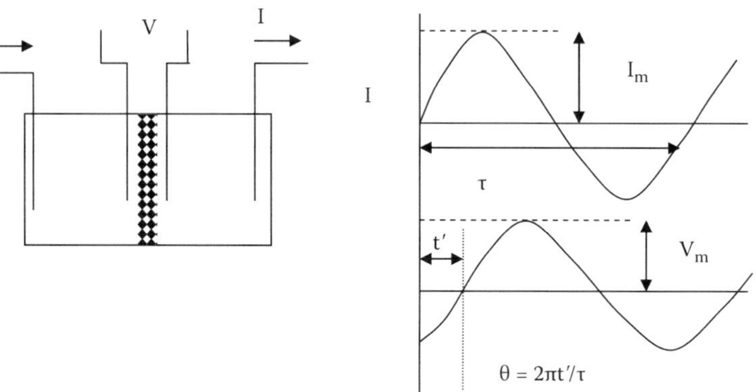

FIGURE 25.22
An experimental arrangement for determining impedance of an epithelial membrane (such as frog skin) from the voltage induced by a sinusoidal current.

Then, $R^2 = |Z|^2/(1 + \tan^2\theta)$ (= $\{V(t)/I(t)\}^2$) and $X = R\tan\theta$.

And we can plot an impedance locus (Figure 25.23), which is essentially a plot of real and imaginary parts of impedance at different frequencies (and is very similar to a Nyquist plot in systems analysis: see next section of this chapter).

A semicircle implies a parallel resistor–capacitor combination: the parallel combination $Y = 1/r + j\omega C$ can be transformed to a series combination $Z = R + jX = R - j/(\omega C')$, as illustrated in Figure 25.24.

$$Z = \frac{r}{1+(\omega\tau)^2} - jr\frac{r(\omega\tau)}{1+(\omega\tau^2)}; \quad \tau = rC.$$

Thus,

$$R = \frac{r}{1+(\omega\tau)^2}; \quad X = -r\frac{r(\omega\tau)}{1+(\omega\tau^2)}$$

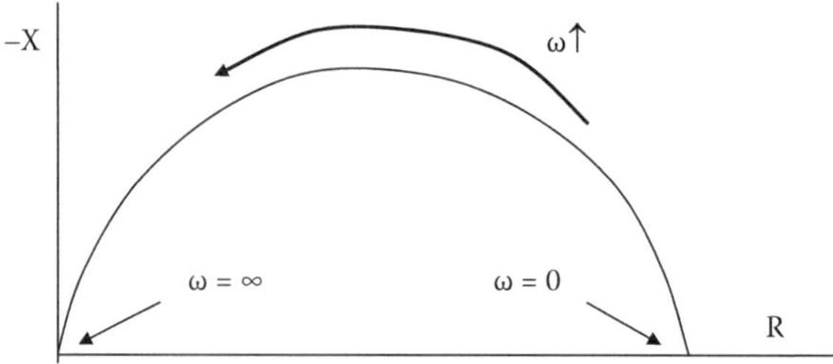

FIGURE 25.23
Impedance locus: the real (R) and imaginary (–X) parts of tissue impedance (Z) plotted for various angular frequencies (ω). A semicircle is characteristic of a simple resistor–capacitor representation.

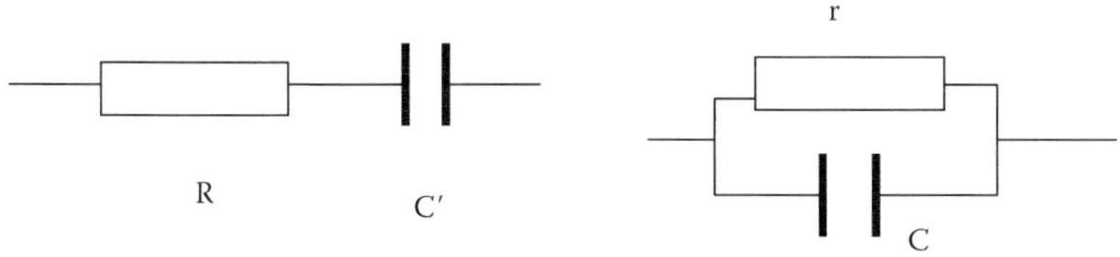

FIGURE 25.24
The two different ways of representing a "leaky capacitor": series representation (L) and parallel representation (R).

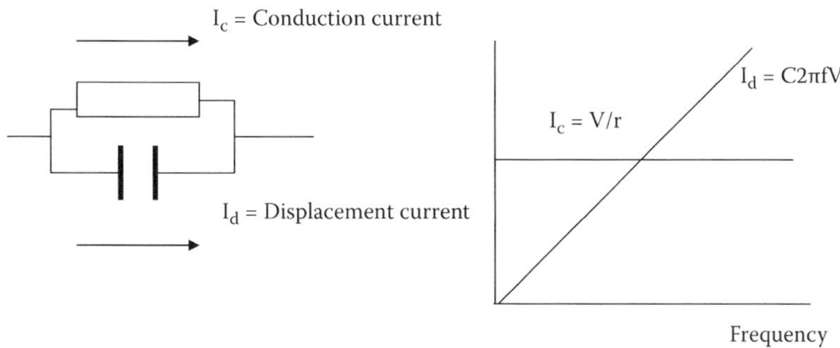

FIGURE 25.25
The variation of conduction and displacement current with frequency in typical tissue. The two are approximately equal at MHz frequencies.

which are the same expressions as for a leaky capacitor. As shown in Figure 25.25, as frequency rises, the amount of current flowing through the capacitor (displacement current) increases, but conduction current through the resistor remains the same.

25.10.5.2 Debye Equations

Many tissues can be more accurately modeled as shown in Figure 25.26—this allows for capacitance at very high frequency:

The admittance is given by

$$Y = j\omega C_\infty + \frac{1}{r_d + \dfrac{1}{j\omega C_d}} = \frac{\omega^2 C_d \tau}{1+(\omega\tau)^2} + j\left[\omega C_\infty + \frac{\omega C_d}{1+(\omega\tau)^2}\right]$$

and we can write this as

$$Y = j\omega K^* C_0 = (K'' + jK')\omega C_0; \quad \text{where } C_0 = \frac{A\varepsilon_0}{d}\text{; the geometric capacitance}.$$

Comparing expressions:

$$K'' = \frac{C_d}{C_0}\frac{\omega\tau}{1+(\omega\tau)^2}.$$

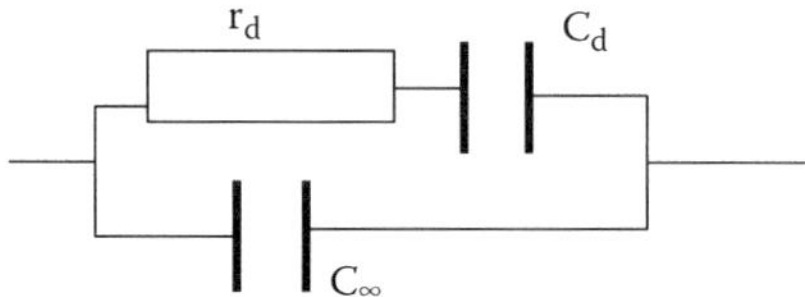

FIGURE 25.26
A more realistic representation of tissue impedance (giving rise to the Debye equations).

Putting

$$\frac{C_d}{C_0} = K_s' - K_\infty'$$

and

$$\frac{C_\infty}{C_0} = K_\infty'$$

$$K' = K_\infty' + \frac{\left(K_s' - K_\infty'\right)}{1+(\omega\tau)^2}; \quad K'' = \frac{\left(K_s' - K_\infty'\right)\omega\tau}{1+(\omega\tau)^2}$$

Or

$$\frac{K' - K_\infty'}{K_s' - K_\infty'} = \frac{1}{1+(\omega\tau)^2}; \quad \frac{K''}{K_s' - K_\infty'} = \frac{\omega\tau}{1+(\omega\tau)^2}$$

where K_∞' and K_0' are the high and low frequency limits, respectively.
Since $K'' = \sigma/(\omega\varepsilon_0)$, we can write:

$$\frac{K''\omega\tau}{K_s' - K_\infty'} = \frac{\sigma\tau}{\varepsilon_0\left(K_s' - K_\infty'\right)} = \frac{(\omega\tau^2)}{\left(1+(\omega\tau)^2\right)}.$$

25.10.5.3 Multiple Relaxation Times

We have assumed that the tissue only has one characteristic time constant τ, but in reality, there will be a range of values τ_i, each with a slightly different K value. Actually, better to think of it as a continuous spectrum of values $G(\tau)d\tau$

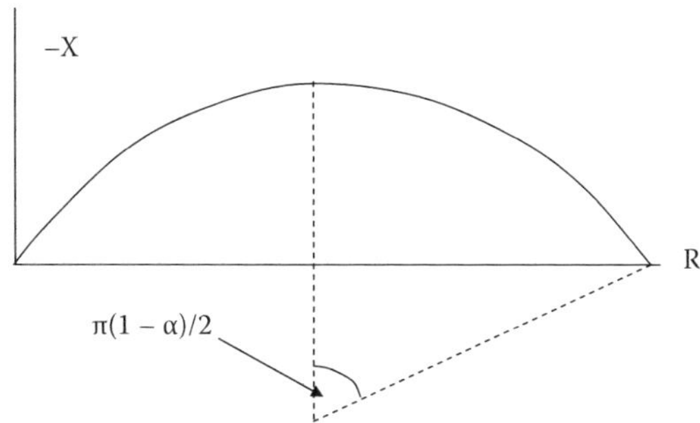

FIGURE 25.27
Most experimental investigations of tissue impedance yield depressed semicircles as shown.

$$K' - K'_\infty = \sum \frac{\Delta K_i}{1 + j\omega\tau_i} \rightarrow \int_0^\infty \frac{G(\tau)\Delta K}{1 + (\omega\tau)^2} d\tau$$

Putting $z = (\tau/\tau_0)$

$$G(z) = \frac{(\sin \alpha\pi)/2\pi}{\cosh\left[(1-\alpha)\ln z\right] - \cos \alpha\pi}; \quad \text{known as the Cole–Cole distribution}$$

$$\text{Or,} \quad K' - K' = \frac{\Delta K}{1 + (j\omega\tau)^{1-\alpha}}$$

which effectively means that the semicircles are depressed below the real axis, which is often what is seen in actual analyses of tissues (see Figure 25.27). A useful resource for data regarding electrical parameters of particular tissues at frequencies up to several GHz can be found at http://niremf.ifac.cnr.it/tissprop/.

25.11 Frequency Analysis of Biological Systems

A powerful method of discovering the characteristic behavior of a biological system is to input standard signals, such as sine waves, and then to observe the output (Figure 25.28: upper). This is essentially a "black box" approach, where no assumptions are initially made about how the system is operating or what the mechanisms are at work in the system. The relationship between output and input is called a "transfer function" which is a concise way of representing this relationship. The best-fit transfer function (TF) for a particular system is "discovered" from experimental data using, for example, what is known as a Bode plot. This type of plot (there are other, related, plots) can also be used to identify the various subcomponents of the TF and can allow the stability of the system to be assessed.

Consider the following "thought" experiment: an isolated pancreas is perfused by a solution in which the glucose concentration varies sinusoidally (Figure 25.28: lower). The secretions from this pancreas enter an experimental animal (whose pancreas has been removed) where the plasma glucose concentration is monitored over several periods of the

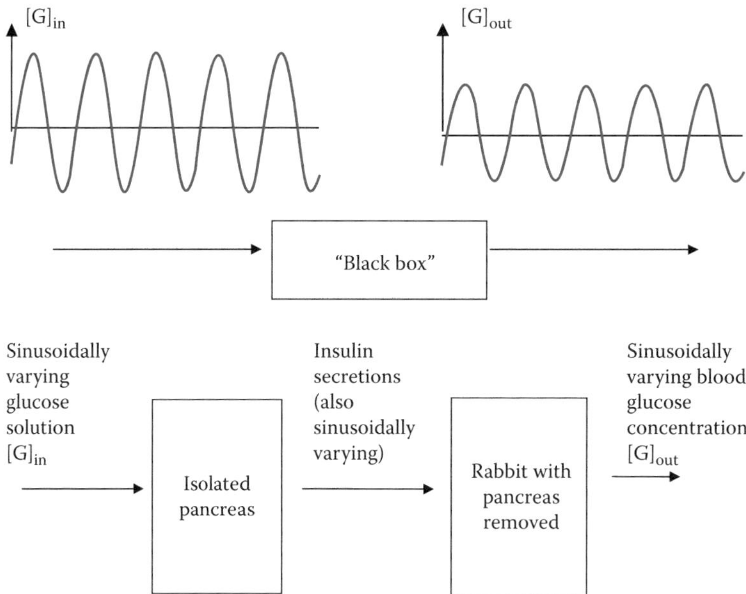

FIGURE 25.28
A "thought" experiment in which an isolated animal pancreas is infused with sinusoidally-varying glucose concentrations, then the resulting insulin secreted from the pancreas infused into an animal in which the pancreas has been removed. The resulting approximately sinusoidal variations in blood glucose in an artery of the animal is then monitored. Top: input and output to the "black box". Lower: the two sub-systems.

sinusoid. The period of input sinusoid is then varied and the resultant change in plasma glucose pattern is then monitored. In general, the changes in plasma glucose are also sinusoidal, but delayed in time in comparison to the sinusoidal changes in insulin (phase shift). In fact, as the insulin concentration rises, the plasma glucose tends to fall, because insulin stimulates cellular glucose uptake from the bloodstream. As the frequency of the insulin oscillations is increased, the amplitude of plasma glucose $[G]_{out}$ oscillations becomes progressively more diminished. This behavior is best characterized by double-log plotting the output amplitude versus 1/period (or frequency). In fact the log Amplitude *RATIO* versus log *ANGULAR* frequency (note that multiplying the $\log_{10}$ angular frequency by 20 gives a decibel (dB) scale: Figure 25.29). In the example given, the input $[G_{in}]$ has a 10 mM

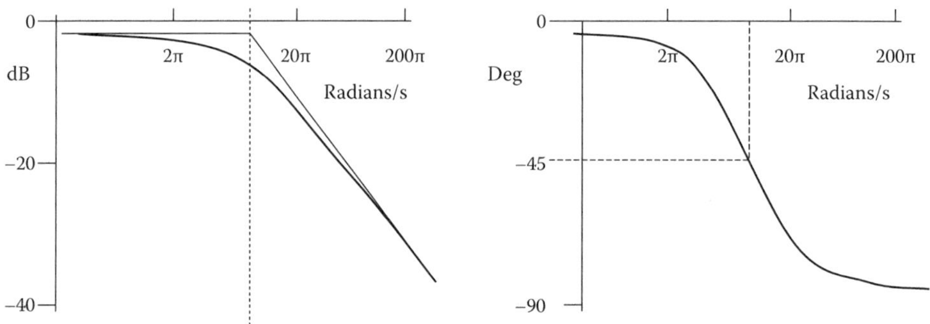

FIGURE 25.29
Left: double-log plot of amplitude ratio versus angular frequency. Right: phase angle versus log angular frequency. Note break frequency at 10π.

amplitude, that is, the steady-state (or resting) level could be 80 mM, but the concentration varies between 70 mM and 90 mM during the sinusoidal oscillations (80 mM ± 10 mM). Note that the amplitudes of the output glucose concentration $[G_{out}]$ gradually diminish, with a linear trend on this log–log plot. The ratio of output to input glucose amplitude ($[G_{out}]/[G_{in}]$) trends to smaller and smaller values as frequency increases. The biological system, that is, the ability of cells to move glucose from the blood stream in response to insulin, is less able to cope with change when this change becomes more rapid. The trend line (or asymptote) has a slope of –1 in logarithmic terms, which is equivalent to –20 dB per decade of frequency. This means that from going from 10 to 100 radians/s (1 $\log_{10}$ unit), the amplitude changes by –20 dB (or 1 $\log_{10}$ unit: refer to the definition of dB). The low frequency asymptote, on the other hand, is parallel to the x-axis. The point where the two asymptotes intersect (Figure 25.29, left) is known is the "break point" and the frequency at which this occurs the "break frequency" (ω_0).

If we now look at the PHASE difference between I/P and O/P (remember, a phase difference of 180° is expected, because this is a negative feedback system, so the O/P waveform is actually inverted before comparing with the I/P, hence as frequency tends to zero the phase now tends to zero, not 180°). As the frequency increases, the O/P lags I/P (hence minus sign), with a high frequency asymptote of –90° (Figure 25.29, right). Note that at the "break point" the phase is –45°. The quantity $1/\omega_0$ is the known as the characteristic time constant of the system. In this particular example (but not in a general sense) at $\omega = 0$, AR = –3dB.

So how can we characterize the system (which in this case is the glucose controller, since in the intact animal with a normally operating pancreas, this will be providing the normal negative feedback to maintain plasma glucose at normal levels)? We can identify a transfer function (TF) from the form of the graphs we have just seen (called a Bode plot). The TF informs us about the nature or characteristics of the system: it is a function of angular frequency $\omega = 2\pi f$, which is actually usually written as the symbol s (where $s = \sigma + j\omega$). Note that, in the present context, we can ignore σ and $j = \sqrt{-1}$, so s is really equivalent to ω.

In this case, TF is given as

$$H(s) = \frac{A}{1 + Ts}.$$

It is not immediately obvious how this TF relates to the Bode plot just described, so some explanation has to be given on why this should be and what the values of A and T are.

First, in the example shown if Figure 25.29, when s (or ω) → 0, A → –3dB (in this example), so A = $\text{antilog}_{10}(-3/20) = 1/\sqrt{2}$ (approximately). Putting s = 0 in the above equation gives $H(0) = 1 \times (1/\sqrt{2})$. Similarly, when $s = \infty$, $H(\infty) = 0$. For other values, taking logs:

$$\log(H) = \log A - \log(1 + Ts) \rightarrow \log A - \log T - \log s, \text{ when } s \text{ is large.}$$

This implies that the slope will be –1 (by looking at d(log(H))/d(log s)).

Note that in this case the break frequency occurs at $s = 10\pi$ and the product $sT = 1$. If we evaluate the above equation at this frequency, $H(10\pi) = A/2 = 1/2 \times (1/\sqrt{2})$.

Also at this frequency $Re(H) = -Im(H)$, so the phase = $\tan^{-1}(-1) = -45°$.

Note that the –1 slope (–20 dB per decade) is indicative of a first-order system. The reason for this is a little complex, but it is sufficient to say that the TF (which gives behavior as a function of frequency) can be transformed into a behavior as a function of time. The symbol s is in fact called the Laplace operator, which indicates, among other things, the types of differential (dy/dt, d^2y/dt^2, …) which appear when this transformation is done. A more detailed description is given in Section 25.11.4. Since there are only terms of the order of s in this TF, this means that it is First Order. Second Order would have s^2 terms, and so on. Putting all this together gives the following TF for this specific system:

$$H(s) = \frac{1}{\sqrt{2}\left(1 + \frac{s}{10\pi}\right)}.$$

A Resistor–Capacitor circuit would behave in the same way, and in this case T = 1/10π = RC. We can say that the current insulin–glucose system has similar characteristics as a RC circuit, and if appropriate, we can assign values to R and C, depending on the physiological interpretation.

In Figure 25.30, we see a depiction of subsystems in series (or cascaded system). The total system TF can be represented as a product of the subsystem TFs (think of amplifiers in series, for example).

$$H_T(s) = H_1(s)\ H_2(s)\ H_3(s).$$

When we take logs (or dB) these then add. In the Bode plot, we just add the log AR or dB contributions of each subsystems by representing each one separately, then adding the log values. In the example we get "break points" at ω = 1/2; 1/10; 1/50 rad/s. Adding the separate TFs together (ignoring the rounded bits near the break points) gets the resultant plot as shown in Figure 25.30.

25.11.1 Analysis of Experimental Data (to "Discover" TF)

A practical way to proceed is to look for segments in the dB versus log ω plot which correspond to slopes of –20, or multiples (or, if the plot is of log AR, slopes of –1, or multiples). The break points are where these lines intersect and the inverse of these will give the characteristic time constants T_1, T_2, T_3. . . . We should now check that phase angle is –45°, at the lowest break point, decreasing by –90° for each extra subunit break points (this may not be exact if there is insufficient differences between break points).

In this case, the TF is

$$H_T(s) = \frac{1}{(1+2s)(1+10s)(1+50s)}.$$

Note that the total phase lag at the high frequency asymptote is 90° + 90° + 90° = 270°.

25.11.2 Other First-Order Systems

In the systems just described the O/P lags the I/P by up to 90° per subsystem. As well as first order lag systems there are first order lead systems, which are similar to inductor-resistor rather than capacitor-resistor combinations. The Bode plot for such systems is similar to that shown in Figure 25.29, except that both parts are reflected about the frequency axis (the high-frequency asymptotic slope is +20 dB per decade). The TF is of the form:

$$H(s) = A'\ (1 + T's)$$

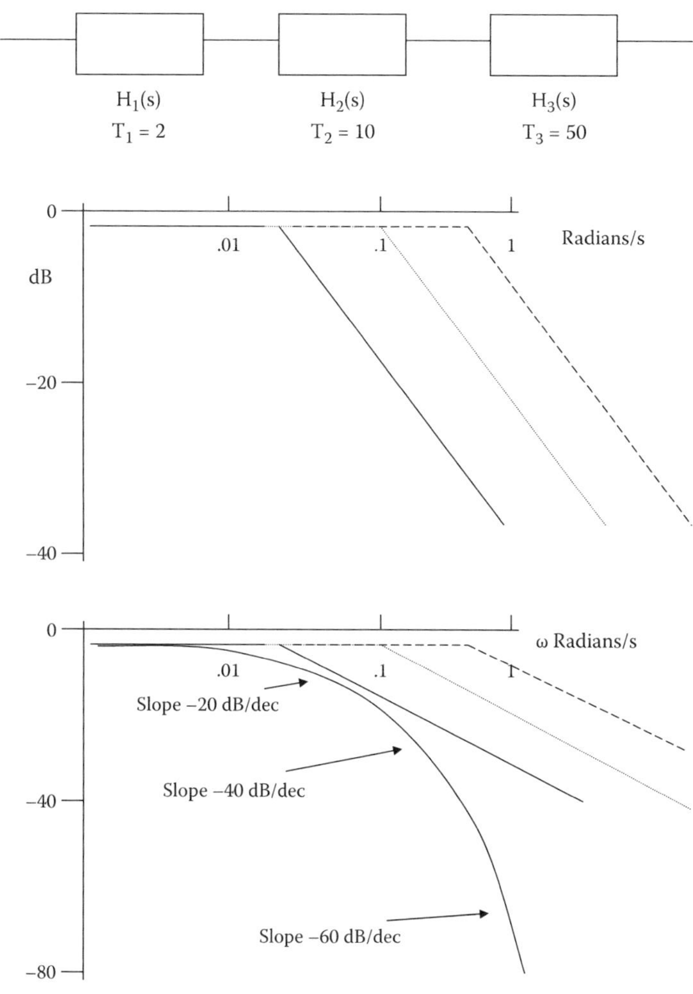

FIGURE 25.30
Top: block diagram of 3-subunit system, each first order. Middle: AR plots (in dB) of individual subunits. Bottom: combined AR plot obtained by adding the contributions from the 3 subsystems (note change of scale).

where in electrical terms, $T' = L/R$. If we have a 1st order lag and a 1st order lead systems cascaded, the TF would be

$$H(s) = \frac{A'(1+T's)}{A(1+Ts)}.$$

The high frequency asymptotic slope is 0, that is, parallel to the frequency axis.

Mathematically, there is interest in the values of s for which either the numerator or the denominator becomes zero. For example, when $s = -1/T'$ or $-1/T$ the numerator and the denominator, respectively, become zero. The first value of s is known as a "zero" and the second the "pole," respectively of the TF (note when the denominator becomes zero the TF becomes infinite. A study of where the poles and zeros occur for a particular system and how these vary with altered system parameters is known as a "root locus plot," and is useful for investigating stability, among other things.

25.11.3 Second-Order Systems

Second order systems have a term s^2 in the TF, which in the time domain refers to a second differential. A series combination of a resistor, a capacitor, and an inductor (where the developed voltage is related to the current, the integral of the current, and the differential of the current, respectively) is a simple second order system. If instead of current we consider the electrical charge in the circuit (current = dq/dt, where q is the charge in the circuit in coulombs) we get

$$V(t) = R\, i(t) + L\, di/dt + 1/C \int i(t')dt'.$$

Differentiating and rearranging gives:

$$LCd^2i/dt^2 + RCdi/dt + 1 = 0.$$

Or, in terms of the Laplace Transform operator s (= jω), and remembering that if $i = i_m \sin\omega t$, $di/dt = \omega i$ and $d^2i/dt^2 = \omega^2 i$

$$-LC\omega^2 + RCj\omega + 1 = LCs^2 + RCs + 1 = 0.$$

This represents a relationship between intput and output voltages (see Figure 25.16, with the special case of G = ∞), with the ratio of voltages across C and the RLC combination being a TF. Going to the frequency domain (and including a constant gain A):

$$H(s) = \frac{A}{LC\, s^2 + RC\, s + 1}$$

$$H(s) = \frac{A}{(1/\omega_0^2)s^2 + (2\zeta/\omega_0)s + 1}.$$

Here, w_0 (= $1/\sqrt{(LC)}$) is the natural frequency of the system and ζ(= $(R/2)\sqrt{(C/L)}$) is the relative damping factor. Sometimes it is possible to factorize the denominator s to give:

$$H(s) = \frac{A}{(1 + Ts)(1 + T's)} = \frac{A}{1 + (T + T')s + TT's^2}.$$

By the same logic as previously, if s >> ω_0 and taking logs of both sides, the limiting slope will now be –2 (or –40 dB per decade). The phase now goes through –90° at the break point and –180° at high frequencies. Depending on the value of RC (or $2\zeta/\omega_0$) there will a resonant peak at or just below the break point. An example is shown in Figure 25.31, with the break point at ω = 10π and a constant gain of –14 dB (or 0.04).

$$H(s) = \frac{0.04}{(1/100\pi^2)s^2 + (1/100\pi)s + 1}.$$

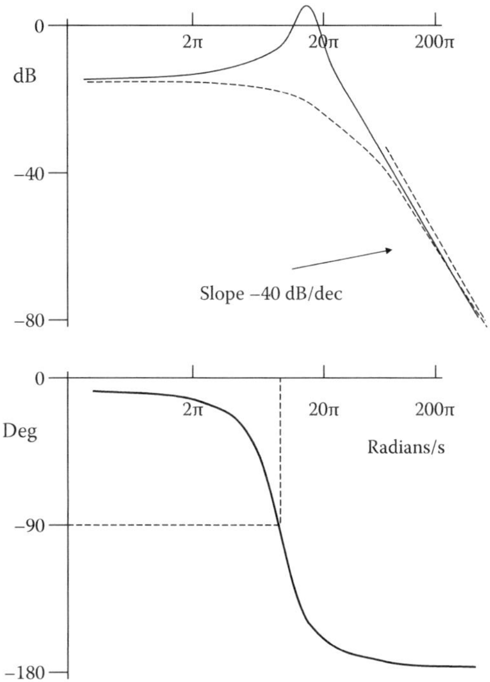

FIGURE 25.31
A Bode diagram of a second order system. Here the characteristic time constant is 10π s and the low frequency gain is 0.04.

The Bode plot of these resonant systems is best visualized by the use of MATLAB code. In earlier versions of MATLAB the coefficients in s had to be entered as separate numerator and denominator vectors as follows (example for $\omega_0 = 1$ and $\zeta = 0.1$ [i.e., RC = 0.05])

```
>> num=1;
>> den=[1 .05 1];
>> Hs = tf(num, den);
>> f = 0:0.01:2;
>> w = 2*pi*f;
>> bode(Hs, w);
```

The Bode plot is shown in Figure 25.32 (top). More recent MATLAB versions permit entry of the TF in s notation as below:

```
>> s = tf('s');  Hs = 1/(s^2+0.05*s+1);
>> f = 0:0.01:2;
>> w = 2*pi*f;
>> bode (Hs, w)
```

Increasing the relative damping factor by 10 then 100 will show how the plot progressively shows less of a resonance behavior, becoming more like a "double pole" with a break point at $\omega_0 = 1$.

The Nyquist plot (which can be seen in the lower part of Figure 25.32) combines AR and phase data in a single diagram. AR is represented by the length of a radial vector from the

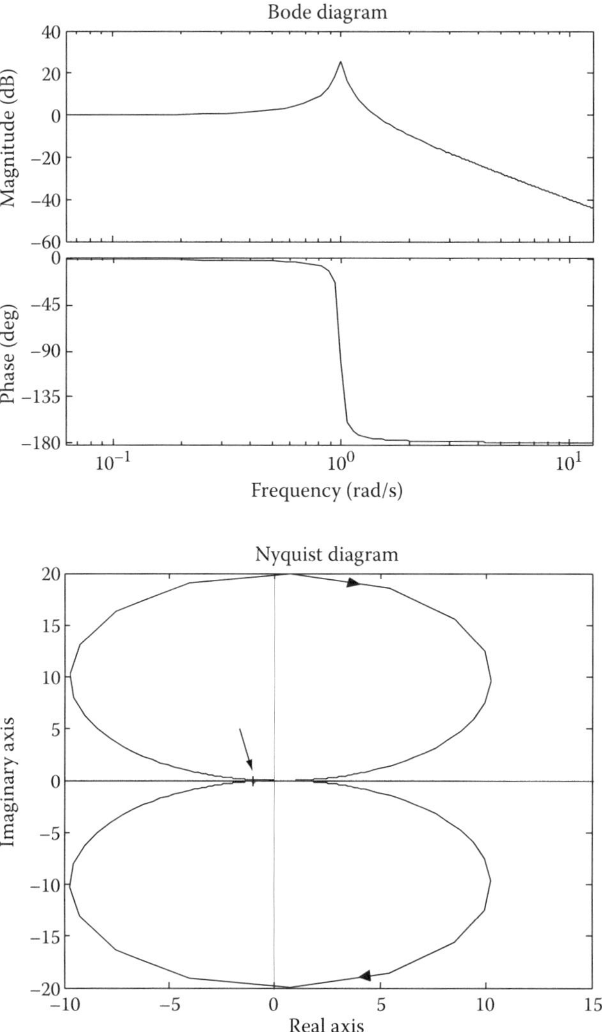

FIGURE 25.32
The MATLAB plot obtained by entering the code in the text. The parameters are $\omega_0 = 1$ and $\zeta = 0.1$. Top: Bode plot; Bottom: Nyquist diagram. The arrow indicates the '–1' point.

center of the diagram and the phase by the angle of that vector with the +x axis. MATLAB gives two loops, in both the upper and lower two quadrants. Since the phase change in this case is a lag, it is only the lower two quadrants we are concerned with. Note that AR is not converted to dB or to logs. The MATLAB diagram is obtained by changing the word "bode" to "nyquist" in the above code. The red cross in the diagram is an important landmark: it represents the point (–1, 0). It represents a point where the phase lag is 180°. If the AR is greater than unity at that phase shift a closed loop controller will become unstable (this is because negative feedback subtracts the feedback signal from the reference signal [see the earlier section of this chapter]). A 180° shift will make the feedback positive and will have a greater amplitude than the initial signal, leading to uncontrolled oscillations. In this example there is a margin of at least 40 dB before this will happen. As an exercise, add a constant gain (numerator of 10^5, say) and see what happens to both the Bode and Nyquist plots. Actually, the chief source of instabilities is dead time or transportation lag (also called pure time delay). If the lag is L units of time, this adds $\phi = \omega L$ radians or $\phi = 360\omega L/2\pi$ degrees to the phase but leaves the AR unchanged. This may seem counter-intuitive at first, but if you imagine a train of say 10 sinusoidal oscillations, with a period of 2 seconds per oscillation, if the entire train were delayed by 6 seconds, this would represent 6/2 = 3 oscillations or 360° × 3 = 1080° of phase. Since the radian frequency is $\omega = 2\pi/T = \pi$ radians/s, the 6 second delay corresponds to $\phi = \omega L = \pi \times 6$ radians, or 1080°, as before.

Remember that the TF is complex, with a magnitude equal to the AR and a phase given by arctan (Im(TF)/Re(TF)), where Im and Re represent the imaginary and real parts of the TF. Adding a factor $e^{-j\omega L}$ or e^{-sL} to the TF effectively adds ωL to the phase, while leaving the AR unchanged. The reason for this is easy to understand by considering Euler's formula $e^{-j\omega L} = \cos(\omega L) - j\sin(\omega L)$, then it follows that $\phi = \arctan(\mathrm{Im(TF)}/\mathrm{Re(TF)}) = \arctan(-\sin(\omega L)/\cos(\omega L)) = -\omega L$. The minus sign indicates phase delay.

As an example, the effect of adding a delay of 1 second to the second order TF considered above is given by the following MATLAB code.

```
>> s = tf('s');  Hs = exp(-s)/(s^2+0.05*s+1);
>> f = 0:0.001:2;
>> w = 2*pi*f;
>> nyquist (Hs, w)
```

Both the Bode and Nyquist plots are shown in Figure 25.33, but the latter is particularly of interest, because now the plot encircles the "–1" point, indicating instability. This behavior is actually observed in physiological systems. For example, in a classic series of experiments on the influence of cerebral blood pressure on overall pressure regulation, Sagawa et al. (1964) (quoted in Sagawa, 1971) showed that for low levels of perfusion pressure the Nyquist plot showed curves encircling the –1 point and the closed-loop behavior showing uncontrollable oscillations in systemic pressure. This behavior is shown in Figure 25.34.

WORKED EXAMPLE

The Bode diagram shown in Figure 25.35 is taken from an experiment in which an isolated carotid artery is filled with a fluid in which the pressure varies sinusoidally. The other baroreceptor nerves are cut and the systemic pressure sinusoidal variations are monitored. Magnitude and phase of output compared to input sinusoidal variations are shown.

(a) Explain how this constitutes an "opened" closed loop control system.
(b) Determine the approximate low frequency gain (N.B. dB = 20 log10(AR)).
(c) What is the order of this system and the characteristic time constant(s)?
(d) Is there transportation lag in the system, and if so, what is its magnitude?
(e) Write down the Transfer Function of the system.
(f) What gain would need to be added to the loop in order to make the closed loop behavior unstable?

Answers

(a) Normally the baroreceptors would be feeding back pressure information directly to the arterioles via the CNS. In this situation this feedback information is externally controlled.

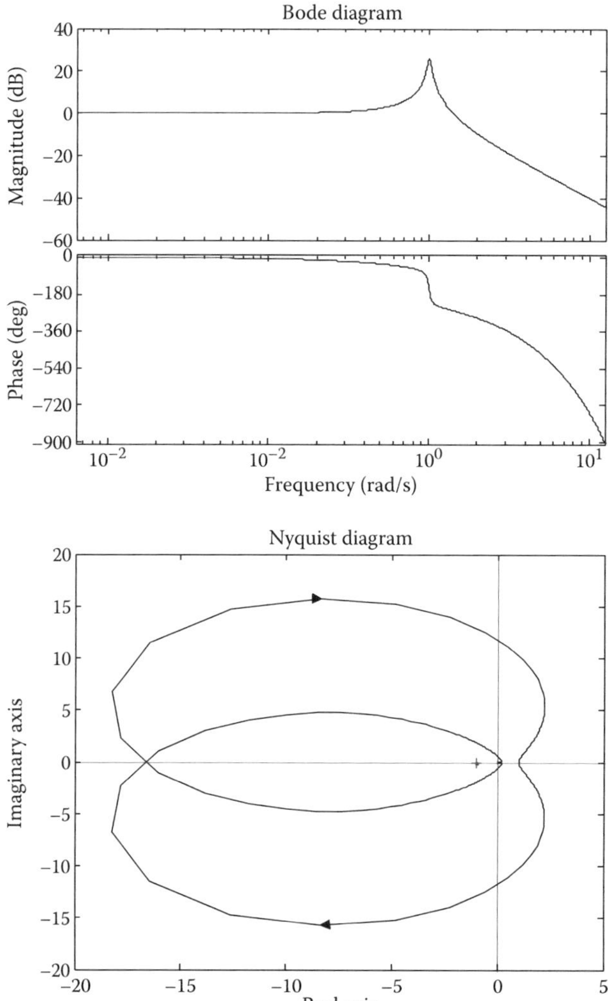

FIGURE 25.33
As for Figure 25.32, but with an added dead time of 1 s.

(b) The gain at the left hand end of the diagram is –6 dB, by interpolation, corresponding to $10^{-6/20} = 0.5$.

(c) Since the slope of the right hand portion is –20 dB/decade, it is a 1st order system and the break point is at 10 rad/s, thus, $T = 1/\omega_0 = 0.1$ s.

(d) The asymptotic lag for a 1st order system would be 90°. In this case, we have a lag of 315° at 10^3 rad/s, an extra 225° or 3.9 radians. The lag is thus, $3.9/10^3 = 3.9 \times 10^{-3}$ s.

(e) The transfer function is thus:

$$\frac{0.5\exp(-3.9\times10^{-3}s)}{(1+0.1\,s)}$$

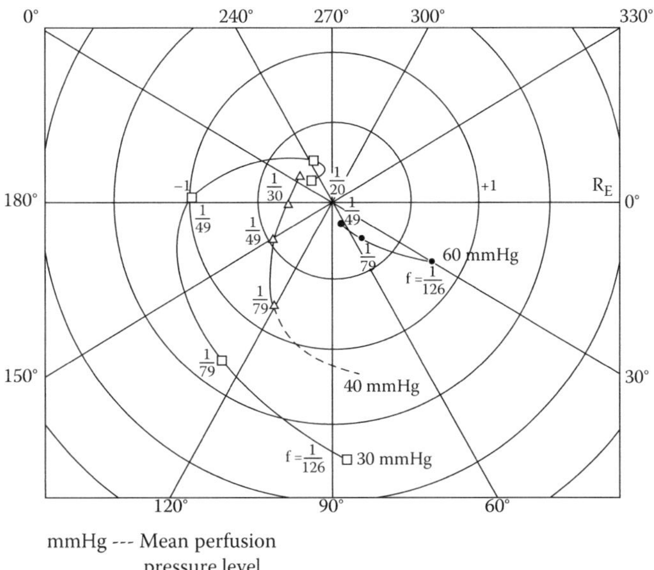

FIGURE 25.34
Experimental data from an open-loop experiment where the head of a donor dog was perfused at various mean pressures and sinusoidal variations superimposed. The variations in pressure in the recipient dog circulation were then plotted as a Nyquist plot as shown. Note that for 30 mmHg perfusion pressure the plot encircles the "–1" point (see Sagawa, 1972).

(f) The lag is 180° at 4×10^2 rad/s, approximately. Here there is a gain margin of around 38 dB before the AR would be positive in dB (that is, greater then unity). Thus, an extra gain of $10^{38/20} = 79$ would need to be added before the system would become unstable.

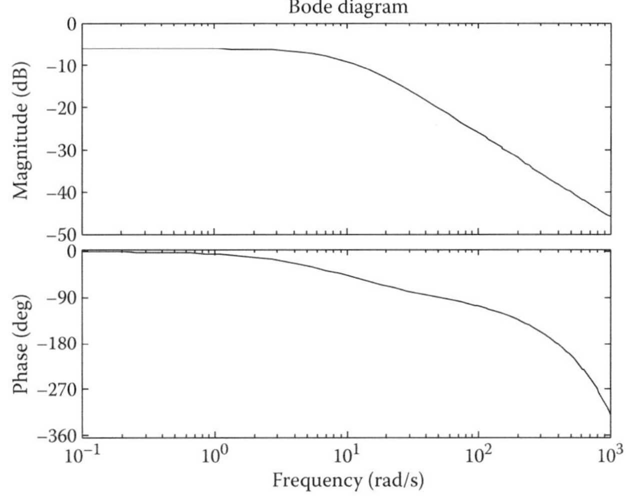

FIGURE 25.35
Worked example question on Bode analysis.

25.11.4 Theoretical Derivation of Transfer Function Forms (Can Be Omitted on First Reading)

So far, the TFs have been introduced without much explanation on why they take on the form they do (although some appeal to analogous electrical systems has been made). The next section will give a brief, but more formal introduction to how TFs are derived.

First, it has not been emphasized that in this type of approach, preliminary investigations need to be done to ensure that the biological system in question is linear and time-invariant. If so, it can be probed by inputting small sinusoidal variations and noting the output resulting from these. For each frequency of the input signal a corresponding amplitude and phase of the output is measured. A plot of $\log_{10}$ Amplitude Ratio versus $\log_{10}$ angular frequency together with relative phase versus $\log_{10}$ angular frequency is called a Bode plot. The Transfer Function (T.F.) of an unknown system (treated as a "black box") can be identified from features of this Bode plot. Since the plot is of logarithms, multiplicative factors in the T.F. appear as additions in the Bode plot. Knowledge of how simple systems appear in a Bode plot allows easy analysis of more complex systems. In turn, analysis of these systems by Laplace transformation reduces differential equations to algebraic expressions.

If we neglect for the moment the question of transportation lag (i.e., deadtime) in biological systems, the general expression

$$\sum a_i y^{[i]} = \sum b_i x^{[i]}$$

describes the relationship between output y and input x of the system. Here $y^{[i]}$ and $x^{[i]}$ refers to the ith derivative of y and x with respect to time. It should be noted that the Laplace transform of an ith derivative is

$$\int y^{[i]} e^{-st}\,dt = s^i Y(s) + y^{[i-1]}(0)$$

where $Y(s) = \int y(t)e^{-st}dt$, s being the Laplace operator $s = \sigma - j\omega$ and $y^{[i-1]}(0)$ is the value of $y_{[i-1]}(t)$ at $t = 0$. We can usually arrange for these initial values to be zero, so the generalized T.F. becomes:

$$H(s) = Y(s)/X(s) = \sum b_i s^i / \sum a_i s^i$$

which as a ratio of two polynomials can be factorized to give:

$$H(s) = K\prod_{i=1}^{m-a}(s+z_i) \bigg/ \left(s^{(b-a)}\prod_{i=1}^{n-b}(s+p_i)\right),$$

where the z_i and p_i are known as the zeros and poles of the expression, respectively. An alternative and equivalent form is

$$H(s) = k\prod_{i=1}^{m-a}(1+T_i s) \bigg/ \left(s^{(b-a)}\prod_{i=1}^{n-b}(1+T_i' s)\right),$$

where the T_i and T_i' are the time constants of the system and $k \neq K$.

Most biological systems are a combination of zeroth, first and second order systems, which will now be described, with their Bode equivalents.

25.11.4.1 Zero Order Systems

$$a_0 y(t) = b_0 x(t): H(j\omega) = k, \text{ where } k = b_0/a_0.$$

In this case, the relationship between y and x is a constant gain, with no frequency dependence (in reality there are no zero-order systems, but many behave as such in the practical range of frequencies of interest).

25.11.4.2 First Order Pole

$$a_i y'(t) + a_0 y(t) = b_0 x(t) \quad \text{or}$$

$$Ty' + y = kx, \text{ where } T = a_1/a_0 \text{ and } k = b_0/a_0, \text{ as before.}$$

Taking Laplace transforms:

$$T\int y'e^{-st}\,dt + \int ye^{-st}\,dt = k\int xe^{-st}\,dt$$

$$T(sY(s) + y(0)) + Y(s) = kX(s), \text{ and if we arrange for } y(0) = 0$$

$$Y(s)\{Ts + 1\} = X(s), \text{ so } H(s) = Y(s)/X(s) = 1/\{1 + Ts\}.$$

If we substitute $j\omega$ for s, it is quite easy to show that the magnitude of $H(j\omega) = k/\sqrt{(1 + (\omega T)^2)}$ and the phase by $\tan^{-1}(\omega T)$. We can see for $\omega << 1/T$, magnitude of $H(j\omega) = k$ (since $\omega T << 1$) and for $\omega >> T$ we get $H(j\omega) = 1/\omega T$. If we plot this on a log–log graph, the slope will be –1, or –20 dB per decade. Note that when $\omega = 1/T$ we get mag $H(\omega) = 1/\sqrt{2}$ and phase is $\tan^{-1}(-1)$ or 45° of lag. For small and large values of ω the lag tends to 0 and –90°s, respectively. The Bode plot thus looks as shown in Figure 25.29. Note that putting $K = k/T$ and $p = 1/T$ we get the alternative form of $H(s) = K/(s + p)$, where p is the pole value in rad/unit time. In the frequency domain, the point $\omega = 1/T$ is known as the "break point" (it has the same value as the pole if real, but the pole can be complex).

A resistor-capacitor circuit is a simple example of a simple pole model, where T = RC.

25.11.4.3 Simple Zero

Similar to the above, $a_0 y(t) = b_i x'(t) + b_0 x(t)$ or

$$y = k(Tx' + x),$$

where $T = b_1/b_0$ and $k = b_0/a_0$.

Thus, $H(j\omega) = k(1 + Ts) = K(s + z)$, where $z = 1/T$.

This behaves similarly to the pole, except that there is an upward slope of 20 dB/decade above the break point $\omega = 1/T$.

An inductor-resistor circuit is a simple example of a zero, with T = L/R. Note that in this case the output *leads* the input in phase.

25.11.4.4 Second Order Systems: Double Poles

These can be modeled as

$$a_2y'' + a_1y' + a_0y = b_0x$$

and putting $k = b_0/a_0$; $\omega_n = \sqrt{(a_0/a_2)}$; $\zeta = a_1/(2\sqrt{(a_0a_2)})$ and taking Laplace transforms we get

$$H(s) = H(j\omega) = \frac{k}{(s/\omega_n)^2 + 2\zeta(s/\omega_n) + 1} = \frac{k}{(T_1s+1)(T_2s+1)}$$

$$\text{Mag of } H(j\omega) = \frac{k}{\sqrt{\{(1-(\omega/\omega_n)^2)^2 + (2\zeta\omega/\omega_n)^2\}}} = \frac{k}{\sqrt{\{(1-\omega^2(T_1T_2))^2 + ((T_1+T_2)\omega)^2\}}}$$

and

$$\tan\varphi = \frac{2\zeta}{(\omega/\omega_n) - (\omega_n/\omega)}.$$

This models a series RLC resonant circuit, where the relevant 2nd order DE is

$$LCq'' + RCq' + q = CV$$

where charging potential can be considered as an input, charge on the capacitor as output.

Similarly, the equation describing the motion of a pointer of an analog voltmeter is

$$I\theta'' + f\theta' + k\theta = k\theta_i,$$

where I, f, and k are inertia, viscous friction and stiffness, respectively.

25.11.4.5 Transportation Lag

Suppose an impulse input gives an impulse output delayed by time L, we can write the output as

$$y(t) = \delta(t - L)$$

$$Y(s) = \int \delta(t-L)e^{-st}\,dt = e^{-sL}\int \delta(t-L)\,dt = e^{-sL}$$

and

$$X(s) = \int \delta(t)e^{-st}\,dt = 1$$

so $H(s) = e^{-sL}$, which has an amplitude of 1, but a phase *lag* of $L\omega$ radians in addition to any other lags or leads.

Tutorial Questions

Systems Analysis

25.1 A hormone, erythropoietin, formed in the kidney, promotes the production of red cells and the bone marrow. Formation of erythropoietin is stimulated by a fall in oxygen level in the blood. These effects can be represented by the following equations where:

N = number of red cells/unit volume

E = blood erythropoietin concentration

O = amount of oxygen in blood/unit volume of blood

$$N = k_1 E \quad \text{(bone marrow red cell production)}$$

$$O = k_2 N \quad \text{(blood oxygen)}$$

$$E = k_3/O \quad \text{(kidney response to } O_2 \text{ level)}$$

(a) Sketch these equations in the form of the following block diagram, and explain how these relationships constitute a control system (Figure 25.36).

(b) Reduce the blocks to two, a controlling system and a controlled system, explaining this.

(c) Derive an expression for open loop gain on the basis of this, comment on whether the body of a climber is able to compensate for blood loss (i.e., hemorrhage) better on top of a mountain than at sea level.

(N.B.: Assume that blood O_2 content is directly related to the O_2 partial pressure in the air)

25.2 The amount of substance produced in a biological system is given by x where

$$x = \sqrt{(y + 16)}.$$

The neural control y which is fed back in response to the substance is given by $y = -3x^2$.

(a) By substitution find the operating point and comment on its value.

(b) Calculate the open loop gain at this operating point and state whether the system is one of negative or positive feedback.

25.3 The following is a schematic diagram of the arterial blood pressure control system: The following approximate relationships exist between the variables in the control system (p is arterial pressure, $k_1 - k_6$ are constants) (Figure 25.37).

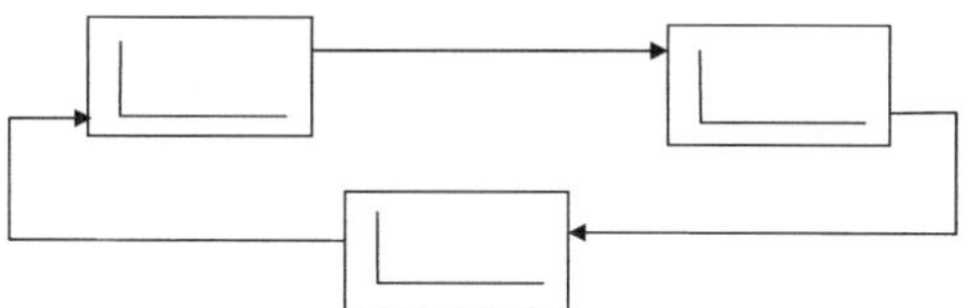

FIGURE 25.36
See Tutorial Question 25.1.

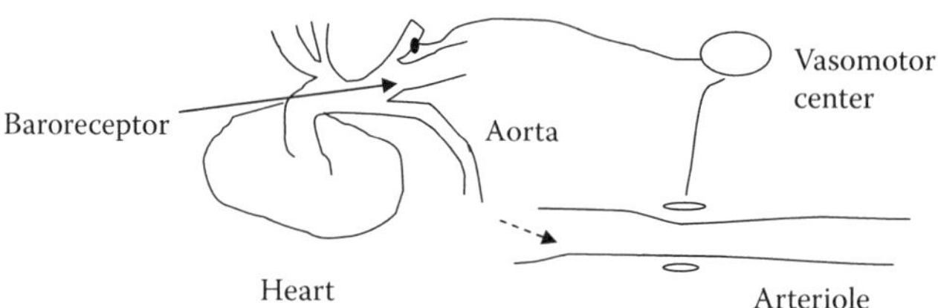

FIGURE 25.37
See Tutorial Question 25.3.

Baroreceptor nerve firing frequency (f_1): $f_1 = k_1p + k_2$

Firing frequency of nerves leading to pressure vessels (f_2): $f_2 = k_3/f_1$

Radius of pressure vessels (r): $r = k_4/f_2$

Resistance of pressure vessels (R): $R = k_5/r^4$

Arterial pressure (fluid analog of Ohm's law): $p = k_6R$

(a) Express these relationships in the form of a block diagram.

(b) Explain how the control system consists of a negative feedback (you may wish to refer to a "symbol-and-arrow" diagram.

(c) Divide the loop into a "controlling" and a "controlled" system (i.e., what is being controlled and what is doing the controlling) with a brief explanation.

(d) Obtain an expression for the open loop gain of the system in terms of the ks and arterial pressure.

(e) Comment on whether individuals with high blood pressure (Hypertension) have a better control of blood pressure than normal.

(f) How can the open loop behavior of this control system be investigated in experimental animals?

25.4 The osmolarity of extracellular fluid is controlled by a system which can be simplified into four blocks, whose steady-state characteristics can be represented an follows:

Supraoptic nuclei: $\ln(f) = Ax + B$ (f is neural frequency)

Neurohypophysis: $y = C\ln(f)$ (x is plasma osmolarity)

Distal tubules and collecting ducts: $z = Dy + E$ (y in rate of ADH secretion)

Extracellular fluid: $x = F/z$ (z is rate of H_2O reabsorption by the kidney)

A–F are constants (in appropriate units). Derive an expression for open-loop gain for the system. Is ECF osmolarity controlled better for states of diuresis or antidiuresis on the basis of this expression?

25.5 In the pupil diameter regulating system of the eye, equations of the following form can be used to model the component blocks in the system:

Retina: $f_a = f_a'(1 - \exp(-k\Phi))$

CNS: $f_e = Bf_a$

Pupil sphincter: $A = A_0\exp(-Kf_e) + A'$

Light flux: $\Phi = AL$

where Φ is the light flux, A is the pupil area and L the light intensity. f_a and f_e are the afferent and efferent nerve impulse frequencies, respectively (these are nerves from the retina and to the sphincter muscle, respectively). A_0 and A' are the maximum and minimum pupil areas, f_a' the maximum afferent nerve impulse frequency and B, K, and k are constants.

(a) Draw (i) a block diagram and (ii) a symbol and arrow diagram for information flow in the system.

(b) Derive an expression for the system open loop gain.

How is this altered by increasing (i) L (ii) A_0?

25.6 Pupil area A is related (via neural feedback mechanisms) to the flux Φ of light entering it by the following empirical relationships:

$$A = 8(5 + 3\sqrt{\Phi})/(1 + 4\sqrt{\Phi}) \qquad (A \text{ in mm}^2;\ \Phi \text{ in mW}).$$

The flux of light entering the pupil is given by

$$\Phi = LA$$

where L is the incident light intensity in mW/mm^2.

(a) Draw a symbol and arrow diagram for this feedback control mechanism.

(b) Obtain an expression for open loop gain. What are the numerical values of this for pupil areas of (i) 4 mm^2 and (ii) 40 mm^2 for L = 0.01 mW/mm^2?

(c) What do these values indicate about pupillary reaction to light?

Compartmental Systems

25.7 100 mg of a foreign substance is injected into the blood at t = 0; blood samples are taken at hourly intervals for the first 6 h then at 12 h intervals after that. The blood concentrations are given below

Time (h)	1	2	3	4	5	6	12	30	42
Conc. (mg/L)	14.3	12.9	12.4	10.4	9.6	9.0	7.0	6.9	6.9

Assuming a simple two compartment system in which the substance is not metabolized or destroyed, calculate:

(a) Total volume of both compartments

(b) Blood volume

(c) Volume of the other compartment

(d) Rate of transfer between compartments (in appropriate units).

Formula:

$$C(t) - C(\infty) = (C(0) - C(\infty)) \exp(-q(1/V_1 + 1/V_2)\, t)$$

25.8 The following is a plot of ln {C(t) – C(∞)} of a substance in plasma, $C_1(\infty)$ being 20 mg/I (Figure 25.38). Assuming a two-compartment system (the plasma forming compartment 1) and given that the original amount added to the plasma was 120 mg (previous concentration being zero) calculate:

(a) The volume of each compartment

(b) The volume flow rate q between compartments, given that the equation below holds

$$C(t) - C(\infty) = (C(0) - C(\infty)) \exp(-q(1/V_1 + 1/V_2)\, t)$$

25.9 An intra-arterial injection of dye (120 mg) is carried out at t = 0. The dye equilibrates with the extracellular fluid (ECF) so that after a few hours the concentration in both the blood and ECF is the same 6.3 mg/I. Assuming a simple two-compartment system:

(a) Estimate the ECF volume, assuming no dye in the ECF for t < 0

(b) With the help of the equations given above, determine k_{12} and k_{21} explaining their significance, from the data below:

t (min)	20	40	60	80	120	160	200
Blood conc. (mg/L)	16.5	12.1	9.3	7.9	6.8	6.4	6.3

(c) Sketch the form of the concentration vs. time graphs for dye in the blood and ECF if the dye were slowly accumulated by the body cells.

HINT: You will need to estimate blood volume by extrapolation to t = 0.

Equations; $A_i(t)$ is the amount of dye in compartment i at time t (similarly $C_i(t)$ for concentrations) and V_i is compartment volume. k_{il} refers to transfer from I to j.

Frequency Analysis

25.10 (a) Discover the transfer function of the opened closed-loop physiological control system whose Bode data are shown (Figure 25.39).

(b) Sketch the Nyquist equivalent of this Bode plot.

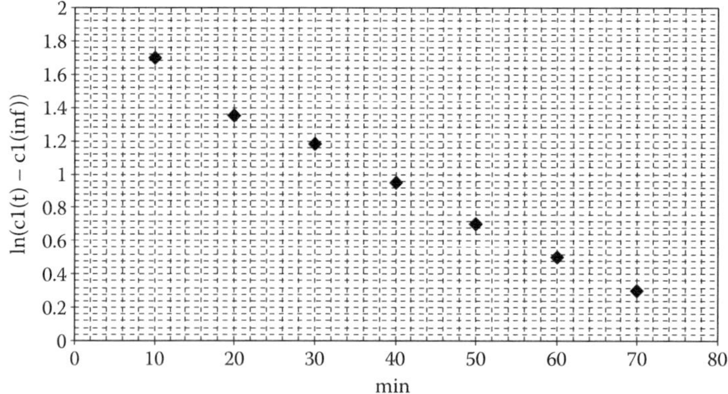

FIGURE 25.38
See Tutorial Question 25.8.

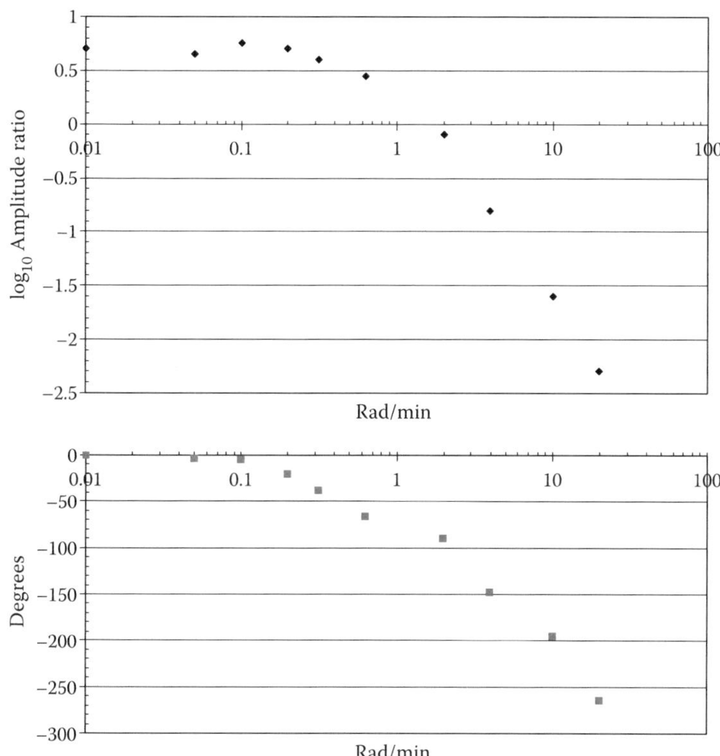

FIGURE 25.39
See Tutorial Question 25.10.

(c) Comment on the behavior of the system if the open-loop system were closed electronically by inserting an amplifier of gain = 10.

25.11 A Bode plot of the open loop analysis of the carotid sinus system is shown below (Figure 25.40):

(a) Is there evidence of transportation lag in the system? If so, estimate its magnitude.

(b) Write down the frequency domain expression for the transfer function.

(c) Is there any evidence of instability in this system? Show reasoning for your answer.

25.12 The following data were obtained by noting the magnitude and phase of systemic pressure variation, in response to small amplitude sinusoidal variations in perfusion pressure:

Period (s)	2000	1000	500	200	100	50	20	10	5
Amplitude Ratio	5.37	4.79	5.01	3.98	2.14	0.87	0.35	0.17	0.06
Phase (deg)	–2	–18	–24	–45	–67	–92	–119	–157	–197

"Discover" the transfer function for this open loop system.

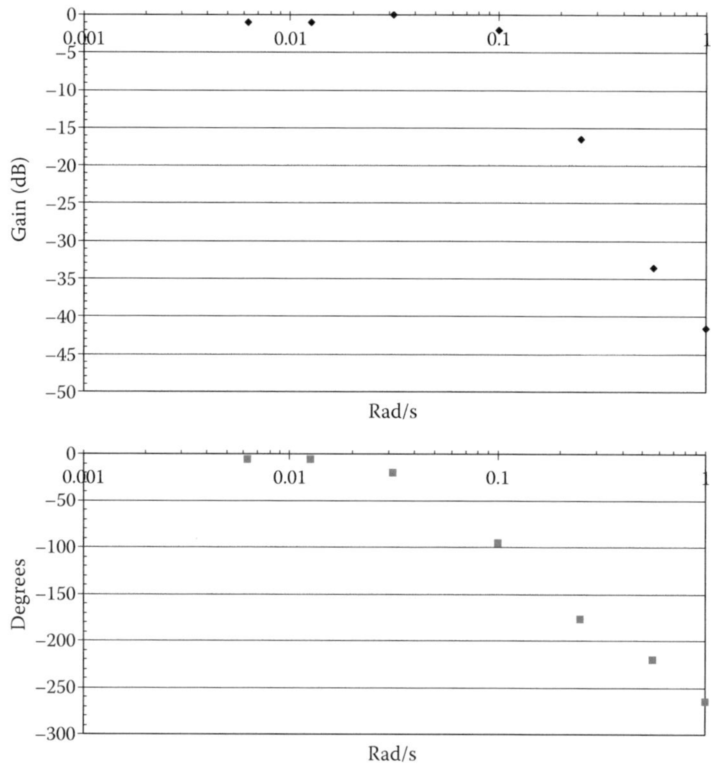

FIGURE 25.40
See Tutorial Question 25.11.

Show that if an extra (constant) gain of approximately 12 were to be introduced into the loop, the closed-loop behavior would become unstable.

25.13 In an isolated segment of carotid artery the pressure was maintained at 100 ± 3 mmHg, at various periods of oscillation. The corresponding pressure amplitudes and phase lags in the aorta are given below:

Period (s)	160	80	30	10	4	1.8	1.0 (seconds)
Press. ampl.	2.97	2.91	3.05	1.69	0.42	0.067	0.017 (mmHg)
Phase lag	4	8	18	98	175	223	265

(i) Plot these results as a Bode plot and discover the transfer function, bearing in mind that a pure time delay is usually involved.

(ii) At a maintained pressure of 35 ± 3 mmHg, the corresponding aortic pressure amplitudes were approximately ten times those given above (i.e., each entry ×10). Comment on the expected closed-loop behavior at this arterial pressure.

25.14 Part of the duodenum is isolated and infused with glucose solution using an infusion pump. The concentration of this solution varies sinusoidally as (1.0 + 0.1 sin ωt) g/L. The blood concentration of glucose is continuously monitored and is found to be of the form (0.8 + A sin(ωt + ϕ)) g/L, where A and ϕ are given below as a function of ω.

ω (rads/min)	0.01	0.03	0.1	0.3	1.0	3.0	10	30	100
A (g/L)	0.79	0.68	0.76	0.45	0.25	0.10	0.03	0.01	0.003
ϕ (lag, in°)	–6	11	21	40	67	73	78	84	86

(a) Express these results in the form of a transfer function.

(b) Explain why this experiment does not really represent the open-loop behavior of the closed-loop blood glucose control system.

(c) Explain how altered transport delays (dead time) or gain might lead to closed-loop instability in the system.

25.15 (a) The iris muscle of the eye forms part of a closed-loop feedback system for regulating the amount of light incident on the retina. Describe these and the other parts of the loop with the aid of a block diagram how can the closed loop system be opened experimentally to determine the (open-loop) transfer function?

(b) Discover the (open-loop) transfer function for a system yielding the following data in open-loop configuration:

Frequency (Hz)	0.001	0.002	0.005	0.016	0.038	0.090	0.160
Amplitude Ratio	0.940	0.960	1.030	0.750	0.150	0.021	0.007
Phase shift (Degrees)	–2	–2	–18	–98	–176	–23.8	–263

(c) If a constant gain of 20 dB were added to the system, would the closed-loop behavior be stable? (Give reasons.)

References

Cuthbertson KS, Chay TR. 1991. Modelling receptor-controlled intracellular calcium oscillators. *Cell Calcium*. 12: 97–109.

Hobbie RK, Roth BJ. 2007. *Intermediate Physics for Medicine and Biology*, 4th Edition. Springer, New York.

Keener J, Sneyd J. 1998. *Mathematical Physiology*. Springer, New York.

Khoo MCK. 2000. *Physiological Control Systems*. Wiley Interscience, Piscataway, NJ, USA.

Lighthill MJ. 1986. *An Informal Introduction to Theoretical Fluid Mechanics*. Cambridge University Press, Cambridge, UK.

Milsum JH. 1966. *Biological Control Systems Analysis*. McGraw Hill, New York.

Ramo S. Whinnery JR, Van Duzer T. 1994. *Fields and Waves in Communications Electronics*, 3rd Edition. John Wiley & Sons, New York.

Riggs DS. 1970. *Control Theory and Physiological Feedback Mechanisms*. Williams & Wilkins, Philadelphia, PA, USA.

Sagawa K. 1971. Use of control theory and systems analysis. In: *Cardiovascular Fluid Dynamics. Volume 1*. DH Bergel (Ed.). Academic Press. New York.

Thompson CJ, Yang, Wood AW. 2001. A mathematical model for the mammalian melatonin rhythm. *Physica A* 296:293–306.

Bibliography

Akay M (ed). 2006. *Wiley Encyclopedia of Biomedical Engineering*, 6-Volume Set. Wiley, Hoboken NJ, USA.

Bronzino JD. 2006. *The Biomedical Engineering Handbook*, 3rd Edition. CRC Press/Tayor & Francis, Boca Raton, FL, USA.

Enderle JD, Blanchard SM, Bronzino JD. 2005. *Introduction to Biomedical Engineering*, 2nd Edition. Elsevier, Burlington, MA, USA.

Fung YC. 1984. *Biodynamics: Circulation*. Springer, (and Fung's other works).

Webster JG (ed.). 2006. *Encyclopedia of Medical Devices and Instrumentation*, 2nd Edition, Wiley, Hoboken NJ, USA.

26

Biomechanics and Biomaterials

Andrew W. Wood

CONTENTS

26.1 Mechanical Properties of Tissue

Biological tissue can be analyzed in a similar manner to nonliving materials, but it is important to preserve its viability as much as possible, so that the measurements carried out in a test-rig in a laboratory reflect the behavior of the tissue in its native state. Invertebrate tissue is easier to test, because it remains viable for longer periods at laboratory temperatures. Most vertebrate tissues need to be maintained at 37°C. All tissue needs to be kept moist with a saline (Ringer's) solution. Important parameters are *stress* (measured in pascal, Pa) and *strain* (which is dimensionless).

26.1.1 Stress in One Dimension

Stress is force per unit area and can be tensile, compressive and shear, as shown below, with the force directions and the relevant area indicated (Figure 26.1). Thus, for shearing stress (which will tend to make the object crack in the direction of the force rather than perpendicular to it) the area is the top or bottom surface rather than the side surface.

Some objects will cleave under a large tensile or shearing force, others deform then restore when the force is removed (elastic behavior), yet others deform and then do not

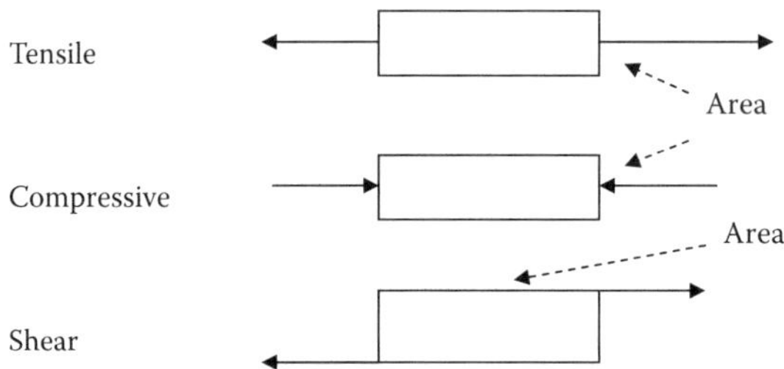

FIGURE 26.1
The difference between the three types of stress which may be experienced by an object, in this case a solid rectangle. The arrows indicate the forces (usually measured in N), which have to be divided by the surface areas indicated (in m²) to calculate the stress (in Pa). Note that in the case of shear stress, the relevant area is of the top or bottom face.

restore (i.e., they show flow-plastic behavior). Compressional stress can cause objects to crumble above a certain threshold.

26.1.2 Wall Stress in Arteries, Heart, etc.

In considering possible rupture of arteries or other tubular structures in the body, it is important to estimate the amount of stress in the wall material.

It can be shown that, for a cylinder, if the thickness of the wall is h, the stress in the wall (which is directed circumferentially) S is given by

$$hS = p_i r_i - p_o r_o;$$

where $p_{i,o}$ and $r_{i,o}$ are the inside and outside pressure and radius, respectively (Figure 26.2).

$$\text{Put } p_i - p_o = p, \quad \text{if } r_i \approx r_o = r,$$

$$hS = pr = T \qquad \text{(known as the law of Laplace)}$$

where T is wall tension (Pa/m). Think of the tension in the wall of a garden hosepipe—a large diameter hose will need to be made out of stronger material to contain a given pressure without bulging. For a sphere, the formula is slightly different, by a factor or 2, hence

$$2pr = T:$$

Two balloons, one large, the other small, when joined have the smaller emptying into the larger one, because if the tension in the balloon material is the same in both, the one with

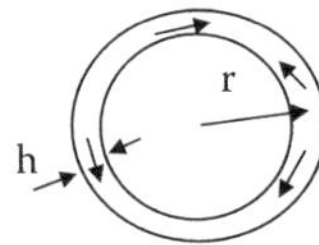

FIGURE 26.2
The law of Laplace in a flexible tube, showing the stress within the wall (thickness h) which depends on the tube radius r as well as the excess pressure inside the tube.

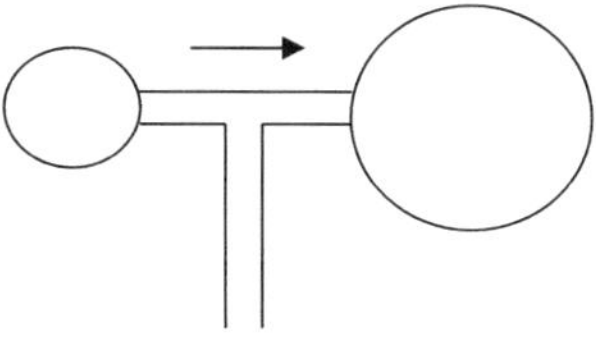

FIGURE 26.3
For a fixed wall tension, the pressure inside smaller interconnecting flexible spheres (such as lung alveoli) will be greater than that in larger ones (via the Laplace law), so the smaller ones will tend to collapse, unless there is a compensatory adjustment in wall tension.

the smaller radius has the higher pressure. In premature infants, surfactant has not been produced in sufficient enough quantities to enable larger alveolar sacs walls to produce higher tension than smaller ones, so the smaller alveoli collapse, giving rise to a much reduced effective area for gas diffusion (Figure 26.3).

26.1.3 Stresses in Three Dimensions

So far, we have only been considering forces applied in one direction, parallel to the long axis of the tissue. Usually (as is the case of the thighbone), the situation is a great deal more complicated. The tensile and compressive stresses can be represented in the three directions as S_{xx}, S_{yy}, and S_{zz}.

The shear stresses are also a little more complicated: they can be represented as S_{xy} S_{xz} and so on—in the case of S_{xy}, x is *normal* to the surface across which the shear force acts and y is the direction of the shearing force (Figure 26.4, upper).

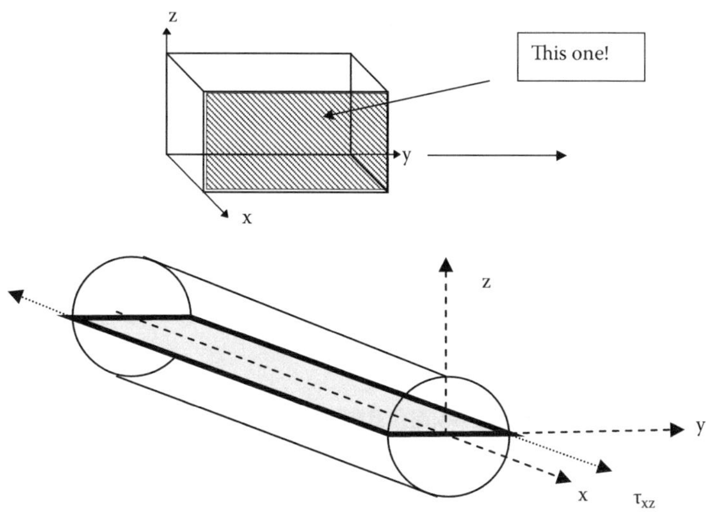

FIGURE 26.4
Upper: see text. When defining shear stresses, it is important to be clear which surface is referred to in subscript notations. Lower: planar analysis where there is an axis of symmetry, such as in a long bone.

26.1.4 Strain

Strain is the fractional deformation. For tension or compression $\varepsilon = \Delta x/x_o$, where x_o is the initial length and Δx is the amount of elongation or diminution.

For moderate stresses, many objects obey Hooke's law, or

$$S_{xx} = E_{xx}\varepsilon_{xx},$$

where E is the Young's modulus in that particular direction (in general it will be different in different directions). The Young's modulus for bone is about 1/10 that of steel.

The relationship between shear stress and strain is as follows:

$$S_{xy} = G_{xy}\tan\theta,$$

where θ is the angle of deformation and for small angles, $\tan\theta \approx \theta$.

26.1.5 Viscosity

Viscosity is fluid resistance to flow due to shearing forces, usually given as rate of shear:

$$du(z)/dz,$$

where u(z) is velocity as a function of distance z (see Chapter 9).

In a Newtonian fluid the stress is proportional to rate of shear, i.e.,

$$S_{zx} = \mu\ du(z)/dz;$$

where μ is the coefficient of viscosity (in Pa s). Note that the direction of the stress is parallel to the streamlines in the fluid (think of two plates with butter in between, one plate is stationary, the other moved at constant velocity with constant force).

In the study of viscoelasticity, which will be considered below, a different measure of viscosity, the stress per unit rate of strain, is used (η has the same units as μ):

$$S_{xx} = \eta\ d\varepsilon_{xx}/dt.$$

In this case, the stress and strain are in the same direction, as they are when pulling on a rubber band. Here we consider the rate of extension, rather than the value itself.

26.1.6 Bulk Modulus

The bulk modulus K is how much fractional decrease in pressure p is produced per fractional increase in volume (dV/V_0)

$$K = -dp/(dV/V_0).$$

For an elastic solid (rather than a gas), it is easier to think of the inverse, for example if a block of rubber were to be put into a pressurized vessel, it would decrease in volume.

26.1.7 Poisson's Ratio

As an object elongates in one direction it gets thinner in orthogonal directions, i.e.,

$$\nu = -\varepsilon_{yy}/\varepsilon_{xx} \approx \varepsilon_{zz}/\varepsilon_{xx} = 0.19 \text{ for steel, } 0.12 - 0.63 \text{ for bone.}$$

The various moduli are interrelated:

$$E = 2G(1 + \nu) = 3K(1 - 2\nu) \approx 3G(1 - G/(3K)).$$

If the material is nonisotropic (i.e., properties are different in different directions, such as in muscle and bone) these relationships are more complicated.

26.1.8 Matrix Analysis of Stress–Strain Relationships

We can combine tensile and shear stress–strain relationships by writing the following general expression:

$$S_i = c_{ij}\varepsilon_j$$

where the c_{ij} represent the influence of a strain in one direction with stresses in other directions. In general, we have three each of normal and shear stresses and a similar number of strains. This is simplified if there exists an axis of symmetry (such as the long axis of bone) or if the analysis can be restricted to a single plane (such as the mid-plane of the tibia, say). These analyses are approximate, but often useful to perform prior to a full analysis. Matrix-based programs such as MATLAB greatly facilitate analysis.

As an example of a planar analysis, consider the plane shown in the lower part of Figure 26.4. Representing the normal stresses along the x and y axis as σ_x, σ_y, and the corresponding strains as ε_1, ε_2, and the single shear stress and strain as τ_{xy} and γ_{xz}, we can write:

$$\begin{vmatrix} \sigma_x \\ \sigma_y \\ \tau_{xz} \end{vmatrix} = \begin{vmatrix} c_{xx} & c_{xy} & 0 \\ c_{xy} & c_{yy} & 0 \\ 0 & 0 & c_{xz} \end{vmatrix} \begin{vmatrix} \varepsilon_x \\ \varepsilon_y \\ \gamma_{xz} \end{vmatrix}.$$

Thus, four coefficients c_{xx}, c_{yy}, c_{xy}, and c_{xz} (shear modulus) are required to fully specify the relationships. Remember, too, that the moduli are related via the Poisson's ratio.

26.1.9 Hysteresis

For many biological materials, the stress for increasing strain is different from stress for decreasing strain: (in the lungs a greater pressure is required to produce a certain lung volume as volume is increased compared to when the volume is decreased; the stress in artery walls is greater for increasing strain than for decreasing strain). The area of the loop equals work done in overcoming viscoelasticity (Figure 26.5).

26.1.10 Viscoelasticity

In some tissues (e.g., muscles), the stress depends on the rate of strain as well as the strain (see discussion of viscosity above). This is exploited in car shock absorbers, which damp

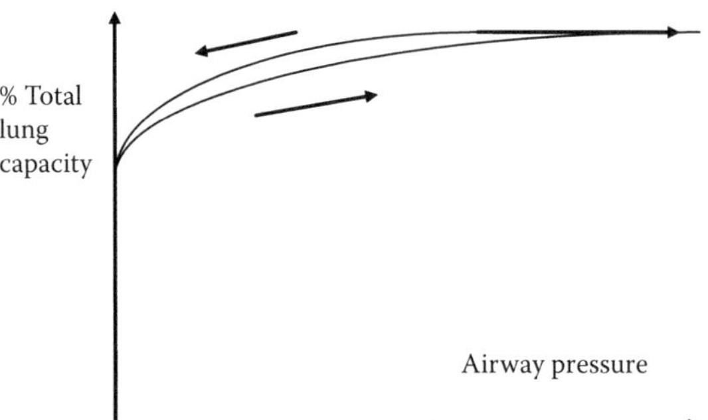

FIGURE 26.5
Hysteresis in lung mechanics. At a given lung volume, the pressure is much higher on inflation than deflation of the lung. The enclosed area is a measure of the work done during the breathing cycle.

out the effects of some of the sudden changes in position, but slow changes do not meet so much resistance. Shock absorber units consist of combinations of springs and pistons in oil, commonly known as dashpots. The oil serves to damp the sudden movements and to inhibit the movement of the spring. The elastic (or spring) elements of muscle tissue are the tendons supporting connective tissue (epimysium, perimysium, endomysium). The contractile, sliding, elements (corresponding to the oil in the dashpot) are the myofibrils.

The simplest model has these two elements in series (the Maxwell model shown in Figure 26.6). The equations are

$$\text{Spring: } \sigma_S = E\varepsilon_S$$

$$\text{Dashpot: } \sigma_D = \eta d\varepsilon_D/dt.$$

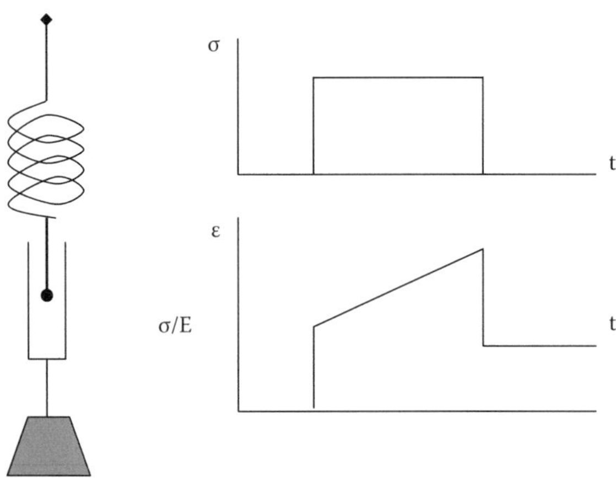

FIGURE 26.6
Maxwell model of a viscoelastic material (dashpot and spring in series). The Creep test (or isotonic test) involves a weight attached to the end of the object (constant stress), which then elongates (strain). When the weight is removed there will be a partial return to initial length as the spring contracts to its initial state.

Total stress $\sigma = \sigma_S = \sigma_D$ (because they are in series).

The rates of strain add: i.e., $d\varepsilon/dt = d\varepsilon_S/dt + d\varepsilon_D/dt$.

The characteristics of materials, including bio-materials, are commonly investigated by at least two tests.

The *creep test* consists of hanging a weight on the material at t = 0 (to produce an initial stress σ_0) and then measuring the strain with time before removing the weight again (this, in physiological terms, is isotonic extension).

$$\sigma_0 = E\varepsilon_{S0} = \eta\, d\varepsilon_D/dt.$$

Thus, $d\varepsilon/dt = d(\sigma_0/E)/dt + d\varepsilon_D/dt = 0 + \sigma_0/\eta$.

Integrating:

$$\varepsilon = \sigma_0 t/\eta + \text{const} = (\sigma_0/E)[1 + t/\tau],$$

where $\tau = \eta/E$.

In the *relaxation test* the object is stretched (providing constant strain) and then the stress $\sigma(t)$ is monitored (using a force transducer).

After initial stress $d\varepsilon/dt = 0$, so $d\varepsilon_S/dt = -d\varepsilon_D/dt$, i.e., $d(\sigma/E)/dt = -\sigma/\eta$.

Thus, $d\sigma/\sigma = -(E/\eta)dt$, which integrates to give

$$\sigma = \sigma_0 e - t/\tau, \text{ where } \tau = \eta/E, \text{ as previously.}$$

Figure 26.6 is a diagram of the behavior of Maxwell model in the "creep test": as mentioned, this is accomplished by hanging a weight then later removing it: elongation (strain) is measured. Figure 26.7 shows the Maxwell Model behavior in the relaxation test. The

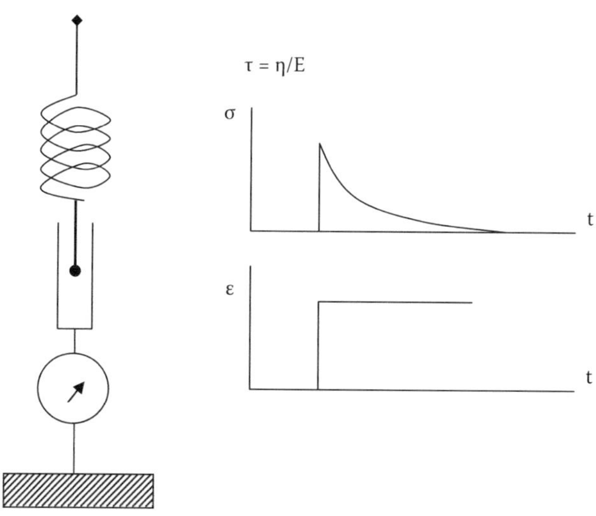

FIGURE 26.7
Maxwell model. Relaxation test (isometric) where the material is pulled down and then held (constant strain). The measurement in this case is of tension (stress) as the spring contracts to its initial state by pulling up the dashpot piston, the stress reduces.

end of the material is pulled down then held: the tension (or stress) in the material is then measured under these conditions of constant length (isometric). Note that at $t = \tau$, the value of σ is 37% of the initial value: this often simplifies analysis.

The *Voigt* model has the spring and dashpot in parallel. The *Kelvin* or hybrid model is favored in biomechanical work: it is similar to the Voigt model, except that the parallel element (PE) is in parallel with both the Series Element (SE) and dashpot (now called the Contractile Element, CE).

The tendon material is represented by a series element SE, and the connective tissue by a parallel element PE. The stress balance equation (representing SE and PE by subscripts S and P, respectively) is

$$\sigma + (\eta/E_S)d\sigma/dt = E_P(\varepsilon + \eta(1/E_S + 1/E_P)d\varepsilon/dt).$$

The progression of strain with time in the creep test ($d\sigma/dt = 0$) is given by the formula

$$\varepsilon(t) = (\sigma_0/E_P)\,[1 - \{E_S/(E_S + E_P)\}\exp(-t/\tau_1)];$$

where $\tau_1 = \eta(E_S + E_P)/(E_S E_P)$.

In the stress relaxation test ($d\varepsilon/dt = 0$),

$$\sigma(t) = \varepsilon_0(E_S\exp(-t/\tau_0) + E_P),$$

where $\tau_0 = \eta/E_S$.

Figures 26.8 and 26.9 are diagrams of the behavior of the Kelvin Model in the creep and the relaxation test, respectively.

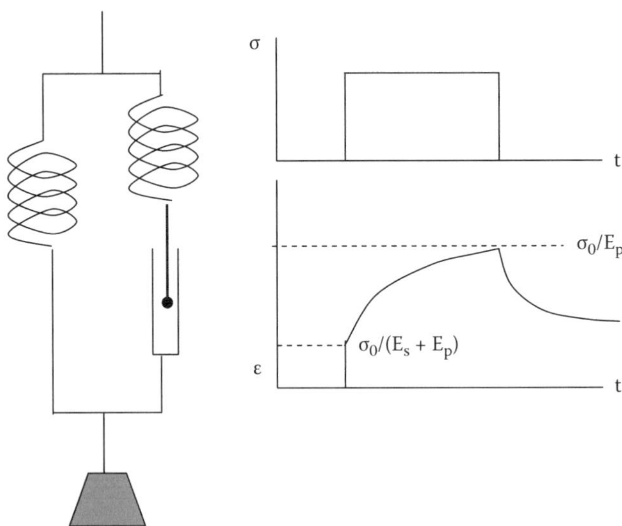

FIGURE 26.8
The Kelvin model, with a parallel elastic component. Creep test as in Figure 26.6.

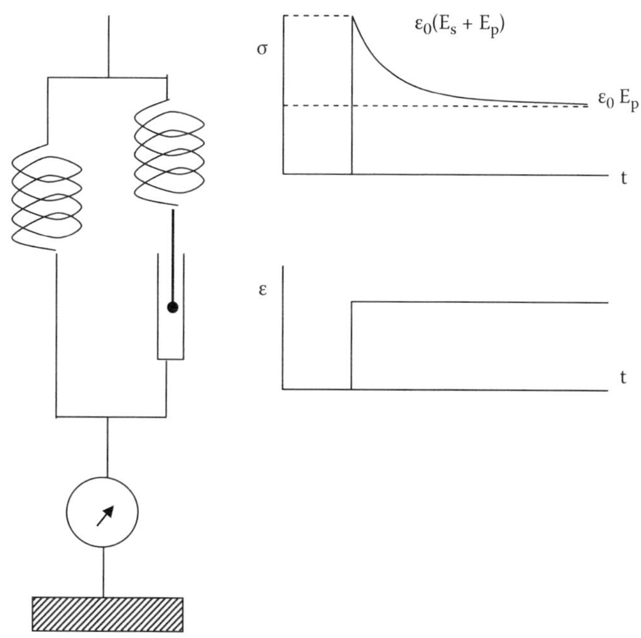

FIGURE 26.9
The Kelvin model Relaxation Test: compare with Figure 26.7.

WORKED EXAMPLE

A strip of human cartilage 30 mm long increases to 34 mm, 5 minutes (effectively $t \to \infty$) after 100 g mass is added to the bottom edge. The strip is 3 mm wide and 1.5 mm thick and is approximately rectangular in cross section.

(a) Estimate the Young's (elastic) modulus of the parallel element.
(b) If the series element elastic modulus has the same value as the parallel element, what is the length of the strip immediately after the mass is added to it (and before further extension has occurred)?
(c) If the viscosity is 20 MPa s, at what time will the length of the strip be 33 mm?

Answer

This is an example of a "creep test" so we can examine how the cartilage extends with time, using $\varepsilon(t) = (\sigma_0/E_P)[1 - \{E_S/(E_S + E_P)\}\exp(-t/\tau_1)]$. If we put $t = \infty$, then $\varepsilon(\infty) = \sigma_0/E_P$, or $E_p = \sigma_0/\varepsilon(\infty)$. σ_0 is force/area $= 0.1 \times 9.81/(3 \times 10^{-3} \times 1.5 \times 10^{-3}) = 0.22$ MPa and $\varepsilon(\infty)$ is fractional increase in length $= 4/30 = 0.133$. So $E_p = 0.22/0.133 = 1.6$ MPa.

For the next part, we put $t = 0$ and note that $\varepsilon(0) = (\sigma_0/E_P)[1 - \{E_S/(E_S + E_P)] = (\sigma_0/(E_s + E_p)) = 0.22 \times 10^6/(2 \times 1.6 \times 10^6) = 0.069$. So the actual initial extension will be $0.069 \times 30 = 2$ mm.

Finally, we need to work out a value for t at which $\varepsilon(t) = 3/30 = 0.1 = (4/30)[1 - 1/2 \times \exp(-t/\tau_1)]$. Rearranging gives $\exp(-t/\tau_1) = 2(1 - ¾) = 0.5$. Taking logs of both sides, $-t/\tau_1 = \ln(0.5) = -0.693$. Now, since $\tau_1 = \eta(E_S + E_P)/(E_S E_P) = 20 \times 10^6 \times (3.2 \times 10^6)/(1.6 \times 10^6)^2 = 25$ s. Thus, the time $t = 25 \times 0.693 = 17$ s.

26.2 Biomaterials

26.2.1 Tensile and Compressive Stress and Strain

In investigating natural materials and artificial materials designed to replace them, data on mechanical properties needs to be collected. For example, in the case of bone, a sample can be taken and exposed to both compressive and tensile stress whilst observing changes in dimension. In the simplest form of analysis a weight can be hung on one end of the bone and the minute changes in length measured using a sensitive instrument (in Figure 26.10, this is a traveling microscope). The strain is the fractional increase in length and the stress (in Pa) the mass times gravitational acceleration divided by the cross-sectional area of the bone. The behavior is shown on the upper diagram of Figure 26.10, with a transition at a certain stress (the Yield Stress, σ_{yield}) from an Elastic Region (where the bone returns to its original length) and above this, a plastic region, where deformation of the bone occurs and the initial length is not restored, because of fundamental irreversible changes in structure (due to microfractures or other disruptions). Eventually, the bone will fracture (Ultimate Stress σ_{ult}). Similarly, under compressive stress there will be a limit (σ_{ult}) where the bone crumbles. For bone, σ_{ult} has values of 0.09 and 0.18 for tension and compression, respectively. The Young's modulus is given by tan ψ, and is around 14 GPa for bone and 200 GPa for steel. If the bone is repeatedly loaded and unloaded beyond σ_{yield}, then the initial strain

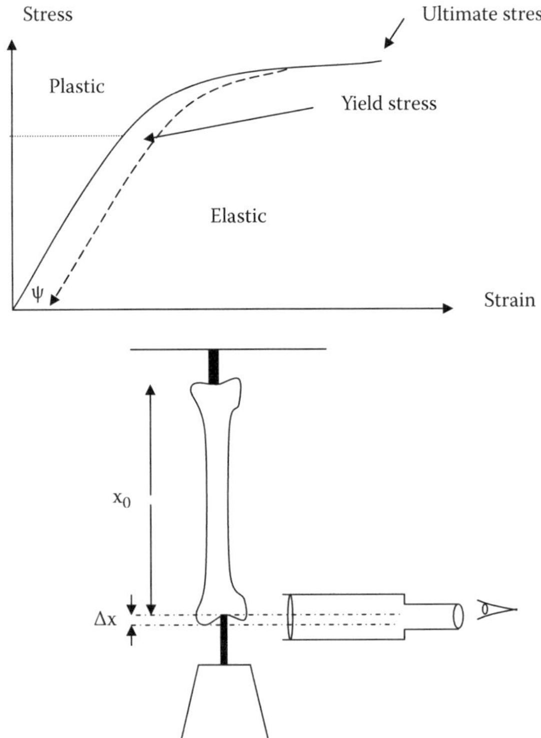

FIGURE 26.10
Investigating the viscoelastic properties of bone. The lower diagram shows the method for measuring the minute elongation caused by the addition of the mass shown. Top diagram shows that above a certain yield stress the bone begins to show plastic behavior (due to microcracking).

increases each time (see dotted line in Figure 26.10) until eventually the bone will fracture. Bone has an interesting structure, consisting of a succession of concentric lamellae (or layers) around a central blood vessel (the Haversian Canal), with each layer separated from the next by longitudinal collagen fibers (which are organic) alternating with inorganic hydroxyapatite crystals ($Ca_{10}(PO_4)_6 2H_2O$), with dimensions 4 × 20 × 50 nm (the long axis parallel to the long axis of the bone). This is often referred to just as apatite. It is soluble in acid (such as 0.5 M HCl) and treatment of bone with acid will cause the bone to be more plastic in its behavior (and the Young's modulus to decline markedly). With aging, the σ_{ult} value declines and bone thus becomes more brittle.

26.2.2 Artificial Materials

The function of an organ or of parts of particular tissue can be performed by a *prosthesis* in the event of disease that is unresponsive or unavailable to treatment. A prosthesis can be used either to replace or to supplement the functions of living tissue. Examples include heart valves, bone implants, breast implants, artificial limbs, and so on. By their nature, prostheses are constructed of materials foreign to the body (although this is not always the case, as in grafts and transplants), and suitable materials are termed *biomaterials.* The essence of successful biomaterials is that they function correctly in contact with biological tissue over many years. Their ability to do so is termed *biocompatibility,* of which there are a number of components. The first of these is mechanical strength and elasticity. The following table gives some values of mechanical properties of tissues compared with those of common materials used in prostheses.

Values (GPa)	Young's Modulus	Compressive σ_{ult}	Tensile σ_{ult}
Bone (wet)	15.2	0.15	0.09
Stainless steel	193		0.54
Ti-Al_6-V (porous)	27		0.14
UHMWPE*	0.7	0.5	0.05

*Ultrahigh molecular weight polyethylene.

Although the values are not identical, the materials in general behave adequately, but there are occasions where this difference becomes significant, particularly where effects such as metal fatigue are involved. Second, biomaterials need to exhibit adequate resistance to corrosion, since tissue naturally attempts to break down foreign material, including metals. This is particularly important in the case of implants which are designed to last for decades, such as artificial hip and knee joints and indwelling pacemakers. Active devices (such as pacemakers and other stimulators) need an adequate power supply, since it is not convenient to have to surgically change batteries. Some of the newer batteries offer a recharge facility using external power input (such as wireless radiofrequency links) and even nonrechargeable batteries are now designed to operate for over ten or so years, by a combination of newer battery technology and smaller current demand of integrated circuits.

In addition to the need for a biomaterial adequate to the task, the other aspect is whether or not the biological system itself is adversely affected. Materials are selected on the basis of minimal tissue response (in terms of immune rejection, ingestion or inflammation) and of minimal blood response (in terms of thrombus formation). The latter can be accomplished by ensuring a smooth surface with positive electrical charge.

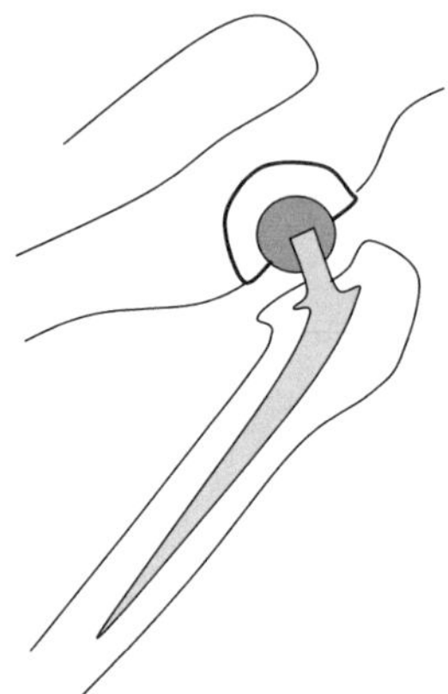

FIGURE 26.11
Hip prosthesis: polyethylene acetabular cup, ceramic ball, and titanium stem and head (Charnley design).

Typical artificial materials to supplement or replace function of living tissue include metals, polymers, ceramics, and carbon fibers. For example, a hip prosthesis can consist of a Teflon-lined metal alloy cup and a metal alloy ball which can be coated with sintered glass (Figure 26.11). The assembly is held into the surgically-shaped bone by bone cement, which is usually poly-methyl-methacrylate (PMMA—liquid perspex or lucite). There is a vast array of alloys used for the metallic parts, such as Co-Ni-Cr-Mo and Ti-Al_4-V alloys as well as pure titanium. Porous or roughened surfaces are important to encourage cell adhesion. There are special cell-adhesion molecules (CAMs) which are transmembrane receptors at the cell surface. These can be sensitive to calcium concentration and have an important role in the binding of prostheses into surrounding tissue. Semiconductor material (porous silicon nitride) is now increasingly finding uses as an implant because of its similarities in structure to bone and because of the availability of a wide range of manufacturing techniques carried over from the electronics industry. Conductive organic polymers are also being trialled because of the possibility of using the conductive properties in the restoration of nerve or muscle function.

A number of natural materials, such as mussel polyphenolic protein glue, liposomes, enzymes, receptors, bacterially-derived membranes in transplants and explants are increasingly being exploited. There is on-going research into the implantation of laboratory grown organs (in which, for example, urinary bladder cells are allowed to grow on an inert carbon fiber bladder-shaped matrix in tissue culture for later surgical replacement of the patient's own bladder). There are prototype heart valves using the same technique: the patient's own endothelial cells are allowed to grow on a valve-shaped matrix.

Tutorial Questions

1. (a) What are the essential requirements to ensure successful bone implantation (for the hip joint, for example)?

   (b) A 3 kg weight is applied to a strip of heart muscle 50 mm long, 10 mm wide and 2 mm thick. The initial elongation is 10 mm, but then after a further 100 seconds the elongation is 16.3 mm, then stabilizing at 20 mm after several minutes.

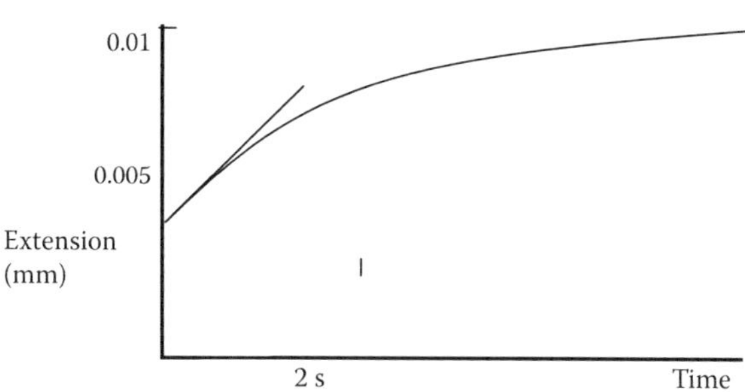

FIGURE 26.12
See Tutorial Question 2.

Starting from this length, the muscle is then attached to a force transducer (the weight having been removed). The force was observed to fall to a constant value, taking 50 s to fall to 37% of the difference between the initial and final values.

Estimate the elastic and plastic (or viscous) moduli for this system, assuming the Kelvin model holds.

2. A new material is to be used as a replacement for broken chordae tendinae (which tether the heart valve leaflets).
   (a) Describe the tests which would be carried out to determine its (i) Yield stress (ii) Poisson ratio.
   (b) A sample of this material is 3 cm long and 2 mm² in cross section. Hanging a 100g weight produces an extension of 0.01 mm after 1 minute (see Figure 26.12).
      (i) Show that the elastic modulus E_p is 1.47 GPa.
      (ii) If the extension varies with time as shown in Figure 26.2, show that $E_s = E_p$ and further determine the value of η.
3. Explain the terms in the following planar analysis of stresses on bone

$$\begin{vmatrix} \sigma_1 \\ \sigma_3 \\ \tau_{12} \end{vmatrix} = \begin{vmatrix} c_{11} & c_{13} & 0 \\ c_{13} & c_{33} & 0 \\ 0 & 0 & c_{12} \end{vmatrix} \cdot \begin{vmatrix} \varepsilon_1 \\ \varepsilon_3 \\ \gamma_{12} \end{vmatrix}$$

A sample of ligament is elongated by 1% and held there. An initial stress of 0.2 GPa is recorded, falling eventually to 0.15 GPa. The time constant of this process is 4 s. Determine E_S, E_P, and η, assuming a Kelvin model.

Bibliography

Ethier CR, Simmons CA. 2007. *Introductory Biomechanics*. Cambridge University Press, Cambridge.
Fung YC. 1990. *Biomechanics: Motion, Flow, Stress, and Growth*. New York: Springer.
Fung YC. 1993. *Biomechanics: Mechanical Properties of Living Tissues*, 2nd Edition. New York: Springer-Verlag.

27

Biosensors and Emerging Technologies

Andrew W. Wood

CONTENTS

27.1 Biosensors

27.1.1 What Is a Biosensor?

These are biological sensors that have (1) a biological recognition component (selectivity) and (2) a supporting structure in intimate contact with a biological component.

They are usually miniature and include monitoring of blood gas and biochemical concentrations: also physical quantities such as force, shear, potential, etc. They can be divided into those that measure electrical, and those that measure optical properties. We will deal with electrical properties first.

27.2 pH Meters and CO_2 Electrodes

The earliest biosensors developed were those for measuring blood gases and pH. These are miniature versions of the types of electrodes used for measuring these quantities extracorporeally. The CO_2 electrode is itself based on the pH meter whose operation will now be explained.

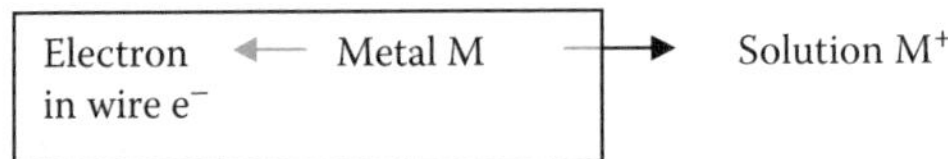

FIGURE 27.1
Representation of metal–solution interface. The rectangle represents the metal electrode in equilibrium with metal ions in solution, with electrons ready to flow in an external circuit via a connecting wire.

At a metal–solution interface there is an equilibrium reaction between metal ions in solution and the metal of the electrode itself (Figure 27.1).

$$M \leftrightarrow ze^- + M^{z+}.$$

Because this reaction represents separation of charge, an "electrical double layer" (a layer of + charges and a layer of – charges) develops, whose potential difference V across is given (see Chapter 3) by

$$V = V_0 + (RT/(zF)) \ln a_M$$

where V_0 is a standard chemical potential that depends on the metal, R is the gas constant, T is the absolute temperature, z is the valence of the metal, F is Faraday's constant, and a_M is the activity of the ion M^{z+}. The activity is more or less equal to the concentration for dilute solutions, but there could be a slight error in making this approximation. If there are two electrodes dipping into separate solutions where the activities are a_{M1} and a_{M2}, and these solutions are joined in some way, the pd between the two electrodes will be

$$\Delta V = (RT/zF)(\ln a_{M1} - \ln a_{M2}) = (RT/zF) \ln (a_{H1}/a_{H2}).$$

The technique is to join the two solutions so that only one specific ion can get from one to the other. Thin glass membranes can be made specific to particular ions, such as H^+, Na^+, etc. Since pH is the negative $\log_{10}$ of $[H^+]$, H^+ selective glass is the basis of a pH meter (Figure 27.2).

$$\Delta V = (RT/zF)(\ln a_{H1} - a_{H2}) = 58 (\log a_{H1} - \log a_{H2}) \text{ (in mV) for pH.}$$

Since $pH = -\log_{10} a_H$, we get

$$\Delta V = \text{constant}_1 + \text{constant}_2 \times pH, \text{ where constant}_2 \text{ is 58 mV at approximately 18°C.}$$

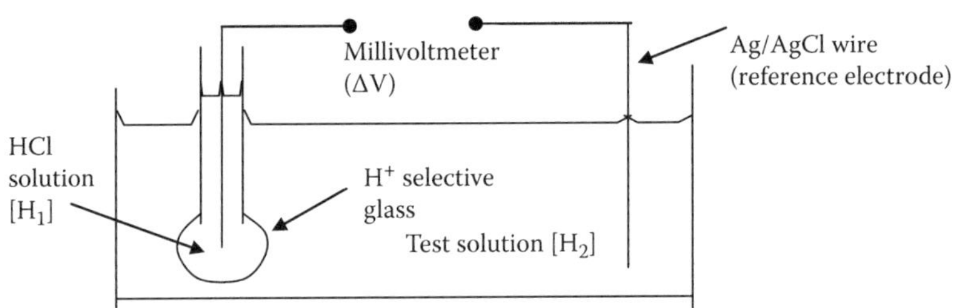

FIGURE 27.2
The pH meter. The electrode on the left is essentially a hydrogen electrode, because of the H^+ selective glass membrane, and the PD is measured relative to a reference electrode on the right.

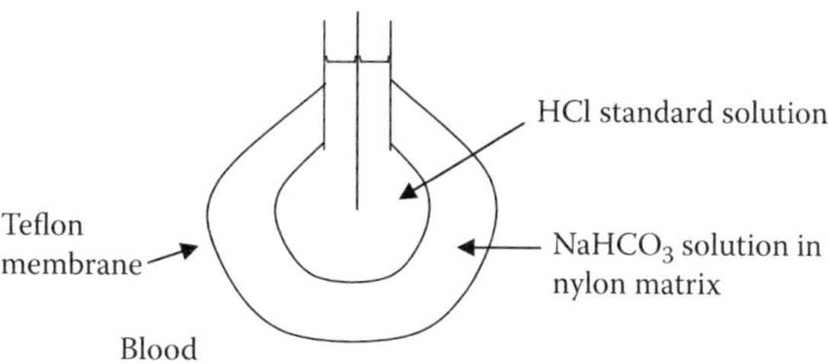

FIGURE 27.3
The CO_2 electrode, which is essentially a pH meter electrode surrounded by an immobilized $NaHCO_3$ solution, which reacts with dissolved CO_2 such that pH changes reflect the [CO_2] in blood or other test solutions.

The older method for measuring blood pH is to pass it through a glass capillary, where the surrounding sleeve contained standard H^+-buffered solution.

The CO_2 electrode is basically a pH electrode with the bulb enclosed by a CO_2 permeable Teflon membrane (Figure 27.3): As CO_2 diffuses through the Teflon, it reacts with water to produce the following equilibrium:

$$CO_2 + H_2O \leftrightarrow H_2CO_3 \leftrightarrow H^+ + HCO_3^-.$$

Thus, if blood [CO_2] (i.e., partial pressure, or pCO_2) rises, [H^+] will rise by a similar amount, thus changing ΔV as before. This is called a voltamic technique.

27.3 Blood (or Other Tissue Fluid) Oxygen

The O_2 electrode, or Clark electrode, is a polarographic (i.e., current measuring) technique. It makes use of the fact that if a reaction is occurring at an electrode, this will limit the actual electrical current that the electrode can draw, since if the number of charge carriers being produced can only be done so in a diffusion-limited process, this, and the concentration of these carriers, will limit the charge transfer, hence current. This "limiting current density" J_{lim} (see Chapter 3) is given by the following:

$$J_{lim} = zFDc/\lambda$$

where z is valence, F is Faraday's constant, D is diffusion coefficient, c is concentration (of O_2 for example), and λ is the diffusion layer thickness.

It is found experimentally that 0.6 V is sufficient for the current to reach its limiting value (Figure 27.4). The Clark electrode consists of a polythene (O_2 selective) membrane and a limiting current passed between platinum and Ag/AgCl electrodes at constant voltage (Figure 27.5). The electrode reaction is

$$O_2 + 2e^- + 2H^+ \leftrightarrow H_2O_2 \text{ or alternatively } O_2 + 4e^- + 2H_2O \leftrightarrow 4OH^-.$$

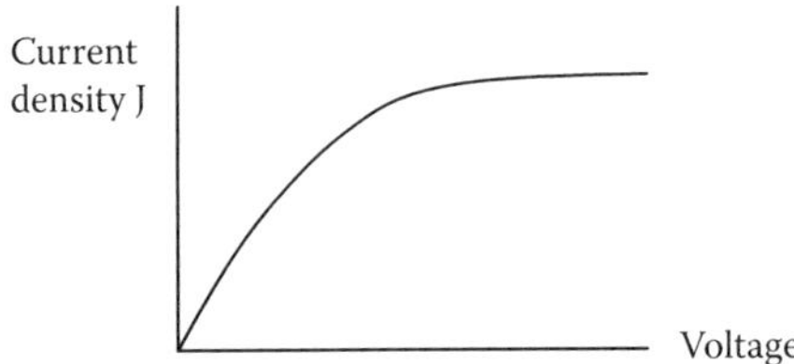

FIGURE 27.4
Current density (J) drawn from an electrode as a function of voltage. The value of J saturates above a certain voltage. This represents a situation in which ions are not able to diffuse away (or toward) the electrode sufficiently quickly to allow further increases in J. This "limiting current density" JL is related to the concentration of critical ions in the immediate vicinity of the electrode surface, thus JL can be used to measure this concentration.

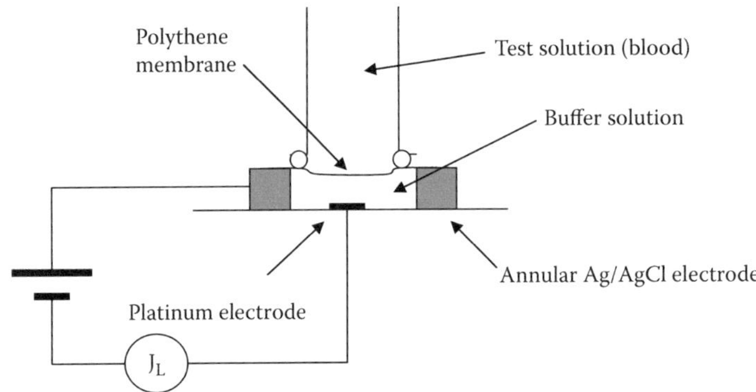

FIGURE 27.5
The Clark O_2 electrode. The limiting current density J_L is a measure of dissolved O_2 in the test solution.

The transcutaneous pO_2 sensor incorporates a heating coil, which causes O_2 to cross the skin from the underlying arterial blood, and this O_2 will then occupy the space between a platinum and an Ag electrode.

By incorporating the enzyme glucose oxidase (GOD), the rate of O_2 consumption is also a measure of glucose present in the sample.

27.3.1 Field-Effect Transistor-Based Biosensors

In a FET, the drain current I_D is very sensitive to gate voltage V_G ($I_D = k(V_G - V_G')^2$ over a particular range). The trick here is to make V_G depend on an electrochemical equilibrium (Figure 27.6).

$$V_G = V_G' - V_{REF} + V_0 + (RT/(zF))\ln a_i'.$$

The chemically sensitive membrane is created via the deposition of ion or other molecule-specific membranes at the gate of an insulated gate FET (or IGFET) to produce an ISFET (ion-specific FET) or ENFET (enzyme FET). This allows miniaturization, but membrane stability remains a problem. Some analysates with their specific selective membranes for their measurement are as follows:

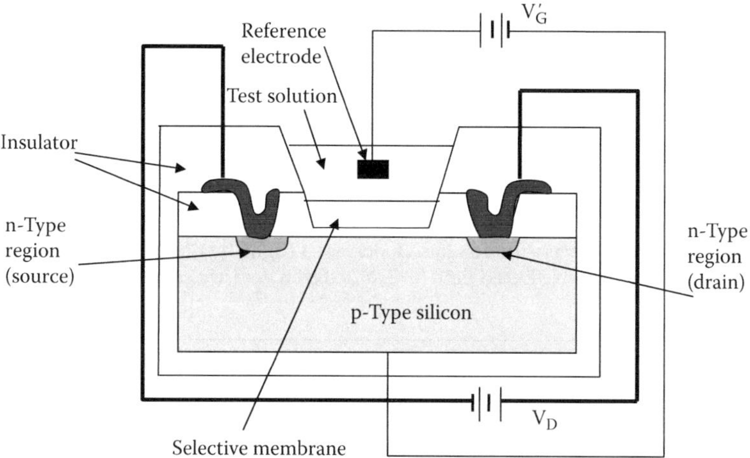

FIGURE 27.6
A CHEMFET. The effective drain voltage V_G is V'_G minus the electrochemical potential developed across the membrane that is specifically selective to the ion whose concentration is to be measured. The value of V_G controls the source–drain current, which can be calibrated in terms of ion concentration. (Diagram redrawn from Turner, A.P.F., Karube, I., and Wilson, G.S., *Biosensors*, 1987, by permission of Oxford University Press.)

(a) pH: layer of SiO_2, Si_3N_4, Al_2O_3, ZrO_2, Ta_2O_5
(b) Br^-: layer of AgBr
(c) Na^+: Al or borosilicate glass
(d) K^+: >50 μm solvent-casted PVC membrane with valinomycin (ionophore)
(e) Ca^{2+}: as above for K^+, but with tHODPP as ionophore
(f) Cl^-, I^-, CN^- : Ag salts in polymer matrix (i.e., Ag/AgCl)

Some enzymes used in ENFETs (with product whose concentration is monitored in brackets) are

(a) penicillinase (H^+)
(b) urease (NH_4^+)
(c) glucose oxidase (via H^+)
(d) acetylcholinesterase (H^+)

Antibody binding of antigen can give rise to a PD, which is exploited in immunochemically sensitive PETs. Gas concentrations (such as H_2) can be detected using a 200-nm layer of palladium on to the SiO_2.

27.4 Optical Biosensors

Most are related in some way to the Beer–Lambert law, which is that at a given wavelength λ, the intensity of light I decreases exponentially within a uniform sample, such that

FIGURE 27.7
Fiber-optic measurement of oxygen saturation in a blood vessel (right) by illuminating red cells with a mixture of light at 650 nm and 805 nm. The reflected light is recorded by a detector as shown. See also Chapter 11.

$$I = I_0 \exp(-\alpha(\lambda)C\Delta x)$$

where C is the concentration of a particular absorbant, $\alpha(\lambda)$ is the absorptivity at wavelength λ, and I_0 and I are the light intensities at the entrance and exit, respectively, of a slab of material Δx thick (see also Section 11.7). Often the light is delivered and collected using two optical fiber bundles in a catheter, with an LED or laser diode as a source and as a detector a photoresistor or photodiode.

An example is oxygen saturation using two wavelengths, red (650 nm) and infrared (805 nm). At 805 nm the $\alpha(\lambda)$ for oxygenated and reduced forms of hemoglobin is the same and at 650 nm they are maximally different. For reflectance oximetry,

$$\text{oxygen saturation (\%)} = k_1 - k_2(I_{805}/I_{650}) \text{ (Figure 27.7).}$$

In evanescent wave spectroscopy, an evanescent wave from the exposed core of the optical fiber penetrates less than a micron into surrounding tissue. If antibodies are immobilized on to the core and antigen is made to react with this, subsequently added fluorescent antibody (2) will attach to antigen already immobilized on antibody (1) (Figure 27.8).

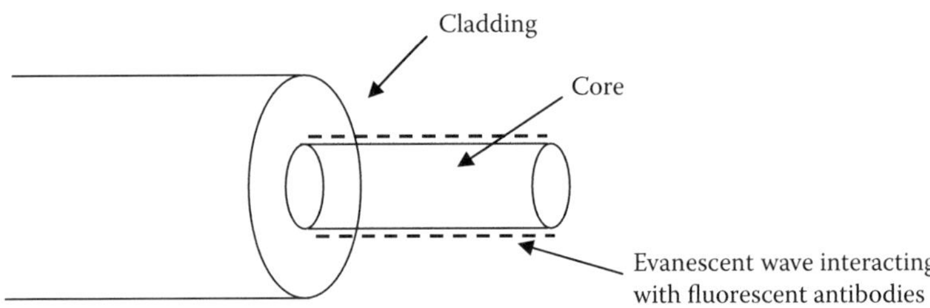

FIGURE 27.8
Fiber-optic-based biosensor, involving interaction of evanescent wave with surface adsorbed fluorescent antibodies.

27.5 Emerging Technologies

This section will outline some of the biomedical technologies that are in some sense emerging, that is, their use in medical diagnosis or therapy has yet to be fully exploited. Since this topic, by its nature, will rapidly become superseded by new developments, the descriptions will be brief and students are encouraged to read further to gain up-to-the minute information on particular technologies. The examples presented are highly selective, and it is hoped to add to this list via supplemental webpage material.

27.5.1 Optical Tweezers and Scissors

A light beam carries with it a momentum that for each photon is h/λ, where h is Planck's constant (6.63×10^{-34} J s) and λ is the wavelength, which for red light is 600 nm. If the photon flux in a light beam is, for example, 10^{26} photons/m^2/s, then the momentum of a beam of 1 mm^2 area is $10^{20} \times 6.63 \times 10^{-34}/(600 \times 10^{-9}) \approx 10^{-7}$ J/m^3. A related quantity is the "radiation pressure" of a beam of electromagnetic radiation, given by 2I/c for complete reflection, where I is the intensity of the beam in W/m^2 and c is the velocity of light. For a class IV laser, with a beam intensity of 1 W/mm^2, the radiation pressure will be 0.3×10^{-8} N/mm^2. This may appear small, but the gravitational down-thrust on a red blood cell is given by $\Delta\rho(4/3)\pi r^3 g$, or a few pN. If the beam strikes a surface at an angle, the radiation pressure is $2I \cos\theta/c$, so as the angle of incidence increases, the radiation pressure falls.

A cell membrane, because of the different refractive index compared to the bathing solution, will tend to refract a laser beam impinging on it. In the left side of Figure 27.9, with a parallel laser beam impinging on the left side of the cell, the cell is seen to gain a lateral component of momentum. It will tend to move to the left, because the light will gain an overall momentum to the right because of the way the different paths are refracted. Similarly, when a focused beam impinges centrally on the cell, the refraction of the cell gives a net downward momentum to the laser beam and hence the cell will tend to move up. These techniques work well for particles around 10 μm in size. As the size increases, so does the laser power needed, leading to greater cell damage. In general, by altering the focusing of the beam and the direction at which it impinges on the cell, the cell can be maneuvered at will. Since the momentum imparted represents a force on the cell if it is attached to other cells, the laser beam can also be used to detach particular cells and thus act as "scissors." By using a ring-shaped laser beam (the so-called Laguerre–Gaussian profile, with a helical wavefront), a twisting motion can be applied to cells. This "optical spanner" makes use of the fact that if the phase of the wavefront changes by 360° m times per rotation, the beam will have angular momentum of $mh/(2\pi)$ per photon. A "quarter-wave plate" can control the rate of rotation.

27.5.2 Electroporation

This technique uses intense pulses of electricity to "punch" holes in cell membranes. These holes or pores last for a few tens of milliseconds and are approximately 40 nm in diameter (Pakhomov, Miklavcic, and Markov 2010). This size is large enough to allow macromolecules (such as genes, drugs, metabolites, antibodies, or molecular probes) to permeate the membrane. The usual arrangement for delivering these pulses is to manoeuvre the cell between miniature plates within the solution (Figure 27.10).

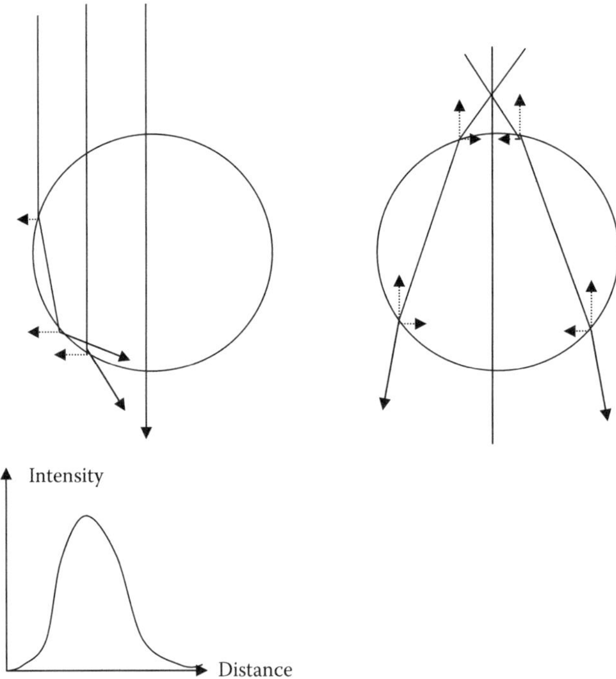

FIGURE 27.9
Optical tweezers. Left: parallel beam producing a movement of the cell into the beam. Right: focused beam producing a movement of the cell into the beam axis. The small dotted arrows indicate the force acting on the cell due to the change in light momentum in the opposite direction.

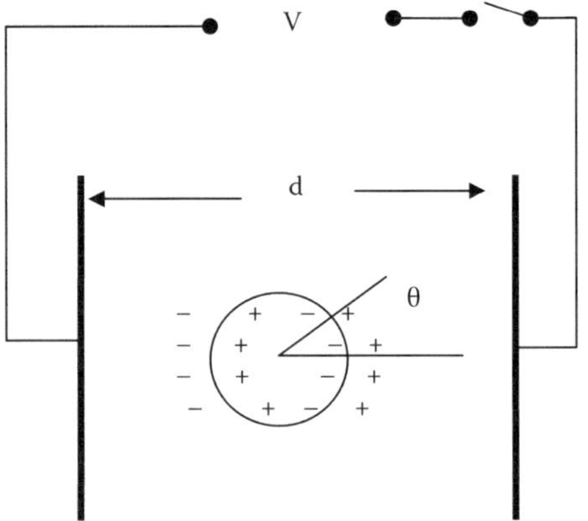

FIGURE 27.10
Electroporation of a cell membrane via a short pulse from a voltage source of around 10 MV/m.

Consider a spherical cell, radius a, subjected to a uniform field $E_0 = V/d$, where V is the voltage between the two plates (in solution) distance d apart. The voltage is applied as a pulse a few ms long. The change in transmembrane potential ($\Delta\Psi$) is given by

$$\Delta\Psi = 1.5\ f\ a\ E_o \cos\ (1 - \exp(-t/\tau)),$$

where

$$f = [1 + a\ G_m(r_i + r_e)]^{-1}$$
$$\tau = f\ a\ C_m(r_i + r_e/2)$$

where r_i and r_e are resistivities of internal and external solution (Ωm) with values of 2 and 0.2, respectively, C_m is membrane capacitance (0.01 F/m^2), and G_m is membrane conductance (normally a few tens of S/m^2, but a great deal larger when the pores open up).

To get electroporation occurring, E_0 is determined experimentally to be 10 kV/m, which will produce $\Delta\Psi$ of 0.75 V from the above equation. In fact, G_m is a function of angle, so parts of the cells will have $\Delta\Psi$ above this threshold, and the other parts will not. In fact, around 0.1% of membrane area will have pores opened up by this procedure. If the gap is 5 mm, the voltage pulse needs to be 500 V. One of the difficulties is to deliver a pulse that retains its "square" waveform. One way of achieving this is to use a circuit known as Blumlein pulse-forming network. This is a coaxial line based design that ensures that shape is maintained. There has been a trend in recent years of using nanosecond duration pulses of greater strength (around 10 MV/m) to produce more predictable pore diameters. The circuits to produce these pulses are understandably more complex. In addition to electroporation, these pulses are also used to electrically trigger intracellular calcium release, shrink tumors, cause block of action potential propagation in nerves, activate platelets, and cause release of growth factors for accelerated wound healing.

A variant of electroporation is electrofusion, which causes two dissimilar cells to fuse together as a single cell, a hybrid. Nonspatially uniform electric fields are used to cause cells to adhere in a "pearl chain" (due to induced attractive charges). A short electric field pulse is then delivered to cause the membranes of adjacent cells to fuse. Hybridomas, or cells created by fusing an immortalized cell line (such as a myeloma) with an antibody-producing cell (such as a B-cell), are useful in the production of monoclonal antibodies, the basis of immunologically based pharmaceuticals.

27.5.3 Atomic Force Microscopy (AFM)

This technique allows the mapping of surfaces on a molecular or atomic scale. It can also be adapted to measure interatomic forces. It forms one class of a number of imaging techniques known collectively as scanning probe microscopy. The principle is fairly straightforward. A silicon or silicon nitride probe with a fine (approximately 1 nm radius) tip is connected to a flexible cantilever. The molecular or atomic forces between an adjacent surface and the tip cause the cantilever to bend according to the nearness of surface features to the probe (hence AFM). A laser beam, reflecting off the cantilever, indicates the position of the probe (Figure 27.11). The tip is then moved across the surface in a raster pattern to build up a 2D image of the height of molecular features on the surface (surface topography). Since this dragging movement would tend to cause damage to the surface, a number of noncontact techniques have been devised. One consists of driving

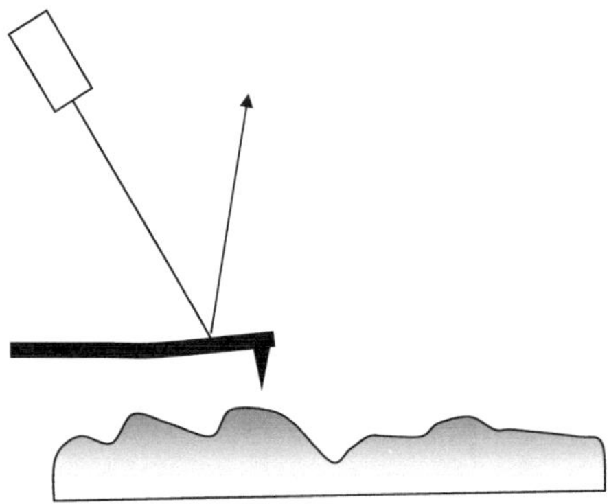

FIGURE 27.11
The principle of the atomic force microscope. A cantilever attached to a nanometer radius probe is repelled by the surface and thus moves in a vertical direction as it is scanned over the surface. A reflected laser beam indicates the position of the probe.

the cantilever to oscillate at just above its natural frequency. The nearness of the probe to the surface alters the natural frequency and hence the amplitude of this oscillation, which can be calibrated in terms of distance. Another approach is to use a "tapping mode," whereby the oscillation amplitude is much greater, and is at the natural frequency. Since the atomic forces are very nonlinear, the interaction is dominated by the force experienced when the oscillation is at its lowest point. There is thus a very small variation in oscillation amplitude with the height of features on the surface. In fact, a servo-system (see Chapter 25) drives piezoelectric elements to keep the tip in such a position that the natural frequency oscillations are maintained constant, thus following the height of the surface features. Since the tip only comes close to the surface intermittently, the damage is further reduced.

For biological work, the surface can be in a wet environment (that is, covered by a layer of water). Although this degrades the performance of the AFM, it provides useful information on viral and bacterial cell surface topographies and other biologically relevant quantities. A number of modes are available, such as force spectroscopy, lateral force measurements, nanomanipulation, and the measurement of "spring constants."

In some applications with conducting surfaces, a current is passed between the probe and the surface. The current for a fixed voltage is strongly dependant on probe–surface distance, so again, using piezoelectric actuators, resistance can be kept constant to measure height.

27.5.4 Carbon Nanotubes

Carbon atoms join in a hexagonal pattern. Graphite is an irregular 2D hexagonal structure and is fairly soft, whereas diamond, which is the hardest known substance, consists of a 3D array. Recent advances in arc discharge and laser or chemical ablation techniques have allowed specific shapes of 2D hexagonal array to be synthesized, in the form of rolled up tubes, either with a single or multiple walled structure. These have several useful properties, including electrical conductivity and semiconductivity, wave guiding properties, and optical properties. The Young's modulus is an order of magnitude higher than steel and the tensile strength around two orders of magnitude higher. There are increasing applications of carbon nanotubes in the area of biosensors, making the use of specific electrode properties.

27.5.5 Quantum Dots

These are highly fluorescent fragments of semiconductor material such as cadmium selenide, cadmium sulfide, indium arsenide, or indium phosphide. Typically, these are between 5 nm and 100 nm in diameter and unlike dye molecules, fluorescence is excited by a broad band of excitation wavelengths. The color of the fluorescence depends on the size, with larger wavelengths associated with larger sizes. In terms of cell biology, the challenge is to insert these into the cytoplasm. There are two main ways of doing this: first, coating with lipids, lipid nanoparticles, or lipid-soluble materials (such as ZnS), and second, the use of electroporation. Unfortunately, there are concerns of the toxicity of entrapped quantum dots to cells, but this may be used to advantage in the ablation of tumors.

27.5.6 Gene Chip Readers (Microarray Scanners)

Molecular biology has been increasingly moving toward the simultaneous analysis of hundreds if not thousands of endpoints. This has led to the increased prominence of the discipline "bioinformatics," which relates to organizing the vast amounts of data arising from the study of gene sequences (genomics) the associated proteins expressed (proteomics) along with signaling pathways (metabolomics). The genes with are active in the genome give rise to mRNA strands in the cytoplasm. These can be profiled by harvesting and then amplification (by the use of a technique called real-time polymerase chain reaction) to give analyzable quantities. These products are then joined to fluorescent markers and then allowed to mix with a "library" of short synthetic nucleotide sequences. Those that associate ("hybridize") with the unknown RNA will have the complementary base pairs and thus the sequence of the unknown can be identified by those locations where there is a good fluorescent signal. A typical array will be up to 10^6 spots, each of area 30 × 30 μm. Laser-jet technology is used to "spot" the molecules of known sequence onto the array.

27.5.7 Micromachining and Microelectromechanical Systems

These technologies exploit the techniques used in the manufacture of integrated circuits to produce nanometer (nm) sized machines, including micromotors, microbalances, pumps, switches, and gearwheels. Various layers can be built up successively on to a silicon substrate, with etching techniques used to remove unwanted regions by the use of masks (photolithography). The main concern in ultraprecision design is to ensure that the devices are not overconstrained, that is, the principles of kinematic design are observed. These principles relate to the three translational (x, y, z) and three rotational (around each axis) degrees of freedom an unconstrained object will have. If the object is only to have one degree of freedom (rotation around the x axis, say) then there must only be five points of constraint. Any more than that will give rise to overconstraint and the danger of binding. As a simple example, the use of five small spheres with point contacts is sufficient to give a precision movement in one direction, in this case the y direction. A force has to be applied in the directions shown to hold the carriage in place (see Figure 27.12).

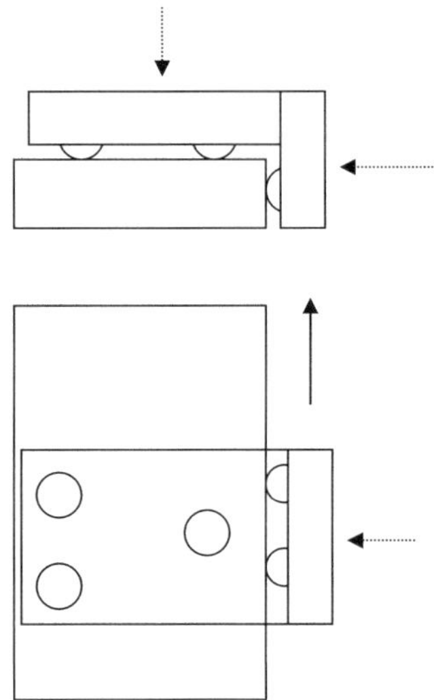

FIGURE 27.12
Illustrates the need for no more than five contact points in order to produce a single degree of freedom sliding movement. Five spheres (miniature ball-bearings) are positioned as shown. Force needs to be applied in the directions shown by the dotted arrows to allow a remaining degree of freedom, translation in the direction shown by the full arrow. (Redrawn from Smith, ST, and Chetwynd, DG, *Foundations of Ultraprecision Mechanism Design*, Gordon & Breach, 1994.)

27.5.8 Neuroengineering

This is concerned with the interfacing of neurons to silicon circuits. The circuits can be used to stimulate the neurons and also to record from the neurons. This is an extension of the principles of the multielectrode array described in Chapter 3. The silicon can be treated with neuropeptides to encourage growth of neurons in particular directions.

Bibliography

Jaiswal JK, Goldman ER, Mattoussi H, Simon SM. 2004. Use of quantum dots for live cell imaging. *Nat Methods* 1:73.

Pakhomov AG, Miklavcic D, Markov MS. 2010. *Advanced Electroporation Techniques in Biology and Medicine*. CRC Press, Boca Raton, FL, USA.

Smith ST, Chetwynd DG. 1994. *Foundations of Ultraprecision Mechanism Design*. Gordon & Breach.

Turner APF, Karube I, Wilson GS. 1987. *Biosensors*. Oxford University Press, Oxford.

Answers to Questions/Problems

Chapter 1

1: 1.6×10^8; **2**: 7.2×10^8 (assuming 30 cm circumference).

Chapter 3

1: 30, 46 mV; **2**: 1.8×10^{-6} mA/cm^2 (assume $\alpha = 0.5$); **3**: 5×10^{-12} mA, 4.8 μA; **4**: 16 kΩ; **5**: 0.5, 0.78 μA/cm^2; **6**: 0.1 V, 1.1 V; **7**: 3.64 mA/cm^2, 21 Ω, 0.8 Ω.

Chapter 5

1: 1.6×10^{-9} m/s, 5.2 and 7.7×10^{-8} m^2/s/V; **2**: 0.15 m/m^4, 10^{-3} m/s; **3**: 109 pm; **4**: 0, –13.1 mV, + 40 mV, 0.40 MPa; **5**: 90 mV, 2.4 MPa; **6**: steady-state, equilibrium, (text), ion fluxes and concentrations equal; **7**: –46.6 (–58 + 11.4) mV; **8**: 249 mOsm/L; **9**: 122, 82 mM, 5 mV, 8.78 kg; **10**: 26.8 kJ; **11**: 341 mM, 615 mOsm/L, 3.2 mV, + 8; **12**: 0.01, 6.5×10^{-9}, 6.5×10^{-11} m/s; **13**: 1.2×10^{-8} m/s, 105 mV (outside positive); **14**: 9×10^{-9} m/s; **15**: 87 mV, K^+ and Cl^-; **16**: 3.4×10^{-11}, 3.8×10^{-9}, 5.2×10^{-9} m/s, 24.4 mV, 58, –27.7, –24.7 mV; **17**: 2.22×10^{-6} mol/m^2/s, **18**: 0.057 S/m^2; **19**: 98.7%, 250 pmol/m^2/s, 29 S/m^2, 20%; **20**: 280, 300 nmol/m^2/s, 1.05, 6.61 S/m^2, 100 nmol/m^2/s, 1.5 S/m^2; **21**: 3,630 nmol/m^2/s, 520 nmol/m^2/s, 2 S/m^2, 5.8 S/m^2, 515 nmol/m^2/s.

Chapter 6

1: 0.01 s, 6.93 ms, 50 mV; **2**: 4 mm, 1.5 μs, 16 μm; **3**: 20 V, 2 ms; **4**: 1.4, 1.4 mA/cm^2, 109, 83 S/m^2, 0.89, 1.79 ms; **5**: 58 mV, 29 mS/cm^2.

Chapter 7

1: 7.32×10^{-6} J; **2**: 80,000, 26×10^{12}; **3**: 650; **4**: 148; **5**: 32, 411; **6**: 66 MPa; **7**: 10^6, 500, 40 N, 0.78 mN; **8**: 1.7 kW; **9**: 14 cm; **10**: 70; **11**: Myasthenia gravis; **12**: Amplitude— large muscle fibers, duration—motor unit with many fibers; **13**: 27.5 W; **14**: 25.6 kg; **15**: 0.043 m/s; **16**: 5 cm/s, 70 g, 14 g, 1 cm/s; **17**: 2/7 V_{max}, 2/21 W_{max}.

Chapter 8

1: 450 m^3/s; **2**: 0.62 Atm; **3**: 247 Pa m.

Chapter 9

1: 10.5 μm; **2**: 1.6×10^{-3} Pa s; **3**: 2.74 L/m, 682 mL/min, Re ~ 2760; **4**: 838, 52% rise; **5**: 3.2 kPa; **6**: 0.40 m/s, 4.1 mL/sec, no; **7**: 10.2 kPa; **8**: 9.6 kPa, 0.314×10^{-6} m^3/s, viscosity not accounted for.

Chapter 10

1: 5.6 L/m, 0.16 J, 13% approximately; **2**: 130 cm/sec, turbulent, 0.10 m; **3**: 5, 6.7 m/s, 0.35, 0.72 MPa; **4**: 2.7×10^{-5}, 9×10^{-6} Pa^{-1}, 0.74 MPa.

Chapter 11

1: 2.1 L/min, 2.5 L; **2**: 10.3 L/min, 1.2 L; **3**: 2.69 L/min; **4**: 4.7 L/min; **5**: 4.9 L/min; **6**: 9.4 mL/sec; **7**: 196 mL/sec; **8**: 4.1, 17.8 L/min, 74% fall; **9**: 0.13 m/s; **10**: 80 mm^2; **11**: 6.9 Hz, no, too wide; **12**: 26 Hz, yes, increase ζ; **13**: 25 Hz, 0.16, overshoot and "ringing," no.

Chapter 12

(MCQ): 1. (c); 2. (b); 3. (b); 4. (b); 5. (e); 6. (b); 7. (d); 8. (d); 9. (a); 10. (a).

Chapter 13

1: 125 mL/min, 625 mL/min, 0.2, 350 mg/min, –.86 mL/min; **2**: 1.67×10^4, 1.13×10^4 cm^2; **3**: 4.5 mg/L, 3 m^2, 90 mL/min; **4**: 82%; **5**: 67%, 80%; **6**: ; **7**: 121, 619 mL/min, 0.195, 696, 1244 mL/min; **8**: 110, 670 mL/min; **9**: 107, 442 mL/min, 0.24, 2.58, –1.66 mL/min; **10**: 141 mL/min, 0.197, 1.95 mL/min.

Chapter 14

1: 0.78 m^2; **2**: yes (195 mL/min at 5 L/min, more at 2 L/min).

Chapter 15

1. (b); **2**. (b); **3**. Frontal, parietal, occipital, and temporal lobes; **4**. The limbic lobe; **5**. No; **6**. Gray matter; **7**. (c); **8**. (a); **9**. The horizontal, coronal, and sagittal planes; **10**. (b) Cross-section; **11**. (d); **12**. Traveling away from a particular structure—output; **13**. (a); **14**. Memory; **15**. Metencephalon; **16**. Midbrain; **17**. Midbrain, pons, and medulla oblongata; **18**. Diencephalon; **19**. Ectoderm; **20**. Neural induction; **21**. Neurulation; **22**. (b); **23**. Gyri; **24**. Sulci; **25**. (b); **26**. Frontal and parietal lobes; **27**. Primary visual cortex; **28**. (d); **29**. (a); **30**. False; **31**. Median brain section; **32**. (e); **33**. (c); **34**. Primary visual processing; **35**. Emotions, learning, and memory; **36**. Parietal lobe; **37**. (b); **38**. (a); **39**. (c); **40**. Visual, auditory, somatic sensory, motor, olfactory, and gustatory cortical areas; **41**. (d); **42**. Three; **43**. (b); **44**. (a); **45**. (c); **46**. (d); **47**. (c); **48**. Deep cerebellar nuclei; **49**. Cerebellar peduncles; **50**. (b); **51**. (a); **52**. 31 pairs; **53**. (b); **54**. (c); **55**. (a); **56**. Posterior, anterior, and lateral white columns; **57**. (b); **58**. (d); **59**. (a); **60**. (c); **61**. (b); **62**. (b); **63**. (a); **64**. (e); **65**. (a); **66**. Midbrain; **67**. (c); **68**. (b); **69**. Thalamus and hypothalamus; **70**. (d); **71**. (b); **72**. (b); **73**. (e); **74**. (a); **75**. (a); **76**. (c); **77**. 12; **78**. (d); **79**. (a); **80**. (d); **81**. (a); **82**. Input nuclei, output nuclei, intrinsic nuclei; **83**. (c); **84**. (b); **85**. (d); **86**. Six; **87**. (c); **88**. (a); **89**. (b); **90**. (b); **91**. The dura mater, arachnoid mater, and pia mater; **92**. Cerebrospinal fluid (CSF); **93**. (d); **94**. (b); **95**. Two lateral ventricles, third ventricle, and fourth ventricle; **96**. (d); **97**. (c); **98**. The blood-CSF (cerebrospinal fluid) barrier and the blood-ECF (extracellular fluid) barrier; **99**. (a); **100**. Middle cerebral artery (MCA).

Chapter 16

1: a; **2**: 52 units; **3**: 7.6 bits; **4**: 1.9°

Chapter 18

(MCQ): 1. (b); **2**. (c); **3**. (b); **4**. (d); **5**. (b); **6**. (c); **7**. (a); **8**. (c); **9**. (a); **10**. (c) **1:** Discuss hair cell function in cupola and the utricle and saccule. Tests include caloric testing and rotational testing to elicit the nystagmus response. **2**: Reference should be made to the vestibular–ocular pathways and the neuroanatomical associations. (See Figures 18.7 and 18.18). **3**: The types of semicircular canals are associated with the animal's orientation in 3D space and their niche in nature.

Chapter 19

1: Between 5 and 10 cm^2; **2**: Cribriform plate of the ethmoid bone; **3**: Three; **4**: At least 5; **5**: 5 layers; **6**: Afferent and efferent (to and from the brain). The sensory flow takes the primary olfactory signals to the brain from the bulb and the efferent (reciprocal) signals are thought to provide feedback control of the bulb by the rest of the brain; **7**: Around every eight weeks; **8**: When many signals synapse onto a smaller number of target neurones, the output is effectively the sum of all inputs in the temporal sense. Providing the information doesn't collide in time on the target neurone or neurones, the target cell still can process all the input information. This can then be sent to areas of the brain where there are an excess of cells available to process all the information; **9**: Action potentials cannot be summed in amplitude, they are all of the same size in any given neurone. If it is necessary to summate signals to determine if they exceed a threshold, then a mechanism for handling proportional signals is required (such as on the cell body before the axon hillock where the influence of very many dendrites can be summed). **10**: Currently, –70 mV; **11**: Three, not counting the olfactory nerve cells; **12**: 25; **13**: Cerebellum; **14**: 80%; **15**: Chemosensory event-related potential; **16**: Sweet, sour, acid, bitter, salt, and umami; **17**: Gustducin, or a G protein coupled receptor; **18**: The glutamate receptor; **19**: 6-*n*-Propylthiouracil, PROP; **20**: Unusually large numbers of fungiform papillae.

Chapter 20

1. When $T = 0.1$ s the sampled sequence is periodic with period $p = 10$ samples. The corresponding temporal period is 1 s.

When $T = 1.1$ s the sequence is also periodic with period 10 samples. The corresponding temporal period is 11 s.

3. $F_n = \dfrac{\sin(2n\pi\tau/T)}{n\pi} = \dfrac{2\tau}{T}\operatorname{sinc}(2n\pi\tau/T)$ where $\operatorname{sinc} x \equiv \dfrac{\sin x}{x}$.

5. $\operatorname{sinc}(\Omega t) \leftrightarrow \dfrac{\pi}{\Omega} P_\Omega(\omega)$.

6. Nyquist rate = 400 Hz.

Nyquist frequency = 190 Hz.

Antialiasing filter cutoff = 190 Hz.

The highest undistorted frequency = 180 Hz.

8. Aliased spectrum (heavy line).

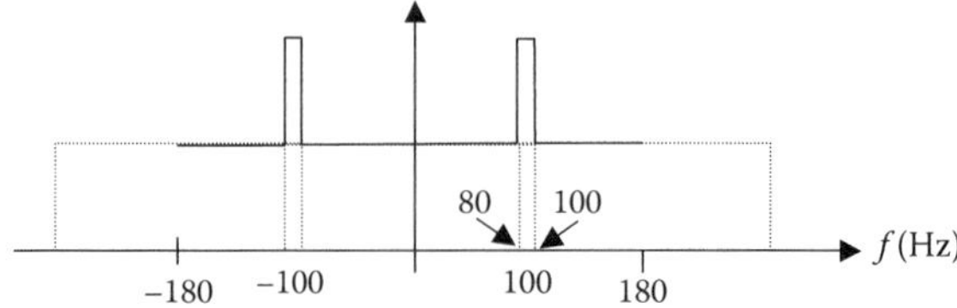

The spectrum of the reconstructed signal:

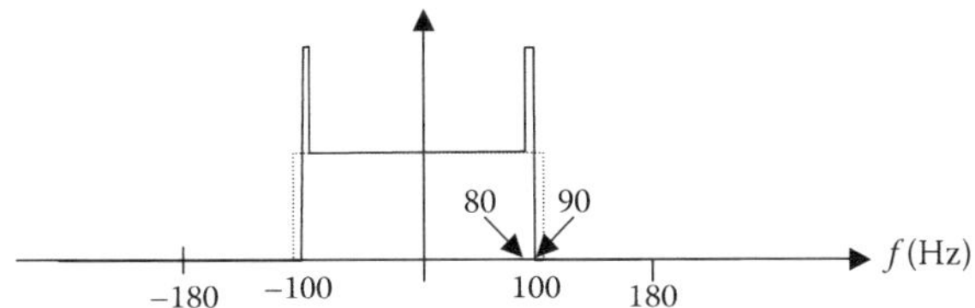

10. (i) The cut-off frequency is $\omega_c = \dfrac{\pi}{T}$.

(ii) $E[x(t)] = 0$ for all t.

(iii) $C_{xx}(t_1, t_2) = \sigma^2 \operatorname{sinc}\left(\dfrac{\pi}{T}(t_2 - t_1)\right) = \sigma^2 \operatorname{sinc}\left(\dfrac{\pi}{T}\tau\right)$

(iv) $x_{rms}^2 = E[x(t)x(t)] = C_m(t,t) = \sigma^2$ so $x_{rms} = \sigma$.

11. Fraction of time = 3.17×10^{-5}.

12. $p(x) = \begin{cases} 1; & 0 \le x \le 1 \\ 0; & \text{otherwise.} \end{cases}$

14. (i) Finite energy
(ii) Finite power
(iii) Finite energy
(iv) Neither
(v) Finite energy
(vi) Finite energy

15. $X(\omega) = -1 + \dfrac{1}{1 - a\exp(j\omega)} + \dfrac{1}{1 - a\exp(-j\omega)} = \dfrac{1 - a^2}{1 + a^2 - 2a\cos(\omega)}$.

22. $\Omega_{eq} = \dfrac{\pi}{2\tau}$.

23. $X(m) = \left\{\exp\left(-j\dfrac{2\pi m}{8}\right) + \exp\left(j\dfrac{2\pi m}{8}\right)\right\} = 2\cos\left(\dfrac{2\pi m}{8}\right); \quad m = 0, 1, 2, \cdots, 7$

Or

$X(m) = 0.25 \times [2, \sqrt{2}, 0, -\sqrt{2}, -2, -\sqrt{2}, 0, \sqrt{2}]$.

24. Aliasing errors appear at 30 Hz with the same amplitude as the interference. The attenuation required is $A \approx 1.7 + 6.02 \times B$ dB = $1.7 + 6.02 \times 12 \approx 74$ dB.

25. $w(t) \leftrightarrow W(\omega) = 1.08T \operatorname{sinc}(\omega T) + 0.46T\left[\operatorname{sinc}(\omega T - \pi) + \operatorname{sinc}(\omega T + \pi)\right]$

or $W(\omega) = \left(\dfrac{0.62(\omega T)^2 - 1.08\pi^2}{(\omega T)^2 - \pi^2}\right) T \operatorname{sinc}(\omega T)$.

26. The required number of trials = 1000.

The moving window length is 1 s.

28. $$H(\omega) = \frac{V_{out}(\omega)}{V_{in}(\omega)} = \frac{1}{1 + j\omega RC}\ ;\ h(t) = \frac{1}{RC} U(t) \exp\left(-\frac{t}{RC}\right).$$

29. $$A(\omega) = |H(\omega)| = \frac{1}{2}(1 + \cos\omega).$$

Chapter 21

Multiple Choice Question Answers: 1 (iii), 2 (iv), 3 (iii), 4 (i), 5 (iv), 6 (ii), 7 (i), 8 (i), 9 (iv).

Questions to research:

1: See Table 21.2 and note solutions; **2**: Cost effective, noninvasive, standard neurological instrument and technique; **3**: Digital storage, computational analysis, visualization of data; **4**: Nerve conduction velocity, eg., motor artifact associated with Parkinson's disease; **5**: See Section 21.2 "Origins of the EEG" for discussion.

Chapter 23

1: 8, 4, 6, 2.5; **2**: 2700 m/s, 5.7 μV; **3**: 0.69, 0.33, 0.01 cm; **6**: 77 μs, 0.98 mV, 0.7 dB/cm tissue, **8**: 1.3 mm, 0.35; **10**: LCLCL, LCLCL, LLLCL, LCLCL, LCLCL.

Chapter 24

1: 109 nM; **3**: 40 nM.

Chapter 25

1: $k_1k_2k_3/O^2$, yes; **2**: 2, –12, 3; **3**: $-4k_5k_6k_1k_3^4/(k_4^4(kp + k_2)^5)$; **4**: DCAF/$z^2$; **5**: kKBA$_0$ f'_a Lexp(–($k\phi$ + Kf_e)); **6**: 68L[(1 + 4$\sqrt{\phi}$)^2$\sqrt{\phi}$]$^{-1}$, 1.05, 0.08; **7**: 14.5, 5.7, 8.6 L, 0.93 L/h; **8**: 1.5, 4.5 L, 0.026 L/min; **9**: 14.3 L, 0.0074, 0.023 min^{-1}; **10**: 5exp(–0.083s)/(1 + 1.1s)^2; **11**: π/2, exp(–πs/2)/(1 + 10s)^2; **12**: 5exp(–1.5s)/(1 + 32s); **13**: exp(–0.4s)/(1 + 1.6s)^2; **14:** 8/(1 + 2.5s); 25.15: exp(–1.4s)/(1 + 11s)^2, unstable.

Chapter 26

1: 3.68, 3.68 MPa, 184 MPa s; **2**: 1.47 GPa s; **3**: 5, 15 GPa, 20 GPa s.

Index

Page numbers followed by f and t indicate figures and tables, respectively.

B

N

O

S

W

X

Y

Z

For Product Safety Concerns and Information please contact our EU representativc GPSR@taylorandfrancis.com Taylor & Francis Verlag GmbH, Kaufingerstraße 24, 80331 München, Germany